2002

Sharks, Skates, and Rays
THE BIOLOGY OF ELASMOBRANCH FISHES

Sharks, Skates, and Rays

THE BIOLOGY OF ELASMOBRANCH FISHES

EDITED BY
WILLIAM C. HAMLETT

THE JOHNS HOPKINS UNIVERSITY PRESS
Baltimore and London

© 1999 The Johns Hopkins University Press
All rights reserved. Published 1999
Printed in the United States of America on acid-free paper

9 8 7 6 5 4 3 2 1

The Johns Hopkins University Press
2715 North Charles Street
Baltimore, Maryland 21218-4363
www.press.jhu.edu

Library of Congress Cataloging-in-Publication Data will be found at the end of this book.
A catalog record for this book is available from the British Library.

ISBN 0-8018-6048-2

CONTENTS

	Preface	vii
CHAPTER 1	*Systematics and Body Form* LEONARD J. V. COMPAGNO	1
CHAPTER 2	*Integumentary System and Teeth* NORMAN E. KEMP	43
CHAPTER 3	*Endoskeleton* LEONARD J. V. COMPAGNO	69
CHAPTER 4	*Muscular System:* *Gross Anatomy and Functional Morphology of Muscles* KAREL F. LIEM AND ADAM P. SUMMERS	93
CHAPTER 5	*Muscular System:* *Microscopical Anatomy, Physiology, and Biochemistry* *of Elasmobranch Muscle Fibers* QUENTIN BONE	115
CHAPTER 6	*Digestive System* SUSANNE HOLMGREN AND STEFAN NILSSON	144

CHAPTER 7	*Respiratory System* PATRICK J. BUTLER	174
CHAPTER 8	*Circulatory System:* *Anatomy of the Peripheral Circulatory System* RAMÓN MUÑOZ-CHÁPULI	198
CHAPTER 9	*Circulatory System:* *Distinctive Attributes of the Circulation* *of Elasmobranch Fish* GEOFFREY H. SATCHELL	218
CHAPTER 10	*Heart* BRUNO TOTA	238
CHAPTER 11	*Nervous System* MICHAEL H. HOFMANN	273
CHAPTER 12	*Special Senses* HORST BLECKMANN AND MICHAEL H. HOFMANN	300
CHAPTER 13	*Rectal Gland and Volume Homeostasis* KENNETH R. OLSON	329
CHAPTER 14	*Urinary System* ERIC R. LACY AND ENRICO REALE	353
CHAPTER 15	*Female Reproductive System* WILLIAM C. HAMLETT AND THOMAS J. KOOB	398
CHAPTER 16	*Male Reproductive System* WILLIAM C. HAMLETT	444
APPENDIX	*Checklist of Living Elasmobranchs* LEONARD J. V. COMPAGNO	471
	Contributors	499
	Index	501

PREFACE

The formal scientific study of elasmobranchs (sharks, skates, and stingrays) began with Aristotle, who made the first significant observations on the natural history of selachians (Cole 1949). He was aware that some sharks were oviparous and that others were viviparous, with a placenta. Hieronymus Fabricius reviewed Aristotle's observations and added his own (Adelmann 1967), diagramming the ovary, uterus, and embryos in *Galeus laevis*. Belon (1551) knew of the attachment of cartilaginous fishes to the maternal oviduct by means of a "navel string," and Rondelet (1554) made a drawing of such a connection in *Mustelus laevis*. The Danish anatomist Steno (1673) claimed the connection was a functional placenta, and Stefano Lorenzini devoted a monograph to *Torpedo* in 1678, making the first mention of red and white muscle. Malpighi made significant observations on fish reproduction, including notes on the shark placenta (Adelmann 1966), and Müller (1842) also presented details of the development and organization of the placenta. Balfour (1878) concentrated on embryological studies, whereas Ziegler and Ziegler (1892) and Rückert (1899) were concerned with early development. Various others produced ichthyological treatises containing information on elasmobranchs, including Beard (1890), Cuvier and Valenciennes (1828–45), and Dean (1909). The work of Garman (1913), likewise an important contribution to the body of knowledge, has recently been reprinted (Garman 1997) and is available to a new generation of students. In 1932 and 1934 Ranzi published his important synthesis on reproductive modes in elasmobranchs.

The contribution of Dr. Frank Daniel is of particular importance to elasmobranch biology. Daniel graduated from the University of Chicago in 1906 and received his Ph.D. from the Johns Hopkins University in 1909, with a dissertation on the "Adaptation and Immunity of Lower Organisms to Ethyl Alcohol." In 1910 he accepted a position in the Department of Zoology at the University of California in Berkeley. His career there spanned 32 years and included a tenure as department chairman from 1936 to 1942.

In 1922, Daniel published the first edition of the now-classic *The Elasmobranch Fishes*. This work subsequently went through three editions and has served many generations of biologists well. It represented the ultimate compilation of information on elasmobranch morphology of its time. Few other volumes have been so consistently appreciated by students in the field. It is still sought after through antiquarian book dealers. A second edition of the book appeared in 1928, followed by a third, revised edition in 1934. I hope the present volume will do justice to the classic work that is its predecessor.

Various compilations of individual papers on elasmobranchs have appeared subsequently in the literature. In 1963 Gilbert published *Sharks and Survival*, and in 1967 *Sharks, Skates, and Rays* was published by the Johns Hopkins University Press. The latter volume was a collection of 39 papers presented at a symposium entitled "Current Investigations Dealing with Elasmobranch Biology" that was convened in 1966 at the Lerner Marine Laboratory, Bimini, Bahamas. In 1970 Lineweaver and Backus gave us *The Natural History of Sharks*, followed by Budker's *The Life of Sharks* in 1971. Both of these volumes were well received by the general public. They also presented considerable technical information. In 1976 Thorson edited a volume on the ichthyofauna of Lake Nicaragua including much of his own work on the bull shark, *Carcharhinus leucas*, and the sawfish, *Pristis perotteti*. In 1976 an important symposium on shark biology was organized by Glenn Northcutt and held in New Orleans.

The results were subsequently published in the *American Zoologist* in 1977. A volume dedicated to sensory biology of elasmobranchs appeared in 1978 (Hodgson and Mathewson 1978) and Compagno published the much-referenced *Sharks of the World* in 1984. Elasmobranchs as models in human biology was the topic of a symposium in Brussels, Belgium, in 1986. The results appeared in the *Archives of Biology* (Brussels) in 1987 (Hamlett 1987).

In 1988 two important volumes appeared, Compagno's *Sharks of the Order Carcharhiniformes* and Shuttleworth's *Physiology of Elasmobranch Fishes*. A symposium held in Rome, Italy, in 1988, entitled "Evolutionary and Contemporary Biology of Elasmobranchs," brought together the leading workers on elasmobranch biology. The 20 papers presented at the symposium were published the following year in the *Journal of Experimental Zoology* (Hamlett and Tota 1989). No previous volume has attempted to combine anatomy, including light and electron microscopy, physiology, and biochemistry in a single resource. Three symposia in the early 1990s were eventually published as *Elasmobranchs as Living Resources* (Pratt et al. 1990), *Vision in Elasmobranchs* (Hueter and Cohen 1991), and *The Reproduction and Development of Sharks, Skates, Rays, and Ratfishes* (Demski and Wourms 1993). In 1996 Klimley and Ainley edited a book on the biology of the great white shark, *Carcharodon carcharias*.

In the present volume, we have built on Daniel's work by keeping a systems approach. We have expanded coverage beyond strict anatomy to include much contemporary information on the physiology and biochemistry of this group. I offer my apologies to my systematics, behavior, and fisheries colleagues for not including these important areas, but this volume is purposely restricted to anatomy and physiology from a systems approach. It is intended to serve as a first resource for individuals interested in the field. It is targeted at an audience at the informed graduate student level that has relatively little formal background in the study of elasmobranchs. The other authors and I have striven to create a volume that presents adequate general introductory information for the relative novice but enough current technical citations to serve as a resource of references to the primary literature. Because of space constraints, no historical reviews have been attempted.

I thank the Johns Hopkins University Press for their patience during the genesis of this volume. I am very grateful to my coauthors for tirelessly striving to make my vision a reality. Some of the authors are of emeritus status, and I was lucky enough to cajole them into participating. To them I extend a special thanks. In all cases, the contributors are leading authorities in their respective fields in elasmobranch biology.

My warmest appreciation to my wife, Martha, and our children, Hayes, Kathleen, and Alex, who had to do without Dad on many occasions as this work unfolded.

REFERENCES

Adelmann, H. B. 1966. *Marcello Malpighi and the Evolution of Embryology.* 5 vols. Ithaca, N.Y.: Cornell University Press.
———. 1967. *The Embryological Treatises of Hieronymus Fabricius of Aquapendente.* 2 vols. Ithaca, N.Y.: Cornell University Press.
Balfour, F. M. 1878. *A Monograph on the Development of Elasmobranch Fishes.* London: Macmillan.
Beard, J. 1890. The inter-relationships of the ichthyopsida: a contribution to the morphology of vertebrates. *Anat. Anz.* 5: 146–59, 179–88.
Belon, P. 1551. *L'Histoire Naturelle des Etranges Poissons Marins.* Paris.
Budker, P. 1971. *The Life of Sharks.* New York: Columbia University Press.
Cole, F. J. 1949. *A History of Comparative Anatomy from Aristotle to the Eighteenth Century.* London: Macmillan.
Compagno, L. J. V. 1984. *Sharks of the World.* FAO Species Catalogue, vol. 4, pts. 1 and 2. Rome: United Nations Food and Agriculture Organization.
———. 1988. *Sharks of the Order Carcharhiniformes.* Princeton, N.J.: Princeton University Press.
Cuvier, G., and A. Valenciennes. 1828–45. *Histoire Naturelle des Poissons,* vol. 1. Paris.
Daniel, J. F. 1922. *The Elasmobranch Fishes.* Berkeley: University of California Press.
Dean, B. 1909. Studies on fossil fishes (sharks, chimaeroids, and arthrodires). *Mem. Am. Mus. Nat. Hist.* 9(5): 211–87.
Demski, L. S., and J. P. Wourms, eds. 1993. *The Reproduction and Development of Sharks, Skates, Rays, and Ratfishes.* Boston: Kluwer.
Garman, S. 1913. *The Plagiostomia.* Memoirs of the Museum of Comparative Zoology, vol. 36. Cambridge, Mass.: Harvard College.
Garman, S. 1997. *The Plagiostomia (Sharks, Skates and Rays).* Los Angeles: Benthic Press.
Gilbert, P. W., ed. 1963. *Sharks and Survival.* Boston: D. C. Heath.
Gilbert, P. W., R. F. Mathewson, and D. P. Rall, eds. 1967. *Sharks, Skates, and Rays.* Baltimore: Johns Hopkins University Press.
Hamlett, W. C., ed. 1987. Elasmobranchs as models in human biology. *Arch. Biol.* (Brussels) 98: 131–260.
Hamlett, W. C., and B. Tota, eds. 1989. Evolutionary and contemporary biology of elasmobranchs. *J. Exp. Zool. Suppl.* 2: 1–198.
Hodgson, E. S., and R. F. Mathewson, eds. 1978. *Sensory Biology of Sharks, Skates and Rays.* Arlington, Va.: Office of Naval Research.
Hueter, R. E., and J. L. Cohen, eds. 1991. Symposium. Vision in elasmobranchs: a comparative and ecological perspective. *J. Exp. Zool. Suppl.* 5: 1–182.
Klimley, A. P., and D. G. Ainley, eds. 1996. *Great White Sharks, The Biology of Carcharodon carcharias.* New York: Academic Press.
Lineweaver, T. H., and R. H. Backus. 1970. *The Natural History of Sharks.* London: Ebenezer Bayliss and Son Ltd.
Lorenzini, S. 1678. *Osservazioni ontorno alle torpedini.* Florence.
Müller, J. 1842. Über den glatten Hai des Aristoteles und über die Verschiedenheiten unter den Haifischen und Rochen in der Entwickelung des Eies. *Abhandl. Akad. Wiss.* 27: 187–257.
Northcutt, R. G., ed. 1977. Recent advances in the biology of sharks. *Am. Zool.* 17: 287–515.
Pratt, H. L., S. H. Gruber, and T. Taniuchi, eds. 1990. *Elasmobranchs as Living Resources: Advances in the Biology, Ecology, Systematics, and the Status of the Fisheries.* NOAA Technical Report NMFS 90.
Ranzi, S. 1932. Le basi fisio-morfologische dello sviluppo embrionale dei Selachi, pt. 1. *Pubb. Staz. Zool.* (Naples) 13: 209–90.
Ranzi, S. 1934. Le basi fisio-morfologische dello sviluppo embrionale dei Selachi, pt. 2, E 3. *Pubb. Staz. Zool.* (Naples) 13: 331–437.
Rondelet, G. 1554. *De Piscibus Marinus.* Lyons: Mattias Bonhomme.
Rückert, J. 1899. Die erste Entwicklung des Eies der Elasmobranchier. In *Festschrift zum siebenzig-*

sten Geburtstag von Carl von Kupffer, 581–704. Jena: Gustav Fischer Verlag.

Shuttleworth, R., ed. 1988. *Physiology of Elasmobranch Fishes.* New York: Springer-Verlag.

Steno, N. 1673. Observationes anatomicae spectantes ova viviparorum. *Acta Med. Hafniensia* 2.

Thorson, T., ed. 1976. *Investigations of the Ichthyofauna of Nicaraguan Lakes.* Lincoln: School of Life Sciences, University of Nebraska.

Ziegler, H. E., and F. Ziegler. 1892. Beiträge zur Entwickelungsgeschichte von *Torpedo. Arch. Mikr. Anat.* 39: 56–102.

Sharks, Skates, and Rays
THE BIOLOGY OF ELASMOBRANCH FISHES

LEONARD J. V. COMPAGNO

CHAPTER 1

Systematics and Body Form

The class Chondrichthyes (cartilaginous fishes) is a major group of aquatic, gill-breathing, finned, fishlike jawed (gnathostome) vertebrates that includes the living elasmobranchs (sharks and rays) and the living holocephalans (chimaeras and elephant fishes), plus numerous extinct taxa extending back through more than 400 million years of geologic time.

Living members of the class Chondrichthyes can be characterized by having the following features:

- Simple endoskeletons of calcified cartilage.
- Paired primary upper jaws (palatoquadrates) and lower jaws (Meckel's cartilages).
- Jointed hyoid and branchial arches.
- Four to seven separate internal and external gill openings.
- No lungs or swim bladders.
- A simple boxlike neurocranium housing or supporting the brain and cephalic sense organs.
- A vertebral column with a notochord secondarily supported by calcified vertebral centra (except in chimaeras).
- Paired pectoral and pelvic fins with endoskeletal girdles.
- Unpaired dorsal and anal fins (secondarily lost in some groups).
- Precaudal fins supported internally by proximal cartilaginous basals and distal radials.
- Paired copulatory organs (mixopterygia or claspers) as rearward extensions of the basal skeleton of the pelvic fins.

- A caudal fin supported along its length by the vertebral column, usually with expanded neural and hemal arch elements serving as fin supports.
- Fin support supplemented by elastic connective tissue rays or ceratotrichia, but without the jointed bony fin rays (lepidotrichia) of bony fishes (class Osteichthyes).
- A dermal skeleton or external covering of small dermal denticles or placoid scales (toothlike structures with enameloid crowns and dentine bases), but without the external dermal bones found in bony fishes.
- Teeth in replicating rows (tooth families) of functional and replacement teeth transverse on the jaws and replaced from inside the mouth (fused into ever-growing tooth plates in living chimaeras and in various fossil chondrichthyans).
- Usually a preoral snout with the nostrils on the ventral surface.
- Nostrils with single (not dual) apertures that are subdivided by cartilage-supported folds of skin.

Cartilaginous fishes primitively are egg-layers (oviparous), and about a third of the living species deposit large eggs enclosed in egg cases, which sit on the bottom until they hatch. It may take several months for the embryo to develop, exhaust its yolk supply, and hatch from the egg case as a miniature adult. About two-thirds of the species of living cartilaginous fishes (sharks and rays only) have changed from oviparity to various forms of live-bearing or viviparity, with ovoviviparity or yolk-sac viviparity the simplest and most wide-ranging type among the various live-bearing elasmobranchs (Compagno 1990a).

The living elasmobranchs

The elasmobranchs (sharks and rays) are the dominant living group of the class Chondrichthyes. Of approximately 60 families, 185 genera, and 929–1164 species of living cartilaginous fishes (estimate from November 4, 1998), approximately 96% of the species are elasmobranchs (sharks and rays) and 4% are chimaeras. The term *elasmobranch* (Gr. *elasmos*, plate, and *branchios*, gill), refers to the five to seven separate gill openings and platelike interbranchial septa and holobranchs of these fishes, with a soft gill cover or operculum lacking (present in chimaeras). Living elasmobranchs also have the upper jaws separate from, not fused with, the neurocranium and usually capable of protrusion from the mouth; the jaws supported by a modified dorsal hyoid arch element (hyomandibula), connecting them and the ventral hyoid elements (ceratohyoids and basihyal) to the neurocranium (otic capsule); tooth rows not fused into three pairs of ever-growing plates; an erect first dorsal fin that cannot fold backward; urogenital and rectal organs discharging into a common cloaca; and no accessory claspers on the head (frontal tentaculum) or in front of the pelvic fins (prepelvic tentacula) in males.

Living sharks and rays are only a small part of the vast adaptive radiation of cartilaginous fishes that includes many extinct higher groups and over 3000 extinct species. They belong to a major subgroup of elasmobranchs, the cohort Euselachii (true sharks), that stems from the radiation of cartilaginous fishes in the Devonian and which also includes the extinct protoselachians (the archaic euselachian sharks, including the hybodont, ctenacanth, and xenacanth sharks). Euselachians primitively have a sharklike form, two spined dorsal fins, an anal fin, and three basal cartilages in the pectoral fins. The living sharks and rays, and their immediate fossil relatives, form a subgroup of Euselachii, the subcohort Neoselachii or modern elasmobranchs, with well-developed vertebral centra (absent in other euselachians) as their most prominent feature. The neoselachians arose from within the Euselachii during the Paleozoic era but began a major adaptive radiation (the "neoselachian revolution") during the Mesozoic era while all other elasmobranch groups became extinct and were replaced by ecologically diverse neoselachians, including those highly successful benthic specialists, the rays. This neoselachian revolution survived the Cretaceous–Tertiary extinction event with continuing high diversity despite loss of large-bodied trophic specialists (apical predators and pelagic durophages), and during the Cenozoic the marine fauna recuperated to produce the living elasmobranch fauna as we know it. Living sharks are far more diverse in supraspecific taxa (genera and higher) than rays (which are derived from sharks), but there are more species of rays than sharks (about 56% versus 44% of total elasmobranch species).

External body form of living elasmobranchs (neoselachians)

Living sharks and rays vary considerably in body form, but there are common elements of topography in generalized living sharks (Fig. 1-1) that can be more or less easily recognized in the rays. The body of a living shark can be delimited into a head (from snout to gill region), a trunk (pectoral girdle to vent), and a tail (including the precaudal tail from vent to caudal origins and the caudal fin). Within the rays, the trunk and tail are functionally supplemented and ultimately replaced as locomotory organs by the enlarged pectoral fins, which form a unified pectoral disk with the head and trunk in all rays except the sawfishes (Pristidae) and sharkfin guitarfishes (Rhinidae).

The external surface of most sharks and rays is more or less covered with dermal denticles or placoid scales, although some sharks and rays are naked on their ventral surfaces and a few batoids are naked or mostly so both above and below. Denticles are small (mostly less than 2 mm long) and toothlike in structure, with enameloid-covered exposed crowns and dentine bases rooted in the skin. They vary greatly in form and size and can develop into enlarged specialized derivatives such as fin spines on the dorsal fins of various sharks and the first, Jurassic rays, rostral teeth in sawsharks and sawfishes, clasper spines, the sting of stingrays and their relatives, and enlarged thorns or bucklers on the dorsal surfaces of many rays and the bramble shark (*Echinorhinus brucus*). Denticles of neoselachians are periodically replaced by being shed and having new denticles erupting through the skin, but some denticle derivatives such as fin spines, the rostral teeth of sawfish, and some thorns and bucklers grow by periodic addition of dentine to their proximal ends and peripheries.

The head can be subdivided into three regions from anterior to posterior: the snout or rostrum in front of the eyes and mouth; the orbital region, including the eyes and mouth; and the hyobranchial region, including the spiracles and gills and terminating at the pectoral girdle. Various paired structures are prominent features of the head: the eyes, spiracles (the external openings of prebranchial respiratory passages between jaws and hyoid arches), external gill openings, nostrils (on the underside of the head or, in a few sharks, nearly terminal on the head), and the endolymphatic pores (on the dorsal surface of head just behind the eyes and communicating with the inner ear).

The eyes have upper and lower eyelids (the upper eyelids are lost in rays) and are situated laterally or dorsolaterally on the head. The eyes are large and rather well developed in most living sharks and rays but are greatly reduced in two genera of blind torpedo rays (*Benthobatis* and *Typhlonarke*). A movable lower eyelid (nictitating lower eyelid) is developed in all requiem sharks (carcharhinoids) and serves to cover the eye, but it is only analogous to the true nictitating membrane of land vertebrates (tetrapods). The collared carpet sharks (family Parascylliidae) have movable upper eyelids.

Spiracles are present in most sharks and rays as external openings of the hyoidian gill pocket between the mandibular (jaw) and hyoid (tongue) visceral arches. Spiracles are usually present in living elasmobranchs. They are usually large in most shark groups and in all batoids, but they are variably reduced in many carcharhinoid, lamnoid, orectoloboid, and hexanchoid sharks and entirely lost in some carcharhinoids, lamnoids, and hexanchoids. The enlarged spiracles of some sharks and many rays have transverse dermal spiracular valves, supported by cartilaginous cores. These sharks and rays with enlarged spiracles can

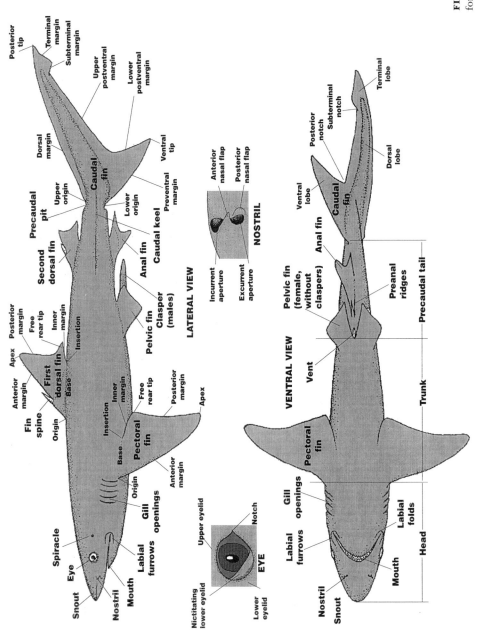

FIGURE 1-1 External body form of living elasmobranchs.

use them as alternate entrances for the respiratory current when the mouth is occluded by the substrate, and the valves may function to prevent water from flowing back out the spiracles during the respiratory cycle. External gill openings vary in number from five to seven and differ greatly in size and shape, from almost porelike in certain sharks and rays to the great arcuate flaps and vast openings that virtually encircle the head of the basking shark (*Cetorhinus maximus*). The frilled sharks (*Chlamydoselachus*) are unique among living elasmobranchs in having the first branchial septa connected ventrally across the throat.

The paired nostrils are well developed in living elasmobranchs and have a single external aperture that is divided by a transverse pair of flaps or valves into anteriorly facing incurrent apertures and posteriorly facing excurrent apertures. Water moves into the nasal organs through the incurrent apertures, through the nasal cavities with their rows of sensory lamellae (schneiderian folds), and out the excurrent apertures, as the shark or ray moves forward. The nostrils are partly covered by anterior nasal flaps, narrow to broad horizontal lobes that extend from the anterior edges of the nostrils under their cavities. The anterior nasal flaps may be elongated or partially differentiated into sensory barbels that are trailed on the substrate and apparently are used to detect food. In many benthic sharks and rays, water pumped into the mouth and through the gills in respiration also helps to draw water through the nostrils when the animal is sitting on the bottom or moving above it. The nostrils of such sharks and rays have their excurrent apertures connected to the upper lip by nasoral grooves, troughs in the upper lips that draw water into the mouth from the nostrils. In elasmobranchs with nasoral grooves, the anterior nasal flaps are usually expanded posteriorly to the front of the mouth and medially to form a broad nasal curtain that forms the floor of the nasoral grooves and help direct the flow of water from the nostrils to the mouth.

The mouth is usually ventral on the head, well behind the snout tip, and has the lips differentiated into distinct labial folds and labial furrows at its corners that are supported by internal labial cartilages. In elasmobranchs with large labial folds and furrows, the labial folds expand laterally when the mouth is open and may serve to restrict the mouth aperture and facilitate suction feeding. The upper and lower lips are often low and inconspicuous but are enlarged and morphologically elaborated in certain squaloid sharks and in some batoids.

All living elasmobranchs have rows of oral teeth along the edges of the upper and lower jaws, which are replaced from inside the mouth and are continually lost at their junction with the outer skin of the lips. Oral teeth vary tremendously in size, form, and function in elasmobranchs, with crushing, cutting, and impaling the primary tooth functions, which are sometimes combined in different sectors of the same dentition. The individual teeth of some scyliorhinid and proscylliid cat sharks and dwarf squaloids are as small as sand grains, while the largest individual teeth (over 50 mm high) of living elasmobranchs are found in the white sharks (*Carcharodon*) and makos (*Isurus*). Most rays have small teeth, but some of the eagle rays (*Aetobatus*) have large platelike teeth, formed from the lateral fusion of several tooth rows. The giant megatooth shark (*Carcharodon megalodon*) is a recently extinct relative of the white shark that had the largest individual teeth of any elasmobranch (with the largest upper anterior teeth over 160 mm high and 130 mm wide). An individual oral tooth consists of an exposed, enameloid-covered distal crown (that part of the tooth that functions in prey capture or processing) and a skin-covered dentine root, which anchors the tooth in the thick skin or dental lamina covering the jaw. Many shark and ray teeth have pointed, sharp-edged projections, cusps and cusplets, on their crowns that serve to impale prey. Serrations may develop on the compressed edges of the crowns and facilitate cutting of large food items.

Just inside the mouth and behind the functional tooth rows in the upper and lower jaws is a deep groove, in the dental lamina, supported by a groove, or shelf, in the jaw cartilage, inside of which the teeth develop. The lamina and jaw itself may be elaborated to form pockets or cavities for each tooth row in those sharks with large teeth. In the upper jaw, just behind the dental groove, there is a broad free fold, the maxillary valve, that extends transversely or in a broad or narrow arc across the front of the palate and serves as an antibackflow valve to limit water flow forward as the elasmobranch respires and compresses its pharynx to drive water into the gills. The palate, behind the maxillary valve, is usually flat and extends posteriorly to form the roof of the pharynx or orobranchial cavity. Ventrally and behind the dental groove is the mouth floor, which in sharks has a deep arcuate pocket or groove that defines the anterior base of the tongue (which is supported internally by the basihyal and ceratohyal cartilages of the hyoid arch). In sharks the tongue is more or less prominent (and is enormous in the megamouth shark, *Megachasma pelagios*), but it is reduced or absent in rays. The floor of the pharynx behind the mouth and the dorsal palate extends posteriorly to the esophagus, and is bounded laterally by the internal gill openings and gill arches.

The pharynx may have papillae on its internal surfaces, which may be large, elevated, and prominent on the anterior floor in various rays. The internal gill openings of sharks and rays may have gill raker papillae, lobate structures along the inner margins of the gill arches mesial to the gill filaments that block food items from entering the gills. These are usually made of dermis and epidermis and are often enlarged and strengthened by a cartilaginous core. In the few filter-feeding elasmobranchs (whale shark, *Rhincodon typus*; megamouth shark, *Megachasma pelagios*; basking shark, *Cetorhinus maximus*; and devil rays, family Mobulidae), the internal gill apertures are more or less restricted by gill filters that serve to collect small food organisms, functioning as the primary organs for food uptake in place of the teeth. The gill filters are cartilage-cored dermal filter plates with screenlike distal margins in the whale shark and devil rays, enlarged and densely packed gill raker papillae in the megamouth shark, and hard gill rakers formed from modified dermal denticles in the basking shark.

The very short to greatly elongated snout, with the nostrils on its ventral or anteroventral surface, houses electrosensitive ampullae of Lorenzini, with groups of ampullae pores on the snout, and the anterior loops of the cephalic lateral line canals. Ampullae and cephalic canals are also present on the orbital and hyobranchial head. In sharks, the canals continue posteriorly as a lateral line along the trunk, tail, and caudal base on each side. Some rays have the cephalic lateral line canals expanded posteriorly onto the pectoral disk. Some deepwater sharks and rays have greatly enlarged snouts, apparently enhancing their sensory function, while the saw sharks and sawfish evolved greatly enlarged, flattened snouts both as sensors and as feeding tools that can be swung laterally to impale small organisms on their rostral teeth.

The trunk encloses the body cavity and viscera and serves as a base for the paired fins, the pectoral and pelvic fins, which originate on its ventrolateral or lateral surfaces. Although certain fossil cartilaginous fishes may have lacked pelvic fins, all living sharks and rays have functional pectoral and pelvic fins of varied shapes, but they are usually broad-based and more or less paddlelike. The pectoral and pelvic fins have broad, anteroposteriorly elongated bases that anchor the fin to the trunk; anterior margins or leading edges that are more or less transverse to the trunk; apices or distal fin tips; posterior margins or trailing edges; and movable posterior lobes consisting of rear apices or free rear tips and anteromedial inner margins connecting the posterior margins with the fin bases. The posterior lobes may function like control surfaces on an aircraft, with maneuvering and braking functions.

The claspers or mixopterygia, the paired reproductive organs found in all living neoselachians, originate on the rear bases of the pelvic fins. Claspers vary greatly in structure among living elasmobranchs but generally have a dorsal or medial clasper groove, an anterior opening, or apopyle, and a posterior opening, or hypopyle, which is confluent with an expanded posterior area, or clasper glans.

The opening of the cloaca, or vent, discharges between the pelvic fin bases. The cloaca has openings for the urogenital ducts, rectum, and abdominal pores, which connect the cloaca with the body cavity. In male neoselachians the urogenital papilla extrudes spermatozoa into the cloaca, which then enter the clasper apopyles and clasper grooves, which fill with sperm. Intromission is accomplished by insertion of a clasper glans into the female's cloaca and injection of the sperm contained in its groove through the hypopyle and glans.

The unpaired fins are located on the median dorsal and median ventral lines of the trunk and tail and include the dorsal, anal, and caudal fins. The first dorsal fin, with or without an anterior, cylindrical-conical, spikelike fin spine, is a broad, compressed fin located on the middorsal surface of the trunk in sharks, with a narrow longitudinal base that varies in position from over the pectoral fin bases to over the pelvic bases. In rays it varies from over the pelvic bases to far rearward on the middle or distal end of the precaudal tail. A few sharks and many rays have lost the first dorsal fin or have it greatly reduced. The second dorsal fin is often similar in shape to the first dorsal, may also have an anterior spine, and is based on the middorsal surface of the trunk or precaudal tail. The second dorsal fin varies in position from over the pelvic bases to near the tail tip. It is apparently always present in sharks but is lost in some skates and many stingrays. Dorsal fin spines often have a posterior groove, which may be partially filled with tissue that is mildly toxic. Many dogfish sharks (squaloids) and all bullhead sharks (heterodontoids) have fin spines, but they are absent in all other sharks and in all living rays.

The anal fin is longitudinally based on the midventral surface of the precaudal tail and is often similar in shape to the dorsal fins (except for lacking a spine). It is present in most living sharks but is absent in the squaloids, squatinoids, pristiophoroids, and in all rays.

The precaudal tail is stout and muscular in sharks and in many rays and, in conjunction with the trunk (sharks and sharklike rays) and caudal fin, serves as their primary organ of locomotion. In skates and stingrays the tail and caudal fin are more or less reduced, commensurate with expansion of the pectoral disk and incorporation of the trunk within it, and may serve other roles such as defense against predators. Many skates have precaudal tails studded with sharp thorns and enclosing small electric organs. Live skates may raise their spiny tails vertically when disturbed and may use them for self-defense. Stingrays have barbed stings with venomous sheaths on the dorsomedial surface of the precaudal tail and can drive the spines into other animals that harass or disturb them. The precaudal tail may have lateral ridges or basal keels that stiffen and streamline it laterally, giving it a lenticular cross section. Precaudal pits are deep transverse, crescentic, or longitudinal grooves of uncertain function at the extreme posterior end of the precaudal tail at the origin of the caudal fin of various active, fast-swimming sharks; they may be related to dorsoventral flexing of the caudal fin or to lateral tail hydrodynamics.

The caudal fin varies tremendously in shape and size among sharks and rays, from crescentic in fast-swimming sharks to elongated and ribbonlike in some rays. In the thresher sharks (family Alopiidae) and zebra shark (*Stegostoma fasciatum*) the caudal fin is about as long as the rest of the shark. In threshers the caudal functions as an accessory food-collecting organ and is used as a whip or flail to herd and strike small, schooling prey. The caudal fin of living elasmobranchs is expanded dorsally, ventrally, and partly behind the posterior end of the vertebral column, which extends nearly to the caudal tip. The dorsal fold of the caudal is the epaxial lobe (above the vertebral column or caudal axis), the ventral fold is the hypaxial lobe (below the caudal axis), and the rear, partly postvertebral part the terminal lobe. The terminal lobe is defined anteriorly by the ventral subterminal notch in most sharks. This is absent in almost all rays except the sharklike knifetooth sawfish (*Anoxypristis*) and may have separate ventral subterminal and terminal margins. The epaxial lobe varies in height in different elasmobranchs, it tends to be reduced in those elasmobranchs with the caudal vertebral axis tilted posterodorsally above the line of the vertebral axis of the trunk and precaudal tail (heterocercal caudal fin). In sharks and rays with strongly heterocercal caudals, the epaxial lobe has a simple dorsal margin extending from the anterior end of the caudal to the tip of the terminal lobe. The hypaxial lobe in such elasmobranchs is more or less differentiated into an anterior preventral margin, a postventral margin (between the preventral margin and the subterminal notch when the notch is present or between the preventral margin and tail tip when the notch is absent), and a variably developed ventral lobe delimited by the preventral and postventral margins. In many sharks the postventral margin is divided into upper and lower sections by a posterior notch or fork. Some living elasmobranchs have the caudal vertebral axis horizontal, in line with the precaudal vertebral axis, and have the epaxial and hypaxial lobes more or less symmetrical and the epaxial lobe well developed (diphycercal caudal fin). In the angel sharks (*Squatina*), the caudal vertebral axis is ventroposteriorly tilted relative to the precaudal vertebral axis (hypocercal caudal fin), and the epaxial lobe is much larger than the hypaxial lobe. In these sharks the epaxial lobe is differentiated into predorsal and postdorsal margins, a dorsal lobe analogous to the ventral lobe, and a weak dorsal subterminal notch, while the hypaxial lobe is simple; the angel shark caudal fin thus resembles an upside-down heterocercal caudal fin from an ordinary shark.

Systematics of elasmobranchs

It has been noted that there are at least as many classifications (and lately, phyletic or cladistic analyses) of cartilaginous fishes as authors who have published them (Compagno 1973).

With the "cladistic revolution" continuing apace and with much renewed interest in the evolution and interrelationships of living elasmobranchs and other cartilaginous fishes, the higher classification of chondrichthyans is currently in a state of flux. Also, fisheries exploration over the past few decades has revealed hitherto unknown diversity of elasmobranchs in sketchily known areas and habitats, particularly the Indo-Australian archipelago and the continental and insular slopes of many areas. This has resulted in the discovery of many new species of elasmobranchs but relatively few new higher taxa. There is considerable agreement with the generic and specific composition of familial or lower taxa in many higher groups of sharks and rays. There is considerable disagreement, however, about the scope and definition of higher, ordinal-level groups of living neoselachians, and there is little agreement on their interrelationships. Chondrichthyan systematics, classification, and phyletics are undergoing a period of "interesting times" for systematists, which may not help the nonsystematist with an interest in elasmobranchs.

Most current workers in chondrichthyan systematics agree that living chondrichthyans are a monophyletic group. The concept of the neat and broad separation of the living neoselachians and holocephalians being exemplary of a long-standing and traditional elasmobranch-holocephalan dichotomy is confounded in part by continuing discoveries and the refining of knowledge of Paleozoic cartilaginous fishes that are difficult to place in the traditional scheme. These problems call into question the validity of including all sharklike fishes in a common "elasmobranch" taxon. It has been suggested that some early palaeoselachian ("cladodont") sharks, such as the famous Devonian shark genus *Cladoselache* and the symmoriioids may not form a monophyletic group with all other elasmobranchs in apposition to the chimaeroids but may be sister groups of the euselachians plus chimaeroids (Maisey 1984, 1986, 1989). This is not accepted by all or even by most workers, however. Some early sharklike and batoidlike fishes, including the edestoids and petalodonts, may not be elasmobranchs as argued by Zangerl (1973, 1981) but instead may be part of a chimaeroid sister group of "paraselachians" (Lund 1985, 1986, 1989, 1990). Although the euselachians, including the aberrant xenacanths (Young 1982; Gaudin 1991), and the neoselachians may be a monophyletic group and the neoselachians are by general consensus a monophyletic group (Cappetta 1987; Compagno 1977, 1988; Gaudin 1991) derived from the protoselachian euselachians, it is debatable which group of protoselachians is closest to the Neoselachii, with authors split on derived Mesozoic hybodonts, on all hybodonts, and on terminal "ctenacanths" (Schaeffer and Williams 1977; Maisey 1975, 1984, 1986, 1989; Compagno 1977, 1988; Gaudin 1991) as possible sister groups of neoselachians.

The interrelationships of living neoselachians and scope, rank, and composition of ordinal-level groups within the Neoselachii are subject to a number of recent interpretations (Maisey 1984, 1986, 1989; Shirai 1992, 1996; Nishida 1990; Compagno 1973, 1977, 1984, 1988; de Carvalho 1996; McEachran et al. 1996) that cannot be discussed at length here for lack of space, but which are strongly divergent. The following high-level phenetic classification of Chondrichthyes is based on Compagno (1973, 1990b) and is used here as an interim treatment and a framework for the lower-level diversity of living elasmobranchs considered below, pending resolution of some of the problems.

>   Class Chondrichthyes: Cartilaginous fishes.
>> Subclass Elasmobranchii: Sharklike fishes.
>>> Cohort Euselachii: Ctenacanth, hybodont, xenacanth, and modern sharks.
>>>> Subcohort Neoselachii: Modern sharks and rays.
>>>>> Living sharks:
>>>>>> Superorder Squalomorphii: Squalomorph sharks.
>>>>>>> Order Hexanchiformes: Cow and frilled sharks.
>>>>>>> Order Squaliformes: Dogfish sharks.
>>>>>>> Order Pristiophoriformes: Saw sharks.
>>>>>> Superorder Squatinomorphii: Angel sharks.
>>>>>>> Order Squatiniformes: Angel sharks.
>>>>>> Superorder Galeomorphii: Galeomorph sharks.
>>>>>>> Order Heterodontiformes: Bullhead sharks.
>>>>>>> Order Orectolobiformes: Carpet sharks.
>>>>>>> Order Lamniformes: Mackerel sharks.
>>>>>>> Order Carcharhiniformes: Ground sharks.
>>>>> Living rays:
>>>>>> Superorder Rajomorphii (Batoidea): Rays.
>>>>>>> Order Pristiformes: Sawfish.
>>>>>>> Order Rhinobatiformes: Guitarfishes.
>>>>>>> Order Torpediniformes: Electric rays.
>>>>>>> Order Rajiformes: Skates.
>>>>>>> Order Myliobatiformes: Stingrays.
>> Subclass Holocephalii (Subterbranchialia): Chimaeroids.

Problem areas in neoselachian classification and phylogeny include:

1. The validity of the shark group Squalomorphii as a monophyletic group, with some authors including the Squatinomorphii and Rajomorphii, or Batoidea, within the squalomorph assemblage.
2. The position of the Hexanchiformes relative to other sharks and the inclusion of the frilled shark (*Chlamydoselachus*) with the cow sharks (Hexanchidae).
3. The validity of Squaliformes as a monophyletic taxon. Shirai (1992, 1996) divides it into four different orders (Echinorhiniformes, Dalatiiformes, Centrophoriformes, and Squaliformes) for several clades. De Carvalho (1996) recognizes the Squaliformes and Echinorhiniformes.

4. The position of the Squatiniformes within the Neoselachii.
5. The position and ranking of the Rajomorphii or batoids. The saw sharks group (Pristiophoriformes) is the likely sister group of Rajomorphii, but relationships of Squatiniformes and squaloids to the batoids are less certain.
6. Within the Rajomorphii, the validity of the Rhinobatiformes as a monophyletic group. Nishida (1990) suggests that the group is paraphyletic and that Rhinidae is the primitive sister group of all living batoids except the pristoids.
7. The validity of the Galeomorphii, including the inclusion of Heterodontiformes within it.
8. Differing interpretations of the relationships of the four galeomorph orders Heterodontiformes, Orectolobiformes, Lamniformes, and Carcharhiniformes.

Diversity of living elasmobranchs

The remainder of this chapter is a review of the living orders and families of neoselachians, with definitions of the taxa. A checklist of the living, described species of elasmobranchs is presented as an appendix. A supraordinal classification is not defined here because of the various problems with existing schemes, including the one presented above. Instead, for surveying lower-level diversity, the ordinal groups of living elasmobranchs are divided into sharks and rays, with problems on the validity of certain orders including the Hexanchiformes, Squaliformes, and Rhinobatiformes to be taken up elsewhere.

Living sharks

The group of living sharks comprises neoselachians with the upper and lower eyelids intact; ceratohyals large and connected with hyomandibulae, not functionally replaced by pseudohyoids; palatoquadrates articulating with the neurocranium; neurocranium with an occipital half-centrum; no chondrified antorbital cartilages on the cranial nasal capsules; pectoral fins not fused to the sides of the head over the external gill openings; pectoral fins with expanded propterygia, these as large or larger than the mesopterygia; pectoral radials not highly jointed and branched distally, with radial musculature and radials not displacing ceratotrichia from the fin webs; suprascapulae of pectoral girdle separate from vertebral column and not articulating with it; and no cervical or cervicothoracic synarcual of fused precaudal vertebrae. There are eight orders of living sharks.

ORDER HEXANCHIFORMES:
COW AND FRILLED SHARKS

The hexanchiformes are sharks with a single spineless dorsal fin; an anal fin; preoral snout short to moderately long and conical or rounded, not greatly elongated, depressed, sawlike, and without rostral barbels; six or seven pairs of gill openings; spiracles reduced or absent; labial furrows, folds, and cartilages reduced or absent; mouth virtually terminal or subterminal, long, and extending behind the eyes; nostrils transverse or diagonal with lateral incurrent apertures and medial excurrent apertures; nostrils not connected to mouth by nasoral grooves; no circumnarial grooves or nasal barbels; eyes lateral, without subocular ridges; no nictitating eyelids; teeth without basal ledges; trunk cylindrical or elongated; caudal peduncle without lateral ridges; no predorsal medial line of enlarged denticles or thorns; pectoral fins small, without triangular lobes expanded over the gill openings; pelvic claspers without lateral spines; the caudal fin heterocercal; valvular intestine with conicospiral or spiral-ring valve; and ovoviviparous reproduction (aplacental viviparity). There are two families.

Family Chlamydoselachidae: Frilled sharks

The Chlamydoselachidae (Fig. 1-2) are sharks with elongated, somewhat compressed bodies, terminal mouths on a snakelike head, and tricuspidate teeth in both jaws. The family comprises one genus and one or more species.

Family Hexanchidae: Cow sharks

The Hexanchidae (Fig. 1-2) are sharks with stout to slender, cylindrical bodies, subterminal

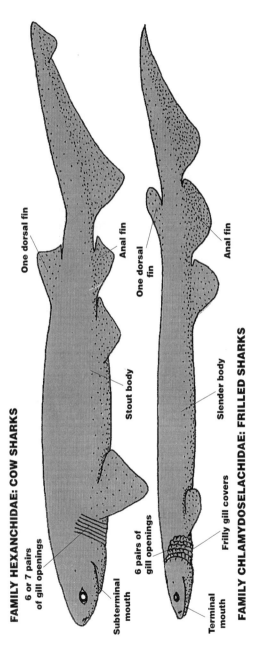

FIGURE 1-2 Order Hexanchiformes: cow and frilled sharks.

mouths below flattened conical or rounded snouts, enlarged sawlike teeth in the lower jaw, and small cuspidate grasping teeth in the upper jaw. There are three genera and four species.

ORDER SQUALIFORMES:
DOGFISH SHARKS

The order Squaliformes comprises sharks with two spined dorsal fins, the spines secondarily lost in some Somniosidae and almost all Dalatiidae (except on the first dorsal of *Squaliolus*); no anal fin; preoral snout short to moderately long and conical or rounded, not greatly elongated, depressed, sawlike, and without rostral barbels; five pairs of gill openings; spiracles small to large, with valve present or absent; labial furrows, folds, and cartilages prominent; mouth usually subterminal but secondarily terminal in *Trigonognathus*, long to short but extending below or behind the eyes; nostrils transverse or diagonal with lateral incurrent apertures and medial excurrent apertures; nostrils not connected to mouth by nasoral grooves; no circumnarial grooves or nasal barbels; eyes lateral, without subocular ridges; no nictitating eyelids; teeth without basal ledges; trunk cylindrical, slightly depressed, or somewhat compressed; caudal peduncle with or without lateral ridges; no predorsal medial line of enlarged denticles or thorns; pectoral fins small, without triangular lobes expanded over the gill openings; pelvic claspers usually with lateral spines; the caudal fin heterocercal to diphycercal; valvular intestine with conicospiral valve; and ovoviviparous reproduction (aplacental viviparity). The classification of squaloids is currently in flux, with the validity of Squaliformes as a monophyletic taxon in question (Shirai, 1992, 1996, whose familial arrangement of squaloids is mostly followed here).

Family Echinorhinidae: Bramble sharks
The bramble sharks (Fig. 1-3) are characterized by a broad and flat head; flat snout broadly rounded in dorsoventral view; nostrils with simple anterior nasal flap; nostrils wide-spaced, internarial width much greater than nostril width; mouth broadly arched, elongated, lips not papillose; labial furrows short, confined to mouth corners and falling well behind level of eyes, not encircling mouth and not elongated posteriorly into postoral grooves; teeth similar in both jaws, these compressed, low-crowned, broad, and bladelike, and with an oblique cusp and one or more cusplets on either side of the cusp in adults; adjacent teeth not imbricated; tooth rows 21–28/20–29, uppers about as numerous as lowers; spiracles very small, far behind eyes; fifth gill opening much larger than first four; trunk broad and cylindrical, abdomen with weak ridges; caudal peduncle compressed and very short, without lateral keels; no photophores (luminous organs); denticles large, sessile, with flat base and median spine, enlarged to form flat plates in *Echinorhinus brucus*; pectoral fins low, broadly rounded to angular, with basal cartilages fused to form a single basipterygium, but with a long metapterygial axis attached to it; pelvic fins larger than pectoral or dorsal fins, separated from caudal fin by a narrow space; claspers with a lateral spine only; dorsal fins small, equal-sized, spineless, and rounded-angular, with convex posterior margins; first dorsal fin far posterior, base over the pelvic bases; and second dorsal base partially over pelvic bases; interdorsal space shorter than dorsal bases; and caudal fin heterocercal, without terminal notch or ventral lobe. Adults range from 1 to 4 m or more in length. There are a single genus and two species.

Family Squalidae: Dogfish sharks
The Squalidae (Fig. 1-3) are sharks with a head moderately broad and somewhat flattened; snout flat and broadly rounded, narrowly rounded, or angular in dorsoventral view; nostrils with medial process on anterior nasal flap, in some species more or less expanded into a barbel; nostrils wide-spaced, internarial width greater than nostril width; mouth nearly transverse and short, lips not papillose; labial furrows short, confined to mouth corners but under posterior corner of eyes, not encircling mouth but elongated posteriorly into postoral grooves; teeth similar in both jaws but somewhat larger in lower jaws, teeth of both jaws slightly compressed, low-crowned, bladelike, and with an oblique cusp and a dis-

tal blade; adjacent teeth imbricated; tooth rows 24–29/20–26, uppers slightly more numerous than lowers; spiracles large, close behind eyes; fifth gill opening not enlarged compared to first four; trunk stout to slender and cylindrical or somewhat compressed, abdomen without ridges; caudal peduncle cylindrical and elongated, with lateral keels and precaudal pits present or absent; no photophores; denticles moderate-sized, pedicellate, with flattened, narrow to broad keeled crowns, slender pedicels, and low bases; pectoral fins high, angular, and falcate, free rear tips not elongated; pectoral basal cartilages separate, mesopterygium not fused to its proximal radials, metapterygial axis moderately long;

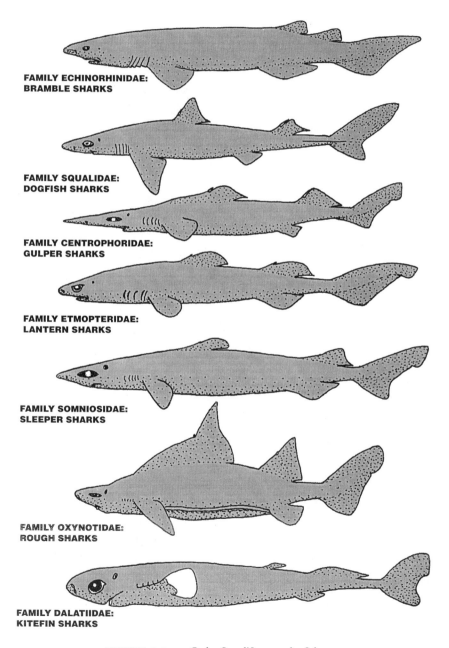

FIGURE 1-3 Order Squaliformes: dogfishes.

pelvic fins smaller than pectoral or first dorsal fin, but subequal to or smaller than second dorsal fin; pelvic-caudal space very long, several times pelvic bases; claspers with a lateral spine and a medial hooklike process; dorsal fins large, falcate-angular, with strongly concave posterior margins and with stout ungrooved spines; dorsal fins subequal (*Cirrhigaleus*) or first dorsal much larger than second (*Squalus*); first dorsal fin over pectoral-pelvic space and well anterior to pelvics, closer to pectoral bases; second dorsal base behind pelvic bases; interdorsal space much longer than dorsal bases; and caudal fin heterocercal, without terminal notch but with a strong ventral lobe. Adults range from less than 1 m to less than 2 m long. There are 2 genera and 11 valid species.

Family Centrophoridae: Gulper sharks
The Centrophoridae (Fig. 1-3) are sharks with a moderately broad to narrow and somewhat flattened head; snout flat and narrowly rounded to elongate-rounded in dorsoventral view; nostrils with simple anterior nasal flaps; nostrils wide-spaced, internarial width greater than nostril width; mouth nearly transverse and short, lips not papillose; labial furrows short to moderately long, confined to mouth corners but under posterior corner of eyes, not encircling mouth but elongated posteriorly into postoral grooves and (in *Deania*) anteromedial preoral grooves; teeth dissimilar in upper and lower jaws, much larger in lower jaws, teeth of both jaws slightly compressed, high-crowned, and bladelike; uppers narrower than lowers, with a narrow, erect to oblique, broad-based cusp; lower teeth with an oblique to semierect cusp and a distal blade; adjacent teeth imbricated; tooth rows 26–42/26–35, with uppers slightly more numerous than lowers; spiracles large, close behind eyes; fifth gill opening not enlarged compared to first four; trunk stout to slender and cylindrical or slightly compressed, abdomen without ridges; caudal peduncle slightly compressed, short to moderately elongated, and without lateral keels or precaudal pits; no photophores; denticles moderate-sized, sessile or pedicellate, the latter with flattened, narrow to broad keeled or forklike crowns, elongated pedicels and low bases; pectoral fins low, angular, and not falcate, free rear tips short to greatly elongated; pectoral basal cartilages with separate propterygium but fused mesopterygium and metapterygium, the mesopterygial radials not fused to the mesometapterygium, metapterygial axis moderate to greatly elongated; pelvic fins smaller than pectoral or first dorsal fin, but subequal to or smaller than second dorsal fin; pelvic-caudal space moderately long, about twice pelvic bases; claspers with a lateral spine; dorsal fins large, angular, broad, not falcate, and with strong grooved spines, both with straight to weakly concave posterior margins; first dorsal usually larger than second, though occasionally subequal in size; first dorsal fin over pectoral-pelvic space and well anterior to pelvics, closer to pectoral bases; second dorsal base partly over or just behind pelvic bases; interdorsal space longer than or subequal to dorsal bases; and caudal fin heterocercal, with strong terminal notch and a weak to moderate ventral lobe. Adults range from less than 1 m to less than 2 m long. There are 2 genera and approximately 14 species.

Family Etmopteridae: Lantern sharks
The Etmopteridae (Fig. 1-3) are characterized by a head moderately broad to narrow and somewhat flattened or cylindrical; snout flat to conical and narrowly to broadly rounded, undulated, or distally truncated in dorsoventral view; nostrils with simple anterior nasal flaps; nostrils wide-spaced, internarial width greater than or subequal to nostril width; mouth arched and long to transverse and short, lips not papillose; labial furrows rudimentary to moderately elongated, confined to mouth corners and under or posterior to eyes, not encircling mouth, variably elongated posteriorly into postoral grooves or not; teeth either similar in upper and lower jaws or not, when dissimilar larger in lower jaw than upper; some species with nonimbricate cuspidate upper and lower teeth that have or lack lateral cusplets; lower teeth enlarged and bladelike in *Etmopterus* and to a lesser degree in *Miroscyllium*; adjacent teeth nonimbricated, or lowers only imbricated; tooth rows 15–68/15–68; spiracles moderate to large, close behind eyes; fifth gill

opening not enlarged compared to first four but gill openings may increase in size from front to back; trunk stout to slender and cylindrical, abdomen without ridges; caudal peduncle cylindrical or slightly compressed, short to moderately elongated, and without lateral keels or precaudal pits; photophores (luminous organs) present on the lower surfaces of many species; denticles moderate-sized, sessile or erect, the latter with pedicellate, spikelike hooked crowns on low bases; pectoral fins low, rounded-angular, and not falcate, free rear tips not elongated; pectoral basal cartilages free, but with mesopterygial radials fused to the mesopterygium, metapterygial axis greatly elongated; pelvic fins subequal to or larger than pectoral fins and subequal to or larger than first dorsal fin, but subequal to or smaller than second dorsal fin; pelvic-caudal space elongated or short, from over twice to about equal to pelvic bases; claspers with medial and lateral spines; dorsal fins small, rounded-angular, broad, not falcate, and with strong to small grooved spines, both fins with straight to weakly concave posterior margins (except some *Etmopterus* with deeply concave second dorsal posterior margins); first dorsal usually smaller than or subequal to second; first dorsal fin over pectoral-pelvic space and well anterior to pelvics, closer to pectoral bases; second dorsal base partly over or just behind pelvic bases; interdorsal space usually longer than dorsal bases; and caudal fin heterocercal, with strong to weak terminal notch and a weak to moderate ventral lobe. Adults are less than 1 m long, with many small or dwarf species adult at less than 0.5 m. There are five genera and over 45 species.

Family Somniosidae: Sleeper sharks
The Somniosidae (Fig. 1-3) have a head moderately broad and somewhat flattened or conical; snout flat and narrowly rounded to elongate-rounded in dorsoventral view; nostrils with simple anterior nasal flaps; nostrils widespaced, internarial width greater than nostril width; mouth nearly transverse and short to broadly arcuate and moderately long, lips not papillose; labial furrows short to moderately long, confined to mouth corners or partly encircling mouth, under posterior corner of eyes, elongated posteriorly into postoral grooves and (in *Centroscymnus crepidater*) anteromedial preoral grooves; teeth dissimilar in upper and lower jaws, much larger in lower jaw, upper teeth small, with narrow cusps and no cusplets or blades; lower teeth broad, bladelike, and with an erect to oblique cusp and a distal blade; adjacent teeth imbricated in lower jaw but not upper jaw; tooth rows 30–70/31–68, with uppers slightly more numerous than lowers or vice-versa; spiracles large, close behind eyes; fifth gill opening not enlarged compared to first four; trunk stout to slender and cylindrical or slightly compressed, abdomen with ridges in some species, uncertain in others; caudal peduncle slightly compressed, short to moderately elongated, and without lateral keels or precaudal pits (some *Somniosus* with a keel on the caudal base); no photophores; denticles moderate-sized and pedicellate, the crowns flattened, round, keeled or smooth, or narrow, thornlike, and hooked, connected by pedicels to low bases; pectoral fins low, rounded-angular, and not falcate, free rear tips not greatly elongated; pectoral basal cartilages with propterygium fused to mesopterygium but with metapterygium free, the mesopterygial radials not fused to the promesopterygium, metapterygial axis moderately elongated; pelvic fins subequal or larger than pectoral and dorsal fins; pelvic-caudal space usually short and between one and two times pelvic bases; claspers with a lateral spine; dorsal fins small or moderate-sized, rounded-angular, broad, not falcate, and with short, strong to weak grooved spines present or absent, both dorsals with convex to weakly concave posterior margins; first dorsal variably subequal, smaller, or larger than second; first dorsal fin over pectoral-pelvic space and well anterior to pelvics, usually closer to pectoral bases but sometimes closer to pelvics; second dorsal base partly over pelvic bases; interdorsal space longer than second dorsal base; and caudal fin heterocercal, with strong terminal notch and a weak to strong ventral lobe. These sharks are moderate-sized to huge, with adults ranging from less than 1 m to over 5 m long. There are 4 or 5 genera and 15 or more species.

Family Oxynotidae: Rough sharks
The Oxynotidae (Fig. 1-3) are characterized by a head moderately broad and somewhat flattened; snout flat, thick, and bluntly rounded or rounded-angular in dorsoventral view; nostrils with broad anterior nasal flaps; nostrils close-spaced, internarial width less than nostril width; mouth nearly transverse and short, lips papillose; labial furrows elongated, under posterior corner of eyes, and encircling mouth, elongated posteriorly into postoral grooves; teeth dissimilar in upper and lower jaws, much larger in lower jaw, upper teeth small, with narrow cusps and no cusplets or blades; lower teeth very broad, bladelike, and with an erect to oblique cusp and a distal blade; adjacent teeth imbricated in lower jaw but not upper jaw; tooth rows less than 12/12; spiracles large, close behind eyes; fifth gill opening not enlarged compared to first four; trunk stout, elevated, and strongly compressed, with a triangular cross-section, abdomen with prominent lateral ridges; caudal peduncle slightly compressed, short, and without lateral keels or precaudal pits; no photophores; denticles large and pedicellate, the crowns flattened, keeled, and cuspidate, connected by pedicels to low bases and forming a rough surface; pectoral fins high, distally lanceolate or leaf-shaped, but not falcate, free rear tips not elongated; pectoral basal cartilages with propterygium fused to mesopterygium but with metapterygium free, the mesopterygial radials not fused to the promesopterygium, metapterygial axis moderately elongated; pelvic fins smaller than the pectoral and dorsal fins; pelvic–caudal space short and subequal to pelvic bases; claspers with a lateral spine; dorsal fins very large, broad-based, high, and triangular or falcate, with large conical spines mostly buried in the fins, both dorsals with straight to (usually) deeply concave posterior margins; first dorsal larger than second; first dorsal fin far anterior, over pectoral bases and anterior third of pectoral–pelvic space and far anterior to pelvics, far in front of pelvic bases; second dorsal base mostly over pelvic bases; interdorsal space longer or shorter than second dorsal base; and caudal fin heterocercal, with strong terminal notch and a weak ventral lobe. Adults range from less than 1 m to less than 2 m long. The family comprises a single genus and five or six species.

Family Dalatiidae: Kitefin sharks
The Dalatiidae (Fig. 1-3) have a head that is narrow and rounded-conical; snout conical and narrowly rounded to elongate-rounded in dorsoventral view; nostrils with simple anterior nasal flaps; nostrils fairly wide-spaced, internarial width greater than or subequal to nostril width; mouth nearly transverse and short, lips thin to fleshy, fringed and pleated in *Dalatias*; labial furrows short to moderately long, confined to mouth corners and under posterior corner of eyes, not or only partly encircling mouth, elongated posteriorly into postoral grooves and anteromedial preoral grooves; teeth dissimilar in upper and lower jaws, much larger in lower jaw, upper teeth small, with narrow cusps and no cusplets or blades; lower teeth broad, bladelike, and with an erect to oblique cusp and usually a distal blade; adjacent teeth imbricated in lower jaw but not upper jaw; tooth rows 16–37/17–34; spiracles large, close behind eyes; fifth gill opening usually not enlarged compared to first four (except in *Euprotomicroides*); trunk slender to stout and cylindrical or slightly compressed (*Euprotomicroides*), abdomen without ridges; caudal peduncle cylindrical, short to moderately elongated, and without or without lateral keels; photophores (luminous organs) present on the skin of some genera (*Isistius*), specialized luminous glands on the shoulders (*Mollisquama*) or in the cloaca (*Euprotomicroides*) of some genera; denticles mostly sessile but pedicellate in some genera, the latter with lanceolate crowns and short pedicels; pectoral fins rounded-angular or lanceolate, but not falcate, free rear tips not greatly elongated; pectoral basal cartilages fused into a single narrow basipterygium, with a short metapterygial axis, mesopterygial radials not fused to the basipterygium; pelvic fins subequal or smaller than pectoral fins, variably smaller to larger than dorsal fins; pelvic-

caudal space short to elongate and between one to over two times pelvic bases; claspers with or without a lateral spine; dorsal fins small or moderate-sized, rounded-angular, broad, not falcate, and without spines in most genera (a small spine present on first dorsal of *Squaliolus*), both dorsals with convex to weakly concave posterior margins; first dorsal usually smaller than second; first dorsal fin over pectoral-pelvic space and usually well anterior to pelvics (partly over pelvics in *Isistius*), varying from closer to pectoral bases to closer to pelvics; second dorsal base over or behind pelvic bases; interdorsal space longer or somewhat shorter than second dorsal base; and caudal fin heterocercal, with strong to weak terminal notch and a weak to strong ventral lobe. Adults range from less than 1 m to nearly 2 m long, but most adults are less than 0.5 m. There are seven genera and nine species.

ORDER PRISTIOPHORIFORMES: SAW SHARKS

The saw sharks (Fig. 1-4) are sharks with two spineless dorsal fins (vestigial spines present internally); no anal fin; preoral snout greatly elongated, depressed, sawlike, and with enlarged lateral saw teeth; snout with unique rostral barbels on the underside between tip and nostrils; five or six pairs of gill openings; spiracles large, with valve present; labial furrows, folds, and cartilages reduced; mouth subterminal, short but behind the eyes; nostrils longitudinal, with anterior incurrent apertures and posterior excurrent apertures, not connected to mouth by nasoral grooves; no circumnarial grooves or nasal barbels; eyes dorsolateral, above prominent subocular ridges; no nictitating eyelids; teeth with basal ledges; trunk cylindrical but head depressed; caudal peduncle with lateral ridges extending onto precaudal tail; no predorsal medial line of enlarged denticles or thorns; pectoral fins moderately large, without triangular lobes expanded over the gill openings; pelvic claspers with lateral spines; caudal fin barely heterocercal; valvular intestine with conicospiral valve; and ovoviviparous reproduction (aplacental viviparity). There is a single family.

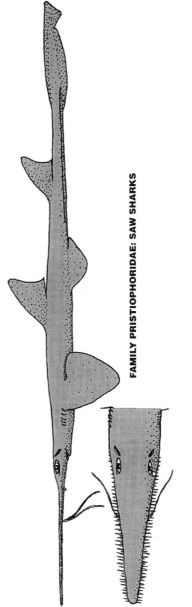

FIGURE 1-4 Order Pristiophoriformes: saw sharks.

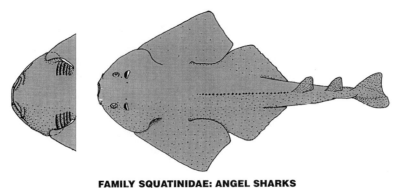

FAMILY SQUATINIDAE: ANGEL SHARKS

FIGURE 1-5   Order Squatiniformes: angel sharks.

Family Pristiophoridae: Sawsharks
The family Pristiophoridae consists of two living genera and five or more species.

ORDER SQUATINIFORMES:
ANGEL SHARKS
The Squatiniformes are raylike sharks with two spineless dorsal fins; no anal fin; preoral snout very short, truncate, and not sawlike, without rostral barbels; five pairs of gill openings; spiracles large but without valves; labial furrows, folds, and cartilages very large; mouth virtually terminal, moderately long and extending behind the eyes; nostrils diagonal, with anterolateral incurrent apertures and posteromedial excurrent apertures, not connected to mouth by nasoral grooves; circumnarial grooves present but weak nasal barbels also present; eyes dorsolateral, above prominent subocular ridges; no nictitating eyelids; teeth without basal ledges; trunk depressed, broad; caudal peduncle with lateral keels; a predorsal medial line of enlarged denticles or thorns present; pectoral fins large, with triangular lobes expanded over the gill openings; pelvic claspers with lateral spines; caudal fin hypocercal, with ventral lobe longer than the dorsal; intestine with conicospiral valve; and ovoviviparous reproduction (aplacental viviparity). The order contains a single family.

Family Squatinidae: Angel sharks
There are one living genus and at least 15 species in the family Squatindae (Fig. 1-5).

ORDER HETERODONTIFORMES:
BULLHEAD SHARKS
The Heterodontiformes (Fig. 1-6) are sharks with two spined dorsal fins; anal fin present; preoral snout very short, not sawlike and with-

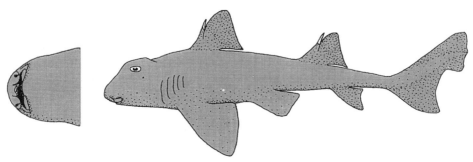

FAMILY HETERODONTIDAE: BULLHEAD SHARKS

FIGURE 1-6   Order Heterodontiformes: bullhead sharks.

out rostral barbels; five pairs of gill openings; spiracles small, without valve; labial furrows, folds, and cartilages large; mouth virtually terminal, short and ending in front of eyes; nostrils diagonal, with anterolateral incurrent apertures and posteromedial excurrent apertures; nostrils connected to mouth by prominent nasoral grooves; circumnarial grooves present but no nasal barbels; eyes dorsolateral, above prominent subocular ridges; no nictitating eyelids; teeth with basal ledges; trunk cylindrical; caudal peduncle without lateral ridges or keels; no predorsal medial line of enlarged denticles or thorns; pectoral fins moderately large, without triangular lobes expanded over the gill openings; pelvic claspers with lateral spines; the caudal fin heterocercal; valvular intestine with conicospiral valve; and oviparous reproduction, with unique screw-shaped egg cases. The order contains a single family.

Family Heterodontidae:
Bullhead or horn sharks
There are one genus and at least eight species.

ORDER LAMNIFORMES:
MACKEREL SHARKS
The Lamniformes are sharks with two spineless dorsal fins; anal fin present; preoral snout long and bladelike or conical to broadly rounded, not sawlike and without rostral barbels; five pairs of gill openings; spiracles small to absent, without valve; labial furrows, folds, and cartilages reduced or absent; mouth subterminal in most genera but terminal in *Megachasma*, long and extending behind eyes; nostrils diagonal, with anterolateral incurrent apertures and posteromedial excurrent apertures; no nasoral grooves, circumnarial grooves, or nasal barbels; eyes dorsolateral to lateral, with or without weak subocular ridges; no nictitating eyelids; teeth with basal ledges reduced or absent; trunk cylindrical; caudal peduncle with or without lateral keels; no predorsal medial line of enlarged denticles or thorns; pectoral fins small to large, without triangular lobes expanded over the gill openings; pelvic claspers with lateral spines; the cau-

dal fin heterocercal; valvular intestine with ring valve; and ovoviviparous reproduction, with uterine cannibalism (egg eating, or oophagy, in most genera and also predation on sibling fetuses, or adelphophagy, in *Carcharias taurus*). The order contains seven very well-defined families.

Family Mitsukurinidae: Goblin sharks
The goblin sharks (Fig. 1-7) are lamnoid sharks with elongated, flattened, bladelike snouts; small eyes; gill openings narrow, not expanded dorsally, fifth anterior to pectoral fin bases; internal gill openings without gill rakers; mouth large, subterminal; enlarged anterior teeth at front of mouth, less than 60 rows of teeth in either jaw; trunk compressed, slender, soft and flabby; dorsal fins small, equal-sized, low and rounded; pectoral fins small and broadly rounded, smaller than pelvic fins; anal fin larger than dorsal fins, broadly rounded; second dorsal and anal fins with broad bases; caudal fin weakly heterocercal, shorter than rest of shark, ventral lobe absent; and caudal peduncle compressed and without lateral keels or precaudal pits. There are a single genus and one species.

Family Odontaspididae: Sand tiger sharks
The Odontaspididae (Fig. 1-7) are lamnoid sharks with short or moderately long, conical or slightly depressed snouts; small to large eyes; gill openings narrow, not expanded dorsally, fifth anterior to pectoral fin bases; internal gill openings without gill rakers; mouth large, subterminal; enlarged anterior teeth at front of mouth, less than 60 rows of teeth in either jaw; trunk compressed-cylindrical, stout, and firm; dorsal fins large, equal-sized or first larger than second, angular; pectoral fins moderately large and angular, larger than pelvic fins; anal fin as large as second dorsal fin or smaller; second dorsal and anal fins with broad bases; caudal fin heterocercal, shorter than rest of shark, ventral lobe short; and caudal peduncle compressed and without lateral keels, but with an upper precaudal pit. There are two genera and three or four species.

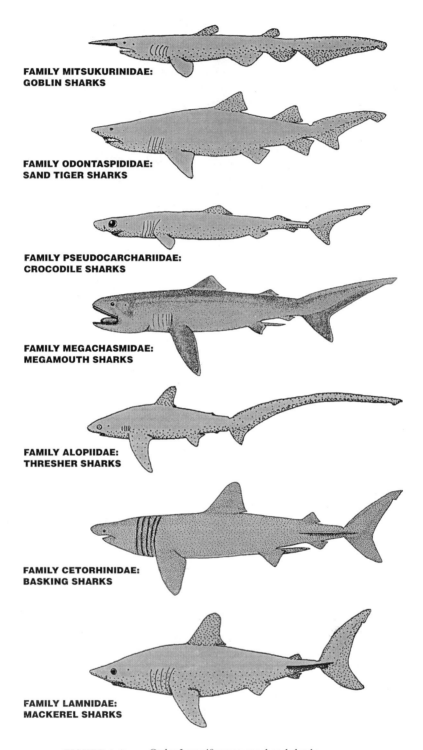

FIGURE 1-7   Order Lamniformes: mackerel sharks.

Family Pseudocarchariidae: Crocodile sharks
The Pseudocarchariidae (Fig. 1-7) are lamnoid sharks with moderately long, conical snouts; very large eyes; wide gill openings, expanded dorsally, with fifth anterior to pectoral fin bases; internal gill openings without gill rakers; mouth large, subterminal; enlarged anterior teeth at front of mouth, less than 40 rows of teeth in either jaw; trunk cylindrical and slender; dorsal fins small, first larger than second, angular; pectoral fins small and angular, larger than pelvic fins; anal smaller than second dorsal fin; second dorsal broad-based but anal fin with narrow, pivoting base; caudal fin heterocercal, shorter than rest of shark, ventral lobe moderately long; and caudal peduncle slightly depressed and with low lateral keels and both upper and lower precaudal pits. The family contains a single genus and species.

Family Megachasmidae: Megamouth sharks
The Megachasmidae (Fig 1-7) are lamnoid sharks with short, broadly rounded, slightly depressed snouts; moderately large eyes; gill openings narrow, not expanded dorsally, fifth over pectoral fin bases; internal gill openings with papillose, denticle-covered gill rakers; mouth large, terminal; no enlarged anterior teeth at front of mouth, over 100 rows of teeth in either jaw; trunk compressed-cylindrical, stout, and flabby; dorsal fins moderately large, first larger than second, angular; pectoral fins large and angular, much larger than pelvic fins; anal fin smaller than second dorsal fin; second dorsal and anal fins with broad bases; caudal fin heterocercal, shorter than rest of shark, ventral lobe long; and caudal peduncle compressed and without lateral keels, but with both upper and lower precaudal pits. There are a single genus and one species.

Family Alopiidae: Thresher sharks
The Alopiidae (Fig. 1-7) are lamnoid sharks with short or moderately long, conical snouts; large to very large eyes; gill openings narrow, not expanded dorsally, fifth over pectoral fin bases; internal gill openings without gill rakers; mouth small, subterminal; moderately large anterior teeth at front of mouth, less than 60 rows of teeth in either jaw; trunk cylindrical, moderately stout, and firm; first dorsal fin very large, second greatly reduced, both angular; pectoral fins very large and angular, larger than pelvic fins; anal fin slightly larger than second dorsal fin, both with narrow, pivotable bases; caudal fin heterocercal, about as long as rest of shark, ventral lobe moderately long; and caudal peduncle compressed and without lateral keels, but with upper and lower precaudal pits present. There are a single genus and three or four species.

Family Cetorhinidae: Basking sharks
The Cetorhinidae (Fig. 1-7) are lamnoid sharks with moderately long, conical or ventrally hooked snouts; small eyes; gill openings very wide, expanded dorsally and ventrally, five anterior to pectoral fin bases; internal gill openings with gill rakers formed from modified dermal denticles, not papillose; mouth large, subterminal; no enlarged anterior teeth at front of mouth, over 200 rows of teeth in either jaw; trunk spindle-shaped, stout, and firm; first dorsal fin large, second reduced, both angular; pectoral fins large and angular, larger than pelvic fins; anal fin subequal to second dorsal fin, both fins with broad bases; caudal fin heterocercal, shorter than rest of shark, and crescent-shaped, with a long ventral lobe; and caudal peduncle depressed, with strong lateral keels and upper and lower precaudal pits. There are a single genus and one species.

Family Lamnidae: Mackerel sharks
The Lamnidae (Fig. 1-7) are lamnoid sharks with moderately long, conical snouts; small eyes; gill openings wide, expanded dorsally on head, fifth anterior to pectoral fin bases; internal gill openings without gill rakers; mouth large, subterminal; enlarged anterior teeth at front of mouth, less than 40 rows of teeth in either jaw; trunk spindle-shaped, stout, and firm; first dorsal fin large, second greatly reduced, both angular; pectoral fins large and angular, larger than pelvic fins; anal fin slightly larger than second dorsal fin, both fins with narrow, pivotable bases; caudal fin het-

erocercal, shorter than rest of shark, and crescent-shaped, with a long ventral lobe; and caudal peduncle depressed, with strong lateral keels and upper and lower precaudal pits. There are three genera and five species.

ORDER ORECTOLOBIFORMES: CARPET SHARKS

The order Orectolobiformes comprises sharks with two spineless dorsal fins; anal fin present; preoral snout terminal or subterminal, short to truncated, not sawlike, and without rostral barbels; five pairs of gill openings; spiracles small to very large, without valve; labial furrows, folds, and cartilages large; mouth short and ending in front of eyes; nostrils diagonal, with anterolateral incurrent apertures and posteromedial excurrent apertures; nostrils connected to mouth by prominent nasoral grooves; circumnarial grooves present or absent; nasal barbels present; eyes dorsolateral to lateral, with or without subocular ridges; no true nictitating eyelids, but some genera with shallow subocular pouches below eyes; teeth with basal ledges, apparently reduced in *Rhincodon*; trunk cylindrical to greatly depressed; caudal peduncle with or without lateral ridges; no predorsal medial line of enlarged denticles or thorns; pectoral fins small to large, without triangular lobes expanded over the gill openings; pelvic claspers with or without lateral spines; the caudal fin heterocercal or diphycercal; valvular intestine with conicospiral or ring valve; and ovoviviparous or oviparous reproduction, but oviparous species do not have screw-shaped egg cases. There are seven families.

Family Parascylliidae: Collared carpet sharks
The Parascylliidae (Fig. 1-8) are carpet sharks with narrow, flattened heads, without lateral skin folds; snout broadly rounded to slightly pointed; eyes dorsolateral on head, with subocular pockets and ridges; spiracles minute; gill openings small, fifth overlaps fourth; internal gill openings without filter screens; nostrils with short, pointed barbels and circumnarial grooves; mouth small, subterminal, arcuate, and without symphysial groove on chin; teeth small, not enlarged and fanglike, less than 60 rows in either jaw; trunk cylindrical or slightly depressed, without ridges on sides; precaudal tail longer than body; dorsal fins equal-sized, first dorsal behind pelvic bases; pectoral fins small, subequal to pelvic fins; anal fin angular, smaller than second dorsal and well separated from caudal base; caudal fin weakly heterocercal, much shorter than rest of shark, with strong subterminal notch but no ventral lobe; caudal peduncle without lateral keels or precaudal pits; valvular intestine with conicospiral valve; and color pattern with spots and saddles. Adults are less than 1 m in total length. There are two genera and at least six species.

Family Brachaeluridae: Blind sharks
The Brachaeluridae (Fig. 1-8) are carpet sharks with broad, flattened heads, without lateral skin folds; snout broadly rounded; eyes dorsolateral on head, with subocular pockets and ridges; spiracles large; gill openings small, fifth not overlapping fourth; internal gill openings without filter screens; nostrils with long, pointed barbels and circumnarial grooves; mouth small, subterminal, transverse, and with a symphysial groove on chin; teeth small, not enlarged and fanglike, less than 40 rows in either jaw; trunk cylindrical or slightly depressed, without ridges on sides; precaudal tail shorter than body; dorsal fins equal-sized or first slightly larger than second, first dorsal origin over pelvic bases; pectoral fins small, subequal to pelvic fins; anal fin angular, smaller than second dorsal and separated by a narrow notch or short space from caudal base; caudal fin weakly heterocercal, much shorter than rest of shark, with strong subterminal notch but no ventral lobe; caudal peduncle without lateral keels or precaudal pits; valvular intestine with spiral ring valve; and color pattern with spots and saddles or plain. Adults less than 1.3 m in total length. There are two genera (sometimes synonymized) and two species.

Family Orectolobidae: Wobbegongs
The Orectolobidae (Fig. 1-8) includes carpet sharks with very broad, flattened heads, with

few to numerous lateral skin folds; snout truncated; eyes dorsolateral on head, with subocular pockets and ridges; spiracles very large; gill openings small, fifth does not overlap fourth; internal gill openings without filter screens; nostrils with long, variably branched barbels and circumnarial grooves; mouth large, nearly terminal, subarcuate, and with a prominent symphysial groove on chin; teeth enlarged and fanglike at symphysis but small at ends of dental bands, less than 30 rows in either jaw; trunk depressed, without ridges on sides; precaudal tail shorter than body; dorsal fins equal-sized, first dorsal with origin over

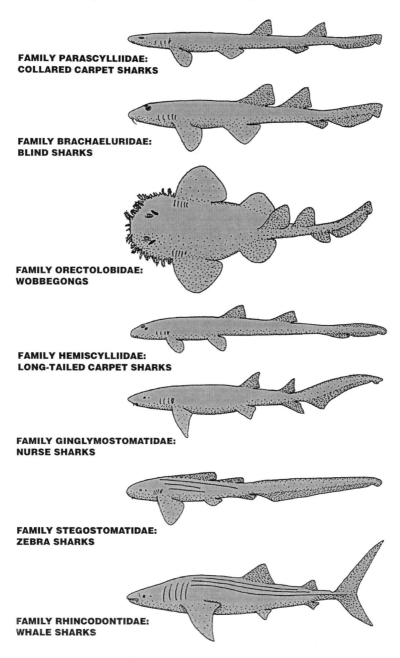

FIGURE 1-8   Order Orectolobiformes: carpet sharks.

pelvic bases; pectoral fins large, subequal to or larger than pelvic fins; anal fin subangular, smaller than second dorsal and separated by a narrow notch from caudal base; caudal fin weakly heterocercal, much shorter than rest of shark, with strong subterminal notch but no ventral lobe; caudal peduncle without lateral keels or precaudal pits; valvular intestine with ring valve; and color pattern with spots, blotches, and saddles. Adults are between 1 and 4 m in total length. There are three genera (sometimes synonymized) and six or more species.

Family Hemiscylliidae:
Long-tailed carpet sharks
The Hemiscylliidae (Fig. 1-8) are carpet sharks with narrow, cylindrical or flattened heads, without lateral skin folds; snout broadly rounded to slightly pointed; eyes dorsolateral on head, with subocular ridges but without subocular pockets; spiracles large; gill openings small, fifth sometimes overlaps fourth; internal gill openings without filter screens; nostrils with short to elongated, pointed barbels and circumnarial grooves; mouth small, subterminal, transverse, and without symphysial groove on chin; teeth small, not enlarged and fanglike, less than 40 rows in either jaw; trunk cylindrical or slightly depressed, with or without ridges on sides; precaudal tail longer than body; dorsal fins equal-sized, first dorsal origin over or just behind pelvic bases; pectoral fins small, subequal to or slightly larger than pelvic fins; anal fin broadly arcuate, subequal to or smaller than second dorsal and separated from caudal base by a shallow notch; caudal fin weakly heterocercal, much shorter than rest of shark, with strong subterminal notch but no ventral lobe; caudal peduncle without lateral keels or precaudal pits; valvular intestine with ring valve; and color pattern with spots, blotches, and saddles. Adults are usually less than 1 m in total length. There are two genera and at least 12 species.

Family Ginglymostomatidae: Nurse sharks
The Ginglymostomatidae (Fig. 1-8) are carpet sharks with broad flattened heads, without lateral skin folds; snout broadly rounded; eyes dorsolateral or lateral on head, with or without subocular ridges but with no subocular pockets; spiracles small; gill openings small, fifth overlaps fourth; internal gill openings without filter screens; nostrils with short to elongated, pointed barbels but no circumnarial grooves; mouth small, subterminal, transverse, and without symphysial groove on chin; teeth small, not enlarged and fanglike, less than 40 rows in either jaw; trunk cylindrical or moderately depressed, without ridges on sides; precaudal tail shorter than body; dorsal fins equal-sized or first slightly larger than second, first dorsal origin over or just anterior to pelvic bases; pectoral fins moderately large, larger than pelvic fins; anal fin subangular, subequal to second dorsal and separated from caudal base by a narrow space; caudal fin heterocercal, much shorter than rest of shark, with strong subterminal notch but ventral lobe very small or absent; caudal peduncle without lateral keels or precaudal pits; valvular intestine with ring valve; and color pattern plain, dark spots present in young of *Ginglymostoma*. Adults are between 2 and 4 m in *Nebrius* and *Ginglymostoma* but less than 1 m in total length in *Pseudoginglymostoma*. There are three genera, each with a single species.

Family Stegostomatidae: Zebra sharks
The Stegostomatidae (Fig. 1-8) are carpet sharks with broad, flattened heads, without lateral skin folds; snout broadly rounded; eyes lateral on head, without subocular ridges or pockets; spiracles large; gill openings small, fifth overlaps fourth; internal gill openings without filter screens; nostrils with short pointed barbels but without circumnarial grooves; mouth moderately large, subterminal, transverse, and without symphysial groove on chin; teeth small, not enlarged and fanglike, less than 40 rows in either jaw; trunk cylindrical, with ridges on sides; precaudal tail shorter than body; first dorsal fin much larger than second, with origin well anterior to pelvic bases; pectoral fins large, much larger than pelvic fins; anal fin angular, somewhat larger than second dorsal and separated from caudal base by a shallow notch or narrow space; caudal fin het-

erocercal, greatly elongated and about as long as rest of shark, with strong subterminal notch but no ventral lobe; caudal peduncle without lateral keels or precaudal pits; valvular intestine with ring valve; and color pattern with dark spots or dark saddles. Adults are less than 4 m in total length. There are a single genus and one species.

Family Rhincodontidae: Whale sharks
The Rhincodontidae (Fig. 1-8) are carpet sharks with broad, flattened heads, without lateral skin folds; snout truncated; eyes lateral on head, without subocular ridges or pockets; spiracles small; gill openings very large, fifth not overlapping fourth; internal gill openings with filter screens of crossbars supporting a spongy web of denticle-studded skin; nostrils with rudimentary barbels and no circumnarial grooves; mouth very broad, terminal, and transverse, without symphysial groove on chin; teeth minute, not fanglike, over 300 rows in either jaw in adults and subadults; trunk cylindrical or slightly depressed, with prominent ridges on sides; precaudal tail shorter than body; first dorsal fin much larger than second and with its origin in front of pelvic bases; pectoral fins very large, much larger than pelvic fins; anal fin angular, subequal to second dorsal and separated from caudal base by a broad space; caudal fin strongly heterocercal, shorter than rest of shark, with rudimentary subterminal notch (at tip) and strong ventral lobe; caudal peduncle with strong lateral keels and precaudal pits; valvular intestine with ring valve; and color pattern of light spots and lines on dark background. Adults are over 5 m in total length, making them the largest living fishlike vertebrates. There are a single genus and species.

ORDER CARCHARHINIFORMES: GROUND SHARKS
The Carcharhiniformes are sharks with two spineless dorsal fins (first dorsal lost in *Pentanchus*); anal fin present; preoral snout short to moderately long, not bladelike, sawlike, and without rostral barbels; five pairs of gill openings; spiracles moderately large to absent, without valve; labial furrows, folds, and cartilages large to reduced or absent; mouth subterminal, long and extending behind eyes; nostrils diagonal, with anterolateral incurrent apertures and posteromedial excurrent apertures; nasoral grooves primitively absent but secondarily developed in a few genera; no circumnarial grooves or nasal barbels; eyes dorsolateral to lateral, with or without weak to strong subocular ridges; true nictitating eyelids with postocular muscles present; teeth with basal ledges strong but secondarily reduced or absent; trunk cylindrical or moderately depressed or compressed; caudal peduncle with or without lateral keels; no predorsal medial line of enlarged denticles or thorns; pectoral fins small to large, without triangular lobes expanded over the gill openings; pelvic claspers without lateral spines; the caudal fin heterocercal; and valvular intestine with conicospiral or spiral ring valve (some *Apristurus* species). The mode of reproduction may be oviparous (but without screw-shaped egg cases) or ovoviviparous, usually without uterine cannibalism (but oophagy secondarily developed in *Pseudotriakis*), or with placental viviparity. There are eight families.

Family Scyliorhinidae: Cat sharks
The Scyliorhinidae (Fig. 1-9) are carcharhinoid sharks with the head not expanded into lateral blades; snout short to elongated, bell-shaped in dorsoventral view or not; eyes elongated, dorsolateral or lateral on head, with or without subocular ridges; no deep grooves or troughs on subocular ridges in front of eyes; nostrils with anterior nasal flaps formed into barbels in *Poroderma* but not in other genera; spiracles moderately large; gill arches with or without gill raker papillae; mouth arcuate or angular, with or without oral papillae; labial furrows varying from very long and about half the mouth width to absent; teeth small, 40–111 rows in either jaw; posterior teeth on dental bands comblike or not; snout–vent length greater, equal, or less than vent–caudal tip; first dorsal fin posteriorly situated over or behind pelvic bases; first dorsal variably higher, lower,

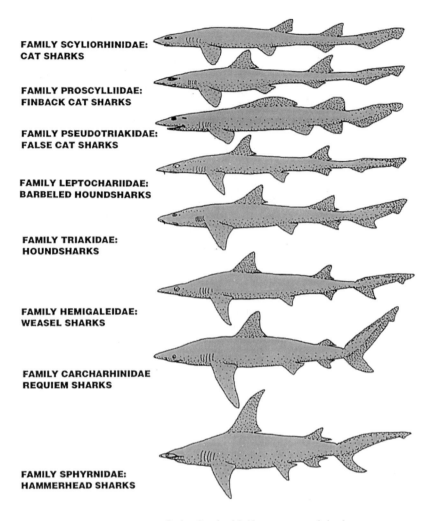

FIGURE 1-9 Order Carcharhiniformes: ground sharks.

or subequal in height to second dorsal, first dorsal length usually less than half of dorsal caudal margin; caudal peduncle without precaudal pits; caudal fin without an undulated dorsal margin, ventral caudal lobe weak or absent; valvular intestine with conicospiral valve of 5–21 turns; neurocranium with or without supraorbital crests; and color pattern varies from highly variegated with spots, blotches, and saddles to plain. Development is usually oviparous, with spindle-shaped egg cases, but some species are ovoviviparous. Adults are usually less than 1 m in total length. This is the largest family of living sharks, with at least 15 genera and over 100 species.

Family Proscylliidae: Finback cat sharks
The Proscylliidae (Fig. 1-9) are carcharhinoid sharks with head not expanded into lateral blades; snout short, not bell-shaped in dorsoventral view; eyes elongated, dorsolateral on head, with subocular ridges; no deep grooves or troughs on subocular ridges in front of eyes; nostrils with anterior nasal flaps not formed as barbels; spiracles moderately large; internal gill openings with gill raker papillae; mouth angular or arcuate, with oral papillae; labial furrows very short, confined to mouth corners, or absent; teeth small, between 46 and 88 rows in either jaw; posterior teeth on dental bands comb-shaped; snout–vent

length less than vent–caudal tip; first dorsal fin over pectoral-pelvic space but midbase closer to pelvics than pectorals; first dorsal subequal in height to second dorsal, first dorsal length less than half of dorsal caudal margin; caudal peduncle without precaudal pits; caudal fin without an undulated dorsal margin, ventral caudal lobe absent; valvular intestine with conicospiral valve of six to ten turns; neurocranium with supraorbital crests; and color pattern variegated with spots, blotches, and obscure saddles. Development is oviparous, with spindle-shaped egg cases, or ovoviviparous. Adults are less than 1 m in total length. There are three genera and five or more species.

Family Pseudotriakidae: False cat sharks
The Pseudotriakidae (Fig. 1-9) are carcharhinoid sharks with head not expanded into lateral blades; snout elongated, bell-shaped in dorsoventral view; eyes elongated, dorsolateral on head, with subocular ridges; deep grooves or troughs present on subocular ridges in front of eyes; nostrils with anterior nasal flaps not formed as barbels; spiracles moderately large; internal gill openings without gill raker papillae; mouth angular, without oral papillae; labial furrows very short, confined to mouth corners; teeth very small, 110–335 rows in either jaw; posterior teeth on dental bands comblike; snout–vent length greater than vent–caudal tip; first dorsal fin over pectoral-pelvic space, midbase closer to pectorals than pelvics; first dorsal lower than second dorsal, first dorsal length varies from half to greater than length of dorsal caudal margin; caudal peduncle without precaudal pits; caudal fin without an undulated dorsal margin, ventral caudal lobe weak or absent; valvular intestine with conicospiral valve of 11–17 turns; neurocranium with supraorbital crests; and color pattern plain, no prominent markings. Development is ovoviviparous. Adults vary from less than 1 m in total length to almost 3 m. There are two described genera and at least two species.

Family Leptochariidae: Barbeled houndsharks
The Leptochariidae (Fig. 1-9) are carcharhinoid sharks with head not expanded into lateral blades; snout short, not bell-shaped in dorsoventral view; eyes horizontally oval, lateral on head, without subocular ridges; no deep grooves or troughs in front of eyes; nostrils with anterior nasal flaps formed as slender barbels; spiracles minute; internal gill openings without gill raker papillae; mouth arcuate, without oral papillae; labial furrows very long; teeth small, 60 rows or less in either jaw; posterior teeth on dental bands not comb-shaped; snout–vent length subequal to space from vent to caudal tip; first dorsal fin over pectoral-pelvic space but midbase slightly closer to pectorals than pelvics; first dorsal subequal in height to second dorsal, first dorsal length less than two-thirds of dorsal caudal margin; caudal peduncle without precaudal pits; caudal fin without an undulated dorsal margin, ventral caudal lobe weak; valvular intestine with conicospiral valve of 14–16 turns; neurocranium without supraorbital crests; and color pattern plain. Development is viviparous, with a yolk sac placenta. Adults are less than 1 m in total length. There are a single genus and one species.

Family Triakidae: Houndsharks
The Triakidae (Fig. 1-9) are carcharhinoid sharks with head not expanded into lateral blades; snout short to long, not bell-shaped in dorsoventral view (except for *Gogolia*); eyes horizontally oval, lateral or dorsolateral on head, with or without subocular ridges; no deep grooves or troughs in front of eyes; nostrils with anterior nasal flaps formed as slender barbels in *Furgaleus* but not in other genera; spiracles moderate-sized; internal gill openings without gill raker papillae; mouth arcuate or angular, without oral papillae; labial furrows short to moderately long; teeth small, 18–106 rows in either jaw; posterior teeth on dental bands not comb-shaped; snout–vent length varying from slightly shorter to slightly longer than space from vent to caudal tip; first dorsal fin over pectoral-pelvic space but midbase usually closer to pectorals than pelvics but slightly closer to pelvics in a few species; first dorsal subequal in height or higher than

the second dorsal, first dorsal length subequal (*Gogolia*) or less than two-thirds of dorsal caudal margin; caudal peduncle without precaudal pits; caudal fin without an undulated dorsal margin, ventral caudal lobe absent to strong; valvular intestine with conicospiral valve of 4–11 turns; neurocranium with supraorbital crests; and color pattern plain or with spots, saddles, and blotches. Development is ovoviviparous or viviparous (placental viviparous). Adults range from less than 1 m to slightly less than 2 m in total length. There are 9 genera and at least 36 species.

Family Hemigaleidae: Weasel sharks
The Hemigaleidae (Fig. 1-9) are carcharhinoid sharks with head not expanded into lateral blades; snout moderately long, not bell-shaped in dorsoventral view; eyes horizontally oval, lateral on head, without subocular ridges; no deep grooves or troughs in front of eyes; nostrils with anterior nasal flaps not formed as slender barbels; spiracles small; internal gill openings without gill raker papillae; mouth arcuate, without oral papillae; labial furrows moderately long; teeth small to large (*Hemipristis*), 25–43 rows in either jaw; posterior teeth on dental bands not comb-shaped; snout–vent length slightly shorter or subequal to space from vent to caudal tip; first dorsal fin over pectoral-pelvic space but midbase closer to pectorals than pelvics; first dorsal slightly higher than the second dorsal, first dorsal length less than two-thirds of dorsal caudal margin; caudal peduncle with precaudal pits; caudal fin with an undulated dorsal margin, ventral caudal lobe strong; valvular intestine with conicospiral valve of four to six turns; neurocranium without supraorbital crests; and color pattern plain except for light or dark fin edges or tips. Development is viviparous (placental viviparous). Adults range from less than 1 m to slightly over 2 m in total length. There are four genera and at least six species.

Family Carcharhinidae: Requiem sharks
The Carcharhinidae (Fig. 1-9) are carcharhinoid sharks with head not expanded into lateral blades; snout short to moderately long, not bell-shaped in dorsoventral view; eyes horizontally oval, circular, or vertically oval, lateral on head, without subocular ridges; no deep grooves or troughs in front of eyes; nostrils with anterior nasal flaps not formed as barbels; spiracles usually absent (small spiracles always present in *Galeocerdo*, vestigial spiracles occasionally present in a few other genera); internal gill openings without gill raker papillae (except in *Prionace*); mouth arcuate, without oral papillae; labial furrows moderately long to vestigial; teeth small to large, 18–60 rows (usually less than 35/35) in either jaw; posterior teeth on dental bands not comb-shaped; snout–vent length slightly shorter to slightly longer than space from vent to caudal tip; first dorsal fin over pectoral-pelvic space but with midbase varying from closer to pectorals to closer to the pelvics; first dorsal usually much higher than the second dorsal (except for *Lamiopsis* and *Negaprion*, in which it is subequal in height), first dorsal length less than two-thirds of dorsal caudal margin; caudal peduncle with precaudal pits; caudal fin with an undulated dorsal margin, ventral caudal lobe strong; valvular intestine with scroll valve; neurocranium without supraorbital crests; and color pattern usually plain, with light or dark fin edges or tips in some species (conspicuously barred in *Galeocerdo*). Development is viviparous (placental viviparous) or possibly ovoviviparous (*Galeocerdo*). Adults range from less than 1 m to over 5 m in total length. There are 12 genera and at least 49 species.

Family Sphyrnidae: Hammerhead sharks
The Sphyrnidae (Fig. 1-9) are carcharhinoid sharks with head expanded into prominent lateral blades (cephalofoil); snout moderately long, not bell-shaped in dorsoventral view; eyes circular, lateral on expanded blades of head, without subocular ridges; no deep grooves or troughs in front of eyes; nostrils with anterior nasal flaps not formed as barbels; spiracles absent; internal gill openings without gill raker papillae; mouth arcuate, without oral papillae; labial furrows vestigial or absent; teeth small, 24–37 rows (usually less than 35/35) in either jaw; posterior teeth on dental bands not comb-shaped; snout–vent length slightly shorter than space from vent to caudal

tip; first dorsal fin over pectoral-pelvic space but with midbase varying from equidistant between these fins to closer to pectorals; first dorsal much higher than the second dorsal, first dorsal length less than two-thirds of dorsal caudal margin; caudal peduncle with precaudal pits; caudal fin with an undulated dorsal margin, ventral caudal lobe strong; valvular intestine with scroll valve; neurocranium without supraorbital crests; and color plain. Development is viviparous (placental viviparous). Adults range from less than 1 m to over 5 m in total length. There are two genera and at least eight species.

Rays (Batoids)

The rays are neoselachians with the upper eyelids fused to the eyeball, but with lower eyelids intact; ceratohyals small and reduced or absent, not connected with hyomandibulae and functionally replaced by pseudohyoids developed from hyoid rays; palatoquadrates not articulating with the neurocranium; neurocranium without an occipital half-centrum; chondrified antorbital cartilages present on the cranial nasal capsules; pectoral fins fused to the sides of the head over the external gill openings; pectoral fins with expanded propterygia, which are as large or larger than the mesopterygia; pectoral radials highly jointed and branched distally, with radial musculature and radials displacing or eliminating ceratotrichia from the fin webs; suprascapulae of pectoral girdle articulating with or fused to vertebral column; and a cervical or cervicothoracic synarcual of fused vertebrae behind the neurocranium. Five orders of living rays are often recognized (ranked by some authors as suborders of a single order of Rajiformes or batoids). A sixth group is tentatively recognized here, following Nishida's (1990) separation of Rhinidae from the Rhinobatiformes.

ORDER PRISTIFORMES: SAWFISHES

The Pristiformes are rays with two large falcate or semifalcate, sharklike dorsal fins, the first over or partly anterior to the pelvic fins; preoral snout depressed, greatly elongated, and sawlike, with large lateral teeth; snout with rostral cartilage large, flat, elongated and strong; antorbital cartilages small, posteriorly directed, wedge-shaped and well anterior to propterygia; nostrils small, far separated from each other and well anterior to mouth, without nasoral grooves; anterior nasal flaps separate from each other and not extending posteriorly to mouth; trunk thick and sharklike, covered with denticles in adults; precaudal tail stout and muscular, without lateral electric organs or a sting; no predorsal medial line of enlarged denticles or thorns; males without hooklike alar spines on pectoral fins; pectoral fins moderate-sized, not expanded anterior to eyes and discrete from the head; pectoral radials ending well behind level of nostrils; no pectoral electric organs; pelvic fins not divided into anterior and posterior lobes; origins of pelvic fins anterior to pectoral free rear tips; and caudal fin large and heterocercal, with or without a strong ventral lobe. Mode of reproduction is ovoviviparous. There is a single living family.

Family Pristidae: Modern sawfishes
The Pristidae (Fig. 1-10) comprise two living genera and approximately seven species.

ORDER RHINIFORMES: SHARKFIN GUITARFISHES OR WEDGEFISHES

The Rhiniformes (Fig. 1-11) consist of rays with two large falcate, sharklike dorsal fins, the first varying from partly anterior to opposite the pelvic fin bases; preoral snout elongated, depressed, or short and bluntly rounded but not sawlike; snout with rostral cartilage large, elongated and strong; antorbital cartilages moderately large, posteriorly directed, wedge-shaped and just anterior to propterygia; nostrils large, well separated from each other and anterior to mouth, without nasoral grooves; anterior nasal flaps separate from each other, ending well anterior to mouth; trunk thick and sharklike, covered with denticles in adults; precaudal tail very stout and muscular, without lateral electric organs or a sting; a predorsal medial line of enlarged denticles or thorns variably developed; males without hooklike alar spines on pectoral fins;

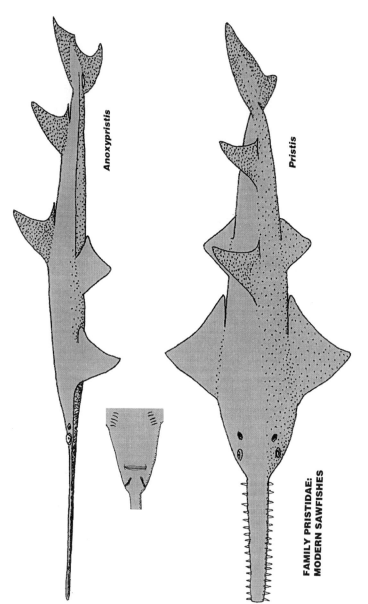

FIGURE 1-10 Order Pristiformes: sawfish.

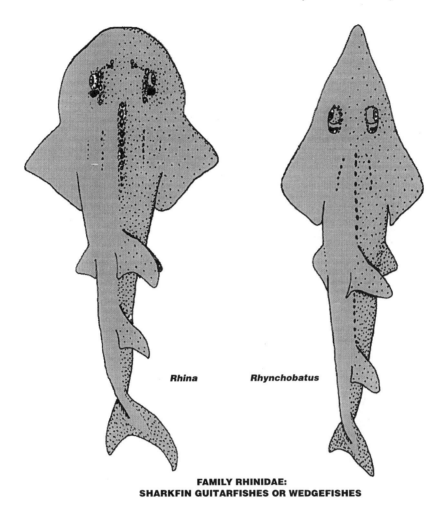

**FAMILY RHINIDAE:
SHARKFIN GUITARFISHES OR WEDGEFISHES**

FIGURE 1-11   Order Rhiniformes: sharkfin guitarfishes or wedgefishes.

pectoral fins moderately large, not expanded anterior to eyes and discrete from the head; pectoral radials ending behind level of nostrils; no pectoral electric organs; pelvic fins not divided into anterior and posterior lobes, origins of pelvic fins just behind pectoral free rear tips; and caudal fin large and heterocercal, with strong ventral lobe. Mode of reproduction is ovoviviparous. There is a single family, usually placed in the Rhinobatiformes but not close to them according to Nishida (1990). McEachran et al. (1996) split this group into two families, Rhinidae and Rhynchobatidae, and placed them in separate orders.

Family Rhinidae:
Sharkfin guitarfishes or wedgefishes
The Rhinidae comprise two genera and five or more species.

ORDER RHINOBATIFORMES:
GUITARFISHES
The Rhinobatiformes (Fig. 1-12) are rays with two moderate-sized rounded-angular (but not falcate) dorsal fins, the first well posterior to the pelvic fin bases; preoral snout elongated or broadly rounded, depressed, but not sawlike; snout with rostral cartilage large, flat, elongated and strong in Rhinobatidae but partially reduced in Platyrhinidae; antorbital

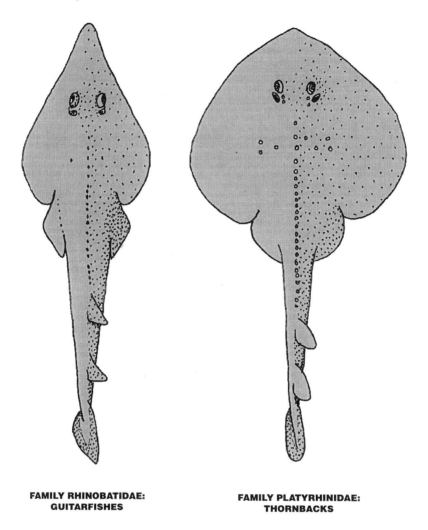

FAMILY RHINOBATIDAE: GUITARFISHES

FAMILY PLATYRHINIDAE: THORNBACKS

FIGURE 1-12 Order Rhinobatiformes: guitarfishes.

cartilages small, posteriorly directed, wedge-shaped and distally attached to propterygia; nostrils varying from small, far separated from each other and well anterior to mouth to close to each other and to mouth, with or without nasoral grooves; anterior nasal flaps separate from each other or fused, ending well anterior to mouth or overlapping it (*Trygonorrhina*); trunk flattened, covered with denticles in adults; precaudal tail fairly stout and muscular, without lateral electric organs or a sting; a predorsal medial line of enlarged denticles or thorns variably developed; males without hooklike alar spines on pectoral fins; pectoral fins moderate-sized to large, not expanded anterior to eyes, more or less fused with the head, and forming an angular to broadly rounded pectoral disk; pectoral radials extending to level of nostrils (Rhinobatidae) or reaching snout tip (Platyrhinidae); no pectoral electric organs; pelvic fins not divided into anterior and posterior lobes, origins of pelvic fins anterior to pectoral free rear tips; and caudal fin moderately large and weakly heterocercal, without a ventral lobe. Their mode of reproduction is ovoviviparous. Two families are generally recognized, which are of uncertain relationship and which probably form a paraphyletic assemblage (Nishida 1990; McEachran et al. 1996).

Family Rhinobatidae: Guitarfishes
The Rhinobatidae (Fig. 1-12) are guitarfishes with narrowly to broadly angular snouts; large angular or rounded pectoral fins forming a wedge-shaped pectoral disk with head, with the anterior pectoral radials extending to the level of the nostrils; small thorns around eyes and along midline of body and on shoulders; fairly stout precaudal tails; first dorsal origin behind pelvic rear tips but closer to pelvic bases than caudal origin. There are 4 genera and at least 37 species.

Family Platyrhinidae: Thornback rays
The Platyrhinidae (Fig. 1-12) are guitarfishes with broadly rounded snouts; pectoral fins, head, and trunk forming a large subcircular or heart-shaped pectoral disk with anterior pectoral radials extending to snout tip; one to three rows of small to large thorns on disk and predorsal tail; slender precaudal tails; and first dorsal far behind pelvic rear tips, with midbase closer to caudal origin than pelvic bases. There are three genera and about five species. McEachran et al. (1996) split the Platyrhinidae into two families, Platyrhinidae and Zanobatidae, and ranked them in separate suborders within the Order Myliobatiformes.

ORDER TORPEDINIFORMES:
ELECTRIC RAYS
The Torpediniformes are rays with two moderately large rounded-angular (not falcate) dorsal fins, the first over, partly anterior, or partly posterior to the pelvic fin bases where present, but second dorsal or both dorsal fins are lost in some genera of Narkidae; preoral snout moderately long, broadly rounded or truncated, and not sawlike; snout with rostral cartilage variably developed, large and broad in Narcinidae but more or less reduced in other families; antorbital cartilages large, anteriorly directed, fan-shaped and anterior to and free from propterygia; nostrils moderately large, close to each other and to mouth, with anterior nasal flaps forming a nasal curtain and with broad nasoral grooves; body thick and depressed, without denticles; precaudal tail stout and muscular, without lateral electric organs or a sting; no predorsal medial line of enlarged denticles or thorns; males without hooklike alar spines on pectoral fins; pectoral fins enlarged, expanded anteriorly to eyes and forming a prominent pectoral disk; pectoral radials extending in front of level of nostrils but not reaching snout tip; prominent paired pectoral electric organs present in disk; pelvic fins divided into distinct anterior and posterior lobes (except in *Typhlonarke*); origins of pelvic fins anterior to pectoral free rear tips; and caudal fin large and more or less heterocercal, with or without a low ventral lobe. The mode of reproduction is ovoviviparous. There are four well-defined families.

Family Narcinidae: Numbfishes
The Narcinidae (Fig. 1-13) are electric rays with broadly rounded snouts; cranial rostrum broad and trough-shaped; disk circular, ovate, or rounded-angular; jaws stout, elongated and highly protrusible; labial cartilages strong; mouth transverse, usually with a deep peripheral groove; teeth monocuspidate, exposed in broad bands on the outer surfaces of the jaws; two large, equal-sized dorsal fins; precaudal tail long and stout; and large caudal fin. There are 4 genera and approximately 21 species.

Family Narkidae: Sleeper rays
The Narkidae (Fig. 1-13) are electric rays with broadly rounded snouts; cranial rostrum reduced to a narrow medial rod; disk circular, ovate, or rounded-angular; jaws stout, short, and not highly protrusible; labial cartilages strong; mouth transverse, with a shallow peripheral groove; teeth monocuspidate, not exposed in broad bands on the outer surfaces of the jaws; two large, equal-sized dorsal fins in *Heteronarce*, but with only one dorsal in *Crassinarke*, *Narke*, and *Typhlonarke* or with no dorsals (*Temera*); precaudal tail moderately long and stout; and large caudal fin. There are four or five genera and at least ten species.

Family Hypnidae: Coffin rays
The Hypnidae (Fig. 1-13) are electric rays with truncate snouts; cranial rostrum reduced to two lateral cartilages that articulate to cranium; disk pear-shaped; jaws very slender, elongated and highly protrusible; labial cartilages absent;

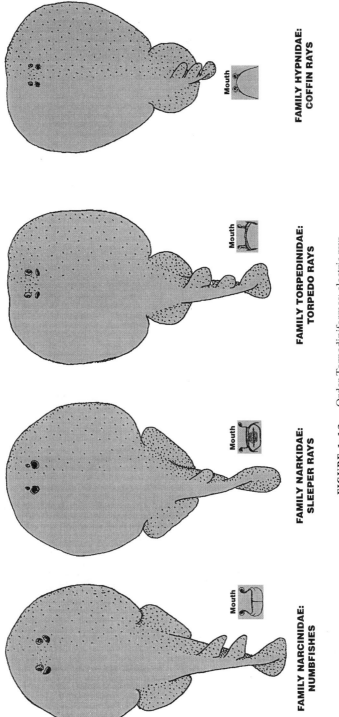

FIGURE 1-13 Order Torpediniformes: electric rays.

mouth broadly arcuate, with paired longitudinal grooves defining its corners; teeth tricuspidate, not exposed in broad bands on the outer surfaces of the jaws; two small dorsal fins, the first larger than the second; precaudal tail very short and reduced; and small caudal fin. There are a single genus and one species.

Family Torpedinidae: Torpedo rays
The Torpedinidae (Fig. 1-13) are electric rays with truncate or emarginate snouts; cranial rostrum reduced to two lateral cartilages that are fused with cranium; disk rounded-subquadrate; jaws very slender, elongated and highly protrusible; labial cartilages absent; mouth broadly arcuate, with paired longitudinal grooves defining its corners; teeth monocuspidate, not exposed in broad bands on the outer surfaces of the jaws; two moderately large dorsal fins, the first larger than the second; precaudal tail moderately long and stout; and large caudal fin. There is a single genus with at least 22 species.

ORDER RAJIFORMES: SKATES

The order Rajiformes comprises rays with two small rounded-angular dorsal fins located far behind pelvic fins on the terminal half of the tail, sometimes one or both dorsals absent; preoral snout moderately long, broadly angular or rounded, and not sawlike; snout with rostral cartilage large, elongated, and strong in many species but variably reduced in others; antorbital cartilages small, wedge-shaped, posteriorly directed, and distally attached to propterygia; nostrils moderately large, close to each other and to mouth, with anterior nasal flaps not forming a nasal curtain; nasoral grooves present; body depressed, usually with denticles and larger thorns but naked (except for alar spines) in *Anacanthobatis*; precaudal tail slender to almost whiplike, with paired lateral electric organs in many species but without a sting; a predorsal medial line of enlarged denticles or thorns on the precaudal tails of most species, extending onto the back in many; males with hooklike alar spines on pectoral fins; pectoral fins enlarged, expanded anteriorly to eyes and forming a prominent pectoral disk; pectoral radials extending in front of nostrils and in some species reaching snout tip; no electric organs in pectoral disk; pelvic fins divided into distinct anterior and posterior lobes; origins of pelvic fins anterior to pectoral free rear tips; and caudal fin small or vestigial, without a ventral lobe. The mode of reproduction is oviparous in all species in which it is known, the egg cases pillow-shaped with a horn at each corner. Family arrangement follows Ishihara's (1990) interpretation of McEachran and Miyake's (1986, 1990) cladistic analyses of the rajoids. McEachran and his coworkers, including McEachran et al. (1996) and McEachran and Dunn (1998) retain all skates in a single family Rajidae. There are three families.

Family Arhynchobatidae: Softnose skates
The Arhynchobatidae (Fig. 1-14) are skates with rostrum more or less reduced, snout soft and flexible (except *Rioraja*, with a stout rostral cartilage); basihyal cartilage with lateral projections; scapulocoracoid with stout or slender anterior bridge (lost in a few genera); clasper glans not greatly expandable, without rhipidion or shield; second and third clasper ventral terminal cartilage spoon-shaped, without a sharp lateral edge and not forming an external clasper shield; clasper dorsal terminal cartilages arranged in parallel; and pelvic fins bilobate but anterior lobe not separated from posterior and not leglike. There are 11 genera and over 81 species.

Family Rajidae: Skates
The Rajidae (Fig. 1-14) are skates with rostrum usually not reduced, snout stiff (secondarily reduced in *Breviraja, Gurgesiella,* and *Neoraja*); basihyal cartilage without lateral projections; scapulocoracoid without anterior bridge; clasper glans greatly expandable, with a rhipidion and shield; clasper ventral terminal cartilage not spoon-shaped, without a sharp lateral edge that forms the shield; second and third clasper dorsal terminal cartilages arranged in series; and pelvic fins bilobate but anterior lobe not separated from posterior and not leglike. There are at least 15 genera and over 136 species, making this the largest chondrichthyan family.

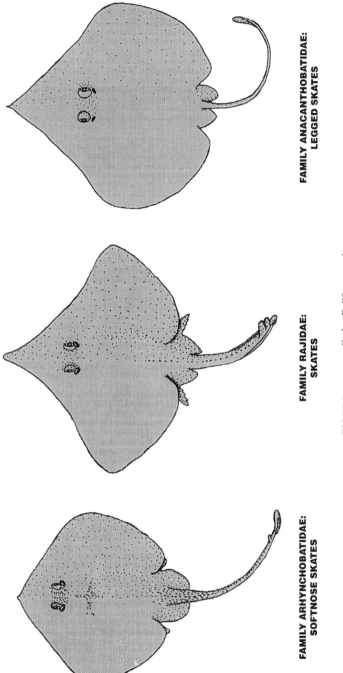

FIGURE 1-14 Order Rajiformes: skates.

Family Anacanthobatidae: Legged skates
The Anacanthobatidae (Fig. 1-14) are skates with rostrum usually not reduced, snout stiff or semiflexible (*Anacanthobatis*); basihyal cartilage without lateral projections; scapulocoracoid without anterior bridge; clasper glans greatly expandable, with a rhipidion and shield; clasper ventral terminal cartilage not spoon-shaped, without a sharp lateral edge that forms the shield; second and third clasper dorsal terminal cartilages arranged in series; pelvic fins with anterior lobes separated from posterior and formed as leglike structures, with a distinct "foot," "knee," and "calf." There are 2 well-defined genera with at least 17 species.

ORDER MYLIOBATIFORMES: STINGRAYS
The order Myliobatiformes comprises rays with one small to moderate-sized rounded or rounded-angular dorsal fin over the pelvic fins or behind them on the anterior half of the tail, but dorsal fin lost in many species; preoral snout moderately long, broadly angular, rounded or bilobate, but not sawlike; snout without a rostral cartilage; antorbital cartilages small, posteriorly directed, wedge-shaped and distally attached to propterygia; nostrils moderately large, close to each other and to mouth, with anterior nasal flaps forming a broad nasal curtain; nasoral grooves present; body depressed, usually with denticles and sometimes larger thorns on dorsal surface but naked in some species; precaudal tail moderately slender and shorter than disk to elongated and whiplike, with no lateral electric organs but usually with a barbed venomous sting on dorsal surface; a predorsal medial line or row of enlarged denticles or thorns on the precaudal tails and backs of some species, but absent in others; males without hooklike alar spines on pectoral fins; pectoral fins enlarged, expanded anteriorly to eyes and forming a prominent pectoral disk; pectoral radials extending to snout tip; no electric organs in pectoral disk; pelvic fins not divided into anterior and posterior lobes; origins of pelvic fins anterior to pectoral free rear tips; and caudal fin either small but well developed, reduced to narrow longitudinal caudal folds, or absent, but without a ventral lobe when present. The mode of reproduction is ovoviviparous as far as known, but with the yolk supply of fetus supplemented by uterine "milk" (histotroph). There are possibly nine or ten families. Phyletic studies of this undoubtedly monophyletic group have produced varying and disparate classifications (Nishida 1990; Lovejoy 1996; McEachran et al. 1996).

Family Plesiobatidae: Giant stingarees
The Plesiobatidae (Fig. 1-15) are stingrays with broadly angular, long snouts, confluent with rest of disk; pectoral disk rounded-rhomboidal in dorsoventral view, slightly narrower than long, disk relatively flat in lateral view and with head not elevated; eyes dorsolateral on head; nasal curtain not reaching mouth; mouth small and subterminal; spiracles close behind eyes; five pairs of small gill openings; internal gill arches without filter plates or ridges; teeth small, not forming flat crushing plates; disk covered with small denticles; pelvic girdle with low blunt medial prepelvic process; dorsal fin absent; tail short, fairly slender, tail including caudal fin slightly shorter than snout–vent length, not whiplike; sting large, functional, and located well behind pelvic fins and about at midlength of the tail; and caudal fin present, moderately large, elongated and prominent, extending to tip of tail. Adults attain a length of nearly 3 m. There are a single genus and one species. This stingray was formerly included in the family Urolophidae but was placed in its own family by Nishida (1990).

Family Hexatrygonidae: Sixgill stingrays
The Hexatrygonidae (Fig. 1-15) are stingrays with broadly angular, extremely long, distally lobate snouts, confluent with rest of disk; pectoral disk rounded-rhomboidal in dorsoventral view, narrower than long, relatively flat in lateral view and with head not elevated; eyes dorsolateral on head; nasal curtain not reaching mouth; mouth small and subterminal; spiracles well separated from eyes; six pairs of small gill openings; internal gill arches without filter plates or ridges; teeth small, not forming flat crushing plates; disk naked; pelvic girdle with low blunt medial prepelvic process; dorsal fin absent; tail short, fairly slender, tail

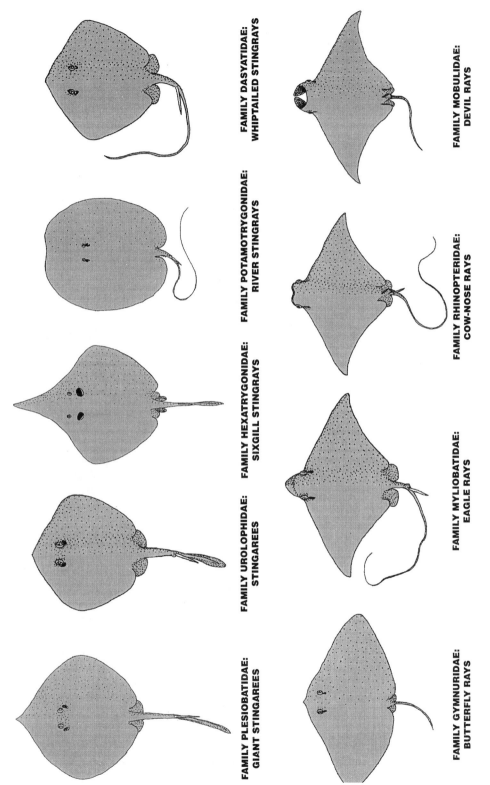

FIGURE 1-15 Order Myliobatiformes: stingrays.

including caudal fin slightly shorter than snout–vent length, not whiplike; sting large, functional, and located well behind pelvic fins and about at midlength of the tail; and caudal fin present, moderately large, elongated and prominent, extending to tip of tail. Adults range between 1 and 2 m long. There are a single genus and possibly only one species.

Family Urolophidae: Stingarees
The Urolophidae (Fig. 1-15) are stingrays with broadly angular or bluntly rounded, short to moderately long snouts, confluent with rest of disk; pectoral disk circular, oval, rounded-rhomboidal or angular in dorsoventral view, slightly wider than long to slightly narrower than long, relatively flat in lateral view, with head not elevated; eyes dorsolateral on head; nasal curtain reaching mouth; mouth small and subterminal; spiracles close behind eyes; five pairs of small gill openings; internal gill arches without filter plates or ridges; teeth small, not forming flat crushing plates; disk smooth or with small denticles or small thorns; pelvic girdle with low medial prepelvic process or not; dorsal fin present or absent, where present small and located well behind the pelvic rear tips and just anterior to the sting; tail short, stout to fairly slender, tail including caudal fin slightly longer to somewhat shorter than snout–vent length, not whiplike; sting large, functional, and located far behind pelvic fins and about at midlength of the tail; and caudal fin present, moderately large, elongated and prominent, extending to tip of tail. Adults are mostly less than 1 m long. There are 4 genera and at least 36 species. McEachran et al. (1996) place American taxa usually grouped in Urolophidae (2 genera and 15 species) in a separate family Urotrygonidae.

Family Potamotrygonidae: River stingrays
The Potamotrygonidae (Fig. 1-15) are stingrays with bluntly rounded or truncate, short to moderately long snouts, confluent with rest of disk; pectoral disk circular or longitudinally oval in dorsoventral view, slightly wider than long to slightly narrower than long, relatively flat in lateral view, with head not elevated; eyes dorsolateral on head; nasal curtain reaching mouth; mouth small and subterminal; spiracles close behind eyes; five pairs of small gill openings; internal gill arches without filter plates or ridges; teeth small to moderately large, not forming flat crushing plates; disk with small denticles or small to large thorns; pelvic girdle with high, acute medial prepelvic process; dorsal fin absent; tail short to moderately long, stout to slender basally but more or less whiplike distal to sting, tail much longer to somewhat shorter than snout–vent length; sting large, functional, and located behind pelvic fins; and caudal fin reduced to elongated dorsal and ventral fin folds that extend to tip of tail or not. Adults are mostly less than 1 m long but some possibly attain 2 m. There are at least 6 genera and 20 species of river stingrays in South America. Following phyletic work by Lovejoy (1996), McEachran et al. (1996) add the Eastern Hemisphere marine dasyatid genus *Taeniura* (three species) and two marine American species of the dasyatid genus *Himantura* to this family.

Family Dasyatidae: Whiptail stingrays
The Dasyatidae (Fig. 1-15) are stingrays with bluntly rounded, angular, or medially elongate, short to long snouts, confluent with rest of disk; pectoral disk circular, longitudinally oval, rounded-rhomboidal, or bluntly rhomboidal in dorsoventral view, varying from slightly wider than long (width less than 1.5 times length) to slightly narrower than long, relatively flat in lateral view and with head not elevated; eyes dorsolateral on head; nasal curtain reaching mouth; mouth small and subterminal; spiracles close behind eyes; five pairs of small gill openings; internal gill arches without filter plates or ridges; teeth small to moderately large, not forming flat crushing plates; disk naked or with denticles or small to large thorns; pelvic girdle with low, blunt medial prepelvic process or none; dorsal fin absent; tail short to greatly elongate, stout to slender basally but more or less whiplike distal to sting, tail over twice as long to somewhat shorter than snout–vent length; sting usually large, functional, and located behind pelvic fins,

greatly reduced or lost in *Urogymnus*; and caudal fin variably reduced to elongated dorsal and ventral fin folds, to a ventral fold only, or completely lost; when present ventral fin folds extend to tail tip or not. Adults vary from less than 1 m long to over 4 m long. There are 5 or 6 genera and over 62 species.

Family Gymnuridae: Butterfly rays
The Gymnuridae (Fig. 1-15) are stingrays with bluntly angular, short snouts, confluent with rest of disk; pectoral disk bluntly rhomboidal and laterally expanded in dorsoventral view, much wider (over 1.5 times) than long, relatively flat in lateral view and with head not elevated; eyes dorsolateral on head; nasal curtain reaching mouth; mouth moderately large and subterminal; spiracles close behind eyes; five pairs of small gill openings; internal gill arches without filter plates or ridges; teeth small and cuspidate, not forming flat crushing plates; disk naked or with small denticles; pelvic girdle with low, blunt medial prepelvic process; dorsal fin small or absent, where present just behind pelvics; tail short, slender basally and distally whiplike, tail shorter than snout–vent length; sting small or absent, when present possibly functional and located just behind pelvic fins; caudal fin reduced to very low dorsal and ventral ridges that may extend to tail tip; adults vary from less than 0.5 m to over 2 m long. There are 2 genera, possibly synonyms, and 12 species.

Family Myliobatidae: Eagle rays
The Myliobatidae (Fig. 1-15) are stingrays with bluntly rounded, short, thick preorbital snouts (rostral fins), separated by lateral notches from rest of disk; postrostral pectoral disk sharply rhomboidal, laterally expanded and "angular or falcate in dorsoventral view, much wider (over 1.5 times) than long, relatively high in lateral view and with prespiracular head elevated; eyes more or less lateral on head; nasal curtain reaching mouth; mouth small and subterminal; spiracles close behind eyes; five pairs of small gill openings; internal gill arches with a row of short dermal ridges or papillae, but no filter plates; teeth large, flat-crowned, and forming flat crushing plates, medial row very wide in both jaws; disk naked or with small denticles; pelvic girdle with short, blunt medial prepelvic process; dorsal fin present, moderately large and over or just behind pelvics; tail elongated, fairly slender basally and distally whiplike, tail longer than snout–vent length; sting large, reduced, or absent, when present apparently functional but close behind pelvic fins and less effective as a weapon than in some other stingrays; and caudal fin absent or greatly reduced to very low dorsal and ventral ridges that do not extend to tail tip. Adults vary from less than 1 m to over 4 m long. There are 4 genera and at least 18 species.

Family Rhinopteridae: Cow-nose rays
The Rhinopteridae (Fig. 1-15) are stingrays with short, bilobate preorbital snouts (rostral fins), with prominent medial notch indenting snout and forehead, separated by lateral notches from rest of disk; postrostral pectoral disk sharply rhomboidal and laterally expanded and falcate in dorsoventral view, much wider (over one and one-half times) than long, relatively high in lateral view and with prespiracular head elevated; eyes lateral on head; nasal curtain reaching mouth; mouth small and subterminal; spiracles close behind eyes; five pairs of small gill openings; internal gill arches with a row of long low dermal ridges and papillae, but no filter plates; teeth large, flat-crowned, and forming flat crushing plates, medial and adjacent lateral rows very wide in both jaws; disk naked or with small denticles; pelvic girdle with moderately long, broadly acute medial prepelvic process; dorsal fin present, moderately large and over pelvics; tail elongated, fairly slender basally and distally whiplike, tail longer than snout–vent length; sting large, apparently functional but close behind pelvic fins and less effective as a weapon than in some other stingrays; and caudal fin absent. Adults usually range between 1 and 2 m long but may possibly be larger. There are a single genus and 11 nominal species, which according to Schwartz (1990) represent five largely allopatric species.

Family Mobulidae: Devil rays

The Mobulidae (Fig. 1-15) are stingrays with long, bilobate preorbital rostral horns, with broad space separating them medially and broad lateral notches separating them from rest of disk; postrostral pectoral disk sharply rhomboidal, laterally expanded and more or less falcate in dorsoventral view, much wider (over 1.5 times) than long, relatively high in lateral view and with prespiracular head elevated; eyes lateral on head; nasal curtain reaching mouth; mouth large and subterminal or terminal; spiracles behind eyes and more or less separated from them; five pairs of large gill openings; internal gill arches with a row of large angular filter plates with cartilaginous cores and distal lobular processes; teeth very small, vestigial, cuspidate or flat-crowned, not forming flat crushing plates; disk naked or with small denticles; pelvic girdle with moderately long, acute medial prepelvic process; dorsal fin present, moderately large and over pelvics; tail elongated, fairly slender basally and distally whiplike, tail shorter or longer than snout–vent length; sting small and possibly nonfunctional, rudimentary, or absent; and caudal fin absent. Adults vary from under 1 m to over 5 m long. They are circumglobal, dwelling in warm temperate and tropical seas, continental and insular shelves and oceans. These batoids are morphologically the most evolved and derived of living elasmobranchs. There are two well-defined genera and at least ten species.

REFERENCES

Cappetta, H. 1987. Chondrichthyes II. Mesozoic and Cenozoic Elasmobranchii. In *Handbook of Paleoichthyology*, vol. 3B, ed. H.-P. Schultze, 1–193. Stuttgart: Gustav Fischer Verlag.

Compagno, L. J. V. 1973. Interrelationships of living elasmobranchs. *Zool. J. Linn. Soc. Lond.* 53(Supp. 1): 15–61.

———. 1977. Phyletic relationships of living sharks and rays. *Am. Zool.* 17 (2): 303–22.

———. 1984. *Sharks of the World*. FAO Species Catalogue. Rome: United Nations Food and Agriculture Organization.

———. 1988. *Sharks of the Order Carcharhiniformes*. Princeton: Princeton University Press.

———. 1990a. Alternate life history styles of cartilaginous fishes in time and space. *Environ. Biol. Fishes* 28(1–4): 33–75.

———. 1990b. Evolution and diversity of sharks. *Underwater Nat., Bull. Am. Littor. Soc.* 19–20(4/1): 15–22.

De Carvalho, M. R. 1996. Higher-level elasmobranch phylogeny, basal squaleans, and paraphyly. In *Interrelationships of Fishes*, ed. M. L. J. Stiassny, L. R. Parenti, and G. D. Johnson, 35–62. San Diego: Academic Press.

Gaudin, T. J. 1991. A re-examination of elasmobranch monophyly and chondrichthyan phylogeny. *N. Jahrb. Geol. Palaontol. Abhandl.* 182(2): 133–60.

Ishihara, H. 1990. The skates and rays of the western North Pacific: an overview of their fisheries, utilization, and classification. In *Elasmobranchs as Living Resources: Advances in the Biology, Ecology, Systematics, and Status of the Fisheries*, ed. H. L. Pratt, Jr., S. H. Gruber, and T. Taniuchi. *NOAA Tech. Rep.* 90: 485–97.

Lovejoy, N. R. 1996. Systematics of myliobatoid elasmobranchs: with emphasis on the phylogeny and historical biogeography of neotropical freshwater stingrays (Potamotrygonidae: Rajiformes). *Zool. J. Linn. Soc.* 117(3): 207–57.

Lund, R. 1985. Stethacanthid elasmobranch remains from the Bear Gulch Limestone Namurian E-2B of Montana, U.S.A. *Am. Mus. Novit.* 2828: 1–24.

———. 1986. The diversity and relationships of the Holocephali. *Indo-Pacific Fish Biology: Proceedings of the Second International Conference on Indo-Pacific Fishes*, ed. T. Uyeno, T. Aria, T. Taniuchi, and K. Matsuura, 97–106. Tokyo: Ichthyological Society of Japan.

———. 1989. New petalodonts (Chondrichthyes) from the Upper Mississippian Bear Gulch Limestone (Namurian E2B) of Montana. *J. Vert. Paleontol.* 9(3): 350–68.

———. 1990. Shadows in time. A capsule history of sharks. *Underwater Nat., Bull. Am. Littor. Soc.* 19-20(4/1): 23–8.

McEachran, J. D., and K. A. Dunn. 1998. Phylogenetic analysis of skates, a morphologically conservative clade of elasmobranchs (Chondrichthyes: Rajidae). *Copeia* 1998(2): 271–90.

McEachran, J. D., and T. Miyake. 1986. Interrelationships within a putative monophyletic group of skates (Chondrichthyes, Rajoidei, Rajini). *Indo-Pacific Fish Biology: Proceedings of the Second International Coference on Indo-Pacific Fishes*, ed. T. Uyemo, T. Aria, T. Taniuichi, and K. Matsuura, 218–90. Tokyo: Ichthyological Society of Japan.

———. 1990. Phylogenetic interrelationships of skates: a working hypothesis (Chondrichthyes, Rajoidei). In *Elasmobranchs as Living Resources: Advances in the Biology, Ecology, Systematics, and Status of the Fisheries,* ed. H. L. Pratt, Jr., S. H. Gruber, and T. Taniuchi. *NOAA Tech. Rep.* 90: 285–304.

McEachran, J. D., K. A. Dunn, and T. Miyake. 1996. Interrelationships of the batoid fishes (Chondrichthyes: Batoidei). In *Interrelationships of Fishes,* ed. M. L. J. Stiassny, L. R. Parenti, and G. D. Johnson, 63–84. San Diego: Academic Press.

Maisey, J. G. 1975. The interrelationships of phalacanthous selachians. *N. Jahrb. Geol. Palaeontol.* 9: 553–67.

———. 1984. Chondrichthyian phylogeny: a look at the evidence. *J. Vert. Paleontol.* 4(3): 359–71.

———. 1986. Heads and tails: a chordate phylogeny. *Cladistics* 2(3): 201–56.

———. 1989. *Hamiltonichthys mapesi* g. & sp. nov. (Chondrichthyes; Elasmobranchii), from the Upper Pennsylvanian of Kansas. *Am. Mus. Novit.* 2931: 1–42.

Nishida, K. 1990. Phylogeny of the Suborder Myliobatidoidei. *Mem. Fac. Fish. Hokkaido U.* 37(1.2): 1–108.

Schaeffer, B., and M. Williams. 1977. Relationships of fossil and living elasmobranchs. *Am. Zool.* 17: 293–302.

Shirai, S. 1992. *Squalean Phylogeny. A New Framework of "Squaloid" Sharks and Related Taxa.* Sapporo: Hokkaido University Press.

———. 1996. Phylogenetic interrelationships of neoselachians (Chondrichthyes, Euselachii). In *Interrelationships of Fishes,* ed. M. L. J. Stiassny, L. R. Parenti, and G. D. Johnson, 9–34. San Diego: Academic Press.

Young, G. C. 1982. Devonian sharks from south-eastern Australia and Antarctica. *Palaeontology* 25: 817–43.

Zangerl, R. 1973. Interrelationships of early chondrichthyians. *Zool. J. Linn. Soc. Lond.* 53 (Suppl. 1): 1–14.

———. 1981. Chondrichthyes: I. Paleozoic Elasmobranchii. In *Handbook of Paleoichthyology,* vol. 3A, ed. H.-P. Schultze, 1–115. Stuttgart: Gustav Fischer Verlag.

NORMAN E. KEMP

CHAPTER 2

Integumentary System and Teeth

Countless students of general zoology or comparative anatomy have dissected the dogfish shark (Fig. 2-1A) as a model for the basic patterns from which the organ systems of higher vertebrates have evolved (Gans and Parsons 1964; Walker and Homberger 1992). The integument, or skin, is the outermost of the organ systems that make up the vertebrate body. It functions as the outer protective barrier that separates the animal from its environment (Bereiter-Hahn et al. 1986). Vertebrate skin is composed of two layers: an outer layer of stratified epithelium, the epidermis, and an underlying layer of connective tissue, the dermis. Moss (1972) has described the dermis as "a uniquely vertebrate structure," much more complex than the relatively simple subepidermal layer of invertebrates. In the protochordate amphioxus, or lancelet, for example, the "dermis" is a thin, fibrous layer underlain by a gelatinous layer containing collagenous fibers. The dermis of the cyclostome *Petromyzon* is markedly thicker, containing bundles of coarse collagenous fibers as well as fine preelastic fibers. Elastic fibers are first seen in the still thicker dermis of the elasmobranchs, and in these fishes the dermis has two distinct layers: a vascularized outer layer of loose connective tissue, the stratum vasculare, and an inner compact layer of dense connective tissue containing orthogonally aligned collagen fibers, the stratum compactum.

Looking at the external surface of a shark, it is apparent that the skin is covered with an investiture of scales (Fig. 2-1B). In elasmobranchs these flat, nonoverlapping units are called placoid scales. The name of a rough-

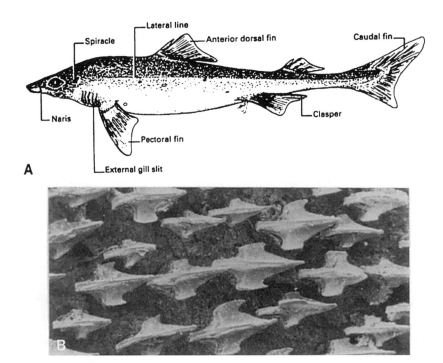

FIGURE 2-1 External morphology of the dogfish, *Squalus acanthias*. **A:** Lateral view, showing external openings (naris, spiracle, gill slits), lateral line, unpaired dorsal fins with fin spines, caudal fin, paired pectoral and pelvic fins, and male pelvic clasper. **B:** Scanning electron micrograph of body scales of a 69-cm specimen. (×45). (A, from Walker and Homberger 1992; B, from Reif 1985a.)

surfaced kind of leather, shagreen, is also commonly used for the dried and processed rough skin of certain sharks and rays. Because the scales are individual toothlike appendages embedded in the skin, they are aptly called dermal denticles, and some comparative anatomists (Maderson 1972) prefer that connotation for elasmobranch scales. Teleost fishes, reptiles, and birds have scales that differ from those of elasmobranchs by developing as folds of the skin, and the term *scale* in a restricted sense may be used for that type of dermal outgrowth. In this discussion we will use the historical terminology, which includes dermal denticles in the general category of dermal scales.

Abruptly around the margin of a shark's mouth the mantle of head skin scales ends and teeth begin. Within the mouth the teeth are aligned in multiple rows behind the outer row, gradually diminishing in size and degree of mineralization toward the interior. In the popular mind the teeth of sharks are awesome killing weapons, a conception shockingly conveyed in Peter Benchley's (1974) gripping novel *Jaws* about a predatory great white shark. From an evolutionary standpoint, teeth are actually modified scales; hence it is instructive to go back in time to the Paleozoic era for information about the origin of vertebrate scales and how they were adapted as the basis for tooth evolution.

EVOLUTION OF SCALES AND TEETH

There are no fossil records to mark the progress of the earliest stages of vertebrate phylogeny. Not until the genesis of mineralization in the skin and endoskeleton was it possible for fossilization to preserve direct evidence about the morphology of early Paleozoic vertebrates. From the investigations of paleontologists, comparative anatomists, and histologists we now know that all four of the principal types of vertebrate hard tissue—

calcified cartilage, bone, dentine, and enamel (enameloid) had already been invented by the most primitive known fishes, the ostracoderms (Kemp 1984; Smith and Hall 1990). In all of these sclerified tissues the mineral component is in the form of crystallites of hydroxyapatite, $3(Ca_3PO_4)_2 \cdot Ca(OH)_2$, deposited in an organic matrix. The basic mechanisms of tissue mineralization in the skeleton, skin, and teeth have remained remarkably conservative throughout vertebrate evolution despite the modulations that have introduced obvious extensive changes in organ morphology.

The earliest known vertebrate fossils are fragments of the dermal skeleton of a primitive agnathan fish, *Anatolepis*, found in late Cambrian and early Ordovician deposits in the Deadwood Formation of northeastern Wyoming in North America and also in Greenland and Spitzbergen (Smith and Hall 1990). The genus is recognized by the characteristic pattern of its flattened, nonoverlapping dermal denticles attached to a continuous underlying bony plate. For 100 million years throughout the Ordovician and into the Silurian periods, the ostracoderms continued to evolve as jawless fish with dermal armor before gnathostomes appeared in the fossil record. The dermal armor evolved into two principal patterns in the Ostracodermi: (1) the "heterostracan" type in the diplorhinid order Heterostraci as well as in the monorhinid orders Osteostraci, Galeaspida, and Anaspida, with denticles (odontodes) attached to plates, tesserae, or scalar sheets of bone deeply embedded in the dermis; and (2) the "thelodont" type in the diplorhinid order Thelodonti, with individual dermal denticles less deeply embedded in the dermis and not attached to an underlying bony plate. These types were the forerunners for the subsequent evolution of vertebrate scales and teeth. Smith and Hall (1990) believe that the thelodont pattern (Fig. 2-2) is more primitive and that chondrichthyan placoid scales (Fig. 2-2) have retained a thelodontlike pattern because it is a character plesiomorphic for vertebrate scales. It appears that the heterostracan pattern (Fig. 2-2) was adopted as the basis for scale evolution in all other groups of fish, including the extinct classes Acanthodii and Placodermi and the persisting class of bony fishes, the Osteichthyes (Fig. 2-2).

Peyer (1968) has concluded "that the transition from agnathous to gnathostomatous organization represented one of the most significant events in the phylogenetic history of vertebrates." Jawed vertebrates, the gnathostomes, first appeared as fossils in the Lower Silurian together with ostracoderms extant at that time (Romer 1966). The innovation of jaws made possible the dental lamina, which develops in a fold of the oral mucosa, bringing the potential for scale development to the expanded mouth. In this environment "scale primordia" had the space and nutritional requirements for growth as teeth (Fig. 2-2). Ingress of the dental lamina may have occurred synchronously with jaw evolution, but Reif (1982) favors the hypothesis that jaws evolved first without dentinous teeth, as in the extinct class Placodermi.

Earliest elasmobranchs were the Paleoselachii (Compagno 1991). *Cladoselache*, a primitive shark, discovered in the Cleveland Shale of Devonian age, had a streamlined body, dermal denticles, and teeth of the cladodont variety (Fig. 2-3) with a high, central pointed cusp and paired lateral cusps. Another order, the freshwater Pleurocanthodii, had pleurocanth teeth with two elongate lateral cusps and a smaller central cusp. Palaeoselachian sharks had become extinct by the end of the Permian to Triassic transition, marking the onset of the Mesozoic era.

An intermediate stage in selachian evolution was represented by the Protoselachii, which included sharks with hybodont teeth (Fig. 2-3). This group separated from the palaeoselachians in the late Paleozoic, and although decimated in the Permian to Triassic transition, they flourished throughout the Mesozoic Era. In the Triassic genus, *Hybodus*, the front teeth were sharp-cusped, but more posterior teeth had lower cusps sometimes reduced to low, rounded crowns. In the midline of both upper and lower jaws, teeth of some hybodont species were retained to form

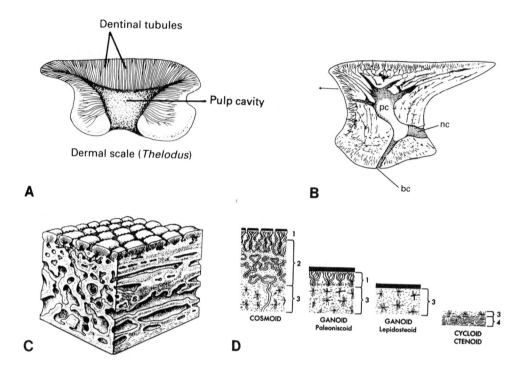

FIGURE 2-2 Evolution of scales in fishes. **A:** Thelodont type of placoid scale in the ostracoderm *Thelodus*, showing pulp cavity, dentinal tubules, and basal canal. **B:** Placoid scale of *Flegestolepis grossi*, a Paleozoic edestid shark, showing similarity to thelodont type. In addition to pulp cavity (pc) and basal canal (bc), neck canal (nc) had evolved and an enameloid layer covered the dentine. **C:** Heterostracan type of scale in the heterostracan genus *Psammosteus*, showing multiple odontodes surmounting a thick base of vascular and laminated bone. **D:** An evolutionary series of scale types in osteichthyan fishes, illustrating probable descent from the heterostracan type. Cosmoid scales of fossil fishes had an outer dentinous layer, cosmine (1), consisting of odontodes capped by enameloid (black layer) underlain by vascular (2) and laminated (3) layers of bone. Paleoniscid scales have lost the layer of vascular bone but retain the dentinous layer (1) and laminated bony layer (3); in addition these scales have a thick ganoid layer. Lepidosteoid ganoid scales have lost the dentinous layer. Cycloid and ctenoid elasmoid scales in teleosts have lost enameloid and dentinous components of scales, and have a fibrillary plate (4), which may be partially mineralized as laminated bone (3). (A, from Wolff 1991; B, after Karatajute-Talimaa, from Smith and Hall 1990; C, from Ørvig 1967; D, from Kent 1973.)

a symphysial whorl rather than being successively shed after attaining maturity, as in the genus *Helicoprion*, for example. Hybodonts have been extinct since the Cretaceous period but have been succeeded by the most recent sharks, the Neoselachii. This group arose during the Jurassic period and by the end of the Mesozoic era had radiated into most of the existing families of sharks. Modern sharks are classified in three principal suborders, the Squalomorphii, Squatinomorphii, and Galeomorphii (see chap. 1). *Heterodontus*, the Port Jackson shark, is considered by some authors to be a surviving member of the hybodont group; others classify it with the galeoids (Young 1981). Its teeth are of two types: sharp-pointed anteriorly and flat plates posteriorly.

Neoselachian sharks have teeth modified for various types of feeding behavior, including piercing (impaling), cutting, or crushing (Fig. 2-3). The sand tiger shark, *Carcharias taurus* (Young 1981; Compagno 1984), has fanglike teeth with a long central cusp flanked by two or more pairs of short lateral cusps, useful for seizing prey. Teeth of the blunt-nose

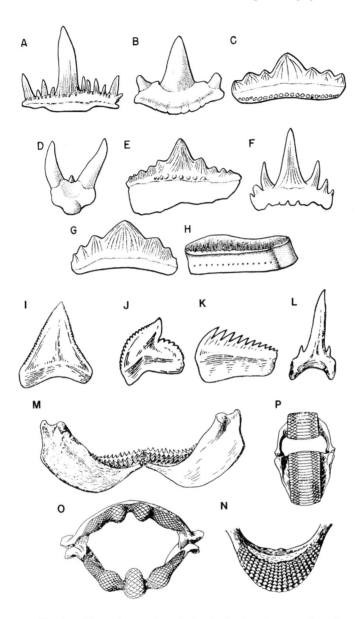

FIGURE 2-3 Drawings illustrating tooth evolution in fossil and recent elasmobranchs. **A–D:** Fossil cladodont-level teeth, showing high central cusp and shorter lateral cusps in *Cladodus* (**A, B**), *Protacrodus* (**C**), or reduced central cusp with elongated lateral cusps in *Xenacanthus* (**D**). **E–H:** Fossil hybodont-level teeth, showing tendency toward reduction of cusp height in *Hybodus* (**E,F**), *Acrodus* (**G**), and *Asterocanthus* (**H**). **I–P:** Neoselachian and batoid teeth, showing adaptations for cutting, piercing, seizing, and crushing. *Carcharodon* (**I**), *Galeocerdo* (**J**), and *Hexanchus* (**K**) have cutting teeth, *Carcharias* (**L**) has fanglike, piercing teeth. The blue shark *Prionace* (**M**) and the nurse shark *Ginglymostoma* (**N**) have pointed teeth for seizing, and the guitar fish *Rhinobatos* (**O**) and eagle ray *Myliobatis* (**P**) have flattened teeth for crushing. (A–H, from Schaeffer 1977; I–P, from Young 1981.)

sixgill shark, *Hexanchus griseus*, have a saw-toothed edge with cusps in a progression of sizes along the cutting edge. The great white shark, *Carcharodon carcharias*, has large triangular teeth with sharp, serrated edges, superb weapons for cutting. The asymmetrical teeth of the tiger shark, *Galeocerdo cuvieri*, are similarly efficient. Another carcharhinid shark, the blue shark, *Prionace glauca*, has triangular teeth, used for capturing fish and squid primarily. In the nurse shark, *Ginglymostoma cirratum*, the teeth are relatively broad with a short central cusp and a pair of shorter lateral cusps on either side.

Rays (Batoidea) first appeared as flattened, bottom-dwelling species, branching from protoselachian stock in the late Jurassic. By the end of the Cretaceous most of the modern skates and rays were already present. Their teeth are typically flat plates, adapted for crushing their molluscan food. The guitarfish, *Rhinobatos* (Fig. 2-3), has pavement-type dentition in both jaws. The pavement pattern is especially pronounced in the eagle stingray, *Myliobatis*, which has a dermal pattern of wide central plates bordered laterally by rows of small plates. Chimaeroids (Holocephali) have three pairs of broad tooth plates rather than individual teeth, two pairs in the upper jaw and a single pair in the lower jaw. This group is thought to be descended from primitive chondrichthyan stock extending back at least to the Devonian period. Some of their early forebears, a group of Palaeozoic sharks with slow tooth replacement rates, the Bradyodonti, also had teeth in the form of tooth plates.

The skin

The skin of vertebrates serves a variety of functions, including sensory reception, secretion, excretion, respiration, heat regulation, locomotion, coloration, digestion, nutrition, mineral deposition, and water balance (Van Oosten 1957; Whitear 1986). Because the skin encases the body tightly, it helps to maintain body form. Its tough, leathery consistency provides protection against mechanical abrasion. Skin of some shark species is used for making shoe leather. Although the epidermis of terrestrial vertebrates typically has an outer layer of dead keratinized cells, the stratum corneum, the entire thickness of the epidermis in most fish consists of living cells. The mantle of nonliving placoid scales protects the epidermis of sharks, but rays and chimaerids have reduced numbers of scales and gain protection instead from a slimy mucus covering, the cuticle (Whitear 1986). Shark embryos are also protected by mucus, but scales displace mucoid cells during body growth so that most adult sharks have only a thin coat of mucus that remains under the scales. The bramble shark, *Echinorhinus brucus*, however, has a mucus coat several millimeters thick (Reif 1985a).

DIFFERENTIATION AND HISTOLOGY

The skin consists of an outer epithelial layer, the epidermis, and an underlying connective tissue layer, the dermis or corium. Between these multicellular layers is a thin noncellular boundary layer, the basement membrane or basal lamina. Epidermis differentiates from embryonic ectoderm, whereas the dermis is a derivative of paraxial mesoderm (dermatome region) in trunk and tail and of head mesenchyme derived from neural crest. Some workers have speculated that mesoderm of the embryonic somatopleure (body wall ectoderm plus underlying somatic mesoderm) may also contribute mesenchyme for the dermis (Kresja 1979). Neural crest contributes ectomesenchyme for head skin, for the dentine of dermal scales and teeth, and for the chromatophores throughout the body.

Differentiation of the skin in sharks is illustrated in Figure 2-4A. At an early stage the skin consists of epidermis and dermatome overlying the myotome, primordium of body wall muscle. Later the epidermis becomes two-layered with an outer periderm and a basal stratum germinativum, while the dermatome proliferates as dermal mesenchyme. Scales begin to develop as dermal papillae and overlying cells of the stratum germinativum bulge outward. Subsequently the epidermis thickens by proliferation from basal cells, and unicellular gland cells differentiate. Scales increase in height and differentiate with an inner dental papilla and layer of odontoblasts surrounded by columnar ameloblasts. Odontoblastic and ameloblastic activity result in deposition of mineralized layers of dentine and enameloid. Vascular and laminated (fibrous) layers and pigment cells (chromatophores) differentiate in the dermis. In mature skin the fully developed scales project through the surface. Histological features of the skin in a leopard shark, *Triakis*, are illustrated in Figure 2-4B, reproduced from Daniel (1934). In the epidermis the stratum germinativum is overlain by several generations of common epithelial cells (terminology of Whitear 1986). Mucus gland cells are present and the primordium of a scale protrudes from the dermis. Below the epidermal basement membrane the dermis is organized as an outer

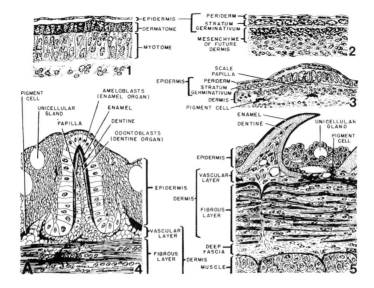

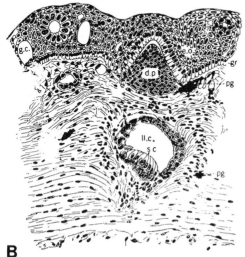

FIGURE 2-4 **A:** Differentiation of the skin in sharks. Early stage (1): Epidermis and dermatome are overlying myotome. Later stage (2): Periderm and stratum germinativum of epidermis; mesenchyme of dermis are seen. Scale primordium (3): Dermal papilla and adjoining columnar cells of stratum germinativum are seen bulging outward. Histogenesis of skin layers (4): Thickened epidermis is shown with common epithelial cells and unicellular gland cells; peripheral cells of developing scale are ameloblasts. Dermal components of scale are the dermal papilla and layer of odontoblasts. Mineralized layers of dentine and enamel (enameloid) are developing between ameloblasts and odontoblasts. Dermis has outer vascular layer containing pigment cells, and deeper fibrous (laminated) layer. Mature skin (5): Epidermal and dermal layers are fully differentiated above deep fascia (connective tissue) and muscle. Erupted scale is capped by exposed enamel layer.
B: Section of integument of leopard shark *Triakis semifasciatus*. d.p. = Dermal papilla; e.o. = enamel organ; g.c. = gland cell; gr = germinative layer of epidermis; ll.c. = lateral line canal; n. = nerve branch; pg = pigment cell; sc = sensory column in lateral line canal. (A, from Nelsen 1953; B, from Daniel 1934.)

layer of loose connective tissue and a deeper layer of compact connective tissue. Blood vessels, nerves, and pigment cells accompany the connective tissue fibroblasts.

EPIDERMIS AND ITS DERIVATIVES

Specialized epidermal cells include the unicellular glands, which may be mucous or granular (Whitear 1986). Multicellular epidermal glands occur in some species. The highly venomous glands embedded in the dermis of paired longitudinal grooves on the ventral side of the long spines of the stingray, *Dasyatis*, and the eagle ray, *Myliobatis* (Bertin 1958b; Quay 1972), are in this category. Males of both elasmobranchs and holocephalans have multicellular, mucus-secreting glands at the base of the claspers (Daniel 1934). Specialized luminescent organs called photophores occur in the skin of some deepwater elasmobranchs, such as *Spinax*, *Etmopterus*, *Isistius*, and *Euprotomicrus* (Daniel 1934; Bertin 1958b; Reif 1985a). In the embryo at neurula and tailbud stages, the epidermis develops a series of basal epidermal thickenings called epidermal placodes. These invaginate and differentiate as the specialized epithelium of head sense organs and of the cutaneous lateral line sense organs (Nelsen 1953; Budker 1958). Paired olfactory, lens, and acousticolateralis (octavolateralis) placodes develop alongside the rudimentary brain, and epibranchial placodes develop above the gill arches. Olfactory placodes produce the nasal epithelium; lens placodes form the eye lenses after optic lobes from the brain make contact with the epidermis; and the acoustic components of the acousticolateralis (octavolateralis) placodes produce the epithelium of the labyrinth of the ear. Lateralis components grow out anteriorly and posteriorly to form the lateral line canals of the head, trunk, and tail. Mechanoreceptive neuromasts develop within these canals and also in individual pit organs open to the exterior through small pores. Electroreceptive neuromasts develop in the ampullae of Lorenzini opening by relatively large pores on the head.

DERMIS

In fishes and larval amphibians, the dermis begins to develop when a layer of orthogonally aligned fibroblasts and collagen fibers differentiates in contact with basal lamina underlying the epidermis. As development proceeds, the primordial dermis delaminates (Kresja 1979) and becomes the deep layer of the dermis, the stratum compactum. Loose connective tissue then proliferates between epidermis and stratum compactum to form the outer layer of the dermis, the stratum vasculare. Blood vessels, nerve endings, and neural crest cells invade the stratum vasculare and spread throughout the dermis.

Neural crest cells play a major role in the differentiation of the dermis. They are the source of the head mesenchyme, and they provide mesenchyme for the development of scales and chromatophores throughout the body. Dermal denticles develop as a result of inductive interaction between dermal papillae of neural-crest–derived mesenchyme and overlying cells of the epidermal stratum germinativum, a relationship that Moss (1968, 1972) has named the "epidermal coparticipation hypothesis." Dermal papilla produces dentine and pulp. Epidermal ameloblasts contribute matrix for the outer enameloid cap of the scale. Mesenchyme at the base of the scale differentiates as the bony base embedded in the stratum vasculare.

SKIN COLORATION

With the possible exception of the pigmented layer of the retina, pigment cells in vertebrates are derived from the neural crest. Prospective chromatophores migrate ventrolaterally in dermal mesenchyme and differentiate as melanophores, xanthophores, erythrophores, and leucophores (iridophores) within the dermis (Bertin 1958a; Taylor and Bagnara 1972). Some chromatophores localize along the underside of the basal lamina, but they may also occur at deeper levels, including within the stratum compactum. Melanophores may invade the epidermis in some vertebrates, including humans, and here are called melanocytes. Melanophores contain the pigment melanin in specialized organelles called melanosomes. Melanins are classified as brown to black eumelanins and yellow to red phaeomelanins. Xanthophores and erythrophores contain yellow to red carotenoid or pteridine pigments. Iridophores contain crystals of guanine, hypoxanthine, or adenine, or combinations of these purines, which reflect light so that the skin appears white or silvery. Work on the distribution of pigment cells in the skin of teleost fishes and amphibians has established that the spots or stripes characteristic of some species are the result of aggregation of chromatophores in distinctive assemblages. Bagnara et al. (1968) have defined a "chromatophore unit" consisting of a basket of melanophores enclosing an assemblage of xanthophores or erythrophores.

Coloration among elasmobranchs is variable, but the colors tend to be muted in comparison to the bright hues of yellow, red, and blue seen in some teleost fish (Daniel 1934). Most sharks are shades of brown, gray, or black dorsally and laterally. Often the venter is white or a contrasting dark shade. Although some sharks may include yellow, red, or blue in their color pattern, these colors are more frequently seen in skates and rays. Many elasmobranchs have characteristic markings contrasting with the background color. These may be in the form of spots, stripes, bars, blotches, or a network of lines (reticulations).

Among sharks (Compagno 1984), the great white shark, *Carcharodon carcharias*, is an example of a species that is uniformly dark gray dorsally and white ventrally. The carcharinid blue shark, *Prionace glauca*, is dark blue above, bright blue laterally, and white below. The tiger shark, *Galeocerdo cuvier*, is gray with dorsal black spots and parallel vertical bars, which are prominent in the young but fade in adults. The leopard shark, *Triakis semifaciatus*, is gray or bronze-gray above and white below with a striking pattern of broad, black saddle marks and scattered large black spots. The ornate angel shark, *Squatina tergocellata*, has prominent large brown ocelli (eye spots) both on the body and the greatly expanded pectoral and pelvic fins. The Japanese wobbegong, *Orectolobus japonicus*, has a series of dark rectangular dorsal saddles enclosing irregular light spots, interspersed with light intersaddle areas containing dark, reticulated lines. The whale shark, *Rhiniodon typus*, is marked with a checkerboard pattern of vertically and horizontally aligned light stripes on a dark background with light spots in the dark areas.

Among batoids, the dorsal surface is generally brown, gray, black, or yellowish, but red, blue, and green colors occur in some species. The spotted skate, *Raja montagui*, is yellow with an abundance of small dark spots. The thornback ray, *Raja clavata*, is gray or brown with lighter patches and has a large black-and-yellow marbled spot on each broad pectoral fin. The manta ray, *Manta birostris*, is dark brown or black above, white below. In manta rays, the occurrence of distinct color patterns unique to individuals on both the dorsal and ventral surfaces is well known. The marbled electric ray, *Torpedo marmorata*, is light or dark brown with darker marbling. Among stingrays, the common stingray, *Dasyatis pastinaca*, is gray, brown, reddish or olive-green, and the young may have white spots. The blue stingray, *Dasyatis violacea*, is gray or brown with a violet tinge. The red stingray, *Dasyatis akajei*, and the sepia stingray, *Urolophus auranticus*, are common Asian species (Sasagawa and Akai 1992). The chimaeroid, *Chimaera colliei*, is brown dorsally with a series of large white spots extending longitudinally in the brown area. Laterally the spotting pattern continues against a background of metallic shades of brown, yellow, and blue. Ventrally the skin is lighter and unspotted (Dean 1906).

An especially intriguing topic in vertebrate development is the relationship between brain and skin in differentiation of the lateral eyes. Optic lobes evaginate bilaterally from the diencephalon of the brain and induce formation of the lens from epidermis. Subsequently the lens becomes detached from the epidermis and enclosed within the optic cup. The cornea is a transparent region of modified skin over the anterior chamber of the eye. Less well known is the vertebrate "third eye" (Eakin 1973), which develops from the epiphysis, a middorsal evagination of the diencephalon. The outer end of the epiphysis (pineal) in vertebrates from cyclostomes to reptiles develops as a vesicular structure containing light receptive cells in a retina similar to that in the lateral eyes. Transparent skin over the third eye of the lizard Sceloporus functions as a cornea, covered by a transparent interparietal scale. The epiphysis in the dogfish *Scyliorhinus canicula* is a long tube up to 16 mm long, only the terminal 1 mm of which is the light-sensitive end vesicle (Rüdeberg 1969). A pigment-free patch of skin over the region of the epiphysis has been reported for some cyclostomes, fishes, frogs, and lizards. Gruber et al. (1975) have discovered a region of reduced opacity functioning as a "window" for light transmission in the chondrocranium of *Negaprion brevirostris*, as well as in the bull shark, *Carcharhinus leucas*, and the smooth dogfish, *Mustelus canis*. Most birds and all investigated mammals have lost the neurosensory function of their epiphyseal (pineal) outgrowths but their pineals are active as endocrine organs.

Elasmobranchs, like many teleost fishes, have the ability to undergo color changes in response to light stimuli. Wilson and Dodd (1973) have shown that lateral eyes are the primary receptors for mediating such changes but that the pineal end vesicle also may affect skin color. Light stimuli cause the pituitary to release melanophore-stimulating hormone (MSH) into the bloodstream. Absorbed by the melanophores in the dermis, the hormone causes their dispersion and the skin darkens. Blinding in *Scyliorhinus canicula* caused a rapid darkening response followed by gradual paling; darkening the surroundings of blinded fish caused slight further paling, indicating a nonvisual effect. Pinealectomy abolished the influence of the pineal on skin color. Experiments on juvenile hammerhead sharks, *Sphyrna lewini*, by Lowe and Goodman-Lowe (1996) have demonstrated that exposure to bright sunlight in a shallow pond resulted in their "tanning." Light tan pups collected from deep murky water in Kaneohe Bay, Hawaii, became dark brown to black after exposures of 21 days and 215 days. Sections of skin showed that the change was due to increased melanin in the dermis.

Placoid scales (dermal denticles)

FORM AND DISTRIBUTION

Scales of some type are present in the skin of all groups of extant fishes except the cyclostomes. Agassiz (1833–44) classified modern fish into four groups according to their scale types, namely, placoid, ganoid, cycloid, and ctenoid. As other comparative anatomical features became known, later systematists tended to minimize the importance of the exoskeleton as a basis for fish taxonomy. Renewed interest in scalation, however, was aroused by the investigations of Goodrich (1904, 1907) on dermal scales and finrays. His evaluations of paleontological evidence were a valuable supplement to his own comparative anatomical and histological studies.

Placoid scales (dermal denticles) are a characteristic of the skin of elasmobranchs and holocephalans. They usually cover the skin of selachians, are distributed discontinuously in batoids, and are scarce in chimaerids (Bertin 1958c; Schaeffer 1977; Reif 1985a). Typically they consist of (1) a basal plate embedded in the dermis, (2) a pedicel that rises from the base and forms a neck connecting with the crown, and (3) the exposed outer portion, the crown (Applegate 1967). Observed outwardly, the scales assume a variety of shapes and sizes depending on the species. All of these varieties can be derived by geometrical transformation of the simple quadrilateral type (Fig. 2-5A) seen in body scales of *Heterodontus* (Reif 1985a). They may be relatively flat, often with ridges and spines running anteroposteriorly. A common type of denticle in many sharks is broad, ridged, and tricuspidate, as in the blackmouth cat shark, *Galeus melanostomus* (Fig. 2-5B). Such scales have a prominent main longitudinal ridge terminating at the central cusp and two lateral ridges extending to the lateral cusps. The valleys between ridges provide a streamlined surface that directs laminar flow of water, thereby reducing drag during active swimming (Reif 1985a,b). Enlarged denticles occur on the upper side of the pectoral fin in many species and may form a crest along the dorsal side of the caudal fin or the ventral side of the caudal peduncle. Differing shapes of the body denticles in *Heptanchus maculatus* have been described by Daniel (1934). The denticles surrounding the pores of the neurosensory pit organs are arranged in a special pattern, usually with an anterior and posterior twin pair around the pore (Budker 1958).

In the bramble shark, *Echinorhinus brucus*, the crowns of the widely scattered round denticles are cone-shaped, described as thorn- or bucklerlike (Compagno 1984). Similarly, in some rays the scales are pointed thorns or enlarged to form shield-shaped bucklers, as in the skate *Raja clavata*. Denticles in the gulper shark, *Centrophorus granulosus*, are rhomboidal, with the crowns sessile on the basal plates. In a related species, *Centrophorus lusitanicus*, the denticles are elongated into a single cusp directed posteriorly; such scales are called lanceolate. The lined lantern shark, *Etmopterus bullisi*, has denticles bearing slender, conical hooks at their posterior tips. The denticles of the sawback angel shark, *Squatina aculeata*, are pyramidal with three strong ridges. An unusual type of denticle occurs in the long-snout dogfish, *Daenia quadrispinosum*. In this species the scales are called pitchfork denticles because the high pedicel terminates in four tines.

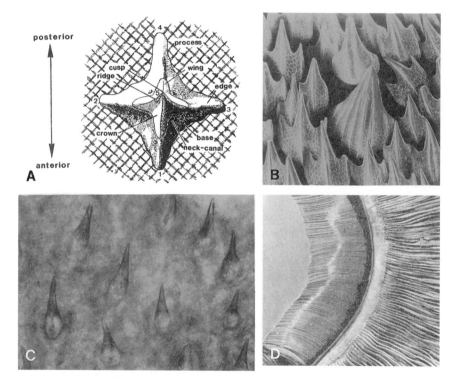

FIGURE 2-5 Scale types in elasmobranchs. **A:** Diagram showing the basic quadrilateral form of shark scales with base embedded in the dermis and exposed crown extended as ridges and cusps. **B:** Scanning electron micrograph showing chiefly triradiate body scales of a 67-cm specimen of the blackmouth cat shark, *Galeus melanostomous*; large scale with multiple lateral ridges borders a pit organ (×74). **C:** Mucous membrane denticles in the oral mucosa of the skate *Raja clavata*. Denticles project posteriorly (×95). **D:** Gill rakers project as a double row from the inner (concave) side of a gill arch; gill filaments project from the convex side of the arch. (A, B, from Reif 1985; C, from Peyer 1968; D, from Arambourg and Bertin 1958.)

The fin spines of sharks and holocephalans are considered to be highly modified dermal scales although, unlike the dermal denticles, they continue to grow throughout life (Reif 1982). Maisey (1979) considers fin spines, scales, and teeth as related differentiation products of the interaction between a dermal papilla and its capping epidermis. The poisonous spines on the tails of stingrays are also thought to be modified scales and like the latter do not grow after reaching a definitive size. Sawfishes of the batoid family Pristidae, for example, *Pristis*, have an extended rostrum equipped with lateral "teeth," which are highly modified scales. The saws may reach 1 ft in width and 6 ft in length in large pristids, and in such fish the "teeth" may be up to 4 in long. They remain constant in number and continue to grow. In contrast, the rostral "teeth" on the rostrum of saw sharks in the selachian family Pristiophoridae, *Pristiophorus* and *Pliotrema*, for example, are nongrowing and are replaced by others of larger size as ontogeny progresses.

Stomodaeal denticles and gill rakers

When the transition from agnathan to gnathostome vertebrates occurred in the Silurian period, the capacity to form denticles was carried into the mouth cavity. Evolution of teeth within a dental lamina was not the only consequence of this ingression. Actually the entire oral mucosa retained the ability to form denticles similar to those in the skin. Within the mouth and pharynx they are known as stomodaeal denticles (Fig. 2-5C). Fahrenholz (1915) distinguished between two parts of the foregut, that portion of the digestive tract within the head. The forward part in his terminology is the *Munddarm*, or mouth gut; the rear part is the *Kiemendarm*, or gill gut. The whole region is the *Mund-Kiemendarm*. In 42 species he examined, distribution of stomodaeal denticles was variable, but he found that they might occur over the entire oropharyngeal mucosa. Since the denticles usually project only slightly through the mucosa, Fahrenholz considered it unlikely that they serve any mechanical function in food processing. A remarkable modification for feeding, however, has occurred in the evolution of the largest species of existing sharks, the giant whale shark, *Rhiniodon typus*, and the large basking shark, *Cetorhinus maximus*. Despite their huge size, these fish do not rely on their relatively small teeth to capture food. Instead they are filter feeders. Slender filaments projecting inward from the pharyngeal mucosa covering the inner side of the gill arches are called gill rakers. In *Rhiniodon* and *Cetorhinus* the gill rakers are greatly expanded (Fig. 2-5D) and overlap to form an efficient screen for filtering small planktonic food organisms from water taken in through the mouth and discharged through the gill slits (Arambourg and Bertin 1958). In *Cetorhinus* the gill rakers are probably shed and replaced annually during a feeding pause in winter (Parker and Boesman 1954; Reif 1985a).

HISTOLOGY

The monumental study by Hertwig (1874) on the development of shark scales and teeth clearly established the homology of these structures. Figure 2-6, from Bertin's (1958c) review, is a drawing first published by Gnadeberg (1926), based on one of Hertwig's original illustrations. It shows that the basal plate of the denticle is situated in the stratum vasculare of the dermis, underlain by the stratum compactum. The crown contains a pulp cavity of connective tissue, surrounded by a relatively thick layer of dentine and a thin layer of enamel (enameloid). Klaatsch (1890) described the progress of development of placoid scales in a series of sequential steps, summarized by Bertin (1958c): (1) Connective tissue cells including presumptive odontoblasts accumulate beneath the epidermis. (2) A dermal papilla pushes outward, capped by basal epidermal cells. (3) The base of the papilla sinks downward into the dermis. Odontoblasts become aligned around the periphery of the pulp tissue and secrete a matrix of dentine externally. As the dentine layer thickens, odontoblastic processes from the odontoblasts remain embedded in the mineralized matrix,

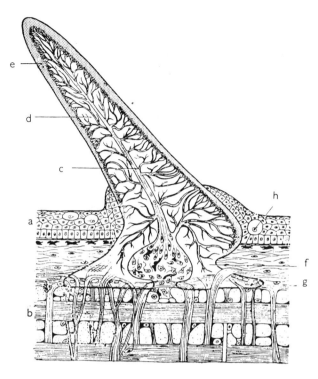

FIGURE 2-6 Diagram of shark placoid scale, showing basal plate and neck in dermis and crown projecting through and above epidermis. a = Epidermis; b = dermis; c = dental papilla; d = dentine; e = enamel (enameloid); f = neck; g = basal plate. (From Bertin 1958c.)

while cell bodies of the odontoblasts remain at the periphery of the pulp in a cellular layer underlying the dentine. (4) Centripetal deposition of the dentinal ivory does not hinder ameloblasts of the epidermal enamel organ from depositing enamel centrifugally outside of the dentine. (5) An outgrowth of connective tissue from the base of the denticle forms a basal plate, which grows peripherally and becomes calcified. Thus the scale body is of mixed origin: dentine distally and calcified fibrous matrix (acellular bone) basally. (6) In the final stage, the scale pierces through the epidermis, thus becoming visible and assuming its role of protecting the integument.

Scales of elasmobranchs are described as nongrowing, because their span of growth is limited. During the life of an individual, scales grow to a definitive size dependent on the size of the animal. Thereafter they are rejected and replaced by scales of larger size. This renewal process is continually repeated, but there is no synchronous pattern of molting in selachians. Individual denticles are rejected and replaced independently of their neighbors.

Budker (1938) has described the decline and removal of ageing denticles. If one examines the skin of a shark, it is apparent that some of the denticles appear white. These are in the process of self-destruction. The process starts by disaggregation of the basal plate and thinning of the dentine layer in the neck. Unmineralized connective tissue increasingly occupies the interior of the scale so that it loses firm anchorage to the dermis. Eventually the enamel layer is resorbed, and the void left by the resorbed scale is rapidly filled by the extension of neighboring scales. Amoeboid connective tissue cells participating in denticle resorption are called odontoclasts (Bertin 1958c; Applegate 1967).

According to the "lepidomorial" theory of Ørvig (1951) and Stensiö (1961), growing scales of elasmobranchs begin their development as a hypothetical unit called a *lepidomorium*, consisting of a crown and base with one neck canal and one basal canal (Reif 1982). A scale that continues to grow by addition of successive increments is called a cyclomorial scale. If growth is abbreviated so that the scale grows by a single morphogenetic step rather than a succession of steps, the scale is called a synchronomorial (nongrowing) scale. By these criteria, the placoid scales of elasmobranchs are synchronomorial, whereas the continually growing elasmoid scales of teleost fish are cyclomorial. Reif (1982) argues that there is no actual evidence for the transition from cyclomorial to synchronomorial in any elasmobranch lineage. He states that in a skeleton of growing scales "new elements are welded *to* existing scales, and in a skeleton of nongrowing scales new elements are inserted *between* older scales as separate scales." Reif rejects the lepidomorial theory as a basis for explaining the differing shapes of elasmobranch scales. According to his "odontode regulation" theory, "complicated shapes of scales result from differentiation of the scale germ rather than from a synchronomorial morphogeny." Variations in the regulation of differentiative potential can account for the variation in numbers of cusps or the variable shapes and sizes of scales developing from individual primordia.

Dermal fin rays

Dermal denticles cover the entire surface of the fins in sharks, and beyond the radials the fins are supported internally by collagenous horny fin rays, which Goodrich (1906) named "ceratotrichia." These rays, described by Goodrich as "fibrous, flexible, unjointed, rarely branched, cylindrical rods," develop in the dermis underlying the epidermis on both sides of the median and paired fins. The tapering proximal ends of the ceratotrichia overlap the distal

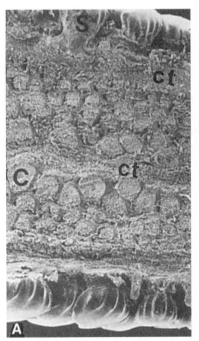

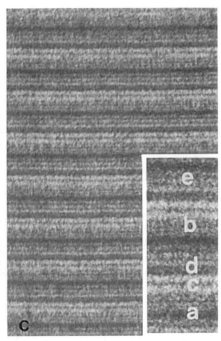

FIGURE 2-7 Ceratotrichia of shark fins. **A:** Scanning electron micrograph of the cross-sectioned surface of a block of tissue excised from the anterodorsal fin of the lemon shark, *Negaprion brevirostris*, showing scales (S), bilateral rows of ceratotrichia (C), and zones of dermal connective tissue (ct) centrally and subepidermally (×45). **B:** Transmission electron micrograph of a longitudinal section of fin dermis, showing narrow conventional collagen fibers adjacent to a wider incipient stage of a ceratotrichium (C) (×26,640). **C:** Longitudinal section of portion of a wide ceratotrichium, showing banding pattern. Insert shows the a, b, c, d, and e bands characteristic of collagen fibers (×175,000; insert ×304,685). (A–C, from Kemp 1977.)

ends of the radial cartilages and their distal ends extend to the edge of the fin. They develop in the dermal connective tissue by polymerization of collagen fibrils (Kemp 1977). Ceratotrichia develop first in a single layer underlying the epidermis, but new generations continue to develop subepidermally so that a multilayered assemblage of these fin rays is present under the epidermis in mature fish.

A section from the anterodorsal fin of the lemon shark, *Negaprion brevirostris* (Fig. 2-7A), shows several rows of ceratotrichia on either side of the central zone of connective tissue. Those deepest in the ceratotrichial layer are larger in diameter than those nearer to the epidermis. A longitudinal section showing conventional collagen fibers adjacent to a newly forming ceratotrichium in the connective tissue of the fin (Fig. 2-7B) demonstrates that both fiber types are crossbanded with a periodicity of about 64 nm. The accompanying section of a portion of a large ceratotrichium (Fig. 2-7C) shows details of its banding pattern. An enlarged view of this pattern (Fig. 2-7C insert) permits identification of the same bands (a, b, c, d, e) designated by Hodge (1967) in mammalian collagen fibers. Clearly the ceratotrichia grow in diameter and length by incorporation of collagen fibers (Kemp and Park 1970).

Fin spines

Single, stout fin spines are present anterior to the dorsal fins of living sharks in the families Heterodontidae, Squalidae, and rarely, in the Echinorhinidae (Maisey 1979). Similar fin spines, however, were present in many fossil elasmobranchs and placoderms. Maisey has studied the development of fin spines in *Squalus acanthias* (Fig. 2-8A,B) and *Heterodontus*. Like the dermal denticles, a fin spine develops from a dermal papilla covered by epidermal epithelium. Anteriorly and laterally the epidermis differentiates into outer and inner epithelial layers. Posteriorly the epidermis thickens as the primordium of glandular tissue at the base of the spine. Projecting into the interior of the dermal papilla is a cartilaginous core, and for this reason the spine develops differently from a dermal denticle. Dentine develops from the dermal papilla, and the inner epithelial layer functions as an enamel organ. The dentine separates into two layers called the trunk and mantle layers respectively. The primordium of the trunk layer starts as a rear plate (*hintere Faserplatte*)

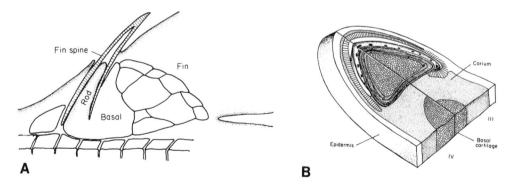

FIGURE 2-8 Diagrams illustrating structure and development of fin spines in the dogfish *Squalus acanthias*. **A:** Fin spine of dorsal fin supported basally by a rod of cartilage projecting from basal cartilage, which abuts the vertebral column. **B:** Three-dimensional model of a section through the base of a fin spine, showing relationship to its triangular central rod of cartilage and the fin basal cartilage posteriorly. Immediately around the central cartilage is the connective tissue of the pulp cavity, which underlies two layers of dentine (inner trunk and outer mantle layers). Capping the spine anteriorly is a layer of enameloid (white with dark inclusions). Columnar epithelium of epidermis is the stratum germinativum for enameloid development. Dermal connective tissue (corium) fills the gap between basal cartilage and fin spine. (From Maisey 1979.)

of mesenchyme, which elongates basally and also grows anteriorly around the cartilage so that it forms a complete ring surrounding an inner segregated portion of the pulp cavity enclosing the cartilage. Before the ring closes, it is open to the rest of the pulp cavity through an anterior fissure. Around the outside of the more peripheral part of the pulp cavity the primordium of the mantle layer develops as the fore plate (*vordere Faserplatte*). It spreads over the anterior margin and sides of the spine under the enamel organ. Both trunk and mantle layers consist of dentine and eventually become fused so that only the central pulp cavity remains around the basal cartilage and distal to it. The fin spines of *Squalus* are pigmented over their exposed crowns but unpigmented over their bases embedded in the skin. Glandular tissue differentiating from the epidermis posterior to the base of the spine is generally found to be atrophied in full-grown adults (Maisey 1979). The mature fin spines of elasmobranchs and chimaerids lack the outer enameloid layer characteristic of dermal denticles.

Teeth of elasmobranchs

Except for the cornified teeth of lampreys and amphibian larvae, the teeth of vertebrates always have dentine as a principal component; hence Waldeyer (1871) named them *Dentinzähne* (dentine teeth) to distinguish them from the nondentinous teeth of cyclostomes and invertebrates. Dentine is the mineralized tissue surrounding the interior pulp cavity within the crown of the tooth and is usually capped by a hypermineralized layer of enamel (enameloid). In elasmobranchs the base of the tooth embedded in the lamina propria of the oral mucosa is composed of acellular bone. It has been suggested that this "bone of attachment" (Smith and Hall 1990) may be homologous to the cement surrounding the dentine of the embedded tooth base of mammalian teeth.

With respect to the origin of the dental lamina, Reif (1982) concludes that there is no indication that it arose independently in the Chondrichthyes and Teleostomi, and thus "one has to assume that it evolved in the common ancestors of both groups after the line leading to the Placodermi had branched off." A common characteristic of chondrichthyans and osteichthyans is that their teeth are polyphyodont, that is, they develop in rows of tooth families within which older teeth function for a time and then are replaced by new teeth in each tooth family. Elasmobranchii and Holocephali have placoid dermal scales and similar fin spines. They are clearly related and have been grouped together as Elasmobranchiomorphi (Jarvik 1980). Yet the teeth in these two groups are markedly different. In elasmobranchs the teeth are unquestionably modified scales, whereas holocephalans have large tooth plates instead of individual teeth. Recent chimaerids have two pairs of plates in the upper jaw and one pair in the lower. According to one theory, chimaerid tooth plates were derived from an originally polyphyodont condition. A recent investigation on development of the tooth plates in *Callorhynchus milii* (Didier et al. 1994) supports this position by demonstrating that each tooth has an oral and an aboral territory representing the fusion of two members of a reduced tooth family. The plates appear to be derived by concrescence of originally separate tooth primordia.

DISTRIBUTION AND
RATE OF REPLACEMENT

The open mouth of a shark presents an array of shiny white teeth set in serried rank and file in both upper (palatoquadrate) and lower (Meckel's cartilage) jaws. A photograph of the lower jaw of the spiny dogfish, *Squalus acanthias* (Fig. 2-9A), shows six ranks extending from the lingual to the labial side of the jaw. In this specimen there are about 24 files (cross rows), 12 on either side of the jaw symphysis in the midline. Crowns of the teeth are directed posteriad and their tips point in opposite directions on left and right sides. Functional teeth are erect at the outer margin of the jaw. Replacement teeth are recumbent against the jaw surface.

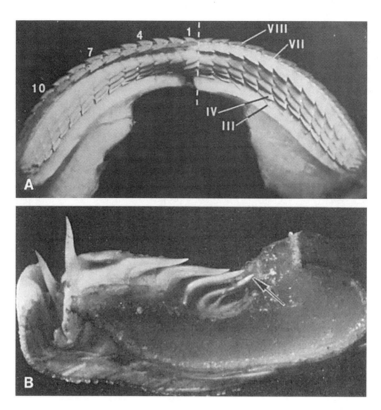

FIGURE 2-9 **A:** Photograph of teeth in the lower jaw of *Squalus acanthias*, showing symmetrical distribution on either side of the central symphysis. Erect functional teeth are in rank VIII along the jaw margin. Files of replacement teeth in ranks III to VII are aligned posteriorly (×1.75). **B:** Lateral view of block of tissue excised from the lower jaw of the shark *Carcharhinus menisorrah*, showing a file of teeth at successive stages of growth. White mineralized tip of tooth (*arrow*) protrudes from tooth follicle posteriorly. Most anterior tooth in trough between lip and jaw has moved forward from the functional position prior to being sloughed. Jaw cartilage is surrounded by calcified plaques, the tesserae (×5.0). (A, from Samuel et al. 1983; B, from Kemp and Park 1974.)

Elasmobranch teeth are lyodont (Moss 1967), meaning that they are not ankylosed to the jaw cartilages. The teeth are anchored in the connective tissue covering the jaw cartilage but are continually being pulled forward as they mature and their anchoring tissue migrates anteriad (James 1953). Functional teeth eventually become detached and are replaced by the generation of teeth next in line behind them. Rates of replacement, that is, the time for one tooth row to be replaced by the next row, have been determined for a number of species (tabulated in Luer et al. 1990). Expressed as time per row, the replacement rate has been reported as varying in different species from 8–10 days/row in *Negaprion bre-virostris* to 5 weeks/row in *Scyliorhinus canicula*. Luer et al. calculated the replacement rate for teeth in three juvenile specimens of the nurse shark over a 3-year period. During summer months when water temperatures were 27°–29°C the rate was 9–21 days/row; in winter months at 19°–22°C the rate was 51–70 days/row. Moss (1967) has shown that tooth size in *Negaprion brevirostris* increases regularly with increase in body length. This relationship has led Luer et al. to theorize that "if tooth size does in fact increase as a shark grows, then tooth replacement rates should change in order to keep pace with changing growth rates. The most rapid rates of tooth replacement for a given species of shark should occur during its juvenile years when growth rates are most rapid. As a shark begins to mature sexually and its growth rate decreases, a corresponding decrease in tooth replacement rates should prevail."

DEVELOPMENT OF DENTINE AND ENAMELOID

Teeth begin their development in the dental lamina, a fold of ectodermally derived oral epithelium underlain by mesenchyme, extending posteriad on the buccal side of the jaw cartilage (Reif 1980). A section of the lower jaw from buccal to labial side shows a succession of developmental stages in a tooth cross row (Fig. 2-9B), starting with tooth buds at the inner end of the lamina and continuing through successive stages of mineralization to the terminal functional stage. Structurally the functional teeth are like the dermal denticles of the skin, except for their larger size. The illustration shows the mineralized prisms of calcified cartilage, the tesserae (Applegate 1967; Kemp and Westrin 1979), which are characteristic of the endoskeleton of the Chondrichthyes. Teeth are anchored in the perichondral connective tissue investing the jaw cartilage.

Shark teeth develop more or less synchronously in successive ranks around the contours of both upper and lower jaws. Early tooth buds of first- and second-rank teeth at the inner (buccal) end of the dental lamina (Fig. 2-10A) develop as progressively larger, dome-shaped dental papillae of mesenchyme surrounded by inner dental epithelium (i.d.e.) of the bilayered enamel organ. Growing teeth move anteriad and heighten to cone-shaped form as the dental lamina shifts toward the outer side of the jaw. Already in a low cone stage, extracellular matrix begins to accumulate at the tooth tip between ameloblasts and odontoblasts. This is the precursor of the enameloid matrix, which in sharks and other elasmobranchs begins to mineralize before dentine.

The tip of a tooth in the third successional rank in a specimen of *Carcharhinus menisorrah* is illustrated in Figure 2-10B. Here enameloid matrix forms a prominent layer between ameloblasts and odontoblasts. In this demineralized and stained section, dot-shaped or elongated sections of odontoblastic processes extend outward into the enameloid matrix. Bases of third- and fourth-rank teeth are seen in Figure 2-10C. In the third-rank tooth the enameloid extends to the base, but no dentine layer is yet present. The more elongated tooth in rank four has a relatively thin layer of dentine under the enameloid, broadening to a thicker layer more basally.

The sequence of enameloid deposition preceding dentine mineralization is shown dramatically in a microradiograph published by Grady (1970), which distinguishes the hypermineralized enameloid from the less-dense dentine (Fig. 2-11A). This picture reveals that both enameloid and dentine layers increase in thickness as the teeth move forward. Enameloid growth terminates early and limits overall tooth size, whereas dentine continues to increase in thickness at the expense of the soft tissue of the pulp cavity. An enlarged view of the tooth tip in a mineralizing tooth (Fig. 2-11B) shows the connective tissue of the pulp cavity surrounded by an unmineralized matrix of predentine underlying a thick layer of dentine. Cell bodies of odontoblasts remain at the periphery of the pulp chamber, which becomes progressively narrower. Elongate odontoblastic processes extend through predentine and dentine layers and continue into the enameloid layer. In shark teeth the odontoblastic processes reach nearly to the layer of ameloblasts, whereas in tetrapod teeth the odontoblastic processes do not extend beyond the dentino-enamel border. Teeth attain full growth before reaching the margin of the jaw so that several ranks may be fully developed before reaching the outer functional position. Details of the process of mineralization in the teeth can be visualized by electron microscopy. Garant (1970) has demonstrated that the dentine of shark teeth mineralizes as it does in mammals by deposition of needlelike hydroxyapatite crystallites in a matrix containing collagen fibers (Fig. 2-11C). Predentine, an unmineralized

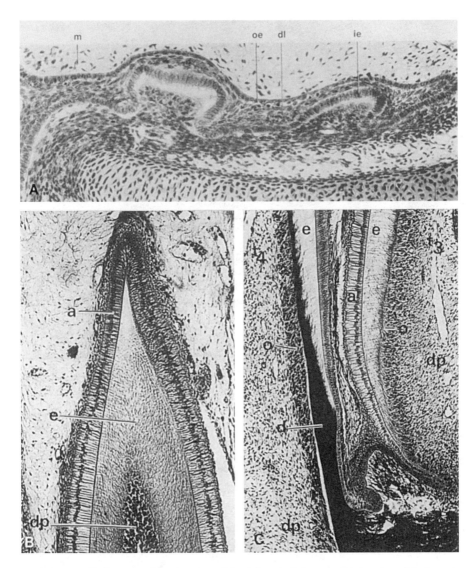

FIGURE 2-10 Early stages of development of sharks' teeth. **A:** Longitudinal section of the lower jaw of a *Squalus acanthias* embryo about 100 mm long. Two earliest tooth buds at the posterior end of the dental lamina are illustrated (×171). ca = Jaw cartilage; dl = dental lamina; ie = inner enamel epithelium; oe = outer enamel epithelium; m = mesenchyme of oral mucosa. **B:** Section of the crown of a demineralized third-rank tooth of *Carcharhinus menisorrah*, showing the enameloid matrix at early stage of mineralization (Heidenhain's azan, ×84). a = Ameloblast layer; dp = dental papilla; e = enameloid matrix. **C:** Section showing the bases of adjacent third- and fourth-rank teeth in *C. menisorrah*. Demineralized matrix of enameloid stains lightly. Dentine staining dark blue first appears in the fourth-rank tooth under enameloid (Heidenhain's azan, ×84). a = Ameloblast layer; d = dentine; dp = dental papilla; e = enameloid; o = odontoblast layer. (A, from Peyer 1968; B, C, from Kemp and Park 1974.)

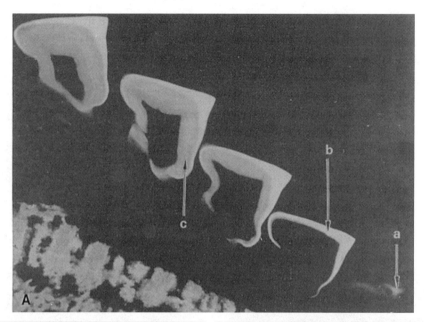

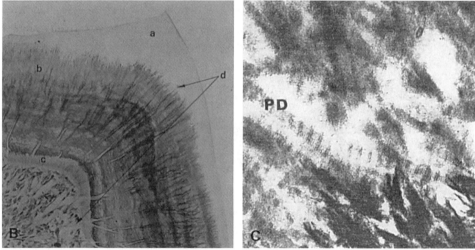

FIGURE 2-11 Mineralization of dentine and enameloid in sharks' teeth. **A:** Microradiograph of a longitudinal section of the lower jaw and mineralizing teeth of *Squalus acanthias*. Mineralized tesserae around the jaw cartilage are seen at lower left. Tooth beginning to mineralize in enameloid at tip is at lower right (a). The next tooth in line has a mineralized enameloid layer all around its surface (b). The next tooth shows a thicker enameloid layer underlain by a less opaque, gray-tinted dentine layer. The dentine layer (c) thickens in the next two teeth and the pulp cavity diminishes in size (×11.2). **B:** Section of the crown of a tooth with well-developed dentine layer in *S. acanthias*. Dental papilla with outer layer of odontoblasts is at lower left (×170). a = Enameloid; b = dentine; c = predentine; d = dentinal tubules extending from odontoblasts through predentine and dentine and into enameloid layer. **C:** Electron micrograph of a section through the border region between predentine (PD) and dentine. Crossbanded collagen fibers occupy the predentine matrix and continue into the dentine layer, which becomes mineralized by deposition of needlelike crystallites in masses among the collagen fibers (×4788). (A, B, from Grady 1970; C, from Garant 1970.)

layer between odontoblasts and dentine, is a zone of collagenous matrix in which collagen fibers develop but remain free of associated crystallites.

The pattern of mineralization in the enameloid layer is distinctly different from that in dentine. Garant (1970) demonstrated that hydroxyapatite crystallites of the enamel layer in *Squalus acanthias* become much larger than those of the dentine layer. Furthermore, the enamel crystallites first form within hollow organic membranes that he called "saccules." The enamel crystallites assume the typical hexagonal shape of apatite crystals as they grow, attaining widths of 200–1000 Å. Collectively the enameline crystallites become packed into zones that Garant called "palisades," separated by unmineralized partitions that he interpreted as "thick collagen bundles in apparent continuity with collagen fibers of the dental pulp." Between the crystallite-rich palisadal zones and the basal lamina of the ameloblasts is a "crystallite-poor" zone containing scattered crystallites in a predominately unmineralized matrix in continuity with the interpalisadal partitions. Garant concluded that "there are clearly two distinct mineralized tissues in the teeth of *Squalus*. The outer layer in no way resembles dentin, nor can it be classified, as others have been, as a modified dentin." Because of its large crystallites and noncollagenous matrix, he concluded that the outer layer could best be termed an "enameloid layer."

Shark ameloblasts lack the apical Tomes processes characteristic of mammalian ameloblasts. These processes control the direction of the crystallites in the alternating prisms and interprisms of mammalian teeth (Moss 1977). Furthermore, the basal lamina is absent from the apical surfaces of Tomes processes during ameloblastic secretion. In shark teeth the basal lamina remains largely intact during amelogenesis, but electron microscopy provides evidence (Fig. 2-12A,B) that ameloblastic secretory products are delivered to the lamina lucida and it is likely that they contribute to the enameloid matrix (Garant 1970; Kemp and Park 1974; Nanci et al. 1983; Kemp 1985).

Nanci and co-workers have demonstrated that granular material secreted by ameloblasts in *Squalus acanthias* accumulates as "amorphous globular material" in masses underlying the basal lamina and that ameloblastic processes free of basal lamina project between such masses. They conclude that "changes in the basal lamina together with the intra-cellular organization of ameloblasts in shark, suggest a merocrine secretion of enameloid constituents occurring along the apical and lateral cell surfaces."

Ameloblasts appear to be the source of the organic matrix that becomes mineralized in both shark enameloid and mammalian enamel. A number of workers have observed that enameloid crystallites in elasmobranch teeth first form within tubular sheaths (Fig. 2-13A). Garant (1970) called them "saccules" and considered them an ectodermal derivative. Kemp and Park (1974) and Kemp (1985) have called them "enameline fibrils (tubules)." Sasagawa (1991) names them "tubular vesicles" and interprets them as a product of odontoblastic rather than ameloblastic secretion. The large, elongate crystallites that develop in shark tooth enameloid are hexagonal (Fig. 2-13B) (Garant 1970). In addition to these tubules, within which enameloid crystallites are seeded, the organic matrix of the enameloid layer in the interpalisadal regions and in the crystallite-poor zone external to the palisades contains nonmineralizing fibrils (Fig. 2-13C,D). Kemp and Park (1974) distinguished these as narrow "unit fibrils" and larger, banded fibers called "giant fibers" with a banding periodicity of 14.5 nm in enameloid of the gray reef shark, *Carcharhinus menisorrah*. Similar large fibers in the lemon shark, *Negaprion brevirostris*, were measured with a periodicity of 17.9 nm (Kemp 1985). Sasagawa (1991) has reported banded fibers with a periodicity of 15 nm in the sharks *Triakis scyllia* and *Heterodontus japonicus* and also fibers with a banding periodicity of 60 nm. He interprets both types of fibers as varieties of collagen and originating from odontoblasts. Undoubtedly the 60-nm fibers are collagenous and the 15-nm fibers may be a short-period polymer of collagen similar

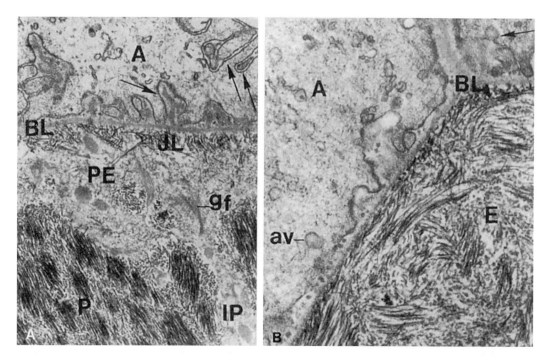

FIGURE 2-12 **A:** The ameloblast-enameloid border in the lemon shark, *Negaprion brevirostris*. Ameloblasts (A) have interdigitating lateral borders (*arrows*). Enameloid crystallites within the enameloid matrix are deposited in zones called "palisades" (P), which are separated by interpalisadal (IP) zones of unmineralized tissue continuous with the crystallite-poor preenameline (PE) zone between ameloblasts and palisades. A juxtalaminar layer (JL) of newly formed crystallites is present just below the basal lamina (BL) of the ameloblasts (×21,510). Gf = giant fiber. **B:** Enlarged view of ameloblast-enameloid border, showing ameloblastic vesicles (av) vesicle at *arrow* about to secrete its contents into lamina lucida of basal lamina (×27,870). A = Ameloblast; E = enameloid. (From Kemp 1985.)

to the type observed by Gross (1956), developing in vitro in cultures of depolymerized collagen. Prostak and Skobe (1988) have theorized that the giant 17-nm fibers become disaggregated and repolymerized as the tubular vesicles that surround enameloid crystallites. Thus they, too, consider the tubular vesicles a product of odontoblasts. Goto (1987) has suggested that the mineralizing saccules are derived from ameloblasts but that the giant fibers are an odontoblastic product.

Similarity of shark enameloid crystallites to those of human enamel, which have been described as "flattened hexagons with a mean thickness of 350 Å and a mean width of 1000 Å" (Garant 1970), raises the question: Is shark enameloid a product of ectodermal ameloblasts like that of humans, or is it a modified type of dentine produced by odontoblasts? Moss (1968) concluded that "shark enamel is ectodermal in origin" and has expressed the consensus (1977) that "all vertebrate teeth form as the result of homologous processes, the mutually inductive interaction between an overlying ectoderm and an underlying core of neural crest derived (mesenchymal) tissues." Whether the enamel layer is homologous in all toothed vertebrates, however, has been a subject of controversy for over a century. Tetrapods, including amphibians, reptiles, and mammals, are known to have ectodermally derived enamel, but it has been argued by some workers that fish teeth do not have "true enamel" of this type (reviews in Moss 1977; Kemp 1985).

Shark teeth differ from tetrapod teeth in three respects: (1) in sharks the enameloid layer begins to mineralize before the start of

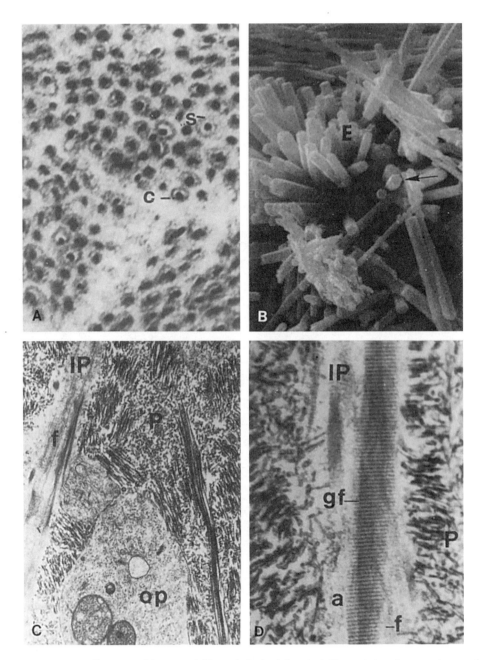

FIGURE 2-13 Structure of the enameloid matrix in sharks' teeth. **A:** Transmission electron micrograph (TEM) showing enameloid crystallites (c) developing within tubular sheaths (s) of organic matrix in tooth of *Carcharhinus menisorrah* (×183,330). **B:** Scanning electron micrograph (SEM) of large enameloid crystallites in a well-developed tooth of the whitetip shark, *Triaenodon obesus*, showing elongated and hexagonal (*arrow*) shape (×10,000). **C:** Enlarged view of the enameloid matrix in *Negaprion brevirostris* tooth, showing a portion of a palisade (P) packed with enameloid crystallites and enclosing a section of an odontoblastic process (op) extending into it. The adjoining interpalisadal tissue (IP) contains fibrils (f) (×17,195). **D:** Enlarged view of interpalisadal region, showing granular amorphous matrix (a), which organizes into fine fibrils (f). The latter then become incorporated into giant fibers (gf) with a banding periodicity of 17.9 nm (×55,850). (From Kemp 1984, 1985.)

dentine mineralization, whereas the chronology of enamel versus dentine mineralization is reversed in tetrapods; (2) shark enameloid is aprismatic, in contrast to the prismatic pattern in "true enamel"; (3) odontoblastic processes extend into the enameloid layer in shark teeth but are absent in tetrapod enamel. The histological distinction that shark enameloid is traversed by odontoblastic processes justifies differentiating it from "true enamel," but the name does not necessarily imply that the mode of mineralization of large, hexagonal crystallites differs in elasmobranch enameloid and tetrapod enamel (Kemp 1984, 1985).

Electron microscopic evidence that the mineralizing matrix of the cap layer of shark teeth is a product of ameloblasts has been supported in recent years by biochemical evidence that ameloblasts of sharks synthesize mammalian-type enamel proteins. Herold et al. (1980) first showed that antisera to fetal calf enamel matrix proteins (amelogenins) cross-reacted with antigens in cells of the i.d.e. and in the enamel matrix of a broad range of other vertebrates, including the cow, mouse, pig, tokay, gecko, salamander, cod, and spiny dogfish shark. Their method was to incubate sections of tooth tissue with rabbit antiserum to calf tooth protein, followed by incubation with goat antiserum to rabbit immunoglobulin conjugated to fluoroscein isothiocyanate. Slavkin et al. (1983) confirmed the interclass cross-reactivity of antisera to mammalian amelogenins, and extensive studies by Slavkin and his collaborators (Slavkin et al. 1983; Samuel et al. 1983) have demonstrated that both the teeth and placoid scales of sharks secrete mammalian-type matrix proteins into their enameloid matrices. From such studies, Slavkin et al. (1983) have postulated "that the synthesis and formation of selachian enameloid in the spiny dogfish are homologous with mammalian secretory amelogenesis."

Enamel proteins consist of two families of acidic glycoproteins called amelogenins and enamelins (Termine et al. 1980). Fetal bovine enamel contains amelogenins ranging in size from 10 to 30 kilodaltons (kDa), constituting 80% of the soluble proteins, and also enamelins from 40 to 70 kDa representing 5–20% of the total protein (Slavkin et al. 1983). The latter authors have suggested that absence of amelogenins in shark enameloid may be related to the fact that enamel in elasmobranchs is aprismatic rather than prismatic. They have further suggested that enamelins are the prevalent type of enameline protein in aquatic vertebrates, whereas evolution has selected the amelogenins as the predominant type in terrestrial vertebrates.

REFERENCES

Agassiz, L. 1833–44. *Recherches sur les poissons fossiles.* 5 vols. Neuchâtel.

Applegate, S. P. 1967. A survey of shark hard parts. In *Sharks, Skates, and Rays,* ed. P. W. Gilbert, R. F. Mathewson, and D. P. Rall, 37–67. Baltimore: Johns Hopkins University Press.

Arambourg, C., and L. Bertin. 1958. Sous-classe des sélaciens (Selachii). In *Traité de Zoologie. Anatomie—Systematique Biologie,* ed. P. P. Grassé, vol. 13, pt. 3, 2016–56. Paris: Masson.

Bagnara, J. T., J. D. Taylor, and M. E. Hadley. 1968. The dermal chromatophore unit. *J. Cell Biol.* 38: 67–79.

Benchley, P. 1974. *Jaws.* New York: Doubleday.

Bereiter-Hahn, J., A. G. Matoltsy, and K. S. Richards, eds. 1986. *Biology of the Integument,* vol. 2: *Vertebrates.* Berlin: Springer-Verlag.

Bertin, L. 1958a. Peau et pigmentation. In *Traité de Zoologie. Anatomie—Systematique Biologie,* ed. P. P. Grassé, vol. 13, pt. 1, 433–58. Paris: Masson.

———. 1958b. Glandes cutanés et organes lumineux. In *Traité de Zoologie. Anatomie—Systematique Biologie,* ed. P. P. Grassé, vol. 13, pt. 1, 459–81. Paris: Masson.

———. 1958c. Denticules cutanés et dents. In *Traité de Zoologie. Anatomie—Systematique Biologie,* ed. P. P. Grassé, vol. 13, pt. 1, 505–31. Paris: Masson.

Budker, P. 1938. Les cryptes sensorielles et les denticules cutanés des Plagiostomes. *Ann. Inst. Oceanogr.* 18: 207–88.

———. 1958. Les organes sensoriels cutanés des sélachiens. In *Traité de Zoologie. Anatomie—Systematique Biologie,* ed. P. P. Grassé, vol. 13, pt. 2, 1033–62. Paris: Masson.

Compagno, L. J. V. 1984. *Sharks of the World.* FAO Species Catalogue. Rome: United Nations Food and Agriculture Organization.

———. 1991. The evolution and diversity of sharks. In *Discovering Sharks,* ed. S. H. Gruber, 15–22. Highland, N.J.: American Littoral Society.

Daniel, J. F. 1934. *The Elasmobranch Fishes*, 3d ed. Berkeley: University of California Press.

Dean, B. 1906. *Chimaeroid Fishes and Their Development*. Publication no. 32. Washington, D.C.: Carnegie Institution Washington.

Didier, D. A., B. J. Stahl, and R. Zangerl. 1994. Development and growth of compound tooth plates in *Callorhinchus milii* (Chondrichthyes Holocephali). *J. Morphol.* 222: 73–89.

Eakin, R. M. 1973. *The Third Eye*. Berkeley: University of California Press.

Fahrenholz, C. 1915. Über die Verbreitung von Zahnbildung und Sinnesorganen im Vorderdarm der Selachier und ihre phylogenetische Beurteilung. *Jena Z. Naturwiss.* 53: 389–444.

Gans, C., and T. S. Parsons. 1964. *A Photographic Atlas of Shark Anatomy. The Gross Morphology of Squalus acanthias*. Chicago: University of Chicago Press.

Garant, P. R. 1970. An electron microscopic study of the crystal-matrix relationship in the teeth of the dogfish *Squalus acanthias* L. *J. Ultrastruct. Res.* 30: 441–9.

Gnadeberg, W. 1926. Untersuchungen über den Bau der Placoidschuppen und der Zähne der Selachier. *Jena Z. Med. Naturwiss.* 42: 473–500.

Goodrich, E. S. 1904. On the dermal fin-rays of fishes—living and extinct. *Q. J. Microsc. Sci.* 47: 465–522.

———. 1906. Notes on the development, structure, and origin of the median and paired fins of fish. *Q. J. Microsc. Sci.* 50: 333–76.

———. 1907. On the scales of fish, living and extinct, and their importance in classification. *Proc. Zool. Soc. Lond.* 1907: 751–74.

Goto, M. 1987. Origin and evolution of tooth enamel. In *Tooth Enamel: Its Formation, Structure, Composition and Evolution*, ed. S. Suga, 222–33 (in Japanese). Tokyo: Quintessence.

Grady, J. E. 1970. Tooth development in sharks. *Arch. Oral Biol.* 15: 613–19.

Gross, J. 1956. The behavior of collagen units as a model in morphogenesis. *J. Biophys. Biochem. Cytol. Suppl.* 2: 261–74.

Gruber, S. H., D. I. Hamasaki, and B. L. Davis. 1975. Window to the epiphysis in sharks. *Copeia* 1975: 378–80.

Herold, R. C., H. T. Graves, and P. Christner. 1980. Immunohistochemical localization of amelogenins in enameloid of lower vertebrate teeth. *Science* 207: 1357–58.

Hertwig, o. 1874. Ueber Bau und Entwickelung der Placoidschuppen und der Zähne der Selachier. *Z. Naturwiss.* 8: 331–404.

Hodge, A. J. 1967. Structure at the electron microscopic level. In *Treatise on Collagen*, vol. 1: *Chemistry of Collagen*, ed. G. N. Ramachandran, 185–205. New York: Academic Press.

James, W. W. 1953. The succession of teeth in elasmobranchs. *Proc. Zool. Soc. Lond.* 123: 419–74.

Jarvik, E. 1980. *Basic Structure and Evolution of Vertebrates*, 2 vols. New York: Academic Press.

Kemp, N. E. 1977. Banding pattern and fibrillogenesis of ceratotrichia in shark fins. *J. Morphol.* 154: 187–204.

———. 1984. Organic matrices and mineral crystallites in vertebrate scales, teeth, and skeletons. *Am. Zool.* 24: 965–76.

———. 1985. Ameloblastic secretion and calcification of the enamel layer in shark teeth. *J. Morphol.* 184: 215–30.

Kemp, N. E., and J. H. Park. 1970. Regeneration of lepidotrichia and actinotrichia in the tailfin of the teleost *Tilapia mossambica*. *Dev. Biol.* 22: 321–42.

———. 1974. Ultrastructure of the enamel layer in developing teeth of the shark *Carcharhinus menisorrah*. *Arch. Oral Biol.* 19: 633–44.

Kemp, N. E., and S. K. Westrin. 1979. Ultrastructure of calcified cartilage in the endoskeletal tesserae of sharks. *J. Morphol.* 160: 75–102.

Kent, G. C. 1973. *Comparative Anatomy of the Vertebrates*, 3d ed. St. Louis: Mosby.

Klaatsch, H. 1890. Zur Morphologie der Fischschuppen und zur Geschichte der Hartsubstanzgewebe. *Morphol. Jahrb.* 16: 97–202.

Kresja, R. J. 1979. The comparative anatomy of the integumental skeleton. In *Hyman's Comparative Anatomy*, 3d ed., ed. M. J. Wake, 112–91. Chicago: University of Chicago Press.

Lowe, C., and G. Goodman-Lowe. 1996. Suntanning in hammerhead sharks. *Nature* 383: 677.

Luer, C. A., P. C. Blum, and P. W. Gilbert. 1990. Rates of tooth replacement in the nurse shark, *Ginglymostoma cirratum*. *Copeia* 1990: 182–90.

Maderson, P. F. A. 1972. When? why? and how?: some speculations on the evolution of the vertebrate integument. *Am. Zool.* 12: 159–71.

Maisey, J. G. 1979. Finspine morphogenesis in squalid and heterodontid sharks. *Zool. J. Linn. Soc.* 66: 161–83.

Moss, S. A. 1967. Tooth replacement in the lemon shark, *Negaprion brevirostris*. In *Sharks, Skates, and Rays*, ed. P. W. Gilbert, R. F. Mathewson, and D. P. Rall, 319–29. Baltimore: Johns Hopkins University Press.

Moss, M. L. 1968. Bone, dentin, and enamel and the evolution of vertebrates. In *Biology of the Mouth*, ed. P. Person, 37–65. AAAS Publication no. 89. Washington, D.C.: American Association for the Advancement of Science.

———. 1972. The vertebrate dermis and the integumental skeleton. *Am. Zool.* 12: 27–34.

———. 1977. Skeletal tissues in sharks. *Am. Zool.* 17: 335–42.

Nanci, A., P. Bringas Jr., N. Samuel, and H. C. Slavkin. 1983. Selachian tooth development: III. Ultrastructural features of secretory amelogenesis in *Squalus acanthias*. *J. Craniofac. Genet. Dev. Biol.* 3: 53–73.

Nelsen, O. E. 1953. *Comparative Embryology of the Vertebrates*. New York: Blakiston.
Ørvig, T. 1951. Histologic studies of placoderms and fossil elasmobranchs: I. The endoskeleton, with remarks on the hard tissues of lower vertebrates in general. *Arch. Zool.* 2: 321–454.
———. 1967. Phylogeny of tooth tissues: evolution of some calcified tissues in early vertebrates. In *Structural and Chemical Organization of Teeth*, ed. A. E. W. Miles, vol. 1, 45–110. New York: Academic Press.
Parker, H. W., and M. Boesman. 1954. The basking shark *Cetorhinus* in winter. *Proc. Zool. Soc. Lond.* 124: 185–94.
Peyer, B. 1968. *Comparative Odontology*. Translated and edited by R. Zangerl. Chicago: University of Chicago Press.
Prostak, K. S., and Z. Skobe. 1988. Ultrastructure of odontogenic cells during enameloid matrix synthesis in tooth buds from an elasmobranch, *Raja erinacea*. *Am. J. Anat.* 182: 59–72.
Quay, W. B. 1972. Integument and environment: glandular composition, function, and evolution. *Am. Zool.* 12: 95–108.
Reif, W.-E. 1980. Development of dentition and dermal skeleton in embryonic *Scyliorhinus canicula*. *J. Morphol.* 166: 275–88.
———. 1982. Evolution of dermal skeleton and dentition in vertebrates. The odontode regulation theory. *Evol. Biol.* 15: 287–368.
———. 1985a. Squamation and ecology of sharks. *Cour. Forsch.-Inst. Senckenberg* 78: 1–255.
———. 1985b. Morphology and hydrodynamic effects of the scales of fast swimming sharks. *Fortschr. Zool.* 30: 483–85.
Romer, A. S. 1966. *Vertebrate Paleontology*, 3d ed. Chicago: University of Chicago Press.
Rüdeberg, C. 1969. Light and electron microscopic studies on the pineal organ of the dogfish, *Scyliorhinus canicula* L. *Z. Zellforsch.* 96: 548–81.
Samuel, N., P. Bringas Jr., V. Santos, A. Nanci, and H. C. Slavkin. 1983. Selachian tooth development: I. Histogenesis, morphogenesis, and anatomical features of *Squalus acanthias*. *J. Craniofac. Genet. Dev. Biol.* 3: 29–41.
Sasagawa, I. 1991. The initial mineralization during tooth development in sharks. In *Mechanisms and Phylogeny of Mineralization in Biological Systems*, ed. S. Suga and H. Nakahara, 199–203. Berlin: Springer-Verlag.
Sasagawa, I., and J. Akai. 1992. The fine structure of the enameloid matrix and initial mineralization during tooth development in the sting rays, *Dasyatis akajei* and *Urolophus auranticus*. *J. Electron Microsc.* 41: 242–52.
Schaeffer, B. 1977. The dermal skeleton in fishes. *Linn. Soc. Symp.* 4: 25–52.
Slavkin, H. C., N. Samuel, P. Bringas, Jr., A. Nanci, and V. Santos. 1983. Selachian tooth development: II. Immunolocalization of amelogenin polypeptides in epithelium during secretory amelogenesis in *Squalus acanthias*. *J. Craniofac. Genet. Dev. Biol.* 3: 43–52.
Smith, M. M., and B. K. Hall. 1990. Development and evolutionary origins of vertebrate skeletogenic and odontogenic tissues. *Biol. Rev.* 65: 277–373.
Stensiö, E. A. 1961. *Permian Vertebrates: Geology of the Arctic*. Toronto: University of Toronto Press.
Taylor, J. D., and J. T. Bagnara. 1972. Dermal chromatophores. *Am. Zool.* 12: 43–62.
Termine, J. D., A. B. Belcourt, P. J. Christner, K. M. Conn, and M. U. Nylen. 1980. Properties of dissociately extracted fetal tooth matrix proteins: I. Principal molecular species in developing bovine enamel. *J. Biol. Chem.* 255: 9760–68.
Van Oosten, J. 1957. The skin and scales. In *The Physiology of Fish*, ed. M. E. Brown, vol. 1, 207–44. New York: Academic Press.
Waldeyer, W. 1871. Bau und Entwicklung der Zähne. In *Handbuch der Lehre von den Geweben des Menschen und der Thiere*, ed. S. Stricker, 333–54. Leipzig: Wilhelm Engelmann.
Walker, W. F., and D. G. Homberger. 1992. *Vertebrate Dissection*, 8th ed. Philadelphia: W. B. Saunders.
Whitear, M. 1986. The skin of fishes including cyclostomes. In *Biology of the Integument*, vol. 2: *Vertebrates*, ed. J. Bereiter-Hahn, A. C. Matoltsy, and K. S. Richards, 8–64. Berlin: Springer-Verlag.
Wilson, J. F., and J. M. Dodd. 1973. The role of the pineal complex and lateral eyes in the colour change response of the dogfish. *J. Endocrinol.* 58:591–98.
Wolff, R. G. 1991. *Functional Chordate Anatomy*. Lexington, Mass.: D. C. Heath.
Young, J. Z. 1981. *The Life of Vertebrates*, 3d ed. Oxford: Clarendon Press.

LEONARD J. V. COMPAGNO

CHAPTER 3

Endoskeleton

The skeletal system, or skeleton, of elasmobranch fishes is deceptively simple compared to that of bony fishes. It consists of a minimalist *endoskeleton*, the internal skeleton of the head, body, tail, and fins; and a relatively inconspicuous *exoskeleton*, the external protective skeleton of the skin, which includes the placoid scales or dermal denticles and their derivatives. The skeleton serves to protect the skin, to support and protect the various organs, and to anchor the musculature. Terminology for the elasmobranch skeleton used here follows Gegenbaur (1865), Jungersen (1899), Huber (1901), Allis (1913, 1914, 1923), Garman (1913), Daniel (1928), White (1930, 1937), Holmgren (1941), Applegate (1967), McEachran and Compagno (1979), Compagno and Roberts (1982), and Compagno (1988).

Endoskeleton

The *endoskeleton* (Fig. 3-1), or internal skeleton, includes the *axial skeleton* with skull, or cranium, and the *vertebral column*, and the *appendicular skeleton*, which supports the fins. The endoskeleton consists primarily of hyaline and calcified cartilage. The cartilaginous skeleton of embryonic elasmobranchs persists into adulthood and is not supplemented or replaced by plates of endochondral bone as in other gnathostomes. In most elasmobranchs the endoskeleton is strengthened by calcified cartilage, which commonly forms as layers of calcified polygons of hydroxyapatite, or

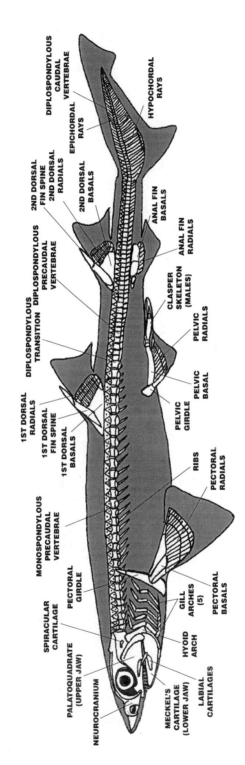

FIGURE 3-1   An idealized neoselachian shark skeleton in lateral view.

tesserae (Ørvig 1951; Applegate 1967), forming a hard tilelike pavement on the surface of skeletal structures. Vertebral centra can have more compact and dense calcification that partially or completely permeates the entire centrum. Various structures on the axial and appendicular skeleton can become regularly hypercalcified in some elasmobranchs (primarily inshore and pelagic sharks), with extremely dense calcification and modification of shape (Compagno 1988) comparable to hyperostosis in bony fishes (Smith-Vaniz et al. 1995). Hypercalcified masses occur on various structures of the neurocranium and on the hyoid arch, extrabranchial cartilages, the pectoral and pelvic fin girdles and fin skeletons, the anal and dorsal fin radials, the hemal arches of the caudal fin, and the vertebral centra. Calcification of the endoskeleton is best developed in inshore demersal elasmobranchs and in mature individuals. Late fetal elasmobranchs may have only the vertebral centra (if present) calcified, while juvenile elasmobranchs are often less calcified than adults. Deep-slope elasmobranchs and some oceanic and inshore elasmobranchs can have endoskeletal calcification more or less reduced, which can include loss of calcified vertebral centra.

AXIAL SKELETON AND CRANIUM

The axial skeleton includes the cranium or skull, which supports the head and protects the brain and sense organs, and the vertebral column, which in neoselachian elasmobranchs is a segmented jointed structure of discrete vertebrae that supports the body, tail, and caudal fin. The cranium, syncranium (Gregory 1933), or skull of elasmobranchs includes the neurocranium and the splanchnocranium, or visceral arches.

Neurocranium

The neurocranium, chondrocranium, chondroneurocranium, or endocranium of elasmobranchs is a box-shaped unitary cartilaginous structure anterior to the vertebral column that is the largest component of the elasmobranch skull (Figs. 3-2, 3-3, 3-4). The neurocranium consists of an axial and medial braincase enclosing and protecting the brain in an internal cerebral cavity and lateral sense capsules for the nasal organs, eyes, and inner ears. Lateral line canals of the head and clusters of ampullae of Lorenzini are also supported by the neurocranium, and the components of the splanchnocranium are closely associated and partly supported by it. The neurocranial description and terminology is mostly derived from the detailed review by Compagno (1988).

Gegenbaur (1865) divided the neurocranium into four external regions from anterior to posterior: (1) the ethmoidal region, including the rostrum and nasal capsules; (2) the orbital region, including the orbits and the basal plate and cranial roof between them; (3) the otic region, with the otic capsules and the cranial roof and basal plate between them; (4) the occipital region, the posterior end of the cranium. Compagno (1988) used a modified system to divide the neurocranium into seven structural areas: (1) the rostrum, including the rostral cartilages and associated structures; (2) the nasal capsules, including the internasal plate between them; (3) the cranial roof, the dorsal cover of the cerebral cavity from the anterior fontanelle to the parietal fossa; (4) the basal plate, the medial and ventral floor of the cerebral cavity, extending from the nasal capsules to the occipital centrum; (5) the orbits, including the supraorbital crests and suborbital shelves, the lateral eye cavities that are separated from one another by the anterior part of the basal plate and cranial roof; (6) the otic capsules, paired containers for the inner ears separated by the rear ends of the basal plate and cranial roof and articulated with the hyomandibulae of the hyoid arches through their hyomandibular facets; (7) the occiput, including the occipital condyles and foramen magnum.

The rostrum is the most anterior part of the neurocranium and supports the prenasal snout, including its lateral line canals and ampullae. The rostrum (when present) partly encloses the precerebral fossa anterior to the tough membrane that closes the anterior

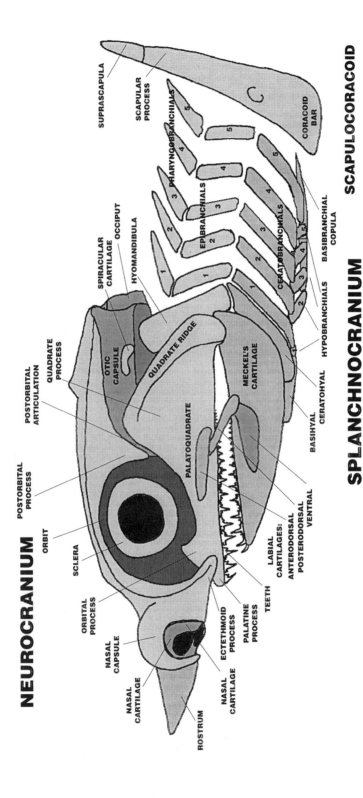

FIGURE 3-2 The cranium of an idealized neoselachian shark in lateral view, with scapulocoracoid also present.

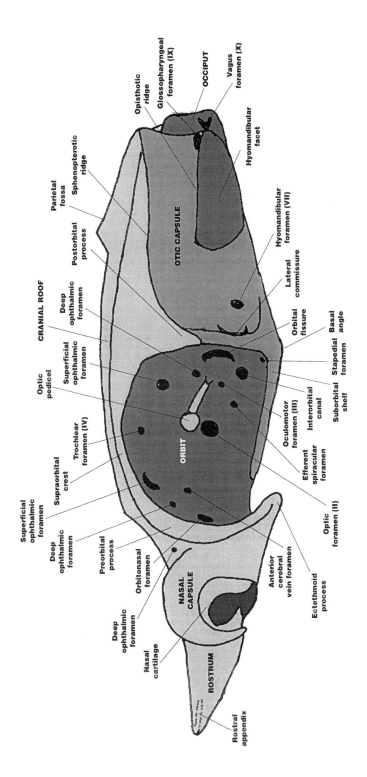

FIGURE 3-3 The neurocranium of an idealized neoselachian shark in lateral view.

fontanelle and entrance to the cerebral cavity enclosing the brain. The rostrum is a dorsally open, lateroventrally walled trough or scoop in squalomorph sharks, squatinoids, and many batoids. It is tripodal in lamnoid and carcharhinoid sharks but greatly reduced to a medial bar in orectoloboid sharks. The rostrum is a flattened blade with the precerebral fossa enclosed within it in saw sharks (pristiophoroids) and sawfish (pristoids). The rostrum is more or less reduced to a slender rod in many skates (rajoids) and narkid electric rays and is absent in posthatchling bullhead sharks (heterodontoids) and in all myliobatoid rays. The tripodal rostrum of lamnoids and carcharhinoids consists of a rodlike medial rostral cartilage originating on the anteroventral end of the internasal septum and a pair of rodlike

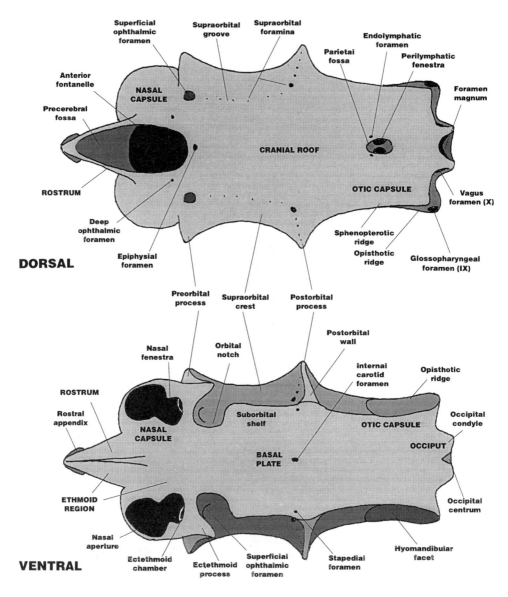

FIGURE 3-4 The neurocranium of an idealized neoselachian shark in dorsal view (*above*) and ventral view (*below*).

lateral rostral cartilages that arise from the dorsal roofs of the nasal capsules or the anteroventral edges of the supraorbital crests. These are expanded anteriorly and either articulate anteriorly or are fused together as a rostral node. The anterior end of the rostrum often has lateral projections, or rostral appendices, which can become greatly expanded in skates.

Posterior to and often somewhat ventral to the rostrum are the paired, hollow nasal capsules, which enclose the olfactory organs in nasal cavities. The nasal capsules are subspherical in many sharks but vary tremendously in shape in different elasmobranch groups. Variants include the trumpet-shaped capsules of orectoloboids and heterodontoids, the laterally expanded capsules of many batoids, and the blade-shaped capsules of sphyrnid sharks. Each nasal capsule is partially or completely open ventrally, with an anterolateral nasal aperture through which the external incurrent and excurrent apertures of the nostrils funnel water into and out of the cavity of each olfactory organ. The aperture and the nasal flaps are reinforced by a ring-shaped nasal cartilage, which is sometimes fused to the floor of the capsule. Many elasmobranchs have a horizontal subnasal plate, or floor of the nasal capsule, that bounds the nasal aperture, but this is absent in some sharks and batoids. There often is a nasal fontanelle or additional aperture posterior and medial to the nasal aperture but closed by a membrane. This may be broadly confluent to the nasal aperture or separated from it by a bar formed from the subnasal plate. The medial wall of each nasal capsule is formed by the internasal septum, which varies from a narrow, compressed high plate to a low, flat, expanded depressed plate, depending on the size and medial separation of the nasal capsules.

The anterior wall of the nasal capsule curves dorsally to form the roof of the nasal capsule and merges posteriorly into the preorbital wall. The preorbital wall forms the posterior boundary of the nasal capsule and the anterior wall of the orbit with the preorbital process and supraorbital crest. Medially the roof of the nasal capsule merges with the internasal septum. The nasal and cerebral cavities are connected by the olfactory canal, a broad posteromedial or medial passage through the preorbital wall. The olfactory canal contains the olfactory bulb and olfactory tract of the brain that connect with the olfactory organ inside the nasal cavity. On the preorbital wall below the olfactory canal is a small to large and variably formed pit, the ectethmoid chamber or fossa, which is the anterior opening of the orbitonasal canal into the nasal cavity. The orbitonasal canal penetrates the preorbital wall to end in the orbitonasal foramen on the anterioventral surface of the orbit and encloses the orbitonasal vein, which drains the nasal sinus around the olfactory organ into the orbital sinus around the eyeball. The deep ophthalmic nerve (ophthalmicus profundus nerve) of the trigeminal, or fifth cranial, nerve enters the nasal capsule from the orbit through a foramen in the preorbital wall just below the preorbital process, crosses the nasal cavity below the roof of the capsule, and leaves the cavity through another foramen that leads to the dorsal surface of the nasal capsule. In some sharks the ventral surfaces of the nasal capsules have a deep embayment, or subethmoid fossa, between them, into which fit the palatine processes of the palatoquadrates, or upper jaws. In certain carcharhinoid sharks there are articular surfaces, or ectethmoid condyles, that abut the orbital processes of the palatoquadrates. In hexanchoid and some squaloid sharks the ventrolateral surfaces of the nasal capsules and the ventral postorbital wall are expanded posteroventrolaterally into ectethmoid processes. In saw sharks and batoids, separate antorbital cartilages are attached to the ventrolateral or anterolateral surfaces of the nasal capsules and support the lateral edge or front of the head (saw sharks, sawfishes, shark-fin guitarfishes, and torpedo rays) or are attached to the pectoral propterygia. In batoids there are prominent antorbital condyles on the nasal capsules for articulation of the antorbital cartilages.

The cranial roof begins anteriorly on the dorsomedial surface of the neurocranium at the anterior fontanelle, the anterior and exterior opening of the cerebral cavity, which is covered by a tough connective tissue membrane. The fontanelle varies tremendously in shape, from a narrow vertical slit to a broad transversely oval aperture, and ranges from being almost horizontal and parallel to the basal plate to perpendicular and vertical. In some sharks and many batoids the fontanelle continues posteriorly to occupy most or all of the cranial roof anterior to the parietal fossa, and in other elasmobranchs there may be separate small frontal fenestrae or large frontal fontanelles in the cranial roof posterior to the anterior fontanelle. The pineal body generally is found in the membrane of the anterior fontanelle near the dorsal lip of the fontanelle, but in some sharks it has a distinct epiphysial notch in the dorsal lip or an epiphysial foramen through the cranial roof just behind the fontanelle. The cranial roof varies from strongly concave to broadly arched or domed in shape. The cranial roof is bounded posteriorly by a parietal fossa, a shallow or deep depression between the otic capsules that houses an anterior pair of endolymphatic foramina for the endolymphatic ducts from the sacculus of the middle ear and a posterior pair of perilymphatic fenestrae opening into the perilymphatic spaces surrounding the auditory organs. In elasmobranchs with broad neurocrania the parietal fossa tends to subdivide laterally, with an endolymphatic foramen and perilymphatic fenestra in a depression on each side separated by a broad space.

The floor of the neurocranium behind the nasal capsule is the basal plate, which varies from flat to broadly arched and angled and extends to the otic capsules and occiput. There may be a distinct basal angle in the plate, extending ventrally below and sometimes just behind the orbit. The basal plate is in proximity to and is penetrated by the main ventral arteries supplying the brain and orbits (terminology from O'Donoghue and Abbott 1928; Daniel 1928; Corrington 1930). The paired efferent hyoidian arteries arise from the dorsal ends of the pretrematic branches of the first collector loops of the efferent branchial arteries and extend anteromedially along the basal plate from the hyomandibular facets, anastomose with the paired dorsal aortas that extend anterolaterally from the medial dorsal aorta, give rise to the stapedial arteries that enter the orbit through a stapedial foramen at the junction of basal plate and suborbital shelf (when the latter is present), and as the internal carotid arteries arch medially to enter the cranial cavity through paired internal carotid foramina or a single foramen. The arrangement of paired internal carotid foramina near the midline of the basal plate or a single medial foramen is very common in living sharks and rays. Stapedial foramina or fenestrae are present in many living elasmobranchs but are absent in some squalomorph sharks and in batoids that lack suborbital shelves. There are several sharks and rays in which the carotid foramina are more laterally situated on the basal plate or even have moved laterally through the stapedial foramina onto the sides of the orbits. The stapedial foramina are greatly enlarged in some carcharhinoid and lamnoid sharks into stapedial fenestrae and house greatly convoluted and elongated arterial masses or retia. In certain carcharhinoid sharks (Hemigaleidae) the efferent hyoidian arteries and paired dorsal aortas are partially buried in the basal plate.

The orbital region is the most complex part of the neurocranium. It contains the eyeballs, their muscles, the orbital sinuses, and most of the cranial nerves that extend from the brain and most of the blood vessels that drain or supply the cranial cavity. The orbits, paired cavities on either side of the cranial cavity and separated by the cranial roof and basal plate, are each delimited anteriorly by the preorbital wall and preorbital process, dorsally by the supraorbital crest (except when secondarily lost), posteriorly by the postorbital process and otic capsule, ventrally by the suborbital shelf, and medially by the medial wall of the orbit. The orbits in lateral view are usually horizontally oval or rectangular, but can be pear- or gourd-shaped or almost circular. The medial wall of the orbit is penetrated by foramina for the optic (second), oculomotor (third), and trochlear (fourth) cranial nerves, the superficial ophthalmic branch of the trigeminal (fifth) and facial (seventh) nerves, the deep ophthalmic branch of the trigeminal nerve, and by the orbital fissure, a large foramen for the hyomandibular, palatine, and buccal branches of the facial nerve, the maxillary and mandibular branches of the trigeminal nerve, and the abducens (sixth) cranial nerve. In many elasmobranchs the superficial ophthalmic foramen through the medial wall is not separate from the orbital fissure, and the hyomandibular branch may have a separate foramen posterior to the orbital fissure and separated from it by a bar of cartilage, or prefacial commissure. The superficial ophthalmic branch leaves the orbit through the preorbital canal at the anterodorsal junction of the orbital wall, preorbital process, and supraorbital crest. The deep ophthalmic nerve leaves the orbit anteriorly through a foramen in the preorbital wall into the nasal capsule.

The space between the orbit and the eyeball is filled by the large venous orbital sinus, which receives blood from the anterior orbitonasal foramen, the foramen of the anterior cerebral vein, and the interorbital canal, and in turn is drained by the large lateral head vein under the postorbital process and through the postorbital wall and an opening, or lateral commissure (where present). Arterial foramina entering the medial wall of the orbit include the stapedial foramen and the foramen for the efferent spiracular artery. The preorbital process is a flat, posteriorventrally curved plate, with its base on the dorsal edge of the preorbital wall from the preorbital canal ventrally to above the basal plate and, in some sharks, it extends posteroventrally onto the lateral edge of the basal plate. Its base is often traversed by the canal for the deep ophthalmic nerve. The supraorbital crest is an arched horizontal plate that extends posteriorly along the dorsal edge of the medial orbital wall from the preorbital process to the postorbital process, with its base continuous with the orbital wall and cranial roof. The crest is penetrated medially by supraorbital foramina for tiny branches of the superficial ophthalmic branch of the facial nerve, which supply the supraorbital lateral line canal that runs in the supraorbital groove on the dorsal surface of the crest at its base. The postorbital process is continuous anteriorly with the supraorbital crest and arises on the anterior end of the otic capsule. From the underside of the postorbital process a postorbital wall extends to the basal plate in squatinoids and is penetrated by the lateral commissure, but this is variably reduced in other sharks and in batoids and the commissure may be present or absent. Supraorbital crests are present in sharks and batoids, with the exception of a few lamnoids and orectoloboids, many carcharhinoids, and all electric rays; crestless taxa may or may not have isolated preorbital and postorbital processes.

The suborbital shelf is a horizontal plate on the ventral junction of the orbital wall and basal plate that is the floor of the orbit. It runs from the nasal capsule to the otic capsule and is penetrated posteriorly by the stapedial foramen and sometimes laterally by a notch, foramen, or fenestra for the palatine branch of the facial nerve. The suborbital shelf is well developed in galeomorph and squatinoid sharks but is variably reduced or absent in squalomorph sharks and in batoids. In sharks with a prominent suborbital shelf, this is often more or less indented immediately behind the nasal capsules to form an orbital notch for the articulation of the orbital process of the palatoquadrate or at least for the attachment of the process to the cranium by the ethmopalatine ligament. The edges of the notch are often thickened as an articular surface for the orbital process. The optic pedicel is a shaft of cartilage that extends from the medial wall of the orbit in front of the orbital fissure to the medial surface of the eyeball. It may have a plate-shaped terminal end that abuts the eyeball. The optic pedicel is absent in a few sharks.

Although not developmentally part of the neurocranium, the eyeballs are sited within the orbits and have their own skeleton. The eyeball is generally reinforced by a chondrified sclera or sclerotic coat, a hollow ball of cartilage that is often subspherical but can be horizontally oval or flattened-oval. The chondrified sclera has a large lateral aperture for the iris and pupil and a medial foramen for the optic nerve. The sclera can be very thin in some deepwater sharks and may not be chondrified, but it can be heavily chondrified in large lamnoid sharks and gigantic and baseball-sized in the big-eyed thresher (*Alopias superciliosus*).

The paired otic capsules are complex boxlike structures containing the inner ears. In most elasmobranchs a horizontal sphenopterotic ridge arises at the rear of the postorbital process and extends along the dorsolateral edge of each otic capsule to either end at the occiput or terminate in a distinct dorsolateral projection, or pterotic process, which anchors the dorsal myotomes of the epibranchial region. Above the sphenopterotic ridge, on the dorsal surface of the otic capsule, are often a pair of distinct low ridges, beneath which are the anterior and posterior vertical semicircular canals. These ridges converge posteromedially and anteromedially to the parietal fossa. Beneath and medial to the sphenopterotic ridge is the horizontal semicircular canal. A horizontal channel for the lateral head vein, the postorbital groove occurs below the sphenopterotic ridge on the lateral surface of the capsule. The opisthotic ridge defines the dorsal rim of the hyomandibular facet below the postorbital groove and terminates posteriorly in the glossopharyngeal foramen for the posterior exit of the main trunk of the glossopharyngeal, or ninth cranial, nerve. The hyomandibular facet is a shallow lateral depression on the otic capsule, usually defined below by a low horizontal ridge between the suborbital shelf and the occiput (absent in some squalomorphs). It is usually horizontally oval but can be round, wedge-shaped, elongated, or bilobate.

The occiput occurs at the posterior convergence of the otic capsules, cranial roof, and basal plate. The large foramen magnum for the spinal cord is placed in the center of the occiput. Below the foramen magnum in sharks is a circular occipital centrum, or hemicentrum, formed by the posterior half of a calcified double cone that is imbedded in the basal plate and that articulates with the first cervical centrum of the vertebral column. On either side of the occipital centrum is an occipital condyle, which articulates with the basiventral of the first cervical vertebra. The large vagus foramen for the main branch of the vagus, or tenth cranial, nerve occurs lateral to each occipital condyle. The occipital crest is a low longitudinal ridge that extends between the parietal fossa and the dorsal surface of the foramen magnum. In living batoids the occipital hemicentrum is lost from the occiput, but a similar hemicentrum occurs partway along the length of the cervicothoracic synarcual and marks the start of synarcual centra.

Splanchnocranium

The splanchnocranium (see Fig. 3-2), or visceral skeleton, includes a set of bilaterally symmetrical, paired structures that support the jaws, tongue, gills, and pharynx. The splanchnocranium includes seven to nine visceral arches: the mandibular arch, or jaws; the hyoid arch, or tongue and jaw supports, with associated hyoid rays, and five to seven branchial arches and branchial rays to support the gill holobranchs. The mandibular arch and hyoid arch are closely associated with and attached to the neurocranium, whereas the dorsal elements of the branchial arches are close behind its basal surface. In addition, there are paired dorsal and ventral labial cartilages and extravisceral cartilages (extrahyoid and extrabranchial cartilages) that are superficial to the visceral arches proper but also support the mouth corners and hyoid and branchial septa.

The mandibular arch of elasmobranchs includes the paired upper jaw cartilages, or palatoquadrates, and the paired lower jaw cartilages, or Meckel's cartilages. The mandibular arch of elasmobranchs is homologous to the embryonic palatoquadrates and Meckel's cartilages of other jawed vertebrates and are *primary jaws*. Elasmobranchs differ from bony fishes and tetrapods in not partially or completely replacing their primary jaws with secondary jaws of dermal and endochondral bone. In most sharks and all batoids the anteriormedial ends of the palatoquadrates and Meckel's cartilages have upper and lower symphysial articulations (which are fused in some batoids). The palatoquadrates and Meckel's cartilages of either side meet one another posterolaterally in double-jointed mandibular articulations. The jaws vary enormously in shape and orientation in elasmobranchs and can be highly specialized in macropredatory and durophagous species. The jaws are generally positioned below and behind the ethmoid region of the neurocranium, although a few sharks have terminal mouths and jaws that extend under the nasal capsules and rostrum. The Meckel's cartilages usually extend more posteriorly and are more mesially situated than the palatoquadrates, and the anteromedial ends of the Meckel's are usually well posterior to the anteromedial ends of the palatoquadrates. Elasmobranch jaws generally have an overbite and are usually subterminal on the head. The palatoquadrates have short to long, stout to slender, anteromesial palatine processes that articulate with or are positioned behind the posterior part of the nasal capsules.

Sharks usually have low and padlike to high and conical orbital processes that may articulate with the orbital notches of the neurocranium or with grooves on the orbital basal plate and are connected to the neurocranium by short to greatly elongated ethmopalatine ligaments. These structures are absent and apparently lost in batoids, many of which have more protrusible jaws than most sharks. In certain deepwater lamnoid and carcharhinoid sharks (the goblin shark and some pentanchine cat sharks) the orbital processes are reduced to attachment surfaces for long, slinglike, elastic ethmopalatine ligaments, and the jaws are highly protrusible. The palatoquadrates of sharks and batoids have low to high, posterior quadrate processes that either articulate with the postorbital processes of the neurocranium or are well separated from them. Shark palatoquadrates are primitively hatchet-shaped, with elevated quadrate processes that have prominent quadrate ridges, but the quadrate processes are low and the ridges are reduced or absent in many sharks and in batoids. Meckel's cartilages are generally in the form of curved anteromedially pointed wedges in sharks, but they vary enormously in shape. The palatoquadrates and Meckel's cartilages of many elasmobranchs have dental grooves on the lingual or inner mesial faces for the dental membrane and teeth. These dental grooves can be elaborated into individual slots for each row of teeth in sharks with large teeth, and there can also be distinct swollen protrusions or dental bullae on the outside, or labial faces, of the palatoquadrates and Meckel's cartilages to accommodate the developing teeth.

The palatoquadrates of sharks are suspended directly from the neurocranium by ligaments and other connective tissue from the ethmoid, orbital, and otic regions and by the preorbital and palatoquadrate levator muscles, and indirectly by the hyomandibulae and the hyomandibular levator and first dorsal constrictor muscles of the hyoid arch. The integument enclosing the head contributes to jaw suspension and attaches to the palatoquadrates at the labial edge of the dental bands and upper lips. In sharks, the Meckel's cartilages are supported by their articulation with the palatoquadrates, by the hyomandibulae and ceratohyals, by the adductor mandibulae muscles, and by the integument. In batoids, the hyomandibulae become the sole hyoid-arch components involved with jaw suspension through degeneration or loss of the ventral hyoid elements. Direct palatoquadrate articulation with the cranium in most living sharks is anterior and ethmoorbital, with the orbital processes and palatine processes abutting the nasal capsules and orbits. Some squalomorph sharks have the jaws greatly shortened for accommodating a scissorslike slicing dentition and the orbital processes are far posterior in the orbits, while hexanchoids and certain lamnoids have loose to tight postorbital articulations of the quadrate processes of the palatoquadrates with the postorbital processes of the neurocranium. Postorbital articulations are thought to be primitive in neoselachian sharks and may be so for hexanchoids at least. In batoids the orbital processes are lost; the palatine processes are only loosely associated with the nasal capsules; the jaws are essentially slung under the neurocranium on the hyomandibulae, jaw and hyoid musculature, and integument; and the jaws are highly protrusible ventrally. In some myliobatoids, jaw mobility is lost with development of short, heavy jaws, massive tooth plates, and durophagy (eagle and cow-nose rays) or of elongated jaws adapted to plankton feeding (devil rays).

Labial cartilages are paired cartilages external to the mandibular arch that support the mouth corners. They are small to large, flat to cylindrical cartilages buried in the skin superficial to the jaws and jaw muscles and at the mouth angles. They form the cores of the superficial labial folds and are usually delimited posteriorly and dorsoventrally by the labial grooves. There are primitively an anterodorsal and posterodorsal labial cartilage on each side of the upper jaw and a ventral labial cartilage on each side of the lower jaw in living sharks. Some sharks with long mouths and wide gapes have the labial cartilages reduced to one on a side or have lost them completely, while saw sharks and most batoids have them variably reduced or modified. In sharks and certain electric rays with large labial cartilages and short mouths, the labial cartilages limit the gape and, with the labial folds, form a short funnel when the mouth is open. This may facilitate suction-feeding. Labial cartilages have sometimes been interpreted as premandibular visceral arches, but they are best regarded as superficial structures that not homologous to the visceral arches (Goodrich 1930) but are possibly homologous to the extravisceral cartilages (Daniel 1928).

The hyoid arch (Fig. 3-5) in living sharks consists of the paired hyomandibulae, or dorsolateral elements (epihyals), of the hyoid arch, which have their proximal ends attached to the otic capsules of the chondrocranium and their distal ends to the posterior ends of the jaw cartilages and to the paired ventrolateral elements, or ceratohyals. The ceratohyals in turn attach to a mesial and possibly compound element, the basihyal, which forms the core of the tongue and is analogous and in part homologous to the basibranchials of the gill skeleton. The hyoid arch runs posteriomesial to the mandibular arch and has the hyoid gill pocket and spiracle between it and the mandibular arch on each side. The hyoid arch serves in jaw suspension, pharyngeal movements, including mouth opening and closing, support for the tongue, spiracular operation, and support for the hyoid gill septum, or hyoid operculum. The ceratohyal and hyomandibula has several hyoid rays, pointed or forked and sometimes conjoined cartilages that extend into the hyoid gill septum and help to support it and the pretrematic hemibranch of the first branchial pocket behind the hyoid arch. In many elasmobranchs with large, functional spiracles there is a spiracular valve, which is supported by a spiracular cartilage just behind the spiracular aperture and between the hyoid and mandibular. Spiracular cartilages may also be present in sharks with small spiracles. In batoids the hyoid arch is more or less modified as the basihyal and ceratohyals are disconnected and separated ventrally from the hyomandibulae and become more or less degenerate or lost. The hyoid rays of batoids have apparently coalesced behind the hyomandibula in the hyoid septum to form a dorsal element, or dorsal pseudohyoid, analogous to the hyomandibula and first epibranchial in series with it and replacing the nonsuspensory functions of the hyomandibula, and a ventral element below it, the ventral pseudohyoid, analogous to the true ceratohyal and first ceratobranchial in series with it and replacing the ceratohyal functionally. The reduced true ceratohyal, where present in batoids, is connected ventrally with the ventral pseudohyoid, and the suspensory hyomandibula is connected posteriodorsally with the dorsal pseudohyoid.

There are five to seven pairs of branchial arches, or gill arches (see Fig. 3-5), for support of the gills. Primitively, sharks and batoids have five branchial arches, but the frilled and six-gill sharks, the six-gilled saw shark, and the sixgill stingray have added an additional sixth arch, while there are two genera of hexanchoid sharks with seven gill arches. The branchial arches include two sets of connected and paired dorsal elements,

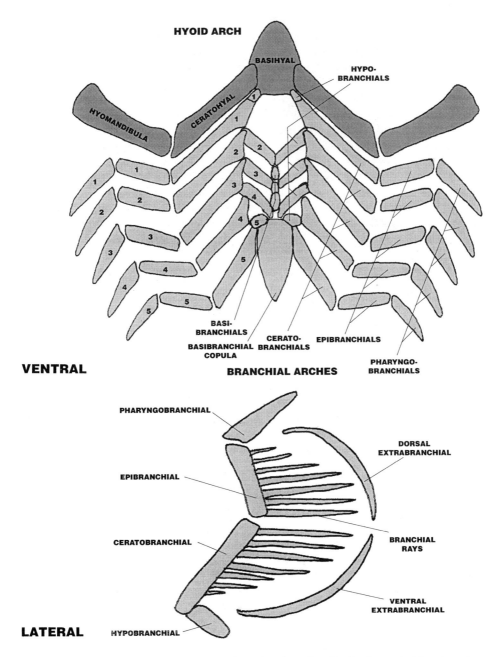

FIGURE 3-5 The splanchnocranium or visceral arches of an idealized neoselachian shark in exploded ventral view (*above*) (hyoid arch elements darker) and a gill arch in exploded lateral view to show branchial rays and extrabranchial cartilages (*below*).

the dorsomedial pharyngobranchials against the roof of the pharynx and the dorsolateral epibranchials (both equivalent to the hyomandibula), supporting the dorsal bases of the gill holobranchs; and two sets of connected ventral elements, the ventrolateral ceratobranchials (equivalent to the ceratohyal) supporting the ventral bases of the gill holobranchs and the variably developed hypobranchials connecting to the ventromedial unpaired elements, or basibranchials, and to the basihyal (first hypobranchials, when present). The hypobranchials and basibranchials, along with the expanded ventral ends of the ceratobranchials, form the basibranchial skeleton of the floor of the posthyoid pharynx. In hexanchoid sharks the pharyngobranchials and epibranchials are separate elements on all gill arches, but in other sharks and rays the last and next to last pharyngobranchials and the last epibranchial are fused to form a single pharyngoepibranchial element. It is unknown whether the condition in hexanchoids is primitive or neotenic, as the three elements fuse in the latter stages of development of the gill arch skeleton in other elasmobranchs. Anterior basibranchials are variably developed, with small unpaired anterior basibranchials often between the anterior hypobranchials in primitive sharks, but these are variably lost in many sharks and in batoids.

A small to large, flat posterior medial element, or basibranchial copula, is well developed in sharks and rays and forms part of the dorsal roof of the pericardial cavity, with the heart below it. More posterior hypobranchials often articulate directly with the branchial copula, and the first and last hypobranchials may be lost. In primitive sharks (including hexanchoids and bramble sharks) the basibranchial skeleton is closely adjacent and posterior to the hyoid arch, but in many sharks a broad gap is present between the hypobranchials and the basihyal, with anterior basibranchials reduced or lost and hypobranchials directed posteriorly and articulating directly with the branchial copula. This may be an adaptation for a possible increase in mobility between the hyoid and branchial arches. In a few sharks there are a pair of small cartilages articulating on the basibranchial copula behind the gill arches that may be rudiments of an additional gill arch (Daniel 1928). The first through next-to-last gill arches have prominent branchial rays attached to the epibranchials and ceratobranchials that stiffen the gill septa, but the last gill arch lacks rays and is closely attached to the scapulocoracoid (shoulder girdle) and in batoids articulates with it on each side by a condyle on the scapulocoracoid. Branchial rays are generally elongated and tapering in sharks and many batoids, but in some guitarfishes and electric rays they are short and have circular, expanded, plate-like tips.

Extravisceral cartilages (sometimes termed extravisceral arches) are paired, elongated cartilages external and circumferential to the hyobranchial arches and rays that function as a framework to support the distal hyoid and branchial septa. In sharks there is only a dorsal extrahyoid cartilage supporting the hyoid septum, but the branchial arches from the first through next to last usually have both a dorsal extrabranchial and a ventral extrabranchial. For each gill arch with extrabranchials, the dorsal and proximal end of each dorsal extrabranchial is platelike and anchored in tissue near the corresponding pharyngobranchial of the arch; likewise, the expanded ventral end of each ventral extrabranchial is anchored near the ventral ceratobranchial. With the extrabranchials, the branchial skeleton forms a bilobate hoop to support and tauten the branchial septum. In batoids the extrabranchials are more platelike and support the bases of the branchial septa without extending into the distal margin; apparently, with movement of gill apertures ventrally in batoids and with the propterygia partially or entirely surrounding the branchial chambers, there is no need for lateral support for the hyobranchial septa. In batoids there are additional opercular cartilages that strengthen the branchial septa just in front of the gill openings and that may be derived from the extrabranchials.

VERTEBRAL COLUMN

The vertebral column of a typical postnatal living shark consists of a string of spool-like or hourglasslike calcified bodies, the vertebral centra, extending from the occiput of the neurocranium to the distal web of the caudal fin. Dorsal to the centra is a covered archlike structure, or neural arch, with a central cavity that contains the spinal cord. Each vertebral centrum, neural arch segment, and associated structures constitutes a vertebra: (1) the centrum itself, with bilaterally symmetrical cartilages attached to it; (2) a pair of wedge-shaped or rectangular cartilaginous plates, or basidorsals, of the neural arch, which surmount each centrum and are penetrated on each side by a foramen for the ventral root of a spinal nerve; (3) a pair of basiventrals below the centrum, and on monospondylous precaudal vertebrae of the trunk and branchial region, a rib attached to the basiventral of each side. Ribs are well developed and elongated in many elasmobranchs, including sawfishes and guitarfishes, and are morphologically dorsal ribs that extend into the horizontal septum of the myomeres (Goodrich 1930). Ribs are reduced or lost in certain lamnoid sharks and in some batoids. The basidorsals of each vertebra are connected by an arched interdorsal cartilage, which is penetrated by a foramen for the dorsal root of a spinal nerve and completes the roof of the neural arch, while the basiventrals of each vertebra are connected by a small interventral cartilage. Some elasmobranchs, including saw sharks, angel sharks, and many batoids, have the dorsal surfaces of the neural arches greatly elevated and partitioned into discrete neural spines. Living elasmobranchs have as few as 60 to as many as 477 vertebrae.

All living elasmobranchs have a segmented vertebral column, with the embryonic notochord and notochordal sheath interrupted and subdivided by the calcified centra, which develop in the notochordal sheath. Calcification patterns of vertebral centra have been important in the classification of elasmobranchs and are discussed by Hasse (1879–85), Regan (1906), Ridewood (1921), Daniel (1928), Goodrich (1930), White (1937), Applegate (1967), and Compagno (1988). In elasmobranchs each centrum has a calcified double cone as a core, with the apices of the two cones connected under the middle of the basidorsals. The cones differentiate in ontogeny as a part of the process of constriction of the notochord and its sheath beneath each pair of basidorsals. The degree of constriction of the notochord and its sheath and the narrowness of the passage of the notochord through the apices of the cones is widely variable in living elasmobranchs and tends to be less developed in deepwater species. No living elasmobranch has a primitively unconstricted notochord and notochordal sheath as in primitive living bony fishes (sturgeons, paddlefish, lungfish, and coelacanths), chimaeroids, and many fossil nonneoselachian elasmobranchs.

Many sharks have the body of the centrum between the basidorsals and basiventrals and surrounding the double cone constituted of hyaline cartilage, which may be displaced by calcifications from the periphery of the centrum or from the double cone. In many sharks there is always a noncalcified area of cartilage extending inward to the double cone from the basidorsals and basiventrals, but in some sharks and batoids these are filled with solid calcification or calcified rings, or annuli. In the vertebral centra of many sharks are four areas of calcification termed *intermedialia* that are present between the basidorsals and basiventrals. One of these intermedialia is present between the two basidorsals on the dorsal surface of the centrum, one occurs on the ventral surface between the basiventrals, and a pair occurs on each side of the centrum between a basidorsal and basiventral. The intermedialia may have solid calcification, thin radial plates, or radii, or annular plates, or annuli. Calcification patterns are apparently size-, habitat-, and activity-related. Deepwater species generally have reduced calcification in their vertebral columns and may lack calcified intermedialia, radii, or annuli or can even lose the calcified double cones of their centra.

The vertebral column in living sharks can be divided into three major parts: (1) the monospondylous precaudal vertebrae of the branchial head and trunk, (2) the diplospondylous precaudal vertebrae of the precaudal tail, and (3) the diplospondylous caudal vertebrae of the caudal fin. Monospondylous precaudal vertebrae extend from the occiput to the diplospondylous transition, which is usually over the pelvic girdle but sometimes is anterior or well posterior to it. From the transition, the diplospondylous precaudal vertebrae extend posteriorly to the base of the caudal fin and continue as diplospondylous caudal vertebrae with expanded neural and hemal arches that support the caudal fin. According to Ridewood (1899), Sécerov (1911), and Goodrich (1930) diplospondyly in elasmobranchs involves the division of vertebral basals and centra, in the tail during embryonic growth to produce two vertebrae for every myomere. The transition to diplospondyly is marked in most sharks and rays by a sudden decrease in length of the vertebral centra and associated basidorsals, with a pair of dorsal and ventral nerve foramina on the basidorsals and interdorsals of every other centrum. The first few diplospondylous centra are usually over half as long as the longest preceding monospondylous centra, but relative length varies with numbers of vertebral centra, size of vertebrae, and commonly the maximum size of the species.

Large sharks and sharklike rays tend to have more vertebrae than smaller ones, and in those species with numerous short centra, the diplospondylous precaudal centra are scarcely shorter than the longest monospondylous centra. Smaller sharks with large centra may have the longest monospondylous centra hypercalcified and over three times the length of the anteriormost diplospondylous centra. The diplospondylous transition commonly occurs opposite the hemal arch transition, where trunk vertebrae with ribs and laterally directed basi- and interventrals change to precaudal tail vertebrae without ribs, and with the caudal artery (rearward extention of the dorsal aorta) and caudal vein enclosed in a hemal canal formed by ventrally expanded hemal arches composed of the basi- and interventrals and having foramina for veins and arteries of the tail. However, some elasmobranchs have the two transitional areas offset, with the hemal arch transition anterior or posterior to the diplospondylous transition. A few thresher sharks have the hemal transition extending far anterior to just behind the cranial occiput.

In batoids the vertebral column is variably elaborated compared with that of sharks and becomes less flexible and quite rigid anteriorly to support the greatly expanded pectoral fins. In all living batoids the monospondylous precaudal vertebrae just posterior to the neurocranium are fused into a rigid tube, or cervicothoracic synarcual, that extends over the branchial region of the head and may include 8 to 45 vertebrae. In sawfishes the cervicothoracic synarcual is very short and ends well in front of the scapulocoracoid, but in other batoids it ends just in front (narkid electric rays) or just behind the scapulocoracoid (most other batoids). The cervicothoracic synarcual usually has a series of entire vertebral centra embedded in its posterior third or half, preceded by a hemicentrum similar to the occipital one of sharks, but in the anterior half of the synarcual, centra are absent. In saw sharks there are a series of abruptly widened unfused cervical vertebrae just behind the occiput, which correspond to the cervicothoracic synarcual of batoids, and in angel sharks a few cervical vertebrae are fused to one another and form an incipient synarcual. Most batoids have free monospondylous and diplospondylous vertebrae behind the cervicothoracic synarcual, but in myliobatoids there is a second, thoracocolumbar synarcual in the monospondylous vertebrae that may have a dozen or more vertebrae and either abuts the cervicothoracic synarcual or is separated from it by one or more free intermediate vertebrae. In most rays separate diplospondylous vertebral centra extend to the tip of the caudal fin skeleton, as in sharks, but in some myliobatoids with whiplike tails centra may be absent behind or even in front of the tail sting and the vertebral column has a secondarily unsegmented notochord extending to the tail tip.

APPENDICULAR SKELETON AND FINS

The appendicular skeleton of living elasmobranchs includes the girdles and skeletons of the paired fins (pectorals and pelvics) and the skeletons of the precaudal unpaired fins (dorsal and anal fins). The pectoral, pelvic, dorsal, and anal fins are supported by cartilaginous basals and radials in the fin bases that form their fin skeletons, but the caudal fin is supported by caudal vertebrae. The fin webs are supported by a core of ceratotrichia, thin, radially elongated elastic connective tissue fibers. Unfortunately, prepared ceratotrichia are the most important ingredient of shark-fin soup. Ceratotrichia are present in the fins of more sharklike batoids but are greatly reduced or lost and functionally replaced by radials in those batoids with greatly expanded pectoral disks and reduced tails and caudal and dorsal fins. Fins of many elasmobranchs are primitively aplesodic, with radials confined to the fin base; but the pectoral fins, the first dorsal fins, and even the pelvic fins of various sharks are plesodic, with the radials elongated and expanded into the ceratotrichia to stiffen the fin webs (Compagno 1988).

Paired fins and girdles

The paired fins (Fig. 3-6) include the anterior pectoral fins at the junction between head and trunk, and the posterior pelvic fins at the junction between trunk and tail. These are supported by the pectoral girdle, or scapulocoracoid, and the pelvic girdle, or puboischiadic bar, which anchor the fins in the body wall.

The pectoral girdle, or scapulocoracoid, is inserted in the body wall just posterior to the branchial arches. The scapulocoracoid is generally U-shaped in sharks and arches dorsally and laterally to the vertebral column. In batoids the scapulocoracoid is more flattened and more expanded anteroposteriorly and is attached to the vertebral column dorsally. The scapulocoracoid consists of a ventral and transverse coracoid bar, a dorsolateral scapular process on each side, and in some species a separate, articulated suprascapula atop the scapular process. In sharks the scapular processes and suprascapulae are directed posteromediodorsally, but in batoids the scapular processes are directed mesially toward the vertebral column and the superscapulae are articulated with or fuse to the neural arches of the vertebral column. In some batoids, including sawfish and some electric rays, the paired superscapulae articulate with the vertebral column behind the cervicothoracic synarcual, but in most batoids they articulate with the synarcual itself. In most myliobatoids the superscapulae are fused to the sides of the synarcual to form sockets, or cotyles, for articulation of the scapular condyles on the dorsomedial tips of the scapular processes, but in devil rays the scapular processes fuse to the superscapulae and synarcual. The coracoid bar in a few primitive living sharks (hexanchoids, bramble sharks) is a paired structure that has a median joint at the midline and may have a separate median sternal cartilage, but it is fused at the midline in most living sharks and all batoids. In many batoids there is a branchial condyle on the anterior face of the coracoid bar on each side for articulation of the ceratobranchial of the last gill arch. The lateral face of the scapulocoracoid at the junction of the coracoid bar and scapular processes has an articular surface for the pectoral basal cartilages and foramina for blood vessels and nerves. In sharks the articular surface may be a simple and irregular ridge or basicondyle, but in angel sharks there is a discrete anteriolateral promesocondyle for the articulation of the propterygium and mesopterygium (pectoral basals) and a posterior metacondyle for the metapterygium. In saw sharks and batoids there are three discrete articular condyles: the procondyle for the propterygium, the mesocondyle for the mesopterygium, and the metacondyle for the metapterygium. Additional condyles or articular ridges may form on the side of the scapulocoracoid between the meso- and metacondyles in batoids with elongated scapulocoracoids for articulation of accessory or neopterygial basals and radials.

Fenestrae or foramina on the lateral face of the scapulocoracoid of primitive batoids

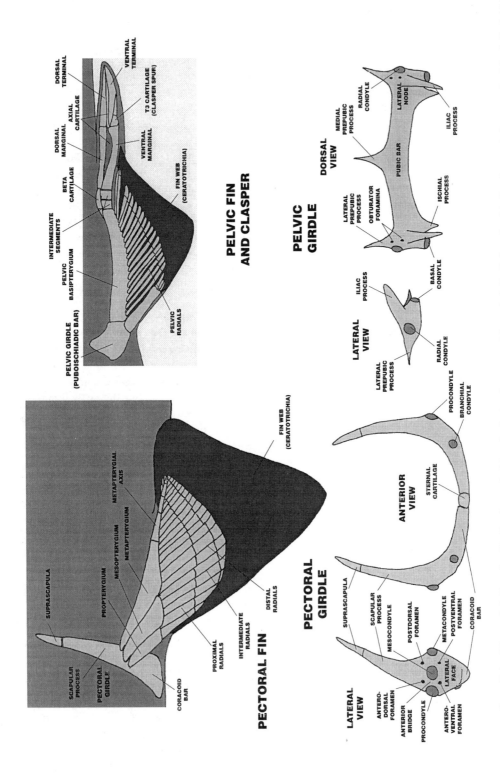

FIGURE 3-6 Paired fins and girdles of an idealized neoselachian shark, showing pectoral fin and girdle (*left*) and pelvic fin, clasper skeleton, and pelvic girdle (*right*).

include anterodorsal and anteroventral foramina or fenestrae between the pro- and mesocondyles and postdorsal and postventral foramina or fenestrae between the meso- and metacondyles. The anterior bridge is a horizontal bar separating the pro- and mesocondyles laterally, and dorsoventrally it separates the anterodorsal and anteroventral fenestra. The anterior bridge is lost in many skates.

The pectoral fins of living sharks and rays primitively have three cartilaginous basal cartilages running from anterior to posterior within the fin, the propterygium, mesopterygium, and metapterygium. The posterior end of the mesopterygium is extended by a short to long metapterygial axis with one or more segments extending into and supporting the posterior lobes of the pectoral fins. The axis is well developed in primitive sharks and is greatly elongated in batoids but is reduced in sharks that have radially elongated, paddle-shaped metapterygia. The pectoral basals are variably fused to each other in some sharks, which can be a case of the propterygium fused to the mesopterygium, the mesopterygium fused to the metapterygium, or in dalatiid and echinorhinid sharks, all three basals fused into a single basipterygium. In batoids the propterygium is more or less expanded anteriorly and is segmented into a propterygial axis. The pectoral basals are distally joined to slender transverse or diagonal segmented cartilages, or radials that support the fin base and may extend into its distal web. In most living sharks radials are mostly trisegmental, but in some sharks with plesodic or semiplesodic pectorals radial segments are increased from 7 to 12, while living batoids have 9 to 30 or more segments. The distal radials of some sharks are irregularly forked, but in batoids several distal rows of radial segments are singly or doubly split longitudinally. In plesodic sharks elongated pectoral radials augment and stiffen the ceratotrichia-supported pectoral fin web. However, in batoids the elongated, slender, distally split, multisegmented pectoral radials and associated dorsal and ventral radial musculature functionally replace the ceratotrichia, which are more or less reduced to a thin distal fringe. The batoid pectoral fin is adapted as a highly flexible and mobile thruster or propellor, which can produce highly controlled sine waves or dorsoventral flapping for locomotion.

The pelvic girdle, or puboischiadic bar, of living elasmobranchs consists of a transverse flattened or cylindrical plate, the pubic bar, connecting the paired lateral nodes for articulation of the skeleton of each pelvic fin. The pubic bar can be straight or arched anteriorly or posteriorly. There may be a medial prepubic process extending anteriorly from the midline of the pubic bar (some myliobatoids), and paired lateral prepubic processes extending anteriorly from the lateral nodes. The lateral nodes usually have vertical, dorsal, or posterodorsal projections, or iliac processes and obturator foramina penetrating the nodes for nerves and blood vessels. There is often a radial condyle on the lateral or posterolateral surface of each lateral node for articulation of the first, compound pelvic radial, a basal condyle on the posterior surface for articulation of the pelvic basipterygium, and a posterior ischial process medial to the posterior condyle and directed posteriorly.

The pelvic fins are attached basally and anteriorly to the lateral nodes of the puboischiadic bar. Usually there is a large anterior compound radial that articulates with the anterior condyle and supports the anterior lobe of the pelvic fins. This seems to be formed by two or three or more radials, by reference to the distal radial segments attached to it. In some sharks with broad lateral nodes and an interspace between the radial and basal condyles, several pelvic radials may articulate directly with the lateral node between the condyles. In sharks and batoids where the two condyles are closely adjacent, the compound radial and basipterygium may be the only structures bearing radials. Most of the pelvic radials articulate with a longitudinal platelike pelvic basal cartilage, or basipterygium, extending posteriorly across the pelvic fin base from an articulation with the posterior condyle. The basipterygium is often curved posterior-medially and its medial face is concave, but in some sharks it is nearly straight. In female elasmobranchs the basipterygium has attached to its posterior end a basipterygial axis analogous to the metapterygial axis of the pectoral fin; this has one to several short segments and is shorter than the basipterygium proper. The basipterygial radials are transverse, diagonally posterolateral, or almost parallel to the axis of the basipterygium. Sharks generally have aplesodic pelvic fins that support the fin base only, except for threshers, which have expanded distal radials that extend into and stiffen the fin web without replacing the ceratotrichia. Shark pelvic radials have two to five radial segments. Batoids have plesodic pelvic fins that parallel the pectoral fins in morphology, with five to nine radial segments, forked or divided distal segments, increased flexibility, reduction of ceratotrichia, and expansion of radial muscles into the fins to enhance their mobility and flexibility. In skates and some electric rays the anterior lobe of the pectoral fin is highly mobile and supported by the anterior compound radial and some basipterygial radials. In legged skates (Anacanthobatidae) and New Zealand legged torpedos (*Typhlonarke*) the anterior lobe is completely separate from the posterior lobe on each pelvic fin and forms a "leg" for walking on the bottom. In legged skates the leg is supported by the compound radial and one basipterygial radial, with a radial-free gap on the basipterygium between them and the radials supporting the posterior lobes. The leg resembles a tetrapod hind limb in being divided into a proximal, medial, and distal part corresponding to a thigh, calf, and foot.

The skeleton of the claspers, or mixopterygia, the paired copulatory organs found in the males of all living elasmobranchs, is apparently formed from the basipterygial axis of the pelvic fin skeleton (and possibly the posteriormost basipterygial radials) and attaches directly to the rear of the basipterygium. A basic elasmobranch clasper skeleton includes one to three intermediate segments, stem-joints, or basal segments (usually one or two) between the basipterygium and the longitudinal cartilage of the clasper, the axial cartilage, or appendix-stem. The beta cartilage is a dorsal cartilage above the intermediate segments that reinforces the joints between the basipterygium and axial cartilage. The axial cartilage is primitively cylindrical and can extend to the posterior tip of the clasper, but it is flattened in many sharks. There is usually a curved dorsal plate, or dorsal marginal cartilage, that originates dorsomedially on the axial cartilage and extends laterally and posteriorly to the clasper glans, the open posterolateral or posterodorsal area of the clasper. A ventral marginal cartilage originates ventromedially on the axial cartilage and parallels the dorsal marginal cartilage. The axial cartilage and marginal cartilages form the clasper shaft, which partially or completely encloses the clasper groove that conducts seminal fluid from the anterior aperture, or apopyle, of the clasper (near the cloaca and urogenital papilla) to the posterior aperture, or hypopyle, at the anterior end of the clasper glans. The area of the glans is variably supported by the posterior end of the axial cartilage, the end-style, and usually has a separate dorsolateral cartilage, or dorsal terminal, extending rearward from the dorsal marginal and paralleling the end-style, as well as a separate ventrolateral cartilage, or ventral terminal, that arises from the ventral marginal and extends rearward in parallel and below the dorsal terminal. Many sharks and rays have an accessory terminal cartilage that often bears a hardened external structure, a clasper spine or clasper spur. Terminal and marginal cartilages are greatly reduced in certain sharks (some hexanchoids and squaloids), but conversely, the basic dorsal and ventral terminals, end-style, and accessory terminal of the clasper glans of many sharks and batoids are elaborated and supplemented in some squaloid and carcharhinoid sharks and especially in skates by several additional cartilages and complex external structures.

Unpaired fins and supports

The unpaired fins (Fig. 3-7) include the dorsal fins, the anal fin, and the caudal fin. The dorsal and anal fins of living sharks have single

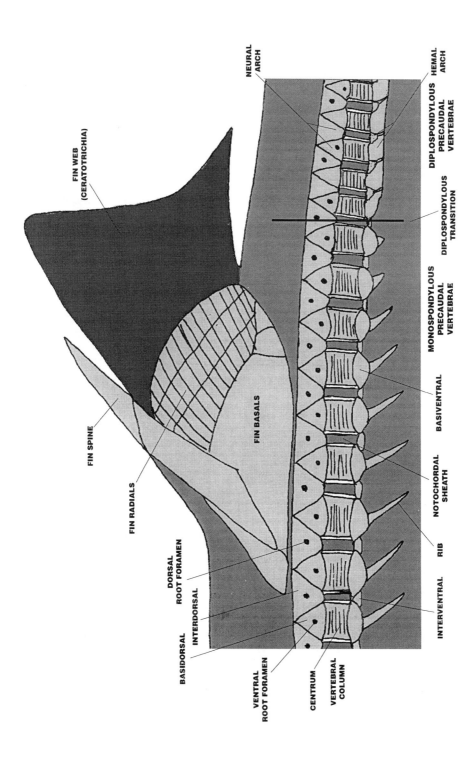

FIGURE 3-7 Neoselochian first dorsal fin and section of vertebral column in lateral view, showing fin skeleton, diplospondylous transition, and hemal arch transition.

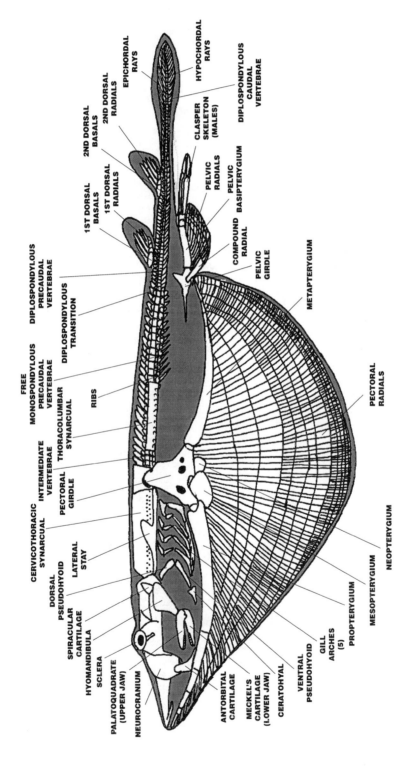

FIGURE 3-8 An idealized batoid skeleton in lateral view. Compare with Figure 3-1 and note attachments of neurocranium, cervicothoracic synarcual, pectoral girdles, and pectoral propterygium to form a strong support for the pectoral disk. Also note subdivision of vertebral column and separation of jaws and hyomandibulae from the rest of the splanchnocranium.

or segmented basal cartilages anchoring them in the midline of the trunk or tail and are joined to distal radials that support the fin base and sometimes the web. Carcharhinoid, lamnoid, and orectoloboid sharks have dorsal fin skeletons with platelike basals reduced or absent; the basals in these sharks are segmented and continuous with the more distal radials. Dorsal fin radials may be unsegmented or divided into two to four or more segments. Dorsal fin spines on the anterior bases of the dorsal fins are are present on the first dorsal fin of the dalatiid shark *Squaliolus* and on both dorsal fins of heterodontid sharks and many squaloids but are absent from most living sharks and all living batoids (Fig. 3-8). Internal, vestigial dorsal fin spines may be present in pristiophorid sharks. Dorsal fin spines are anchored on the basal cartilages and are supported by a cartilage extending from the basal cartilage into the spine base. The first dorsal fin is plesodic or semiplesodic in some lamnoid sharks, but ceratotrichia are well developed in the distal fin webs. Both dorsal fins are aplesodic in most sharks and in sharklike batoids and have well-developed ceratotrichia, which are variably reduced in batoids with small dorsal fins.

The caudal fin is an unpaired medial fin like the dorsal and anal fins and has its distal fin web supported by ceratotrichia, but in living elasmobranchs the caudal fin skeleton is part of the axial skeleton and vertebral column and does not have separate basals and radials as do other unpaired fins. The caudal fin base is supported by expanded and dorsoventrally elongated neural and hemal arch cartilages of the vertebral column, the epichordal and hypochordal rays. The hypochordal rays sometimes have a few distal segments. The caudal fin web in sharks and in sharklike rays with large caudal fins is supported only by ceratotrichia, which are the only support of the well-developed ventral lobes of those elasmobranchs with lunate caudal fins. However, in many Paleozoic chondrichthyans there are true epichordal and hypochordal radials in addition to expanded neural and hemal arch cartilages, and in forms with lunate caudal fins, expanded radials support the ventral caudal lobe. Caudal fin supports are greatly reduced in many batoids with small caudal fins, and in those with vestigial caudal fins (longitudinal folds) or no caudals and whiplike tails, the vertebral column is not dorsoventrally expanded and can be simple, tubelike, and lacking vertebral centra.

Exoskeleton

Unlike that of most jawless and jawed fish groups, the *exoskeleton*, or external skeleton, of living elasmobranchs is relatively inconspicuous and does not form large, external bony plates or imbricated rows of large bony scales on the head and body. The exoskeleton consists of dermal denticles (placoid scales) and their derivatives. Dermal denticles usually cover the head, trunk, tail, fins, and orobranchial cavity of elasmobranchs, although denticles can be absent from certain of these structures in various sharks and batoids, while some batoids completely lack denticles. Possible and probable derivatives of dermal denticles include thorns, rostral sawteeth, fin spines, stings, clasper spines, and gill raker denticles. Oral teeth are probably derived from dermal denticles and are also part of the exoskeleton. Specialized oral teeth with linear replacement have been characteristic of cartilaginous fishes since the Devonian, although they could have developed from unspecialized small oral teeth that were shed and replaced like denticles and were essentially denticles that functioned as primitive teeth.

REFERENCES

Allis, E. P., Jr. 1913. The homologies of the ethmoidal region of the selachian skull. *Anat. Anz.* 44: 323–24.

———. 1914. The pituitary fossa and trigeminofacialis chamber in selachians. *Anat. Anz.* 46: 225–53.

———. 1923. The cranial anatomy of Chlamydoselachus anguineus. *Acta. Zool.* (Stockh.) 4: 123–21.

Applegate, S. P. 1967. A survey of shark hard parts. In *Sharks, Skates, and Rays,* ed. P. W. Gilbert, R. F. Mathewson, and D. P. Rall, 37–67. Baltimore: Johns Hopkins University Press.

Compagno, L. J. V. 1988. *Sharks of the Order Carcharhiniformes.* Princeton, N.J.: Princeton University Press.

Compagno, L. J. V., and T. R. Roberts. 1982. Freshwater stingrays (Dasyatidae) of Southeast Asia and New Guinea, with description of a new species of *Himantura* and reports of unidentified species. *Envir. Biol. Fish* 7: 321–39.

Corrington, J. D. 1930. Morphology of the anterior arteries in sharks. *Acta. Zool.* (Stockh.) 11: 185–261.

Daniel, J. F. 1928. *The Elasmobranch Fishes,* 2d ed. Berkeley: University of California Press.

Garman, S. 1913. *The Plagiostomia.* Memoirs of the Museum of Comparative Zoology, vol. 36. Cambridge, Mass.: Harvard College.

Gegenbaur, C. 1865. *Untersuchungen zur Vergleichenden Anatomie der Wirbelthiere,* vol. 2, pts. 1, 2. Leipzig: Wilhelm Engelmann.

Goodrich, E. S. 1930. *Studies on the Structure and Development of Vertebrates.* London: Macmillan.

Gregory, W. K. 1933. Fish skulls, a study of the evolution of natural mechanisms. *Trans. Am. Philos. Soc.* 23: 75–481.

Hasse, J. C. F. 1879–85. *Das natürliche System der Elasmobranchier auf Grundlage des Baues und der Entwicklung ihrer Wirbelsäule. Eine morphologische und paläontologische Studie.*

Holmgren, N. 1941. Studies on the head in fishes. Embryological, morphological, and phylogenetical researches: II. Comparative anatomy of the adult selachian skull, with remarks on the dorsal fins in sharks. *Acta. Zool.* (Stockh.) 22: 1–100.

Huber, O. 1901. Die Kopulationsglieder der Selachier. *Z. Wiss. Zool.* 70: 592–674.

Jungersen, H. F. E. 1899. On the appendices genitales in the Greenland shark, *Somniosus microcephalus* (Bl. Schn.), and other selachians. In *The Danish Ingolf Expedition,* vol. 2, pt. 2. Copenhagen: Bianco Luno.

McEachran, J. D., and Compagno, L. J. V. 1979. A further description of *Gurgesiella furvescens* with comments on the interrelationships of Gurgesiellidae and Pseudorajidae (Pisces: Rajoidei). *Bull. Mar. Sci.* 29(4): 530–53.

O'Donoghue, C. H., and E. Abbott. 1928. The blood vascular system of the spiny dogfish, *Squalus acanthias* Linne, and *Squalus sucklii,* Gill. *Trans. R. Soc. (Edinb.)* 55: 823–90.

Ørvig, T. 1951. Histologic studies of placoderms and fossil elasmobranchs: I. The endoskeleton, with remarks on the hard tissues of lower vertebrates in general. *Arch. Zool.* 2: 321–454.

Regan, C. T. 1906. A classification of the selachian fishes. *Proc. Zool. Soc. Lond.* 1906: 722–58.

Ridewood, W. 1899. Some observations on the caudal diplospondyly of sharks. *J. Linn. Soc.* (Zool.) 27: 46–59.

———. 1921. On the calcification of the vertebra centra in sharks and rays. *Philos. Trans. R. Soc. Lond.* 210B: 311–407.

Sécerov, S. 1911. Über die Entstehung der Diplospondylie der Selachier. *Arb. Zool. Inst. Univ. Wien Zool. Sta. Trieste* 29: 1–28.

Smith-Vaniz, L., S. Kaufmann, and J. Glowacki. 1995. Species-specific patterns of hyperostosis in marine teleost fishes. *Mar. Biol.* 121: 573–80.

White, E. G. 1930. The whale shark, *Rhiniodon typus.* Description of the skeletal parts and classification based on the Marathon specimen captured in 1923. *Bull. Am. Mus. Nat. Hist.* 61: 129–60.

———. 1937. Interrelationships of the elasmobranchs with a key to the order Galea. *Bull. Am. Mus. Nat. Hist.* 74: 25–138.

KAREL F. LIEM
ADAM P. SUMMERS

CHAPTER 4

Muscular System

GROSS ANATOMY AND
FUNCTIONAL MORPHOLOGY
OF MUSCLES

The gross anatomy of the muscles of the shark, usually represented by *Squalus acanthias*, has been the object of dissections by countless students from all over the world. There are many descriptions of elasmobranch musculature in the literature, ranging from detailed examinations of some region (e.g., Luther 1909), to comparative studies (e.g., Daniel 1934), to comprehensive descriptions of the entire musculature aimed a the comparative anatomy student (e.g., Gilbert 1973; Walker and Homberger 1992). Recent studies of the function of shark muscles have also included useful comparative anatomy (e.g., Wu 1994; Motta et al. 1997; Wilga and Motta 1998). In this chapter we rely solely on the most comprehensive study of the myology of *Squalus acanthias*, that of Marinelli and Strenger (1959). We have adopted Marinelli and Strenger's findings as a necessary prelude to future functional studies. To complete the gross anatomy we include a section on recent findings of muscle function as they relate to feeding, respiration, and locomotion. An important caveat is that the musculature of *Squalus*, particularly in the cranial region, is not necessarily representative of the rest of the elasmobranchs. A complete treatment of the variation in musculature would constitute a volume by itself.

A regional grouping of the muscles (e.g., those of the pectoral fin or the trunk) or a functional grouping (e.g., adductors or abductors) is helpful in dissection or functional analysis, but such groupings often include muscles of different phylogenetic origins. To trace the anatomical evolution of the muscular system meaningfully, we group muscles ac-

TABLE 4-1	GROUPS OF SOMATIC MUSCLES
EXTRINSIC OCULAR MUSCLES	
First two somitomeres (oculomotor nerve, III)	Dorsal rectus Ventral rectus Medial rectus Ventral rectus
Third somitomere (trochlear nerve, IV)	Dorsal oblique
Fifth somitomere (abducens nerve, VI)	Lateral rectus
BRANCHIOMERIC MUSCLES	
Mandibular muscles (trigeminal nerve, V)	Adductor mandibulae Levator palatoquadrati Spiracularis Preorbitalis Intermandibularis Ventral superficial constrictor
Hyoid muscles (facial nerve, VII)	Levator hyomandibulae Dorsal deep constrictors Ventral deep constrictors Ventral superficial constrictors Interhyoideus
Branchiomeric muscles of remaining arches (glossopharyngeal nerve, IX; vagus nerve, X)	Lateral interarcual Dorsal interarcual Cucullaris (trapezius) Interarcualis Adductor branchialis Superficial constrictors Interbranchial
EPIBRANCHIAL MUSCLES	
Epibranchial muscles (dorsal rami of anterior spinal nerves)	Anterior epibranchialis muscles inserting on skull
HYPOBRANCHIAL MUSCLES	
Prehyoid muscles (ventral rami of anterior spinal nerves)	Coracomandibularis
Posthyoid muscles (ventral rami of anterior spinal nerves)	Coracoarcualis (rectus cervicis) Coracobranchialis
TRUNK MUSCLES	
Epaxial group (dorsal rami of spinal nerves)	Epaxialis muscle complex
Hypaxial group (ventral rami of spinal nerves)	Hypaxialis muscle complex
APPENDICULAR AND CLASPER MUSCLES	
Pectoral and pelvic muscles (ventral rami of spinal nerves)	
Pectoral dorsal group	Pterygoideus dorsalis
Pectoral ventral group	Pterygoideus ventralis
Pelvic dorsal group	Dorsal superficial pterygoideus Dorsal deep pterygoideus
Pelvic ventral group	Ventral pterygoideus Pelvicobasalis Superficial mixipodii Medial mixipodii Dorsal mixipodii Ventral mixipodii

cording to their embryonic origin and nerve supply. We consider muscles with common development and nerve supply homologous. Such a consistent grouping can be recognized and used more effectively in comparisons between evolutionary lineages (Table 4-1).

Descriptive anatomy

EXTRINSIC OCULAR MUSCLES

The first three and fifth somitomeres give rise to the somatic extrinsic ocular muscles that move the eyeball. So, the most rostral somatic muscles in *Squalus acanthias*, as in most other vertebrates, are represented by six strap-shaped parallel fibered muscles that insert musculously on the scleral wall of the eyeball. They effect eyeball movements as the moving fish continues to look at a fixed object, or as the animal shifts its field of vision. On the anteromedial wall of the orbit two *oblique muscles* originate and extend obliquely caudad to insert on the dorsal and ventral surfaces of the eyeball (Fig. 4-1). The superior (or dorsal) oblique is innervated by the trochlear nerve (IV) and the inferior (or ventral) oblique is innervated by the oculomotor nerve (III). Four *recti muscles* fan out from the posteromedial wall of the orbit to insert on the eyeball. They are named for their points of insertion: superior rectus, inferior rectus, medial rectus, and lateral rectus. The lateral rectus is innervated by the abducens nerve (VI), while the other recti muscles are innervated by the oculomotor nerve (III). The superficial ophthalmic and lateral branches of the trigeminal (V) nerve pass dorsal to the superior rectus and superior oblique muscles, while infraorbital and maxillary branches of the trigeminal pass ventral to inferior oblique and inferior and lateral recti muscles (see Fig. 4-1).

BRANCHIOMERIC MUSCLES

The largest cranial muscle is the massive *adductor mandibulae* muscle, which exhibits very complex pinnation and occupies the angle of the jaws. One can recognize four subdivisions, all innervated by the trigeminal (V) nerve, in which different fiber directions indicate that each is active and efficient at a different stage of jaw closure. The most anterior

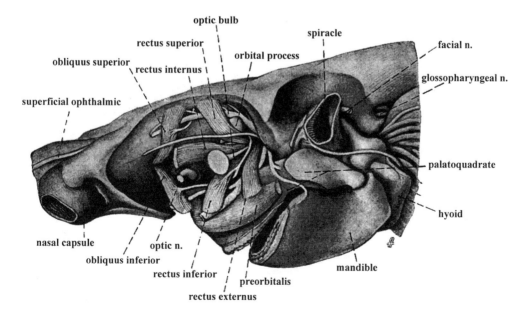

FIGURE 4-1 *Squalus acanthias* L. Lateral view of orbital region of the skull with ocular muscles, major nerves, and jaws. (From Marinelli and Strenger 1959.)

subdivision, or head, originates from the palatoquadrate between the attachments of the levator palatoquadrati and spiracularis muscles to insert on the ventral surface of the mandible. Posterior to this preorbital head, the adductor mandibulae muscle fibers of the second head wrap around the mandible, originating from a restricted region on the lateral surface of the palatoquadrate just medial to the preorbital portion. The remainder of the adductor mandibulae complex is composed of two extremely large parts. The anterior part is triangular in cross-section (Fig. 4-2), originating from the lateral surface of the posterior wing of the palatoquadrate and inserting on the lateral surface of the posterior quarter of the mandible. Finally, the most posterior portion of the adductor mandibulae originates from the most posterior portion of the lateral surface of a special process of the palatoquadrate. Its fibers form a large, complex, concentrated structure from which muscle fibers run posteroventrally to wrap around and attach to the posterior corner of the mandible. Surprisingly, the four heads of the adductor mandibulae lack tendons even though they are clearly separate, with fibers running in different directions and different angles (see Fig. 4-2).

The *levator palatoquadrati* muscle (Figs. 4-2 and 4-3) is located just posterior to the postorbital process and is innervated by the trigeminal (V) nerve. It originates from a very shallow fossa in the dorsal rim of the otic capsule and inserts on the palatoquadrate just medial to the adductor mandibulae process. Both origin and insertion are muscular and the fibers are by and large parallel. The function of this muscle is to retract the palatoquadrate after prey prehension.

Just posterior to the levator palatoquadrati and anterior to the external spiracular opening is the small *spiracularis* muscle, which closes the spiracle and is also innervated by the trigeminal (V) (see Figs. 4-2 and 4-3). Thus it is closely associated with the cranial side of the spiracle. The fibers partially encircle the spiracular cartilage, originating from the lateral wall of the otic capsule and inserting both on the spiracular cartilage and on the palatoquadrate just posterior to the site of insertion of the levator palatoquadrati.

The spindle-shaped *preorbital* muscle takes its origin musculously from the underside of the chondrocranium. The muscle fibers of the spindle-shaped belly converge on a strong tendon, which attaches to the connective tissue at the jaw joint just cranial to the adductor mandibulae muscle (see Fig. 4-2). Topographically the muscle is situated just ventral to the eyeball and is innervated by the trigeminal (V) nerve. Functionally this muscle is quite variable, serving to retract or protract the palatoquadrate, depending on the site of origin and the morphology of the rostral cartilage.

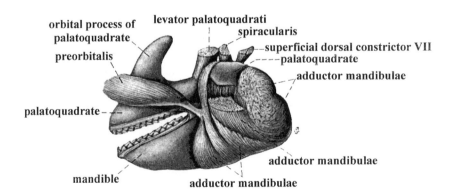

FIGURE 4-2   *Squalus acanthias* L. Lateral view of the left side of the jaws with associated musculature. (From Marinelli and Strenger 1959.)

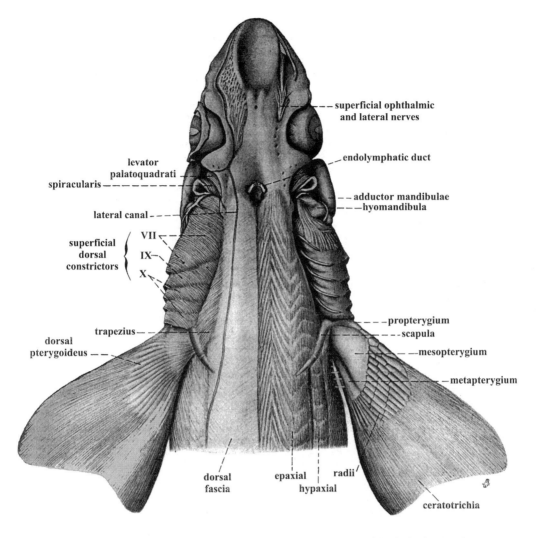

FIGURE 4-3 *Squalus acanthias* L. Dorsal view of the anterior portion of the body showing the superficial fascia on the left and a deeper dissection on the right. (From Marinelli and Strenger 1959.)

The ventral, transverse, thin *intermandibularis* muscle runs between the two rami of the mandible (Fig. 4-4). The muscle merges insensibly with the ventral superficial constrictors of the gill pouches. Therefore it is often considered a component of the constrictor series and named as such (e.g., Marinelli and Strenger 1959). The muscle fibers insert on the median longitudinal tendinous raphe, which divides the muscle in right and left halves. The intermandibularis is innervated dually by branches of both the trigeminal (V) and facial (VII) nerves. The fibers of this muscle run in a pattern that is oriented slightly obliquely. The muscle compresses the anterior pharyngeal cavity and the buccal cavity during feeding and respiration.

Just dorsal to the intermandibularis muscle runs a second thin transverse muscle, the *interhyoideus*, which is innervated purely by the facial (VII) nerve and is technically a hyoid rather than a mandibular muscle. However, its intimate proximity with the intermandibularis warrants its discussion at this juncture. The interhyoideus does not reach as far anteriorly as the intermandibularis, while its fibers

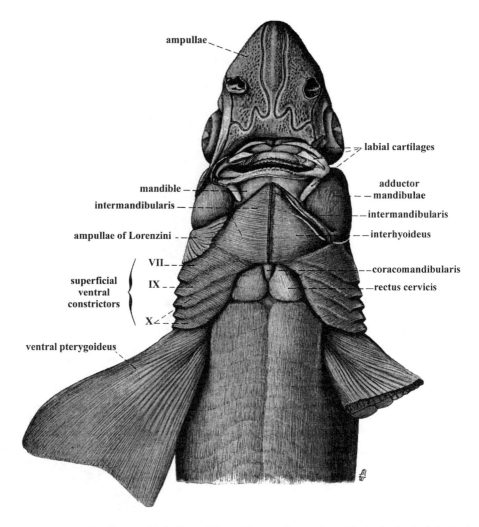

FIGURE 4-4 *Squalus acanthias* L. Ventral view of the anterior portion of the body. On the left side the intermandibularis has been removed to show the interhyoideus. (From Marinelli and Strenger 1959.)

show a more transverse orientation. Unlike the intermandibularis, the origin of the interhyoideus does not include the mandibular rami but is restricted to the ceratohyal. Caudally the interhyoideus merges insensibly with the ventral hyoid constrictor (see Fig. 4-4).

Among the hyoid muscles innervated by the facial nerve (VII) are the segmental superficial *constrictor* muscles associated with the gill slits (Figs. 4-3, 4-4, and 4-5). They are thin sheets of muscle, whose dorsal and ventral components do not differ from one another except in their topography. However, the most anterior of the superficial constrictors is in-

nervated by the facial (VII) nerve, whereas the subsequent superficial constrictors are actually receiving their innervation from branches of the glossopharyngeal (IX) and vagus (X) nerves (see Figs. 4-3 and 4-4).

The *levator hyomandibulae* is a large muscle, innervated by the facial (VII) nerve, that occupies the space between the first typical gill slit and the spiracle (Fig. 4-5). Its origin is from the lateral surface of the otic capsule and the fibers run anteroventrally, in a parallel fashion, to insert mainly on the lateral surface of the hyomandibular cartilage. The muscle passes medial to the adductor mandibulae

Muscles: Gross Anatomy and Functional Morphology   99

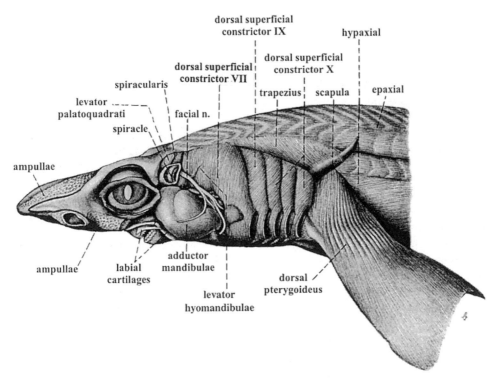

FIGURE 4-5 *Squalus acanthias* L. Lateral view of the anterior portion of the body showing superficial musculature. (From Marinelli and Strenger 1959.)

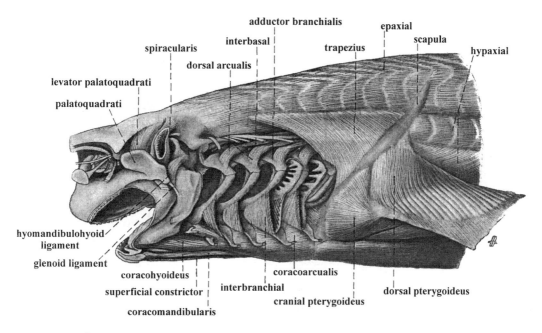

FIGURE 4-6 *Squalus acanthias* L. Lateral view of deep dissection of anterior portion of the body after removal of left gill pouches and mandible. (From Marinelli and Strenger 1959.)

complex and just anterior to the first constrictor located between the first typical gill slit and the spiracle (see Figs. 4-5 and 4-6).

It is generally accepted that the levator muscles of the five branchial arches behind the hyomandibula have united with one another to form a large triangular muscle, the *cucullaris* (*trapezius*), which lies dorsal to the superficial constrictors (see Figs. 4-5 and -6). The fibers of the cucullaris originate from the dorsal fascia of the cranial part of the epaxial body musculature and extend ventrocaudally to insert on the epibranchial of the last gill arch as well as on the scapular portion of the pectoral girdle. The anatomy suggests that this muscle pulls each gill arch up, through its action on each epibranchial cartilage, and it also protracts the pectoral girdle.

Four small lateral *interarcual* muscles are straplike, with parallel fibers that run anteroventrally, originating from the pharyngobranchial cartilages of the gill arches and inserting on the corresponding epibranchial cartilage (see Fig. 4-6). Since the interarcual muscles insert on epibranchial cartilages in series, they are thought to represent specialized portions of an originally single levator group.

Dorsal to the interarcual muscles run small ribbonlike muscles in a more longitudinal direction. These muscles extend between the pharyngobranchial cartilages of the gill arches and are called the *dorsal interarcuals* (*interbasilar muscles*) (see Fig. 4-6). These muscles are hypothesized to draw the branchial arches together and compress the pharynx.

In the angle between each epibranchial and ceratobranchial cartilage is a distinct short muscle strategically situated to make the angle more acute: the *adductor branchialis* muscle (see Fig. 4-6). Each arch possesses such a muscle, which plays an important role in compressing the pharynx by reducing its volume.

The deepest part of the constrictor musculature is closely associated with the interbranchial septum and is called the *interbranchial* muscle (see Fig. 4-6). Interbranchial muscles have circularly arranged fibers that interconnect the gill rays.

HYPOBRANCHIAL MUSCLES

The hypaxial musculature of the body extends forward from the pectoral region toward the

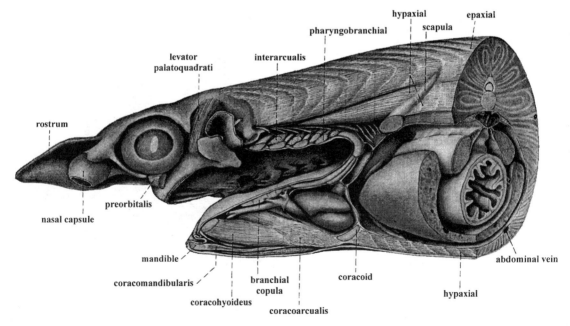

FIGURE 4-7 *Squalus acanthias* L. Lateral view of dissected anterior portion of the body after removal of the left visceral apparatus and pectoral girdle. (From Marinelli and Strenger 1959.)

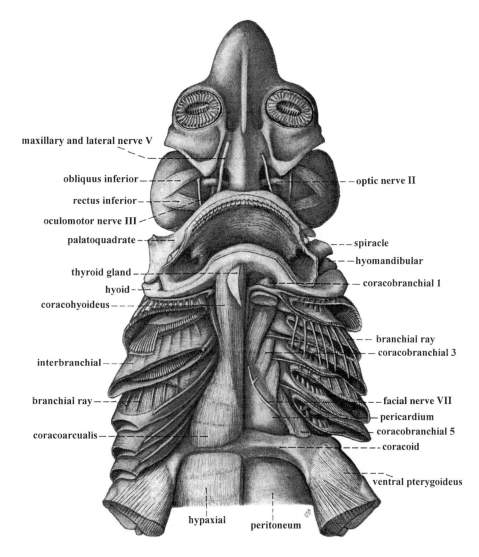

FIGURE 4-8 *Squalus acanthias* L. Ventral view of deep dissection of the anterior portion of the body, after removal of the mandible, the coracomandibular muscle, and the hypobranchial musculature originating from the left pectoral girdle. At the right, gill rays of the septum of the left second septum have been dissected. (From Marinelli and Strenger 1959.)

tip of the mandible. These head muscles are supplied by spinal nerves rather than cranial nerves, indicating that they are derived from anterior myotomes of the trunk.

The *coracomandibular* muscle originates posteriorly from fascia associated with the rectus cervicis in the area of the coracoid portion of the pectoral girdle and inserts on the tip of the mandible (Fig. 4-7). The fibers run parallel and are grouped into a roughly cylindrical shape.

It is situated just dorsal to the intermandibularis and interhyoideus muscles.

Dorsal to the coracomandibularis muscle is a complex of long cylindrical muscles, which originate from the pectoral girdle and insert on the hyoid arch. This complex is generally referred to as the *rectus cervicis* muscle. However, it can be considered two pairs of muscles, since the fibers of the posterior half converge toward the midline, while those of the anterior half extend longitudinally to the hyoid arch (Figs. 4-7 and 4-8). The posterior pair is often called the *coracoarcualis* while the anterior pair is referred to as the *coracohyoideus* (see Figs. 4-7 and 4-8).

Dorsal to the rectus cervicis is a group of segmental straplike muscles, the *coracobranchials*, which originate from the fascia of the ventral wall of the pericardium, the dorsal fascia of the rectus cervicis, and the coracoid bar. From this broad origin, the coracobranchial muscles insert on the ventral parts of the gill arches (see Fig. 4-8). Each coracobranchial muscle is covered with a sheet of connective tissue.

EPIBRANCHIAL MUSCLES

Anteriorly and dorsally the myotomes of the trunk extend forward to insert on the dorsal and posterior parts of the chondrocranium (see Figs. 4-3 and 4-7). This massive muscle is called the *epaxial* muscle and is innervated by dorsal rami of the anterior spinal nerves. It plays a major role in head lifting of sharks during feeding and locomotion. The muscles are similar in architecture to the myomeres of bony fishes.

Before moving on to the muscles of the trunk, a few words on the musculature of batoids are in order. These flattened elasmobranchs make up nearly half the diverse number elasmobranchs extant and have many unique and interesting morphological features. The cranial musculature of batoids is grossly similar to that of selachians with a few marked exceptions. The batoids all possess a levator and a depressor rostri muscle, which are

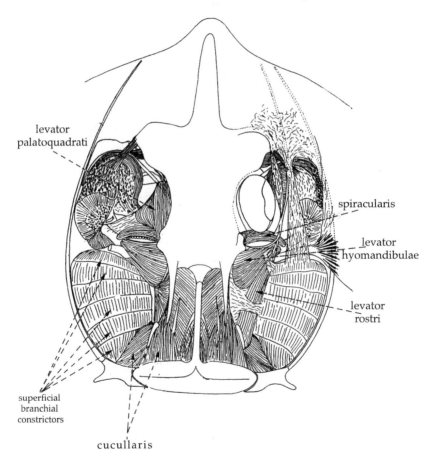

FIGURE 4-9 *Raja erinacea* Mitchill. Dorsal view of a superficial dissection of the cranial musculature showing the arrangement of the branchiomeric and constrictor muscles. (From Marion 1905.)

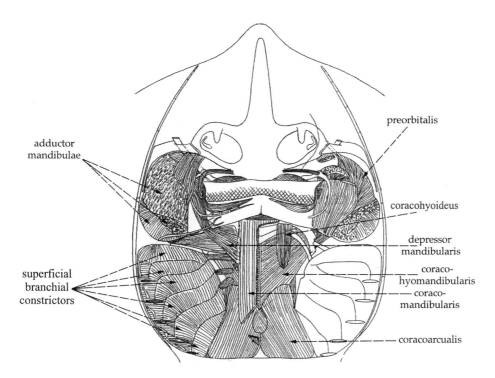

FIGURE 4-10 *Raja erinacea* Mitchill. Ventral view of the superficial musculature showing the arrangement of the branchiomeric and constrictor muscles. (From Marion 1905.)

the most superficial muscles of the dorsal and ventral sides respectively. These muscles serve to elevate and depress the snout and are generally quite thin. The sawfish has exceptionally large depressor and levator rostri muscles, which occupy the entire lateral aspect of the head. They are presumed to function in the slashing of the saw from side to side as this fish feeds. In addition to these muscles the origin or insertion of many of the cranial muscles is quite different in batoids. The muscles of the branchial arches are very well treated in a study by Miyake et al. (1992). The effect of the ventrally directed mouth on the arrangement of the muscles can be seen in Figures 4-9 and 4-10.

TRUNK MUSCLES

As in the vast majority of fishes, the trunk and tail musculature of *Squalus* is represented by the axial group of somatic muscles, forward extensions of which penetrate the head: the epibranchial muscles dorsally and the hypobranchial muscles ventrally. In this, *Squalus* can be taken as representative of the majority of selachians. However, batoids have very much reduced trunk musculature, the majority of the body muscle being greatly expanded pectoral muscles.

Many of the fibers of the segmental trunk and tail musculature insert on the well-developed connective tissue of the dermis, which acts as a tendon during swimming. Each segment, or *myomere*, has differentiated from the embryonic myotome and is innervated by branches of the dorsal and ventral rami of the spinal nerves. Adjacent myomeres are separated by distinct connective tissue sheets, the *myosepta*, which have connections with the dermis and thus act as transmitters of force. The myomeres and associated myosepta are bent in a complex zigzag fashion (Figs. 4-11 and 4-12). A longitudinal sheet of connective tissue, the *horizontal skeletogenous septum*, situated deep to the lateral line, subdivides each myomere. The dorsal portions of the myomeres form the *epaxial* musculature, which is innervated by the dorsal rami of the spinal nerves; the ventral portions are innervated by the ventral rami of the spinal nerves and are called the *hypaxial* musculature. The most lateral longitudinal bundles of muscle fibers, dorsal and ventral to the horizontal skeletogenous septum, are red muscle fibers with aerobic metabolism. These fibers are presumably used by the swimming fish at normal cruising speeds. Each myomere consists of several V-like folds, which are cone-shaped and extend further anteriorly as well as posteriorly than the folds do at the body surface (see Fig. 4-10). The cone-within-a-cone pattern of the folds can be seen in a cross section of the tail (see Figs. 4-11 and 4-12). When cross-sectioned, the overlapping V-shaped folds of several myomeres appear as a series of nearly concentric rings (see Fig. 4-11).

PECTORAL AND PELVIC FIN MUSCULATURE

The appendicular muscles consist of a single dorsal and single ventral mass. We recognize the *ventral pterygoideus* muscle, which originates from the ventral third of the scapular part of the pectoral girdle. It runs posterolaterally and subdivides into small segments that insert on the ventral surfaces of the pectoral radials. It contributes to the bulk of the muscular shoulder (see Figs. 4-3 and 4-4). The dorsal muscle mass is the *dorsal pterygoideus* muscle (the levator or abductor), which can be subdivided into superficial and dorsal parts. The superficial part originates from the fascia covering the trunk musculature, while the deep part takes its origin from the dorsal half of the scapular part of the girdle. The insertion of the deep part is more distal on the radials than that of the superficial part. The dorsal pterygoideus muscle (see Figs. 4-5 and 4-6) is considered the abductor of the fin, whereas the ventral is an adductor, even though this hypothesis has not been confirmed electromyographically. The ventral and dorsal pterygoideus muscles form the body of batoid fishes. The arrangement of the muscles is similar to that of selachians; however, the extent, both of the fin and the musculature, is greatly enlarged.

In the pelvic girdle the ventral muscle mass is more complex than the dorsal muscle mass (Fig. 4-13). Dorsally, the *pterygoideus* muscle originates from the elongate basal cartilaginous bar, to divide into numerous small segments inserting on the dorsal surfaces of the pelvic radials near their articulation with the ceratotrichia. In a mirror image, the *ventral pterygoideus* muscles also originate from the pelvic and metapterygoid cartilage to insert as spindle-shaped subdivisions on the ventral surfaces of the pelvic radii. The *pelvicobasal* muscle occupies the area medial to the pelvic metapterygoid cartilage. Electromyographic

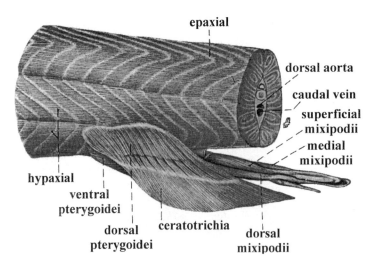

FIGURE 4-11 *Squalus acanthias* L. Lateral view of the trunk and tail musculature, including pelvic fin and clasper. (From Marinelli and Strenger 1959.)

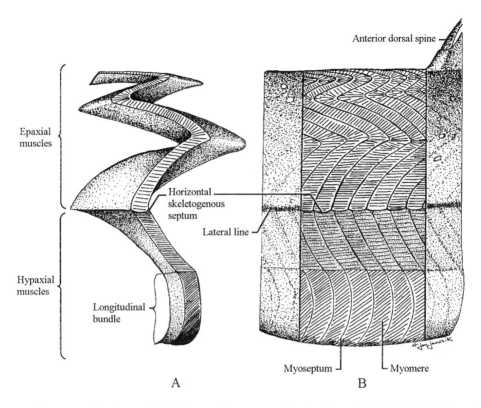

FIGURE 4-12 *Squalus acanthias* L. Diagram of a myomere. Anterior is toward the left. (After Walker and Homberger 1992.)

evidence for the functions is lacking, though it is thought that the dorsal pterygoideus has the capacity to lift and spread the pelvic fin, while the ventral pterygoideus accomplishes the opposite.

Male chondrichthians are characterized by the presence of specialized copulation organs, the *claspers*. In selachians the clasper has a well-developed *saccus glandulosus* (clasper siphon), which runs in close association with muscles ventromedially throughout the length of the clasper. It emerges in a groove on the dorsal side of the clasper. Each clasper possesses a set of four clasper muscles: the superficial, dorsal, medial, and ventral *mixipodial* muscles (see Fig. 4-13). The *ventral mixipodial* muscle originates together with, and distal to, the pelvicobasal muscle to insert with fascia on the medial mixipodial muscle. This medial mixipodial muscle takes its origin from the dorsal surface of the medial part of the metapterygium and inserts on a terminal cartilage. Dorsal to the medial clasper muscle is the *superficial mixipodial* muscle, which originates from the dorsal aspect of the metapterygoid cartilage and inserts on a terminal cartilage. Finally, the *dorsal mixipodial* muscle originates from the outer edge of the metapterygial cartilage. Its muscle fibers run parallel to the terminal end of the basal mixipodial cartilage. In between the dorsal and superficial mixipodial muscles is lodged the clasper siphon. The combined action of the four clasper muscles is to flex the clasper anteriorly (Gilbert and Heath 1972) without rotating it. Elastic recoil of the bent structure returns the entire complex to resting position.

MUSCULATURE OF UNPAIRED FINS

The muscle masses of the median and caudal fins exhibit a primitive, simple condition. They consist of rows of parallel fibers on either side

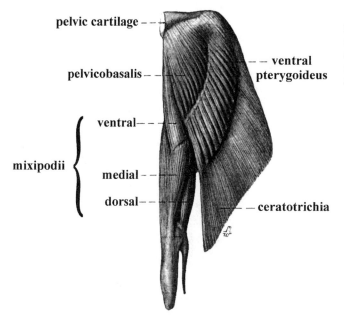

FIGURE 4-13 *Squalus acanthias* L. Ventral aspect of the musculature of the pelvic girdle of the male. (From Marinelli and Strenger 1959.)

of the cartilaginous elements embedded in the median septum separating the left and right trunk musculature. The insertion is on the radial cartilages supporting the fins, from which the long, flexible, unjointed ceratotrichia originate.

SOME KEY LIGAMENTS

Well-differentiated ligaments not only connect different cartilages but also function as force transmitters. We follow the description of Marinelli and Strenger (1959) for the major ligaments found in the head of *Squalus*. For a more comprehensive description of cephalic ligaments in the lemon shark, *Negaprion brevirostris*, the reader is referred to Motta and Wilga (1995). A strong ligament between upper and lower jaws is the elongate *quadratomandibular ligament*, which runs from the medial aspect of the mandible, just anterior to the jaw joint, to the medial aspect of the palatoquadrate about halfway along its length (Fig. 4-14). This ligament ensures that movements between palatoquadrate and mandible are linked and coordinated. A second, short, stout ligament, the *quadratohyomandibular ligament*, connects the anteroventral corner of the hyomandibula with the posterodorsal corner of the palatoquadrate (see Fig. 4-14). This ligament plays a role in transmitting rotational movements of the hyomandibula to the palatoquadrate, a linkage that helps effect jaw protrusion. Behind the quadratohyomandibular ligament is the short *mandibulohyoid ligament*, which is attached to the ventrolateral corner of the hyoid and the medial surface of the mandibular cartilage just posterior and ventral to the attachment of the quadratomandibular ligament (see Fig. 4-14). The mandibulohyoid ligament transmits forces and movements of the hyoid to the mandible. A *glenoid ligament* bridges the gap between mandible and palatoquadrate and runs over the capsule of the quadratomandibular joint. It also connects the hyoid with the mandible. The mandibulohyoid and glenoid ligaments transmit hyoid movements to the mandible. A strong enforcement in the connective tissue connects the orbital process of the palatoquadrate to the chondrocranium. The connection is distinct even though the structure is not given a name by Marinelli and Strenger (1959). Other authors call this ligament the *ethmopalatine ligament*, and in the lemon shark it originates from the preorbital wall of the chondrocranium and inserts on the orbital process of

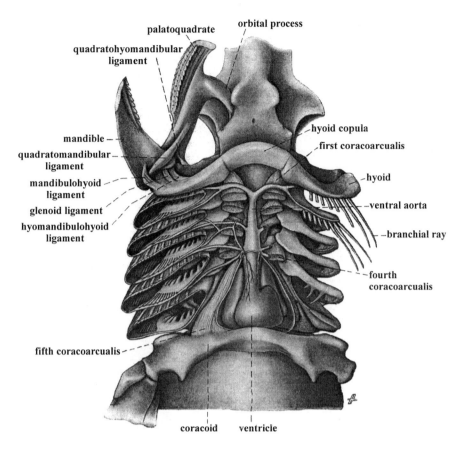

FIGURE 4-14 *Squalus acanthias* L. Ventral view of the head after removal of the mandibular, hyoid, and hypobranchial musculature. Key ligaments are shown on the right. (From Marinelli and Strenger 1959.)

the palatoquadrate. It plays a significant role in retracting the palatoquadrate during prey capture and preventing the dislocation of the palatoquadrate (Motta and Wilga 1995).

The *hyomandibulo-hyoid ligament* (called hyomandibular-ceratohyal by Motta and Wilga) originates dorsally from the hyomandibulo-ceratohyal capsule sheath. It wraps around the sheath and runs ventromedially to insert on the anterodorsal aspect of the distal end of the ceratohyal.

Functional morphology

Before the widespread use of technology, determinations of the function of muscles and the kinetics of the mechanical systems were based on manipulation and dissection of dead specimens (Daniel 1934; Luther 1909; Moss 1977; Springer 1961). Recently, techniques such as cineradiography, computer axial tomography (CAT) scans, high-speed video, and electromyography have led to interesting new models of muscle function as well as support of some well-established hypotheses (e.g., Frazzetta 1994; Wu 1994). The greatest advances in understanding function have involved experimentation and observation of free-swimming elasmobranchs in natural and captive environments. Three mechanical systems that have received significant attention from functional morphologists are the feeding apparatus, respiratory mechanism, and locomotor system.

The function of the cranial muscles during feeding in elasmobranchs has been studied for over 80 years (Frazzetta and Prange 1987; Haller 1926; Luther 1909; Moss 1972, 1977; Motta et al. 1997; Shirai and Nakaya 1992; Springer 1961; Tanaka 1973; Tricas and McCosker 1984; Wu 1994). This interest can be ascribed in part to fascination with a voracious predator, but the unique jaw suspension and the phylogenetic position of elasmobranchs have also played a role (for a review see Frazzetta 1994). Several factors have made the study of muscle function difficult. The suspension of the upper and lower jaws from the chondrocranium by the hyomandibula (hyostyly) adds an element of structural complexity not seen in other vertebrates. The relatively loose mandibular symphysis allows extensive medial and lateral movements of the jaws in many elasmobranch taxa, making three-dimensional analysis of the feeding mechanism more important than in most other aquatic vertebrates. There are very few well-defined tendons in the cranial region, as most muscles originate and insert very broadly on irregularly shaped skeletal elements or other muscles. This complicates the process of generating a model by obscuring the line of action of the muscle and the stability of origin or insertion. In spite of these difficulties, a great deal of progress has been made.

The kinematics of feeding have been studied in several species of unrestrained free-swimming sharks (Ferry-Graham 1997; Frazzetta and Prange 1987; Motta et al. 1997; Tricas and McCosker 1984; Wu 1994). These studies have involved standard and high-speed film and video, cineradiography, and CAT scans. Though several different prey capture behaviors have been described, the sequence of events resembles that of other aquatic vertebrates (Lauder and Shaffer 1986; Reilly and Lauder 1990). A preparatory phase characterized by tightly closing the mouth precedes the rapid opening of the mouth, accompanied by lifting of the head and retraction of the lower jaw. Peak gape precedes the peak hyoid protraction. At some point in the gape cycle the upper jaw may be protruded anteroventrally. Mouth closure is followed by hyoid retraction as water drawn in through the mouth is pushed out of the oropharyngeal cavity, primarily through the gill slits.

The muscular basis of these kinematic events has been determined through dissection, manipulation, electrical stimulation, and electromyography. The epaxial and the cucullaris muscles lift the head, while the hypobranchial, the coracomandibularis, coracohyoideus, and coracoarcualis muscles open the jaw. The coracohyoideus and the coracobranchialis also retract the ceratohyal and depress the floor of the oropharyngeal chamber. The mouth is closed by contraction of the adductor mandibulae complex, and the jaws and hyoid are returned to their resting position by the levator hyomandibulae and, in some species, the preorbitalis muscles (Wilga and Motta 1998).

The upper jaw is protruded to various degrees and at various times during feeding, depending on the taxon and the prey capture method. Many different functions for upper jaw protrusion during prey capture have been proposed, including: (1) rotation of the upper teeth into a functional position (Frazzetta and Prange 1987); (2) increasing the depth of cut of the upper teeth so that chunks of large prey can be removed (Moss 1977); (3) rounding the mouth for more efficient suction feeding (Wu 1994); (4) the simultaneous impact of upper and lower teeth on the prey (Frazzetta and Prange 1987); and (5) decreasing the gape cycle by bringing the upper jaw toward the lower during closure (Motta et al. 1997). Data suggest that jaw protrusion varies both within and among taxa. This explains the several different models of the protrusion mechanism that have been proposed and supported with experimentation.

The protrusion mechanism of orectolobid sharks is particularly well understood (Wu 1994). The orectolobids are suction feeders with a relatively small oral opening and well-developed labial cartilages, which form the round gape. In these sharks a complex linkage between the ceratohyal, hyomandibula, palatoquadrate, and Meckel's cartilage causes the jaws to protrude. During mouth opening, the ceratohyal is retracted by the coracohyoideus, which presses the dorsal part of the ceratohyal into the hyomandibula protracting it anteriorly. As it moves forward, the hyomandibula protrudes the upper and lower jaws by pushing on the mandibular knob of Meckel's cartilage (Fig. 4-15). A second mechanism acts simultaneously by essentially lengthening the forward sweep of the jaws. This is caused by the contraction of the interhyoideus, which narrows the span of the ceratohyals. Because the jaws form a V shape and the base of the V is being narrowed, the apex moves anteriorly. It should be noted that these two mechanisms of protrusion depend on the retraction of the hyoid and lowering of the mandible and thus cannot be decoupled from mouth opening. This explanation does not apply to sharks that can protrude the jaws with the mouth entirely open or closed.

The protrusion mechanism of carcharhinid and lamnid sharks is also well understood and seems more flexible than that of orectolobids (Frazzetta 1994; Motta et al. 1997; Tricas and McCosker 1984). The preorbitalis muscle, which may function as a weak jaw protruder in orectolobids (Wu 1994), and the levator palatoquadrati muscles provide the protruding force in carcharhinids.

In the retracted state, the palatoquadrate in carcharhinids and lamnids is braced against the orbital notch of the chondrocranium. The preorbitalis muscle has two heads: the dorsal head originates from the adductor mandibulae and inserts on the palatoquadrate; the ventral head originates in the nasal region and inserts on the adductor mandibulae. When the dorsal preorbitalis muscle contracts, it draws the orbital process of the palatoquadrate out of the orbital notch. The subsequent contraction of the ventral head of the preorbitalis and the levator palatoquadrati muscles force the palatoquadrate to slide anteroventrally on the anterior wall of the orbital notch. The adductor mandibulae muscle is also an important protrusion muscle when the mouth is open. When the mouth is held open, either by the action of the hypobranchial muscles or by inertia, the contraction of the adductor mandibulae will decrease the gape both by lifting the lower jaws and protruding the upper jaw (Motta et al. 1997). A recent study of protrusion in squaloid sharks shows that the mechanism is similar to that of carcharhinids. The primary difference is that the preorbitalis muscle, by virtue of a more posterior origin, functions

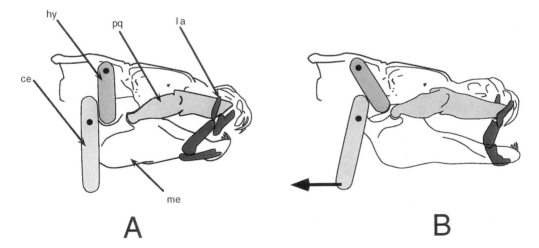

FIGURE 4-15 Schematic representation of the ceratohyal mediated jaw protrusion mechanism in orectolobid sharks. **A:** The skeletal elements in the resting position: ce = ceratohyal; hy = hyomandibula; la = labial cartilages; me = Meckel's cartilage; pq = palatoquadrate. **B:** The position of the skeletal elements after the ceratohyal has been drawn posteroventrally. The ceratohyal acts on the hyomandibula to push the jaws anteriorly. (After Wu 1994.)

as a retractor of the upper jaw rather than a protractor (Wilga and Motta 1998).

Two final aspects of the feeding mechanism require further investigation. First, it has been proposed that the attitude of the teeth is under active muscular control (Frazzetta 1994). The intermandibularis muscle might serve to erect the lower jaw teeth, but this has not been confirmed, nor has another muscle been proposed to erect the teeth of the upper jaw. Second, the adductor mandibulae is clearly a multiheaded muscle; however, the broad insertions and origins and the convoluted nature of the fiber pathways have hampered understanding of the precise functions of the different heads. Since the anatomy of the several heads varies significantly across elasmobranch taxa, the precise determination of the function of the several parts of this muscle will reveal important biomechanical as well as evolutionary trends.

RESPIRATION

Two modes of respiration occur in elasmobranchs. Many fast and steady swimmers, and all filter feeders, use ram ventilation to force water through the gills. By the mouth's being opened with the rectus cervicis, coracobranchial, and hypaxial muscles during swimming, the gills are exposed to a continuous flow of oxygenated water. Other sharks, and nearly all batoids, use cranial muscles to actively pump water through the gills. The functional morphology of the respiratory mechanism as determined by electromyography and pressure transducers is similar to that of bony fishes. Respiration is powered by two pumping chambers, which can generate a constant, unidirectional flow over the gill filaments (Hughes 1960; Hughes and Ballintijn 1965).

The respiratory apparatus consists of a medial oropharyngeal chamber and two laterally located parabranchial chambers. The parabranchial chambers contain the gill filaments, and are bounded medially by the branchial arch and laterally by the gill slits. The respiratory cycle starts with a compression phase that decreases the volume of these chambers (Fig. 4-16). Activity in the adductor mandibulae, levator palatoquadrati, and the levator hyomandibulae muscles compresses the oropharyngeal cavity while the spiracularis muscle closes the spiracle at the same time. The parabranchial chambers are constricted dorsoventrally by the combined action of the levator hyomandibulae and the superficial and deep constrictors. Posterior movement of the gill slits relative to the branchial arch, driven by the dorsal interarcuals, also decreases the volume of the parabranchial chambers. During this compression phase water is forced over the gill filaments as it is expelled from the parabranchial chambers through the gill slits. A minor amount of water is expelled through the spiracle before it closes (Hughes 1960; Hughes and Ballintijn 1965).

An expansion phase closely follows the compression phase. Under most circumstances the expansion occurs passively by elastic recoil of the skeletal elements and soft tissue. Water is drawn into the expanding oropharyngeal chamber through both the spiracle and the mouth. Flexible flap valves prevent water from being drawn into the parabranchial chambers through the gill slits. As the parabranchial chambers return to their relaxed, uncompressed state they fill with water that passes over the gill filaments as it flows between the branchial arches from the oropharyngeal chamber. Thus, as in bony fishes, the flow over the respiratory surface is continuous from the beginning of the compression phase to the end of the expansion phase. During expansion there is little or no activity in the hypobranchial muscles. During heavy respiration and feeding these muscles do act to lower the oropharyngeal floor (Hughes and Ballintijn 1965; but see Woskoboinikoff 1932 on hypobranchials).

After the expansion phase there is a period of inactivity, sometimes prolonged, before the next compression phase. During this quiescent time there is no flow of oxygenated water across the gills. The synchronization of the heart with the respiratory rhythm may ensure that blood is only pumped through the gills when there is water flowing across them (Hughes 1972; Satchell 1960).

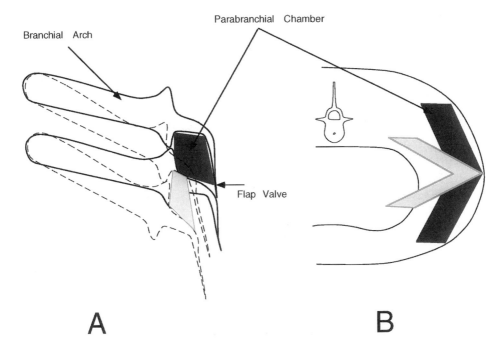

FIGURE 4-16 Illustration of the volume changes in the parabranchial chambers during contraction of the respiratory muscles. A parabranchial chamber is shaded dark gray before contraction and light gray after contraction. **A:** Dorsal view of two branchial arches. The *dotted lines* indicate the position of the arches after contraction. **B:** A cross section through a shark at the level of a parabranchial chamber. The outline of the body is shown before contraction. (After Hughes and Ballintijn 1965.)

The respiratory mechanics of the batoid elasmobranchs are poorly understood, but they are presumably not very different from that of sharks (Hughes 1960). An obvious exception is that the ventrally directed mouth forces the benthic batoids to rely almost entirely on the spiracle as the water intake. Also, when the periphery of a resting stingray is covered entirely with sand, the ventrally directed gill slits cannot function as the outlet. In this case the spiracle is open, at least partially, during both compression and expansion phases.

LOCOMOTION
The sinuous, and seemingly effortless, swimming glide of sharks is a form of axial undulation. The segmented myomeres of sharks are well adapted to produce such an undulation of the trunk and tail. Since the vertebral column prevents the body from shortening, contraction of several myomeres on one side pulls the myosepta toward each other causing curvature. The zigzag foldings (see Fig. 4-12) of the overlapping myomeres allow one myomere to exert a force over a greater length than would be the case if the arrangement of the fibers were conventional. The overlap of myomeres ensures a smooth generation of force. The direction of the muscle fibers is not alike in all parts of a myomere. The most lateral fibers are longitudinal and longer than the obliquely oriented fibers near the middorsal and midventral lines. It is generally accepted that this difference compensates for the varying distance of the fibers from the plane of the bending. Longer fibers near the center lie farther from the sagittal plane and must shorten more than the fibers near the middorsal and midventral lines. The shorter obliquely oriented fibers near the middorsal and midventral lines shorten less or not at all. The superficial fibers of the myomeres insert directly into the dermis, which acts as

an exotendon, transferring forces along the body. As a wave of curvature passes down one side of the body, the exotendon on the opposite convex side is stretched and stores energy. This energy is released and assists bending when a wave of contraction passes down the previously convex side.

Thus, thrust is generated by serial contractions of the trunk myomeres, resulting in an undulatory wave that propagates posteriorly over some portion of the body. The amount of the body that is thrown into waves by these muscular contractions varies from the anguilliform locomotion of *Chlamydoselachus* to the carangiform locomotion of the mackerel sharks (Lamnidae) (Breder 1926). The efficiency of transport goes up as less of the body is thrown into waves; however, at the same time the ability to accelerate is compromised (Webb 1984). Unique properties of the skin of some sharks allow them to circumvent this constraint, enabling them to be good accelerators as well as efficient cruisers. The thick stratum compactum of the dermis is arranged in helically wound layers of collagen fibers. This helical pattern allows the exotendon to stiffen the body as the fish swims faster (Wainwright et al. 1978). As the shark gets stiffer it undulates less of its body and cruises more efficiently.

Heterocercal tails are characteristic of many sharks and are the primary propulsive element in undulatory swimming. The classical model of tail function assumes that the longer and stiffer upper lobe of the tail, reinforced by the vertebral column, exerts a force downward as well as backwards (Fig. 4-17) (Affleck 1950; Aleev 1969; Gray 1933; Grove and Newell 1936). This hypothesis is supported by experiments on detached tails (Alexander 1965). However, observations of free-swimming sharks showed that the smaller lower lobe of the fin was actually leading the stiffer upper lobe as the tail swept from side to side. This would imply that the tail generates nearly pure thrust, with just a slight upward force (Thomson 1976, 1990; Thomson and Simanek 1977). These two models predict very different swimming kinetics. Because the tail is situated well behind the center of gravity, the classical model requires a righting force, provided primarily by the pectoral fins, to

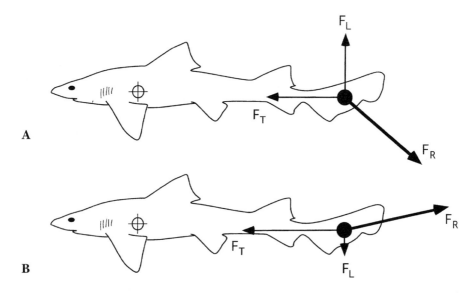

FIGURE 4-17 Thrust production during swimming by a typical shark, showing the direction in which water moved by the tail (F_R), the lift produced (F_L), and the thrust (F_T). **A:** The classical model in which water is pushed downward by the tail. **B:** Thomson's model, in which water is pushed slightly upward by the tail. (After Ferry and Lauder 1996.)

keep the shark's nose from pitching down. Thomson's model requires no such force, and the function of the pectoral fins would presumably be solely depth adjustment. A three-dimensional analysis of the kinematics of the tail of a free-swimming leopard shark, *Triakis semifasciatus*, conclusively demonstrates that the classical model of tail function is correct (Ferry and Lauder 1996). Force from the tail is indeed generated downward and presumably the pectoral fins must counter the resultant lift.

Though studies of swimming in sharks have focused mainly on axial undulation, elasmobranchs exhibit two other forms of locomotion. The batoid fishes use their pectoral fins as wings to "fly" through the water (see Fig. 4-17). Most of the skates and some benthic sharks also "walk" on the bottom using either the pelvic or pectoral fins (Pridmore 1994). During pectoral fin locomotion an undulatory wave is generated by serial contraction of the dorsal and ventral pterygoideus muscles (Rosenberger and Westneat 1997). This results in a wave that passes posteriorly at a rate greater than the speed of travel. The application of aerodynamic theory to the problem of undulatory swimming has been illuminating. For steady flight it is clear that the lowest cost of transport is achieved by having the highest aspect ratio wings. In other words, to swim efficiently, the wingspan should be large and the distance from leading to trailing edge should be small. This is a quite different shape from that of the typical batoid. The reason for the squat wing shape may lie in the unsteady nature of the flapping flight of skates and rays. Thin blade analysis of the unsteady locomotion of a skate, *Raja eglanteria*, shows that the cost of transport decreases with aspect ratio at biologically relevant swimming speeds (Daniel 1988).

REFERENCES

Affleck, R. J. 1950. Some points on the function development and evolution of the tail in fishes. *Proc. Zool. Soc. Lond.* 120: 249–68.

Aleev, Y. G. 1969. *Function and gross morphology in fish.* Jerusalem: Keter Press.

Alexander, R. M. 1965. The lift produced by the heterocercal tails of Selachii. *J. Exp. Biol.* 43: 131–38.

Breder, C. M. J. 1926. The locomotion of fishes. *Zoologica* 4: 159–297.

Daniel, J. F. 1934. *The Elasmobranch Fishes*, 3d. ed. Berkeley: University of California Press.

Daniel, T. D. 1988. Forward flapping flight from flexible fins. *Can. J. Zool.* 66: 630–38.

Ferry, L. A., and G. V. Lauder. 1996. Heterocercal tail function in leopard sharks: a three-dimensional kinematic analysis of two models. *J. Exp. Biol.* 199: 2253–68.

Ferry-Graham, L. A. 1997. Feeding kinematics of juvenile swellsharks. *Cephaloscyllium ventriosum*. *J. Exp. Biol.* 200: 1255–69.

Frazzetta, T. H. 1994. Feeding mechanisms in sharks and other elasmobranchs. *Adv. Comp. Environ. Physiol.* 18: 31–57.

Frazzetta, T. H., and C. D. Prange. 1987. Movements of cephalic components during feeding in some requiem sharks (Carcharhiniformes: Carcharidae). *Copeia.* 1987: 979–93.

Gilbert, P. W., and G. W. Heath. 1972. The clasper-siphon sac mechanism in *Squalus acanthias* and *Mustelus canis*. *Comp. Biochem. Physiol.* 42A: 97–119.

Gilbert, S. G. 1973. *Pictorial Anatomy of the Dogfish.* Seattle: University of Washington Press.

Gray, J. E. 1933. Studies in animal locomotion: I. The movement of fish with special reference to the eel. *J. Exp. Biol.* 10: 88–104.

Grove, A. J., and G. E. Newell. 1936. A mechanical investigation into the effectual action of the caudal fin of some aquatic vertebrates. *Ann. Mag. Nat. Hist.* 10: 280–90.

Haller, G. 1926. Ueber die Entwicklung, den Bau, und die Mechanik des Kieferapparates des Dornhais (*Acanthias vulgaris*). *Z. Mikrosk. Anat. Forsch.* 5: 749–93.

Hughes, G. M. 1960. The mechanism of ventilation in the dogfish and skate. *J. Exp. Biol.* 37: 11–27.

———. 1972. The relationship between cardiac and respiratory rhythms in the dogfish (*Scyliorhinus caniculus* L.). *J. Exp. Biol.* 57: 415–34.

Hughes, G. M., and C. M. Ballintijn. 1965. The muscular basis of the respiratory pumps in the dogfish (*Scyliorhinus canicula* L.). *J. Exp. Biol.* 57: 363–83.

Lauder, G.V., and H. B. Shaffer. 1986. Functional design of the feeding mechanism in lower vertebrates: unidirectional and bidirectional flow systems in the tiger salamander. *Zool. J. Linn. Soc.* 88: 277–90.

Luther, A. 1909. Untersuchungen ueber die vom N. Trigeminus innervierte Muskulatur der Selachier (Haie und Rochen). *Acta Soc. Scient. Fenn.* 36: 1–145.

Marion, G. E. 1905. Mandibular and pharyngeal muscles of *Acanthias* and *Raia. Tufts Coll. Stud.* 2: 1–34.

Marinelli W., and A. Strenger. 1959. *Vergleichende Anatomie und Morphologie der Wirbeltiere: III. Squalus acanthias L.* Vienna: Franz Deuticke Verlag.

Miyake, T., J. D. McEachran, and B. K. Hall. 1992 Edgeworth's legacy of cranial muscle development with an analysis of muscles in the ventral arch region of batoid fishes (Chondrichthyes: Batoidea). *J. Morphol.* 212: 213–56.

Moss, S. A. 1972. The feeding mechanism of sharks of the family Carcharhinidae. *J. Zool. Lond.* 167: 423–36.

———. 1977. Feeding mechanisms in sharks. *Am. Zool.* 17: 355–64.

Motta, P. J., T. Tricas, R. L. Hueter, and A. P. Summers. 1997. The feeding mechanics of the lemon shark, *Negaprion brevirostris. J. Exp. Biol.* 200: 2765–80.

Motta, P. J., and C. D. Wilga. 1995. Anatomy of the feeding apparatus of the lemon shark, *Negaprion brevirostris. J. Morphol.* 226: 309–29.

Pridmore, P. A. 1994. Submerged walking in the epaulette shark *Hemiscyllium ocellatum* (Hemiscyllidae) and its implications for locomotion in rhipidistian fishes and early tetrapods. *Zool. Anal. Complex Syst.* 98: 278–97.

Reilly, S. M., and G. V. Lauder. 1990. The evolution of tetrapod feeding behavior—kinematic homologies in prey transport. *Evolution* 44: 1542–57.

Rosenberger, L., and M. W. Westneat. 1997. Undulatory pectoral fin locomotion in the stingray, *Taeniura lymma. Am. Zool.* 35(5): 77A.

Satchell, G. H. 1960. The reflex coordination of the heartbeat with respiration in the dogfish. *J. Exp. Biol.* 37: 719–31.

Shirai, S., and K. Nakaya. 1992. Functional morphology of the feeding apparatus of the cookie cutter shark, *Isistius brasiliensis. Zool. Sci.* 9: 811–21.

Springer, S. 1961. Dynamics of the feeding mechanism of large galeoid sharks. *Am. Zool.* 1: 183–85.

Tanaka, S. 1973. Suction feeding by the nurse shark. *Copeia* 1973: 606–08.

Thomson, K. S. 1976. On the heterocercal tail in sharks. *Paleobiology* 2: 19–38.

———. 1990. The shape of the shark's tail. *Am. Sci.* 78: 499–501.

Thomson, K. S., and D. E. Simanek. 1977. Body form and locomotion in sharks. *Am. Zool.* 17: 343–54.

Tricas, T., and J. E. McCosker. 1984. Predatory behavior in the white shark (*Carcharodon carcharias*), with notes on its biology. *Proc. Calif. Acad. Sci.* 43: 221–38.

Wainwright, S. A., F. Vosburgh, and J. H. Hebrank. 1978. Shark skin: function in locomotion. *Science* 202: 747–49.

Walker, W. F., and D. Homberger. 1992. *Vertebrate Dissection*, 8th ed. New York: Saunders.

Webb, P. 1984. Form and function in fish swimming. *Sci. Am.,* July, 72–82.

Wilga, C., and P. J. Motta. 1998. Conservation and variation in the feeding mechanisms of the spiny dogfish *Squalus acanthias. J. Exp. Biol.* 201: 1345–58.

Woskoboinikoff, M. M. 1932. Der Apparat der Kiemenatmung bei den Fischen. Ein Versuch der Synthese in der Morphologie. *Zool. Jahrb. Abt.* 2 55: 315–488.

Wu, E. H. 1994. Kinematic analysis of jaw protrusion in orectolobiform sharks: a new mechanism for jaw protrusion in elasmobranchs. *J. Morphol.* 222: 175–90.

CHAPTER 5

Muscular System
MICROSCOPICAL ANATOMY, PHYSIOLOGY, AND BIOCHEMISTRY OF ELASMOBRANCH MUSCLE FIBERS

Very few of the 950 or so species of elasmobranchs have been examined with regard to the structure and operation of their muscle fibers. Indeed, only one small shark, the dogfish *Scyliorhinus canicula*, has been examined in any detail, while fewer aspects of two other small sharks, *Galeus melastomus* and *Etmopterus spinax*, have been studied. It seems from the examination of these three that there are only minor differences among species. However, for obvious reasons, only *small* sharks have been studied, and the difference in scale between dogfishes up to 1 m or so long and the 12-m or more basking shark *Cetorhinus* (in which the myotomal muscle fibers may be no less than 13 cm long) may well suggest differences in innervation pattern and operation. Shark musculature has been reviewed (Bone 1988); nevertheless some new information has appeared since then.

Although batoids outnumber sharks in living species, their muscle fibers have received almost no attention since early work by Lorenzini (1678) and Ranvier (1873). There has been some electromyographic work on stingrays (Droge and Leonard 1983) and a brief note on the innervation of the myotomal fibers (Bone 1964). For this reason, some attention is paid in this chapter to the arrangement and properties of the pectoral muscle fibers of rajids. It is not yet possible to say whether the musculature of different batoids has essentially the same arrangement as in rajids (though this seems likely from cursory studies of other batoids).

Sharks swim by oscillating the body with myotomal muscles, while (most) batoids swim quite differently, using the pectoral fins, and this difference is naturally reflected in the quite different ways in which their locomotor muscle fibers are arranged. Those batoids (like rhinobatids and the electric ray *Torpedo*), which oscillate their tails as they swim, have sharklike caudal myotomal musculature, though the innervation pattern is not precisely the same as in dogfish.

Sharks

MICROSCOPICAL ANATOMY

General

As observed in the previous chapter, there is an obvious division in the myotomal and other muscles between those regions that are white

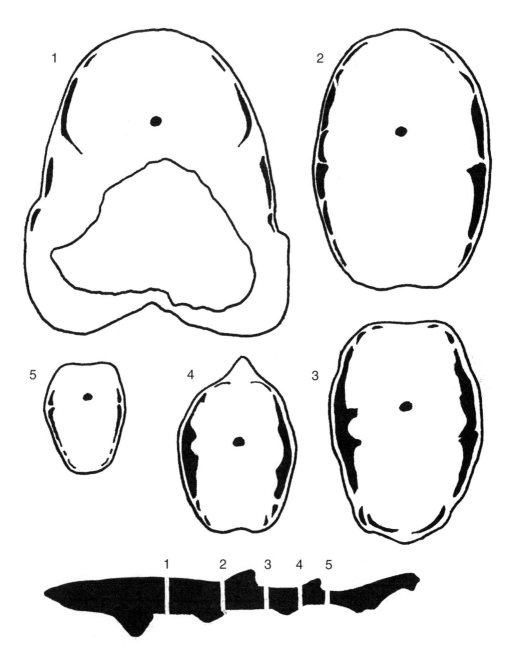

FIGURE 5-1 The proportion of red muscle (*black*) in the myotomes at different positions along the body of the dogfish *Scyliorhinus canicula*. (From Bone 1966.)

or yellowish and those smaller regions conspicuous by their red color. In fact it was Lorenzini (1678), in his dissections of *Torpedo*, who first observed red and white muscles in vertebrates. These differently colored regions of the musculature reflect striking differences in the muscle fibers that compose them, as they do in other fish groups, where there are similar color differences. The difference in color is due not only to the greater vascularity of the red muscles but also to their much higher myoglobin content, as compared with the white muscles (Kryvi et al. 1981). In the majority of sharks, the myotomal locomotor musculature consists of an outer layer of red fibers overlying a much greater mass of white or pale yellow fibers. The proportion of red fibers varies along the length of the shark, as seen in the dogfish *Scyliorhinus canicula* (Fig. 5-1), ranging from less than 5% to 20% or a little more in the caudal region. The more active fish have a relatively greater proportion of red fibers in the caudal region (22% in the wide-ranging oceanic *Prionace glauca*), and in the most active warm-blooded isurids the red muscle mass is infolded into the myotome (Fig. 5-2), just as it is in tunas. Red muscle is also infolded into the myotomes in the sluggish benthic *Squatina*, here unrelated to elevated temperature or activity but presumably simply a consequence of the flattening of this ambush predator.

Fiber structure

The ultrastructure of the fibers composing these red and white zones of the myotome has been examined in *Scyliorhinus*, *Galeus melastomus*, and *Etmopterus spinax*. In *Scyliorhinus* (but not in the other two species), as well as the broad duality of small, red, mitochondria-rich fibers and large-diameter, white, mitochondria-poor fibers (typical of all fish groups), there are small numbers of a superficial fiber type, peculiar to *Scyliorhinus* so far as is known at

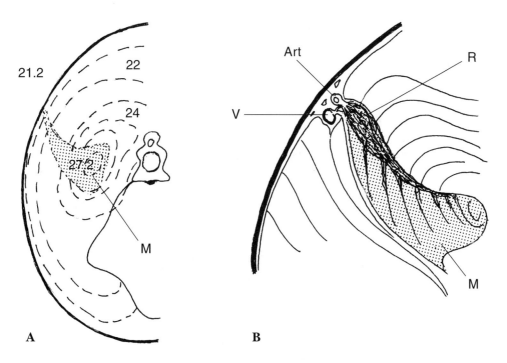

FIGURE 5-2 Internalized red muscle in the warm-blooded shortfin mako shark, *Isurus oxyrhynchus*, and its blood supply. **A:** Transverse section of visceral region showing temperatures measured with thermistor probes. **B:** Red muscle. A = artery; M = muscle; R = rete; V = vein. The *dotted lines* are 1°C isotherms. (After Carey and Teal 1969.)

present. In cryostat sections stained for oxidative mitochondrial enzymes such as lactate or succinic dehydrogenase (LDH, SDH) or for Ca^{2+}-activated myofibrillar adenosine triphosphatase (ATPase), these different fiber types can easily be recognized. In addition, it seems that the red and white fiber types may be further subdivided.

In *Scyliorhinus* (Fig. 5-3), from the outer surface of the myotome inward, there is first a single layer of large-diameter (up to 125 μm) mitochondria-poor fibers. The relative volume density of mitochondria and of the sarcotubular system of these fibers is compared with that of the other two main types in Table 5-1. These superficial fibers are found only along the flank of the fish and do not appear dorsally and ventrally, so they form a long, thin lateral strip. In an immediately postanal myotome, there are only 80–90 superficial fibers, compared with around 8000 red fibers and 11,000 white fibers. They are much larger in diameter than the red fibers, yet they only make up around 0.6% of myotomal cross-sectional area, as compared with 24.4% for the red fibers and 75% for the white fibers. They are absent from other sharks so far examined (including *Prionace* and *Squalus*, as well as *Galeus* and *Etmopterus*) but may well be present in other species. Since they are easy to distinguish even in crude sections of the myotomes viewed under a binocular, search in other species might prove rewarding and might suggest their functional role, at present enigmatic.

Next is the zone of smaller (up to 75 μm) mitochondria-rich red fibers, thickest beside the lateral line and lateral vein and thinning dorsally and ventrally. In cryostat sections stained for SDH and other oxidative enzymes there is a gradation across the red fiber zone, such that the outermost stain most deeply, while the inner ones, adjacent to the outermost fibers of the white zone, stain less deeply. Thus it is reasonable to distinguish inner and outer red fibers, although it is not yet clear if their differences underlie different functional roles, as those between the superficial and red and white fibers evidently do. In Table 5-2, the values for red fibers are for the outer red fibers.

Last, making up the main mass of the myotome, are the larger diameter (up to 200 μm) mitochondria-poor white fibers. Curiously enough, as will be seen below, although these fibers make up some three-fourths of the total myotomal mass, they are only relatively rarely employed by the fish. The white fibers close to the inner red fiber zone are different from those of the main white mass in the myotome, for they contain slightly more mitochondria and are surrounded by a few more capillaries than the inner white fibers.

They have sometimes been termed *intermediate fibers* (e.g., Bone 1966; Kryvi and Totland 1977; Totland et al. 1981), since in several respects they are intermediate between the inner red fibers and the outermost white fibers. For example, microdensitometric measurements of oxidative enzyme content lie between the values for red and white fibers (Kryvi and Totland 1977), as do values for their myoglobin content (Kryvi et al. 1981) and

TABLE 5-1 SOME ULTRASTRUCTURAL CHARACTERISTICS OF THE DIFFERENT FIBER TYPES IN *SCYLIORHINUS*

	Muscle fiber type		
Parameter	SUPERFICIAL	RED	WHITE
Relative volume density of myofibrils	75.7±2.5	62.2±2.2	77.8±1.2
Relative volume density of mitochondria	9.5±1.2	21.3±2.1	5.4±0.7
Relative volume density of sarcotubular system	3.3±0.24	2.0±0.14	5.8±0.4

SOURCE: From Bone et al. (1986).
Values given as means ± S.E.M.

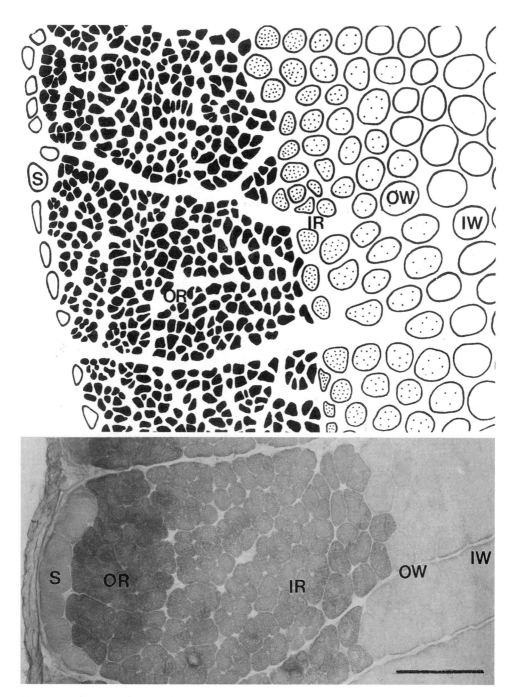

FIGURE 5-3 **Top:** Semischematic transverse section of outer portion of *Scyliorhinus* myotome as seen in sections stained for succinic dehydrogenase, showing the different activities of the various fiber types. S = superficial fibers; OR = outer red fibers; IR = inner red fibers; OW = outer white fibers; IW = inner white fibers. (From Bone 1966.) **Bottom:** Cryostat section stained for maleate dehydrogenase showing the relative enzyme activity (equivalent to mitochondrial content) of the different fiber types. Scale bar: 500 μm.

TABLE 5-2 DIFFERENCES AMONG FIBER TYPES IN SOME SMALL SHARKS

Species	Fiber type	Diameter (μm)[a]	Mitochondrial area in cross section[a] (%)	Lipid content (% wet wt.)[b]	Glycogen content (mg) (% wet wt.)[b]	Myoglobin content (nmolmg^{-1} wet wt.)[c]	Maximum hypothetical diffusion distance[d]
Scyliorhinus	Superficial	120.0	2.3			565	63.3
	Red	61.	24.8	3.17–4.87	1106–1866		27.4
	Intermediate	127.0	—				49.3
	White	152.0	0.99	0.51–0.82	1064		53.5
Galeus	Red	52.3	34.1			170	25.7
	Intermediate	85.1	16.3				66.8
	White	138.0	0.9				168.5
Etmopterus	Red	64.9	30.4				47.5
	Intermediate	78.8	7.2				78.8
	White	113.9	0.5				341.1
Squalus	Red			High[b]			

[a]Totland et al. (1981).
[b]Bone (1966).
[c]Kryvi et al. (1981).
[d]This measure of the maximum hypothetical diffusion distance is based on fiber number and area, divided by capillary number (see Totland et al. 1981: 231).

mitochondrial content (Kryvi 1977), while their lactate dehydrogenase (LDH) isozymes are similar to those of the red fibers rather than the white (Totland et al. 1978). However, it seems preferable to term them *outer white fibers*, since this emphasizes their evident relation with the white fibers in innervation pattern and activation. There are therefore inner and outer white fibers in the myotome, thus making five different fiber types. While these different fiber types can be readily distinguished on the basis of their lipid and SDH content, there are also differences in color, glycogen and lipid content, innervation pattern, vascular bed, mechanical properties, mitochondrial and myoglobin content, enzyme profile, and in electrophysiology (where this last has been examined). Table 5-2 notes some of these differences among fiber types in these small sharks.

It is natural with the large differences in mitochondrial and oxidative enzyme content among these different fiber types that there should also be a large difference in their vascular bed (Fig. 5-4). Table 5-3 shows differences in vascular bed among the fibers of *Scyliorhinus* and several different species of small shark. On the whole, although it is difficult to make comparisons with other groups owing to the different methods used for analyzing vascular supply in different studies, elasmobranch muscles are less well vascularized than those of comparable teleosts.

The development of the different myotomal fiber types during ontogeny has not been followed in any detail. However, in *Scyliorhinus*, in 48-mm embryos, only red and white fibers can be distinguished, while superficial fibers (intermediate in mitochondrial content between the red and white fibers) are present at 94 mm. Later development involves increase in the mitochondrial content of the red fibers and decline in the white fibers so that by 100 mm the fiber types resemble those of the adult (Table 5-4). What is striking is that, in contrast to teleosts, elasmobranch muscle fiber number is fixed early, and increase in myotomal volume is brought about by fiber hypertrophy rather than hyperplasia. It is not uncommon in small sharks, for instance, to find myotomal muscle fibers up to 150 μm in diameter, all forming a uniform population, while in teleosts

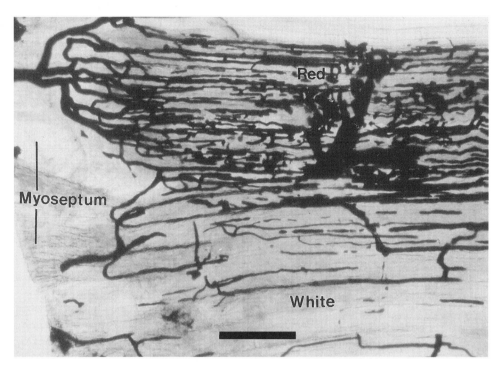

FIGURE 5-4 Vascular bed of outer part of *Scyliorhinus* myotome injected with latex. Red muscle fiber bundle *above*, white *below*, myocommata at *left*. Scale bar: 500 μm.

TABLE 5-3 VASCULAR BED OF THE FIBER TYPES IN THE MYOTOMES OF SOME SMALL SHARKS

Species	Fiber type	Percent of fibers surrounded by 0–6 capillaries							Mean no. capillaries around each fiber (± S.D.)
		0	1	2	3	4	5	6	
Galeus	Red	1	19	34	27	14	5		2.5 ± 1.2
	Intermediate	52	26	17	4	1			0.8 ± 0.9
	White	68	27	4	1				0.4 ± 0.6
Etmopterus	Red	11	48	35	5	1			1.4 ± 0.7
	Intermediate	67	28	5					0.4 ± 0.6
	White	96	4						0.04 ± 0.19
	Superficial	63	18	18	1				1.2 ± 1.3
Scyliorhinus	Red	5	15	26	24	20	7	3	2.8 ± 1.4
	Intermediate	1	9	23	39	18	7	2	3.0 ± 1.2
	White	0	5	2	22	38	31		3.9 ± 1.1

SOURCE: Bone et al. (1986).

TABLE 5-4 PERCENTAGE OF MITOCHONDRIAL VOLUME OF DIFFERENT FIBER TYPES IN *SCYLIORHINUS* AT DIFFERENT GROWTH STAGES

Fiber type	Body length (mm)			
	48	94	100	450
Superficial	—	8.30–3.98	3.27	2.39–0.49
Red	10.14–1.77	13.79–3.12	18.07–3.55	21.55–3.39
White	5.67–2.42	0.81–0.26	1.11–0.29	0.99–0.16

the largest fibers are much smaller, and even in adult teleosts they are surrounded by much smaller, younger fibers.

Motor innervation pattern

Superficial and red myotomal fibers are multiply innervated, receiving large en grappe type terminals at intervals along their length. The axons supplying the red fibers pass onto the fibers from both myoseptal ends; hence each is presumably innervated by axons that pass out of the cord in two adjacent spinal nerves. On both superficial and red fibers, the terminals lie above extensive subjunctional sarcolemmal folds, and in at least one instance, terminals on both fiber types have been traced back to the same axon. Despite their multiterminal innervation (Fig. 5-5A,B) with motor terminals typically some 250 µm apart along the fibers, at least some of the red fibers propagate action potentials. Nothing is known of the membrane properties of the superficial fibers.

The white fibers are innervated in an entirely different manner. Each receives at one myoseptal end, a large basketlike terminal much resembling a hand clasping a newel post (Fig. 5-5C–E). These endings were first observed by Giacomini (1898), who supposed them to be sensory. It was soon suggested that they might be motor, and later work has confirmed this. The fibers are innervated at one end alone, and in a single myotome this may be at either end, so a single myotome is supplied by two adjacent ventral roots. In experiments in which numbers of the white fibers of a single myotome in *Torpedo* were successively impaled to record their action potentials as nerve bundles in the myocommata were stimulated, half the fibers were innervated at the anterior myocommatal junction and half at the posterior (Bone and Murray, unpublished). An ultrastructural investigation of these endings (Bone 1972) showed that there were apparently two kinds of endings over the extensive subjunctional folds at the myoseptal ends of the fibers (Fig. 5-6). One contained electron-lucent 50-nm vesicles typical of cholinergic terminals, while the other contained much larger vesicles, many dense-cored, up to 100 nm in diameter. This rather curious feature is apparently due to the cooperation of two different axons at each nerve terminal. In methylene blue and silver whole-mount preparations, the two axons supplying the terminal can often be traced for long distances independently away from the terminal, and although it is impossible to be certain that they are not branches of the same axon, this is highly improbable. Here, then, in elasmobranchs, there is dual innervation of skeletal muscle fibers, reminiscent of the long-discussed "sympathetic" innervation of the skeletal muscle of higher vertebrates (see Tiegs 1953). There is no evidence at present for the source of origin of these axons, though it seems obvious that one must be equivalent to the usual vertebrate ventral root motor axons. Subsequently, a similar dual innervation of fast fibers was found in a variety of other fish (see Sakharov and Kashapova 1979; Ono 1983). Hence it appears to be the primitive pattern of innervation of fast vertebrate skeletal muscle, as these workers have suggested.

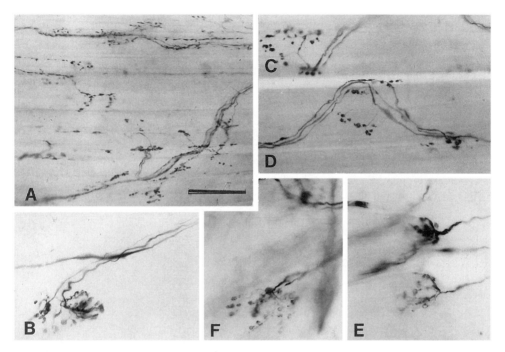

FIGURE 5-5 Innervation of *Scyliorhinus* myotomal fibers visualized with supravital methylene blue staining. **A, C,** and **D**: Multiple innervation of red muscle fibers. **B, E,** and **F**: Basketlike (en panier) terminals at the myoseptal ends of the muscle fibers. In **B** and also partly in **F** note two fibers contributing to end formation. All to same scale; scale bar: 200 μm. (From Bone 1966.)

It is clear from acetylcholinesterase preparations (as well as from the effects of acetylcholine on isolated fiber bundles of both red and white fibers) that acetylcholine is a transmitter at the neuromuscular junctions of red and white fibers. Unfortunately the identity of the second transmitter is unknown, although the presence of dense-cored vesicles suggests the possibility of a neuropeptide. As yet desultory tests with such possible transmitters as γ-aminobutyric acid (GABA), RF amide, and glutamate have yielded negative results.

Sensory innervation

Two types of sensory ending have been observed in elasmobranchs, those in batoids being intramuscular; in sharks, there are coiled corpuscles first described by Wunderer (1908). These lie not among the muscle fibers, but subcutaneously, just superficial to the myocommata (Fig. 5-7). Roberts (1969) has given information about these slowly adapting mechanoreceptors and shown how their properties could be utilized to inform the shark about the frequency and angle of bending of its body. Ultrastructural investigation has shown the endings to be blind whorls and that they are not encapsulated (Bone and Chubb 1975).

ELECTROPHYSIOLOGY

So far as the author is aware, the only study of the cable constants of shark muscle fibers remains that of Stanfield (1972), who used a two-electrode voltage clamp to investigate the conductance properties of the myotomal fibers of *Scyliorhinus*. His results are shown in Table 5-5 and compared with the less detailed observations of Hagiwara and Takahashi (1967) on batoids. Stanfield found that red muscle fiber resting potentials in the Ringer's solution he used were around −70 mV (−71.1 ± 1.2 mV), while the white fibers had higher resting potentials (−85.2 ± 0.4 mV). The white fibers propagated rapid action potentials similar to those seen in the fast twitch muscle

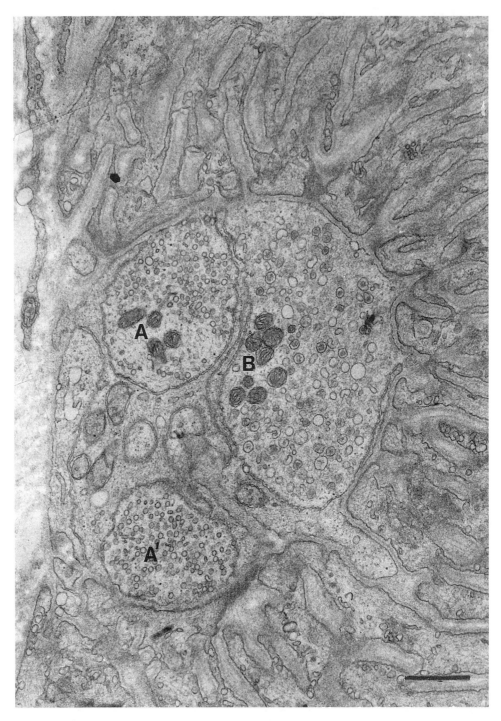

FIGURE 5-6 Ultrastructure of nerve endings on white muscle fiber, showing different vesicle populations within terminals A, A', and in terminal B. Extensive subjunctional folds surround the endings. Scale bar: 1.0 μm. (From Bone 1972.)

Muscles: Microscopical Anatomy, Physiology, Biochemistry

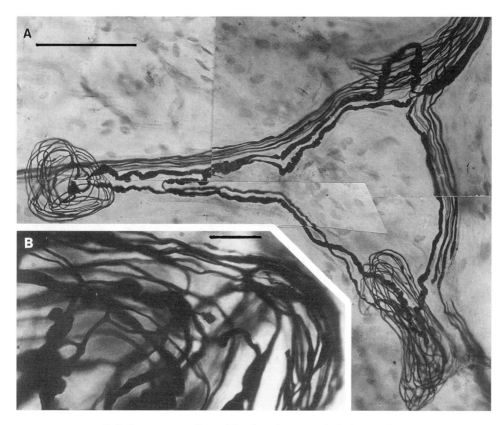

FIGURE 5-7 Coiled sensory endings (Wunderer's corpuscles) from subcutis of *Scyliorhinus*. **A:** General view of silver-impregnated whole mount after removal of surrounding connective tissue. Scale bar: 100 μm. **B:** Part of another ending showing complexity of coiling. Scale bar: 5 μm.

TABLE 5-5   MEMBRANE CONSTANTS OF RED AND WHITE MUSCLE FIBERS IN ELASMOBRANCHS

Species and fiber type	Input resistance R (MΩ)	Space constant λ (mm)	Series internal resistance $r_i \times 10^6$ (cm^{-1})	Series membrane resistance $r_m \times 10^5$ (cm)	Specific internal resistance R_1 (cm)	Specific membrane resistance R_m (cm^2)	τ_m (msec)	C_m (μF cm^{-2})
Scyliorhinus canicula								
White fibers	0.14	2.36	0.62	0.34	108	1588	15.9	10.3
Red fibers	0.75	2.27	6.92	3.44	136	5410	47	10.2
Himantura uarnak								
White fibers	0.36–0.46	3.0–3.5						
Taeniura lymma								
White fibers	0.06–0.08	2.5–3.0						

SOURCE: From Stanfield (1972) and Hagiwara and Takahashi (1967).

fibers of higher vertebrates. Stanfield's most striking finding, however, was that a proportion of the (multiply innervated) red fibers exhibited sufficiently large early inward sodium currents to suggest that they were capable of propagating action potentials. Of 27 fibers he examined, 8 (29%) were of this kind, 6 (22%) had no inward sodium currents when depolarized, and in the remaining 13 (49%) sodium currents were present but too small to be capable of carrying action potentials (Fig. 5-8). On one occasion Stanfield observed an abortive spike from a red fiber. This situation has recently been reinvestigated using

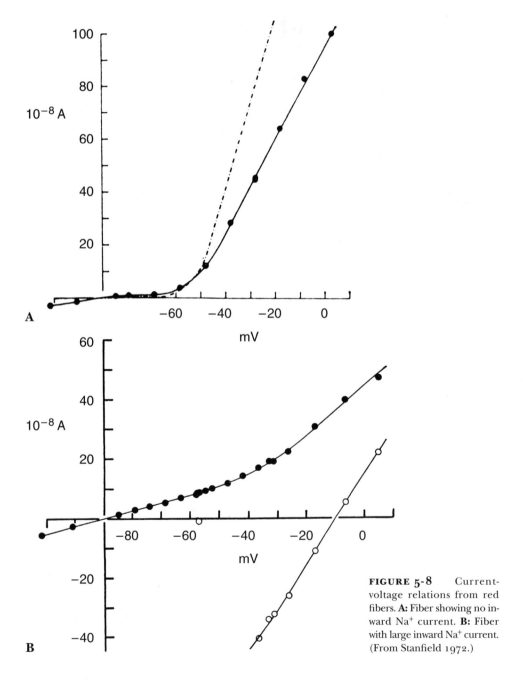

FIGURE 5-8 Current-voltage relations from red fibers. **A:** Fiber showing no inward Na^+ current. **B:** Fiber with large inward Na^+ current. (From Stanfield 1972.)

current injection into red fibers impaled with a recording electrode (Bone et al. 1994). Although action potentials were not evoked in all fibers examined, they were seen in 15 fibers. In this study, mean resting potential of the red fibers was −77.4 mV (S.E.M. 3.7), slightly higher than that observed by Stanfield. They propagated without diminution up to 4.2 mm (almost the entire length of the fiber): Stanfield found the space constant (λ) to be 2.27 mm.

The action potentials were, however, unusual in that they did not overshoot zero potential (mean maximum potential reached was −31.9 mV [S.E.M. 2.4]) and were relatively long and slow (Fig. 5-9). In line with Stanfield's results, many red fibers with apparently normal resting potentials failed to produce action potentials on injection of current. The reason for this curious difference among apparently normal fibers is not known, and even adjacent fibers may behave differently. It will be recalled that adjacent fibers may differ much in their mitochondrial content (between the dark and pale red), and it is conceivable that there may be some correlation between enzyme and membrane channel activity.

Mechanical properties

Although the Levin-Wyman ergometer was first employed to study the viscous elastic properties of the "jaw" muscles of dogfish (Levin and Wyman 1927), until recently the only modern mechanical measurements from myotomal muscle were from chemically "skinned" single muscle fibers (Altringham and Johnston 1982; Bone et al. 1986). These enabled force-velocity curves to be determined, and hence power outputs. As seen in Figure 5-10, the force-velocity relations are typical of vertebrate fibers, and that for the red fibers (Fig. 5-10B) clearly shows the characteristic double hyperbolic relation demonstrated by Edman (1988) in frog fibers. After correction for the difference in myofibrillar content of the different fiber types (the consequence of differences in mitochondrial content), the ratios of the maximum forces produced were 1:1.74:3.55 for superficial:red:white. The relative power outputs (W kg^{-1}) were, in the same order, 1.4, 8.5, and 55. These values alone suffice to indicate that the superficial fibers can play little role in swimming.

More recently, the mechanical properties of the red and white fibers of *Scyliorhinus* have been examined in detail in a series of papers by Curtin and Woledge (1988, 1991, 1993a,b). As a result, the mechanical properties of dogfish myotomal muscle are known in almost as much detail as those of the frog or tortoise.

In their first paper Curtin and Woledge (1988) redetermined the force-velocity relation and power output of the white fibers (using, instead of single skinned fibers, small

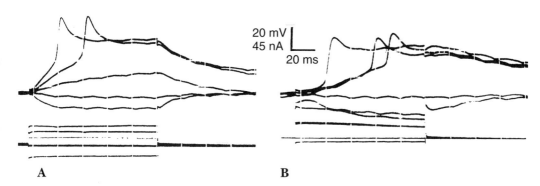

FIGURE 5-9 Nonovershooting action potentials from *Scyliorhinus* red fibers recorded 1.2 mm (*left*) and 4.2 mm (*right*) from current injection electrode. The largest electrode separation was very close to the width of the myotome, that is, to the length of the fibers, and was nearly double the space constant.

bundles of intact fibers stimulated electrically). As expected, they observed higher power outputs than from the earlier skinned fiber preparation. In subsequent studies, they monitored force and heat production simultaneously (Fig. 5-11), and so were able to investigate the efficiency of energy conversion during shortening and during sinusoidal movements (driven by a computer-controlled combined motor and length transducer). Evidently, sinusoidal movements are much closer to the normal operation of the fibers as the fish alternately stretches and contracts its myotomes, and such experiments were designed to test the (reasonable) idea that maximum efficiency occurs at the speeds of contraction found during normal swimming movements. Efficiency may be defined in a number of ways, but Curtin and Woledge take it as the net work output per molecule of ATP used.

Figure 5-12 shows example records of the force produced by red and by white fibers under isometric conditions. Tetanic stimulation of the red fibers gave the same force at 50 and 60 Hz. During slow spinal swimming, when the red fibers alone are employed and the dogfish swims steadily at 35–40 tail beats/min, regular bursts of action potentials could be recorded from spinal motoneurons at 60 Hz and above (Mos et al. 1990), so producing maximum force from the red fibers.

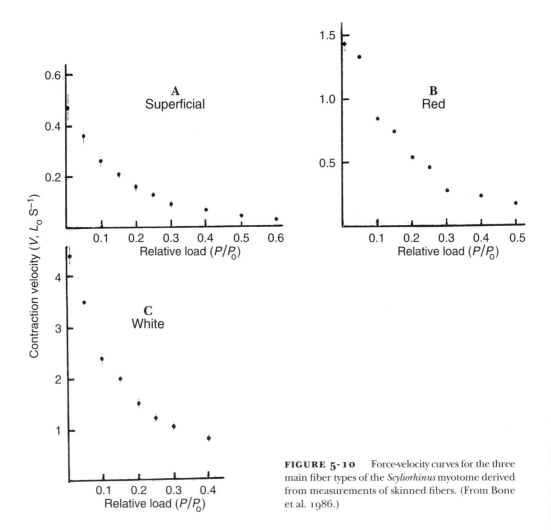

FIGURE 5-10 Force-velocity curves for the three main fiber types of the *Scyliorhinus* myotome derived from measurements of skinned fibers. (From Bone et al. 1986.)

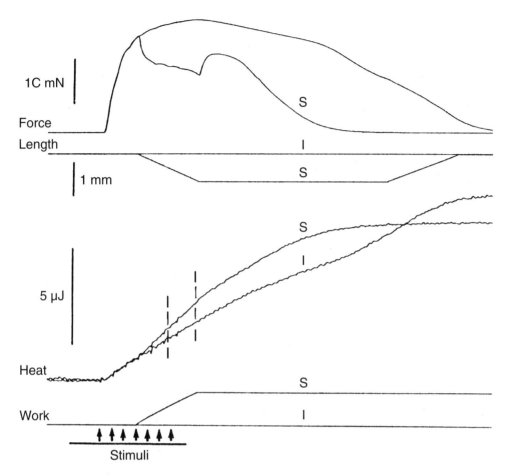

FIGURE 5-11 Sample records of simultaneous force, length, heat, and work measurements from *Scyliorhinus* white fibers stimulated electrically. The records labeled *I* are from a tetanus where length was held constant, and *S* are those from a tetanus with a period of shortening at constant velocity. The heat records came from a thermopile, and rate of heat production was measured between vertical dotted lines. (From Curtin and Woledge 1991.)

The results of their oscillatory studies can be summarized briefly (the original papers should be consulted):

1. Maximum work and efficiency were produced if stimulation began during the stretch phase of the cycle.
2. Maximum power from the red fibers was obtained during sinusoidal movement at just over 1 Hz (rather faster than tail beat frequency during spinal swimming).
3. Maximum power from the white fibers was obtained during similar movement at 3.5 Hz.
4. Maximum efficiency for the red fibers (mean 0.507 ± 0.045, N = 9) was obtained during movements at lower frequency than maximum power (between 0.61 and 0.95 Hz in different preparations).
5. Maximum efficiency for the white fibers (mean value 0.41 ± 0.021, N = 13) was obtained during movements at 2.0–2.5 Hz.
6. In terms of work done per molecule of ATP split, red fibers obtain 29×10^{-21} J, while the white fibers obtain a little less, 23×10^{-21} J. Perhaps this may be due to less energy being used in the red fibers

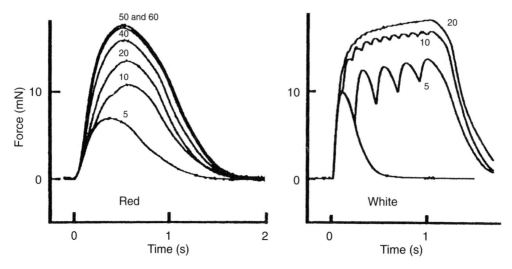

FIGURE 5-12 Records of force produced by *Scyliorhinus* red fiber bundle (*left*) and white fiber bundle (*right*) under isometric conditions. For the red fibers, brief tetani lasted for 0.4 sec at different frequencies (numbers in Hz); for the white fibers, the tetanic stimuli lasted 1.0 sec.

than in the white for Ca^{2+} release from the sarcoplasmic reticulum (SR) rather than turned into work.

The values for the efficiency of chemical conversion considered above are only one of the efficiencies of the multistage process, which have as their product the overall efficiency of the swimming fish. For instance, the efficiency of ATP production per molecule of fuel metabolized is vastly superior in the red muscle fiber to that in the white fibers, which use glycogen anaerobically.

The careful mechanical experiments of Curtin and Woledge suggest several points underlying the way that the dogfish uses its two motor systems, and these are considered in the next section.

Functional role of the different fiber types in the shark myotome

Direct electromyographic evidence (Fig. 5-13) from swimming spinal *Scyliorhinus* as well as studies of metabolites following different swimming regimens has shown that the red fiber portion of the myotome is devoted to sustained cruising swimming, while the much larger white portion is employed only during burst activity (Bone 1966). Subsequent work on teleosts and other fish groups has confirmed that this division of labor is general in fish. The role of the superficial fibers in *Scyliorhinus* is unclear, but their small number, their structure, and their arrangement as a lateral strip have suggested (Bone et al. 1986) that they might be regarded as tonic postural fibers, and this view is not in conflict with the shape of their force-velocity curve, which resembles that of frog tonic fibers, or their low power output. *Scyliorhinus* is a relatively inactive bottom-dwelling fish, which rests with its body slightly curved upward at either end, perhaps to enable it to make an effective first tail stroke when disturbed. The superficial fiber lateral strip perhaps provides the small amount of tension required for maintaining this curvature. It would certainly be worth examining other bottom-dwelling sharks such as Heterodontids and Orectolobids to see if they too possess superficial fibers. None of the more active free-swimming sharks such as *Squalus*, *Galeus*, and *Mustelus* have superficial fibers.

Lamnid sharks have "internalized" red muscle with retial connection to the systemic circulation that enables them to retain metabolic heat in the myotome (Carey and Teal

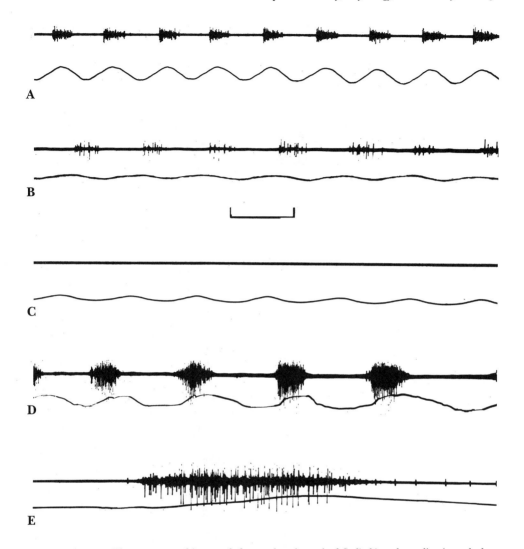

FIGURE 5-13 Electromyographic records from swimming spinal *Scyliorhinus*, lower line in each showing movement of fish. **A** and **B**: Regular slow spinal swimming, electrode tip in red muscle zone. **C**: as in **A** and **B**, but electrode tip in white muscle, showing absence of electrical activity. **D** and **E**: Bursts of potentials from the white muscle during rapid movements. Scale bar represents 2 sec for all except **D**, 4 sec, and **E**, 1 sec.

1969). So far as is known, there are no specializations of the red fibers in warm lamnids, and indeed, Ballantyne et al. (1992) find that mitochondrial substrate preferences and respiration rates are similar to those from dogfish "cold" red muscle fibers.

The main gap in our understanding of the functional role of the different fiber types in the shark myotome (apart from the relatively insignificant and perhaps uncommon superficial fibers) lies in the possible roles of the two types of red and white fiber. Given the red-white dichotomy of function, it would seem reasonable to suppose that when the shark is cruising very slowly, only a proportion of the red fibers are engaged, and that as speed rises toward the maximum sustainable cruising speed, more and more of the red fibers are operated. If this be so, then perhaps the outer red fibers are the first to become active, and the inner red fibers only play a part in higher-speed cruising.

A similar case might be made for supposing that the outer white fibers, with their slight but distinct aerobic capacity (at any rate slight in *Scyliorhinus* and *Galeus*), are employed during swimming above speeds sustainable for long periods, and that the inner white fibers are used only during very rapid-burst escape swimming. Kryvi and Totland (1977) made just this suggestion on the basis of their histochemical studies. At present, so far as I am aware, the experiments to support or demolish this view have yet to be performed. It would be testable by such experiments as those of Curtin and Woledge, who used for their experiments the outer red fibers and the inner white fibers (N. A. Curtin, personal communication, 1995). Curiously enough, the most suggestive evidence that the inner and outer white fibers have different roles is provided not by sharks but by *Torpedo*, where the two types of fiber receive different types of innervation.

BIOCHEMISTRY

Since sharks and batoids are apparently very similar in muscle biochemistry, both groups are considered under this heading to avoid repetition.

The great majority of published work on fish muscle biochemistry has concerned teleosts, and in most respects, the work on elasmobranchs has shown them to be similar to teleosts, with two notable exceptions: (1) the effects of the presence of urea and the other nitrogenous osmolytes on muscle enzyme systems and contractile proteins and (2) the role of ketone bodies in lipid metabolism.

On the whole, apart from these two elasmobranch "specialities," attention has focused on intermediate metabolism and its control and on the scope for activity suggested by the enzyme profiles of the different muscle types. As Dickson et al. (1993) point out, it is no easy matter to measure swimming speeds, metabolic rates, and energy budgets of large, active sharks, or of large rays for that matter, and so biochemical indices of muscle activity are used instead.

In *Scyliorhinus*, the activities of anaerobic glycolytic enzymes in the superficial fibers (the only measure we have of their functional role) are similar to those in the intermediate fibers (see Table 5-1), with the exception of lower SDH activity. Unfortunately, this statement merely hints at the difficulty of inferring function from enzyme profiles, for it does not seem likely that the two fiber types can have the same functional role.

Insofar as the white muscle is concerned, there is agreement that anaerobic glycolysis is the route whereby ATP is provided for the contractile proteins and that there is little gluconeogenesis in the muscle. In *Scyliorhinus*, white muscle activities of the glycolytic rate-limiting enzymes are five times those in the red muscle (Crabtree and Newsholme 1972), while in the little skate (*Raja erinacea*) Moon and Mommsen (1987) find high levels of LDH, phosphofructokinase, glycogen phosphorylase, and pyruvate kinase, similarly indicating high potential for anaerobic glycolysis. Dickson et al. (1993) tabulate the LDH activities of white muscle in the eight elasmobranchs they examined (and data from other authors for five others, including *R. erinacea*), showing a clear relation between activity levels and the general activity of the fish (Fig. 5-14). The extremely active shortfin mako shark (*Isurus oxyrinchus*) had much higher levels than any of the other fish they examined.

In consequence of this anaerobic glycolysis in the white muscle, even short periods of burst activity when it is in use result in large (tenfold) increases in plasma lactate (Zammit and Newsholme 1979; Lowe et al. 1995). In the dusky shark (*Carcharhinus obscurus*) the highest values of plasma lactate were found more than an hour after capture (Cliff and Thurman 1984). The poor vascularity of the white portion of the myotome is evidently at least partially responsible for the slow release of lactate into the circulation.

There is some evidence for a degree of oxidative metabolism in the white muscle of active sharks. Dickson et al. (1993) found citrate synthase activity levels in the white muscle of shortfin mako high enough to suggest the

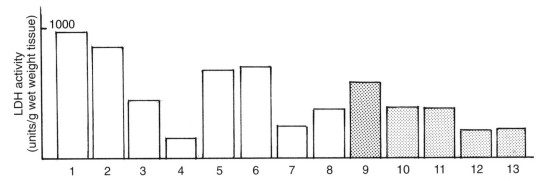

FIGURE 5-14 Lactate dehydrogenase levels in various elasmobranchs. Batoids are shaded; dark shading indicates a batoid that swims with myotomal oscillations. 1 = mako shark; 2 = blacktip shark; 3 = sandbar shark; 4 = blue shark; 5 = leopard shark; 6 = gray smooth hound; 7 = dogfish; 8 = spur dogfish; 9 = shovelnose guitarfish; 10 = thornback ray; 11 = round stingray; 12 = little skate; 13 = California skate. (After Dickson et al. 1993.)

possibility of some aerobic activity, possibly during rapid cruising. So far as I am aware, there is no *direct* evidence in any elasmobranch for the use of the white zone of the myotome during cruise swimming (as there is in some teleosts), but there is no reason that the intermediate fibers might not be employed for this purpose. After over 30 years work on fish muscle on and off, I feel it still seems necessary to warn the reader, as did Austin (1962), against the deeply ingrained worship of tidy-looking dichotomies!

So far as the red muscle is concerned, enzyme profiles in the species examined by Dickson et al. (1993) and in the little skate (Moon and Mommsen 1987) indicate aerobic operation. In view of the mitochondrial content of the fibers and their abundant vascular bed, this is hardly surprising. However, it is still not entirely clear whether different elasmobranchs may preferentially use different "fuels" in the red muscle. Both glycogen and fat decline after prolonged red muscle use during spinal swimming of *Scyliorhinus* (Bone 1966), and both lipid and glycogen are found in red muscle fibers.

Crabtree and Newsholme (1972) examined lipid levels and enzymes in a wide range of vertebrates and invertebrates, finding very low levels of triglyceride and of carnitine palmitoyltransferase in comparison with teleosts and other vertebrates. E. A. Newsholme (personal communication, 1975) queried these low levels in dogfish, but it was not until some years later that Zammit and Newsholme (1979) discovered the importance of ketone body metabolism in dogfish red muscle. It is unclear why dogfish differ from higher animals in lipid metabolism, but these authors suggest that it may relate to the lack of a suitable transport protein for triacylglycerols in elasmobranch blood.

Recent work on cardiac enzymes in freshwater stingrays (Driedzic and de Almeida-Val 1996) has shown the presence of detectable amounts of carnitine palmitoyltransferase. Unfortunately this was not determined in the red muscle, but it is possible that the inability to catabolize fatty acids may not be universal in all elasmobranch species.

Lacking a swim bladder, many sharks rely for static lift mainly on neutral lipids stored in the liver, but there may also be significant amounts in muscle tissue. For example, Van Vleet et al. (1984) found that the muscle of the seven-gill shark (*Heptranchias perlo*) contained no less than 57% of the total body lipid, that is, 14% of the wet weight of the muscle. Since sharks appear capable of regulating their buoyancy, there are evidently interesting problems in lipid metabolism and conservation in both liver and muscle, and lipid metabolism in elasmobranchs seems more complex than in teleosts.

The regulation of the glycolytic metabolism of elasmobranch white muscle is not by circulating catecholamines, but the activity of phosphorylase is directly Ca^{2+}-dependent, so that glycolysis in the white muscle is triggered by Ca^{2+} release from the SR when the muscle propagates action potentials. Battersby et al. (1996) have recently shown (in the spurdog, *Squalus acanthias*) that 3,5,3'-triiodothyronine (T_3) significantly increases pyruvate kinase activity in white muscle in the short term, although several interpretations of the manner in which it does so are possible.

Last, the effects of urea and how they are counteracted has interested several groups. It was early realized that urea (present at half molar levels in marine elasmobranchs) would disrupt proteins in other vertebrates at these concentrations and that at least some elasmobranch enzymes were resistant to urea. Other enzymes were found to require urea for maximum in vitro activity, for example, Mg^{2+}-myofibrillar ATPase. More unexpected, perhaps, than the adaptation of these enzymes for operation in the high-urea elasmobranch environment was the discovery by Yancey and Somero (1979, 1980) that the methylamine compounds such as trimethylamine oxide (TMAO), betaine, and sarcosine, when present in the ratio of 2:1 urea: methylamine compounds, effectively counteracted the perturbing effects of urea on enzymes such as pyruvate kinase. This ratio is that actually found in marine elasmobranchs. Simple experiments on the mechanical responses of elasmobranch skinned fibers by Altringham et al. (1982) showed that maximum isometric tension was depressed by the addition of urea but could be nearly fully restored by addition of TMAO. However, Hand and Somero (1984) found that phosphofructokinase from the thornback ray (*R. clavata*) was rapidly inactivated in vitro by even 50 mM urea below pH 7.0, and this was not counteracted by TMAO. Partial protection against deactivation was afforded by F-actin and muscle thin filaments, to which the enzyme binds, but in vivo the authors suggest that the enzyme may be completely protected by being in a resistant particulate form.

Batoids

GENERAL

In contrast to sharks, batoid locomotor musculature has received little attention, despite the fact that it propels one of the largest and most spectacular fish in the oceans (the manta ray, *Mobula*) and also has a certain culinary interest. Although excellent kinematic records of swimming rays were made more than a century ago by Marey (1894), and their flapping aquatic flight has been analyzed by Daniel (1988), surprisingly little is yet known of the arrangement of the locomotor muscle fibers that move the pectoral fins. Early work by Ranvier (1873) showed that the pectoral musculature contained red and white fiber bundles, with different mechanical properties, but he gave no details of their disposition in the fin. Apart from a very brief description of the pectoral musculature of the electric ray *Typhlonarke* and a transverse section of the pectoral fin of *Dasyatis* figured by Droge and Leonard (1983), there has been no recent work on the structure of the pectoral musculature. Since Ranvier, physiological investigations have been limited to studies of the resting and spike potentials from muscles in the pelvic fins of three stingray species by Hagiwara and Takahashi (1967); a brief report of tetanus fusion frequency of twitch fibers from the pectoral fin of *Leucoraja naevis* (Johnston 1980); and the electromyographic study of stingray swimming by Droge and Leonard (1983). This section is based largely on unpublished work on the pectoral musculature of *Raja montagui, R. clavata*, and *R. microocellata* (Bone and Holst, in preparation).

PECTORAL AND PELVIC FIN STRUCTURE

The fins that propel rajids are built on a repeating series of units, arranged on jointed fin rays that are flexed upward and downward by a rather complex set of muscles, in which there are three different fiber types. Figure 5-15 shows the architecture of the unit of the fin. Most of the fin unit consists of two layers of white fibers dorsally and ventrally in a pinnate arrangement. These insert at their prox-

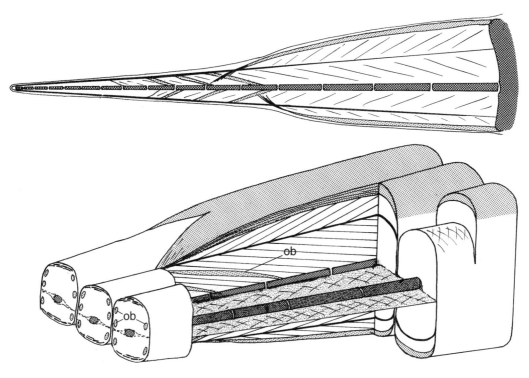

FIGURE 5-15 The architecture of the ray fin. **ABOVE:** Sagittal section of a fin unit showing cartilage (dark stipple) and red fibers (lighter stipple). **BELOW:** Stereodiagram showing the way that the red fibers and oblique red fiber bundles (ob) are arranged around the white fibers. (From Bone and Holst, in preparation.)

imal ends either on the connective tissue sheath surrounding the radial cartilages or, for the outer layer, on a connective tissue tendinous layer, linked ultimately to the girdles. The distal ends of both outer and inner white fibers are joined in a thick tendinous layer, which toward the outer third of the fin forms a long tendon reaching to its outer border. In addition, around the outer (lateral) borders of each fin unit, there are small, flattened, oblique bundles of red fibers, which have the same disposition as the inner white fiber layer and insert proximally on the edges of the radial cartilage sheath and distally on the tendinous sheath formed by the distal end of the white fiber attachments. These small red bundles are visible in living material and conspicuous after SDH staining. Lastly, the outer surface of each unit consists of very long red fibers inserting proximally onto the girdle and wrapping distally around the outer layer of white fibers to end as tendons that join the sheath of the radial cartilages. These long "slips" of red fibers were probably those examined by Ranvier (1873), who demonstrated by perfusion of the vessels that their red coloration was intrinsic to the fibers.

It is notable that the elevator musculature of the fin units is considerably larger in mass than the depressor musculature (see Fig. 5-15). In a medium-sized *Raja clavata*, for example, the dorsal (depressor) muscles of the fin units were only 35–40% of the equivalent ventral elevator muscles. Since the rays examined, like *R. clavata*, are among the densest of all elasmobranchs, it is at first sight paradoxical that such batoids should have more powerful elevator than depressor musculature in the fins, the converse of the pectoral musculature of birds. It is possible that some down force as well as rearward thrust has to be generated by the propulsive waves across the disk of the ray, because in section the ray is convex above and flat beneath, and thus circulation around the body generates lift as it swims forward. In other words, unless down force is generated during the swimming motions of the ray, it would loop the loop during rapid forward movement. No tests have examined this speculation, however.

FIBER STRUCTURE

The white fibers that make up the major part of the fin units vary in length according to their position along the unit and are very large in diameter (up to 340 μm). They are mitochondria-poor and contain little lipid. Their capillary bed is sparse and, presumably, like other fish fast fibers, their metabolism is primarily anaerobic. In contrast, the smaller-diameter red fibers (up to 160 μm) in the outer red layer and in the oblique red fiber bundles have an abundant capillary bed and contain many mitochondria, although few lipid droplets, and are evidently aerobic. However, although in cryostat sections stained for SDH or resin sections stained with p-phenylenediamine most red fibers are darkly stained, owing to their high mitochondrial content, some are lightly stained (Fig. 5-16) and, apart from their smaller size, rather resemble *Scyliorhinus* superficial fibers. Mean diameters for these pale red fibers are similar to those for red fibers, both being about half the mean diameter of the white fibers. The significance of this diversity among the "red" fibers is not known.

Last, there are intermediate fibers that are in some respects intermediate between the red and white fiber types. For example, they are larger in diameter than the red though smaller than the white (up to 200 μm) and contain few mitochondria or lipid droplets. They are found in the oblique red fiber bundles at the edges of the white muscle masses of the fin units and where the superficial red fiber layers abut the underlying white fibers. The histogram of Figure 5-17 shows the relative abundance and size ranges of the fibers in a ventral deep red oblique fiber bundle. At hatching from the egg capsule, although smaller in diameter, all three main fiber types are present; nothing is known of their ontogeny at present.

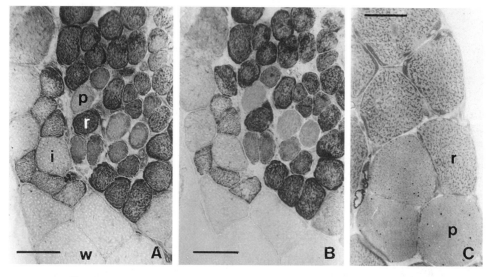

FIGURE 5-16 Transverse sections of fibers from deep oblique fiber bundle. **A** and **B:** Successive cryostat sections stained for (**A**) lipid and (**B**) succinic dehydrogenase. Note wide variation in mitochondrial enzyme and lipid content of the small red (r) and pale red (p) fibers. White fibers (w) are larger than the intermediate fibers (i). Scale bar: 100 μm. **C:** Resin section stained with p = paraphenylenediamine to show difference in mitochondrial content between red (r) and pale red (p) fibers. Scale bar: 50 μm.

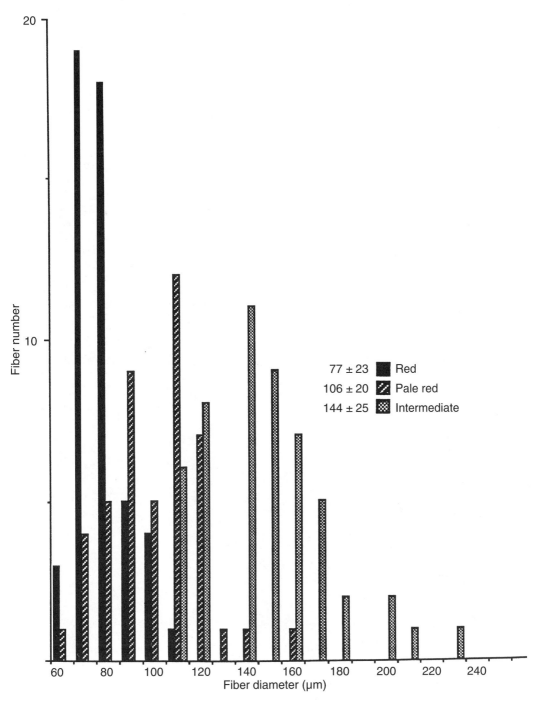

FIGURE 5-17 Histogram of fiber types in ventral deep oblique red fiber bundle. (From Bone and Holst, in preparation.)

MOTOR INNERVATION AND ELECTRICAL PROPERTIES

The white fibers are innervated as in the myotomes of *Scyliorhinus* and *Squalus* (the only two sharks I have examined) by single large end plates (Fig. 5-18B). Unlike the sharks, these are in the middle of the fibers and are derived from a single axon only. Resting potentials are around 80–90 mV (85.8 mV ± 4.9), and current injection evokes rapid, overshooting action potentials (Fig. 5-18A). Curiously, the end plates on the *myotomal* white muscle fibers of rajids are terminal and derived from two separate axons (Fig. 5-18E). Hagiwara and Takahashi (1967) examined the electrical properties of the white fibers of several tropical stingrays (see Table 5-5) and found the resting potential to be mainly determined by the Cl^- concentration difference across the sarcolemma. They later used the white fibers of

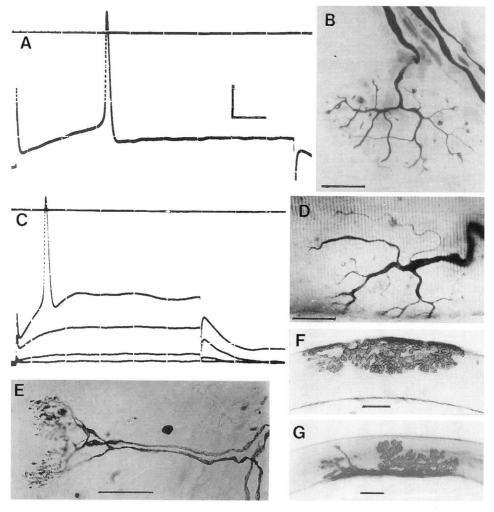

FIGURE 5-18 **A:** White fiber action potential evoked by current injection. Scale bars: 20 mV; 20 msec. **B:** Motor end plate on white fiber (reduced silver impregnation). Scale bar: 20 μm. **C:** Intermediate fiber action potential evoked by current injection. Scale as in A. **D:** One of the motor endings on an intermediate fiber (reduced silver impregnation). Scale bar: 20 μm. **E:** Dual innervation of terminal basket ending on myotomal white fiber (supravital methylene blue staining). **F** and **G:** Endings on intermediate fibers (acetylcholinesterase staining). Scale bars: 100 μm. (E from Bone 1972; remainder from Bone and Holst, in preparation.)

the stingray *Taeniura lymma* as a model system for examining anion permeation (Hagiwara and Takahashi 1974). Apart from these papers, I am not aware of other published work on the electrical properties of batoid muscle fibers. The intermediate fibers are innervated by a small number of large en plaque end formations (Fig. 5-18D, F, and G), either two or three in the oblique bundles, and by three in the longer red fiber layer. The end plates are thus spaced at least 2000 µm apart, sometimes up to 7500 µm apart. Resting potentials were less than those of the white or red fibers (80.13 mV ± 5.4, N = 8). Naturally, they propagate action potentials (Fig. 5-18C) and in such long fibers multiple end plates may well speed activation of the fiber significantly.

The red fibers are innervated by small en grappe endings along their length spaced between 450 µm and 1500 µm apart. Occasional much larger en plaque endings are found on some red fibers (Fig. 5-19). Resting potentials are slightly above those of the white fibers (89.15 mV ± 4.74, N = 20). Overshooting action potentials have not been observed in ray red fibers after injection of current, but just as in *Scyliorhinus* red fibers, current injection evokes smaller potentials up to 30 mV. Although it is not known whether these are propagated, their similarity to those of *Scyliorhinus* and the very large distance between nerve terminals suggest that they probably are.

As in sharks, the myotomal muscle fibers are divided into red and white portions, but their innervation has only been examined in *Torpedo*, where the outer white fibers are innervated by central end plates rather than by end plates at the ends of the fibers (Bone 1964). It is not known whether two axons contribute to these central end plates as they do to the terminal basketlike endings of the inner white fibers.

SENSORY INNERVATION

With the doubtful exception of unconfirmed reports of muscle spindles in some teleost fishes, the elongate beaded endings of large-diameter nerve fibers among muscle fibers are the only known intramuscular stretch receptors in fishes (Fig. 5-20). An ultrastructural study (Bone and Chubb 1975) showed that they were not encapsulated, yet others found that their responses were very similar to those of the (encapsulated) spindles of terrestrial vertebrates. It is noteworthy that in all batoid genera examined (including *Myliobatis, Trygonorrhina, Aptychotrema, Raja, Torpedo,* and *Dasyatis*), these endings are only found among red muscle fibers. Even in each of the

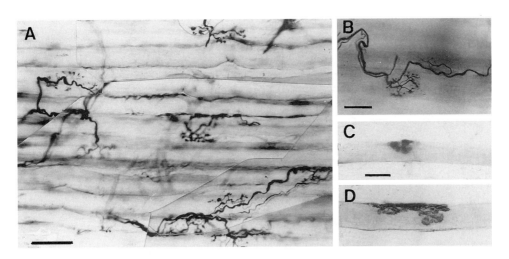

FIGURE 5-19 The multiterminal innervation of red fibers in rays. **A:** General view of silver impregnated whole mount. Scale bar: 200 µm. **B:** Endings of two different sizes above and below axons supplying them. Scale bar: 100 µm. **C** and **D:** Acetylcholinesterase preparations of the two sizes of ending on red fibers. Scale bar: 100 µm.

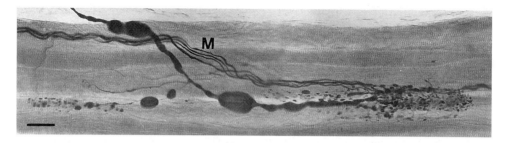

FIGURE 5-20 A small part of a complex sensory ending in the dorsal superficial red muscle sheet of a ray fin unit, showing extensive beaded fibers aligned between the muscle fibers. Silver-impregnated whole-mount preparation. Scale bar: 100 μm.

smallest of the oblique fiber bundles there are one or two such endings. In contrast to sharks, there are no coiled corpuscular endings under the skin, but there are what appear to be modified Pouloumordwinoff endings closely associated with the most superficial muscle fibers.

In addition to the sensory endings associated with the red muscle fibers, there are also simple branching endings, which ramify in the connective tissue of the capsules of the joints between the radial cartilages along the fin unit. These are likely to be proprioceptive endings; similar endings are found in joint capsules in other vertebrate groups.

MECHANICAL PROPERTIES

The mechanical properties of the three different fiber types have only been examined in a preliminary way, using supramaximal stimulation of small bundles of fibers attached to a transducer. The significant features of our preliminary results are: (1) that the white and intermediate fibers differ in contraction speed, white fibers being almost twice the speed of the intermediate fibers, and fused tetani and peak tensions are reached at stimulus frequencies lower than those of the intermediate fibers; and (2) that the red fibers respond to single stimuli with very slow contractions lasting 1 sec or so and to increasing stimulus frequency by

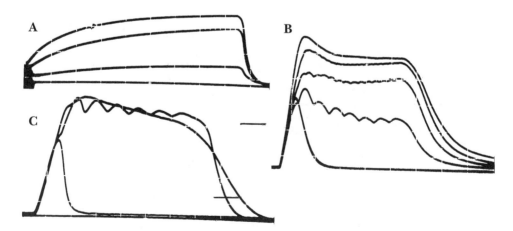

FIGURE 5-21 Mechanical responses of bundles of different ray fiber types evoked by direct electrical stimulation. **A:** Red fibers. Series of slow contractions stimulated at 20, 50, and 90 Hz. Scale bar: 1 sec. **B:** Intermediate fibers. Single twitch and tetani at 15, 30, 50 and 100 Hz. Scale bar: 100 msec. **C:** White fibers. Single twitch and tetani at 20 and 50 Hz. Scale bar: 50 msec. Note difference in speed between intermediate and white fibers. (From Bone and Holst, in preparation.)

slowly rising tension increases, reaching a plateau only after some 6 sec. Maximum tensions are not reached until stimulus frequency exceeds 90 Hz. Figure 5-21 illustrates some of these points. It is interesting that Ranvier (1873) observed (using a stimulating set-up of Marey's [1894]) that ray red fibers produced a continuous tension curve when stimulated repetitively at frequencies that evoked a fused tetanus in the white fibers.

ROLE OF THE DIFFERENT FIBER TYPES IN RAYS

Some of the results of a series of electromyographic studies on rajids are shown in Figure 5-22. During slow cruising swimming only the red fibers of each fin unit are active. If the ray is disturbed during cruising swimming, it accelerates into more rapid swimming, and as it does so, the white fibers of the fin units become active, as they are when captive rays swim vigorously and protrude part of the snout out of the water. These results are as expected from previous work on sharks, namely, the white fibers are active during rapid swimming movements and, when cruising, the ray employs the superficial red fibers. It did not prove possible to record from the oblique red fiber bundles, or from intermediate fibers underlying the dorsal and ventral red muscle strips. As Ranvier suggested over a century ago for the superficial red fibers, "les rouges seraient plutot équilibrateurs ou régulateurs." That is, he supposed them to play a postural or tonic role. Perhaps these small fiber bundles with their stretch receptors aligned along the axes of the white fibers may be concerned with monitoring the rate of bending of the fin during rapid movements, for it would seem that monitoring fin *position* could be better carried out by the receptors within the outer dorsal and ventral red fiber bundles.

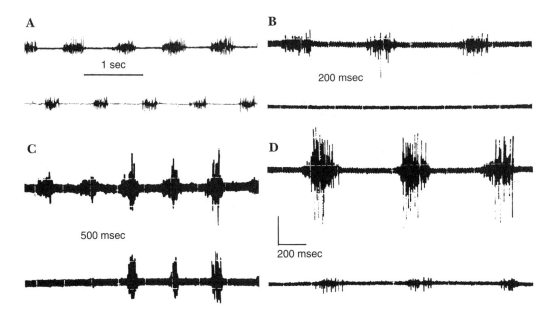

FIGURE 5-22 Electromyograms from fin unit muscles of rays. **A:** Slow swimming. Both electrodes in dorsal superficial red muscle several units apart across fin, showing sequential activity as waves pass a ross fin. Scale bar: 1 sec. **B:** Slow swimming, lower line electrode in dorsal white muscle showing absence of activity. Scale bars: 0.2 sec, 50 V. **C:** Electrodes as in B, cruise swimming on left, with transition to rapid swimming the electrode in the white muscle (*lower line*) shows activity. Scale bars: 0.5 sec, 50 V. **D:** Vertical strong swimming with nose out of the water. Activity in both white (*lower*) and red dorsal muscle of the fin unit. Scale bars: 0.2 sec, 50 V. Note that the signal amplitude does not necessarily reflect the amplitude of the potentials from red or white muscle fibers, simply being related to more or less successful placement of the electrode tip. (From Bone and Holst, in preparation.)

Conclusions

Provided it is assumed that observations on small sharks and rays represent elasmobranchs in general, including large species, our knowledge of the structure of elasmobranch muscle is reasonable. In comparison with higher vertebrates, their fast muscle fibers are extremely large, yet very few workers apart from Hagiwara and Takahashi (1974) have taken advantage of them as experimental models. So far, it seems that their red fibers are unique among vertebrates in propagating nonovershooting action potentials, and rays (perhaps also some sharks) have fast-twitch fibers of different contraction speeds. There is still room for further work on the role of intermediate fibers in swimming, and another area that would seem to offer scope for interesting studies is the control of lipid metabolism in elasmobranchs.

REFERENCES

Altringham, J. D., and I. A. Johnston. 1982. The pCa-tension and force-velocity characteristics of skinned fibers isolated from fish slow and fast muscles. *J. Physiol.* 333: 421–49.

Altringham, J. D., P. H. Yancey, and I. A. Johnston. 1982. The effects of osmoregulatory solutes on tension generation by dogfish skinned muscle fibers. *J. Exp. Biol.* 96: 443–45.

Austin, J. L. 1962. *Sense and Sensibilia*. Reconstructed by G. J. Warnock. London: Oxford University Press.

Ballantyne, J. S., M. E. Chamberlin, and T. D. Singer. 1992. Oxidative metabolism in thermogenic tissues of the swordfish and mako shark. *J. Exp. Zool.* 261: 110–14.

Battersby, B. J., W. J. McFarlane, and J. S. Ballantyne. 1996. Short-term effects of 3,5,3'-triiodothyronine on the intermediary metabolism of the dogfish shark *Squalus acanthias*—evidence from enzyme activities. *J. Exp. Zool.* 274: 157–62.

Bone, Q. 1964. Patterns of muscular innervation in the lower chordates. *Int. Rev. Neurobiol.* 6: 99–147.

———. 1966. On the function of the two types of myotomal muscle fiber in elasmobranch fish. *J. Mar. Biol. Assoc. U.K.* 46: 321–49.

———. 1972. The dogfish neuromuscular junction: dual innervation of vertebrate striated muscle fibers? *J. Cell Sci.* 10: 657–65.

———. 1988. Muscles and locomotion. In *Physiology of Elasmobranch Fishes*, ed. T. J. Shuttleworth, 99–141. New York: Springer-Verlag.

Bone, Q., and A. D. Chubb. 1975. The structure of stretch receptor endings in the fin muscles of rays. *J. Mar. Biol. Assoc. U.K.* 55: 939–43.

Bone, Q., N. A. Curtin, and R. C. Woledge. 1994. Action potentials in red muscle fibers isolated from dogfish. *J. Physiol.* 479: 4–5.

Bone, Q., I. A. Johnston, A. Pulsford, and K. P. Ryan. 1986. Contractile properties and ultrastructure of three types of muscle fiber in the dogfish myotome. *J. Muscle Res. Cell Motil.* 7: 47–56.

Carey, F. G., and J. M. Teal. 1969. Mako and porbeagle: warm-bodied sharks. *Comp. Biochem. Physiol.* 28: 199–204.

Cliff, G., and G. D. Thurman. 1984. Pathological and physiological effects of stress during capture and transport in the juvenile dusky shark, *Carcharhinus obscurus*. *Comp. Biochem. Physiol.* 78A: 167–73.

Crabtree, B., and E. A. Newsholme. 1972. The activities of lipases and carnitine palmitoyltransferase in muscle from vertebrates and invertebrates. *Biochem. J.* 130: 697–705.

Curtin, N. A., and R. C. Woledge. 1988. Power output and force-velocity relationship of live fibers from white myotomal muscle of the dogfish, *Scyliorhinus canicula*. *J. Exp. Biol.* 140: 187–97.

———. 1991. Efficiency of energy conversion during shortening of muscle fibers from the dogfish *Scyliorhinus canicula*. *J. Exp. Biol.* 158: 343–53.

———. 1993a. Efficiency of energy conversion during sinusoidal movement of white muscle fibers from the dogfish *Scyliorhinus canicula*. *J. Exp. Biol.* 183: 137–47.

———. 1993b. Efficiency of energy conversion during sinusoidal movement of red muscle fibers from the dogfish *Scyliorhinus canicula*. *J. Exp. Biol.* 183: 189–206.

Daniel, T. L. 1988. Forward flapping flight from flexible fins. *Can. J. Zool.* 61: 1406–20.

Dickson, K. A., M. O. Gregorio, S. J. Gruber, K. L. Loefler, M. C. Tran, and C. Terrell. 1993. Biochemical indices of aerobic and anaerobic capacity in muscle tissues of California elasmobranch fishes differing in typical activity level. *Mar. Biol.* 117: 185–93.

Driedzic, W. R., and V. M. F. de Almeida-Val. 1996. Enzymes of cardiac energy metabolism in Amazonian teleosts and the freshwater stingray (*Potamotrygon hystrix*). *J. Exp. Zool.* 274: 327–33.

Droge, M. H. and R. B. Leonard. 1983. Swimming pattern in intact and decerebrated stingrays. *J. Neurophysiol.* 50: 162–77.

Edman, P. 1988. Double-hyperbolic force/velocity relation in frog muscle fibers. *J. Physiol.* 404: 301–21.

Giacomini, E. 1898. Sulla maniera onde I nervi si terminano nei miocommi e sulle estremita delle fiber muscolari nei Selacii. *Atti R. Accad. Fisocritica Siena* 4, 10: 560.

Hagiwara, S. and K. Takahashi. 1967. Resting and spike potentials of skeletal muscle fibers of saltwater elasmobranch and teleost fish. *J. Physiol.* 190: 499–518.

———. 1974. Mechanism of anion permeation through the muscle fiber membrane of an elasmobranch fish, *Taeniura lymma. J. Physiol.* 238: 109–27.

Hand, S. C., and G. N. Somero. 1984. Influence of osmolytes, thin filaments, and solubility state on elasmobranch phosphofructokinase in vitro. *J. Exp. Zool.* 231: 297–302.

Johnston, I. A. 1980. Contractile properties of fish fast muscle fibers. *Mar. Biol. Lett.* 1: 323–28.

Kryvi, H. 1977. Ultrastructure of the different fiber types in axial muscles of the sharks *Etmopterus spinax* and *Galeus melastomus. Cell Tissue Res.* 184: 287–300.

Kryvi, H., and G. K. Totland. 1977. Histochemical studies with microphotometric determinations of the lateral muscles in the sharks *Etmopterus spinax* and *Galeus melastomus. J. Mar. Biol. Assoc. U.K.* 57: 261–71.

Kryvi, H., T. Flatmark, and G. K. Totland. 1981. The myoglobin content in red, intermediate, and white fibers of the swimming muscles in three species of shark, a comparative study using high-performance liquid chromatography. *J. Fish Biol.* 18: 331–38.

Levin, A., and J. Wyman. 1927. The viscous elastic properties of muscle. *Proc. R. Soc. Lond. B Biol. Sci.* 101: 218–43.

Lorenzini, S. 1678. Osservazioni intorno alle Torpedini. Firenze: Onofri.

Lowe, T. E., R. M. G. Wells, and J. Baldwin. 1995. Absence of regulated blood-oxygen transport in response to strenuous exercise by the shovelnosed ray, *Rhinobatos typus. Mar. Freshw. Res.* 46: 441–46.

Marey, E. J. 1894. *Le mouvement.* Paris: Masson.

Moon, T. W., and T. P. Mommsen. 1987. Enzymes of intermediary metabolism in tissues of the little skate, *Raja erinacea. J. Exp. Zool.* 244: 9–15.

Mos, W., B. L. Roberts, and R. Williamson. 1990. Activity patterns of motoneurons in the spinal dogfish in relation to changing fictive locomotion. *Proc. R. Soc. Lond. B Biol. Sci.* 330: 329–39.

Ono, R. D. 1983. Dual motor innervation in the axial musculature of fishes. *J. Fish Biol.* 22: 395–408.

Ranvier, L. 1873. Propriétés et structures différentes des muscles rouges et des muscles blancs, chez les Lapins et chez les Raies. *C. R. Acad. Sci.* 77: 1030–34.

Roberts, B. L. 1969. The response of a proprioceptor to the undulatory movements of dogfish. *J. Exp. Biol.* 51: 775–85.

Sakharov, D. A. and L. Kashapova. 1979. The primitive pattern of the vertebrate body muscle innervation: ultrastructural evidence for two synaptic transmitters. *Comp. Biochem. Physiol.* 62A: 771–76.

Stanfield, P. R. 1972. Electrical properties of white and red muscle fibers of the elasmobranch fish *Scyliorhinus canicula. J. Physiol.* 222: 161–86.

Tiegs, O. W. 1953. Innervation of voluntary muscle. *Physiol. Rev.* 33: 90–144.

Totland, G. K., H. Kryvi, and E. Slinde. 1978. LDH isozymes in the axial muscle of the sharks *Galeus melastomus* and *Etmopterus spinax. J. Fish Biol.* 12: 45–50.

Totland, G. K., H. Kryvi, Q. Bone, and P. R. Flood. 1981. Vascularization of the lateral muscle of some elasmobranchiomorph fishes. *J. Fish Biol.* 18: 223–34.

Van Vleet, E. S., S. Candileri, J. McNellie, S. B. Reinhardt, M. E. Conkright, and A. Zwissler. 1984. Neutral lipid components of eleven species of Caribbean sharks. *Comp. Biochem. Physiol.* 79B: 549–54.

Wunderer, H. 1908. Über terminal Körperchen der Anamnien. *Arch. Mikroskop. Anat. Entwicklungmechanik* 71: 504–69.

Yancey, P. H., and G. N. Somero. 1979. Counteraction of urea destabilization of protein structure by methylamine osmoregulatory compounds of elasmobranchs. *Biochem. J.* 183: 317–23.

———. 1980. Methylamine osmoregulatory solutes of elasmobranch fishes counteract urea inhibition of enzymes. *J. Exp. Zool.* 212: 205–13.

Zammit, V. A., and E. A. Newsholme. 1979. Activities of enzymes of fat and ketone body metabolism and effects of starvation on blood concentrations of glucose and fat fuels in teleosts and elasmobranchs. *Biochem. J.* 184: 313–22.

SUSANNE HOLMGREN
STEFAN NILSSON

CHAPTER 6

Digestive System

The digestive system uses mechanical and chemical processes to break down, transport, and absorb ingested food in order to provide nutrients for the living animal. Among vertebrates, the anatomy and the physiological functions of the digestive system follow the same basic plan, but variations are plenty and often may be correlated to the type of food preferred by the animal.

The most common image of a feeding elasmobranch is that of the chasing predator, viciously tearing into swimmers and surfers. However, elasmobranchs, like other fish groups, comprise species representing a wide variety of feeding patterns, ranging from hunting carnivores and omnivorous scavengers to ram-feeding plankton-eaters. Sharks are often opportunistic in their feeding behavior and have a reputation for eating anything that can possibly fit into their mouths, such as fishing gear, clothes, and furniture (!) (e.g., Lineweaver and Backus 1970; Keenleyside 1979; Yano and Musick 1992). On the other hand, whale sharks, basking sharks, and manta rays, which are among the largest fishes in the world, are planktivores (Keenleyside 1979). Skates are often benthic and live on small prey within or close to the bottom.

In general, herbivores and detritus eaters have a longer intestine than carnivores (see Stevens 1988). However, although we find all types of eating habits among elasmobranchs, the elasmobranch intestine is always short. Instead, as is described in more detail below, elasmobranchs use their special feature of a spiral intestine to increase the area (and time) for enzymatic treatment and absorption of the food.

General anatomy

The alimentary canal in vertebrates may be divided into a headgut, a foregut, a midgut, and a hindgut (Bertin 1958a; Stevens 1988). In elasmobranchs the orobranchial cavity forms the headgut. The foregut, by definition, ends with the pyloric sphincter between stomach and intestine and includes the esophagus and the stomach. The midgut comprises the parts of the intestine that absorb carbohydrates, proteins, and fat, while the hindgut consists of the water-absorbing parts. In elasmobranchs, the midgut consists of a proximal intestine (or duodenum) followed by the spiral intestine (ileum). The proximal intestine is usually short, but in some species it can reach considerable length. There is no clear border to the hindgut, which consists of a rectum and terminates with a cloaca (Pernkopf and Lehner 1937; Jacobshagen 1937) (Fig. 6-1).

MOUTH AND PHARYNX

The orobranchial cavity comprises the mouth and pharynx. The mouth and the teeth are the structures that probably most clearly reflect the feeding habits of a particular species. The teeth are mainly used for capture and holding of food, rather than chewing. Disjointing of large prey is effected through tearing by vigorous thrashing of the body.

The pharynx (or branchial cavity) possesses five pairs of lateral gill slits in most species, and six or seven pairs in Hexanchiformes. The spiracles are a paired opening in front of the gill slits. A pharyngeal sieving of particulate food has been observed in the southern stingray, *Dasyatis americana* (Stokes and Holland 1992). The plankton-feeding basking sharks, *Cetorhinus maximus*, and whale sharks, *Rhincodon typus*, have exceptionally large gill slits, large mouth, and long gill rakers, which trap the food (Nelson 1984). In the basking shark, branched papillae, which become progressively larger posteriorly, cover the mucosa of the pharynx and esophagus and aid in the catching of food (Matthews and Parker 1950).

There are no reports on secretion of digestive enzymes in this part of the alimentary canal in elasmobranchs, but glandular cells producing mucus frequently occur. The pharyngeal epithelium of *Mustelus schmitti* secretes sulfated and carboxylated acid mucosubstances (Galindez 1992).

ESOPHAGUS

The esophagus is a short, broad tube between the gill slits and the stomach. There are no "demarcation zones" such as the upper and lower esophageal sphincters in mammals, but the esophageal wall easily can be distinguished histologically from that of the stomach (Fig. 6-2). The muscle layers of the esophagus contain striated muscle. The mucosal layer forms folds, which allow distention during swallowing. There are no enzyme-producing or acid-producing glands in the mucosa, but strongly sulfated mucus is secreted (Galindez 1992). The mucosa is often provided with cone-shaped or branched papillae directed anally. Many species have an "organ of Leydig," consisting of masses of lymphomyeloid tissue in the submucosa (Fänge and Grove 1979; Estecondo et al. 1988).

STOMACH

The stomach is used for storage and initial digestion of the food. The elasmobranch stomach is a large, J-shaped organ with a "descending" cardiac part and an "ascending" pyloric part (see Fig. 6-1). In some species there is a chamberlike enlargement of the pyloric part of the stomach, the bursa entiana, which adds to the storage space (Matthews and Parker 1950). Chimaeroids lack a stomach (Fänge and Grove 1979; Kobegenova 1992).

The cardiac part of the stomach contains simple tubular glands perpendicular to the mucosal surface (see Fig. 6-2). The gland cells are of one type only. They contain acidophilic granules, which indicates that they are oxyntic (acid-secreting) (Hogben 1967a,b). Since there are no other types of gland cells present and since antibodies raised against chicken pepsinogen react with these cells (Yasugi et al. 1988), it may be presumed that the cells are actually oxynticopeptic, producing

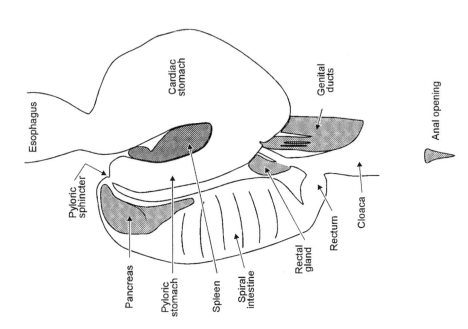
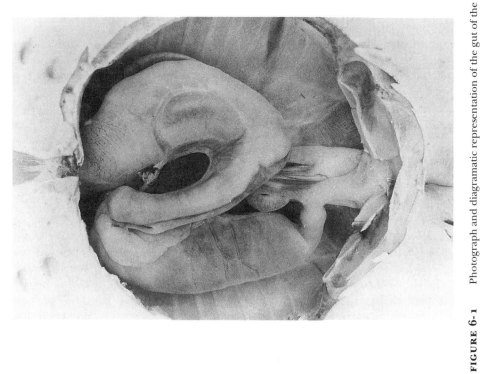

FIGURE 6-1 Photograph and diagramatic representation of the gut of the skate *Raja clavata*. The body wall has been dissected away and the liver removed.

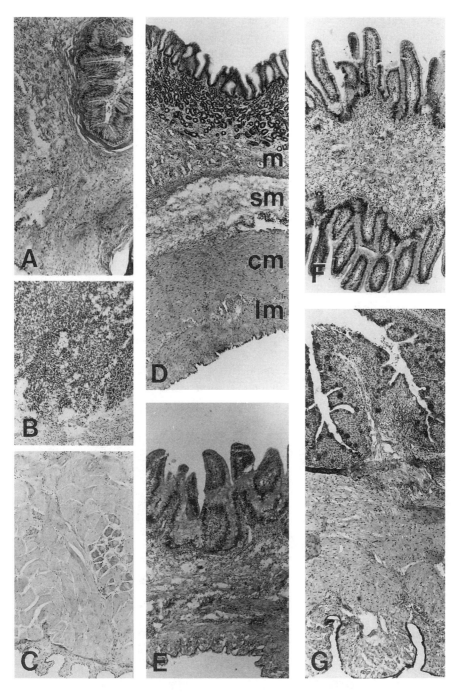

FIGURE 6-2 The morphology of the gut wall of *Raja clavata* at different levels. **A, B,** and **C** are (not continuous) sections from the thick wall of the esophagus: **A:** Mucosa. **B:** Organ of Leydig. **C:** Outer layer of striated muscle. **D:** Cross section of cardiac stomach wall; m = mucosa; sm = submucosa; cm = circular muscle; lm = longitudinal muscle. **E:** Cross section of the wall of the spiral intestine. The same layers as indicated in **D** can be seen. **F:** Cross section of a valve of the spiral intestine. The valve is built of mucosa and submucosa only. **G:** Cross section of the wall of the rectum. The same layers as indicated in **D** can be seen (×40).

both gastric acid and pepsinogen, as in teleosts. This is further confirmed by the ultrastructural similarity of the cells to oxynticopeptic cells in other nonmammalian vertebrates (Rebolledo and Vial 1979).

INTESTINE

Elasmobranchs have a short intestine compared to most other animals (see Fig. 6-1). Instead, enlargement of the surface area is achieved by the presence of a spiral fold of mucosa and submucosa along a major portion of the length of the intestine. The proximal part of the intestine, the duodenum, extends from the pyloric sphincter to the spiral intestine. In most species the duodenum is very short and merely forms a chamber anterior to the first turn of the spiral valve. This is where the bile duct and pancreatic ducts open into the intestine. One or a few large appendages (pyloric ceca) are present in a few species (Pernkopf and Lehner 1937) (Fig. 6-3).

There are large variations in the anatomy of the spiral fold (Fig. 6-4). Basically, four different types have been described: a spiral winding around a central column and attached to the outer wall (Fig. 6-4A); a series of interconnecting cones directed posteriorly (Fig. 6-4B); a series of interconnecting cones directed anteriorly (Fig. 6-4C); and a scroll valve with the central border free and the outer border attached along the length of the intestinal wall (Fig. 6-4D). The number of turns of the spiral reflects the diet and can be as few as two or three, as in the holocephalan *Chimaera monstrosa*, or as many as fifty, as in *Cetorhinus maximus* (Rauthner 1940; Bertin 1958a). The fold delays the passage of digesta, thus increasing the time allowed for digestion at the same time as it provides an increased surface for absorption.

RECTUM AND CLOACA

The spiral intestine tapers off into a distal part of the intestine with a thicker muscular wall and a stratified epithelium, the rectum (see Fig. 6-1). The rectum empties into the cloaca, into which the urinary and genital ducts also open. A particular feature of the elasmobranchs is the rectal gland, which opens

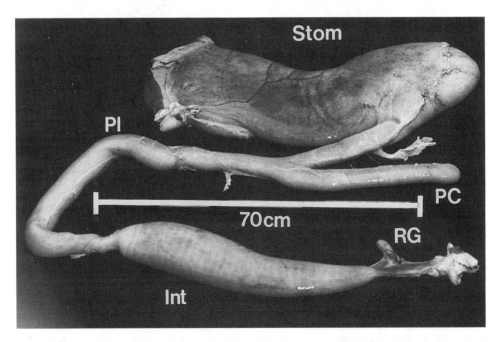

FIGURE 6-3 The gut of *Somniosus microcephalus*. Note the large pyloric cecum and the long proximal intestine (duodenum). Scale bar: 70 cm. Int = spiral intestine; PC = pyloric cecum; PI = proximal intestine; RG = rectal gland; Stom = (cardiac) stomach. (Photograph courtesy of Professor Ragnar Fänge, Göteborg.)

into the intestine-rectum posterior to the spiral fold (see Fig. 6-1). The rectal gland is involved in osmoregulation (see chap. 13).

THE GUT WALL
The arrangement of the layers of the gut wall follows that of fish in general (see Fig. 6-2, also Fänge and Grove 1979), which is also essentially the same as for other vertebrates. Along the length of the alimentary tract, from the stomach to the rectum, there is a coat of smooth muscle, the muscularis externa, beneath the outer epithelial layer, the serosa. The outermost part of this muscle coat consists of one or two thin layers of longitudinally oriented smooth muscle fibers. The inner,

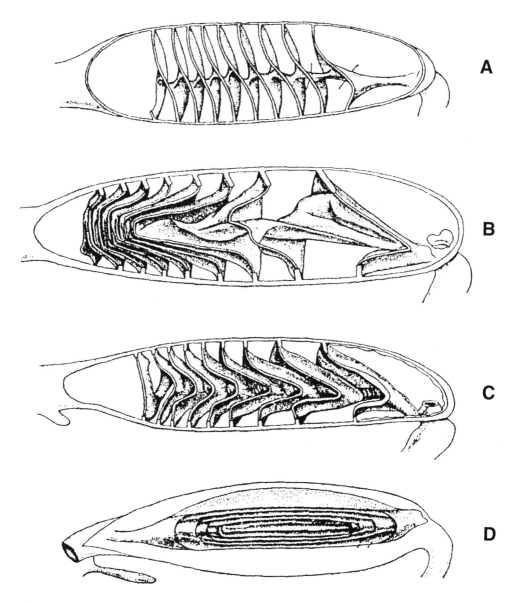

FIGURE 6-4   Examples of the arrangement of the spiral valve of the elasmobranch intestine. Columnar spiral (**A**), funnels pointing backward (**B**), funnels pointing forward (**C**) (all from *Raja* species), and cylindrical valve (scroll valve) from *Zygæna malleus* (**D**). (From Bertin 1958a, after Parker 1885.)

circular, layer varies in thickness. It is comparatively thin in the spiral intestine, thicker in other parts of the intestine and the rectum, and very well developed in the stomach, particularly the pyloric part. The muscle layers of the esophagus are formed from striated muscle fibers. In some species, such as the spiny dogfish, *Squalus acanthias*, the striated muscle layer extends into the upper part of the stomach.

The luminal surface of the whole gastrointestinal canal is lined by a mucosal layer. The mucosal epithelium may be ciliated in the esophagus. The epithelium of the stomach is simple and consists of cylindrical cells that probably secrete mucin. Multicellular glands formed by oxyntic (oxynticopeptic) cells open into foveolae in the stomach mucosa (Hogben 1967a,b). The mucosal epithelium of the intestine is simple and columnar and contains several enterochromaffin cells. The multicellular secreting structures form crypts rather than proper glands as in tetrapods. The mucosa of the posterior intestine is stratified (Bertin 1958a).

Between the mucosa and the muscle layers of the gut wall, there is a thick submucosal layer of connective tissue and larger vessels. A thin muscle layer, the muscularis mucosae, is present below the mucosa. The spiral fold of the intestine is formed by mucosa covering a central layer of submucosa.

VESSELS OF THE GUT

Arterial blood reaches the gut from the dorsal aorta via several arteries (Fig. 6-5). The most anterior, the celiac artery, branches off the dorsal aorta a short distance from the most posterior pair of gill arches and supplies the stomach and liver regions. Several mesen-

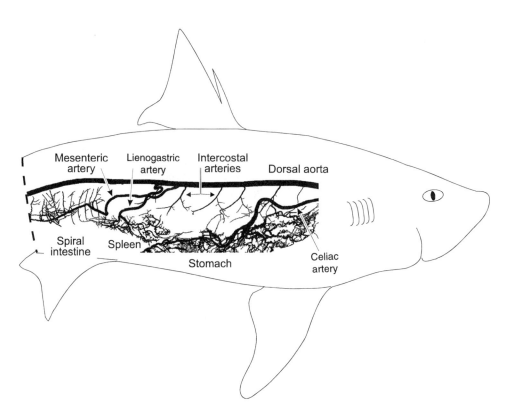

FIGURE 6-5 Scanned image of an acrylic vascular cast of the visceral blood vessels in the spiny dogfish, *Squalus acanthias*. The vasclar beds of the celiac, the lienogastric and the mesenteric arteries are outlined. (Courtesy of Dr. Michael Axelsson, Göteborg.)

teric arteries supplying the midgut and hindgut branch off more posteriorly along the length of the dorsal aorta, the number depending on species (Bertin 1958b; Holmgren et al. 1992a).

Doughnut-shaped sphincters are described around veins in the stomach wall (and elsewhere) of *Raja* (Weinland 1901), but their function is not clear. As in all other vertebrates, the veins of the elasmobranch gut fuse into a hepatic portal system (Bertin 1958b).

Innervation of the gut
EXTRINSIC NERVES
The elasmobranch gut is innervated by extrinsic autonomic nerves (Fig. 6-6). Cranial autonomic ("parasympathetic") nerve fibers run in the intestinal branches of the vagus to the stomach and anterior intestine. These fibers are preganglionic and synapse on ganglion cells within the distal parts of the nerves or within the stomach wall (enteric neurons). Spinal autonomic ("sympathetic") fibers reach the gut via splanchnic nerves. The anterior splanchnic nerve runs along the celiac artery and innervates an area similar to that of the intestinal vagus. It carries postganglionic fibers emanating from ganglion cells in the most anterior pair of paravertebral ganglia, the axillary bodies, and preganglionic fibers from the spinal cord. A midsplanchnic nerve arises from the middle paravertebral ganglia, runs along the anterior mesenteric artery, and innervates most parts of the spiral intestine. Posterior splanchnic nerves, arising from the posterior paravertebral ganglia, innervate the posterior part of the spiral intestine and the rectum.

The fibers reaching the gut may act directly on effector units (muscle, blood vessels, glands) of the gut, but the major influence, particularly of the vagal fibers, is probably via intrinsic nerves of the enteric nervous system. The anatomy and functions of the nervous system of the elasmobranch gut have been extensively reviewed by Müller and Liljestrand (1918), Müller (1920), Young (1933), and Nicol (1952) and more recently by Nilsson (1983), Nilsson and Holmgren (1988), and Jensen and Holmgren (1994).

INTRINSIC NERVES
The intrinsic nerves of the gut form the enteric nervous system. The elasmobranch enteric nervous system follows the general arrangement of all vertebrate groups, but the overall density of ganglion cells, nerve plexuses, and nerve fibers penetrating muscle layers and mucosa is less than in tetrapods (Jacobshagen 1937; Kirtisinghe 1940). A myenteric plexus is present in the thick tunica intermuscularis between the longitudinal, and circular muscle layers. The myenteric plexus is ganglionated and well developed. The innervation of the outer, longitudinal, muscle layer(s) is sparse, with only few nerve fibers penetrating the thin muscle layer, and the main control may be executed by the parts of the myenteric plexus adjacent to the longitudinal muscle. In the circular muscle layer, thick bundles of nerve fibers follow vessels and connective tissue between the muscle bundles. Ganglion cells commonly occur in these bundles.

NEUROTRANSMITTERS
The autonomic neurons function by the release of neurotransmitters. In mammals, each neuron normally contains more than one neurotransmitter, and different combinations of neuropeptides, catecholamines, or acetylcholine, and other nonpeptide transmitters occur. So far, there are no studies of the coexistence of transmitters in elasmobranchs, but the transmitters are very likely to occur in combinations in this group also.

The presence of catecholamines in nerves of the elasmobranch gut has been demonstrated by fluorescence histochemistry in *Squalus acanthias* (Holmgren and Nilsson 1983) (Fig. 6-7). Euler and Fänge (1961), using the same species for biochemical analyses of catecholamine levels in the gut, showed that noradrenaline levels (330 ng/g tissue) were higher than adrenaline levels (30 ng/g tissue).

Cholinergic nerves are difficult to identify histochemically, and this has so far not been done in elasmobranchs, but pharmacological studies suggest a cholinergic control of gastric secretion (Hogben 1967a,b). A

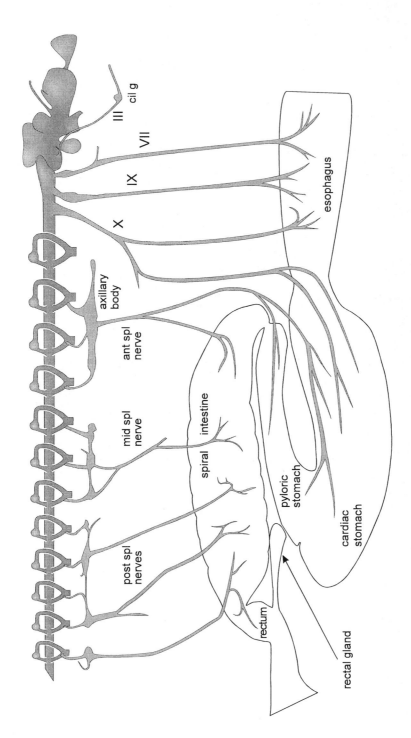

FIGURE 6-6 Summary of the extrinsic autonomic innervation of the elasmobranch gut. Chromaffin tissue is present in all paravertebral ganglia, especially in the axillary bodies. Cil g = ciliary ganglion; Ant, Mid, and Post spl nerve(s) = anterior, middle, and posterior splanchnic nerves, respectively; III = oculomotor nerve; VII = facial nerve; IX = glossopharyngeal nerve; X = vagus nerve. (Redrawn from Nilsson 1983.)

cholinergic innervation of the gut musculature, as seen in other vertebrates, is less certain (cf. Nilsson 1983).

Immunohistochemistry, often in combination with radioimmunoassay, has been used to demonstrate a number of neurotransmitters within the enteric neurons. These include 5-hydroxytryptamine (serotonin) and several neuropeptides (Table 6-1; see Fig. 6-7). In some cases the exact amino acid sequences of

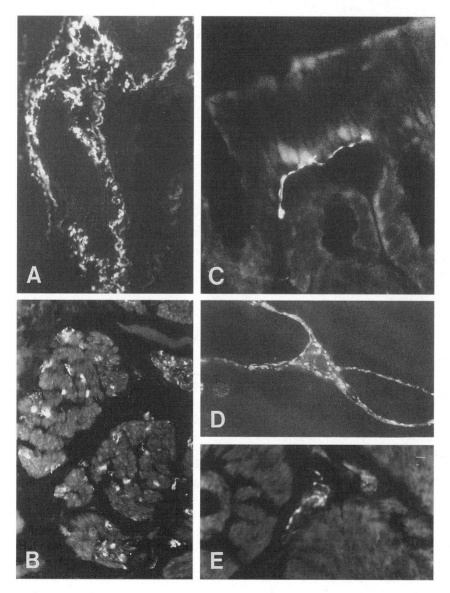

FIGURE 6-7 **A:** Catecholamines visualized by the Falck-Hillarp method in the wall of vessels running in the stomach wall of *Squalus acanthias*. Figures **B–E** show immunohistochemistry of neuropeptide-containing nerves in the gut wall. **B:** Neuropeptide Y–like immunoreactivity in nerve bundle innervation of smooth muscle bundles of the circular muscle layer in *Raja clavata* (×200). **C:** Vasoactive intestinal petpide (VIP)-like immunoreactivity in nerve fiber innervation the mucosa of the cardiac stomach in *Raja clavata* (×400). **D:** VIP-like immunoreactivity in nerve fibers in the myenteric plexus of the spiral intestine of *Squalus acanthias* (×200). **E:** Phe-met-arg-phe amide (FMRF)–like immunoreactivity in nerve fibers innervating circular smooth muscle of the rectum of *Raja clavata* (×200).

TABLE 6-1 LOCALIZATION OF NEUROPEPTIDE-LIKE AND
5-HYDROXYTRYPTAMINE–LIKE IMMUNOREACTIVITIES
IN NERVES OF THE GUT OF ELASMOBRANCH FISH

Species	SP	VIP	BM/CGRP	CCK/G	FMRF	5-HT	NPY	SST	References
Squalus acanthias	S,I,R	S,I,R	S,I,R	S,I,R		S,I,R		S,I,R	1,2,3,4
Squatina aculeata		S,I							5
Scyliorhinus canicula		S,I					S,I		5,6
Scyliorhinus stellaris	S,I,R	S,I,R	S,I,R	S	S,I,R	S		S	7,8,9
Raja clavata	S,R	S,I,R	S,I,R		S		S,I,R		10,11,12
Raja erinacea		S,I,R					S,I,R		13,14
Raja microocellata			S,I,R					S,I,R	11
Raja montaguii		S,R					S,R		11
Raja naevus		S,I,R					S,I,R		11
Raja radiata		I					S,I		10,13
Raja rhina		S,I,R							14
Chimaera monstrosa		I		I					15

SOURCES: (1) Reinecke et al. 1981; (2) Holmgren and Nilsson 1983; (3) El-Salhy 1984; (4) Holmgren 1985; (5) Tagliafierro et al. 1988a; (6) Bjenning et al. 1993; (7) Cimini et al. 1985; (8) Cimini et al. 1989b; (9) Tagliafierro et al. 1988b; (10) Falkmer et al. 1980b; (11) Bjenning and Holmgren 1988; (12) S. Holmgren, unpublished; (13) Bjenning et al. 1989; (14) Bjenning et al. 1991; (15) Yui et al. 1990.

BM/CGRP = bombesin or gastrin-releasing peptide; CCK/G = cholecystokinin or gastrin; FMRF = phe-met-arg-phe amide; 5-HT = 5-hydroxytryptamine (serotonin); I = intestine; NPY = neuropeptide Y; R = rectum; S = stomach; SP = substance P; SST = somatostatin; VIP = vasoactive intestinal peptide.

elasmobranch neuropeptides have been determined, and somatostatin-14 from *Torpedo marmorata* is identical to somatostatin-14 in most other investigated species (Conlon et al. 1985). Usually, however, there are some variations in amino acid sequence of neuropeptides from elasmobranchs compared with other animals. The substitutions are commonly conservative, occur mainly in the less biologically active parts of the molecule, and are more frequent the less related the species are. Thus, the somatostatin-14 in the Pacific ratfish, *Hydrolagus colliei*, has a substitution at position 5, which is considered to have little effect on the biological activity (Conlon 1990). A gastrin-releasing peptide (GRP) from shark (Conlon et al. 1987) and neuropeptide Y (NPY) from dogfish (*Scyliorhinus canicula, Squalus acanthias*) and ray (*Torpedo marmorata*) have been sequenced (Conlon et al. 1991, 1992; Blomqvist et al. 1992; Pan et al. 1992) and are strongly conserved in the C-terminal region, while dogfish vasoactive intestinal polypeptide (VIP), isolated from *Squalus acanthias* and *Scyliorhinus stellaris*, shows a high degree of conservation in the N-terminal part (Dimaline et al. 1987). Two different tachykinins, scyliorhinin I and scyliorhinin II, are isolated from elasmobranchs (Conlon et al. 1986; Conlon and Thim 1988) and resemble physalaemin and neurokinin A, respectively. Receptor binding studies indicate the presence of two types of tachykinin receptors in the gut of *Scyliorhinus canicula*, one binding substance P and being similar to the mammalian neurokinin (NK_1) receptor while the other binds scyliorhinin II and does not resemble any mammalian tachykinin receptor (Van Giersbergen et al. 1991).

Gut endocrine cells and hormones

The so-called pancreatic hormones (insulin, glucagon, pancreatic polypeptide, and somatostatin) often occur in endocrine cells of the gut mucosa as well (Falkmer and Stefan 1978;

Holmgren and Nilsson 1983; El-Salhy 1984; Reinecke et al. 1984; Tagliafierro et al. 1985a; Cimini et al. 1989b; Yui et al. 1990). In addition, immunohistochemistry has shown material similar to neurotransmitters in endocrine cells. These substances include 5-hydroxytryptamine and peptides related to GRP and bombesin, gastrin and cholecystokinin (CCK), neurotensin, neuropeptide Y, substance P, and VIP (Falkmer et al. 1980b; Reinecke et al. 1981; Holmgren and Nilsson 1983; El-Salhy 1984; Holmgren 1985; Cimini et al. 1989a, 1992; Yui et al. 1990). There are occasional reports on the presence of calcitonin-, endorphin-, enkephalin-, phe-met-arg-phe amide (FMRF)–, gastric inhibitory peptide (GIP)–, glicentin-, insulinlike growth factor (IGF)-1–, peptide histidine isoleucine (PHI)–, and peptide YY (PYY)–like peptides and histamine (Falkmer et al. 1980a; El-Salhy 1984; Håkanson et al. 1986; Cimini et al. 1989b; Yui et al. 1990; Reinecke et al. 1992).

As in neurons, the active substances in endocrine cells coexist in certain combinations. There are, however, only a few histological studies of this condition in elasmobranchs (Cimini et al. 1989a, 1992), and no physiological correlations can be drawn.

Gut secretion and digestion

Studies of the functions of the digestive system in elasmobranchs are scarce, and comparisons to other fish and to mammals are necessary for the interpretation of the available scattered information. For detailed recent reviews of gut secretion in mammals, readers are referred, for example, to Debas (1987), Forte and Wolosin (1987), Soll and Berglindh (1987), and Tache (1987). Extensive comparative surveys of studies in nonmammals are given by Smit (1968), Jönsson and Holmgren (1989), and Jönsson (1994).

GASTRIC SECRETION
Elasmobranchs produce large amounts of distinctly acid fluid in association with food intake. The measured pH varies, but values as low as 1.69 in *Scyllium stellare* are reported (Van Her-werden and Ringer 1911). In unfed *Raja*, a gastric pH of 3–3.8 was measured (Babkin et al. 1935a,b). The acidity is at its greatest when the stomach is filled with semidigested crustaceans and may remain high for several days after the stomach is emptied (Sullivan 1905; Dobreff 1927). The gastric fluid in the absence of food is usually less acid or even neutral.

The initial chemical digestion in the stomach is performed by acid proteases. Pepsin is the major acid protease in fish. Four different pepsinogens were isolated from the mucosa of the dogfish *Mustelus canis* (Merret et al. 1969). Pepsinogen is secreted from the cardiac part of the stomach and is converted to pepsin at low pH. Pepsin has been found both in mucosal extracts and in gastric juice from several species of elasmobranchs (Sullivan 1905; Bodansky and Rose 1922; Dobreff 1927; Babkin and Friedman 1934). Chitinases and high chitinolytic activity are present in the gastric mucosa of, for example, *Etmopterus spinax* and *Raja radiata* (Micha et al. 1973; Fänge et al. 1979), which is important for the breakdown of shells of crustaceans and other invertebrates with a chitinous integument.

Mucus is produced by the epithelial cells lining the mucosa. This protects the mucosa against mechanical and chemical damage. The mucus produced in the stomach of *Mustelus schmitti* is a carboxylated mucopolysaccharide, probably a sialomucin (Galindez 1992).

Cellulases aid in the breakdown of wooden material and are present in the gut of several invertebrates. A weak cellulase activity has been demonstrated in the stomach of *Raja clavata*, but no activity was found in a number of other elasmobranch species investigated (Stickney and Shumway 1974; Lindsay and Harris 1980). This occurrence of a cellulase activity in *Raja* (and in some teleost species) has been attributed to ingested bacteria or even ingested cellulases.

Control of gastric secretion
The earlier studies of the control of gastric secretion have yielded confusing and contradictory results. It may be suspected that the

experimental protocols have disturbed the normal processes and have produced atypical responses. However, evidence so far suggests the presence of a basal acid secretion (*Raja* spp., Babkin et al. 1935a,b; *Scyliorhinus*, Hogben 1967a,b).

The secretion is stimulated by cholinergic drugs and to some degree by histamine (Ungar 1935; Hogben 1967a,b) (Fig. 6-8). Histamine is present in endocrine cells located predominantly at the base of the gastric glands and in a few mast cells in *Squalus acanthias* and *Raja clavata* (Håkanson et al. 1986). However, the low sensitivity to histamine and the unusually low transmembrane potential of the gastric mucosa suggest that the mechanisms of secretion of gastric acid in elasmobranchs are different from those in other vertebrates (Hogben 1967a).

Adrenergic drugs and stimulation of the sympathetic nervous system inhibit the acid secretion (Babkin et al. 1935a,b). This has been interpreted as an effect of reduced blood flow through the secreting tissue rather than an effect directly on the gland cells. Catecholamine containing nerve fibers are indeed present in the submucosa of *Squalus acanthias*, where they mainly have a perivascular distribution (Holmgren and Nilsson 1983).

Gastrin is a potent stimulator of acid secretion in mammals, and acid secretion in dogfish is stimulated similarly by porcine gastrin (Vigna 1983). Multiple gastrinlike peptides are present in endocrine cells of both stomach

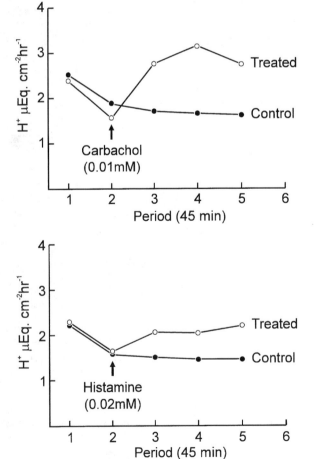

FIGURE 6-8 Effects of carbachol (*top*, means of seven pairs) and histamine (*bottom*, means of eight pairs) on acid secretion by isolated gastric mucosa from *Squalus acanthias*. Acid secretion was measured over five consecutive 45-minute periods. (Redrawn from Hogben 1967a.)

and intestine of *Squalus acanthias* and may provide a hormonal control of the secretion (El-Salhy 1984; Holmgren and Nilsson 1983; Aldman et al. 1989).

No nervous control of gastric secretion in elasmobranchs has been demonstrated so far.

INTESTINAL SECRETION

Secretion of digestive enzymes in the intestine occurs from epithelial cells of the intestinal mucosa and from the exocrine pancreas. Enzymes involved in membrane digestion (α-amylase, saccharase, dipeptidase, alkaline phosphatase) have been investigated in *Squalus acanthias*, *Raja clavata*, and *Dasyatis pastina*. In contrast to other fish and mammals, the proximal to distal decline in enzymatic activity is small, except for alkaline phosphatase. The activity at low temperatures ($0°$–$20°C$) of the enzymes from the dogfish is approximately double that of the thornback ray, which possibly reflects an adaptation to a life in cold waters. The pH-optimum of alkaline phosphatase is in the same range as in other animals, around 10 (Kuz'mina 1990).

PANCREAS

The pancreas is thought to have evolved from a primitive stage in which a discrete exocrine pancreas is absent and the "islet hormones" occur in endocrine cells diffusely spread throughout the intestinal mucosa. A proper pancreas with an acinar exocrine function (an exocrine pancreas) and with islets of endocrine cells occurs for the first time in holocephalans. It is attached to the spleen and connects to the gut via a duct several centimeters long (Falkmer and Van Noorden 1983; Falkmer et al. 1981, 1984). The bilobed pancreas in sharks and rays is also a discrete large gland (see Fig. 6-1) with both exocrine and endocrine functions, but it remains closely associated with the gut wall in its proximal part. The short exocrine duct opens into the proximal part of the intestine (the duodenum). The endocrine cells occur in islets of 5 to 20 cells (Fig. 6-9) (Falkmer and Van Noorden 1983; Falkmer et al. 1981, 1984; Thorndyke and Falkmer 1985).

The exocrine function of the pancreas

The exocrine pancreas (see Fig. 6-9) produces digestive enzymes (in an inactive form) that are released into the duodenum when food enters the intestine. There is no complete survey of the types of enzymes produced by the elasmobranch pancreas. The fragmentary information available confirms the presence of peptidases such as trypsin (or rather, the inactive form, trypsinogen) in *Chimaera monstrosa*, *Ginglymostoma cirratum*, and *Squalus suckleyi* (Nilsson and Fänge 1969; Zendzian and Barnard 1967), elastase (or proelastase) in *Chimaera monstrosa* and *Dasyatis americana* (Nilsson and Fänge 1969; Zendzian and Barnard 1967) and carboxypeptidases in *Squalus acanthias* (Prahl and Neurath 1966a,b). Colipase, which is needed for activation of lipases in the presence of bile salts, is found in both sharks and rays (Sternby et al. 1983), and sequence analysis of colipase from *Squalus acanthias* reveals considerable homology with mammalian colipases in the enzymatically active N-terminal, although the isoelectric point of the enzyme is unusually high (Sternby et al. 1984). Chitinases, which are common in the digestive system of animals that feed on insects or crustaceans, are found at high activity levels in the pancreas of *Chimaera monstrosa* (Fänge et al. 1979).

Secretion of pancreatic juices from the exocrine pancreas could be initiated by introduction of hydrochloric acid into the anterior intestine of *Raja* (Babkin 1929, 1933). This probably involves a hormonal link, as in mammals, where acid in the intestine stimulates release of hormonal secretin and CCK, which in turn stimulate release of thin, watery, bicarbonate-rich pancreatic juice and enzyme-rich pancreatic juice, respectively. Secretin has been extracted from the intestinal wall of elasmobranchs (Bayliss and Starling 1903; Nilsson and Fänge 1969). Endocrine cells containing CCK-like peptides are present in the intestinal wall of elasmobranchs (Holmgren and Nilsson 1983; El-Salhy 1984; Yui et al. 1990).

Scattered NPY- and somatostatin-immunoreactive cells are found in the exocrine tissue

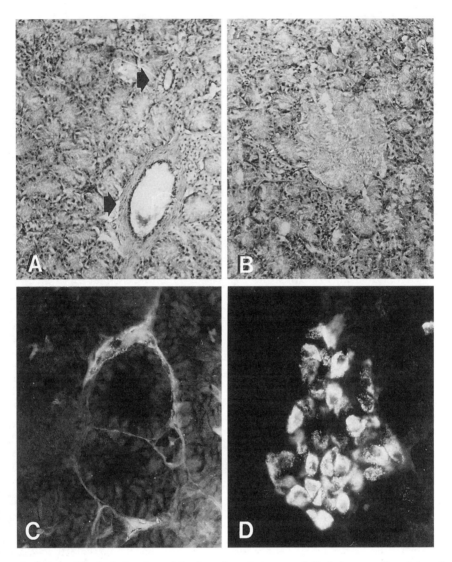

FIGURE 6-9 Histological sections of the elasmobranch pancreas. **A:** Eosin-hematoxylin staining of exocrine tissue from *Raja radiata*. *Upper arrow,* exocrine duct; *lower arrow,* small artery (×100). **B:** Eosin-hematoxylin staining of section showing islet of endocrine tissue within the exocrine pancreas of *Raja radiata* (×100). **C:** Nerve fibers showing galaninlike immunoreactivity surrounding acini of exocrine cells in the pancreas of *Raja radiata* (×250). **D:** Cluster of endocrine cells showing neuropeptide Y (NPY)-like immunoreactivity in the pancreas of *Squalus acanthias* (×250). (Figures courtesy of Dr. Ann-Cathrine Jönsson, Göteborg.)

and along the ducts in several species (Tagliafierro et al. 1985b; Pan et al. 1992), and occasional IGF-1– and gastrin-and-CCK–immunoreactive cells are found along the ducts in *Squalus acanthias* and *Raja* species (Jönsson 1991; Reinecke et al. 1992); these cells may be involved in the local control of the exocrine pancreas. In addition, autonomic nerves innervate the exocrine tissues. Galanin, bombesin-GRP– and VIP-immunoreactive nerves are present in *Squalus acanthias* and *Raja radiata* (Jönsson 1991) and GRP-immunoreactive nerves in *Chimaera monstrosa* (Yui and Fujita 1986). There are no studies of the effects of these peptides in elasmobranchs, but dogfish VIP stimulates exocrine pancreatic secretion in turkey (Dimaline and Thorndyke 1986).

The endocrine function of the pancreas
Clusters of 5 to 20 cells form islets of endocrine tissue within the exocrine pancreas. All the four types of endocrine islet cells that occur in most vertebrates (insulin, glucagon, pancreatic polypeptide, and somatostatin) have been demonstrated in different shark and dogfish species (Stefan et al. 1978; El-Salhy 1984; Tagliafierro et al. 1985b; El-Salhy et al. 1987; Faraldi et al. 1990; Jönsson 1991). With the exception (so far) of pancreatic polypeptide, similar cells have been found in rays and holocephalans (Tagliafierro et al. 1985b; Conlon et al. 1985; Jönsson 1991). There is in most cases no specific regional distribution of the cells within the islets. In addition, FMRF-like, gastrin- or CCK-like, GIP-like, IGF-1–like, NPY-like (see Fig. 6-9), and neurotensinlike immunoreactivities have been reported present in populations of endocrine islet cells of elasmobranchs (El-Salhy 1984; Tagliafierro et al. 1985b; Jönsson 1991; Cimini et al. 1992; Pan et al. 1992; Reinecke et al. 1992). There are, however, no conclusive reports on the possible coexistence of these peptides with each other or with the "classical" islet hormones, and their functions can so far only be guessed from comparison with studies in mammals.

Galanin- and VIP-immunoreactive nerves have been demonstrated in endocrine tissue in *Squalus acanthias* and *Raja radiata* (Jönsson 1991), but there is so far no indication of their function.

LIVER AND GALLBLADDER

The elasmobranch liver is a large organ and fills a considerable portion of the body cavity. Pelagic species have a liver weight ranging between 6.5% and 23% of their body weight, while bottom dwelling species range between 1% and 6% (Bone and Roberts 1969; Rossouw 1987). In the basking shark, *Cetorhinus maximus*, the liver constitutes about a quarter of the body weight and may weigh as much as a ton (Matthews and Parker 1950). Squalene and other lipids accumulate in large fat vacuoles in the liver cells and may constitute as much as 80% of the liver volume (Baldridge 1972). These fats are an important source of fuel for muscle activity (Rossouw 1987) and also affect the buoyancy of the fish. In a description of basking-shark fishery, Morris (Went and O'Suilleabhain 1967) notes that as the liver was cut out "the fish immediately fills with water and loses its buoyance, and the ropes must be slipped with great dexterity or it would bring down the boat."

Squalene has a specific gravity of 0.86 g/cm^3 at 20°C (Tsujimoto 1916), and in pelagic species it may amount to 90% of the total oil content, which gives the fish neutral or almost neutral buoyancy (Bone and Roberts 1969).

Bile secreted from the liver is collected into a large gallbladder. The production of bile by the liver is continuous in the ray *Raja erinacea* and the dogfish *Squalus acanthias*. The amount produced (1–1.5 µl/kg/min) appears to be dependent on the portal vein blood pressure and is approximately 1% of the amount produced by a rat (Boyer et al. 1976). The dogfish gallbladder epithelium functions like that in a mammal, concentrating bile acids in micelles and reducing the levels of Cl^- and HCO_3^-, while the gallbladder epithelium of the ray is less active (Boyer et al. 1976). In contrast to teleosts and all tetrapods, where bile acids form taurine salts, the bile acids from elasmobranchs (and cyclostomes and dipnoans) form alcohol sulfate esters (Haselwood 1968).

Control of the release of bile from the elasmobranch gallbladder by a cholecystokininlike peptide, as occurs in other vertebrates, is likely, since peptides from the gastrin and CCK family contract the gallbladder in *Raja* species (Andrews and Young 1988b). Gastrin- or CCK-related peptides are present in endocrine cells of the gut of *Squalus acanthias* (Holmgren and Nilsson 1983), and CCK-like biological activity of these peptides has been demonstrated in extracts of the gastric mucosa from *Rhinobatos productus* and *Squalus acanthias* (Hansen 1975; Vigna 1979).

Gut absorption

After appropriate digestion, nutrients are absorbed from the intestine via the mucosa by active and passive mechanisms. Extremely little is known about similarities and differences between elasmobranchs and other vertebrate groups in this matter.

Gut motility and food transport

The few studies made on the amount of food consumed in elasmobranchs show that about 1–2% of the body weight is consumed per day, which is in the lower range of food consumption for fish in general (0.1–28%) (Jones and Genn 1977; Medved et al. 1988; Cortés and Gruber 1990). One single meal may be much larger (more than 10% of the body weight in *Scyliorhinus canicula*) but then the emptying of the stomach takes a longer time, corresponding to 1–2% per day (S. Stead and D. J. Grove, personal communication).

Transport of food through the fish gut is dependent on several interacting parameters, such as digestion rate, the coordinated function of the smooth muscle of the gut, and environmental factors such as temperature and stress. Meal size (total size and size of individual food items), digestibility of the food (relative amounts of fat, proteins, and carbohydrates, and presence of integuments and shells, for example) and previous feeding patterns all affect digestion rate (Fänge and Grove 1979). The presence of food in the esophagus and stomach also affect muscle activity, both by the mechanical stretching of the muscle wall and by initiating release of hormones.

GASTRIC EMPTYING TIME

Different methods have been developed to study transport of food through the gastrointestinal canal. Most studies have measured gastric evacuation rather than passage through the whole gut. Among fish, the bulk of studies have been made on teleosts (see Fänge and Grove 1979), and only a few consider elasmobranchs. The most common method has been to feed a group of fish and kill the individuals at different time intervals afterwards to measure the remaining food contents in the stomach. Other approaches have been to make the fish vomit or to evacuate the stomach contents with a pump. The more elegant methods involve roentgenographic studies at different time intervals (Fig. 6-10).

Several mathematical models for gastric evacuation of food in fish have been constructed. There is usually an initial lag phase from the entry of food to the stomach until emptying into the intestine starts. Following this initial delay, the amount of food in the stomach decreases; this decrease may be linear with time or may follow an exponential curve, a square root function, or a logistic growth function (Fänge and Grove 1979; Schurdak and Gruber 1989). The decrease in stomach contents after a meal follows a so-called Gompertz growth curve in the sandbar shark, *Carcharhinus plumbeus* (Medved 1985; Medved et al. 1988) and a linear function in the lemon shark, *Negaprion brevirostris* (Cortés and Gruber 1992). In the juvenile lemon shark, gastric evacuation is exponential (Schurdak and Gruber 1989). The time to total gastric evacuation in these species during the experiments, around 24 hours, is longer than for most teleosts during corresponding circumstances. Similarly, the time for total passage of a meal through the gut of the lemon shark (68–82 hours) is substantially longer than that for most teleosts (Wetherbee et al. 1987), and during laboratory conditions the time for total emptying of the stomach after a large meal may be as much as 10 days in *Scyliorhinus canicula* and up to 7 days in *Raja clavata* (T. de Sousa and D. J. Grove, personal communication).

Effects of temperature

The rate of food processing is dependent on temperature, and an increase in temperature of 10°C has been calculated to increase gastric emptying time by about 2.6 times ($Q_{10} = 2.6$) in fish (Backiel 1971; Jones 1974), provided the increase is within physiological limits. Being poikilothermic, the temperature of most fish depends on the ambient temperature and, not surprisingly, fishes from cold habitats

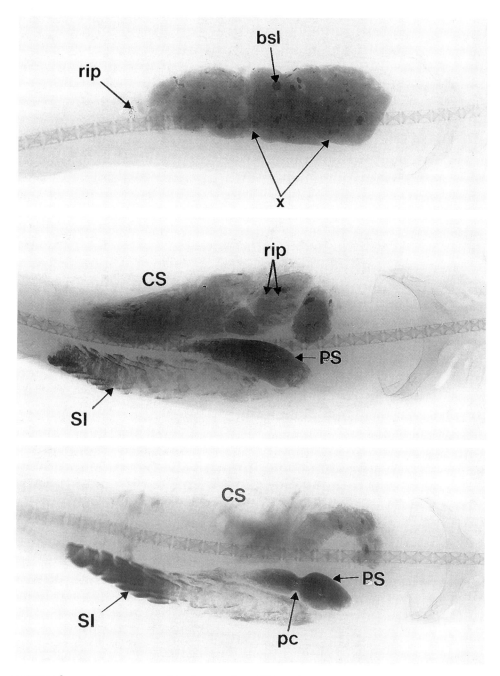

FIGURE 6-10 Roentgenographic photographs at different intervals after feeding the dogfish *Scyliorhinus canicula* an experimental meal of barium sulfate label (bsl) containing radiopaque material. The pictures show that food in the cardiac stomach (CS) is mostly digested in the posterior end, since the food in the anterior part remains relatively intact even after 47 hours of feeding (*middle photograph*). In the *upper photograph*, x indicates contracted areas of the cardiac stomach and rip indicates residual individual particles of label. A peristaltic contraction (pc) is evident in the pyloric stomach (PS), demonstrating intermittent gastric emptying of experimental meals (*lower photograph*). Temperature is 9.5 + 1.5°C. (Photographs courtesy of Dr. David J. Grove, School of Ocean Science, University College of North Wales, Menai Bridge.)

characteristically show longer food processing times than tropical species. However, laminid sharks have the ability to specifically increase the temperature of the gut region in connection with digestion of food, as a result of counter current systems of vessels forming large retia mirabilia in the gut region (Carey et al. 1981). The white shark, *Carcharodon carcharias*, may increase the stomach temperature as much as 7.4°C above the ambient water temperature (McCosker 1987). The reduced digestion time this produces is of importance for an animal that is an intermittent feeder of large meals and that depends on much swimming activity to chase its prey.

THE PHYSIOLOGY OF GUT MOTILITY
The movements of the gut are caused by coordinated contractions of the smooth muscle cells. The muscle layers have an intrinsic contractile activity; stimulatory impulses are initiated in certain areas and spread from there over large areas of muscle. Superimposed on this system is the control exerted by nerves and hormones. Together, the nerves control such actions of the gastrointestinal smooth muscle as peristalsis for propulsion of food, stationary contractions for mixing of stomach contents, receptive relaxation on arrival of food to the stomach, and opening and closing of the pyloric sphincter. The physiology of gut smooth muscle in general has been reviewed by Johnson (1987). Comparative reviews of the control of gut motility are made by Nilsson (1983), Holmgren (1989), and Jensen and Holmgren (1994).

The motility of the elasmobranch stomach, with its well-developed muscular layers, much resembles that of other vertebrate stomachs. Strong contractions alternate with periods of inactivity (Andrews and Young 1993). Considering the sometimes unorthodox feeding habits of elasmobranchs and the narrow pyloric sphincter, it may be fortunate that stomach contents may be ejected by strong retrograde peristalsis (Andrews and Young 1993) or, in other species, by gastric eversion (Keenleyside 1979). This latter behavior has recently been video filmed in a *Raja* species treated with an emetic substance (P. L. R. Andrews and J. Z. Young, personal communication).

The muscle wall of the intestine is thin, and observed spontaneous movements are slow and irregular, involving both elongation and shortening phases (Young 1933, 1983). Contraction waves lasting 1.5–2 minutes are reported in the intestine (Young 1983), while the rectum shows spontaneous contractions at a frequency of one per minute or less (Young 1988). A regular intestinal eversion leading to protrusion of the intestine through the cloaca is an unusual feature that has been observed among species with a scroll valve intestine. The physiological significance of this is unclear and, at least in captivity, this habit has proved to be fatal after attacks by hungry tankmates (Crow et al. 1990, 1991).

Nervous control of gut motility
Several types of nerve fibers are involved in the control of gut motility. Extrinsic nerves comprise fibers running in the vagus and in the splanchnic nerves. These fibers may act directly on the smooth muscle or may affect the action of the intrinsic neuronal plexuses. Intrinsic neurons form plexuses between the muscle layers and in the submucosa and penetrate the muscle layers. Immunohistochemistry has revealed the presence of several putative neurotransmitters in gut neurons of elasmobranchs (see Table 6-1) (Nilsson and Holmgren 1988; Jensen and Holmgren 1994), but it has seldom been established whether these are truly intrinsic or of extrinsic origin.

The vagus. The vagus nerve contains both cholinergic excitatory neurons and noncholinergic inhibitory neurons in most vertebrates. Stimulation of the vagal roots causes inhibition of the spontaneous rhythmic activity of the stomach of the dogfish *Scyliorhinus canicula*, which suggests the presence of inhibitory innervation (Campbell 1975) (Fig. 6-11). This would explain the inhibitory parts of the responses obtained on stimulation of the vagus trunk in several elasmobranch species in earlier studies (Botazzi 1902; Müller and Liljestrand 1918). The presence of an inhibitory innervation was further confirmed by Young (1980, 1983), although the results were less clear because of the experimental design, which recorded events from longitudinal muscle rather than circular muscle.

On the other hand, so far there is no clear evidence of an excitatory innervation in elasmobranchs that is truly vagal. Early investigations report excitatory effects of peripheral vagal stimulation (Botazzi 1902; Müller and Liljestrand 1918; Lutz 1931; Babkin et al. 1935a), but these may possibly be attributed to a so-called rebound contraction following the cessation of an inhibitory stimulation or to stimulation of intrinsic noncholinergic neurons. There is so far no clear pharmacological or histological evidence demonstrating a cholinergic vagal innervation.

Splanchnic nerves. In most vertebrate species, the splanchnic innervation of the gut contains an adrenergic component. Adrenaline stimulates both the stomach and intestine of elas-

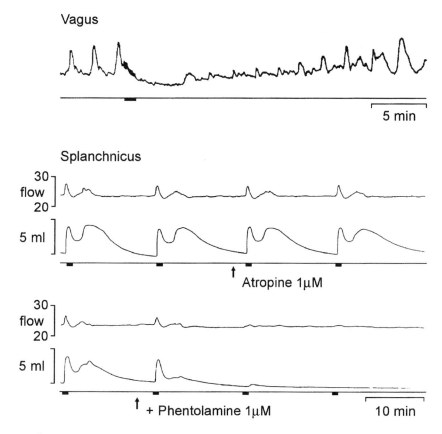

FIGURE 6-11 Effects on stomach motility caused by electrical stimulation of the vagus nerve (*top tracing*) and the splanchnic nerve (*middle* and *lower tracings*). *Upper tracing:* Intracranial stimulation of both vagi was performed in *Scyliorhinus canicula* for 60 sec (marker) with trains of 1-msec pulses (10 V) at 4 Hz. Tracing shows intragastric pressure recorded via a balloon in the fundus. *Lower tracings:* Stimulation of the splanchnic nerve supply to the isolated perfused stomach of *Squalus acanthias* was performed at intervals for 60 sec (markers) with trains of 1-msec pulses (10 V) at 10 Hz. Tracings show perfusion flow expressed as drops per minute, and volume changes of the stomach (fluid displacement) expressed in milliliters. Antagonists were added, as shown by *arrows* in the concentrations indicated, and were present for the rest of the experiment. Note that the effect of splanchnic nerve stimulation is unaffected by atropine but abolished by phentolamine, suggesting that the innervation is adrenergic, acting via α-adrenoceptors. (Upper tracing from Campbell 1975; lower tracings from Nilsson and Holmgren 1983.)

mobranchs to contraction (Nilsson and Holmgren 1983; Young 1980, 1983; Euler and Östlund 1957). This is clearly correlated with adrenergic innervation in *Squalus acanthias*, where numerous catecholamine-containing nerve fibers are present in the splanchnic nerve and stomach wall (Holmgren and Nilsson 1983) and stimulation of the splanchnic nerve causes a contraction that can be blocked by adrenergic antagonists (Nilsson and Holmgren 1983) (see Fig. 6-11). However, some other elasmobranch species may lack adrenergic innervation via the splanchnic nerves. In *Raja* and *Scyliorhinus*, splanchnic stimulation causes inhibition, followed by a rebound contraction, but neither of these effects were affected by adrenergic antagonists (Young 1983; Andrews and Young 1993). The rectum of *Raja* species is initially stimulated and then inhibited by stimulation of the posterior mesenteric nerve (Young 1988).

Excitatory effects of putative transmitters
Spontaneous contractions of the elasmobranch gut have been observed, and contractions may also be elicited by stimulation of the splanchnic nerves and (sometimes) the vagus. Clearly, the contraction caused by stimulation of the splanchnic nerves may in many cases be explained by the action of adrenergic nerves. Dopa and dopamine, which are precursors of adrenaline and may act as transmitters on their own, have stimulatory effects on the electrical and motor activities of the stomach of *Raja radiata*; these effects were not antagonized by α- or β-adrenoceptor blockade (Groisman and Shparkovskii 1989). In addition, several other transmitters have been found using histochemistry on the gut wall (see Table 6-1). Of these, 5-hydroxytryptamine, bombesin, gastrin-and-CCK–related peptides, and some tachykinins have shown excitatory effects in pharmacological experiments and may be involved in the neuronal control of the gut, as transmitters or cotransmitters.

The distribution of 5-hydroxytryptamine–containing nerves in the gut is sparse, but 5-hydroxytryptamine is clearly excitatory on stomach and rectum from both dogfish and skate species (Nilsson and Holmgren 1983; Young 1983, 1988). GRP- or bombesin-containing nerves, on the other hand, are numerous, and bombesin stimulates both basic tone and rhythmic activity of the rectum from *Squalus acanthias* (Lundin et al. 1984) (Fig. 6-12).

Both intestine and rectum of the spiny dogfish are activated by the related peptides CCK, gastrin, and caerulein (Aldman et al. 1989). Similarly, longitudinal muscle preparations of the stomach and intestine from *Raja* species increase tone and rhythmic activity in response to pentagastrin (Andrews and Young 1988a). However, the responses are weak and irregular, and it is possible that the endogenous peptide observed in nerves and endocrine cells is different from the mammalian peptides used for the experiments.

Substance P (a representative for the tachykinin group of peptides) has a direct effect on the gut of *Squalus acanthias* and on *Raja* species (Holmgren 1985; Andrews and Young 1988a). This direct effect of tachykinins occurs in all vertebrates examined to date. However, most other groups show additional indirect actions of substance P on cholinergic or serotonergic pathways. Therefore, the single direct action of substance P on the elasmobranch gut may be a primitive feature shared with lungfishes and holosteans (see Jensen and Holmgren 1994).

There is no evidence for an extrinsic vagal innervation of the elasmobranch gut as in most other vertebrates (see above). However, exogenous acetylcholine causes a contraction of the stomach that is blocked by atropine or hyoscine, suggesting the presence of muscarinic receptors on gastric smooth muscle cells (Nicholls 1934; Euler and Östlund 1957; Nilsson and Holmgren 1983). These receptors may be noninnervated and of little physiological importance, or they may be innervated by enteric cholinergic neurons that are unaffected by the stimulation of extrinsic neurons.

Inhibitory effects of putative transmitters
Vagal stimulation, and in some cases splanchnic nerve stimulation, causes inhibition of gut

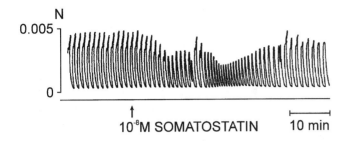
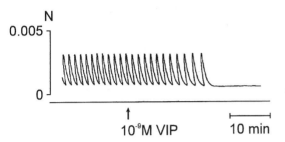
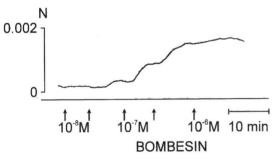
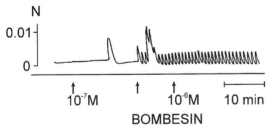

FIGURE 6-12 Motor effects of neuropeptides on the rectum of *Squalus acanthias*. Somatostatin has mixed effects, vasoactive intestinal polypeptide (VIP) is inhibitory, and bombesin is excitatory, increasing either tonus or the frequency of rhythmic contractions. (Reproduced from Nilsson and Holmgren 1988.)

activity. The transmitters mediating these responses have not yet been identified. However, one possible candidate is VIP. VIP often has inhibitory effects on the vertebrate gut and reduces or abolishes the rhythmic activity of the rectum of *Squalus acanthias*. Similar, but weaker, effects are obtained with somatostatin (Lundin et al. 1984). Adenosine triphosphate (ATP) also has been proposed as an inhibitory transmitter of the mammalian gut. In the rectum of *Raja* species it has inhibitory effects followed by a rebound contraction (Young 1988).

Gut circulation

INCREASE OF FLOW TO THE GUT
Processing of food is usually associated with an increased blood flow to the gut of vertebrates (Fara 1984). A number of intrinsic factors

including nervous and hormonal tonus, smooth muscle metabolism, blood oxygen levels, and the compression caused by surrounding muscle affect the flow. The postprandial hyperemia has been verified in teleosts (Axelsson et al. 1989; Axelsson and Fritsche 1991), but so far there are no direct measurements of this phenomenon in an elasmobranch species. Some observations, however, suggest the presence of both hormonal and neuronal control mechanisms. Perivascular fibers containing a bombesinlike peptide innervate submucosal arteries and major arteries to the gut (Holmgren and Nilsson 1983; Tagliafierro et al. 1988a; Bjenning et al. 1990, 1991). In *Scyliorhinus canicula* and *Squatina aculeata* these neurons also contain 5-hydroxytryptamine (Tagliafierro et al. 1988a). The fibers are present in the adventitiomedial border of the vessel wall, which suggests an involvement in the control of the vascular resistance. In accordance with this phenomenon, bombesin increases the flow through the vascularly perfused stomach of *Squalus acanthias* in vitro (Bjenning et al. 1990). However, the increase of the flow to the gut obtained after bombesin injection in vivo, is due to constriction of the somatic vasculature causing a redistribution of blood to the gut (Holmgren et al. 1992b).

NPY similarly produces an increase in the flow to the gut in vivo, but in contrast to the effect of bombesin, this is caused by a reduction in local vascular resistance of the gut (Holmgren et al. 1992b). It may be noted that the effect of NPY is opposite to the effect of adrenaline and noradrenaline (Holmgren et al. 1992a). This finding is in contrast to findings in the mammalian cardiovascular system, where NPY and catecholamines often are synergistic, and in the instances where NPY has been found in vasodilatory nerves, it does not actually cause vasodilation of these blood vessels (Morris and Gibbins 1994).

Substance P or a related tachykinin, possibly one of the scyliorhinins, is present in endocrine cells of the gut wall (El-Salhy 1984; Holmgren 1985; Conlon et al. 1986). So far, no tachykinins have been identified in perivascular neurons of elasmobranchs. However, local blood flow may well be affected by tachykinins released from neurons in the myenteric plexus that diffuse to nearby vasculature. The response to injected substance P in *Squalus acanthias* is an increase in flow to the gut resulting from a decrease in vascular resistance of the gut vessels (Holmgren et al. 1992b; also see Mione et al. 1990; Jensen et al. 1991). This response is similar to that often seen in mammals.

DECREASE OF FLOW TO THE GUT

The blood flow to the gut is rapidly reduced at periods of voluntary swimming or at stress (Holmgren et al. 1992a) (Fig. 6-13). This effect could probably to a large extent be ascribed to an adrenergic mechanism. Adrenaline and noradrenaline contract the major vessels to the gut by an α-adrenoceptor–mediated effect (Nilsson et al. 1975). Adrenergic nerves innervate the vessels of the gut (Nilsson et al. 1975; Holmgren and Nilsson 1983), and the levels of circulating catecholamines (released from chromaffin tissue) increase during voluntary exercise and during stress (Abrahamsson 1979; Opdyke et al. 1982; Butler 1986; Butler et al. 1986). In vivo injections of adrenaline and noradrenaline mimic the effects of exercise or fright. Both catecholamines cause an increase in resistance of the celiac artery vascular bed, adrenaline being the more potent of the two (Holmgren et al. 1992a). The relative importance of the two control systems (adrenergic nerves versus circulating catecholamines) is not clear, but it is possible that the effects of circulating catecholamines are dominating.

VIP nerves innervate the large arteries to the gut in *Squalus acanthias* (Holmgren and Nilsson 1983). Porcine VIP causes a reduction in blood flow to the gut, as a result of an increase in vascular resistance (Holmgren et al. 1992a). This is an unusual effect, since VIP is widely considered to be vasodilatory in the gut. It is notable that in the rectal gland of *Squalus*, VIP causes the same vasodilation combined with increased glandular secre-

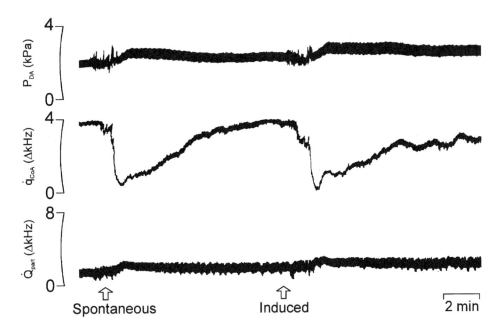

FIGURE 6-13 Recording showing the effects of spontaneous activity and "fright" (tapping the tank) on dorsal aortic blood pressure (P_{DA}), celiac artery blood flow ($\dot{q}_{CoA}$), and cardiac output ($\dot{Q}_{part}$). Blood pressure is expressed in kilopascals (kPa) and flows are shown as unprocessed output from the Doppler flowmeter (kHz Doppler shift). Note the decrease in celiac artery blood flow despite the increase in dorsal aortic blood pressure. (Reproduced from Holmgren et al. 1992a.)

tion, as in mammalian exocrine glands (Thorndyke et al. 1989).

Conclusions

The elasmobranch gut is certainly not more primitive, in the sense that it is less effective, than that of any other vertebrate. It provides the body with nutrients with an efficiency that allows high metabolism, high-speed swimming, breathing, reproduction, and more eating. Elasmobranch species occur as predators in all food niches of the sea, and prey capture is as effective and varying as in any other vertebrate group. Adaptations to the different feeding habits are most obvious in the anatomy of the mouth region and in the surface area of the intestine; herbivores have a larger mucosal surface than carnivores, as in other species.

The basic anatomical structure of the gut follows that of other vertebrates, with only few exceptions. The most striking exception is the spiral intestine, which provides for an enlarged surface area by means of spiral folds of mucosa within a short intestine, rather than by an increase in the actual length of the intestine. The absence of esophageal sphincters, the oxynticopeptic cells in the gastric mucosa, and the presence of secreting cells in folds and crypts of the mucosa rather than in proper glands in the intestine are features shared with other fish species but are different from mammals.

Similarly, the control mechanisms in the gut follow the basic vertebrate plan, but the number of ganglion cells, nerve fibers, and endocrine cells involved in the control of the gut functions is smaller in elasmobranchs than in tetrapods. However, the available information suggests a variety of neurotransmitters and hormones equaling that in other vertebrates, allowing complex and advanced control systems. Conspicuous elasmobranch features are the lack of a cholinergic innervation of the gut smooth muscle and different control mechanisms for acid secretion.

The processing of food takes a longer time in elasmobranchs than in homeotherms. This is partly dependent on the body temperature of the fish but also, in comparison with teleost fish living at similar temperatures, the elasmobranchs are slower. Pelagic, carnivore sharks, in particular, often are intermittent feeders on large prey. These species store energy as low-density fats (such as squalenes) in the large liver, which at the same time plays an important role in the buoyancy of the fish.

All in all, it may be concluded that the elasmobranch gut shows some interesting adaptations and evolutionary variations, but basically the gut functions follow the general vertebrate plan.

ACKNOWLEDGMENTS

We are grateful to Professor Ragnar Fänge, Professor Ian Gibbins, and Dr. David Grove for constructive criticism of the manuscript; to Dr. Paul Andrews, Dr. Michael Axelsson, Professor Ragnar Fänge, Dr. David Grove, and Dr. Ann-Cathrine Jönsson for letting us use their unpublished data and figures; and to Mrs. Christina Hagström for expert help with histochemistry and photography.

REFERENCES

Abrahamsson, T. 1979. Phenylethanolamine-N-methyl transferase (PNMT) activity and catecholamine storage and release from chromaffin tissue of the spiny dogfish, *Squalus acanthias*. *Comp. Biochem. Physiol.* 64C: 169–72.

Aldman, G., A. C. Jönsson, J. Jensen, and S. Holmgren. 1989. Gastrin/CCK-like peptides in the spiny dogfish, *Squalus acanthias*; concentrations and actions in the gut. *Comp. Biochem. Physiol.* 92C: 103–09.

Andrews, P. L. R., and J. Z. Young. 1988a. The effect of peptides on the motility of the stomach, intestine, and rectum in the skate (*Raja*). *Comp. Biochem. Physiol.* 89C: 343–48.

———. 1988b. A pharmacological study of the control of motility in the gall bladder of the skate (*Raja*). *Comp. Biochem. Physiol.* 89C: 349–54.

———. 1993. Gastric motility patterns for digestion and vomiting evoked by sympathetic nerve stimulation and 5-hydroxytryptamine in the dogfish *Scyliorhinus canicula*. *Philos. Trans. R. Soc. Lond.* 342B: 363–80.

Axelsson, M., and R. Fritsche. 1991. Effects of exercise, hypoxia, and feeding on the gastrointestinal blood flow in the Atlantic cod, *Gadus morhua*. *J. Exp. Biol.* 158: 181–91.

Axelsson, M., W. R. Driedzic, A. P. Farrell, and S. Nilsson. 1989. Regulation of cardiac output and gut blood flow in the sea raven, *Hemitripterus americanus*. *Fish Physiol. Biochem.* 6: 315–26.

Babkin, B. P. 1929. Studies on the pancreatic secretion in skates. *Biol. Bull.* 57: 272–91.

———. 1933. Further studies on the pancreatic secretion in skates. *Contrib. Can. Biol. Fish.* 7: 1–9.

Babkin, B. P., and M. H. Friedman. 1934. The relation of the autonomic nervous system to the motility and secretion of the stomach in elasmobranchs. *Am. J. Physiol.* 109: 3.

Babkin, B. P., M. H. F. Friedman, and M. E. MacKay-Sawyer. 1935a. Vagal and sympathetic innervation of the skate. *J. Biol. Board Can.* 1: 239–50.

Babkin, B. P., A. F. Chaisson, and M. H. F. Friedman. 1935b. Factors determinating the course of the gastric secretion in elasmobranchs. *J. Biol. Board Can.* 1: 251–59.

Backiel, T. 1971. Production and food consumption of predatory fish in the Vistula River. *J. Fish Biol.* 3: 369–405.

Baldridge, H. D. 1972. Accumulation and function of liver oil in Florida sharks. *Copeia* 2: 306–25.

Bayliss, W. M., and E. H. Starling. 1903. On the uniformity of the pancreatic mechanism in vertebrata. *J. Physiol. (Lond.)* 29: 174–80.

Bertin, L. 1958a. Appareil digestif. In *Traité de Zoologie*, ed. P. P. Grassé, vol. 13, 1248–302. Paris: Masson.

———. 1958b. Appareil circulatoire. In *Traité de Zoologie*, ed. P. P. Grassé, vol. 13, 1399–458. Paris: Masson.

Bjenning, C., and S. Holmgren. 1988. Neuropeptides in the fish gut. A study of evolutionary trends. *Histochemistry* 88: 155–63.

Bjenning, C., Driedzic, W., and Holmgren, S. 1989. Neuropeptide Y-like immunoreactivity in the cardiovascular nerve plexus of the elasmobranchs *Raja erinacea* and *Raja radiata*. *Cell Tissue Res.* 255: 481–86.

Bjenning, C., A. P. Farrell, and S. Holmgren. 1991. Bombesin-like immunoreactivity in the gut and heart region of elasmobranch fish (skates), and the in vitro effect of bombesin on coronary vessels from the longnose skate, *Raja rhina*. *Regul. Pept.* 35: 207–19.

Bjenning, C., N. Hazon, A. Balasubramaniam, S. Holmgren, and J. M. Conlon. 1993. Distribution and activity of dogfish NPY and peptide YY in the cardiovascular system of the common dogfish. *Am. J. Physiol.* 264: R1119–24.

Bjenning, C., A. C. Jönsson, and S. Holmgren. 1990. Bombesin-like immunoreactive material in the gut, and the effect of bombesin on the stomach circulatory system of an elasmobranch fish, *Squalus acanthias. Regul. Pept.* 28: 57–69.

Blomqvist, A. G., C. Söderberg, I. Lundell, R. J. Milner, and D. Larhammar. 1992. Strong evolutionary conservation of neuropeptide Y: sequences of chicken, goldfish, and *Torpedo marmorata* DNA clones. *Proc. Natl. Acad. Sci. U.S.A.* 89: 2350–54.

Bodansky, M., and W. C. Rose. 1922. Digestion in elasmobranchs and teleosts. *Am. J. Physiol.* 62: 482–87.

Bone, Q., and B. L. Roberts. 1969. The density of elasmobranchs. *J. Mar. Biol. Assoc. U.K.* 49: 913–37.

Botazzi F. 1902. Untersuchungen über das viscerale Nervensystem der Selachier. *Z. Biol.* 43: 372–442.

Boyer, J. L., J. Schwartz, and N. Smith. 1976. Biliary secretion in elasmobranchs: I. Bile collection and composition. *Am. J. Physiol.* 230: 970–73.

Butler, P. J. 1986. Exercise. In *Fish Physiology: Recent Advances*, ed. S. Nilsson and S. Holmgren, 102–18. London: Croom-Helm.

Butler, P. J., J. D. Metcalfe, and S. A. Ginley. 1986. Plasma catecholamines in the lesser spotted dogfish and rainbow trout at rest and during different levels of exercise. *J. Exp. Biol.* 123: 409–21.

Campbell, G. 1975. Inhibitory vagal innervation of the stomach in fish. *Comp. Biochem. Physiol.* 50C: 169–70.

Carey, F. G., J. M. Teal, and J. W. Kanwisher. 1981. The visceral temperature of mackerel sharks (Lamnidae). *Physiol. Zool.* 54: 334–44.

Cimini, V., S. Van Noorden, G. Giordano-Lanza, V. Nardini, G. P. McGregor, S. R. Bloom, and J. M. Polak, 1985. Neuropeptides and 5-HT immunoreactivity in the gastric nerves of the dogfish (*Scyliorhinus stellaris*). *Peptides* 6 (suppl. 3): 373–77.

Cimini, V., S. Van Noorden, and V. Nardini. 1989b. Peptides of the gastrointestinal tract of the dogfish (*Scyliorhinus stellaris*). *J. Exp. Zool. Suppl.* 2: 146–57.

Cimini, V., S. Van Noorden, and J. M. Polak, 1989a. Co-localization of substance P–, bombesin–, and peptide histidine isoleucine (PHI)-like peptides in gut endocrine cells of the dogfish *Scyliorhinus stellaris. Anat. Embryol. (Berl.)* 179: 605–14.

Cimini, V., S. Van Noorden, and M. Sansone. 1992. Neuropeptide Y–like immunoreactivity in the dogfish gastroenteropancreatic tract: light and electron microscopical study. *Gen. Comp. Endocrinol.* 86: 413–23.

Conlon, J. M. 1990. [Ser]-somatostatin-14 isolation from the pancreas of a holocephalan fish, the Pacific ratfish (*Hydrolagus colliei*). *Gen. Comp. Endocrinol.* 80: 314–20.

Conlon, J. M., D. V. Agoston, and L. Thim. 1985. An elasmobranchian somatostatin: primary structure and tissue distribution in *Torpedo marmorata. Gen. Comp. Endocrinol.* 60: 406–13.

Conlon, J. M., A. Balasubramaniam, and N. Hazon. 1991. Structural characterization and biological activity of a neuropeptide Y–related peptide from the dogfish *Scyliorhinus canicula. Endocrinology* 128: 2273–79.

Conlon, J. M., C. Bjenning, and N. Hazon. 1992. Structural characterisation of neuropeptide Y from the brain of the dogfish *Scyliorhinus canicula. Peptides* 13: 493–97.

Conlon, J. M., C. F. Deacon, L. O'Toole, and L. Thim. 1986. Scyliorhinin I and II: two novel tachykinins from the dogfish gut. *FEBS Lett.* 200: 111–16.

Conlon, J. M., I. W. Henderson, and L. Thim. 1987. Gastrin-releasing peptide from the intestine of the elasmobranch fish *Scyliorhinus canicula* (common dogfish). *Gen. Comp. Endocrinol.* 68: 415–20.

Conlon, J. M., and L. Thim. 1988. Isolation of the tachykinin des-Ser1-Pro2-scyliorhinin II from the intestine of the ray, *Torpedo marmorata. Gen. Comp. Endocrinol.* 71: 383–88.

Cortés, E., and S. H. Gruber. 1990. Diet feeding habits and estimates of daily ration of young lemon sharks, *Negaprion brevirostris* (Poey). *Copeia* 1990: 204–18.

———. 1992. Gastric evacuation in the young lemon shark, *Negaprion brevirostris*, under field conditions. *Environ. Biol. Fishes* 35: 205–12.

Crow, G. L., J. A. Brock, J. C. Howe, and B. E. Linnon. 1991. Shark bite wounds of the valvular intestine the cause of an acute mortality syndrome of captive blacktip reef sharks, *Carcharhinus melanopterus. Zool. Biol.* 10: 457–64.

Crow, G. L., J. C. Howe, S. Uchida, S. Kamolnic, M. G. Wisner, and J. N. Caira. 1990. Protrusion of the valvular intestine through the cloaca in sharks of the family Carcharhinidae. *Copeia* 1990: 226–29.

Debas, H. T. 1987. Peripheral regulation of gastric acid secretion. In *Physiology of the Gastrointestinal Tract*, 2d ed., ed. L. R. Johnson, J. Christensen, E. D. Jacobson, M. J. Jackson, and J. H. Walsh, 931–45. New York: Raven Press.

Dimaline, R., and M. C. Thorndyke. 1986. Purification and characterisation of VIP from two species of dogfish. *Peptides* 7: 21–25.

Dimaline, R., J. Young, D. T. Thwaites, C. M. Lee, T. J. Shuttleworth, and M. C. Thorndyke. 1987. A novel vasoactive intestinal peptide (VIP) from elasmobranch intestine has full affinity for mammalian pancreatic VIP receptors. *Biochem. Biophys. Acta* 930: 97–100.

Dobreff, M. 1927. Experimentelle Studien über die vergleichende Physiologie der Verdauung. *Arch. Ges. Physiol.* 217: 221–34.

El-Salhy, M. 1984. Immunocytochemical investigation of the gastro-entero-pancreatic (GEP) neurohormonal peptides in the pancreas and gastrointestinal tract of the dogfish *Squalus acanthias*. *Histochemistry* 80: 193–205.

El-Salhy, M., L. Grimelius, P. C. Emson, and S. Falkmer. 1987. Polypeptide YY– and neuropeptide Y–immunoreactive cells and nerves in the endocrine and exocrine pancreas of some vertebrates: an onto- and phylogenetic study. *Histochem. J.* 19: 111–17.

Estecondo, S., S. M. Codon, and E. J. Galindez. 1988. Anatomical and histological study of the digestive tract of *Mustelus schmitti* (Chondrichthyes, Triakidae). *Physis. Sec. A. Oceanos Org.* 46: 31–41.

Euler, U. S. von., and R. Fänge. 1961. Catecholamines in nerves and organs of *Myxine glutinosa*, *Squalus acanthias*, and *Gadus callarias*. *Gen. Comp. Endocrinol.* 1: 191–94.

Euler, U. S. von., and E. Östlund. 1957. Effects of certain biologically occurring substances on the isolated intestine of fish. *Acta Physiol. Scand.* 38: 364–72.

Falkmer, S., and Y. Stefan. 1978. Pancreatic polypeptide (PP): phylogenetic aspects in gastrointestinal mucosa and endocrine pancreas. *Scand. J. Gastroenterol.* 13: 59.

Falkmer, S., and S. Van Noorden. 1983. Ontogeny and phylogeny of the glucagon cell. *Handb. Exp. Pharmacol.* 66: 81–119.

Falkmer, S., R. E. Carraway, M. El-Salhy, S. O. Emdin, L. Grimelius, J. F. Rehfeld, M. Reinecke, and T. W. Schwartz. 1981. Phylogeny of the gastroenteropancreatic neuroendocrine system. A review. In *Cellular Basis of Chemical Messengers in the Digestive System*, ed. M. Grossman, M. Brazier, and J. Lechago, 21–42. London: Academic Press.

Falkmer, S., R. Ebert, R, Arnold, and W. Creutzfeldt. 1980a. Some phylogenetic aspects on the enteroinsular axis with particular regard to the appearance of the gastric inhibitory polypeptide. *Front. Horm. Res.* 7: 1–6.

Falkmer, S., M. El-Salhy, and M. Titlbach. 1984. Evolution of the endocrine systems in vertebrates: a review with particular reference to the phylogeny and postnatal maturation of the islet parenchyma. In *Evolution and Tumor Pathology of the Neuroendocrine System*, ed. S. Falkmer, R. Håkanson, and F. Sundler, 59–87. Amsterdam: Elsevier.

Falkmer, S., J. Fahrenkrug, J. Alumets, R. Håkanson, and F. Sundler. 1980b. Vasoactive intestinal polypeptide (VIP) in epithelial cells of the gut mucosa of an elasmobranch cartilaginous fish, the ray. *Endocrinol. Japon. S. R.* 27 (suppl. 1): 31–35.

Fänge, R., and D. J. Grove, 1979. Digestion. In *Fish Physiology*, ed. W. S. Hoar, and D. J. Randall, 161–260. London: Academic Press.

Fänge, R., G. Lundblad, J. Lind, and K. Slettengren. 1979. Chitinolytic enzymes in the digestive system of marine fishes. *Mar. Biol.* 53: 317–21.

Fara, J. W. 1984. Postprandial mesenteric hyperemia. In *Physiology of the Intestinal Circulation*, ed. A. Shepherd and D. N. Granger, 99–119. New York: Raven Press.

Faraldi, G., M. Canepa, L. Borgiani, and T. Zanin. 1990. Identification of pancreatic glucagon cells in the cartilaginous fish *Scyliorhinus canicula* by ultrastructural immunocytochemistry. *Basic Appl. Histochem.* 34: 199–208.

Forte, J. G., and J. M. Wolosin. 1987. HCl secretion by the gastric oxyntic cell. In *Physiology of the Gastrointestinal Tract*, 2d ed., ed. L. R. Johnson, J. Christensen, E. D. Jacobson, M. D. Jackson, and J. H. Walsh, 853–63. New York: Raven Press.

Galindez, E. J. 1992. Observations on the mucopolysaccharide histochemistry of the gut of the *Mustelus schmitti* (Chondrichthyes, Triakidae). *Iheringia. Ser. Zool.* 72: 127–34.

Groisman, S. D., and I. A. Shparkovskii. 1989. The effect of dopamine and dopa on the electrical activity of stomach muscles in the skate *Raja radiata* and cod *Gadus morhua*. *Zh. Evol. Biokhim. Fiziol.* 25: 505–11.

Hansen, D. 1975. Evidence of a gastrin-like substance in *Rhinobatos productus*. *Comp. Biochem. Physiol.* 52C: 61–63.

Haselwood, G. A. D. 1968. Evolution and bile salts. In *Handbook of Physiology*, ed. C. F. Code, vol. 5, sect. 6, 2375–90. Washington, D.C.: American Physiological Society.

Håkanson, R., G. Böttcher, E. Ekblad, P. Panula, M. Simonsson, M. Dohlsten, T. Hallberg, and F. Sundler. 1986. Histamine in endocrine cells of the stomach. *Histochemistry* 86: 5–17.

Hogben, C. A. M. 1967a. Response of the isolated dogfish gastric mucosa to histamine. *Proc. Soc. Exp. Biol. Med.* 124: 890–93.

———. 1967b. Secretion of acid by the dogfish *Squalus acanthias*. In *Sharks, Skates and Rays*, ed. P. W. Gilbert, R. F. Mathewson, and D. P. Rall. 299–315. Baltimore: John Hopkins University Press.

Holmgren, S. 1985. Substance P in the gastrointestinal tract of *Squalus acanthias*. *Mol. Physiol.* 8: 119–30.

———. 1989. Gut motility. In *The Comparative Physiology of Regulatory Peptides*, ed. S. Holmgren, 231–55. London: Chapman & Hall.

Holmgren, S., and S. Nilsson. 1983. Bombesin-, gastrin/CCK-, 5-hydoxytryptamine–, neurotensin-, somatostatin-, and VIP-like immunoreactivity and catecholamine fluorescence in the gut of the elasmobranch *Squalus acanthias*. *Cell Tissue Res.* 234: 595–618.

Holmgren, S., M. Axelsson, and A. P. Farrell, 1992a. The effect of catecholamines, substance P, and vasoactive intestinal polypeptide on blood flow to the gut in the dogfish *Squalus acanthias*. *J. Exp. Biol.* 168: 161–75.

———. 1992b. The effects of neuropeptide Y and bombesin on blood flow to the gut in dogfish *Squalus acanthias*. *Regul. Pept.* 40: 169.

Jacobshagen, E. 1937. Darmsystem: IV. Mittel- und Enddarm (Rumpfdarm). In *Handbuch der vergleichenden Anatomie der Wirbeltiere*, ed. L. Bolk, E. Göppert, E. Kallius, and W. Lubosch, 563–724. Berlin: Urban & Schwarzenberg.

Jensen, J., and S. Holmgren. 1994. The gastrointestinal canal. In *Comparative Physiology and Evolution of the Autonomic Nervous System*, ed. S. Holmgren and S. Nilsson, 193–246. Chur, Switzerland: Harwood Academic Publishers.

Jensen, J., M. Axelsson, and S. Holmgren. 1991. Effects of substance P and vasoactive intestinal polypeptide on gastrointestinal blood flow in the Atlantic cod, *Gadus morhua*. *J. Exp. Biol.* 156: 361–73.

Johnson, L. R., ed. 1987. *Physiology of the Gastrointestinal Tract*. New York: Raven Press.

Jones, R. 1974. The rate of elimination of food from the stomachs of haddock, *Melanogrammus aeglefinus*, cod, *Gadus morhua*, and whiting, *Merlangius merlangus*. *J. Cons., Cons. Int. Explor. Mer.* 35: 225–43.

Jones, B., and G. Genn. 1977. Food and feeding of spiny dogfish (*Squalus acanthias*) in British Columbia waters. *J. Fish. Res. Board. Can.* 34: 2006.

Jönsson, A. C. 1991. Regulatory peptides in the pancreas of two species of elasmobranchs and in the Brockmann bodies of four teleost species. *Cell Tissue Res.* 266: 163–72.

———. 1994. Glands. In *Comparative Physiology and Evolution of the Autonomic Nervous System*, ed. S. Holmgren and S. Nilsson, 169–92. Chur, Switzerland: Harwood Academic Publishers.

Jönsson, A. C., and S. Holmgren, 1989. Gut secretion. In *The Comparative Physiology of Regulatory Peptides*, ed. S. Holmgren, 256–71. London: Chapman & Hall.

Keenleyside, M. H. A. 1979. *Diversity and Adaptation in Fish Behavior*. Berlin: Springer-Verlag.

Kirtisinghe, P. 1940. The myenteric nerve plexus in some lower chordates. *Q. J. Microsc. Sci.* 81: 521–39.

Kobegenova, S. S. 1992. Morphology and morphogenesis of the digestive system of some cartilaginous fishes (Chondrichthyes). *Zool. Zh.* 71: 108–22.

Kuz'mina, V. V. 1990. Characteristics of enzymes involved in membrane digestion in elasmobranch fishes. *Zh. Evol. Biokhim. Fiziol.* 26: 161–66.

Lindsay, G. J. H., and J. E. Harris. 1980. Carboxymethylcellulase activity in the digestive tracts of fish. *J. Fish Biol.* 16: 219–33.

Lineweaver, T. H., and R. H. Backus. 1970. *The Natural History of Sharks*. Philadelphia: J. B. Lippincott.

Lundin, K., S. Holmgren, and S. Nilsson. 1984. Peptidergic functions in the dogfish rectum. *Acta Physiol. Scand.* 121: 46A.

Lutz, B. R. 1931. The innervation of the stomach and rectum and the action of adrenaline in elasmobranch fishes. *Biol. Bull.* 61: 93–100.

Matthews, L. H., and H. W. Parker. 1950. Notes on the anatomy and biology of the basking shark (*Cetorhinus maximus*). *Proc. Zool. Soc. Lond.* 120: 535–76.

McCosker, J. E. 1987. The white shark, *Carcharodon carcharias*, has a warm stomach. *Copeia* 1987: 195–97.

Medved, R.J. 1985. Gastric evacuation in the sandbar shark, *Carcharhinus plumbeus*. *J. Fish Biol.* 26: 239–54.

Medved, R. J., C. E. Stillwell, and J. G. Casey. 1988. The rate of food consumption of young sandbar sharks, *Carcharhinus plumbeus*, in Chincoteague Bay, Virginia. *Copeia* 1988: 956–63.

Merret, T. G., E. Bar-Eli, and H. Van Vunakis. 1969. Pepsinogens A, C, and D from the smooth dogfish. *Biochemistry* 8: 3696–702.

Micha, J. C., G. Dandrifosse, and C. Jeuniaux. 1973. Distribution et localisation tissulaire de la synthèse de chitinases chez les vertébrés inférieurs. *Arch. Int. Physiol. Biochim.* 81: 439–51.

Mione, M. C., V. Ralevic, and G. Burnstock. 1990. Peptides and vasomotor mechanisms. *Pharmacol. Ther.* 46: 429–668.

Morris, J. L., and I. L. Gibbins. 1994. Cotransmission and neuromodulation. In *Comparative Physiology and Evolution of the Autonomic Nervous System*, ed. S. Holmgren and S. Nilsson, 33–119. Chur, Switzerland: Harwood Academic Publishers.

Müller, E. 1920. Über die Entwicklung des Sympathicus und des Vagus bei den Selachiern. *Arch. Mikrosk. Anat.* 94: 208.

Müller, E., and F. Liljestrand. 1918. Anatomische und experimentelle Untersuchungen über das autonome Nervensystem der Elasmobranchier nebst Bemerkungen über die Darmnerven bei den Amphibien und Säugetieren. *Arch. Anat.* 1918: 137–72.

Nelson, J. S. 1984. *Fishes of the World*. New York: John Wiley & Sons.

Nicholls, J. V. V. 1934. Reaction of the smooth muscle of the gastro-intestinal tract of the skate to stimulation of autonomic nerves in isolated nerve-muscle preparations. *J. Physiol. (Lond.)* 83: 56–67.

Nicol, J. A. C. 1952. Autonomic nervous systems in lower chordates. *Biol. Rev. Cam. Philos. Soc.* 27: 1–49.

Nilsson, A., and R. Fänge. 1969. Digestive proteases in the holocephalan fish *Chimaera monstrosa* (L.). *Comp. Biochem. Physiol.* 31: 147–65.

Nilsson, S. 1983. *Autonomic Nerve Function in the Vertebrates.* Berlin: Springer-Verlag.

Nilsson, S., and S. Holmgren. 1983. Splanchnic nervous control of the stomach of the spiny dogfish, *Squalus acanthias. Comp. Biochem. Physiol.* 76C: 271–76.

———. 1988. The autonomic nervous system. In *Physiology of Elasmobranch Fishes,* ed. T. J. Shuttleworth, 143–69. Berlin: Springer-Verlag.

Nilsson, S., S. Holmgren, and D. J. Grove. 1975. Effects of drugs and nerve stimulation on the spleen and arteries of two species of dogfish, *Scyliorhinus canicula* and *Squalus acanthias. Acta Physiol. Scand.* 95: 219–30.

Opdyke, D. F., R. G. Carroll, and N. E. Keller. 1982. Catecholamine release and blood pressure changes induced by exercise in dogfish. *Am. J. Physiol.* 242: R306–10.

Pan, J. Z., C. Shaw, D. W. Halton, L. Thim, C. F. Johnston, and K. D. Buchanan. 1992. The primary structure of peptide Y (PY) of the spiny dogfish *Squalus acanthias. Comp. Biochem. Physiol.* 102B: 1–5.

Parker, T. J. 1885. On the intestinal spiral valve in the genus *Raja. Trans. Zool. Soc. Lond.* 11: 49–61.

Pernkopf, E., and J. Lehner. 1937. Vergleichende Beschreibungen des Vorderdarms bei den einzelnen Klassen der Cranioten. In *Handbuch der vergleichenden Anatomie der Wirbeltiere,* ed. L. Bolk, E. Göppert, E. Kallius, and W. Lubosch, 349–476. Berlin: Urban & Schwarzenberg.

Prahl, J. W., and H. Neurath. 1966a. Pancreatic enzymes of the spiny pacific dogfish: I. Cationic chymotrypsinogen and chymotrypsin. *Biochemistry* 5: 2131–46.

———. 1966b. Pancreatic enzymes of the spiny pacific dogfish: II. Procarboxypeptidases B and carboxypeptidase B. *Biochemistry* 5: 4137–45.

Rauthner, M. 1940. Der Intestinaltraktus. In *Dr. H. G. Bronns Klassen und Ordnungen des Tierreichs.* Pt. 1: *Echte Fische,* 657–1050. Leipzig: Akad. Verlagsges.

Rebolledo, I. M., and J. D. Vial. 1979. Fine structure of the oxynticopeptic cell in the gastric glands of an elasmobranch species (*Halaelurus chilensis*). *Anat. Rec.* 193: 805–22.

Reinecke, M., P. Schluter, N. Yanaihara, and W. G. Forssmann. 1981. VIP immunoreactivity in enteric nerves and endocrine cells of the vertebrate gut. *Peptides* 2: 149–56.

Reinecke, M., S. Falkmer, D. Heinrich, K. Almasan, and W. G. Forsmann. 1984. A phylogenetic study on the occurrence of somatostatin immunoreactivity in endocrine cells and enteric nerves of the vertebrate gut. In *Proceedings of the 2nd International Symposium on Somatostatin, 1–3 June, 1985, Athens,* 47–53. Tübingen: Attempto Verlag.

Reinecke, M., K. Drakenberg, S. Falkmer, and V. R. Sara. 1992. Peptides related to insulin-like growth factor 1 in the gastro-entero-pancreatic system of bony and cartilaginous fish. *Regul. Pept.* 37: 155–65.

Rossouw, G. J. 1987. Function of the liver and hepatic lipids of the lesser sand shark, *Rhinobatos annulatus* (Mueller & Henle). *Comp. Biochem. Physiol.* 86B: 785–790.

Schurdak, M. E., and S. H. Gruber. 1989. Gastric evacuation of the lemon shark, *Negaprion brevirostris* (Poey), under controlled conditions. *Exp. Biol. (Berl.)* 48: 77–82.

Smit, H. 1968. Gastric secretion in the lower vertebrates and birds. In *Handbook of Physiology,* ed. C. F. Code, vol. 5, sect. 6, 2791–805. Washington, D.C.: American Physiological Society.

Soll, A. H., and T. Berglindh. 1987. Physiology of isolated gastric glands and parietal cells: receptors and effectors regulation function. In *Physiology of the Gastrointestinal Tract,* 2d ed., ed. L. R. Johnson, J. Christensen, E. D. Jacobson, M. D. Jackson, and J. H. Walsh, 883–909. New York: Raven Press.

Stefan, Y., C. Dufour, and S. Falkmer. 1978. Mise en évidence par immunofluorescence de cellules à polypeptide pancréatique (PP) dans le pancréas et le tube digestif de poissons osseux et cartilagineux. *C. R. Acad. Sci. Paris* 286: 1073–75.

Sternby, B., A. Engström, and U. Hellman. 1984. Purification and characterization of pancreatic colipase from the dogfish, *Squalus acanthias. Biochim. Biophys. Acta* 789: 159–63.

Sternby, B., A. Larsson, and B. Borgström. 1983. Evolutionary studies on pancreatic colipase. *Biochim. Biophys. Acta* 750: 340–45.

Stevens, C. E. 1988. *Comparative Physiology of the Vertebrate Digestive System.* Cambridge: Cambridge University Press.

Stickney, R. R., and S. E. Shumway. 1974. Occurrence of cellulase activity in the stomachs of fishes. *J. Fish Biol.* 6: 779–90.

Stokes, M. D., and N. D. Holland. 1992. Southern stingray *Dasyatis americana* feeding on lancelets *Branchiostoma floridae. J. Fish Biol.* 41: 1043–44.

Sullivan, M. X. 1905. The physiology of the digestive tract of elasmobranchs. *Am. J. Physiol.* 15: 42–45.

Tache, Y. 1987. Central nervous system regulation of gastric acid secretion. In *Physiology of the Gastrointestinal Tract,* 2d ed., ed. L. R. Johnson, J. Christensen, E. D. Jacobson, M. D. Jackson, and J. H. Walsh, 911–30. New York: Raven Press.

Tagliafierro, G., G. Faraldi, and R. Bandelloni. 1985a. Distribution histochemistry and ultrastructure of somatostatin-like immunoreactive cells in the gastroenteric tract of the cartilaginous fish *Scyliorhinus stellaris*. *Histochem. J.* 17: 1033–42.

Tagliafierro, G., E. Bonini, G. Faraldi, L. Farina, and G. G. Rossi. 1988b. Distribution and ontogeny of VIP-like immunoreactivity in the gastro-entero-pancreatic system of a cartilaginous fish, *Scyliorhinus stellaris*. *Cell Tissue Res.* 253: 23–28.

Tagliafierro, G., G. Faraldi, and M. Pestarino. 1985b. Interrelationships between somatostatin-like cells and other endocrine cells in the pancreas of some cartilaginous fish. *Cell. Mol. Biol. (Naisy-le-grand)* 31: 201–07.

Tagliafierro, G., G. Zaccone, E. Bonini, G. Faraldi, L. Farina, S. Fasula, and G. G. Rossi. 1988a. Bombesin-like immunoreactivity in the gastrointestinal tract of some lower vertebrates. *Ann. N. Y. Acad. Sci.* 547: 458–60.

Thorndyke, M. C., and S. Falkmer. 1985. Evolution of gastro-entero-pancreatic endocrine systems in lower vertebrates. In *Evolutionary Biology of Primitive Fishes*, ed. R. E. Foreman, A. Gorbman, J. M. Dodd, and Olsson, 379–400. New York: Plenum Press.

Thorndyke, M. C., J. H. Riddell, D. T. Thwaites, and R. Dimaline. 1989. Vasoactive intestinal polypeptide and its relatives—biochemistry, distribution, and functions. *Biol. Bull.* 177: 183–86.

Tsujimoto, M. 1916. Highly unsaturated hydrocarbon in shark liver oil. *J. Ind. Eng. Chem.* 8: 889–920.

Ungar, C. 1935. Perfusion de l'estomac des Sélaciens: étude pharmacodynamique de la sécrétion gastrique. *C. R. Séances Soc. Biol. Fil.* 119: 859–60.

Van Giersbergen, P. L. M., J. M. Conlon, and S. H. Buck. 1991. Binding sites for tachykinin peptides in the brain and stomach of the dogfish *Scyliorhinus canicula*. *Peptides* 12: 1161–63.

Van Herwerden, M., and W. E. Ringer. 1911. Die Acidität des Magensaftes von *Scyllium stellare*. *Hoppe-Seylers Z. Physiol. Chem.* 75: 290.

Vigna, S. R. 1979. Distinction between cholecystokinin-like and gastrin-like biological activities extracted from gastrointestinal tissues of some lower vertebrates. *Gen. Comp. Endocrinol.* 39: 512–20.

———. 1983. Evolution of endocrine regulation of gastrointestinal function in lower vertebrates. *Am. Zool.* 23: 729–38.

Weinland, E. 1901. Zur Magenverdauung der Haifische. *Z. Biol.* 1: 35–68, 275–94.

Went, A. E. J., and S. O'Suilleabhain. 1967. Fishing for the sunfish or basking shark in Irish waters. *Proc. R. Ir. Acad.* 65: 91–115.

Wetherbee, B. M., S. H. Gruber, and A. L. Ramsey. 1987. X-radiographic observations of food passage through digestive tracts of lemon sharks. *Trans. Am. Fish Soc.* 116: 763–67.

Yano, K., and J. A. Musick. 1992. Comparison of morphometrics of Atlantic and Pacific specimens of the false catshark, *Pseudotriakis microdon*, with notes on stomach contents. *Copeia* 1992: 877–86.

Yasugi, S., T. Matsunaga, and T. Mizuno. 1988. Presence of pepsinogens immunoreactive to antiembryonic chicken pepsinogen antiserum in fish stomachs: possible ancestor molecules of chymosin of higher vertebrates. *Comp. Biochem. Physiol.* 91A: 565–69.

Young, J. Z. 1933. The autonomic nervous system of selachians. *Q. J. Microsc. Sci.* 75: 571–624.

———. 1980. Nervous control of stomach movements in dogfishes and rays. *J. Mar. Biol. Assoc. U.K.* 60: 1–17.

———. 1983. Control of movements of the stomach and spiral intestine of *Raja* and *Scyliorhinus*. *J. Mar. Biol. Assoc. U.K.* 63: 557–74.

———. 1988. Sympathetic innervation of the rectum and bladder of the skate and parallel effects of ATP and adrenalin. *Comp. Biochem. Physiol.* 89C: 101–07.

Yui, R., and T. Fujita. 1986. Immunocytochemical studies on the pancreatic islets of the ratfish, *Chimaera monstrosa*. *Arch. Histol. Japan* 40: 369–77.

Yui, R., M. Shimada, and T. Fujita. 1990. Immunohistochemical studies of peptide- and amine-containing endocrine cells and nerves in the gut and the rectal gland of the ratfish *Chimaera monstrosa*. *Cell Tissue Res.* 260: 193–201.

Zendzian, E., and E. A. Barnard. 1967. Distribution of pancreatic ribonuclease, chymotrypsin, and trypsin in vertebrates. *Arch. Biochem. Biophys.* 122: 699–713.

PATRICK J. BUTLER

CHAPTER 7

Respiratory System

The major function of the respiratory system of any animal is the exchange of the respiratory gases (oxygen and carbon dioxide) between the environment and the metabolizing cells, and in all vertebrates this is achieved in conjunction with the circulatory system. The organs of gas exchange in fishes are the gills. Typically, in elasmobranch fishes, there are five gill arches on each side of the pharynx, the first of which bears only a single row of filaments (hemibranch). Each of the remaining four gill arches consists of a row of filaments on either side (holobranch) of a sheet of muscular and connective tissue, the interbranchial septum (Fig. 7-1), which is supported by lateral rods of cartilage, the gill rays, which radiate out from the skeletal structure of the arch. The tip of each filament is usually free of the interbranchial septum. However, the septum extends beyond the end of the gill filaments to form a flap that covers each gill slit. Thus in elasmobranch fish, there are a number of gill slits (usually five) on each side of the head. In some species, the number of gill slits and gill arches is greater, for example, seven in *Heptanchus*. In addition to the gill slits as such, there is also, at the front end of the pharynx, the remnant of a gill opening known as the spiracle (see Fig. 7-1). This is particularly well developed in the bottom-dwelling skates and rays and in some of the more sluggish sharks. However, it is small in many forms of shark and is absent altogether in the more active genera.

On both the dorsal and ventral surfaces of each gill filament are arranged a row of (secondary) lamellae (Fig. 7-2) and these are the sites

Respiratory System 175

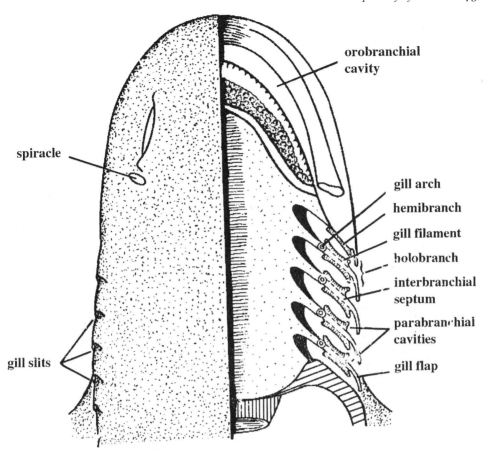

FIGURE 7-1 Head of a sharklike elasmobranch. The right half is sectioned horizontally through the pharynx in order to show the arrangement of the gills. (Modified from Romer 1955.)

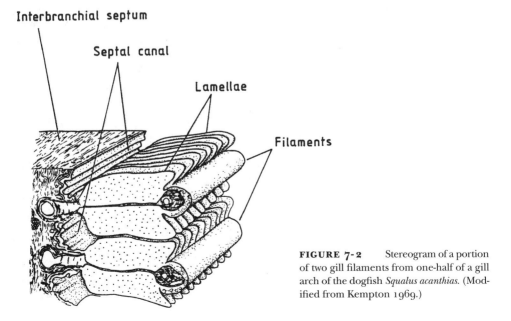

FIGURE 7-2 Stereogram of a portion of two gill filaments from one-half of a gill arch of the dogfish *Squalus acanthias*. (Modified from Kempton 1969.)

of gas exchange. This arrangement of the gills separates the orobranchial cavity from the five parabranchial cavities (see Fig. 7-1), and water is pumped from the former to the latter during the ventilation cycle. This chapter considers the ventilation mechanism, the detailed anatomy of the gills, and the factors affecting gas exchange in the gills of elasmobranchs.

Ventilation

The way in which the flow of water is maintained across the gill filaments was determined in an elasmobranch, the dogfish *Scyliorhinus canicula*, by Hughes and Ballintijn in 1965. They simultaneously monitored the movements of the lower jaw and one of the gill slits, the pressures in the oro- and parabranchial cavities and the electrical activity (electromyogram [EMG]) in the muscles that were thought to be involved in ventilation (Fig. 7-3). Water is drawn into the orobranchial cavity via the mouth and via the spiracles. In actively ventilating fish, gill ventilation is mainly achieved by the superficial sheet of constrictor muscles, which contracts against the elastic skeleton of the gills (sometimes known as the branchial basket). These muscles contract in a peristaltic sequence, with the closure of the jaw preceding the main phase of contraction, which travels caudally down the branchial basket. As these muscles contract, the volumes of both the orobranchial and parabranchial cavities are reduced and the water is forced through the gills from the orobranchial cavity to the parabranchial cavities.

The increase in pressure in both cavities is almost simultaneous (see Fig. 7-3). When the constrictor muscles relax, the parabranchial cavities expand passively because of the elasticity of the branchial skeleton. However, since the external gill flaps are closed, water is drawn across the gills from the orobranchial cavity to the parabranchial cavities. During this phase of the ventilatory cycle, the main muscles to be active are the adductor mandibulae (see Fig. 7-3), which function to maintain the position of the mandible and hyoid arch.

These muscles regulate the amount of water that enters the orobranchial cavity by adjusting the degree of opening of the mouth. Thus, an almost continuous flow of water is maintained across the gills by the combined action of orobranchial pressure and of parabranchial suction, although there is no anatomical basis for considering the existence of two separate pumps. Little activity was recorded in the hypobranchial muscles during the normal ventilation cycle, although they may be active during hyperventilation.

In the dogfish *S. canicula*, water that enters the mouth leaves via the three posterior gill slits, while the water that enters via the spiracles leaves by the more anterior gill slits (Hughes 1960), indicating that there is little mixing of water in the orobranchial cavity. This separation of water flows is less marked in the thornback ray (skate), *Raja clavata*. In fact, when this dorsoventrally flattened species is resting on the substratum, water enters via the two dorsally situated spiracles, whereas when the animal is swimming, water also enters through the mouth. The differential pressure between the orobranchial and parabranchial cavities in the dogfish (Fig. 7-4) indicates that parabranchial suction is at least as important as orobranchial pressure in maintaining water flow, whereas in the skate, parabranchial suction may be of more importance (Hughes 1960). The lack of even a slight reversal of the differential pressure across the gills of the skate and the electric ray, *Torpedo marmorata* (Hughes 1960, 1978) suggests that the gill flaps may be controlled actively, rather than passively. Hughes (1960) suggested that such control may be functionally significant in preventing a reversal of the water flow and thus the entry of sand into the branchial apparatus of these bottom-dwelling species.

The spiracles are much reduced in size in some of the sharklike elasmobranchs, and Grigg (1970a) noted that under certain circumstances, the Port Jackson shark, *Heterodontus portusjacksoni*, takes in water through the first gill slit. He suggested that this phenomenon may be related to the fact that this

Respiratory System 177

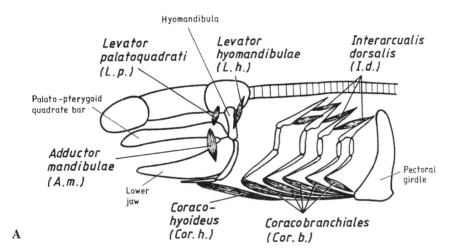

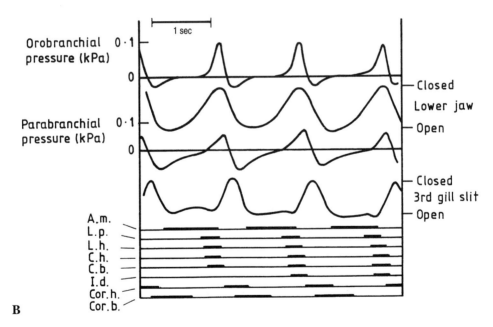

FIGURE 7-3 **A:** Diagram of the anterior skeleton and main ventilatory muscles of the dogfish *Scyliorhinus canicula* (the superficial constrictor muscles have been omitted for clarity). **B:** Diagram of the pressures in the orobranchial and third parabranchial cavities, movements of the lower jaw and the third gill slit, and the main phase of activity of the ventilatory muscles shown in A, plus the superficial constrictor muscles, constrictor hyoideus (C.h.) and constrictor brachialis (C.b.). (Modified from Hughes and Ballintijn 1965.)

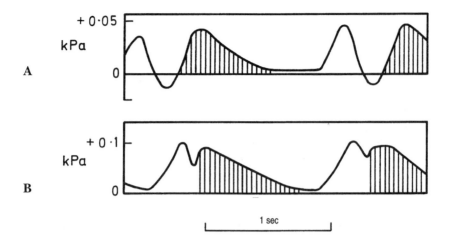

FIGURE 7-4 Diagram of the difference in pressure between the orobranchial and third parabranchial cavities of (**A**) the dogfish, *S. canicula*, and (**B**) the thornback ray, *R. clavata*. A positive pressure indicates that pressure in the orobranchial cavity is greater than that in the parabranchial cavity. The phase when parabranchial suction predominates has been shaded. (Modified from Hughes 1960.)

sluggish animal, like the rays and skates, crushes its food (molluscs and crustaceans) in its mouth, thus preventing the entry of water through the mouth. Unlike the rays and skates, it is unable to use its spiracles, so it uses its first gill slits instead. In more active sharks, such as the leopard shark, *Triakis semifasciata*, the mouth and gill flaps are held open during normal swimming so that water passes in a continuous stream over the gills (Hughes 1960). This is known as ram ventilation, and the only branchial muscles to be active under these conditions may be the adductor mandibulae (Hughes and Ballintijn 1965). When the fish is not swimming, it appears to ventilate its gills in much the same way as the dogfish.

There has been some discussion concerning the pathway taken by the water current as it flows across the gills from the orobranchial cavity to the parabranchial cavities. It has been suggested that the presence of the interbranchial septum prevents the water passing freely between the lamellae from their outer to inner edges and thus traveling in the opposite direction to the flow of blood through them, as in teleosts (Hughes and Shelton 1962; Piiper and Schumann 1967). It has since been demonstrated that, at least in *S. acanthias* and the Port Jackson shark, there is a channel, the septal canal, between the inner edges of the lamellae and the interbranchial septum (Kempton 1969; Grigg 1970b; De Vries and De Jager 1984). This arrangement does allow the free passage of water between the lamellae (Fig. 7-5). It has been noted for the Endeavour dogfish, *Centrophorus scalpratus*, that the basal 20 or so pairs of lamellae of every filament are covered by a sheet of tissue formed from the gill arch (Cooke 1980), and it is concluded that this canopy must prevent water from flowing over the lamellae that it covers.

Gill morphology

In terms of gas exchange, the lamellae are the functional elements of the gills. Each lamella consists of a double sheet of epithelium, beneath which is a well-developed basement membrane. The space between the epithelial layers is divided into a number of channels by pillar cells, and blood cells fill the channels (Fig. 7-6). In a study on a number of species of elasmobranchs, Hughes and Wright (1970) found that the epithelium consists of a double

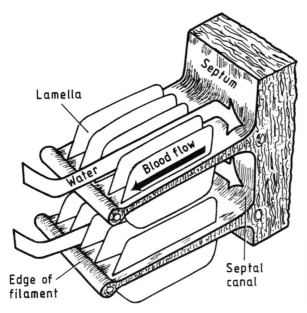

FIGURE 7-5 Schematic diagram showing the proposed direction of water flow between the lamellae and into the septal canals in the Port Jackson shark. (Redrawn from Grigg 1970a.)

layer of cells, with the outer membrane of the external layer forming numerous microvilli. Flanges of the pillar cells form the linings of the blood channels. Both chloride-secreting cells and mucus cells are present in the lamellae, and both types of cells are more numerous on their filament edge than on their free edge. Thus, the respiratory gases must pass through two epithelial cells, the basement membrane, and the flange of the pillar cell, as they move between the blood and the water. The blood channels are small, which means that the red blood cells are close to the flanges of the pillar cells and that the diffusion distance within the blood is very small.

Measurements of the distance between the water and blood across the lamellae (water-blood distance) are sometimes made of the overall distance from the tips of the microvilli to the inner membrane of the flange of the pillar cell and sometimes from the base of the microvilli, which is the minimum distance (Hughes and Wright 1970). These authors present data using both methods applied to electron micrographs obtained from the lamellae of a number of species of elasmobranchs, and two interesting features emerge. First, the distance is longer in elasmobranchs than in teleosts, and second, there is a large range in the measurements from any one species. For example, in the dogfish, *S. canicula*, measurements to the tips of the microvilli range from 5.2 μm to 19.1 μm, with an arithmetic mean for about 50 measurements of 11.3 μm. It was also noted that there is a tendency for the mean thickness to decrease from more active fish such as *S. acanthias* (10.1 μm) to the less active ray, *R. montagui* (4.8 μm). This is rather strange, particularly in view of the fact that the most active teleosts, such as tuna, have the smallest diffusion distances (Hughes 1970).

The earliest measurements of the water-blood distance were made perpendicularly to the lamellar surface at different parts of the blood channel (see Hughes and Morgan 1973 for discussion). A problem with this approach is that the investigator is never certain that the sections are cut precisely perpendicular to the surface. In later studies, the measurement was made of the thinnest part of the blood-water barrier in any section and then Hughes (1972a) measured the harmonic mean thickness of the lamellae from the gills of tench (Fig. 7-7). This method, which is a modification of that developed by Weibel (1971) for the mammalian lung, takes many possible

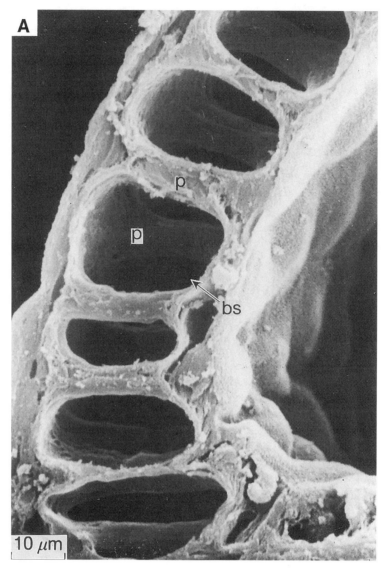

FIGURE 7-6 **A:** Scanning electron micrograph of a transverse section through the lamella from the gill of the dogfish *S. canicula*. bs = Blood space; p = pillar cell. (Metcalfe and Butler 1986.) **B:** Diagram of transverse section through the lamella of a dogfish showing the details of the cell layers, the basal channels, and the marginal channel. (Based on Randall 1982.)

diffusional pathways into account and not just the shortest. As such, it might be expected that the mean harmonic thickness would be greater than the thickness measured by the other two methods. However, Hughes (1972a) found that there was little difference in the values obtained by the three different methods.

Hughes et al. (1986) found that the water-blood distance varied little with body mass in *S. canicula*.

The other important morphological feature of the lamellae in terms of gas exchange is their surface area. As might be imagined, this is not very easy to measure, as the shape and di-

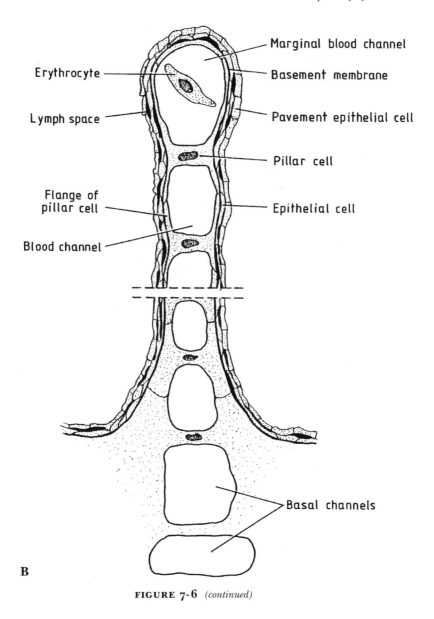

FIGURE 7-6 *(continued)*

mensions of the lamellae vary with their position along each filament (Fig. 7-8). Lamellae at the tip of the filaments are shorter and higher than those at the base and the surface area of individual lamellae seems to be maximal in the middle portion of each filament (Hughes et al. 1986). Another problem is the shrinkage of tissue during its fixation and subsequent storage in alcohol. Despite these problems, Hughes et al. (1986) measured total lamellar area in the dogfish *Scyliorhinus stellaris* and found that it does not increase in direct proportion to body mass, but rather to (body mass)$^{0.78}$. They also point out that their data

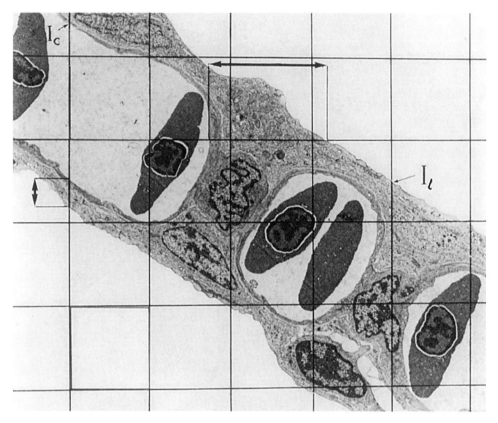

FIGURE 7-7 Electron micrograph of a lamella of a tench with a superimposed grid. The *thick arrows* give two examples of the distances measured during the determination of mean harmonic thickness of the blood-water barrier. The *small arrows* indicate intercepts of the surface of the lamella (I_l) and of the surface of a blood channel (I_c). (Hughes 1972a.)

represent the minimal values and could underestimate by as much as 30%. The allometric formula that they produce is

Total gas exchange area (cm)^2 =
1350 (body mass in kg)$^{0.78}$

This means that the area of the gas exchange surface is relatively smaller in larger fish and is different from the situation described by Hughes (1972a, 1977, 1978) for other species of elasmobranchs, where the mass exponent is close to 1. The physiological significance of this difference is unclear.

Gill circulation

For a more complete analysis of gas exchange, it is necessary to consider the circulatory system through the gills. Each gill (holobranch) receives mixed venous blood direct from the heart via its afferent branchial artery (aba). The innominate arteries, which divide to produce the first two pairs of afferent arteries, receive approximately 33% of the cardiac output in a number of species of elasmobranchs and in the leopard shark, the third and fourth afferent arteries receive approximately 25% each, with the remaining 17% going to the fifth aba (Lai et al. 1989). In a 5.9-kg leopard shark, there is approximately 25 mL of blood in the secondary lamellae, which is approximately 6% of its total blood volume (Bhargava et al. 1992). Within each gill there there are two distinct yet extensively interconnected blood pathways. One appears to supply blood to the lamellae and has become known as

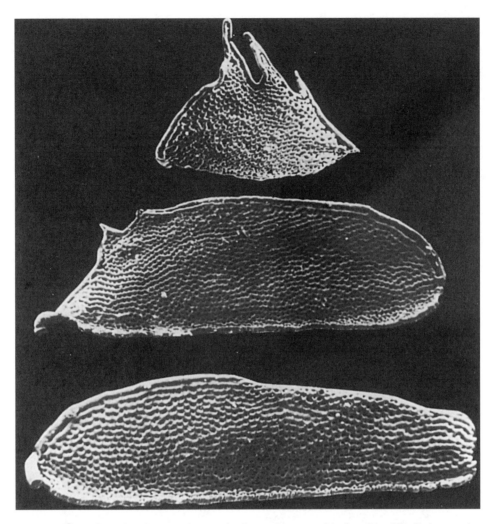

FIGURE 7-8 Scanning electron micrograph of corrosion casts of the lamellae of the Endeavour dogfish showing the variation in their shape along the length of a filament. **Top:** lamella from the tip of a filament. **Middle:** lamella from midway along a filament. **Bottom:** lamella from near the base of a filament. (Cooke 1980.)

the "respiratory" blood pathway; the other is called the "nonrespiratory" blood pathway and probably has a mixed nutritive and associated venous drainage function. For the sake of clarity, these two blood pathways are described separately.

THE "RESPIRATORY" BLOOD PATHWAY
Each gill filament receives blood from the afferent branchial artery (Fig. 7-9) via an afferent filament artery, which may itself divide and supply several adjacent filaments. The afferent filament artery passes through the tissues of the interbranchial septum and joins the filament about one-third of the way along its length, where it divides into two branches. One of these branches supplies the distal two thirds of the filament, while a recurrent branch supplies the basal one-third. These two branches can functionally be considered as one continuous afferent filament artery.

Adjacent to the afferent filament artery on the filament side lies the corpus cavernosum (Figs. 7-9, 7-10, and 7-11), which is an

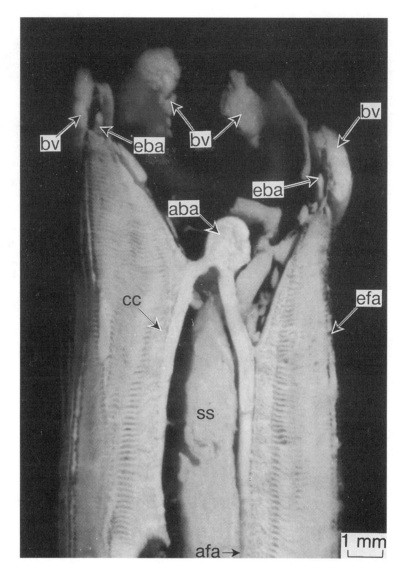

FIGURE 7-9 Light micrograph of a transverse section through a latex cast of a gill of the dogfish *S. canicula*, showing the general form of the vascular anatomy. aba = Afferent branchial artery; afa = afferent filament artery; bv = branchial vein; cc = corpus cavernosum; eba = efferent branchial artery; efa = efferent filament artery; ss = septal sinus. (From Metcalfe and Butler 1986.)

irregular structure almost uniquely confined among fishes to the elasmobranchs. This runs the entire length of the filament and receives blood from the afferent filament artery via numerous connections. Wright (1973) reports that in *S. canicula* these junctions possess smooth muscle sphincters, but this has not been confirmed in more recent studies of the species (Dunel and Laurent 1980) or in *S. acanthias* (De Vries and De Jager 1984). Toward the tip of each filament, as it separates from the interbranchial septum, the distinction between a separate afferent filament artery and corpus cavernosum becomes less clear, and they appear to merge into a single structure in both *S. acanthias* (De Vries and De Jager 1984) and *S. canicula* (Metcalfe and Butler 1986).

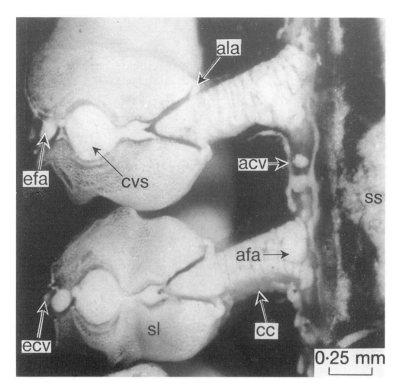

FIGURE 7-10 Light micrograph of a transverse section through a latex cast of two consecutive filaments of a gill of the dogfish *S. canicula*. acv = Afferent companion vessel; afa = afferent filament artery; ala = afferent lamellar arteriole; cc = corpus cavernosum; cvs = central venous sinus; ecv = efferent companion vessel; efa = efferent filament artery; sl = secondary lamella; ss = septal sinus. (From Metcalfe and Butler 1986.)

The separation of the corpus cavernosum into lateral and medial parts (Olson and Kent 1980) does not appear to be functionally meaningful (De Vries and De Jager 1984). The physiological role of the corpus cavernosum is not fully understood, and it seems likely that it may perform three important functions. Acrivo (1935) suggested that it is the site of erythrocyte destruction and, although Kempton (1969) could find no evidence for this, Wright (1973) describes endothelial cells that appear to be phagocytic. He has also suggested that the corpus cavernosum may act as a pulse-smoothing capacitance vessel, and De Vries and De Jager (1984) support this idea.

Cooke (1980) suggested that the corpus cavernosum may act as a hydraulic skeleton, providing support for the gill filaments, since the skeletal cartilage usually observed in the gill filaments of teleosts is absent from those of elasmobranchs. This is consistent with the observation of De Vries and De Jager (1984) that artificially applied ventral aortic pressures, at the top of the physiological range, cause the tips of filaments to move into the position normally observed in the live animal. It is probably of functional significance, therefore, that the corpus cavernosum occupies the same position in the gill filaments of elasmobranchs as that occupied by the skeletal cartilage in the gill filaments of teleosts.

The lamellae receive blood from the corpus cavernosum via afferent lamellar arterioles, and the blood flows through the lamellae in the channels formed by the pillar cells. Casts of the lamellar blood channels indicate that they form an interconnecting network, which creates a near sheet of blood, which is "perforated" by the pillar cells (see Fig. 7-8). The channel at the outer edge is larger than

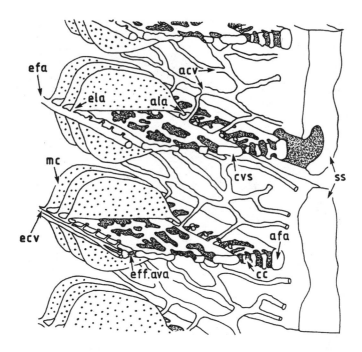

FIGURE 7-11 Semidiagrammatic stereogram of the vascular anatomy of a gill of the dogfish S. *canicula*. It illustrates portions of two consecutive gill filaments from one hemibranch. afa = Afferent filament artery; ala = afferent lamellar arteriole; acu = afferent companion vessel; cc = corpus cavernosum; ecv = efferent companion vessel; efa = efferent filament artery; eff. ava = efferent arteriovenous anastomosis; ela = efferent lamellar arteriole; mc = marginal channel; ss = septal sinus. (From Metcalfe and Butler 1986.)

the others (Kempton 1969; Wright 1973) and in the Endeavour dogfish, at least, it is somewhat separate from the others (Cooke 1980). This author also noted that the basal three or four blood channels of each lamella are buried below the epithelium of the filament (see Fig. 7-6B), as has been reported for teleosts (Kuhn and Koecke 1956).

Blood, now oxygenated, leaves the lamellae via efferent lamellar arterioles and enters the efferent filament artery (see Fig. 7-11). Both the afferent and efferent lamellar arterioles are reported to possess smooth muscle sphincters (Wright 1973; Dunel and Laurent 1980). From the efferent filament arteries, blood enters the appropriate efferent branchial artery, of which there are two in each holobranch; one receives blood from filaments on its anterior surface and the other receives blood from filaments on the posterior surface (see Fig. 7-9). The efferent branchial arteries enter the dorsal aorta, which distributes blood to the systemic circulation. In *S. canicula*, smooth muscle sphincters have been observed in the efferent filament artery, just prior to its junction with the efferent branchial artery. Such sphincters do not exist, however, in *Raja clavata* (Dunel and Laurent 1980).

THE "NON-RESPIRATORY"
BLOOD PATHWAY

In addition to the vascular pathways described above, which are directly concerned with blood flow to and from the gas-exchange surfaces, there is also an extensive vacular network that consists of a number of components and that appears not to be involved in gas exchange.

As well as supplying blood to the lamellae via the corpus cavernosum and afferent lamellar arterioles, the afferent filament artery also gives rise to a network of blood vessels within the interbranchial septum. These vessels extend for a short distance on both the ventral and dorsal sides of the afferent filament artery at its more distal end before it separates from the interbranchial septum. As the gill filaments separate from the interbranchial septum, this vascular network becomes more extensive and joins with similar networks from adjacent filaments in such a way that it is continuous throughout the distal portion of the interbranchial septum (Fig. 7-12). The function of this network is not fully understood, but it must presumably, at least in part, serve a nutritive function for the tissues of the interbranchial septum.

Within the central part of the gill filament, between the corpus cavernosum and the efferent filament artery, lies an extensive central venous sinus (Figs. 7-10, 7-11, and 7-13). In a number of species, such as *S. acanthias*, *C. scalpratus*, and *R. clavata*, the central venous sinus receives both pre- and postlamellar blood via anastomoses with the corpus cavernosum (effectively the afferent filament artery) and the efferent filament artery, respectively (Olson and Kent 1980; De Vries and De Jager 1984; Cooke 1980; Dunel and Laurent 1980), but in other species, such as *S. canicula* and *R. erinacea* (Metcalfe and Butler 1986; Dunel and Laurent 1980; Olson and Kent 1980), the central venous sinus receives only postlamellar blood from the efferent filament artery (Fig. 7-11).

Blood that has entered the central venous sinus drains both into the large blood sinuses within the interbranchial septum via vessels that have variously been termed "afferent companion vessels" (see Figs. 7-10, 7-11, and 7-13) (Cooke 1980; Metcalfe and Butler 1986) or a "venous web" (De Vries and De Jager 1984) and into efferent companion vessels that run

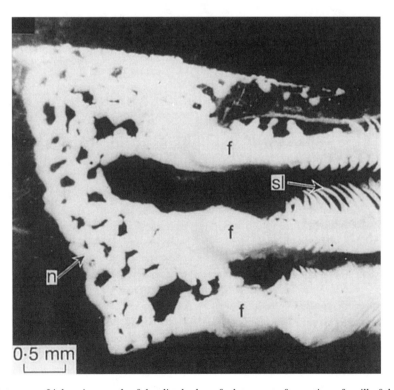

FIGURE 7-12 Light micrograph of the distal edge of a latex cast of a portion of a gill of the dogfish *S. canicula*, showing the network of blood vessels (n) that originates from the distal end of each afferent filament artery. The major part of the gill filaments has been dissected away. f = Gill filament; sl = secondary lamella. (From Metcalfe and Butler 1986.)

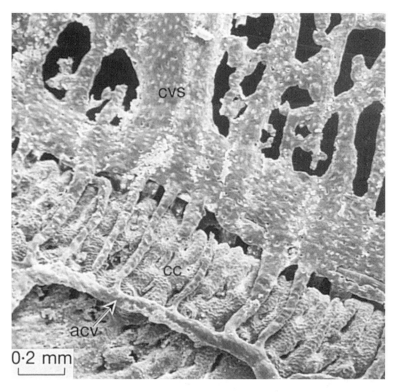

FIGURE 7-13 Scanning electron micrograph of a methyl methacrylate cast of a gill filament of the dogfish *S. canicula*, showing the connections between the central venous sinus (cvs) and the afferent companion vessels (acv). cc = Corpus cavernosum. (From Metcalfe and Butler 1986.)

parallel to the efferent filament artery (see Figs. 7-10 and 7-11) and empty into subepithelial sinuses in the gill arch. The efferent companion vessels are also reported to receive a small blood supply directly from the efferent filament artery. The venous sinuses of the interbranchial septum and the gill arch drain dorsally into the anterior cardinal sinus and ventrally into the inferior jugular sinuses.

Although the functional role of the central venous sinus has yet to be resolved, it is now generally agreed (Cooke 1980; De Vries and De Jager 1984; Metcalfe and Butler 1986) (Fig. 7-14) that even in those elasmobranch species where it connects with both the afferent and the efferent sides of the filament circulation, it does not constitute a route by which blood may bypass the exchange surface of the lamellae as suggested by Steen and Kruysse (1964) for teleosts. Since the central venous sinus is directly connected to the large venous sinuses of the gill arch and of the interbranchial septum, it must operate at the pressures within the venous system and, as these are below those of the efferent filament artery, its functioning as a lamellar bypass is precluded. Although efferent arteriovenous anastomoses will allow oxygenated, post-lamellar blood to be diverted via the branchial venous system back to the heart, it seems unlikely that in elasmobranchs this is an important route for supplying cardiac muscle with oxygen, since there is frequently a well-developed coronary blood supply (Tota et al. 1983).

As blood flows through the gills, blood pressure is reduced by some 25–30% from prebranchial values of 3–5 kPa to postbranchial values of between 2 and 4 kPa. It is presumably because of these relatively low postbranchial blood pressures that the arteries that branch off the dorsal aorta and supply the skeletal muscles and viscera are valved close to their origin. These valves prevent any backflow of blood that would otherwise occur when blood pressure in the peripheral arteries rises as a consequence of the compression of the blood vessels in the postpelvic trunk during locomotion.

Analysis of gas exchange in relation to the morphology of the gills

Gas exchange across any respiratory surface is by *diffusion* and is, therefore, governed by Fick's law of diffusion:

$$\dot{M} = \frac{KA \, \Delta P}{X} \qquad (7\text{-}1)$$

Where $\dot{M}$ = amount of gas diffusing per unit time (mmol sec^{-1})

K = Krogh's diffusion constant (mmol sec^{-1} mm^{-1} kPa^{-1})

A = area of exchange surface (mm^2)

ΔP = difference in partial pressure of gas on either side of the exchange surface (kPa)

X = thickness of the exchange surface or water-blood distance (mm)

The conditions that are necessary to ensure that diffusion proceeds adequately are maintained by the activity of the respiratory and cardiovascular systems, that is, by *convection*. These systems together maintain ΔP at as high a level as possible and, at the same time, transport oxygen to the gills (ventilation) and then to the tissues (circulation). The transport of carbon dioxide occurs in the opposite direction. These processes can be described by Fick's laws of convection, as shown:

For the respiratory system:

$$\dot{M} = \dot{V}_w \, (C_i - C_e) = \dot{V}_w \cdot \beta_w (P_i - P_e) \qquad (7\text{-}2)$$

And for the circulatory system:

$$\dot{M} = \dot{V}_b \, (C_a - C_{\bar{v}}) = \dot{V}_b \cdot \beta_b (P_a - P_{\bar{v}}) \qquad (7\text{-}3)$$

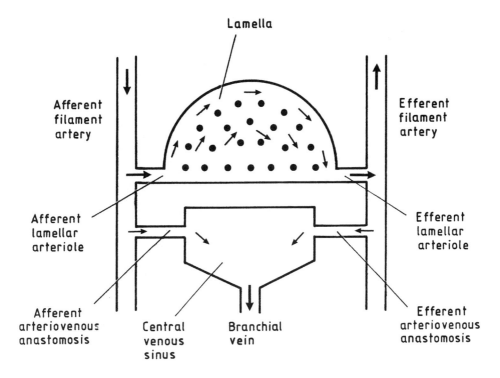

FIGURE 7-14 Diagram to show the anastomoses between the central venous sinus and the afferent and efferent filament arteries in the gills of elasmobranchs. (From Butler and Metcalfe 1983.)

Where

$\dot{V}_w$ and $\dot{V}_b$ = volumes of water and blood, respectively, passing over and through the gills in unit time (1 min^{-1})

C_i and C_e = concentrations of gas in inspired and expired water, respectively (mmol L^{-1})

C_a and $C_{\bar{v}}$ = concentrations of gas in arterial and mixed venous blood, respectively (mmol L^{-1})

P_i and P_e = partial pressures of gas in inspired and expired water, respectively (kPa)

P_a and $P_{\bar{v}}$ = partial pressures of gas in arterial and mixed venous blood, respectively (kPa)

β_w and β_b = capacitance coefficients of gas in water and blood, respectively (mmol L^{-1} kPa^{-1})

The effectiveness of the gills in exchanging gases depends on a number of factors, and these include: the resistance to diffusion that exists at the gills, in the water, and in the blood; the relative direction in which the blood and water flow; the proportion of each fluid that is effectively not involved in gas exchange, thus forming shunts; and unequal distribution of water and blood to the gills.

Gas exchange across the gills of fishes has been analyzed in terms of three different conductances, which can be derived from equations (7-1), (7-2), and (7-3) (Piiper and Scheid 1975; Scheid and Piiper 1976):

1. *Diffusion conductance,*

$$G_{diff} = \frac{\dot{M}}{\Delta P} = \frac{KA}{x}; \text{ see equation (7-1)}$$

2. *Ventilation conductance,*

$$G_{vent} = \dot{V}_w \cdot \beta_w = \frac{\dot{M}}{P_i - P_e}; \text{ see equation (7-2)}$$

3. *Perfusion conductance,*

$$G_{perf} = \dot{V}_b \cdot \beta_b = \frac{\dot{M}}{P_a - P_{\bar{v}}}; \text{ see equation (7-3)}$$

RESISTANCE TO DIFFUSION

Diffusion conductance, also known as the diffusion capacity (D) or transfer factor (T), is clearly related to the effective surface area of the gills (A) and to the thickness of the water-blood distance (x) as well as to Krogh's diffusion constant. It is the rate at which a gas (usually oxygen) will diffuse across the exchange surface for a given difference in partial pressure of that gas across the exchanger. The reciprocal of diffusion conductance, $1/G_{diff}$ gives an indication of the overall resistance to gas exchange across the gills (Fig. 7-15C).

Using morphometric data and K for oxygen in tissue, Hughes (1972a), in his review paper, gives a value for D for *S. stellaris* of 0.0272 mL min^{-1} kg^{-1} mmHg^{-1} (= 9.1 μmol min^{-1} kg^{-1} kPa^{-1}) whereas Hughes et al. (1986) give a value, for the same species, of approximately 4.1 μmol min^{-1} mmHg^{-1} (= 30.7 μmol min^{-1} kPa^{-1}). It is not clear whether this is for a 1 kg fish. What these authors do point out, however, is that these estimates do not take into account the resistance to gas exchange that resides in the water between the lamellae. This is important, because this factor has been shown to be a substantial proportion of the overall resistance to diffusion (Hills and Hughes, 1970; Scheid and Piiper 1971, 1976). For *S. stellaris* weighing 1.6 kg, the mean distance between the lamellae, that is, the thickness of the water between the lamellae, at 83 μm, is more than eight times the harmonic mean of the water-blood distance (Hughes et al. 1986).

G_{diff} has also been determined physiologically, basically by using the expression $\dot{M}/\Delta P$. Randall et al. (1967), working on the rainbow trout, calculated ΔP as:

$$\frac{P_iO_2 + P_eO_2}{2} - \frac{P_aO_2 + P_{\bar{v}}O_2}{2} \quad (7\text{-}4)$$

This assumes that the oxygen equilibrium curve is linear rather than sigmoid and that the oxygen partial pressure profiles in both water and blood are linear along the lamellae (see Fig. 7-15). The latter is only the case if $G_{vent} = G_{perf}$ (Piiper and Scheid 1982) and other methods for calculating ΔP have been proposed should this not be the case (see Butler and Metcalfe 1988 for details). It seems

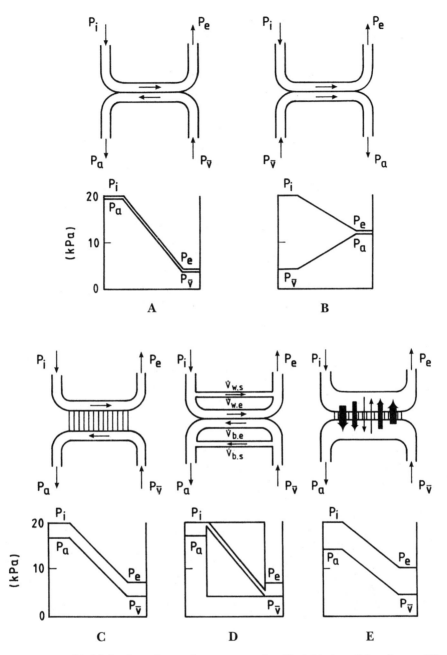

FIGURE 7-15 Models for the exchange of oxygen across the gills. **A:** Ideal conditions (i.e., no diffusion limitation and $G_{vent} = G_{perf}$) with countercurrent flow of blood and water. Lengths of arrows in the exchange unit indicate relative magnitude of G_{vent} or G_{perf}. **B:** Ideal conditions with concurrent flow. **C:** Countercurrent with diffusion resistance. **D:** Countercurrent with blood and water shunts. In this case, some blood and water are thought of as equilibrating completely and, therefore, as being effective ($\dot{V}_{b.e}$ and $\dot{V}_{w.e}$), while the rest of the blood and water, which undergo no gas exchange, can be thought of as being ineffective and constituting shunts ($\dot{V}_{b.s}$ and $\dot{V}_{w.s}$). **E:** countercurrent with combination of diffusion resistance and shunts. In this case, the *vertical arrows* in the exchange unit represent gas exchange by diffusion, with the thickness indicating the degree of equilibration over the distance denoted by the length. Beyond a certain distance, no gas exchange occurs within the time that the liquids are in the exchange unit.

The graphs indicate changes in the partial pressure of oxygen (PO_2) in water and in blood under each of the above conditions. P_i and P_e are partial pressures in inspired and expired water, respectively, while P_a and $P_{\bar{v}}$ are partial pressures in arterial and mixed venous blood, respectively. (Butler and Metcalfe 1988.)

that for elasmobranchs, G_{vent} is close enough to G_{perf} for equation (7-4) to give similar values of ΔP to those obtained by the other methods (Piiper and Baumgarten-Schumann 1968; Piiper et al. 1977; Short et al. 1979). Physiological values of G_{diff} for elasmobranchs vary between 2.75 and 5.25 µmol min^{-1}kg^{-1} kPa^{-1} (Butler and Metcalfe 1988).

The disparity between the morphometric and physiological values of G_{diff} is, at least partly, due to the influence of the water between the lamellae. Using physiological data on gas transfer across the gills and morphometric data on the gills themselves, Scheid and Piiper (1976), have estimated that in the resting dogfish, *S. stellaris*, 42% of the total resistance to O_2 transfer resides in the interlamellar water and 49% in the water-blood barrier. The remainder (9%) may be accounted for by factors such as diffusion and chemical reaction within the blood, cyclic variations of water and blood flows, and shunting of water and/or blood. In a more recent study, Piiper et al. (1986) found that the resistance to diffusion in the water is some 85% greater than that across the water-blood barrier in resting *S. stellaris*, and when this and other factors are taken into account, there is closer agreement between the morphometric and physiological values of G_{diff} (4.68 versus 2.85 µmol min^{-1}kg^{-1}kPa^{-1}, respectively). In swimming animals, the values are even closer, at 5.31 and 5.78 µmol min^{-1} kg^{-1} kPa^{-1}, respectively.

Using the modern technique of digital radiographic imaging of blood circulation through the gills of the leopard shark, Bhargava et al. (1992) have demonstrated that it takes, on average, 6.5 seconds for blood to travel through the lamellae. These authors were able to perform time domain analysis of O_2 diffusion and hemoglobin-O_2 reaction kinetics; they concluded that the relatively long transit time is indicative of a large diffusion limitation and a relatively slow rate of binding between hemoglobin and O_2.

A problem with the physiological assessment of G_{diff} in dogfish is that the measurement of P_e involves the attachment of collecting devices around the gill slits, a procedure that appears to prevent the fish from increasing ventilation sufficiently to cope with hypoxia (see Butler and Taylor 1975; Short et al. 1979). In an attempt to circumvent this problem, Metcalfe (1983) has shown that G_{diff} for oxygen can be reliably determined in *S. canicula* from measurements of $\dot{M}O_2$, P_aO_2 and P_iO_2, provided that $G_{perf} = G_{vent}$.

DIRECTIONS OF WATER AND BLOOD FLOWS

If there is no diffusion barrier between blood and water and if $G_{vent} = G_{perf}$, with water and blood flowing in opposite directions (countercurrent), it is theoretically possible that P_aO_2 would equal P_iO_2 and that P_eO_2 would equal P_vO_2 (see Fig. 7-15A). This situation has never been found to occur in any aquatic animal; however, the fact that P_aO_2 can be > P_eO_2 in a number of elasmobranchs (Baumgarten-Schumann and Piiper 1968; Hanson and Johansen 1970; Grigg and Read 1971; Short et al. 1979) indicates that there is not a parallel (cocurrent) flow arrangement (see Fig. 7-15B). The presence of the septal canal between the inner edges of the lamellae and the interbranchial septum does provide a route for water to flow over the exchange area in the opposite direction to that of the blood, and it is now generally accepted that such a countercurrent arrangement does exist in elasmobranchs (Piiper and Scheid 1984). It is perhaps worth pointing out here that because of the approximate ten times difference between βO_2 in blood and in water, $\dot{V}_w / \dot{V}_b$ is usually about 10.

SHUNTS

Some water may not pass through the sieve produced by the lamellae and therefore not come into contact with the gas exchange surfaces of the gills. This represents an anatomical, or true, shunt. A similar anatomical shunt exists when water passes between lamellae that are not perfused. In resting rainbow trout in well aerated water, some 40% of the lamellae are not perfused with blood (Booth 1978) and a similar situation may exist in elasmobranchs. For water that does flow between the lamellae that are perfused, the extent of its participation in gas exchange will, because of the resistance to diffusion of oxygen through water, depend on its distance from the lamellar surface. Another way to look at the effect of diffusion resistance in water is to think in terms of a shunt.

For water more than a given distance away, the diffusion resistance of the water may be too high for any significant gas exchange to occur during its transit time through the gills (see Fig. 7-15D). Such water can be considered to constitute a shunt (Piiper and Baumgarten-Schumann 1968). Water participating in gas exchange, but not equilibrating completely, can be considered as consisting of a portion that equilibrates completely and of another component that takes no part in gas exchange at all, that is, it is functionally similar to the water that completely bypasses the gills or is too far away to take part in gas exchange. Thus, total water flow ($\dot{V}_{w \cdot t}$) can be divided into ineffective or shunted water ($\dot{V}_{w \cdot s}$) and effective water ($\dot{V}_{w \cdot e}$). Thus, in otherwise ideal conditions, the deviation of P_e from its ideal value ($P_{\bar{v}}$) can be attributed to the fraction of total water that is effectively shunted (Piiper and Scheid 1984).

Abnormally high ventilation rates, possibly as a result of the animals being disturbed, would reduce the contact time of the water with the lamellae, thus increasing the ineffective or shunted water and maybe also increasing the proportion of water bypassing the lamellae (true shunt). This could explain why P_aO_2 is not always greater than P_eO_2 in a number of species of elasmobranchs (Lenfant and Johansen 1966; Baumgarten-Schumann and Piiper 1968; Hanson and Johansen 1970; Short et al. 1979). Even in the absence of any diffusion barrier, P_e will not equilibrate to $P_{\bar{v}}$ if G_{vent} is greater than G_{perf} (Scheid and Piiper 1976).

There is no anatomical evidence for the existence of direct connections between afferent and efferent filament arteries in elasmobranchs, which would allow blood from the ventral aorta to bypass the lamellae and enter the dorsal aorta (Metcalfe and Butler 1986). It is possible, however, that anatomical (true) shunts do occur in elasmobranch gills via the basal three or four blood channels in the lamellae (see Fig. 7-6B), which are so far from the lamellar surface (Cooke 1980) as to make it unlikely that the blood within them takes part in gas exchange (cf. Tuurala et al. 1984 for teleosts). Also, in the Endeavour dogfish, approximately 20% of the total gas exchange area is not ventilated as a result of the branchial canopy, which covers the basal 20 or so pairs of lamellae of every filament (Cooke 1980). Whether blood actually perfuses these lamellae, thus constituting an anatomical shunt, remains to be seen. Cooke (1980) also states that this situation is not unique to this species and quotes five other antipodean species of elasmobranchs in which it has been observed. In *S. acanthias*, however, the branchial canopy covers only 4% of the proximal lamellae, and these are not thought to be in a dead space (De Vries and De Jager 1984).

Besides these possible true shunts, it is also conceivable that effective shunting of blood could occur within the lamellae in a fashion similar to that described above for water passing between the lamellae (Piiper and Baumgarten-Schumann 1968). As with water, these shunts would incorporate diffusion resistances within the blood (Fig. 7-15D) and, under otherwise ideal conditions, deviation of P_a from P_i can be attributed to the fraction of total blood flow that is effectively shunted. Also, excessively high rates of perfusion could increase the effective shunt and any true shunt that existed on the blood side of the exchanger so that, even in the absence of any diffusion limitation, P_a would not equilibrate with P_i (Piiper and Scheid 1984).

In reality and in addition to anatomical shunts, the limitations of the gas exchange system consist of a combination of diffusion resistance and physiological shunts (see

Fig. 7-15E), and there is a continuous transition from one to the other.

UNEQUAL DISTRIBUTION OF WATER AND BLOOD FLOWS

If some lamellae are perfused but not ventilated, or vice versa, there is a blood or a water shunt. Between these two extremes and the optimum ratio between ventilation and perfusion, that is, when $V_w \cdot \beta_w = V_b \cdot \beta_b$, there are intermediate conditions where some lamellae may be underperfused and others may be underventilated (Fig. 7-16). Such inequalities of water and blood flows between different lamellae can reduce the overall effectiveness of gas exchange, but the extent to which this occurs depends on the overall ratio of $G_{vent}:G_{perf}$ (see Piiper and Scheid 1984). Data obtained from the dogfish $S.$ $suckleyi$ and the stingray $Dasyatis$ $sabina$ suggest that elasmobranchs have little control over the distribution of water to various gill arches if some of them are deprived of blood, although they do seem to be able to readjust blood flow in response to cessation of water flow to some gill arches (Cameron et al. 1971). The latter may be the direct result of hypoxia causing local constriction of the blood vessels in the unventilated gills (Satchell 1960).

The pulsatile nature of the flows of water and blood through the lamellae may produce effects that are similar to those of unequal distribution of water and blood. This would particularly be so if slow oscillations of the

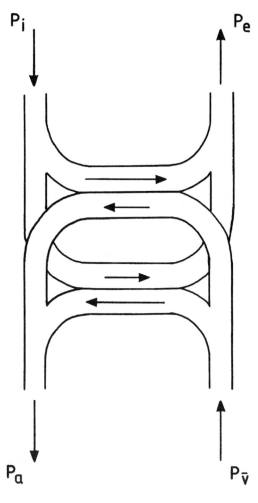

FIGURE 7-16 Model showing unequal distribution of water and blood among parallel gas exchange units, that is, G_{vent} is greater than or less than G_{perf} in different units. Lengths of *arrows* in the exchange units indicate the relative magnitude of G_{vent} or G_{perf}. P_a = partial pressure of gas in arterial blood; P_e = partial pressure of gas in expired water; P_i = partial pressure of gas in inspired water; $P_{\bar{v}}$ = partial pressure of gas in mixed venous blood. (After Piiper and Scheid 1984.)

two fluids are out of phase (Piiper and Scheid 1984). The data on synchrony between ventilation and heartbeat from restrained dogfish are inconsistent. In *Squalus lebruni* and *Mustelus antarcticus*, the heart tends to beat during a particular phase of the respiratory cycle (Satchell 1960), whereas in *S. canicula*, there is no lasting synchrony between the ventilation and cardiac cycles (Taylor and Butler 1971; Hughes 1972b). Later studies on unrestrained and undisturbed electric ray and *S. canicula* in well-aerated water have demonstrated a clear 1:1 synchrony between the two rhythms (Taylor 1985), which may maximize the effectiveness of gas exchange in the resting fish.

Conclusions

This chapter presents the current state of knowledge regarding the respiratory system of elasmobranchs in a general sense. The majority of physiological studies have been carried out on small, sedentary species. Size constraints have limited studies to species that can conveniently serve as laboratory models. Availability of animals has also dictated the choice of model species. The spiny dogfish, *Squalus acanthias*, and the little skate, *Raja erinacea*, are the most commonly studied species in the United States. The Port Jackson shark, *Heterodontus portusjacksoni*, is commonly used in Australia and New Zealand. In Europe, the dogfishes *Scyliorhinus stellaris* and *S. canicula* are widely found in the Atlantic and Mediterranean.

While these species can be regarded as typical elasmobranchs, there are notable exceptions, such as lamnid sharks, which exhibit regional heterothermy. There is a dearth of physiological data regarding the respiratory system in other types of elasmobranchs. Further studies of other species with different life history characteristics will increase our understanding of elasmobranchs as a whole.

REFERENCES

Acrivo, C. 1935. Sur l'organisation et la structure du corps caverneux chez *Scyllium canicula* Cuv. *Bull. Histol. Appl. Physiol. Pathol.* 12: 362–72.

Baumgarten-Schumann, D., and J. Piiper. 1968. Gas exchange in the gills of resting unanesthetized dogfish (*Scyliorhinus stellaris*). *Respir. Physiol.* 5: 317–25.

Bhargava, V., N. C. Lai, B. Graham, S. C. Hempleman, and R. Shabetai. 1992. Digital image analysis of shark gills: modelling of oxygen transfer in the domain of time. *Am. J. Physiol.* 263: R741–46.

Booth, J. H. 1978. The distribution of blood flow in the gills of fish: application of a new technique to rainbow trout (*Salmo gairdneri*). *J. Exp. Biol.* 73: 119–29.

Butler, P. J., and J. D. Metcalfe. 1983. Control of respiration and circulation. In *Control Processes in Fish Physiology*, ed. J. C. Rankin, T. J. Pitcher, and R. T. Duggan, 41–65. London: Croom-Helm.

Butler, P. J., and J. D. Metcalfe. 1988. Cardiovascular and respiratory systems. In *Physiology of Elasmobranch Fishes*, ed. T. J. Shuttleworth, 1–47. New York: Springer-Verlag.

Butler, P. J., and E. W. Taylor. 1975. The effect of progressive hypoxia on respiration in the dogfish (*Scyliorhinus canicula*) at different seasonal temperatures. *J. Exp. Biol.* 63: 117–30.

Cameron, J. N., D. J. Randall, and J. C. Davis. 1971. Regulation of the ventilation-perfusion ratio in the gills of *Dasyatis sabina* and *Squalus suckleyi*. *Comp. Biochem. Physiol.* 39A: 505–19.

Cooke, I. R. C. 1980. Functional aspects of the morphology and vascular anatomy of the gills of the Endeavour dogfish, *Centrophorus scalpratus* (McCulloch) (Elasmobranchii: Squalidae). *Zoomorphologie* 94: 167–83.

De Vries, R., and De Jager, S. 1984. The gill in the spiny dogfish, *Squalus acanthias*: respiratory and nonrespiratory function. *Am. J. Anat.* 169: 1–29.

Dunel, S., and P. Laurent. 1980. Functional organisation of the gill vasculature in different classes of fish. In *Epithelial Transport in the Lower Vertebrates*, ed. B. Lahlou, 37–58. Cambridge: Cambridge University Press.

Grigg, G. C. 1970a. Water flow through the gills of Port Jackson sharks. *J. Exp. Biol.* 52: 565–68.

———. 1970b. Use of the first gill slits for water intake in a shark. *J. Exp. Biol.* 52: 569–74.

Grigg, G. C., and B. Read. 1971. Gill function in an elasmobranch. *Z. Vergl. Physiol.* 73: 439–51.

Hanson, D., and K. Johansen. 1970. Relationship of gill ventilation and perfusion in Pacific dogfish, *Squalus suckleyi*. *J. Fish. Res. Board Can.* 27: 551–64.

Hills, B. A., and G. M. Hughes. 1970. A dimensional analysis of oxygen transfer in the fish gill. *Respir. Physiol.* 9: 126–40.

Hughes, G. M. 1960. The mechanism of gill ventilation in the dogfish and skate. *J. Exp. Biol.* 37: 11–27.

———. 1970. Morphological measurements on the gills of fishes in relation to their respiratory function. *Folia Morphol. (Wars.)* 18: 78–95.

———. 1972a. Morphometrics of fish gills. *Respir. Physiol.* 14: 1–25.

———. 1972b. The relationship between cardiac and respiratory rhythms in the dogfish, *Scyliorhinus canicula* L. *J. Exp. Biol.* 57: 415–34.

———. 1977. Dimensions and the respiration of lower vertebrates. In *Scale Effects in Animal Locomotion*, ed. T. J. Pedley, 57–81. London: Academic Press.

———. 1978. On the respiration of *Torpedo marmorata*. *J. Exp. Biol.* 73: 85–88.

Hughes, G. M., and C. M. Ballintijn. 1965. The muscular basis of the respiratory pumps in the dogfish (*Scyliorhinus canicula*). *J. Exp. Biol.* 43: 363–83.

Hughes, G. M., and M. Morgan. 1973. The structure of fish gills in relation to their respiratory function. *Biol. Rev.* 48: 419–75.

Hughes, G. M., S. F. Perry, and J. Piiper. 1986. Morphometry of the gills of the elasmobranch *Scyliorhinus stellaris* in relation to body size. *J. Exp. Biol.* 121: 27–42.

Hughes, G. M., and G. Shelton. 1962. Respiratory mechanisms and their nervous control in fish. In *Advances in Comparative Physiology and Biochemistry*, vol. 1, ed. O. E. Lowenstein, 275–364. New York: Academic Press.

Hughes, G. M., and D. E. Wright. 1970. A comparative study of the ultrastructure of the water-blood pathway in the secondary lamellae of teleost and elasmobranch fishes—benthic forms. *Z. Zellforsch.* 104: 478–93.

Kempton, R. T. 1969. Morphological features of functional significance in the gills of the spiny dogfish *Squalus acanthias*. *Biol. Bull.* 136: 226–40.

Kuhn, O., and H. V. Koecke. 1956. Histologische und zytologische Veränderungen der Fischkieme nach Einwirkung im Wasser enthaltener schädigender Substanzen. *Z. Zellforsch.* 43: 611–43.

Lai, N. C., J. B. Graham, V. Bhargava, W. R. Lowell, and R. Shabetai. 1989. Branchial blood flow distribution in the blue shark (*Prionace glauca*) and the leopard shark (*Triakis semifasciata*). *Exp. Biol.* 48: 273–78.

Lenfant, C., and K. Johansen. 1966. Respiratory function in the elasmobranch *Squalus suckleyi* G. *Respir. Physiol.* 1: 13–29.

Metcalfe, J. 1983. A reappraisal of the estimation of the oxygen partial pressure difference between blood and water across the gills of the dogfish (*Scyliorhinus canicula*). *J. Physiol. (Lond.)* 338: 54–55P.

Metcalfe, J. D., and P. J. Butler. 1986. The functional anatomy of the gills of the dogfish (*Scyliorhinus canicula*). *J. Zool.* 208: 519–30.

Olson, K. R., and B. Kent. 1980. The microvasculature of the elasmobranch gill. *Cell Tissue Res.* 209: 49–63.

Piiper, J., and D. Baumgarten-Schumann. 1968. Effectiveness of O_2 and CO_2 exchange in the gills of the dogfish (*Scyliorhinus stellaris*). *Respir. Physiol.* 5: 338–49.

Piiper, J., M. Meyer, H. Worth, and H. Willmer. 1977. Respiration and circulation during swimming activity in the dogfish (*Scyliorhinus stellaris*). *Respir. Physiol.* 30: 221–39.

Piiper, J., and P. Scheid. 1975. Gas transport efficacy of gills, lungs and skin: theory and experimental data. *Respir. Physiol.* 23: 209–21.

———. 1982. Physical principles of respiratory gas exchange in fish gills. In *Gills*, ed. D. F. Houlihan, C. Rankin, and T. J. Shuttleworth, 45–61. Cambridge: Cambridge University Press.

———. 1984. Model analysis of gas transfer in fish gills. In *Fish Physiology*, vol. 10A, ed. W. S. Hoar and D. J. Randall, 229–62. New York: Academic Press.

Piiper, J., P. Scheid, S. F. Perry, and G. M. Hughes. 1986. Effective and morphometric oxygen-diffusing capacity of the gills of the elasmobranch *Scyliorhinus stellaris*. *J. Exp. Biol.* 123: 27–41.

Piiper, J., and D. Schumann. 1967. Efficiency of O_2 exchange in the gills of the dogfish, *Scyliorhinus stellaris*. *Respir. Physiol.* 2: 135–48.

Randall, D. J. 1982. The control of respiration and circulation in fish during exercise and hypoxia. *J. Exp. Biol.* 100: 275–88.

Randall, D. J., G. F. Holeton, and E. D. Stevens. 1967. The exchange of oxygen and carbon dioxide across the gills of rainbow trout. *J. Exp. Biol.* 46: 339–48.

Romer, A. S. 1955. *The Vertebrate Body*. Philadelphia: W. B. Saunders.

Satchell, G. H. 1960. The reflex co-ordination of the heart beat with respiration in the dogfish. *J. Exp. Biol.* 237: 719–31.

Scheid, P., and J. Piiper. 1971. Theoretical analysis of respiratory gas equilibrium in water passing through fish gills. *Respir. Physiol.* 13: 305–18.

———. 1976. Quantitative functional analysis of branchial gas transfer: theory and application to *Scyliorhinus stellaris* (Elasmobranchii). In *Respiration in Amphibious Vertebrates*, ed. G. Hughes, 17–38. London: Academic Press.

Short, S., E. W. Taylor, and P. J. Butler. 1979. The effectiveness of oxygen transfer during normoxia and hypoxia in the dogfish (*Scyliorhinus canicula* L.) before and after cardiac vagotomy. *J. Comp. Physiol.* 132: 289–95.

Steen, B., and A. Kruysse. 1964. The respiratory function of teleostan gills. *Comp. Biochem. Physiol.* 12:127–42.

Taylor, E. W. 1985. Control and co-ordination of gill ventilation and perfusion. *Symp. Soc. Exp. Biol.* 39: 123–61.

Taylor, E. W., and P. J. Butler. 1971. Some observation on the relationship between heartbeat and respiratory movements in the dogfish (*Scyliorhinus canicula* L.). *Comp. Biochem. Physiol.* 39A: 297–305.

Tota, B., V. Cimini, G. Salvatore, and G. Zummo. 1983. Comparative study of the arterial and lacunary systems of the ventricular myocardium of elasmobranch and teleost fishes. *Am. J. Anat.* 167: 15–32.

Tuurala, H., P. Part, M. Nikinmaa, and A. Soivio. 1984. The basal channels of secondary lamellae in *Salmo gairdneri* gills—a non-respiratory shunt. *Comp. Biochem. Physiol.* 79A: 35–39.

Weibel, W. R. 1971. Morphometric estimation of pulmonary diffusion capacity: I. Model and method. *Respir. Physiol.* 11: 5–75.

Wright, D. E. 1973. The structure of the gills of the elasmobranch *Scyliorhinus canicula. Z. Zellforsch.* 144: 489–509.

RAMÓN MUÑOZ-CHÁPULI

CHAPTER 8

Circulatory System
ANATOMY OF THE PERIPHERAL CIRCULATORY SYSTEM

The basic blueprint of the vertebrate circulatory system consists of (1) the heart, a muscular pump ventral and posterior to the pharynx, (2) a system of conduits for the distribution of the blood to the branchial arches, (3) a system of efferent vessels that collect the blood from the branchial arches and join in a dorsal vessel, the aorta, (4) a set of somatic and visceral branches from the aorta and the efferent vessels, which supply all the body, and (5) a set of somatic and visceral veins returning the blood to the heart, sometimes passing through another organ, such as the liver or the kidneys.

The circulatory system of the elasmobranchs is, possibly, one of the closest to a basic plan among gnathostomes. Teleost fishes have secondary (dermal) jaws and a highly modified hyoid arch that involve important changes in the anatomy of the circulatory system. On the other hand, lungfishes and tetrapods have a pulmonary circulation and in the latter, the system of branchial arches, still present in the embryo, is extensively rearranged for exclusive air-breathing.

There are excellent descriptions in classical literature about the anatomy of the circulatory system in elasmobranchs. Of note are the papers by Parker (1886) on *Mustelus,* Carazzi (1905) on *Cetorhinus* and other species, Daniel (1922) on *Heptanchus maculatus* (= *Notorynchus cepedianus*) and other species, O'Donoghue and Abbott (1928) on *Squalus,* and Marples (1936) on *Squatina.* These studies, however, are lacking in a wide comparative scope that allows an insight into the main changes in the anatomy of the circulatory system throughout elasmobranch evolution.

For this chapter we have used the information collected from the classical sources named previously as well as our own observations obtained by corrosion-casting techniques on a number of elasmobranch species. These observations show that some specific features of the circulatory system are shared by representatives of some higher taxa of elasmobranchs, but it would be necessary to study more material in order to assess the usefulness of these features for taxonomy and phylogenetic analysis.

The methods used to obtain the casts that illustrate this chapter have been described in Muñoz-Chápuli and García Garrido (1986).

Ventral aorta and afferent branchial arteries

The ventral aorta is a large artery that extends forward in the midventral line, from the conus arteriosus of the heart almost to the symphysis of the lower jaw. The ventral aorta supplies the venous blood to the branchial arches through lateral branches called *afferent branchial arteries* (Fig. 8-1).

The most usual arrangement of the afferent branchial arteries is as follows: A common trunk at each side of the anterior end of the ventral aorta forks into the first and second afferent branchial arteries, which pass to the hyoidean and first branchial arches, respectively. The third afferent branchial artery, which passes to the second branchial arch, arises from the ventral aorta, usually behind its midpoint. The fourth and fifth afferent branchial arteries originate close to the proximal part of the ventral aorta, though the distance may vary.

The anatomical arrangement of the afferent branchial arteries is different in some species (see Fig. 8-1). In *Squatina*, a common trunk splits into posterior afferent arteries (see Fig. 8-1). In the Hexanchiforms, the six- and sevengill sharks, there are one or two extra pairs of afferent branchial arteries. These extra afferents seem to originate from the middle part of the ventral aorta, since the proximal and distal pair of afferents show a common origin, as in other elasmobranchs. The third, fourth, and fifth afferent branchials have a common origin from the ventral aorta in *Torpedo*, while in *Raja* they originate successively from a common trunk (see Fig. 8-1). However, the myliobatoids *Dasyatis* and *Mobula* follow the general pattern.

The afferent branchial arteries give off many small afferent arterioles that pass to the

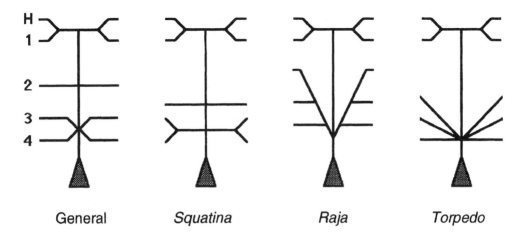

FIGURE 8-1 Diagram of the ventral aorta and afferent branchial arteries. The figure shows the general anatomical arrangement of most elasmobranchs and some variants found in *Squatina*, *Raja*, and *Torpedo*. H = hyoidean afferent; 1–4 = afferent branchial arteries to the branchial arches one to four.

anterior and posterior branchial filaments of the holobranchs or only to the posterior filaments of the hyoidean hemibranch. The afferent arteries taper gradually and terminate in the upper part of the branchial arches. An exception is the hexanchoid *Chlamydoselachus*, whose afferent branchial arteries are joined at their dorsal ends, forming closed loops around the gill clefts (Allis 1911). The afferent arterioles within the branchial filaments break up into a complex network from which the efferent arterioles emerge.

Efferent branchial arteries

Oxygenated blood exits from the gill filaments through efferent arterioles, which connect with a second set of arteries located within the branchial arches. These arteries consist of a pretrematic and a postrematic branch, joined above and below the gill cleft. For that reason they are called the *efferent collector loops* (Fig. 8-2). The arterioles arising from the anterior and posterior filaments of a given holobranch connect, respectively, with the postrematic and pretrematic branches of the two adjacent collector loops.

There is a complete efferent collector loop around each of the four anterior gill clefts (five or six in hexanchoids). However, only the pretrematic branch is present in front of the most posterior gill cleft, since there are no branchial filaments in the last branchial arch. Each postrematic branch is connected with the pretrematic branch of the immediately posterior loop through short longitudinal anastomoses. The number of anastomoses is variable, from 10–20 in hexanchoids to 3–5 in squaloids or only one in most batoids and galeomorph sharks (Fig. 8-3). Some branches arise from these anastomoses in *Dasyatis* and *Torpedo*. In the latter, these branches are large, and supply the electric organ. A second type of anastomosis appears between the ventral ends of the loops and has been mainly observed in squaloids and galeomorph

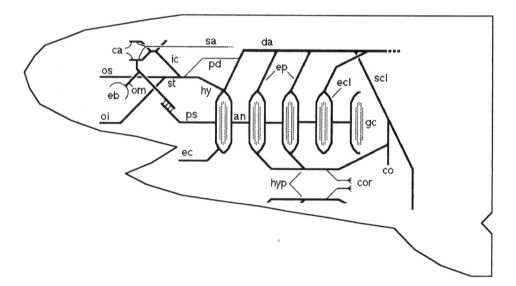

FIGURE 8-2 Diagram of the general pattern of the main cephalic arteries in elasmobranchs. The figure is arranged to show the paired hypobranchial arteries. an = Anastomosing branches between the efferent collector loops; ca = cerebral arteries; co = coracoidean branch of the subclavian; cor = coronary arteries; da = dorsal aorta; eb = eyeball; ec = external carotid artery; ecl = efferent collector loop; ep = efferent epibranchial arteries; gc = gill cleft; hy = hyoidean artery; hyp = hypobranchial arteries; ic = internal carotid artery; oi = inferior orbital or maxillonasal branch of the stapedial artery; om = ophthalmic artery (*Ophthalmica magna*); os = superior orbital branch of the stapedial artery; pd = paired dorsal aorta; ps = pseudobranchial or spiracular artery; sa = spinal artery; scl = subclavian artery; st = stapedial artery.

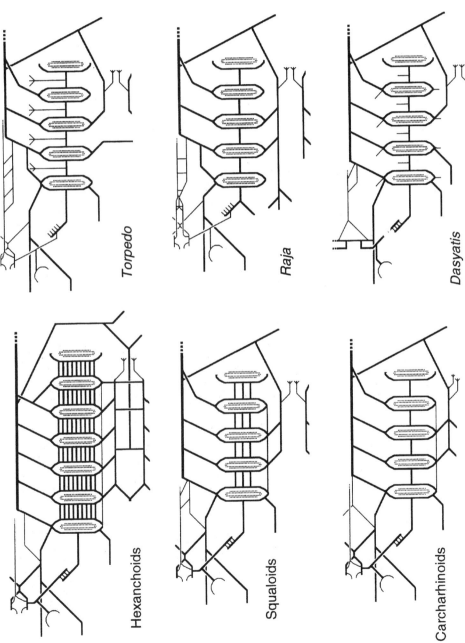

FIGURE 8-3 Diagrammatic comparison of the anatomical arrangements of the main cephalic arteries in representatives of six elasmobranch taxa. The diagram of the hexanchoids is based on the vascular anatomy of *Hexanchus*, *Heptranchias*, and *Notorynchus*. The diagram of squaloids is based on corrosion casts of *Squalus*, *Centrophorus*, and *Oxynotus*. The diagram of carcharhinoids is based on corrosion casts of *Scyliorhinus*, *Galeus*, *Galeorhinus*, *Mustelus*, *Prionace*, and *Carcharhinus*. Note the similarities between the diagrams of *Torpedo* and *Raja*.

sharks. Dorsal anastomoses between the collector loops have only been reported in *Chlamydoselachus* (Allis 1911). Some small branches emerge from both ends of the efferent collector loops in *Dasyatis* and other species to supply the neighboring tissues.

The epibranchial or efferent branchial arteries, which are the dorsal part of the aortic arches, arise from the dorsal end of all the complete collector loops. Therefore, the usual number of epibranchial arteries is four, although this number is larger in the hexanchiforms, with five epibranchial arteries in *Hexanchus* and *Chlamydoselachus* and six in *Heptranchias*. In these species, the last epibranchial artery joins the anterior one (Figs. 8-3 and 8-4). The first epibranchial artery is short in *Raja* and *Torpedo*, and it connects with the second one, leaving only three main epibranchial arteries in the roof of the pharynx (see Fig. 8-3). *Dasyatis* and *Mobula* follow the general pattern (see Fig. 8-3). The phylogenetic significance of these anatomical arrangements is unknown.

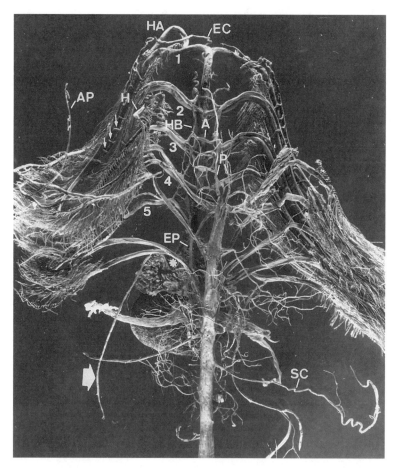

FIGURE 8-4 Corrosion cast of the hexanchoid *Hexanchus griseus*, dorsal view. Note the numerous anastomoses between the efferent collector loops (*small arrows*), the contribution of most efferent collector loops to the hypobranchial arteries (HB), and the connection of the fourth and fifth efferent epibranchial arteries (*asterisk*). An innominate artery arises from the fifth efferent epibranchial artery and runs backwards (*large arrow*). 1–5 = First to fifth afferent branchial arteries; A = anastomosis between the lateral hypobranchial arteries; AP = afferent portion of the pseudobranchial artery; EC = external carotid artery; EP = epigastric artery originating from the hypobranchial arteries; H = origin of the hyoid artery (not perfused); HA = hyoidean afferent branchial artery; P = paired dorsal aortas; SC = subclavian artery.

The epibranchial arteries unite just above the midline of the pharynx to form a large longitudinal vessel, the dorsal aorta. In galeoid sharks and some hexanchoids, such as *Heptranchias* or *Chlamydoselachus*, the dorsal aorta passes forward from the level of the first epibranchial artery and splits into two branches, which connect with the hyoidean artery. In most elasmobranchs, however, there is not a noticeable dorsal aorta before the confluence of the first epibranchial arteries. Instead, there are usually two laterodorsal arteries arising from the first epibranchial artery, or in *Raja* and *Torpedo*, the second one. The paired dorsal aortas can also make a connection with the hyoidean artery, and these vessels often contribute to the blood supply of the central nervous system, as described subsequently. The dorsal aorta passes backward under the vertebral centra, branching off into smaller arteries, is discussed subsequently.

Hypobranchial arteries

The hypobranchial region of the head is usually irrigated by two lateroventral vessels called the *hypobranchial arteries*, which can be joined by transverse anastomosing branches above or

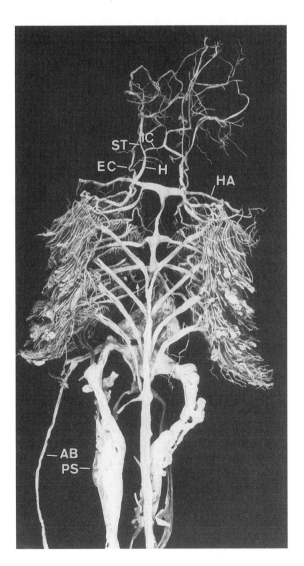

FIGURE 8-5 Corrosion cast of the squaloid *Oxynotus centrina*, dorsal view. The left abdominal (AB) and the posterior cardinal veins were retrogradely perfused from the ventricle. Note the contribution of the third efferent collector loop to the lateral hypobranchial arteries (*arrow*). EC = external carotid artery; H = hyoid artery; HA = hyoidean afferent branchial; IC = internal carotid artery; PS = postcardinal sinus; ST = stapedial artery.

below the ventral aorta. The hypobranchial arteries receive blood from branches arising either from the ventral end of the efferent collector loops (Figs. 8-4 and 8-5), usually the second and third, or from the longitudinal arteries that connect the ventral ends of the collector loops, when these vessels exist. In *Hexanchus*, the hypobranchial arteries receive branches from almost all the complete collector loops (see Fig. 8-4). In most carcharhiniform sharks one or two midventral hypobranchial arteries replace the lateroventral hypobranchial arteries. When there are two midventral hypobranchial arteries, they are arranged above and below the ventral aorta, respectively. In lamniforms there are one midventral and two lateroventral hypobranchial arteries. The latter are very large vessels, which convey most of the blood flow to the digestive tract. This is due to the connection of the hypobranchial arteries to the visceral heat-exchangers (Burne 1923). Caudal to the heart, the hypobranchial arteries break into many small arteries that pervade the cavity of the posterior cardinal sinuses. These heat-exchangers recirculate the heat in the body cavity, avoiding thermal losses through the gills.

The lateroventral hypobranchial arteries pass above the pericardial cavity and usually connect with the coracoidean branch of the subclavian artery (Fig. 8-6). In carcharhiniforms, one or two epigastric arteries, arising either from the midventral hypobranchial arteries or from the posterior branches that supply them, irrigate the pericardium, the esophagus and, in some species, also anastomose with the subclavian arteries. The hypobranchial arteries can also anastomose with the system of abdominal arteries.

The coronary arteries supplying the heart are connected in differing fashions with the hypobranchial arteries or with the branches that supply them. There are usually two main coronary trunks, which are arranged dorsoventrally in the conus arteriosus of the lamniform and in most carcharhiniform sharks and laterally in the other elasmobranchs (De Andrés et al. 1990, 1992; Muñoz-Chápuli et al. 1994). Most often, the lateral coronary trunks turn clockwise along the conus in such a way that the right trunk supplies the dorsal region of the ventricle, while the left trunk supplies the ventral ventricular area (De Andrés et al. 1992). Two posterior coronary arteries arise from the coracoidean branch of the subclavian in *Raja* and *Dasyatis*. They reach the dorsal and posterior region of the ventricle through the lateral margins of the sinus venosus. These vessels have not been reported in other batoids like *Torpedo* (Muñoz-Chápuli et al. 1994).

Cephalic arteries

The cephalic blood supply is mainly furnished by three arteries that originate from the first efferent collector loop, together with branches from the single or paired dorsal aortas. These arteries are the external carotid, the pseudobranchial, and the hyoidean (Fig. 8-2).

The external carotids originate from the ventral ends of the first efferent collector loops. They pass forward and supply the lateroventral part of the head, including the adductor mandibularis muscle. The mandibular branch of the external carotid extends toward the symphysis and supplies the thyroid gland. The skates (*Raja*) lack external carotids, their functions being performed by anterior branches of the hypobranchial arteries and a branch of the pseudobranchial artery (see Fig. 8-3).

The pseudobranchial arteries originate about midway in the pretrematic branches of the first efferent collector loops. Their course can be very tortuous in some species. When the pseudobranch is well developed, the pseudobranchial artery is divided into afferent and efferent portions, connected by the pseudobranchial arterioles. When the pseudobranch is rudimentary, the pseudobranchial artery courses nearly uninterrupted between the mandibular and hyoidean arches. In all the cases, the distal portion of the pseudobranchial artery passes through the cranial wall at the orbitary region and anastomoses with the internal carotid. Before this anastomosis, the pseudobranchial artery gives off the

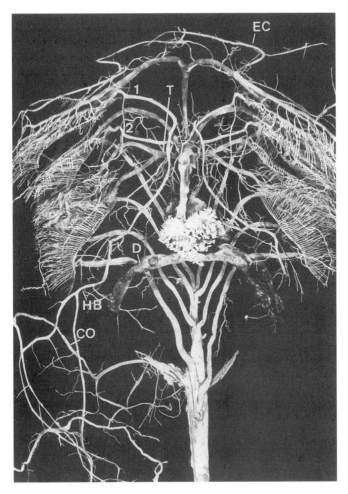

FIGURE 8-6 Corrosion cast of the squatinoid *Squatina squatina*, ventral view. Large hypobranchial arteries (HB) originate from the ventral ends of the first (1) and second (2) efferent collector loops and connect posteriorly with the coracoidean branch (CO) of the subclavians. The *small arrow* points to the right coronary artery, which runs along the large conus arteriosus. D = right duct of Cuvier; EC = external carotid artery; T = common trunk of the third and fourth afferent branchial arteries.

ophthalmic artery, which supplies the inner part of the eyeball (choroid) (Fig. 8-7). The large afferent pseudobranchial artery, in the skates, sends off branches to the adductor mandibularis muscle, which is not supplied by an external carotid. The efferent pseudobranchial portion is comparatively narrow in *Raja* and *Torpedo*.

The hyoidean arteries originate from the dorsal ends of the first efferent collector loops. They pass inward and fork into two important vessels, the orbital or stapedial artery and the internal carotid artery. Near this bifurcation there may be a connection of either the paired dorsal aortas or the anterior branches originating from the bifurcation of the medial dorsal aorta.

The stapedial artery supplies the dorsolateral part of the head, external to the chondrocranium, through three main branches. The superior orbital artery sends twigs to the oculomotor muscles, the levatores of the palatoquadrate, hyomandibular, and the occipital area. Then the superior orbital artery leaves the orbit through the orbitonasal canal and supplies branches to the olfactory organ and

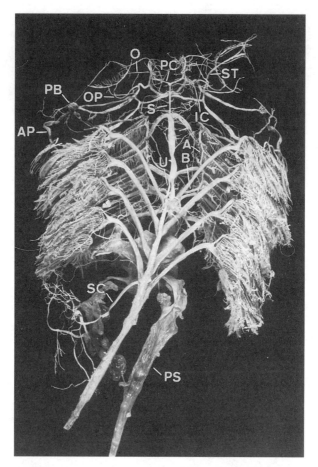

FIGURE 8-7 Corrosion cast of the carcharhinoid *Mustelus mustelus*, dorsal view. Note the unpaired dorsal aorta (U) in front of the connection of the first efferent epibranchial arteries. A = anastomosis joining the ventral ends of two efferent collector loops; AP = afferent portion of the pseudobranchial artery; B = branches contributing to the midventral hypobranchial artery (not visible); IC = internal carotid artery; O = optic artery; OP = ophthhalmic artery, supplying the choroid plexus of the eyeball; PB = pseudobranch; PC = posterior cerebral arteries; PS = postcardinal sinus; S = spinal artery; SC = subclavian artery; ST = stapedial artery.

the rostrum. The inferior orbital or maxillo-nasal artery supplies the roof of the buccal cavity and part of the levator mandibularis. Well-developed stapedial arteries, larger than the internal carotids, are always present in the batoids (Figs. 8-8–8-10).

The internal carotid arteries, originating from the bifurcation of the hyoidean, pass inward and join in the midline to form a common trunk. The union is located either outside or inside the chondrocranium. In the first case, the common trunk enters through a single basilar foramen of the basal plate, and in the second case there are two lateral basal foramina for the internal carotid arteries. Two lateral arteries arise from the common carotid trunk, and they pass forward at the sides of the forebrain. These arteries receive

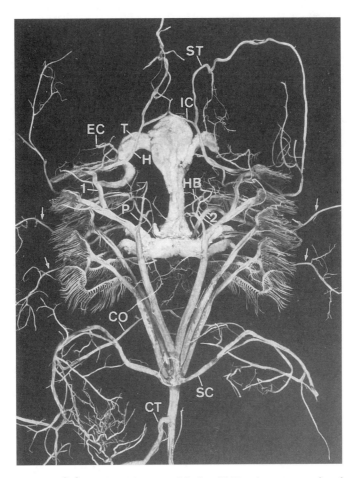

FIGURE 8-8 Corrosion cast of the batoid *Torpedo marmorata*, dorsal view. Note the remarkable diameter of the ventral aorta, especially at the proximal and distal portions, as well as the large common trunk (T) of the hyoidean and first afferent branchial arteries. The second afferent branchial artery (2) shows a common origin with the third and fourth. Another singular feature of the vascular anatomy of *Torpedo* is the presence of branches supplying the electric organs (*arrows*), branches that originate from the anastomoses between the efferent collector loops. The arrangement of the celiac trunk (CT) toward the left also seems exceptional among the elasmobranchs. 1 = First efferent epibranchial connecting with the second one; CO = coracoidean branch of the subclavian artery; EC = external carotid artery; H = hyoid artery; HB = hypobranchial artery; IC = internal carotid artery; P = paired dorsal aortas; SC = subclavian artery; ST = stapedial artery.

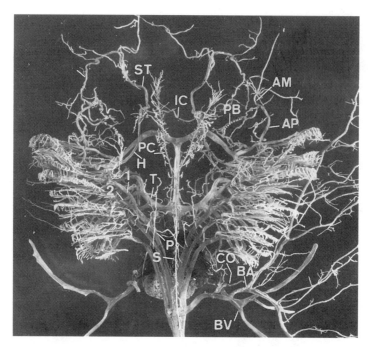

FIGURE 8-9 Corrosion cast of the batoid *Raja clavata*, dorsal view. The posterior cerebral arteries (PC) are well developed, in contrast to the relatively small internal carotid arteries (IC). This is due to the contribution of the paired dorsal aortas (P), which originate from the second efferent epibranchial artery (*arrow*). Note the connection of the first efferent epibranchial (1) to the second one (2). AM = branch of the pseudobranchial artery to the adductor mandibularis muscle; AP = afferent portion of the pseudobranchial artery; BA = brachial artery; BV = brachial vein; CO = coracoidean branch of the subclavian, which gives off a posterior coronary artery; H = hyoid artery; PB = pseudobranch; S = spinal artery; T = common trunk of the second, third, and fourth afferent branchial arteries.

the intracranial portion of the efferent pseudobranchial, and rise to the optic arteries, which run along the optic nerve to the eyeball (see Fig. 8-7), and the anterior, medial, and posterior cerebral arteries, which supply the encephalon. The posterior cerebral arteries anastomose under the hindbrain, giving rise to the basilar artery, which continues backward along the base of the medulla as the spinal artery. Usually, the spinal artery receives tributaries from the first efferent epibranchial artery (the second one in *Raja* and *Torpedo*). This contribution from the second epibranchial to the spinal artery is especially significant in these batoids, probably because of their relatively small internal carotids (see Fig. 8-9). In *Squatina*, three or four pairs of segmentally arranged vessels connect the spinal artery with the paired dorsal aorta and the first efferent epibranchial arteries.

There are other anatomical arrangements of the internal carotid arteries in some elasmobranchs. In the myliobatoids *Dasyatis* and *Mobula*, the paired internal carotids are replaced by a single, transversal artery, which originates at the hyoidean bifurcation. This artery plays the same role as the internal carotids, since it receives the efferent pseudobranchial and gives rise to the cerebral and optic arteries (see Fig. 8-10). In *Isurus*, and probably in other lamnoids (Burne 1923), the internal carotids are small, and most of the

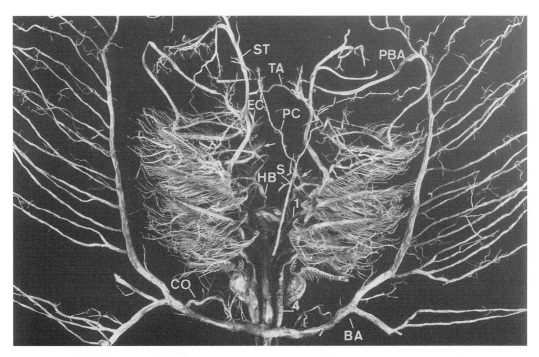

FIGURE 8-10 Corrosion cast of the batoid *Dasyatis pastinaca*, dorsal view. The most remarkable features of this cast are the large brachial arteries (BA) supplying the pectoral fins, the first efferent epibranchial artery (1), which does not connect with the second one, and the transversal anastomosis (TA) between the hyoid arteries, a vessel that replaces the internal carotid arteries. Note also the featherlike extensions of the dorsal and ventral ends of the efferent collector loops (*arrows*) and the long course backward of the fourth efferent epibranchial artery (4). CO = coracoidean branch of the subclavian artery; EC = external carotid artery; HB = lateral hypobranchial artery originating from the third efferent collector loop; PBA = pseudobranchial artery connecting with the vessel labeled as TA; PC = posterior cerebral artery; S = spinal artery; ST = stapedial artery.

blood supply to the head is conveyed by a large pseudobranchial artery (Fig. 8-11). This artery and the hyoid form a complex vascular network bathed by the venous blood of the orbital sinus. It constitutes a cephalic heat exchanger, which keeps the temperature of the brain several degrees above that of the seawater (Block and Carey 1985). The hammerhead sharks (*Sphyrna*) also have vascular coilings formed by both the hyoid and pseudobranchial arteries. The vascular coilings of the hammerhead are also located in the orbital cavity and are percolated by venous blood from the orbital sinus. However, they differ from those of the lamnoid sharks since they are constituted of a few coils of thick and unbranched vessels. Thus, a heat-exchanger function can be discarded for these structures. It has been suggested that they have a function related to the control of the blood pressure (Muñoz-Chápuli and De Andrés 1995). Another interesting feature of the hammerhead is the lack of basilar foramina, the internal carotids entering the chondrocranium through the orbitary wall.

The dorsal aorta and its branches

The dorsal aorta runs along the trunk between the hemal processes of the vertebrae. In its course, the dorsal aorta gives rise to two kinds of branches, namely the *somatic*, which

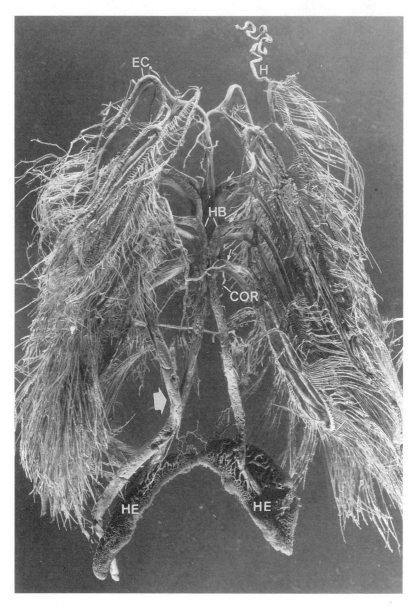

FIGURE 8-11 Corrosion cast of the mako shark *Isurus oxyrinchus*, ventral view. Two large hypobranchial arteries (HB) arise from ventral branches (*small arrows*) from the second, third, and fourth efferent collector loops. These hypobranchial arteries break up into a network of vessels that constitute the visceral heat-exchangers (HE). The *large arrow* points to a thick branch originating in the dorsal end of the third efferent collector loop. Note the remarkable length of the branchial filamentary arterioles. COR = coronary artery; EC = external carotid; H = hyoid artery.

are usually paired and arise laterally, and the visceral branches, which arise ventrally and run inside the mesenteria (Fig. 8-12). Beyond the posterior end of the body cavity, the dorsal aorta continues as the caudal aorta within the hemal canal of the vertebrae. The caudal aorta also branches into somatic arteries, similar to those of the trunk, which exit from the vertebral column through the interhemal spaces.

SOMATIC BRANCHES

The general somatic branches are called the *segmental arteries*. They are variable in number, more abundant in primitive forms, such as the hexanchoids, and considerably reduced in number in batoids. Each segmental artery has several branches. The vertebromuscular branch supplies the dorsal musculature and dorsal fins. The vertebrospinal branch enters the spinal canal and divides into a dorsal and a ventral branch, which supply the medulla. The ventral branch anastomoses with the spinal artery, which runs at the base of the medulla. Each segmental artery also branches to the lateral musculature and then continues as the intercostal arteries, which run under the parietal peritoneum.

The kidneys and gonads are supplied by ventral branches of the segmental arteries (see Fig. 8-12). The oviducal and testicular arteries, derived from one or more pairs of anterior segmental arteries, are especially well developed in the female viviparous sharks, as they convey blood to the uterine wall. In *Dasyatis*, the blood supply to the gonads is

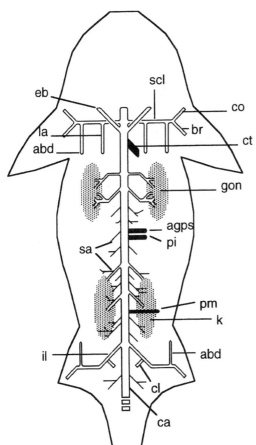

FIGURE 8-12 Diagram of the main aortic branches in elasmobranchs, dorsal view. Somatic arteries are depicted in white, visceral arteries in black. abd = Abdominal arteries; agps = anterior gastropancreaticosplenic artery; br = brachial artery; ca = caudal aorta; cl = cloacal artery; co = coracoidean artery; ct = celiac trunk; eb = most posterior efferent epibranchial artery; gon = gonads and reproductive tract; il = iliac artery; k = kidney; la = lateral artery; pi = posterior intestinal artery; pm = posterior mesenteric artery; sa = segmental arteries; scl = subclavian artery.

furnished by a proximal branch from the subclavian arteries (Daniel 1922). The renal arteries are a set of ventral branches from the posterior segmental arteries.

There are also special somatic branches arising from the aorta. As a general rule, the subclavian arteries originate just in front of the connection of the most posterior efferent epibranchial arteries. They are very large in batoids and squatinoids, since they supply the enlarged pectoral fins (see Figs. 8-6 and 8-8–8-10). The first important branches of the subclavian artery are the anterior dorsolateral and scapular arteries, which supply the area in front and behind the scapula, respectively, including the levators of the branchial arches. Two lateral and longitudinal vessels that run backward along the trunk also originate from the subclavian, the lateral and ventrolateral arteries, being located at the levels indicated by their names. The ventrolateral, or abdominal, artery can eventually reach the level of the pelvic fins to anatomose with the iliac artery (see Fig. 8-12). The subclavian artery finally gives rise to a brachial artery, which passes through a foramen in the pectoral girdle to the pectoral fin. The brachial artery of the batoids divides into anterior and posterior branches, from which many lateral branches originate (see Figs. 8-8–8-10). Another branch from the subclavian is the coracoidean, which runs in a midventral direction, along the margin of the coracoid, and usually anastomoses with branches of the hypobranchial system.

The iliac arteries are posterior somatic branches of the dorsal aorta that carry blood to the cloaca, the pelvic fins, and the claspers. They are also well developed in batoids. In some species, the iliac arteries branch to supply the rectum.

VISCERAL BRANCHES

There are three or four main visceral branches arising from the dorsal aorta, the celiac trunk, and the anterior and posterior mesenteric arteries. The anterior mesenteric artery is very often divided into two arteries from its origin, namely the anterior gastropancreaticosplenic and the posterior intestinal arteries (Figs. 8-12 and 8-13).

The celiac trunk is a very large artery of variable length, which arises just behind the aortic connections of the most posterior efferent epibranchial arteries. It runs through the dorsal mesentery and divides into several branches, namely (1) a short gastrohepatic artery, which soon forks into a hepatic artery and a gastric artery, and (2) a large anterior intestinal artery, which runs through the ventral part of the intestine.

The gastric artery divides into an anterior and a ventral branch. The anterior branch supplies the esophagus and the cardiac region of the stomach. The ventral branch supplies the fundic and pyloric regions of the stomach and sends branches to anastomose with the gastrosplenic artery. The hepatic artery bifurcates into two arteries that carry blood to both hepatic lobes. It has branches to the gall bladder and sometimes to the spleen.

The second division of the celiac trunk, the anterior intestinal artery, arrives at the gut at the level of the duodenum and gives rise to: (1) a posterior gastropancreaticosplenic artery, which supplies the pyloric region of the stomach, the pancreas, and as the posterior gastrosplenic artery, the spleen; (2) the gastroduodenal artery, which has branches to the pylorus and duodenum; (3) the ventral intestinal artery, with annular branches that run along the base of the spiral valve of the intestine. This vessel is absent in the skates (Daniel 1922). According to this author, the anterior intestinal artery also gives rise to the anterior gastropancreaticosplenic artery in *Dasyatis*. After the above-mentioned branchings, the anterior intestinal artery continues within the valvular intestine as the intraintestinal artery.

The anterior mesenteric artery, when present, is only a short branch from the dorsal aorta that divides into an anterior gastropancreaticosplenic (or lienogastric) artery (except in *Dasyatis*) and a posterior intestinal artery. In most elasmobranchs, these arteries arise independently from the dorsal aorta. The former vessel supplies the dorsal part of the cardiac

Circulatory System: Peripheral System Anatomy 213

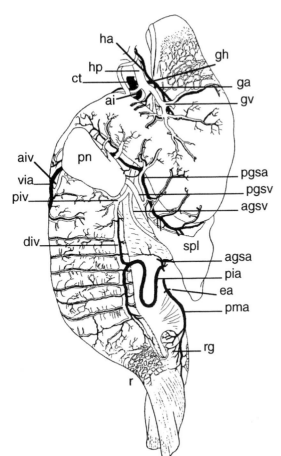

FIGURE 8-13 Visceral vessels in the shark *Heterodontus francisci*, lateral view. Arteries and veins are depicted in black and white, respectively. agsa = Anterior gastrosplenic artery; agsv = anterior gastrosplenic vein; ai = anterior intestinal artery; aiv = anterior intestinal vein; ct = celiac trunk; div = dorsal intestinal vein; ea = epigonal artery; ga = gastric artery; gh = gastrohepatic artery; gv = gastric vein; ha = hepatic artery; hp = hepatic portal vein; pgsa = posterior gastro-pancreaticosplenic artery; pgsv = posterior gastropancreaticosplenic vein; pia = posterior intestinal artery; piv = posterior intestinal vein; pma = posterior mesenteric artery; pn = pancreas; r = rectum; rg = rectal gland; spl = spleen; via = ventral intestinal artery. (Redrawn from Daniel 1922.)

stomach, the dorsal lobes of the pancreas, and the spleen. The posterior intestinal artery joins the intestine at about the middle of its length. This artery passes backward as the dorsal intestinal artery and gives rise to annular branches around the base of the spiral valve. Some of these branches anastomose with similar branchings from the ventral intestinal artery. The dorsal intestinal artery terminates at the base of the rectal gland.

The posterior mesenteric artery arises from the aorta some segments behind the origin of the anterior mesenteric and runs along the anterior margin of the mesorectal mesentery to join the apex of the rectal gland. This artery supplies the epigonal organ, the rectal gland, and then breaks up into a vascular plexus around the rectum.

Venous system

Blood distributed by the arteries is returned to the heart through the veins. The veins of elasmobranchs can be classified as somatic and visceral. We describe the anterior and posterior cardinals, the caudal, the renal portal, the subclavians, the abdominal, and the cutaneous veins between the somatic veins (Fig. 8-14). The suprahepatic, the hepatic portal and its tributaries are the main visceral veins (see Fig. 8-13).

SOMATIC VEINS

Anterior cardinal veins

The anterior cardinal veins are large vessels that originate behind the eyeball, at the orbital sinus, passing backward dorsal to the branchial

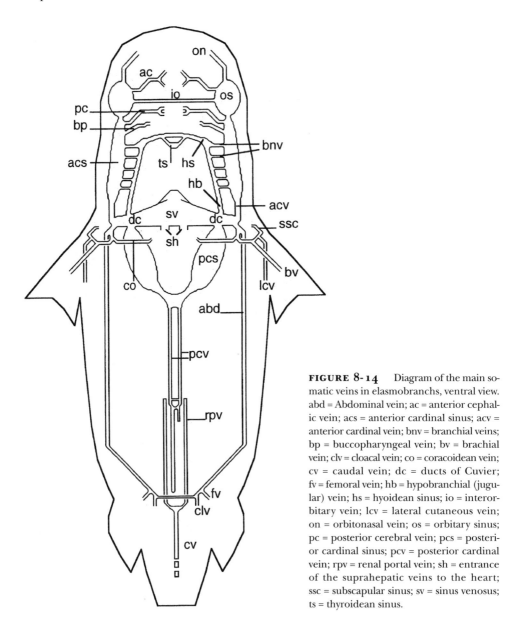

FIGURE 8-14 Diagram of the main somatic veins in elasmobranchs, ventral view. abd = Abdominal vein; ac = anterior cephalic vein; acs = anterior cardinal sinus; acv = anterior cardinal vein; bnv = branchial veins; bp = buccopharyngeal vein; bv = brachial vein; clv = cloacal vein; co = coracoidean vein; cv = caudal vein; dc = ducts of Cuvier; fv = femoral vein; hb = hypobranchial (jugular) vein; hs = hyoidean sinus; io = interorbitary vein; lcv = lateral cutaneous vein; on = orbitonasal vein; os = orbitary sinus; pc = posterior cerebral vein; pcs = posterior cardinal sinus; pcv = posterior cardinal vein; rpv = renal portal vein; sh = entrance of the suprahepatic veins to the heart; ssc = subscapular sinus; sv = sinus venosus; ts = thyroidean sinus.

arches and turning downward and inward, at the level of the scapulocoracoid cartilage, to join the ducts of Cuvier (common cardinal) (see Fig. 8-14). The main veins draining into the orbital sinus are the anterior cerebral, the orbitonasal, and the ophthalmic veins. The anterior cerebral vein drains the prosencephalon and the optic lobes. The orbitonasal and the ophthalmic veins carry blood from the nasal region and the eyeball, respectively. An interorbital vein connects the orbital sinuses, passing through the cranium by way of the interorbital canal.

In the pharyngeal region, the anterior cardinal veins receive the posterior cerebral vein from the hindbrain, the buccopharyngeal vein from the roof of the mouth, and the branchial veins from the hyoidean hemibranch and from all the holobranchs. The branchial veins are located at the branchial arches, in close contact with the afferent arteries. At the ventral part of the pharynx, the branchial veins connect with the hypobranchial veins, two lateroventral vessels that drain the hypobranchial region, the lower jaw, and the thyroid. Although they are connected to the system of anterior cardinals through the branchial veins, they pass backward, traverse the region of the heart, and connect with the ducts of Cuvier near the entrance of the anterior cardinal veins.

Posterior cardinal veins

The posterior cardinal veins are two large vessels located in the trunk, under the vertebral column, at the lateral and ventral sides of the dorsal aorta (see Fig. 8-14). They originate between the kidneys, from which they receive their most posterior tributaries. In their posterior portion, they join in a single vessel in some cases, as in *Scyliorhinus* or *Raja;* in others they extend as separate vessels until the posterior end of the kidney or, more usually, only the right one runs the entire length between the kidneys, while the left one is connected to the right (Daniel 1922). Along their course, the posterior cardinals also receive segmental veins from the body wall and from the genital sinuses. The segmental veins have a similar arrangement to that described for the segmental arteries; dorsal branches drain the spinal cord, the dorsal fins, and the dorsal musculature and ventral branches drain the lateral parts of the body. The posterior cardinals terminate anteriorly in the ducts of Cuvier. The anterior portion of the posterior cardinal veins is enlarged, forming the posterior cardinal sinuses, in the walls of which some chromaffin tissue, secreting catecholamines, is located (see Figs. 8-5 and 8-7).

Caudal vein

The caudal vein is located in the hemal canal of the caudal vertebrae, surrounding the caudal aorta. In its anterior portion, where it reaches the body cavity, it divides into the renal portal veins. All along its course, the caudal vein receives tributaries from the segmental veins of the tail and also from the dorsal and posterior ventral cutaneous veins.

Renal portal veins

The renal portal veins originate from the split of the caudal vein behind the cloaca (see Fig. 8-14). They pass forward at the sides of the cloaca and run along the outer sides of the kidneys, giving rise to branches that merge with the renal tissue. The renal portal veins also receive segmental veins from the body wall.

Subclavian veins

The large subclavian veins are formed by the successive union of the brachial veins from the pectoral fins and the subscapular and coracoidean veins from the dorsal and ventral regions of the pectoral girdle, respectively, and the abdominal veins, which are described in the next section. The subclavian veins drain into the ducts of Cuvier near the entry of the hypobranchial veins (see Figs. 8-9 and 8-14).

Abdominal veins

The abdominal veins are a pair of vessels that extend from the pelvic to the pectoral region at a lateroventral level, just beneath the peritoneum (see Figs. 8-5 and 8-14). The left and right abdominals are joined across the midventral line at the cloacal region. The abdominal veins receive the iliac veins from the pelvic fins, the cloacal veins, and several veins from the lateroventral musculature of the trunk. The coracoidean veins, which drain the ventral region of the pectoral girdle, also connect with the abdominal veins in some cases. The abdominal veins terminate in the subclavian veins.

Cutaneous veins

Four main longitudinal veins run along the subdermal connective tissue of the trunk and tail at middorsal, lateral, and midventral levels.

The dorsal cutaneous vein forms a loop around the dorsal fins, and it also forks to collect the blood from both sides of the upper lobe of the caudal fin. This dorsal vein drains into the lateral cutaneous veins as well as into the caudal vein through a deep branch.

The paired lateral cutaneous veins are located ventral to the lateral line, from the middle region of the caudal fin to the subscapular sinus of the subclavian veins, where they drain (see Fig. 8-14). The lateral cutaneous veins also connect with the system of the abdominal veins via the cloacal veins. They receive numerous branches, segmentally arranged, by way of the subdermal tissue of the sides of the trunk.

The ventral cutaneous vein has two segments, which are located at the midventral line of the trunk and tail, respectively. The tail segment, in a fashion similar to the dorsal cutaneous vein, forms a loop around the anal fin and bifurcates in the region of the lower lobe of the caudal fin. This vein drains into the cloacal veins. The trunk segment of the ventral cutaneous vein runs along the belly and drains into the coracoidean vein.

VISCERAL VEINS

Hepatic portal system

The hepatic portal system is composed of the hepatic portal vein and its tributaries, collecting blood from the digestive tract (see Fig. 8-13). The large hepatic portal vein divides into two branches, one supplying each of the lobes. Both branches break into many sinusoids, which intermingle with the hepatic tissue.

The visceral branches that contribute to the hepatic portal vein are the intraintestinal, anterior intestinal, posterior intestinal, and gastric veins. (1) The intraintestinal vein emerges from the anterior end of the valvular intestine and is sometimes located along the free edge of the spiral valve. (2) The anterior intestinal vein runs along the midventral line of the intestine like the ventral intestinal vein and receives venous branches from the base of the spiral valve. The anterior intestinal vein then leaves the intestine and joins the intraintestinal and the posterior gastropancreaticosplenic veins. The anterior intestinal vein is small in rays (Daniel 1922). (3) The posterior intestinal vein drains the rectal gland and the intestine, like the dorsal intestinal vein. The posterior intestinal vein leaves the intestine before the point at which the posterior intestinal artery joins it. Then this vein receives the anterior gastropancreaticosplenic vein and joins either the anterior intestinal or the gastric vein. (4) The gastric vein or veins drain the cardiac portion of the stomach and join the anterior and posterior intestinal veins. The resulting vessel is the hepatic portal vein.

Suprahepatic veins

The suprahepatic veins are the vessels that collect the blood from the liver and convey it to the heart. At the point where they connect with the heart, through the transverse septum, they are usually two short and very wide vessels, which drain into the posterior side of the sinus venosus. However, in the rays, the suprahepatic veins drain into the ducts of Cuvier.

ACKNOWLEDGMENT

The author thanks L. García Garrido for cooperation in the preparation of the corrosion casts that illustrate this chapter.

REFERENCES

Allis, E. P. 1911. The pseudobranchial and carotid arteries in *Chlamydoselachus anguineus*. Anat. Anz. 39: 511–19.

Block, B. A., and F. G. Carey. 1985. Warm brain and temperatures in sharks. *J. Comp. Physiol.* 156B: 229–36.

Burne, R. H. 1923. Some peculiarities of the blood vascular system of the porbeagle shark (*Lamna cornubica*). *Philos. Trans. R. Soc. Lond.* 212B: 209.

Carazzi, O. 1905. Sul sistema arterioso di *Selache maxima* e di altri squalidi. *Anat. Anz.* 26: 63–96.

Daniel, J. F. 1922. *The Elasmobranch Fishes.* Berkeley: University of California Press.

De Andrés, A. V., R. Muñoz-Chápuli, V. Sans Coma, and L. García Garrido. 1990. Anatomical studies of the coronary system in elasmobranchs: I. Coronary arteries in lamnoid sharks. *Am. J. Anat.* 187: 303–10.

———. 1992. Anatomical studies of the coronary system in elasmobranchs: II. Coronary arteries in hexanchoid, squaloid, and carcharhinoid sharks. *Anat. Rec.* 233: 429–39.

Marples, B. J. 1936. The blood-vascular system of the elasmobranch fish *Squatina squatina. Trans. R. Soc. (Edinb.)* 58: 817–40.

Muñoz-Chápuli, R., and A. V. De Andrés. 1995. Anatomy and histology of the cephalic arterial coilings in hammerhead sharks (genus *Sphyrna*). *Acta Zool. (Stockh.)* 76: 301–05.

Muñoz-Chápuli, R., A. V. De Andrés, and G. Dingerkus. 1994. Coronary artery anatomy and elasmobranch phylogeny. *Acta Zool. (Stockh.)* 75: 249–54.

Muñoz-Chápuli, R., and L. García Garrido. 1986. Cephalic blood vessels in elasmobranchs: anatomy and phylogenetic implications. In *Indo-Pacific Fish Biology: Proceedings of the Second International Conference on Indo-Pacific Fishes*, ed. T. Uyeno, R. Arai, T. Taniuchi, and K. Matsuura, 164–72. Tokyo: Society of Ichthyology of Japan.

O'Donoghue, C. H., and E. Abbott. 1928. The blood vascular system of the spiny dogfish *Squalus acanthias* Linné and *Squalus suckleyi* Gill. *Trans. R. Soc. (Edinb.)* 55: 823–90.

Parker, T. J. 1886. On the blood vessels of *Mustelus antarcticus*. A contribution to the morphology of the vascular system in the Vertebrata. *Philos. Trans. R. Soc. Lond.* 177B: 685–731.

GEOFFREY H. SATCHELL

CHAPTER 9

Circulatory System

DISTINCTIVE ATTRIBUTES OF THE CIRCULATION OF ELASMOBRANCH FISH

There exist a number of general accounts of circulation in fishes (Randall 1970; Randall et al. 1992; Satchell 1971, 1991) in which the particular features of elasmobranchs have inevitably to take second place to discussion of the so much more numerous teleosts. This chapter provides the opportunity to rectify this imbalance and to discuss in greater detail features of the circulation that are peculiar to the cartilaginous fish.

Some general functional similarity between the circulatory systems of elasmobranchs and teleosts is to be expected (Table 9-1).

In both elasmobranchs and teleosts, the cardiac muscle of a single ventricle ejects blood into a leaky elastic reservoir, the ventral aorta and afferent arteries, the walls of which include much elastic tissue and some smooth muscle and collagen. From here it passes through a branching system of blood vessels, the arteries, arterioles, and capillaries, and is returned to the heart in the veins. The shape and magnitude of flow and pressure pulses are in part dictated by the proteins, that is, elastin, collagen, actin, and myosin, of which their walls are made. Any one parameter, such as, blood pressure, measured over several species of elasmobranch is seen to span a range of values (Table 9-2).

Elasmobranchs differ greatly from teleosts in many significant features. Selachians and batoids lack an operculum, and the two hemibranchs of a gill filament are separated by and attached to the septum that separates the gill pouches. They lack a swim bladder. Their body fluids are rendered virtually isosmotic with seawater by the reabsorption of urea, and the recent study by Lacy and Reale (1991) shows that the elasmobranch

TABLE 9-1   SOME CIRCULATORY PARAMETERS OF TWO SPECIES OF DOGFISH, *SCYLIORHINUS CANICULA* AND *S. STELLARIS*, AT 15°–19°C, COMPARED WITH THE RAINBOW TROUT, *ONCORHYNCHUS MYKISS*, AT 14.5°C

Datum	Dogfish	Trout	References
Mass (kg)	0.9 ± 1–0.05	0.1 ± 1–0.72	1, 5
$\dot{V}O_2$ (mL kg/min)	0.79 ± 1–0.08	0.56 ± 1–0.002	1, 5
Mean blood pressure (mmHg)			
Ventral aorta	40.3 ± 1–0.7	31.5 ± 1–1.8	1, 4
Dorsal aorta	33.1 ± 1–1.3	25.4 ± 1–0.6	1, 4
Heart rate (beats/min)	35.3 ± 1–1.8	37.8 ± 1–1.5	1, 5
Stroke volume (mIs/kg/beat)[a]	1.2 ± 1–0.1	0.46 ± 1–0.02	1, 5
Cardiac output, Q (mIs/min/kg)	43.7 ± 1–2.4	17.6 ± 1–1.1	1, 5
Hematocrit (%)	18.4 ± 1–0.9	22.6 ± 1–1.0	1, 5
Arterial blood pH	7.78	7.93	2, 5
Arterial O_2 tension, P_aO_2 (mmHg)	64.8 ± 1–11.3	137.0 ± 1–4.2	1, 5
Venous O_2 tension, P_vO_2 (mmHg)	34.5 ± 1–3.3	33.0 ± 1–6	1, 5
Arterial O_2 content (mmol 1.1)	1.64 ± 1–0.02	4.64 ± 1–0.22	3, 5
Venous O_2 content (mmol 1.1)	0.53 ± 1–0.06	3.17 ± 1–0.3	3, 5

SOURCES: (1) Taylor et al. 1977; (2) Piiper et al. 1972; (3) Piiper et al. 1977; (4) Wood and Shelton 1980; (5) Kiceniuk and Jones 1977.

[a] $I = Q$ blood volume; $\dot{V}O_2$ = oxygen consumption.

kidney tubule is unique among the vertebrates in both gross and fine structure. The exocrine and endocrine tissues of the pancreas are intermixed as in higher vertebrates, rather than separate as in most teleosts (Jönsson 1991), and their spermatozoa have a unique ultrastructure (Matte 1991). Their circulatory systems, too, differ significantly and in this chapter we focus primarily on these differences.

Figures 9-1A and B contrast the arrangement of the five pairs of vessels in a selachian, *Squalus acanthias*, and a batoid, *Torpedo fairchildi*, with that of the four pairs (Figure 9-1C) seen in a teleost, the lingcod, *Ophiodon elongatus*. The most evident difference in batoids is the fusion of the last three afferent branchial arteries to form a posterior innominate and the origin of this vessel close to the base of the aorta.

TABLE 9-2   BLOOD PRESSURES IN ELASMOBRANCH FISH

Species	Ventral aorta (mmHg)	Dorsal aorta (mmHg)	References
Scyliorhinus canicula	40.3 ± 1–0.7	31.5 ± 1–1.8	Taylor et al. 1977
Squalus acanthias	38–28	32–24	Burger and Bradley 1951
Heterodontus portusjacksoni	32–22	21–16	Satchell 1971
Raja binoculata	28–19	21–16	Hanson 1967
Mustelus canis	26–19	—	Sudak 1965

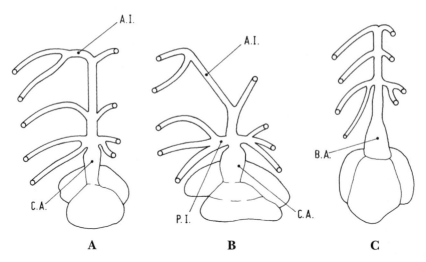

FIGURE 9-1 Afferent branchial arteries and ventral aorta of (**A**) a selachian (*Squalus acanthias*), (**B**) a batoid (*Torpedo fairchildi*), and (**C**) a teleost, the lingcod (*Ophiodon elongatus*). AI = anterior innominate; BA = bulbus arteriosus; OA = conus arteriosus; PI = posterior innominate. (C, Redrawn from Allen 1905.)

REGULATION OF THE VENTILATION:PERFUSION RATIO IN THE GILLS

Cameron et al. (1971) have recorded flow in the branchial arteries of the Pacific dogfish, *Squalus suckleyi*, and the Atlantic stingray, *Dasyatis sabina*. Elasmobranchs, like teleosts, match perfusion to ventilation by locally changing the resistance to blood flow through the gill vessels. When flow through the first three hemibranchs was prevented by occluding the ventral aorta just caudal to the origin of the anterior innominate artery (see Fig. 9-1A), it did not change the distribution of water to each hemibranch. It did, however, increase blood flow to the perfused arches and decrease it to those in which blood flow had been curtailed. The overall oxygen uptake (Vo_2) and arterial and venous oxygen tension (P_ao_2) and P_vo_2 remained unchanged, and the mean ventilation:perfusion ratio ($\dot{V}g:Q$) was preserved. Likewise, sewing the gill slits shut on one side and thereby depriving the gills of a flow of water did not change arterial or venous oxygen saturation. We know that the efferent lamellar arterioles constrict in response to local hypoxia, and this shunts blood away to better-ventilated gills. $\dot{V}g:Q$ ratios are regulated at the level of the gill arch, an arrangement appropriate to the elasmobranchs, in which water can enter the pharynx in varying proportions through these two separate ports. When skates and rays settle on the substrate, the entry of water via the mouth is reduced. The reduction of blood flow through the posterior gills (Fig. 9-2B) thus curtails the flow into the dorsal aorta of blood destined to supply the (now-resting) skeletal muscles. The water entering through the spiracle irrigates the first three hemibranchs, perfused from the anterior innominate, and passes via the afferent and efferent pseudobranchial and hyoidean epibranchial vessels to the optic and cerebral arteries and from these to the brain and eye.

The blood supply to the spiracle

Within the opening of the spiracle in batoid and squaloid elasmobranchs lies the pseudobranch (see Fig. 9-2), a reduced mandibular hemibranch. It still retains something of its gill-like structure, consisting of a group of lamellae supported by branchial rays. Within the lamellae, parallel afferent and efferent vessels are connected by small loops of capillary dimensions. During its development in the embryo, the blood supply to the pseudobranch changes. The ventral connection with the mandibular artery is lost and a new vessel, the afferent pseudobranchial artery, forms. It gives rise to no branches until it enters the pseudobranch. Blood from the small vessels is collected into the efferent pseudobranchial artery, which gives rise to the cerebral artery to the brain and the optic artery to the retina (see Figs. 9-2A and B).

The arrangement of the palatal vessels differs in selachians and batoids. In selachians (see Fig. 9-2A) the dorsal aorta is continued forward rostral to the fusion of the paired efferent branchial arteries from the first gill slits as a very narrow median vessel. This does not divide into paired lateral dorsal aortas until the level of the spiracle. In batoids the flattened body form and the wide spiracle space the paired lateral dorsal aortas apart so that they remain separate as far back as the entrance of the second efferent branchial arteries. As a result, the palate has in it an empty bell-like space free of blood vessels (Fig. 9-2B).

From the top of the first epibranchial loop the hyoidean epibranchial runs diagonally to join, in selachians, the remnant of the lateral dorsal aorta of its side and shortly thereafter gives rise to a commissural vessel, which decussates to join the cerebral artery, not of its side but of the opposite side. Thus blood from the anterior two gill slits of one side can, potentially, pass to the circulation of the opposite olfactory bulb. The implications of this crossed circulation are as yet obscure.

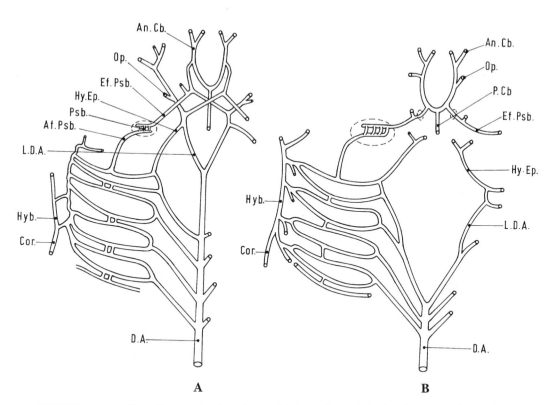

FIGURE 9-2 The epibranchial and cranial arterial circulation and the dorsal aorta of (**A**) a selachian (*Mustelus antarcticus*) and (**B**) a batoid (*Raja radiata*). Af.Psb. = afferent pseudobranchial; An.Cb. = anterior cerebral; Cor. = coronary; D.A. = dorsal aorta, Ef.Psb. = efferent pseudobranchial; Hy.Ep. = hyoidean epibranchial; Hyb. = hypobranchial; L.D.A. = lateral dorsal aorta; Op. = optic; P.Cb. = posterior cerebral; Psb. = pseudobranch in spiracle. (A, redrawn partly from Parker 1886; B, redrawn partly from Allis 1912.)

What, we may ask, is the function of the pseudobranch? It is located directly in the main vascular pathway to the brain and eye; it cannot be involved in gas exchange despite its gill-like structure, for the blood flowing through it (see Figs. 9-2A and B) has already been equilibrated with the incoming water in the secondary lamellae of the first gill. In teleosts pseudobranchial function has been much investigated. There the pseudobranch appears as a sunken or covered vascular mass on the inside of the operculum and is the site of certain receptors. They are known to discharge in response to changes in blood pressure, pH, and $P\text{co}_2$. These nerve endings are too far from the surface to be able readily to respond to changes in the incoming water and they are viewed primarily as vascular proprioceptors, able to monitor the blood passing to the brain.

In most elasmobranchs the location of the pseudobranch in the spiracular canal may imply that its receptors can also monitor the water entering the pharynx. Certainly the pseudobranch is well supplied with receptor endings. Mechanical or chemical stimulation of it evokes the cough or spouting reflex. Electrical recordings from the peripheral cut end of the prespiracular branch of the seventh nerve show bursts of action potentials when the pseudobranch is mechanically stimulated. Electrical stimulation of the central cut end of the nerve evokes cardiac and respiratory inhibition (Satchell 1959). Initially, we may surmise, the spiracle served as a significant portal for inflow into the orobranchial cavity, as it does in batoids today. The most aerated water is likely to lie above and around the front of the fish; farther back it will be mixed with exhaled water. The tendency for the most rostral gill slit to serve this function (Grigg 1970) is seen in *Heterodontus*, a rather isolated and primitive genus. A pattern of ventilation is sometimes seen in which water enters by the first gill slit and augments that entering through the mouth; water leaves via gill slits 2–5. Subsequently, we may suppose that as the spiracle became less important for the intake of water, the pseudobranch persisted because its various receptors are well positioned to monitor the blood passing forward to the brain.

The secondary blood system

In teleost fish we know that a second system of arteries, capillaries, and veins arises from the efferent branchial arteries and dorsal aorta and is in parallel with those of the primary system. At their origin from the primary vessels these secondary arteries are surrounded by a sphincter of smooth muscle, and a fringe of microvilli extends across the openings and holds back a varying proportion of the erythrocytes. The blood in the secondary vessels thus tends to have a lower hematocrit than that in the primary system. Secondary blood vessels are largely absent from the nervous system, skeletal muscle, and the solid organs of the digestive system; they are well represented in epithelia that face into the environmental water, such as those of the gills, the operculum, the skin, and in part, the intestinal lining (Vogel 1985; Satchell 1991). The secondary system is surmised to be involved in ion and water movements.

We do not yet know how the secondary system of elasmobranch fish takes origin, but four prominent, interconnected secondary veins, the dorsal, lateral, and ventral cutaneous vessels are very evident underlying the skin of elasmobranchs and are obviously homologous with those of teleosts. Moreover, the paired lateral cutaneous veins of teleosts lead posteriorly in many genera to a small, two-chambered caudal heart in the tail, powered by skeletal muscle and driven from motoneurons in the tip of the spinal cord; this caudal heart pumps blood from the secondary veins into the caudal vein. In the selachian genera *Cephaloscyllium* and *Mustelus* (Satchell and Weber 1987), an elongate vessel, the caudal sinus, occurs in the ventral lobe of the tail fin (Fig. 9-3). It receives blood from the major cutaneous veins and is compressed against the median fin skeleton by the radial muscles, which serve to deflect the tail when the fish swims. These muscles contract rhythmically

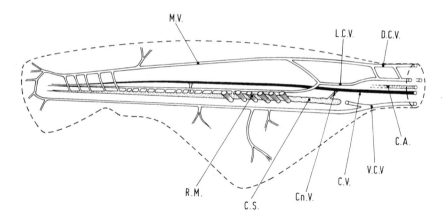

FIGURE 9-3 The caudal heart of the carpet shark, *Cephaloscyllium isabella*. Many of the radial muscles and smaller vessels have been omitted for the sake of clarity. C.A. = caudal artery; Cn.V. = connecting vein; C.S. = (stippled) caudal sinus; C.V. = caudal vein; D.C.V. = dorsal cutaneous vein; L.C.V. = lateral cutaneous vein; M.V. = marginal vein; R.M. radial muscles; V.C.V. = ventral cutaneous vein. (Redrawn from Satchell and Weber 1987.)

in a caudal-rostral sequence when the fish is at rest and blood is pumped from the sinus into the caudal vein.

Paired fins and the lateral abdominal vein

Studies of the heterocercal tail, so characteristic of elasmobranchs, show that its line of thrust is oriented through the center of gravity and the transverse plane of insertion of the pectoral fins (Thomson 1976). The net sinking component at the center of gravity is offset by the planing action of these fins. Adjustments effected by the pterygial muscles of the pectoral and, to a lesser degree, of the pelvic fins are important in vertical mobility. The paired fins of sharks and rays are large and well-muscled structures compared with those of teleosts, in which the swim bladder provides neutral buoyancy.

Paired fins develop from a continuous lateral fin fold in the embryo, and in Upper Devonian sharks such as *Cladoselache* they are still broadly based. In elasmobranchs, but not in teleosts, blood draining from the fins enters a prominent vessel, the lateral abdominal vein. It is a vein of considerable size that runs lateroventrally up each side of the abdominal cavity, immediately beneath the coelomic lining, and enters the sinus venosus. Along its length it receives small factors from the hypaxial muscles and at the end of the abdomen it joins with its fellow of the opposite side above the pelvic bar. At this level it receives the drainage of the pelvic fins and, via a network of small valved veins, the blood from the cloaca and rectum. There are also small connections with the caudal vein.

Flow is forward into the sinus venosus; in *Heterodontus* pressure at the central end is subambient, $-2.2 \pm 1-1.5$ mmHg and is lower than that at the pelvic end, $1.0 \pm 1-0.4$ mmHg; its fluctuations reflect the cyclic aspiration created by ventricular ejection (Birch et al. 1969). This major vein is absent in most bony fish; their relatively small paired fins drain separately into the posterior cardinal sinus or caudal vein.

The venous sinuses and venous capacitance

The concept of venous capacitance grew largely from mammalian studies (Caldini et al. 1974; Green and Jackman 1984). Capacitance defines the total contained volume of the vessels at a given transmural pressure. Veins are mostly thin-walled and compliant and thus have a higher capacitance than arteries. Mammalian circulatory physiologists have developed models with two compartments, that is, vascular beds. The peripheral compartment (P) comprises all the vessels other than those of the viscera, which constitute the splanchnic compartment (S) (Rothe 1983).

Constriction of arterioles lowers the pressure in the capillaries and veins in the downstream vascular bed, which results in the decrease of blood volume there secondary to the passive elastic recoil of the vessel walls. This blood augments the inflow into the heart and is redistributed to the peripheral vascular bed. Reciprocal changes occur when arterioles are relaxed. The volume of blood redistributed is influenced by the compliance of the vessel walls. The time constant (t) of each vascular bed is the product of the sum of the arteriolar and venous resistances (R) and their capacitances (C), expressed as $t = (R_{AS} + R_{VS}) \times C_S$ and $t = (R_{AP} + R_{VP}) \times C_P$. These values have been determined in mammals, and it emerges that t_S is three times t_P. The flow from a compartment (Q_V) is inversely proportional to the time constant (i.e., $Q_V = D_V/t$). When transmural pressure falls, the outflow from a system of semirigid channels is less than that from one with more compliant walls. If $t_S = 3t_P$, the decrease in flow in the splanchnic circulation is only one-third that of the increase in flow in the peripheral vessels.

The compliance and capacitance of elasmobranch vessels have not yet been determined; nevertheless some comments can usefully be made.

First, in a resting supine vertebrate an overall gradient of pressure exists along the veins. The highest pressure occurs at the point where small veins are formed by the coalescence of capillaries; the low end of the gradient is at the entrance to the heart. The elasmobranch venous system is characterized by the development of large venous sinuses centrally located at the places where the veins enter the sinus venosus. The anterior and posterior cardinal sinuses, containing blood from the somatic circulation, are particularly large. Blood from the posterior cardinal sinuses is the vehicle for secreted catecholamines from the axillary bodies and for urotensins from the large caudal neurosecretory cells that project onto diffuse neurohemal areas on the ventral surface of the posterior spinal cord and it is the elasmobranch equivalent of the teleost urophysis. The hepatic sinuses are also extensive. A fingerlike prolongation extends down each liver lobe and from these smaller branches extend into the substance of the liver; they are so finely divided that a cast of the sinuses has the shape of the liver. A hepatic sinus contains the blood that has just come from the hepatic sinusoids and, more remotely, from the large vessels of the intestine, the dorsal, ventral, and intraintestinal veins. It is the vehicle for nutrients from the liver.

Second, in elasmobranchs the gradient of venous pressure is disturbed by the presence of various venous pumps constituting valved reservoirs that are compressed by skeletal muscles; they have been reviewed recently (Satchell 1992). This compression is incidental to their role in locomotion or ventilation. The most important of these is the hemal arch pump, consisting of the intercostal veins draining the segmental axial muscles of the trunk, which are valved at their entrance to the caudal vein. The vascular bed of the muscle is compressed as the wave of muscular contraction passes caudally along the sequence of myotomes. This causes blood to pass into the caudal vein, which is itself sheltered from compression by the cartilaginous surround of the hemal arch.

Among the batoids of various body form, the trunk is always much reduced compared to that of a selachian, and it is no more than a whiplike extension in the myliobatiformes. The importance of the hemal arch pump must here be much reduced. Swimming is effected by rostral to caudal waves of alternate elevation and depression of the enlarged pectoral fins, caused by the alternate contraction of antagonist dorsal elevator and ventral depressor muscles (Bernau et al. 1991). Alexander (1991) gives a good account of the three venous sinuses of the fin, the pro-, meso- and metapterygial veins, which appear to be in a position to be compressed against the pterygial cartilages during swimming.

Third, the presence of these large centrally located venous sinuses facilitates the aspiratory role of the heart of elasmobranchs, whereby ventricular ejection lowers intrapericardial pressure and assists flow into the atrium. In mammals the greatest vascular volume is located in the small veins, and the redistribution from this volume is primarily effected by upstream arteriolar vasomotor activity. It is doubtful if this mechanism occurs in elasmobranchs, in which a greater proportion of the venous blood is located centrally. Central venous pressure and thus capacitance may depend less on innervated arterioles and more on changes in heart rate and stroke volume.

Hepatic vein sphincters

The openings of the hepatic veins into the sinus venosus in elasmobranchs are guarded by sphincters of smooth muscle, which by their constriction can narrow or close these openings. The pericardioperitoneal diaphragm is formed of dense collagenous fibers and must create a radial tension opposing the sphincters. In the spiny dogfish, *Squalus acanthias*, 10 mg/mL of epinephrine causes the sphincter to relax (Johansen and Hanson 1967) and acetylcholine in the same concentration causes its contraction. Relaxation allows transmural pressure in the veins and venous sinuses of the intestine and liver to fall and visceral blood to be transferred into the peripheral circulation. The transfer is thus the result of changes in R_{AS} at both upstream and downstream points, that is, the arterioles that admit blood to the gut and liver vasculatures and the outflow from these through the hepatic veins.

Shark livers synthesize and are rich in wax esters, triacyl glycerols, and alkyl diacyl glycerols; muscle does not synthesize any of these (Sargent et al. 1972; Van Vleet et al. 1984). The alkyl diacyl glycerols are relatively conserved during starvation and the migration associated with reproduction, suggesting that their primary role is to enhance buoyancy. Moreover, if the fish are made artificially heavier (Malins and Barone 1970) the proportion of these relative to the other lipids increases. In contrast, the other two lipids turn over rapidly and in guitarfish (*Rhinobatos annulatus*) are most abundant during peak breeding periods (Rossouw 1987). In *Ginglymostoma* we know that norepinephrine and epinephrine stimulate lipolysis of its liver lipids, converting them to fatty acids (Lipshaw et al. 1972). In vivo studies in *Squalus acanthias* of labeled fatty alcohols show that they are massively oxidized to fatty acids, the bulk of which subsequently appear in skeletal muscle (Sargent et al. 1972). *Ginglymostoma* possesses a fatty acid–binding protein (FABP), the primary structure of which has been studied by Medzihradszky et al. (1992); it shows structural affinities, not with the liver FABPs of higher vertebrates, but with that of human adipose tissue, the lipid stores of which can be similarly mobilized by catecholamines. Shark liver is metabolically different from mammalian liver; it resembles adipose tissue in that under the influence of catecholamines it releases fatty acids into the circulation from its stores of wax esters and triacyl glycerols. This bidirectional flux of fatty acids into and out of the liver stores and into muscle is facilitated by FABP. The rise in blood catecholamines in elasmobranchs during swimming, noted above, thus enriches the blood in the liver sinuses with fatty acids and, by dilating the hepatic sphincters, facilitates the acids' transfer into the peripheral circulation to muscle.

Warm-blooded elasmobranchs

Heat diffuses through tissues more than ten times as rapidly as oxygen, and gills of sufficient area and thinness to enable the blood flowing through them to be oxygenated will ensure that the temperature of the blood (T_{BL}) closely matches that of the water (T_W). Any heat retained in the blood from the aerobic metabolism of the tissues is largely lost.

As a result, tissue temperatures in most elasmobranchs are closely similar to ambient water temperature (T_W). This is well shown in Figure 9-4A; hollow circles plot brain temperature (T_{BR}) against T_W for eight blue sharks, *Prionace glauca*, and one each of *Carcharhinus signatus, C. falciformis, C. obscurus, Negaprion brevirostris*, and *Odontaspis* sp. (Block and Carey 1985).

Certain other sharks, notably the porbeagle (*Lamna nasus*), mako (*Isurus oxyrhynchus*), and the white shark (*Carcharodon carcharias*) have certain parts of the body warmer than the water. In Figure 9-4A the solid circles plot T_{BR} against T_W for mako and porbeagle. The two plots approach each other in water above 26°, but at $T_W = 12°$ the brain is some 6° warmer than the water. In addition, these elevated temperatures are found in the eye, red muscle, and parts of the digestive system, notably the liver, stomach, and spiral valve of the mako and porbeagle.

The anatomical basis of these elevated temperatures are the retia mirabilia. The normal arterial blood supply is rerouted through a network of small arterioles, which come into close association with the venous blood draining from the organ or tissue; heat diffuses into the arterial blood and is carried back to the tissue by counter current flow.

THE SUPRAHEPATIC RETE

Burne (1923) and Carey et al. (1981) describe the suprahepatic rete of the porbeagle. Small arteries, some 80 μm in diameter, are packed at a density of $85/mm^2$ to form two columnlike masses within the hepatic sinus interposed between the pericardial diaphragm and the liver. Paired pericardial arteries arise from the caudal margin of the last hypobranchial vessel and run parallel with the coronary arteries. There are two of these, dorsal and ventral, as in other carchariniform sharks (De Andrés et al. 1992).

In other elasmobranchs the pericardial arteries would be small twiglike vessels but are here massive arteries that supply almost all of the blood to the rostral pole of the rete (RPR). From the caudal end the outflow from the many small arterioles is collected into large paired arteries, from which distributors carry blood to the liver, stomach, intestine, and spleen. The normal arteries supplying these viscera from branches of the dorsal aorta, the celiacomesenteric and lienogastric arteries, are of narrow diameter.

This subversion of some of the hypobranchial blood (see Fig. 9-2A,B) and its passage through what is in effect a heat exchanger enables the inflowing arterial blood, which will be at water temperature, to take up

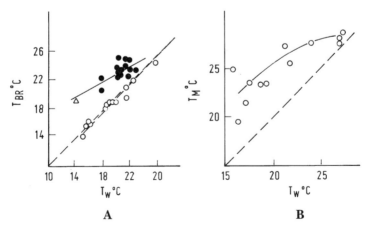

FIGURE 9-4 **A:** Plot of brain temperature (T_{BR}) and water temperature (T_W) of sharks with a carotid rete (*solid circles*) and sharks without one (*hollow circles*). In both A and B the *dashed line* is the plot of water temperature. **B:** Muscle temperature (T_M) of mako caught at various temperatures. (A, redrawn from Block and Carey 1985; B, redrawn from Carey and Teal 1969a.)

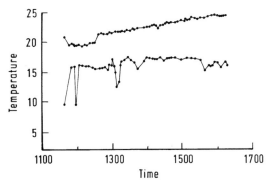

FIGURE 9-5 *Upper line* shows the stomach temperature of the mako shark; *lower line* shows water temperature. When released the shark went briefly below the thermocline, where it encountered 9°C water. (Redrawn from Carey et al. 1981.)

heat from the returning visceral venous blood. Figure 9-5 shows a telemetric record of stomach temperature and water temperature of a mako shark. Shortly after the start of the record, the shark descended briefly through the thermocline; the stomach was some 5°C warmer than the water. The rete itself is cool at the pericardial end and warm at the visceral end. Longer term records suggest that there is some ability to regulate visceral temperature. Carey et al. (1981) note that the two masses of retial arterial vessels do not entirely fill the venous sinus; the irregular spaces between them coalesce to form two veins that at their rostral ends are provided with bands of circular and longitudinal smooth muscle and may provide a mechanism that could allow the warm blood to flow directly into the sinus venosus and thus bypass the rete.

THE RETIAL VESSELS
OF SKELETAL RED MUSCLE

In most selachians, as in teleosts, slow maintained cruising is effected by red muscle, which is spread as a thin fillet immediately beneath the skin at the level of and below the lateral line. Red muscle is characterized by its abundant capillaries, its high content of myoglobin, and its ability to metabolize lipid. Blood flows centrifugally to it in segmental arteries that arise from the dorsal aorta and is returned centripetally in segmental veins that enter the caudal vein or posterior cardinal sinus.

A cross section of a lamnid shark (Fig. 9-6B) shows a different disposition of the red muscle; it is tucked in laterally to form a fillet close to the spine and extends the length of the fish from the pectoral fin to the muscle of the tail. It is thus insulated from the water by the mass of surrounding white muscle. Between it and the lateral margin of the trunk is a slab of retial tissue composed of intermeshed arterial and venous vessels; from a lateral cutaneous artery beneath the skin at the level of the lateral line, numerous small arterioles carry blood centrifugally to the red muscle. Warm blood is returned to a lateral vein through the small venous channels of the rete. The shift of the red muscle toward the center of the body and of its blood supply to a peripheral site provides space to accommodate the rete. The lateral arterial vessel, absent in fish lacking a muscle rete, is derived from the epibranchial vessel of the fourth gill arch. Associated with it is the reduction of the epibranchial vessels and aorta, which would normally supply the myotomal muscles. The epibranchial arteries have a minute lumen. The lumen of the dorsal aorta is only one-third that of a normal shark of equivalent size. The visceral arteries (celiaco-mesenteric and lienogastric) are also reduced.

Figure 9-4B shows a plot of the warmest muscle temperature T_M plotted against water temperature T_W in mako sharks (Carey and Teal 1969a). It resembles that of the brain temperature of mako and porbeagle sharks shown in Figure 9-4A; in cooler water the muscle, like the brain, can be 5–8°C warmer than the water.

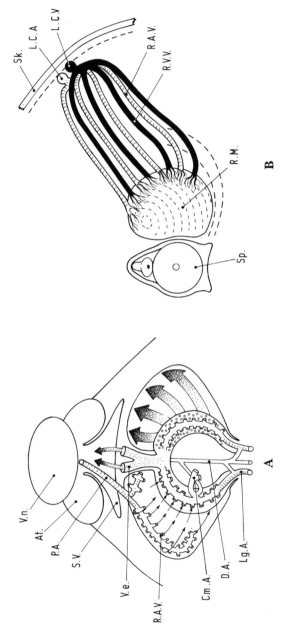

FIGURE 9-6 **A:** The suprahepatic rete of the mako shark, *Isurus oxyrhynchus*. The atrium (At) and ventricle (Vn) have been folded forward to show the rete behind the sinus venosus (S.V.). On the left, the large pericardial arteries (P.A.) bring arterial blood into the retial arterial vessels (R.A.V.); these are collected together to form the celiomesenteric artery (Cm.A.) and lienogastric and spermatic arteries (Lg.A.). The dorsal aorta (D.A.) is very narrow. The right side shows how the venous blood, collected from the sinusoids of the liver, is gathered into the hepatic sinus and flows (*large arrows*) into paired veins (Ve.), which are provided with smooth muscle. **B:** Semidiagramatic drawing of the muscle rete of the porbeagle shark, *Lamna cornubica*. L.C.A. = lateral cutaneous artery; L.C.V. = lateral cutaneous vein; R.A.V. = retial arterial vessel; R.V.V. = retial venous vessel; R.M. = red muscle; Sk. = skin; Sp. = spine. (A, redrawn from Carey et al. 1981; B, redrawn partly from Carey and Teal 1969a.)

THE CAROTID RETE

In the porbeagle the pseudobranch is much reduced and consists of a few big loop vessels. From it the efferent pseudobranchial artery runs forward as a series of ten or so loops. At the point where it enters the orbit and the orbital venous sinus that surrounds the eye, it contributes to a mass of 200 or so small arterial vessels constituting the *pseudobranchial plexus*, part of the carotid rete. These vessels become smaller and finally coalesce to form a stem vessel that passes through the cranial wall and divides into the anterior and posterior cerebral arteries to the brain. In addition, the hyoidean epibranchial, arising from the top of the first epibranchial loop, follows a less looped, twisting course and gives rise to the stapedial artery, which supplies a separate portion of the carotid rete, termed the *stapedial plexus*. From its distal pole arises the optic artery, which supplies the retina. The continuation of the hyoidean efferent artery, termed the *internal carotid artery*, enters the braincase and may provide a bypass to the carotid rete, allowing for some regulation of brain temperature.

Blood in these retial vessels will flow in a rostral direction, countercurrent to the venous blood draining through the orbital sinus into the anterior cardinal sinus. Eye temperatures in mako and porbeagle sharks may be 5°C warmer than the water.

The presence of retia, which permit a localized warm-bloodedness in these elasmobranchs, is paralleled by similar structures in certain large fast-swimming carnivorous teleosts such as tuna, and the topic has been commented on by earlier writers (Carey and Teal 1969b; Carey et al. 1971; Carey 1982; Satchell 1991). Swordfish (*Xiphius gladius*) have an eye heater derived from the external rectus muscle and formed of thermogenic tissue comparable to the brown fat of mammals. In some ways elasmobranch retia are less elaborate; only in those of the red muscle are arterial and venous small vessels arranged in closely packed arrays. The carotid and suprahepatic retia are merely a meshwork of small arterioles tucked into a venous sinus. Moreover they lack thermogenic tissue and the mitochondria of the muscle retia in *Isurus oxyrhinchus* do not show the biochemical specializations of the swordfish thermogenic organ that enables it to oxidize alpha glycerophosphate at a high rate (Ballantyne et al. 1992). Their regionally elevated temperatures are derived from heat conservation rather than locally stimulated thermogenesis.

What are the physiological advantages, we may ask, of this partial and localized warm-bloodedness? The Q_{10} for enzymatic reactions is generally in the range of two to three (Stevens and McLeese 1984), and this is certainly true for the hydrolyses mediated by the proteolytic enzymes. Elevated temperatures increase the rate of digestion; a rise from a visceral temperature of 10°C up to 25°C decreases the time for protein digestion to about one-third in tuna (Carey et al. 1984) and as the stomach is emptied faster, the number of meals per day can be increased.

The power available from vertebrate muscle increases some threefold for each 10°C (Carey and Teal 1969b). The diffusion of oxygen within red muscle fibers is facilitated by the myoglobin they contain; it provides the last link in the supply line for the oxygen-needing contractile machinery. A 10°C rise increases the facilitated diffusion by 400%. Lactate levels in the blood return to normal faster at higher temperature and thus shorten the time between bouts of high activity (Stevens and Carey 1981). Studies on trout show that cooling the brain lengthens the latency and duration of responses to light (Konishi and Hickman 1964).

Hemopoiesis and phagocytosis— the mononuclear phagocytic system

The defense of the tissues against agents that invade the body comprises, in elasmobranchs as in other vertebrates, two basic mechanisms. The agent may be physically engulfed (i.e., by phagocytosis) or it may be agglutinated by immunoglobulins synthesized against it. The cells and organs involved in these functions are here termed the mononuclear phagocytic system (MPS). In higher vertebrates they are closely related to the system of lymphatic capillaries that serve to convey spilt plasma protein back into the central veins. Such a lymphatic system is absent in elasmobranchs, and our understanding of fluid balance across their capillaries is incomplete. For this reason the term *lymphatic system* is best avoided.

In the laboratory, sharks are able to produce circulating antibodies resembling human immune macroglobulins of the immunoglobulin (Ig) M class in response to various antigenic stimuli such as the streptococcal A variant and the PR8 and equine influenza virus (Sigel and Clem 1965). In recent years, great interest has attached to the immunoglobulins of sharks, in the belief that they are of very ancient lineage. As in higher vertebrates, immunoglobulins are composed of two heavy (μ), and two light (κ or λ) chains. The κ and λ chains have constant and variable regions; in the sandbar shark, *Carcharinus plumbeus*, Hohman et al. (1992) have described the sequence of nucleotides and that of the corresponding amino acids of the light chains. The sequence in both the constant (C) and the variable (V) regions suggest that they are λ rather than κ chains. The sequences in *Carcharinus* have more homology with human light chains than with those from the horned shark, *Heterodontus francisci* (Shamblott and Litman 1989). Grogan and Lund (1991) comment on the fact that, despite the phylogenetic distance, factors in shark serum stimulate activity in human lymphocytes and monocytes. However, it has been shown that in sharks, antibody responses are slow to develop. In the demersal shark, *Ginglymostoma cirratum*, the response to equine influenza virus does not peak until 28–40 days (Sigel and Clem 1965). Such slow responses may be of little utility in the short term, and nonspecific defenses may well be more important.

PERIPHERAL BLOOD LEUCOCYTES

In *Scyliorhinus canicula*, peripheral blood leucocytes include four types of granulocyte (G1–G4), monocytes, thrombocytes and lymphocytes (Morrow and Pulsford 1980). G2, constituting 0.70% of the leucocytes, has many features in common with the mammalian neutrophil. G1 (5%), G3 (40%), and G4 (25%) are eosinophils (Hunt and Rowley 1985, 1986). Plasma cells have been described in the blood of *Mustelus canis* (Barnes et al. 1967). In the study by Hunt and Rowley (1986) using a range of particles, it was concluded that of the blood leucocytes, monocytes and thrombocytes but not granulocytes were responsible for clearing the blood.

Monocytes (5%) are released into the blood stream and migrate into the tissues, where they develop into macrophages. These scavengers occur in various situations and are an important part of the nonspecific immune system. In the spleen, some of the macrophages have erythrocyte debris within them and are clearly involved in the destruction of effete erythrocytes. Others are surrounded with a rosette of lymphocytes, which establish cytoplasmic contacts with them; in some, complete cytoplasmic fusion occurs. Pulsford et al. (1982) suggest that some kind of immunological exchange may be occurring; sessile macrophages may be trapping and processing antigens and presenting them to lymphocytes, which then produce antibodies.

In all elasmobranchs and a few teleosts the abdominal cavity communicates with the outside water via the paired abdominal pores, small passages through the body wall anterolateral to the vent (George et al. 1982). These peritoneal surfaces are scavenged by macrophages and small quantities of abdominal fluid containing loaded effete macrophages pass out of them from time to time.

THE CAVERNOUS BODIES

Cavernous bodies (CB) are largely confined to elasmobranchs and provide a further means for nonspecific defense. When *Scyliorhinus canicula* is injected with colloidal carbon, latex beads, or sheep erythrocytes, these are cleared from the circulation in varying degrees (Hunt and Rowley 1986). The CB are intercalated into the arterioarterial pathway in the manner depicted in Figure 9-7. Blood from the afferent filament arteries, which are branches from the main afferent branchial vessels, passes

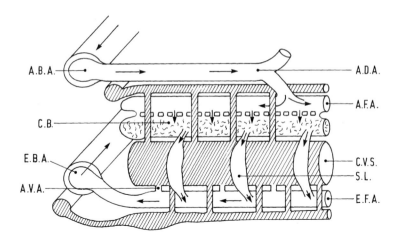

FIGURE 9-7 The cavernous body of the gill of *Urolophus mucosus*, the stingaree. Vessels in the arterio-arterial pathway are shown *clear* and *arrows* show the direction of blood flow; elements of the venous drainage are *cross hatched*. A.B.A. = afferent branchial artery; A.D.A. = afferent distributing artery; A.F.A. = afferent filament artery; A.V.A. = arteriovenous anastomosis; C.B. = cavernous body; E.B.A. = efferent branchial artery; E.F.A. = efferent filament artery; C.V.S. = central venous sinus; S.L. = secondary lamella. (Redrawn from Donald 1989.)

through the meshwork of small vessels of the CB prior to entering the secondary lamellae (Donald 1989). These vessels are spanned by trabeculae with a core of connective tissue and muscle cells, and all surfaces are lined by an endothelium, on which are scattered fixed phagocytic cells, sometimes forming a tightly packed layer; they are rounded and project into the lumen. CB cells are not strictly members of the MPS, for they are derived from endothelial cells and are thus mesenchymal in origin. Most of the CB cells will have taken up carbon within 30 minutes and the blood will have been cleared of particles within 12 hours. They are strategically located to be able to screen the blood entering the gill lamellae and free it of pathogens that may have entered from the gut.

THE THYMUS
The thymus in some species involutes with age and in adult *Scyliorhinus* is largely regressed. In the rays and in *Heterodontus* it persists as a large organ throughout life (Fänge 1987). In *Raja clavata* and *Torpedo marmorata* (Zapata 1980) the thymus is an opaque yellowish organ above the gills on each side and, as in higher vertebrates, consists of a cortex and medulla. The cortex consists of numerous lymphocytes and lymphoblasts, some in mitosis, in a network of highly branched reticular cells, which are joined by desmosomes. Some of the cells contain tonofilaments. Myoid cells, which as in higher vertebrates have an imperfect resemblance to striated muscle, have cytoplasm filled with myofilaments, T tubes, and lipid inclusions. The medulla stains less densely and has reticular fibers and macrophages. It lacks the thymic corpuscles of higher vertebrates, and epithelial cysts are scarce. The authors make no mention of plasma cells.

In higher vertebrates the thymus has a key role in early life as it is the source of the T lymphocytes, which migrate out and are responsible for local immunity. They are distinct from the B lymphocytes derived from the bone marrow; these give rise to plasma cells, which secrete circulating Igs. Whether there is any such division of lymphocyte function in elasmobranchs, which have no bone marrow, is unknown. Zapata (1979) argues that in brown trout, *Salmo trutta*, such a division does exist and that the equivalent of B lymphocytes are generated in the pronephric kidney. In teleosts the pronephric kidney lacks tubules but contains fixed macrophages, reticular cells, and many thin-walled arteries, a histological picture reminiscent of mammalian bone marrow. It is a region of intense hemopoietic activity as seen in the paddle fish *Polyodon spathula* (Georgi and Beedle 1978). In elasmobranchs the pronephric kidney is prominent but its histology has not been studied recently.

In higher vertebrates the thymus is also the source of various thymic peptides, including the hormone thymulin, which are involved in communication between the neuroendocrine and immune systems (Millington and Buckingham 1992); there has not as yet been any systematic search for these in elasmobranchs.

THE SPLEEN

The spleen of elasmobranchs is large; in a 650–950 g *Scyliorhinus canicula* it is elongate (7–9 cm) and weighs 2.5–3.5 g (Pulsford et al. 1982). It is supplied with blood from the celiomesenteric artery and drains into the hepatic portal vein. It consists of masses of white pulp, mainly located around large blood vessels, with red pulp between. Vessels of arteriolar and capillary dimensions penetrate the pulp. Some arterioles give rise to structures termed *ellipsoids*, which consist of a fusiform bunch of capillaries surrounded by a sheath of endothelial cells that are actively phagocytic. A similar structure occurs in the batoid spleen (Zapata 1980). The red pulp contains erythroid tissue. The white pulp contains a few granulocytes and plasma cells and numerous lymphocytes, monocytes, and macrophages. Plasma cells produce antibodies, and Fänge and Mattisson (1981) found large numbers of plasma cells in the spleen of nurse sharks. Macrophages are common within the ellipsoids and lie free within the white pulp. The role of macrophages and lymphocytes in producing antibodies has been mentioned above; the spleen is considered to be the main site of antibody synthesis.

THE ORGAN OF LEYDIG AND THE EPIGONAL ORGAN

In higher vertebrates the bone marrow is the dominant site of hemopoiesis, but in elasmobranchs this is in part taken over by two organs unique to the group, the organ of Leydig and the epigonal organ. The first appears as a large pink or white mass in the wall of the esophagus, extending dorsally and ventrally from the buccal region to the cardia (Fänge 1968). It contains coarse and fine eosinophilic granulocytes, heterophils, and lymphocytes. Fänge (1987) reports that in some genera (*Squalus* and *Etmopterus*) G3 granulocytes are generated in it.

A rather similar collection of granulopoietic cells is found in certain elasmobranchs in the epigonal organ (Fänge 1984). This occurs in the gonadal mesenteries and may be single or paired. In *Cetorhinus maximus* and in certain other large sharks, the epigonal organ is dark red in color, suggesting that some erythropoiesis occurs there. There is a tendency for these two structures to be reciprocally related in size so that sharks like *Prionace glauca* and *Ginglymostoma*, which have little or no Leydig's organ, have large epigonal organs. The angel shark, *Squatina squatina*, has the proportions between the two reversed.

When the spleen of *Scyliorhinus canicula* is removed, the hematocrit falls for 2 weeks but thereafter rises and returns to normal in 3 weeks (Fänge and Johansson-Sjobeck 1975). It is at present uncertain where the new red cells are generated. Elasmobranchs possess less erythropoietic and granulopoietic tissue as a percentage of body weight than do higher vertebrates. In mammals the two together amount to 3% of body weight; in *Scyliorhinus canicula* and *Raja batis* they come to only 0.6–1% (Fänge 1968).

Conclusion

The elasmobranchs arose in the middle Devonian period, some 380 million years ago (Carroll 1988). We know that today the rate of change of their mitochondrial DNA is very slow; their low metabolic rate ensures a smaller flux of oxygen radicals, which are potent intracellular mutagens (Martin et al. 1992). Their circulatory systems bear the stamp of their primitive origins. Yet Table 9-1 bears testimony to how similar their circulatory performance is to that of teleosts, which did not arise until the Triassic or later. There are no values in Table 9-1 that could identify a fish as being specifically an elasmobranch or a teleost. Animals exist as functional wholes, not as collections of systems, and similar end results may be achieved by different means.

Two points require further discussion. First, elasmobranchs differ greatly from teleosts in the extent to which their circulatory systems are under direct autonomic control. In higher vertebrates, direct sympathetic and parasympathetic innervations progressively replace control by circulating humoral agents. Inasmuch as the heart has a direct and powerful vagal control but lacks sympathetic innervation, elasmobranchs lie between agnathans and teleosts. A crucial advance in the evolution of autonomic control was the use of the spinal nerves as pathways along which sympathetic fibers could gain access to effectors such as arterioles, chromatophores, and glands distributed through the tissues and at the body surface. This step is fully achieved in the teleosts but is not yet attained in elasmobranchs. They uniformly lack the gray rami that in all higher vertebrates carry postganglionic unmyelinated sympathetic fibers into the spinal nerves. A simple dissection shows that the spinal sympathetic ganglia are not segmentally arranged and are not connected to form the ladderlike chain so familiar in higher vertebrates (Young 1933). In the abdominal region some postganglionic fibers run directly to the viscera and Bjenning et al. (1989) report that catecholamine-containing nerve endings are largely restricted to the gut vasculature in *Raja erinacea* and *R. radiata*. This primitive feature of the systems involved in circulatory control characterizes Batoidea, Squalomorpha, and Galeomorpha alike. It does not, however, leave the fish totally deprived of the ability to regulate flow through its blood vessels. Elasmobranchs characteristically have axillary bodies, that is, paired concentrations of chromaffin tissue and sympathetic postganglionic neurons, which are located in the dilated rostra portion of each posterior cardinal sinus. Catecholamines are liberated into the blood in the posterior cardinal sinus, aspirated at the next heartbeat into the heart, and passed on through the gills to the general circulation. Butler et al. (1986) have shown in *Scyliorhinus canicula* that even during spontaneous swimming, resting levels of epinephrine increase 3.3 times to 1.9×10^{-8} and norepinephrine 2.3 times to 3.2×10^{-8} mol/L. These levels are, respectively, 80% and 50% of the levels shown by other workers to have the maximum effect of increasing gas exchange at the gills and gas transport to the tissues. In contrast catecholamines in teleost blood do not rise above resting levels until swimming speeds exceed 80% of U_{crit} (critical swimming speed).

Differential effects on tissues of elasmobranchs depend more on end-organ responsiveness to α-, β_1-, and β_2-adrenergic agonists than on specific autonomic innervation. A corollary of this situation is the need to minimize the long-term accumulation of transmitter substances. Elasmobranchs have not been studied in this regard, but Nekvasil and Olson (1986) report that the perfused trout gill removes 60% of a norepinephrine and 47% of an epinephrine pulse in one passage; most of the deamination occurs in the venous drainage of the gill. Blood perfusing this pathway does not pass into the dorsal aorta but returns directly to the heart, in elasmobranchs as in teleosts.

Second, the MPS arose in the vertebrate series in close association with the blood system, for blood provided the MPS's cells with a ready access to the tissues. In the elasmobranchs the various sources of hemopoietic and granulopoetic cells outlined above make strange reading to a mammalian hematologist, accustomed to the generation of these cells in the bone marrow. Early groups of fish such as Ostracoderms, Placoderms, and Arthrodires were bottom-living and had an inflexible bony armor. Elasmobranchs uniformly lack a swim bladder and, with few exceptions, maintain themselves in the water column by continuous swimming. The deposition of mineral salts in bone makes it heavy, and the reversion to an entirely cartilaginous skeleton in elasmobranchs must be regarded as a secondary loss that reduced the work to be done to keep afloat when deposit feeding was abandoned for the pursuit of prey. Cartilage is a strong avascular tissue. It contrasts with bone, which is a very vascular tissue, and endochondral ossification is a vascular event. In limbs, blood vessels from the surrounding tissue penetrate into the cartilage, and a ring of bone indicates the start of the marrow cavity. The relocation of erythropoiesis and granulopoiesis into these cavities is a late event in vertebrate evolution, occurring first in the amphibia. It led to the loss of such outlying sites of hemopoiesis as the organ of Leydig and the epigonal organ.

In conclusion, we may note that the batoids and galeomorphs represent two extremes of modification. Although each group contains various genera, most typically they are represented by the one, flattened, bottom-living fish that feed on mollusks and other demersal animals; the other, the fast-swimming sharks that are predators of the nutritionally dilute open oceans. So disparate are they that earlier writers derived them from separate groups of fossil fish. Rays, it was suggested, were derived from Devonian rhenanid arthrodires, which were flattened bottom-living plagiostomes; *Jagorina* is a well-known example. Sharks, it was held, were closely related to acanthodians, plagiostomes that existed from the upper Silurian to lower Permian. Jarvik (1980) points out that no fossils are known that could link the two groups. Carroll (1988), in a recent cladistic study, argues that all modern elasmobranchs can be traced to a single ancestral group, the Neoselachii, which occurred sometime in the Triassic or perhaps as early as the Permian. Batoids, squalomorphs, and galeomorphs all possess 15 of the 28 derived characters used by this author to construct the cladogram. These are all features of the skeleton and scales apart from one, the bell-like configuration of the palatal blood vessels described above. It is reassuring to find that other features of the circulatory system, such as the presence of CBs in the branchial pathway, the expansion of the central veins to form sinuses close to the heart, and the absence of gray rami in the autonomic nervous system, support such conclusions. Despite their great adaptive differences, batoids and selachians are rightly included in the one group, the elasmobranch.

REFERENCES

Allen, W. F. 1905. The blood-vascular system of the loricati, the mail-cheeked fishes. *Proc. Wash. Acad. Sci.* 7: 27–157.

Allis, E. P. 1912. The branchial, pseudobranchial, and carotid arteries in *Raja radiata*. *Anat. Anz.* 41: 579–89.

Alexander, R. L. 1991. Variation in venous drainage of the pectoral fins of batoids (subclass Elasmobranchii: superorder Batoidea). *J. Morphol.* 209: 1–14.

Ballantyne, J. S., M. E. Chamberlin, and T. D. Singer. 1992. Oxidative metabolism in thermogenic tissues of the swordfish and mako shark. *J. Exp. Zool.* 261: 110–14.

Barnes, S. N., A. L. Bell, and S. Gelfant. 1967. The fixation and fine structure of white blood cells of the smooth dogfish shark, *Mustelus canis. Biol. Bull.* 133: 457.

Bernau, N. A., R. L. Puzdrowski, and R. B. Leonard. 1991. Identification of the midbrain locomotor region and its relation to descending locomotor pathways in the Atlantic stingray *Dasyatis sabina. Brain Res.* 557: 83–94.

Birch, M. P., C. G. Carre, and G. H. Satchell. 1969. Venous return in the trunk of the Port Jackson shark. *J. Zool. (Lond.)* 159: 31–49.

Bjenning, C., W. Driedzic, and S. Holmgren. 1989. Neuropeptide Y-like immunoreactivity in the cardiovascular nerve plexus of the elasmobranchs *Raja erinacea* and *Raja radiata. Cell Tissue Res.* 255: 481–86.

Block, B. A., and F. G. Carey. 1985. Warm brain and eye temperatures in sharks. *J. Comp. Physiol.* 156B: 229–36.

Burger, J. W., and S. E. Bradley. 1951. The general form of the circulation in the dogfish *Squalus acanthias. J. Cell. Comp. Physiol.* 37: 389–402.

Burne, R. H. 1923. Some peculiarities of the blood vascular system of the porbeagle shark (*Lamna cornubica*). *Philos. Trans. R. Soc. Lond.* 212B: 209–57.

Butler, P. J., J. D. Metcalf, and S. A. Ginley. 1986. Plasma catecholamines in the lesser spotted dogfish and rainbow trout at rest and during different levels of exercise. *J. Exp. Biol.* 123: 409–21.

Caldini, P., S. Permutt, J. A. Waddell, and R. L. Riley. 1974. Effect of epinephrine on pressure, flow, and volume relationships in the systemic circulation of dogs. *Circ. Res.* 34: 606–23.

Cameron, J. N., D. J. Randall, and J. C. Davis. 1971. Regulation of the ventilation-perfusion ratio in the gills of *Dasyatis sabina* and *Squalus suckleyi*. *Comp. Biochem. Physiol.* 38A: 505–19.

Carey, F. G. 1982. Warm fish. In *A Companion to Animal Physiology*, ed. C. R. Taylor, K. Johansen, and L. Bolis, 216–33. Cambridge: Cambridge University Press.

Carey, F. G., J. W. Kanwisher, and E. D. Stevens. 1984. Bluefin tuna warm their viscera during digestion. *J. Exp. Biol.* 109: 1–20.

Carey, F. G., and J. M. Teal. 1969a. Mako and porbeagle: warm-bodied sharks. *Comp. Biochem. Physiol.* 28: 199–204.

———. 1969b. Regulation of body temperature by the bluefin tuna. *Comp. Biochem. Physiol.* 28: 205–13.

Carey, F. G., J. M. Teal, and J. W. Kanwisher. 1981. The visceral temperatures of mackerel sharks (Lamnidae). *Physiol. Zool.* 54: 334–44.

Carey, F. G., J. M. Teal, J. W. Kanwisher, K. D. Lawson, and J. S. Beckett. 1971. Warm-bodied fish. *Am. Zool.* 11: 135–43.

Carroll, R. L. 1988. *Vertebrate Paleontology and Evolution*. New York: W. H. Freeman.

De Andrés, A. V., R. Muñoz-Chápuli, V. Sans Coma, and L. García Garrido. 1992. Anatomical study of the coronary system in elasmobranchs: II. Coronary arteries in hexanchoid, squaloid, and carcharinoid sharks. *Anat. Rec.* 233: 429–39.

Donald, J. A. 1989. Vascular anatomy of the gills of the stingarees *Urolophus mucosus* and *U. paucimaculatus* (Urolophidae, Elasmobranchii). *J. Morphol.* 200: 37–46.

Fänge, R. 1968. The formation of eosinophilic granulocytes in the oesophageal lymphomyeloid tissue of the elasmobranchs. *Acta Zool. (Stockh.)* 49: 155–61.

———. 1984. Lymphomyeloid tissues in fish. *Vid. Meddr. Dan. Nat. Foren.* 145: 143–62.

———. 1987. Lymphomyeloid system and blood cell morphology in elasmobranchs. *Arch. Biol. (Brussels)* 98: 187–208.

Fänge, R., and Johansson-Sjobeck. 1975. The effect of splenectomy on the hematology and on the activity of d-aminolevulinic acid dehydratase (ala-D) in hemopoietic tissues of the dogfish, *Scyliorhinus canicula* (Elasmobranchii). *Comp. Biochem. Physiol.* 52A: 577–80.

Fänge, R., and A. Mattisson. 1981. The lymphomyeloid (hemopoietic) system of the Atlantic nurse shark, *Ginglymostoma cirratum*. *Biol. Bull.* 160: 240–49.

George, C. J., A. E. Ellis, and D. W. Bruno. 1982. On resemblance of the abdominal pores in rainbow trout, *Salmo gairdneri* Richardson, and some other salmonid spp. *J. Fish Biol.* 21: 643–47.

Georgi, T. A., and D. Beedle. 1978. The histology of the excretory kidney of the paddle fish (*Polydon spathula*). *J. Fish Biol.* 13: 587–90.

Green, J. F., and A. P. Jackman. 1984. Peripheral limitations to exercises. *Med. Sci. Sports Exerc.* 16: 299–305.

Grigg, G. C. 1970. Use of the first gill slits for water intake in a shark. *J. Exp. Biol.* 52: 569–74.

Grogan, E. D., and R. Lund. 1991. Reactivity of human white blood cells to factors of elasmobranch origin. *Copeia.* 2: 402–8.

Hanson, D. 1967. Cardiovascular dynamics and aspects of gas exchange in Chondrichthyes. Ph.D. diss., University of Washington, Seattle.

Hohman, V. S., S. F. Schluter, and J. J. Marchalonis. 1992. Complete sequence of a cDNA clone specifying sandbar shark immunoglobulin light chain: gene organization and implications for the evolution of light chains. *Proc. Natl. Acad. Sci. U.S.A.* 89: 276–80.

Hunt, T. C., and A. F. Rowley. 1985. Separation of leucocytes in the dogfish (*Scyllorhinus canicula*) using density gradient centrifugation and differential adhesion to glass coverslips. *Cell Tissue Res.* 241: 283–90.

———. 1986. Studies on the reticulo-endothelial system of the dogfish, *Scyliorhinus canicula*. Endocytic activity of fixed cells in the gills and the peripheral blood leucocytes. *Cell Tissue Res.* 244: 215–26.

Jarvik, E. 1980. *Basic Structure and Evolution of Vertebrates*, vol. 1. London: Academic Press.

Johansen, K., and D. Hanson. 1967. Hepatic vein sphincters in elasmobranchs and their significance in controlling hepatic blood flow. *J. Exp. Biol.* 46: 195–203.

Jönsson, A. C. 1991. Regulatory peptides in the pancreas of two species of elasmobranchs and in the Brockman bodies of four teleost species. *Cell Tissue Res.* 266: 163–72.

Kiceniuk, J. W., and D. R. Jones. 1977. The oxygen transport system in trout (*Salmo gairdneri*) during sustained exercise. *J. Exp. Biol.* 69: 247–60.

Konishi, J., and C. P. Hickman. 1964. Temperature acclimation in the central nervous system of rainbow trout (*Salmo gairdneri*). *Comp. Biochem. Physiol.* 13: 433–42

Lacy, E. R., and E. Reale. 1991. The fine structure of the elasmobranch renal tubule: intermediate, distal, and collecting duct segments of the little skate. *Am. J. Anat.* 192: 478–97.

Lipshaw, L. A., G. J. Patent, and P. P. Foa. 1972. Effects of epinephrine and norepinephrine on the hepatic lipids of the nurse shark, *Ginglymostoma cirratum*. *Horm. Metab. Res.* 4: 34–8.

Malins, D. C., and A. Barone. 1970. Glyceryl ether metabolism: regulation of buoyancy in dogfish (*Squalus acanthias*). *Science* 167: 79–80.

Martin, A. P., G. J. P. Naylor, and S. R. Palumbi. 1992. Rates of mitochondrial DNA evolution in sharks are slow compared with mammals. *Nature* 357: 153–55.

Matte, X. 1991. Spermatozoon ultrastructure and its systematic implications in fishes. *Can. J. Zool.* 69: 3038–55.

Medzihradszky, K. F., B. W. Gibson, K. Surinder, Y. Zhonghua, D. Medzihradszky, A. L. Burlingame, and N. M. Bass. 1992. The primary structure of fatty-acid–binding protein from nurse shark liver. Structural and evolutionary relationship to the mammalian fatty-acid–binding protein family. *Eur. J. Biochem.* 203: 327–39.

Millington, G., and J. C. Buckingham. 1992. Thymic peptides and neuroendocrine-immune communication. *J. Endocrinol.* 133: 163–68.

Morrow, W. J. W., and A. Pulsford. 1980. Identification of peripheral blood leucocytes of the dogfish (*Scyliorhinus canicula* L.) by electron microscopy. *J. Fish Biol.* 17: 461–75.

Nekvasil, N. P., and K. R. Olson. 1986. Extraction and metabolism of circulating catecholamines by the trout gill. *Am. J. Physiol.* 250: R526–31.

Parker, T. J. 1886. On the blood-vessels of *Mustelus antarcticus*: a contribution to the morphology of the vascular system in the vertebrata. *Philos. Trans. R. Soc. Lond.* 177B: 685–732.

Piiper, J., M. Meyer, and F. Drees. 1972. Hydrogen ion balance in the elasmobranch *Scyliorhinus stellaris* after exhausting activity. *Respir. Physiol.* 16: 290–303.

Piiper, J., M. Meyer, H. Worth, and H. Willmer. 1977. Respiration and circulation during swimming activity in the dogfish *Scyliorhinus stellaris*. *Respir. Physiol.* 30: 221–39.

Pulsford, A., R. Fänge, and W. J. W. Morrow. 1982. Cell types and interactions in the spleen of the dogfish *Scyliorhinus canicula* L.: an electron microscopic study. *J. Fish Biol.* 21: 649–62.

Randall, D. J. 1970. The circulatory system. In *Fish Physiology*, vol. 4: *The Nervous System, Circulation and Respiration*, ed. W. S. Hoar and D. J. Randall, 133–72. New York: Academic Press.

Randall, D. J., W. S. Hoar, and A. P. Farrell, eds. 1992. *The Heart and Circulation*, vol. 12A: *Fish Physiology*. London: Academic Press.

Rossouw, G. J. 1987. Function of the liver and hepatic lipids of the lesser sand shark, *Rhinobatos annulatus* (Müller & Henle). *Comp. Biochem. Physiol.* 86B: 785–90.

Rothe, C. F. 1983. Venous system: physiology of the capacitance vessels. In *Handbook of Physiology*, ed. J. T. Shepherd and F. M. Abboud, sect. 2, vol. 3, pt. 1, 397–452. Bethesda: American Physiological Society.

Sargent, J. R., R. R. Gatten, and R. McIntosh. 1972. The metabolism of neutral lipids in the spur dogfish, *Squalus acanthias*. *Lipids* 7: 240–45.

Satchell, G. H. 1959. Respiratory reflexes in the dogfish. *J. Exp. Biol.* 36: 62–71.

———. 1971. *Circulation in Fishes*. Cambridge: Cambridge University Press.

———. 1991. *Physiology and Form of Fish Circulation*. Cambridge: Cambridge University Press.

———. 1992. The venous system. In *Fish Physiology*, ed. W. S. Hoar, D. J. Randall, and A. P. Farrell, vol. 12A, 141–83. New York: Academic Press.

Satchell, G. H., and L. J. Weber. 1987. The caudal heart of the carpet shark, *Cephaloscyllium isabella*. *Physiol Zool.* 60: 692–98.

Shamblott, M. J., and G. W. Litman. 1989. Complete nucleotide sequence of primitive vertebrate immunoglobulin light chain genes. *Proc. Natl. Acad. Sci. U.S.A.* 86: 4684–88.

Sigel, M. M., and L. W. Clem. 1965. Antibody response of fish to viral antigens. *Ann. N. Y. Acad. Sci.* 126: 662–77.

Stevens, E. D., and F. G. Carey. 1981. One why of the warmth of warm-bodied fish. *Am. J. Physiol.* 240: R151–55.

Stevens, E. D., and J. M. McLeese. 1984. Why bluefin tuna have warm tummies: temperature effect on trypsin and chymotrypsin. *Am. J. Physiol.* 246: R487–89.

Sudak, F. N. 1965. Intrapericardial and intracardiac pressures and the events of the cardiac cycle in *Mustelus canis*. (Mitchell). *Comp. Biochem. Physiol.* 14: 689–705.

Taylor, E. W., S. Short, and P. J. Butler. 1977. The role of the cardiac vagus in the response of the dogfish *Scyliorhinus canicula* to hypoxia. *J. Exp. Biol.* 70: 57–75.

Thomson, K. S. 1976. On the heterocercal tail in sharks. *Paleobiology* 2: 19–38.

Van Vleet, E. S., S. Candileri, J. McNeillie, S. B. Reinhardt, M. E. Conkright, and A. Zwissler. 1984. Neutral lipid components of eleven species of Caribbean sharks. *Comp. Biochem. Physiol.* 79B: 549–54.

Vogel, W. O. P. 1985. Systemic vascular anastomoses, primary and secondary vessels in fish, and the phylogeny of lymphatics. In *Cardiovascular Shunts*, ed. K. Johansen, and W. W. Burggren, 143–49.

Alfred Benzon Symposium 21. Copenhagen: Munksgaard.

Wood, C. M., and G. Shelton. 1980. Cardiovascular dynamics and adrenergic responses of the rainbow trout in vivo. *J. Exp. Biol.* 87: 247–70.

Young, J. Z. 1933. The autonomic nervous system of selachians. *Qt. J. Microsc. Sci.* 75: 571–624.

Zapata, A. 1979. Ultrastructural study of the teleost fish kidney. *Dev. Comp. Immunol.* 3: 55–65.

———. 1980. Ultrastructure of elasmobranch lymphoid tissue: I. Thymus and spleen. *Dev. Comp. Immunol.* 4: 459–72.

BRUNO TOTA

CHAPTER 10

Heart

With increasing body mass, developing vertebrate embryos soon reach the state at which a circulatory system is a prerequisite for further growth (i.e., for supplying the growing tissues and for removing catabolic material). Accordingly, in elasmobranchs, as in teleosts, birds, and mammals, cardiac development precedes that of other organs (Pelster and Bemis 1991). The importance of the design of the "internal machinery" is well exemplified by early stages of vertebrate cardiogenesis when the dorsal aorta joins the heart and ventral aorta by way of the first pharyngeal arch vessel; from that time, the parts of the system function together. Alteration in flow in one region during the early stages of development influences not only flow in another region, but also morphogenesis of other areas. The diagrams in Fig. 10-1 show that the cardiovascular system is a dynamic closed loop in which the heart pumps blood to the arteries perfusing the microvasculature while the veins return blood to the heart, in such a way that there is always a head of pressure in the arterial tree and the resistances in the vascular network provide sufficient flow for the particular organ facing a particular metabolic demand. The autonomic nervous system plays an important role in this closed-loop system. Sensory inputs from a variety of receptors that monitor the levels of chemical components (chemoreceptors) as well as blood volume and pressure (baroreceptors) inform the cardioregulatory and vasomotor centers of the brain. Information flows from the same centers via efferent autonomic neurons to adjust both vascular

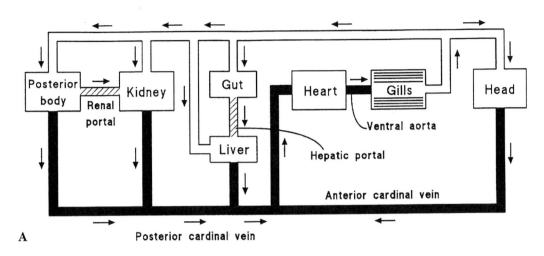

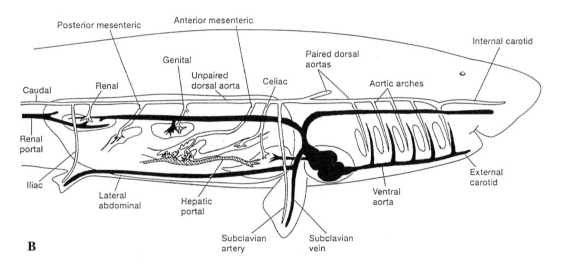

FIGURE 10-1 The basic single circulation pattern of fish in a shark illustrated (**A**) schematically and (**B**) diagramatically. The blood moves from the heart to the gills to the systemic tissues and back to the heart, through which it passes only once during each complete circuit.

resistance, by controlling mainly smooth muscle of the arterioles and precapillary sphincters, and cardiac output, by controlling heart rate and stroke volume. Part of this neuronal integration is achieved solely by the cardiac vagus and allows for the adjustment of cardiovascular activity according to the respiratory pattern of the fish. The heart must adequately perfuse other organs in addition to its own myocardial mass. As discussed later, two routes of cardiac blood supply, lacunary and coronary, have evolved in elasmobranchs to perfuse their well-developed myocardial mass, particularly the ventricle. As a consequence, the cardiac pump and the arterial network on one side and the venous system ending in the sinus venosus on the other side are close interacting subsystems.

General design of the cardiovascular apparatus

As in teleosts and cyclostomes, the elasmobranch cardiovascular system consists of a single closed circuit in which the systemic (branchial) heart pumps deoxygenated venous blood to the gills, from which oxygenated blood is distributed via the dorsal aorta to the body microvasculature. Deoxygenated blood then returns from the periphery to the heart via the venous system (see Fig. 10-1). The heart consists of a caudocranially arranged series of four contractile chambers (sinus venosus, atrium, ventricle, and conus arteriosus) whose walls contain myocardial tissue. This situation contrasts with that found in teleosts, in which smooth muscle replaces myocardial muscle in the bulbus arteriosus and often also in the sinus venosus. Venous blood enters the sinus venosus from the common cardinal veins or ducts of Cuvier and hepatic veins and is pumped first into the atrium, thence to the ventricle, and finally to the ventral aorta via the conus arteriosus (elasmobranchs) or the noncontractile bulbous arteriosus (teleosts). Consequently, the cardiac pump must develop enough kinetic energy to maintain adequate blood flow across the hindrance of several capillary beds in series. The first such location is the gills, where oxygenation of the blood and ion transfer takes place in the lamellae, and then the systemic capillary bed. In mackerel sharks an important arteriole-venule network consists of the retial countercurrent systems that concentrate heat (Block and Carey 1985). Data obtained in small sharks, like the dogfish, have shown that after perfusion through the gills, blood pressure is reduced by some 25–30% from prebranchial values. We shall see that both the myoarchitecture of the ventricular chamber and the structure and function of the conus are well suited for the hemodynamic task of ensuring adequate blood flow to the periphery despite this loss of pressure across the gills.

In comparison with ancestral fishes, in which the arrangement of the heart, dorsal aorta, and branchial vasculature evolved from six gill arches, the cardiovascular system of elasmobranchs, like that of teleosts, has been rearranged with a reduced number of gill arches (Fig. 10-2). However, primary differences between teleosts and elasmobranchs concern the arrangement of afferent and efferent branchial vessels, which in turn are related to basic differences in the mechanism of ventilation (see chap. 9). In contrast with the cyclostomes, in which the circulatory system is partly open, with large blood sinuses, elasmobranchs have remarkably reduced the sinusoidal circulation by a structural arrangement of the circulatory system consisting of closed, narrow vessels, a condition that facilitates higher intravascular pressures, a trend further developed in teleosts.

In elasmobranchs, the very large liver is well suited to function as a reservoir of blood, in that it has an extensive hepatic portal system allowing prolonged transit time of blood through the organ. In dogfish and skates (Johansen and Hanson 1967), prominent sphincters located on the large hepatic veins just before their entrance into the sinus venosus may play a role in control of hepatic blood flow and in adjustment of cardiac output. The presence of these sphincters must be relevant in view of the apparent general lack of venomotor activity observed in elasmobranchs. Of great importance is the development of large, centrally located venous sinuses, such as the anterior and posterior cardinal sinuses and the hepatic sinuses, that facilitate the aspiratory role of the heart (see Fig. 10-2).

Many elasmobranch species are viviparous and some experience an expansion of blood volume related to a highly vascularized placenta (Hamlett 1987) and the hypertrophied uterine mucosa in stingrays (Hamlett et al. 1985, 1996a,b). For example, in the pregnant *Squalus suckleyi*, the blood volume increases from 5% to 10–13% of volume, with no change in the hematocrit (Martin 1950). Although no studies have so far documented the "cost" of viviparity in terms of cardiovascular adjustments, one must expect notable cardiac adaptations in response to such increased hemodynamic loads.

Heart 241

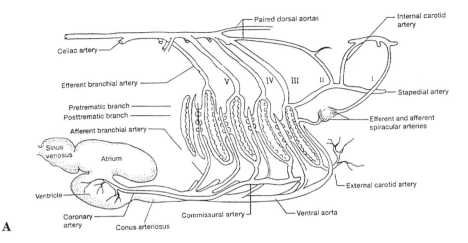

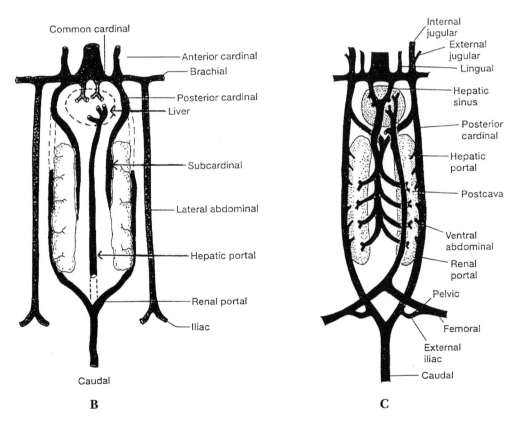

FIGURE 10-2 Adult shark. **A:** Aortic arch derivatives, with roman numerals indicating aortic arches. The section of the aortic arch delivering blood to the gills is the afferent branchial artery, while the efferent branchial artery is the dorsal section carrying blood away. **B:** The basic pattern of major veins showing how the caudal blood from the tail passes through the kidneys, where the caudal vein becomes the renal portal system. From the kidneys, blood is drained by the subcardinal veins. **C:** The large venous sinuses facilitate the aspiratory action of the heart.

Cardiovascular function is affected by buoyancy. Static lift is attained in fish in two main ways: either by storing a sufficient amount of low-density material such as lighter lipids (like squalene) in the tissues or by regulating the mass of gas in a gas-filled organ (the swim bladder). The very large liver of squaloid sharks, which stores enormous amounts of low-density squalene oil, account for almost one quarter of the total body weight, in contrast to 4–6% of the total body weight shown by the livers of most vertebrates, including humans (Bone et al. 1995). As a result, elasmobranchs are nearly neutrally buoyant, while they sink if the liver is removed. Elasmobranchs lack a swim bladder.

Skin morphology is another feature of the shark that can affect circulatory dynamics. As pointed out by Vogel (1985), commenting on Wainwright et al. (1978), data show that sharks utilize a hydroskeleton: "shark skin is sturdier stuff than fish skins in general and has the crossed helical fiber array diagnostic of hydrostatic arrangements." In sharks, muscles attach directly to the skin, which thus acts as both an external pressure-resisting membrane and an external whole-body tendon. Fiber angles vary with location on the fish; the nonelastic skin must transmit the forces generated by the body musculature back toward the tail. During locomotion, the pressure inside the body of a shark rises to as much as 200 kPa, twice atmospheric.

It has been suggested that skeletal muscle tonus and rhythmic swimming movements facilitate venous return, as documented in the Port Jackson shark (Satchell 1991). Blood flow increased 46% in the caudal vein in relation to postabdominal trunk swimming movements. It is not known if or how much the respiratory movements that set up rhythmical pressure changes in the pericardium may affect venous return. The branchial and caudal pumps, as well as the caudal heart of the carpet shark, *Cephaloscyllium isabella*, have been reviewed (Satchell 1991).

A more complex view of the blood circulatory system shared by typical elasmobranchs and water-breathing teleosts considers the circulation as divided into a primary and a secondary vessel network. Common to both circuits are the central cardiovascular apparatus, that is, the cardiac pump, the branchial respiratory vasculature, and the main blood distributor, the dorsal aorta (Vogel 1985; Steffensen and Lomholt 1992).

Location of the heart

The fish heart is always posterior to the gills, being situated in teleosts relatively farther forward in the body than in elasmobranchs. Well protected by the shoulder girdle, the heart is located in the pericardial cavity made up of the membranous pericardial sac, which is more spacious in elasmobranchs than in bony fish (Fig. 10-3). The pericardium consists of two layers: the parietal, or outer, pericardium, which makes up the outer wall of the pericardial cavity, and the inner, visceral pericardium, which reflects over the heart rostrally at the junction of the conus arteriosus with the ventral aorta and caudally at the posterior margin of the sinus venosus. The semirigid parietal pericardium, adherent to surrounding cartilages and muscles, has ventral and lateral surfaces supported by the dense cartilage of the pectoral girdle, while the rostral walls lie against the coracobrachial and depressor muscles, thus acting as a rigid, boxlike structure. This arrangement has functional consequences since it constrains the contractile performance of the heart (suctorial heart), particularly in elasmobranchs, in which the parietal pericardium is more rigid than in teleosts. This anatomical feature may account for some of the functional differences between the heart of elasmobranchs and mammals. In mammals the pericardium is in actual contact with the surface of the heart and moves with it, while in elasmobranchs the pericardium is separated by a large volume of pericardial fluid and does not move during the cardiac cycle (Shabetai et al. 1985; Lai et al. 1989). The visceral pericardium, also called the epicardium, is inseparable from the heart and is in intimate contact with the subjacent myocardium through the subepicardium. Some features of the subepicardium, together with ultrastructural features of the mesothelial cells of the epicardium, suggest an active exchange between the pericardial cavity and the subepicardial compartment in *Scyliorhinus stellaris* (Helle et al. 1983).

During ontogenesis, the pericardial cavity becomes separated from the primary coelomic cavity by the septum transversum and in the adult is closed toward the rear. However, in elasmobranchs, as well as in the cyclostome *Myxine* and in a few primitive bony fish (Acipenseriformes), the pericardial lumen

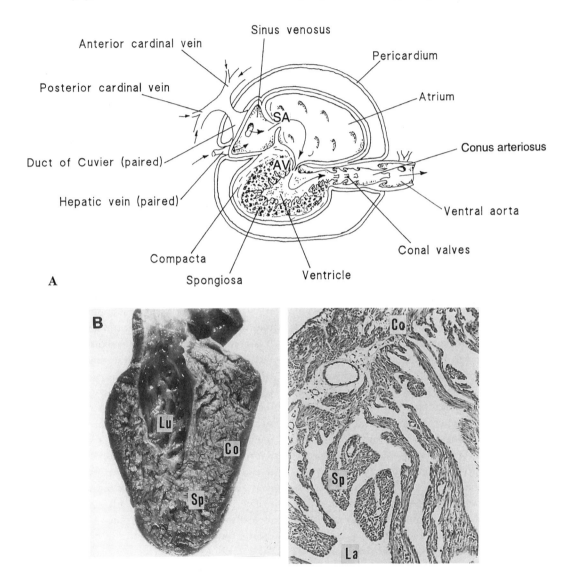

FIGURE 10-3 Location of the elasmobranch heart in the pericardial cavity. **A:** Schematic illustration showing how the outer, or parietal, pericardium reflects over the cardiac chambers, becoming the visceral pericardium, or epicardium. The **S**-shaped arrangement of the cardiac chambers places the thin-walled sinus venosus and atrium dorsal to the ventricle, so that atrial contraction may assist ventricular filling. Note the sinoatrial (SA) and atrioventricular (AV) valves and the mixed type of ventricular myoarchitecture with the trabeculated and compact musculature. **B:** Heart of *Alopias*. Lu = lumen; Co = compacta; Sp = spongiosa. Right panel ×100, reproduced at 70%.

communicates with the peritoneal cavity through the pericardioperitoneal canal. The canal arises dorsally above the atrium, and after running through the septum along the ventral surface of the esophagus, ends with two openings into the peritoneal cavity. It is flaccid and not patent unless the pressure of the pericardial fluid exceeds that in the peritoneal cavity, thus making the delicate walls act like a valve. Pericardial fluid fills the space between the inner and outer pericardial layers. The fluid not only is free of protein, as documented in both *Raja* (Smith 1929) and *Squalus* (White and Satchell 1969 as quoted by Satchell 1971), but also differs in composition from the peritoneal fluid and from plasma by having higher sodium- and potassium-ion and lower urea content in *R. diaphanes* and *R. stabuliforisa* (Maren 1967; Satchell 1991).

Structural design of the heart

The muscular heart is interposed between the two principal parts of the circulation, the high-pressure arterial system and the low-pressure venous system. The position of the heart is reflected in the different structure and muscularity of its chambers, which mirror their diverse hemodynamic loads and functions (see Fig. 10-3). In homiotherm embryos (Clark and Hu 1982) and the skate (*Raja erinacea*) embryo, an increase in body mass is correlated with both arterial blood pressure and heart growth, with older embryos showing not only more myocardiocytes but also a higher concentration of contractile proteins (Pelster and Bemis 1991).

The characteristic manner in which event follows event during every heartbeat is the result of structural features and the specialized pathway followed by the wave of depolarization that activates all myocardial chambers. Specialized myocardial cells (nodal cells) in the most caudal chamber initiate and propagate an action potential so that all cells of the myocardium contract in a caudorostral pattern. The presence of intercalated disks, that is, specialized cell-cell junctions representing regions of low electrical resistance, makes the myocardium act as a functional syncytium. An efficient cardiac pump needs a suitable geometrical arrangement of myocardial fibers and a series of competent one-way valves. When the myocardial fibers contract to compress the blood contained in a given chamber, blood is expelled sequentially in a unidirectional flow from the sinus venosus to the atrium, thence to the ventricle, and finally to the conus arteriosus, from which it is distributed to the ventral aorta. The nerve supply, together with endocrine input, local hormones, and humoral factors released from other cell components of the heart and from the myocardium itself play a complex modulatory role, decreasing or increasing several aspects of cardiac function.

Sinus venosus

As in cyclostomes and teleosts, the sinus venosus is the most caudal and dorsal chamber of the elasmobranch heart. It has a conical shape, with the broader oval base located caudally, where four openings, two lateral from the common cardinal veins and two caudal from the suprahepatic veins, allow systemic venous return. The apex is toward the atrium, to which it is connected through the sinoatrial junction. Venous blood flows to the atrium through the sinoatrial orifice, where a sinoatrial valve prevents backflow during atrial systole. The presence of ligaments between the sinus venosus and the parietal pericardium has not frequently been observed; however Grant and Regnier (1926) described a V-shaped cardiac ligament situated on either side of the sinus venosus in *R. clavata*. Just above the sinoatrial orifice, on the right and left sides, are openings of the right and left coronary veins that drain blood returning from the heart into the sinus.

The sinus venosus is the thinnest-walled and most compliant of the cardiac chambers. The wall is made up of an outer epicardium, the visceral pericardium, and an inner endocardium, which are both monolayered, between which a collagenous matrix with variable amount of myocardial cells is interposed. In *Scyliorhinus*, in contrast to the situation in *Chimaera*, the sinus wall is tightly packed with myocardiocytes. The wall structure, studied in *Squalus acanthias, Etmopterus spinax*, and *Galeus melastomus* (Saetersdal et al. 1975), *S. stellaris* (Helle et al. 1983), and in *Scyliorhinus canicula* (Ramos et al. 1996), is characterized by a collagen-rich subepicardial layer and, in particular, by a subendocardium densely innervated by myelinated and unmyelinated neurons and intramural ganglionic cells. In *S. stellaris* (Helle et al. 1983), neural input via large bundles of granule-containing axons terminating at fenestrated regions of the endocardium suggests a neurohumoral function. The thin processes of the endocardium are often less than 50 nm thick and together with fenestrations may represent sites for transepithelial diffusion. In the three species studied by Saetersdal et al. (1975), an accumulation of granulated processes could be detected in the subendothelial space together with larger endothelial pores (about 0.4–1.6 µm diameter) through which granulated processes protruded into the venous cavity. The presence and distribution of these endocardial pores has been further detailed by the scanning electron microscopic (SEM) study of Muñoz-Chápuli et al. (1995) in the dogfish *S. canicula*. As shown in Figure 10-4, the pores fit into the category of large gaps (several hundreds of nanometers) between endothelial cells and were always found on large bundles protruding into the cardiac lumen, mainly constituted of granule-containing nerve fibers. Interestingly, Muñoz-Chápuli et al. (1994), on the basis of the demonstration of a distinct substance P–like immunoreactivity in these fibers, have suggested that a tachykinin-related peptide, alone or together with other substances, could be released into the lumen from the neuroendocrine cells and fibers through the endocardial pores. Taken together, these findings support the view that the sinus venosus in elasmobranchs can be the site of an important neuroendocrine system and may add a new dimension to the concept of the heart as an endocrine organ.

Myocardial cells are frequently arranged in bundles of few cells of variable size and shape, cylindrical being the most common. Sinusal myocardiocytes are usually larger than atrial or ventricular myocytes, ranging in diameter 7–9 µm in *S. canicula* (Ramos et al. 1996) or 9–15 µm in *S. acanthias*. In *G. melastomus* and *E. spinax* (Saetersdal et al. 1975), nuclei are circular, oval, or pyriform and have a sparse number of myofibrils randomly oriented at the periphery. Sarcomeres have the typical A, I, H, and Z bands with lengths ranging from 1.8 to 2.0 µm in *S. canicula* (Ramos et al. 1996) to 2.4 µm in *S. acanthias, G. melastomus*, and *E. spinax* (Saetersdal et al. 1975). Elliptic or elongated mitochondria are commonly dispersed throughout the cytoplasm, when they are not concentrated in areas like the subsarcolemmal space. The sarcoplasmic reticulum consists of peripheral couplings and subsarcolemmal vesicles. The abundance of these structures, more in the sinusal than in the atrial, suggest myocardiocytes might be related to the ability for spontaneous excitation of the cell membrane, as suggested by Bride (1977). Gap junctions, desmosomes, or specific granules are uncommon in the sinusal myocardiocytes of *S. canicula*, while laterally arranged cell junctions of the intermediate type at the Z-line levels may be present. Intercalated discs are scarce, exhibiting only fasciae adherentes.

Myocardial bundles, surrounded by a basal membrane, are densely innervated. Nerve terminals are present at the periphery and more often in the center of the bundles and are in simultaneous contact with several myocardiocytes, together with some granule-containing unmyelinated neurons. The myelinated axons, all belonging to branches of the vagus nerve, break into a diffuse plexus, a characteristic feature of the sinus venosus. Postganglionic parasympathetic neurons are represented by numerous ganglion cells, which may eventually pass into the atrium via the atrioventricular junction into the ventricle, as reported in *Carcharias sorrah* (Nair 1970). The ganglion cells, which in *Triakis scylla* exhibit a diameter between 8 and 14 µm

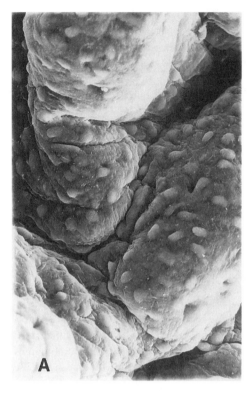
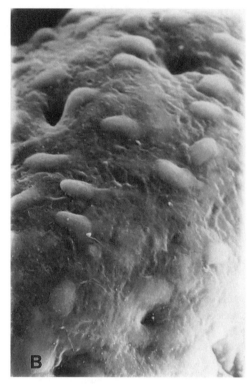

FIGURE 10-4 The endocardial endothelium of the sinus venosus of the dogfish *Scyliorhinus canicula*, as visualized from the luminal side by transmission electron microscopy (TEM). (With kind permission of Dr. R. Muñoz-Chápuli, unpublished material.) **A:** Note the discontinuities of the endothelial cells (×833). **B:** High-magnification view of the endocardial pores (×2000). Reproduced at 80%.

(Yamauchi 1980) are endowed with many large (150–250 nm) membrane-bound granules in their cytoplasm and are surrounded by satellite cells. Occasionally, by formaldehyde-induced fluorescence histochemistry positive for catecholamine-containing neurons, very few fluorescent varicose fibers have been found. As cardiac sympathetic innervation is absent in the elasmobranch heart (see below), these fluorescent neurons have been considered errant adrenergic axons normally distributed to other regions of the vascular system (Nilsson et al. 1975).

The structures of the sinoatrial valve and the sinoatrial junction are of interest for their electrophysiological properties, related to the location of the pacemaker and nodal tissue. The sinoatrial valve, which defines a vertical slotlike communication between the sinus venosus and atrium, consists of two laterally inserted, transversal folds of the cardiac wall that, with the exception of their free margins, are separated by connective tissue. The valve has been studied in detail by Ramos et al. (1996) in *S. canicula*. In this species both folds of the valve, the one facing the sinus venosus and the other facing the atrial chamber, have bundles of myocardial cells. Such bundles are larger than those of the sinusal wall. In the free margin of the valve they join each other, while they are separated in the remaining portion of the folds by collagen-rich connective tissue. The connective tissue also surrounds the sinoatrial junction so that only the free margins of the valve ensure myocardial continuity between the sinusal and atrial chambers. While nerve fibers or terminals are very rare in the atrium, they are abundant in the region of the dorsal and ventral commissures of the valve.

On the whole, sinusal myocardiocytes of elasmobranchs, with their scarce, randomly arranged myofibrils, clear cytoplasm, and rich innervation, fulfill the morphological criteria of nodal tissue. Moreover, the structure of the sinoatrial valve as it appears in S. canicula, with nodal tissue throughout the sinusal face of the folds and the disposition of the ring of connective tissue around the sinoatrial junction and between the double sheet of the valve, is consistent with the hypothesis of impulses being generated in the sinus and conducted into the atrium through the free margin of the valve itself.

The potential contribution of the sinus venosus to atrial filling and to neurohumoral control of the whole heart represent two unexplored aspects for future study.

Atrium

The atrium is the second successive chamber of the heart, lying ventral to the sinus venosus and dorsal to the ventricle. The wall of the chamber may be folded and give the appearance of possessing two rooms (lobes). In some sharks, like *Scyliorhinus*, it may partially infold the ventricle. In contrast with the situation in teleosts, in which the degree of apposition of the atrium to the bulbus arteriosus varies among species, the elasmobranch atrium usually does not show any extensive contact with the conus arteriosus. It has a larger luminal volume than the ventricle, while its wall thickness is much less than that of the very thick walled ventricle but more than that of the sinus venosus. In some elasmobranchs, because of the notable bulk of ventricular musculature, the ventricle is on the left side of the atrium.

The atrioventricular orifice lies in the atrial midline or may be shifted sharply to the left, so communication between the atrium and ventricle is visible in ventral view. The opening is circular or ellipsoid and surrounded by two pocketlike flaps, the atrioventricular valves, projecting into the ventricle with the concavities directed toward the ventricular lumen.

The valves, which prevent the return of blood into the atrium, are supported by both a flat sphincter, made up of circular bands of atrial myocardium surrounding the atrioventricular orifice, and by fine threads, chordae tendineae, anchored to the ventricular wall. The anatomy of the cardiac valvular system, including the atrioventricular valve complex, has recently been studied in *Urolophus jamaicensis* and *Rhizoprionodon terraenovae* (Hamlett et al. 1996b).

The outermost layer of the atrial wall, the epicardium, as evidenced in *S. stellaris*, in *E. spinax*, and in the holocephalan *Chimaera monstrosa* (Berge 1979; Helle et al. 1983), consists of a monolayer of cuboidal cells exhibiting a high degree of cell-cell interdigitation with desmosomes, especially evident at the pericardial side. The cytoplasm of these cells appears densely granulated, with small extensions into the pericardial cavity, prominent rough endoplasmic reticulum, and Golgi complexes. These ultrastructural features may reflect the ability of the epicardium to act both as a secretory layer toward the pericardial cavity and as a protective barrier for subjacent myocardial cells. The subepicardial space is occupied by fibroblasts and collagen fibers.

The innermost lining of the atrium, the endocardium, is made up of a thin (0.1–0.5 μm thick), single layer of electron dense, flat, membrane-bound, granule-rich cells (*Scyliorhinus*, *Chimaera*). The cells have desmosomes at lateral cell-to-cell abutments. A basal lamina is interposed between them and the subendocardial space is occupied by collagen fibers and fibroblasts.

Insinuated between these two layers is the delicate trabeculated myocardium, which projects thin muscle strands (musculi pectinati) inward. By arising from the atrioventricular orifice, these radiating myocardial bundles arch and intersect each other over the roof of the atrium, so that upon contraction, they pull the atrial wall toward the atrioventricular orifice, thus moving the incoming blood into the ventricular cavity.

The connection between the circular bundles of atrial muscle around the atrioventricular orifice and the ventricular muscle is effected by means of an extremely narrow and prolonged strand of myofibers (McWilliam 1885). The muscle strand all around the atrioventricular orifice is prolonged from that musculotubular communication. It is reminiscent of the "canalis auricularis" in the embryo heart, which enters the ventricle at the atrioventricular orifice and penetrates the connective tissue of the atrioventricular junction by means of a remarkably long and slender isthmus of myocardium becoming continuous with the subendocardial trabeculum of the ventricular musculature. Since the time of Gaskell's studies on the turtle heart, the canalis auricularis was considered of particular significance in the electrical activity of the myocardium, being regarded as a functional prolongation of the sinus tissue (Nair 1970). An opposite view, based on morphological investigation of the embryonic dogfish heart, has been presented by De Andrés et al. (1993), who believe the canalis auricularis to be nothing more than the atrioventricular venous ring.

Atrial myofibers, less thick than those in the ventricle and in the conus arteriosus, range from 2 to 3 µm in diameter. Packed in small numbers, they are arranged in trabeculae enclosed by endocardial cells. The sarcolemma is associated with a basal lamina that surrounds bundles of cells in the trabeculae. Sarcomere organization is orthodox, while the sarcoplasmic reticulum both in atrial and ventricular myocytes seems less developed than in those teleost species so far analyzed. In *C. monstrosa* (Berge 1979) subsarcolemmal cisternae are frequently observed. In this species membrane-bound dense bodies are found in atrial myocardiocytes as well as in both the ventricle and conus (Helle et al. 1983).

Electron microscopic, immunohistochemical, and immunocytochemical evidence, complemented by biological data, has recently established that in elasmobranchs the heart is an endocrine organ in the classical sense (Aardal and Helle 1991). Myocytes with secretory granules, as well as myoendocrine cells (i.e., cells characterized by spherical granules with a dense core and a clear halo intimately associated with the Golgi apparatus, intermingled between the myocardiocytes without a preferential localization), have been described in different cardiac regions. They are especially abundant in the atrium of *S. acanthias, S. canicula, R. clavata, C. monstrosa* (Reinecke et al. 1987), *Triakis scyllia* (banded dogfish), and *Narke japonica* (electric ray) (Uemura et al. 1990). These secretory granules contain homologous cardiac hormones of the atrial natriuretic peptide (AP) family. A high-molecular-weight form of C-type natriuretic peptide (CNP), reported to exist only in the brain, has been isolated and characterized from both atrium and ventricle of the dogfish *S. canicula* (Suzuki et al. 1991). C-type natriuretic peptide–like immunoreactivity has been detected in the atrium, and scantly in the ventricle, of *S. acanthias* (Donald et al. 1992). These findings have thus extended to the elasmobranchs the long-held view of the atrium as a specialized volume or stretch receptor chamber well suited for regulating blood volume and pressure. Such sensitivity of the atrium to monitor and adapt to volume and pressure changes has relevance to the reciprocal fashion by which atrial and ventricular functions influence each other.

Ventricle

Providing the major impetus for ejection of blood, the ventricle is the chamber with the largest amount of myocardial tissue and with structural characteristics that best reflect adaptive morphodynamic features.

GROSS MORPHOLOGY

There are two major architectural differences between teleosts and elasmobranchs. First, in elasmobranchs, as well as in chondrostean and dipnoan fishes, the ventricle is connected with the myocardial contractile wall of the conus arteriosus, which in teleosts is replaced by the noncontractile bulbus arteriosus. Second, in teleosts the atrioventricular and bulboventricular orifices are aligned in the midsagittal plane, whereas in elasmobranchs the atrioventricular and conoventricular orifices are arranged lateral to one another. For these reasons the elasmobranch ventricle displays a looped appearance, which may geometrically constrain the overall myoarchitectural design of the chamber and in turn cause differences in the mechanical determinants of ventricular performance and in the directional flow of blood during systole.

While teleosts have a variable tubular, saccular, or pyramidal-shaped ventricle (Santer 1985), elasmobranchs usually have ventricles with the shape of a dorsoventrally flattened pyramid. Several faces are distinguishable: two rhomboidal faces, dorsal and ventral, separated by lateral borders and four vertices: the cranial, two laterals, and the caudal vertex (Sanchez-Quintana and Hurle 1987). When the lateral vertices are less accentuated, as in the dogfish, the ventricle shape is more saccular.

MYOARCHITECTURE

When the ventricle is cut along the sagittal plane, the lack of a spacious central lumen and the high ratio of anastomosing surface area to ventricular volume can be discerned. The prominent trabecular organization of the wall consists of crisscrossed arrays of thin muscle bundles (trabeculae) interlacing in an extremely complex meshwork to give the internal surface a spongy appearance, thence the name *spongiosa*. The trabeculae are lined by a monolayer of endothelial cells (endocardial endothelium) and are supplied by venous blood from the intertrabecular spaces (lacunae) of the ventricular lumen. The trabeculated myocardium is the basic ventricular myoarchitecture in all fish groups. In all elasmobranchs the spongiosa is invested in an outer layer of densely and orderly arranged myocardial bundles, termed *compacta*. This is only present in some species of teleosts associated with high levels of locomotory or metabolic activity (Santer 1985; Tota 1989). A distinct arteriocapillary bed supplies the compacta. Ventricles made of an outer compact layer enveloping the spongiosa are termed "mixed type" hearts. Table 10-1, based on histological and blood supply characteristics of the ventricular wall, summarizes different types of cardiac organization in fishes.

Myocardial fiber architecture has been studied recently in the dogfish (*Galeorhinus galeus*), the blue shark (*Prionace glauca*) and the Atlantic shortfin mako (*Isurus oxyrinchus*) (Sanchez-Quintana and Hurle 1987). In the spongiosa, the innermost subendocardial fibers are grouped into arcuate trabeculae protruding toward the ventricular lumen, and they follow an ordered appearance in a U-shaped direction from the atrioventricular orifice toward the conoventricular orifice, where they are inserted into the fibrous ring. These subendocardial bundles originate at the atrioventricular orifice, where they are continuous with subepicardial bundles of the compacta. Conceivably, this arrangement of the spongiosa, which is not found in teleosts, represents a geometrical determinant of the ventricular mechanics in elasmobranchs.

As visualized by angiocardiographic imaging in the leopard shark, *Triakis semifasciata* (Lai et al. 1990), the spongiosa is compressed against the compacta at the end of diastole. In systole, it appears as an irregular, diffusely opaque structure surrounding the cavity.

The compacta exhibits varying thickness in various species and, in the same species, in relation to age and body size (see Santer 1985 for references). Its relative thickness appears to be related to the number of fiber layers present. In species having a modestly developed compacta, such as the dogfish and the blue shark, only a single fiber layer is present, while two or three layers are observed in other elasmobranchs. The superficial bundle, which at the level of the atrioventricular orifice is continuous with the subendocardial trabeculum, is made up of fibers that are transversely arranged along the ventricular faces. In pyramidal-shaped ventricles these fibers typically encircle the vertices. In the Atlantic mako, in which the compacta accounts for about 30% of the ventricular wall thickness, two main layers are identified. The outermost exhibits fibers arranged as loops with the origins and ends located around the atrioventricular orifice, without inserting in it but continuous

TABLE 10-1 TYPES OF VENTRICULAR ORGANIZATION IN FISH

Type	I (spongy)	II (mixed)	III (mixed)	IV (mixed)
Ventricular myoarchitecture	Spongiosa	Spongiosa, compacta	Spongiosa, compacta (<30%)	Spongiosa, compacta (>30%)
Blood supply	Lacunary (venous blood)	Lacunary (venous blood) + vascular (oxygenated blood)	Lacunary (venous blood) + vascular (oxygenated blood)	Lacunary (venous blood) + vascular (oxygenated blood)
Degree of vascularization	Absent or limited to the epicardium	Coronary vessels limited to the compacta	Vascularization of compacta and spongiosa Thebesian-like shunts	Vascularization of compacta and spongiosa Thebesian-like shunts
Interface	Endocardium	Endocardium + endothelium	Endocardium + endothelium	Endocardium + endothelium
Representative species	*Scorpaena*, *Pleuronectes*, icefish	*Oncorhynchus mykiss*	*Scyliorhinus*, *Torpedo*	Sharks, tuna

SOURCE: After Tota (1989).

Most of the spongy hearts (type I) are typical "venous and avascular hearts" according to Grant and Regnier (1926) (subtype *a*). Other subtypes are related in the presence of epicardial vascularization. Types II–IV are all mixed types of ventricles, with the differences lying in the amount of compacta and the degree of vascularization. The term *interface* is used to indicate the internal boundary, mediating the interaction between myocardial cells and blood. In the spongy heart, the endocardium is the only barrier between the superfusing blood and the myocardium, whereas in the higher types there is, in addition, the vascular endothelium. Whether and how this different endocardial–endothelial composition might exert a different type of control over cardiac performance are completely unknown. In most fish, the coronary artery(ies) originates cranially from the hypobranchial artery. In some species (eel and swordfish) a caudal coronary artery originating from a branch of the dorsal aorta is also present.

with the subendocardial trabeculum. The inner layer of the compacta has fibers arranged in transverse circles, being responsible for the saccular appearance of the ventricle.

There is wide agreement that the amount of compacta, which varies in different species, is related more to the hemodynamic demands of the animal than to its phylogenetic position. When warm-bodied pelagic sharks, such as the great white shark, *Carcharodon carcharias*, and the mako, *I. oxyrinchus*, are compared with ecophysiologically similar ectothermic species, the ventricle weights are the same, but in the former, the compacta is almost twice as large as that of ectothermic species (Emery et al. 1985). Similarly, in teleosts it has been shown that the compacta, when present, increases its thickness in relation to body size and to hemodynamic loads of the animal (Poupa and Lindstrom 1983). From a bioconstructional point of view, compared with the trabecular arrangement of the spongiosa, a compact type of musculature realizes a higher density of properly oriented contractile units (i.e., better economy of space) and thus an improvement of the mechanical efficiency of the ventricular wall. It is not surprising that the compacta is always present in ventricles that function as high-pressure pumps (Tota 1989; Tota and Gattuso 1996).

Less clear is the functional specialization and usefulness of the spongiosa. This type of myocardial organization appears in vertebrates as a basic condition ("initial condition"), ontogenetically, and phylogenetically that has opened the pathway for more complex myoarchitectures. The trabeculum, with its ridges and valleys made of interlacing myocardial bundles, is well suited for retaining venous blood for a relatively long time. This facilitates exchange across the endocardial-endothelial lining and the subjacent trabeculae with the superfusing blood both in systole and diastole. A metabolic adaptation and specialization of the spongiosa to this condition has been suggested (Tota 1983; Driedzic 1992). An important bioconstructional advantage of a ventricular chamber with the wall arranged as thousands of pumps, rather than a single pump made up solely of one compact layer, has been illustrated by Johansen (1965). As compared to a larger ventricle, a small ventricle must contract to a larger size range to eject the same volume, yet for the smaller ventricle a lesser stress (tension [T]) on the myocardial fibers will exist for generating the same internal fluid pressure (P), as dictated by the law of Laplace ($T = PR$, where R is the radius of the chamber). The trabeculated arrangement confers on the ventricle the advantage of functioning both as a single large pump (i.e., the composite ventricular volume) and as many small pumps (lacunae) that are designed for rapid ejection and generation of pressure without a concurrent large tension (stress) on the involved myocardial trabeculae.

ULTRASTRUCTURAL AND FUNCTIONAL ASPECTS

Ultrastructural data cover only a limited number of elasmobranch species; however, when these data are compared with those obtained from bony fish, it appears that ventricle ultrastructure is rather conservative, similarities between teleosts and elasmobranchs far outweighing differences.

Common to all myocardial cells is the peripheral position of myofibrils and the central location of nucleus and mitochondria, absence of T tubules, and similar arrangement of myofilaments, sarcoplasmic reticulum, and couplings. Fish myocardiocytes have smaller diameters than those of homeotherms (Randall 1968; Santer 1985). Elasmobranch ventricular myocardiocytes are usually smaller than those in teleosts, ranging on average from 2.0 to 3.2 nm, as reported in *Torpedo nobiliana*, *Scyliorhinus canicula*, *Scyliorhinus stellaris*, *Squalus acanthias*, *Etmopterus spinax*, and *Chimaera monstrosa* (see Tota 1989 for references). Accordingly, they exhibit a high surface:volume ratio and for their shorter diffusion distances, they may undergo rapid concentration changes in cytoplasmic calcium ions, with a minor contribution by the sarcoplasmic reticulum to electromechanical coupling in *S. acanthias* (Maylie et al. 1980). This makes the shark ventricle somewhat similar to the frog ventricle in that Ca^{2+} for activation of contraction enters the myocytes with membrane depolarization (Nabauer and Morad 1992). In contrast to the frog ventricle, which apparently lacks low-voltage–activated T-type Ca^{2+} channels, in shark ventricular myocytes both T- and L-type Ca^{2+} currents are present with an absolute density greater than those reported in any other vertebrate species. They are considered to contribute to the initiation of action potential and initiation of contraction (Maylie and Morad 1995).

Driedzic and Gesser (1988) have analyzed force-frequency relationships and calcium dependency in ventricle strips from teleost and elasmobranch hearts, and while teleost hearts showed poor calcium storage capabilities, elasmobranch hearts exhibited the classical positive force-frequency response seen in most mammalian, turtle, and frog hearts. These findings are not consistent with the hypothesis that the sarcoplasmic reticulum is better developed in teleost than in elasmobranch hearts (Santer 1985).

One may think of the ventricular myocardium as being "suspended" in extracellular space, the microenvironment on which its homeostasis is dependent. As in mammals, in *S. stellaris* the extracellular spaces comprise the subepicardial, the subendocardial, perivascular, and interstitial compartments, as detailed by Helle et al. (1983). Subepicardial and subendocardial spaces are rich in fibroblasts and collagen. Fibroblasts exhibit very thin processes, often insinuated between two myocardial cells within each trabecula. Simple squamous mesothelium of the ventricular epicardium has ultrastructural features consistent with both secretory and mechanoprotective functions. The endocardium consists of thin, irregularly shaped endothelial cells, about 0.5 µm in diameter, that show slender, long processes by which they connect through interdigitated, loosely coupled membrane areas. Fenestrations, considered sites for transepithelial diffusion, have been described in both *S. canicula* (Ostadal and Schiebler 1970) and *S. stellaris* (Helle et al. 1983). The endothelium, with its basal lamina, covers the trabeculum, thus mediating exchange of materials and information between the luminal blood and subjacent myocardium. In mammals, growing experimental evidence indicates endocardial endothelium is an important modulator of myocardial performance (Brutsaert 1989). Nothing is known of this function in fish. The perivascular compartment reflects the distributional pattern of the intramural vasculature (see discussion of vascularization of the ventricular wall below). Since mixed-type ventricles have both endocardial endothelium and vascular endothelium, myocardial performance may be modulated by a number of humoral factors mediated or released by both types of endothelial cells, such as endothelins, APs, eicosanoids, and nitric oxide (NO). Unfortunately, nothing is known of the humoral heterogeneity of the elasmobranch heart ventricle.

Conus arteriosus

This tubular and most anterior chamber of the elasmobranch heart communicates posteriorly with the ventricle via the ventriculoconal ostium and anteriorly is confluent with the ventral aorta. The conus has been progressively lost in evolution and thus is considered by most authors, starting with Gegenbauer (1866), as a phylogenetically primitive feature. It is only present in elasmobranch, chondrostean, and dipnoan fishes in which the chamber is invested by myocardial muscle. In contrast, in teleosts it is replaced by an elastic chamber, the bulbus arteriosus, endowed with smooth muscle, and thence is outside the heart in strict sense. In elasmobranchs the conus is usually characterized by three longitudinal rows, or ridges, of valve cusps arranged in three to five tiers. A fourth row of very rudimentary valve cusps may be located between the two ventral rows. The three cusps of the most distal tier are by far the larger, with their insertion almost into the ventral aorta, and are the only ones capable of spanning the relaxed conal lumen. The valves appear early in embryonic development, preventing regurgitation of blood into the ventricle during diastole (Pelster and Bemis 1991). O'Donoghue and Abbott (1928) in their study on the macroscopic appearance of the conus in *S. acanthias* and in *S. suckleyi* viewed these distal valves as homologous to the semilunar valves in mammals. The cusps, covered by endothelium and made of fibrous tissue, are inserted by means of chordae tendineae. Most recently, Hamlett et al. (1996b) have presented a light and scanning electron microscopic study of the development and structure of the various valvular structures and their supports (i.e., chordae tendineae) in *U. jamaicensis* and *R. terraenovae*. The important relationships between the conal wall and the coronary arteries and veins are considered below (see blood supply).

In other cardiac chambers, the wall of the conus consists of three layers: the epicardium, myocardium, and endocardium. According to the ultrastructural description given in the dogfish, *S. stellaris*, by Zummo and Farina (1989), the outer layer (visceral pericardium) is constituted of cuboidal mesothelial cells with a subepicardium rich in collagen fibers and blood vessels. The architecture of the middle layer, made of bundles of myocardial cells separated by collagen fibers and capillaries, closely resembles that of the ventricular wall. The arrangement of the myocytes, about 23 μm in diameter, is continuous with those of the ventricular wall and maintained in close contact by desmosomes at the lateral surface and by intercalated disks at the terminal ends. This provides the morphological basis for the electrical and mechanical coupling between the ventricle and conus. Myocytes, particularly those of the dorsal surface of the conal wall, contain dense core vesicles with a clear halo, located both in the Golgi complex and in the interfibrillar and subsarcolemmal regions. Their appearance is similar to the natriuretic cardiac peptide–containing granules described by Reinecke et al. (1987) in the atrial and ventricular myocardiocytes of other elasmobranchs. The endocardial layer, about 0.5 μm thick, consists of a continuous lining of squamous endothelial cells and a subendocardial space rich in collagen fibers. Near the ventricle, on the ventral surface, the endocardial cells are particularly rich in granules resembling those described in the endocardial cells of the ventricle. In the same region, the subendocardium shows an abundant innervation by myelinated and unmyelinated axons. Many of the axonal ends are closely associated with chromaffin cells organized in a glomuslike arrangement and endowed with numerous dense core vesicles of about 150–220 nm in diameter, surrounded by a distinct halo. The authors have suggested that the chromaffin tissue present in the inner layer of the conus may play a neurosecretory function similar to that of the axillary bodies and of the sinus venosus.

Blood supply

The highly vascularized elasmobranch heart is nourished by two ontogenetically and morphologically distinct routes of blood supply, the lacunary and the vascular (Fig. 10-5). The lacunary system, perfused by venous blood from the cardiac lumen, is made of the myriad of intertrabecular spaces, or lacunae, located in the spongy meshwork of the myocardial trabeculum. The cardiac vascular system consists of arterial and venous networks with an interposed terminal bed of capillaries and thebesian vessels. The arteries and the veins constitute the coronary circulation; while the former derive from the hypobranchial arterial system, the veins are only related to the heart chambers. Clearly, an arteriocapillary network represents a hydraulic bioconstructional prerequisite for the development of a compact type of myoarchitecture (compacta) in which the condensation of myocardial bundles results in the obliteration of the intertrabecular spaces.

The development of cardiac vasculature in fish has scarcely been studied; however, available data indicate morphological and developmental stages closely similar to those of birds and mammals. Early nourishment of the growing myocardium is achieved by intertrabecular sinusoids and later by the coronary vessels (Lewis 1904; Raffaele 1904; Nair

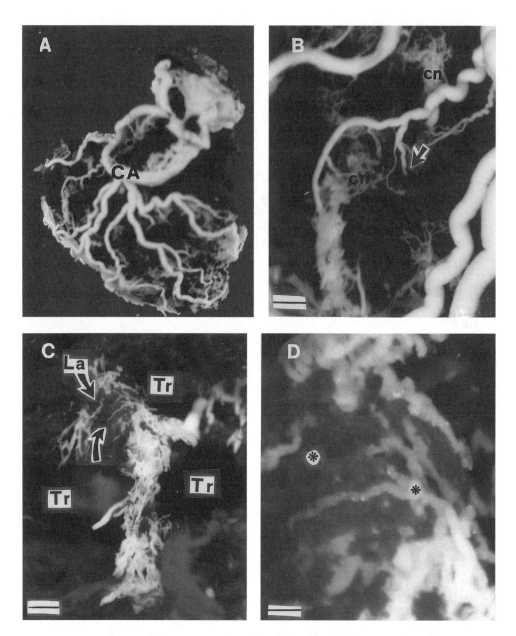

FIGURE 10-5 The arterial coronary supply and its relationships with the venous lacunary system as revealed from macroscopic to microscopic levels by neoprene corrosion casts of the heart of *Torpedo marmorata*. Note that the hollow empty zones correspond to the cardiac trabeculae (Tr) removed by the corrosion process. **A:** General appearance of the extramural coronary system, with the arteries (CA) running along the cardiac wall and branching in a dichotomous pattern, as seen in the homeotherms. **B:** The highly vascularized ventricle, in which the same territory is usually perfused by capillary networks originated from two adjacent arterioles, shows typically tortuous arteries with terminal arborization and capillary network (cn). Small arteriolar branches penetrate the spongiosa perpendicularly (*arrow*) (scale bar: 210 µm). **C:** Three trabeculae (Tr) surrounding a lacunary space (La, *arrow*). Part of the neoprene filling the lacuna was removed to permit visualization of the area outlined by the two arrows where the relationships between the arterial terminal bed and the venous lacunary system are seen (shown at higher magnification in **D**) (scale bar: 200 µm). **D:** The coronary capillaries (*asterisks*) terminating in the lacuna are part of the arterioluminal connections (i.e., thebesian vessels) that characterize the blood supply of the spongiosa (scale bar: 600 µm). (From Tota 1989.)

1970; De Andrés et al. 1993). As in the embryonic heart of homeotherms, cardiac veins develop earlier than coronary arteries. In the dogfish *S. canicula*, the appearance of a diverticulum from the anteroventral region of the sinus venosus, at the level of the posterior limit of the atrioventricular junction, that outlines the future coronary sinus represents the earliest evidence of coronary vessel development (embryo stage: 31 mm of total body length). There then appear the atrioventricular and conoventricular venous rings, into which all the ventricular veins drain, and they are completely developed (embryo stage: 36 mm of total length) before the first appearance of the coronary arteries (in embryos of 40 mm of total length) (De Andrés et al. 1993). Coronary arteries were also described by Lewis (1904) in the ventricle of a 22-mm-total-length *Torpedo* embryo. According to De Andrés et al. (1993) in the dogfish, the first coronary arteries arise as two sprouts from the midventral hypobranchial artery and divide to give rise to four arteries that reach the anteriormost part of the conus beneath the epicardium, when the cortical ventricular myocardium still consists of two myocyte layers in a 40-mm-total-length embryo. In 57 mm-total-length embryos, thin perforating branches originate from the subepicardial arteries around the atrioventricular groove and penetrate into the inner spongy ventricle. Thus, coronary vascularization of the ventricle takes place before myocyte condensation in the subepicardium constitutes the compacta. In contrast, atrial arteries do not appear until posthatching.

CORONARY ARTERIES

A recent basic description of the main coronary trunks and their branching patterns is given by De Andrés et al. (1990, 1992), while the intracardiac distribution of coronary arteries, including their relationship with the lacunary system, has been analyzed by Tota et al. (1983) and Tota (1989). From the phylogenetic survey of De Andrés et al. (1990) in lamnoid, hexanchoid, squaloid, and carcharinoid sharks, two main patterns of coronary arterial distribution are detected. These include the dorsoventral coronary trunk pattern, shared by lamnoid and advanced carcharinoids, and the lateral coronary trunk pattern, shared by squaloids and hexanchoids, with *Scyliorhinus* and *Galeus* having an intermediate vascular arrangement. In the dorsoventral coronary trunk pattern, dorsal and ventral coronary trunks originate from the dorsal and ventral medial hypobranchial arteries, respectively, run along the ventral aorta and enter the conus arteriosus, giving rise to transverse branches that fuse into a vascular ring around the conoaortic limit. In the lateral coronary trunk pattern, right and left coronary trunks originate from the lateral or commisural arteries and then run along the conus. They give rise to several ventricular branches, dorsal and ventral being the most important. Constant anatomical features, such as the position of the conal coronary trunks, together with the origin, trajectory, and relative importance of the ventricular branches have probable taxonomic value. Each coronary artery breaks into a series of smaller vessels that supply both the atrium and ventricle. As stressed by De Andrés et al. (1992), the atrioventricular and conoventricular grooves are preferential places both for the establishment of anastomoses and for branching of main arteries that give rise to atrial vessels and long, perforating branches supplying the spongiosa. At the level of the ventricle, coronaries show two distinct distributional patterns, the extramural (or subepicardial) and intramural (or intramyocardial).

The extramural network consists of subepicardial vessels that, after running along the ventricular surface and often branching in a dichotomous fashion, penetrate the compacta, providing capillary networks of the outer epimyocardial layer. In hearts in which the compacta is highly developed, as in the most active pelagic sharks, extramural arterial vessels, often alternating with venules, give rise to arterioles and capillary networks in a way that resembles the microvascular pattern seen in the subepicardium of the mammalian ventricle.

The intramural network originates from two types of arterial ramifications, the extrahilar and hilar vessels (Tota et al. 1983). Extrahilar vessels are represented by continuations and ramifications of subepicardial branches that penetrate deeply into the ventricular wall. Hilar vessels are those rather straight arteries distributed along the conus that become partly embedded into the conal wall and partly penetrate the ventricular wall at the conoventricular junction. A constant feature in both the compacta and spongiosa is the richness of "collateral" or "anastomotic" vessels that connect many of these arterial branches before giving rise to terminal capillaries. The same area is frequently supplied by extensive capillary networks originating from two adjacent arterioles. Of interest from a functional point of view is the higher intramyocardialization of the ventricular branches and the higher number of intercoronary anastomoses in larger, fast-swimming pelagic species in comparison to sluggish, demersal sharks (Tota et al. 1983; De Andrés et al. 1992). Many of the small arteries and arterioles, frequently originating from distant extramural arterial branches, that maintain their original lengthwise direction within the trabeculum, penetrate the spongiosa. Here, irrespective of trabecular width, arterioles penetrate several trabeculae in a "fingerlike" fashion, giving rise to three-dimensional side branches or to capillary meshworks. This results in the characteristic angioarchitecture of the spongiosa, which appears widely supplied by both the venous lacunary circuit and the oxygenated vascular system (see Fig. 10-5). These two systems are anatomically connected, since a variety of arterioluminal shunts (thebesian vessels) have been described (Parker and Davis 1899; Tota et al. 1983; Tota 1989). These comprise either the capillaries originating from the terminal arterioles that open into the lacunae, that is, arteriocapillary pathways, or arterioles that open directly into the lacunae, that is, arterioluminal pathways. The functional significance of these arteriolacunary shunts is unknown.

CORONARY VEINS

Blood from the heart is returned by coronary veins, but knowledge of their intracardiac microvascular distribution is lacking, since most of the available data derive from old studies, mainly focused on macroscopic appearance and extramural venous patterns (see O'Donoghue and Abbott 1928 for references). In *Mustelus antarcticus*, Parker (1886) described two coronary veins, a right and a left, located one on either side of the furrow between the atrium and the ventricle. Each receives veins from its own side of the ventricle and the conus arteriosus and passes backward to open into the sinus venosus in the immediate vicinity of the sinoatrial ostium. In dogfish *S. acanthias* and *S. suckleyi*, both coronary veins arise on the conus and run parallel with the right and left coronary arteries, respectively. They receive tributary vessels mostly from the ventral and dorsal surfaces of the ventricle; vena cardiaca dextra (i.e., right) and vena cardiaca sinistra (i.e., left) are the larger of these tributaries and enter the right and left coronary veins, respectively, just before the two coronary veins open on the right and on the left of the sinoatrial orifice.

In the atrial wall, a series of thebesian vessels exist between veins, and perhaps between arteries, and the internal cavity of the atrium (O'Donoghue and Abbott 1928). These authors stressed the striking similarities and homology of the elasmobranch coronary arteries and veins to those of mammals, a concept further corroborated by Foxon (1950), while more recently Muñoz-Chápuli and García Garrido (1986) have analyzed the phylogenetic relationships of the coronary arteries in eight orders of elasmobranchs.

Innervation

Innervation of the elasmobranch heart has been reviewed by Laurent et al. (1983), Nilsson (1983), Santer (1985), Nilsson and Holmgren (1992), and Taylor (1992). With rare exceptions (e.g., *Mustelus*), direct or indirect cardiac innervation by spinal autonomic pathways is absent in elasmobranchs, the nerve supply being represented by two distinct pairs of cardiac branches from the vagi, the branchial cardiac branch and visceral cardiac branch. The branchial cardiac branch arises from the postbranchial rami of the fourth vagal branch, while the visceral cardiac branch originates from the visceral branches. Both cardiac branches enter common cardinals and run toward the heart, where they ramify into a dense, interwoven plexus on the sinus venosus, terminating at the sinoatrial junction (Taylor 1992). Vagal stimulation, as well as application of acetylcholine, exerts an inhibitory effect on heart rate. This effect is mediated by muscarinic receptors, since it is abolished by atropine. It has been suggested on the basis of anatomical and electrophysiological data that visceral branches have mostly sensory function, while branchial branches mediate cardioinhibitory effects (Taylor et al. 1977; Short et al. 1977), accounting for the major portion of normoxic vagal tone and reflex bradycardia during hypoxia (Taylor 1992). Vagal branches are entirely constituted of myelinated axons. The absence of unmyelinated fibers is consistent with lack of sympathetic innervation in the elasmobranch heart, with the exception of some adrenergic fibers in the coronary vessels in the conus arteriosus. Bjenning et al. (1989) agree with the absence of physiopharmacological evidence for a direct adrenergic cardiac tone (Randall 1970; Nilsson and Axelsson 1987); however, some degree of adrenergic cardioregulation may be exerted through a specialized neurosecretory mechanism by circulating catecholamines, whose levels are particularly high in venous blood and even increase during hypoxia, (Butler et al. 1978; Farrell and Jones 1992; Randall and Perry 1992). Catecholamines are concentrated and stored in clusters of chromaffin cells in axillary bodies (i.e., anterior chromaffin bodies strategically located within the posterior cardinal sinus) before entering the heart. The axillary bodies, innervated by preganglionic cholinergic axons, under proper stimuli may release catecholamines to the bloodstream entering the sinus venosus, thus affecting cardiac function. Both the sinus venosus and atrium of chimaeroids and selachians possess specialized catecholamine-storing endocardial endothelial cells innervated by cholinergic vagal fibers that can exert adrenergic influence on cardiac performance. Positive chronotropic and inotropic effects mediated by β-adrenoceptor mechanisms have been reported in the isolated heart of *S. acanthias* (Capra and Satchell 1977).

Recent evidence indicates that some purine derivatives (e.g., adenosine), certain amines (e.g., serotonin), and several peptides can be involved in neurohumoral control of the heart (see Nilsson and Holmgren 1992, for review), but nothing is known regarding the occurrence of purinergic, serotoninergic, and peptidergic nerves in the elasmobranch cardiovascular system. P1-purinoceptors, mediating negative chronotropic and inotropic effects, were reported in the atrium but not ventricle of *S. canicula* (Meghji and Burnstock 1984). A rich distribution of nerves showing neuropeptide Y (NPY) immunoreactivity in ventricular endocardium and myocardium, in the conus arteriosus, and in coronary vessels of *R. erinacea* and *R. radiata* have been detected (Bjenning et al. 1989). In the dogfish *S. canicula*, in which only few fibers in the atrium showed NPY-like immunoreactivity, homologous NPY-related peptides and porcine NPY contracted the afferent branchial artery in a direct, endothelium-independent manner (Bjenning et al. 1993a). In the coronary artery of the skate *R. rhina*, porcine NPY potentiated the effect of noradrenaline-induced contraction via an indirect mechanism (Bjenning et al. 1993b). The same group reported a positive bombesinlike immunoreactivity in the sinus venosus, conus arteriosus, and coronary arteries of *R. erinacea* (Bjenning et al. 1991). Notable amounts of angiotensin-converting enzyme (ACE) activity have been detected among other tissues in the heart of *S. acanthias* and *R. erinacea* (Olson 1992). Clearly, there is a need for reevaluation of the nerve supply to the elasmobranch heart, particularly in view of the growing interest in the mechanisms of the neurohumoral and "hormonal" control of the cardiac performance.

Cardiac cycle

As in all vertebrates, the pumping action of the elasmobranch heart depends on the ability of its musculature to undergo spontaneous rhythmic depolarizations and repolarizations that trigger synchronized contractions (systole), during which cardiac chambers propel blood, and relaxations (diastole), during which chambers refill. The sequence of these electrical and mechanical events, called the "cardiac cycle," represents the focal point of cardiac physiology. The cardiac cycle can best be visualized by a series of superimposed curves, called a "Wiggers diagram" after the classic description of the cycle in mammals offered by Carl J. Wiggers. The electrical events occurring within the heart as a whole are monitored by the electrocardiogram (ECG) as a sequence of changes in electrical potential differences recorded by electrodes placed on the surface of the body or, in the case of an intracardiac electrogram, in the cardiac cavities. A detailed analysis of such electrical events in different fishes, including the dogfish *S. stellaris* and *S. canicula*, was reported by Marchetti et al. (1962), who in addition to using bipolar reference peripheral electrodes, performed epicardial electrocardiograms directly on the exposed heart after removing the pericardium. The main results of this study are summarized in Figure 10-6. The mechanical events reflecting the systolic and diastolic phases of a given cardiac chamber are recorded as pressure, flow, and volume changes. The cardiac cycle of the shark heart, as analyzed by Sudak (1965), Satchell and Jones (1967), and Satchell (1970), is illustrated in Figure 10-7.

The first event of the cycle is represented by depolarization of the sinus venosus (sinoatrial junction, according to Marchetti et al. 1962), monitored as the V wave of the ECG, followed by a feeble sinus contraction that can scarcely be detected on the pressure record, with sinus pressure ranging between −2 to −7 cmH_2O (Sudak 1965) or −8 and 1 cmH_2O (Lai et al. 1989a). Subsequent atrial depolarization is signaled by the P wave on the ECG, followed by atrial systole. The peak of atrial systole is the only event of the cycle in which atrial pressure slightly exceeds ventricular pressure, causing a corresponding expansion of both the ventricle and conus.

Ventricular depolarization causes the QRS complex on the ECG, indicating the spreading of the action potential to the whole chamber, with a velocity that in *Heterodontus* is 40–100 cm/sec. An atrioventricular delay, that is, the P–Q interval, allows complete atrial contraction before the onset of ventricular systole. There is lack of information on the occurrence of a specialized atrioventricular conduction pathway, but the possibility that the depolarization wave from the atrium runs throughout the canalis auricularis to the subendocardial trabeculum of the ventricle, thence reaching the compacta, should be experimentally tested. Following the QRS, ventricular systole is signaled by the notable rise of ventricular pressure, which closes the two pocket valves of the atrioventricular orifice and, as ventricular tension increases, forces the blood to flow into the conus. Since the conal myocardium has not yet been activated, the conus arteriosus is passively dilated. The contour of the ventricular pressure curve provides important information about the state of the myocardium. The slope of the ascending limb of the ventricular pressure curve reflects the rate of force developed by the ventricle (change in pressure with time $[=dP/dt]$) and at any given degree of ventricular filling may be used as an index of the initial contraction velocity, hence of contractility. Accordingly, if the fish heart is under adrenergic stimulation or in exercise (Lai et al. 1990), there will be fast rising ventricular pressure, a somewhat reduced end-diastolic pressure, and a brief systole, while an opposite pattern will characterize a hypodynamic heart.

Several authors have stressed the point that ventricular performance in elasmobranchs differs from that of teleosts. In contrast with teleosts and mammals, in the elasmobranch ventricle the isovolumic contraction (i.e., the interval of time between the start of ventricular systole and the opening of the conal valves, when ventricular pressure rises abruptly), in the strict sense, is missing. While in teleosts blood in the ventricle cannot pass the tier of valves in the bulbus arteriosus until ventricular pressure becomes higher than that in the ventral aorta, in the elasmobranch there is no period between closure of the atrioventricular valves and opening of the upper

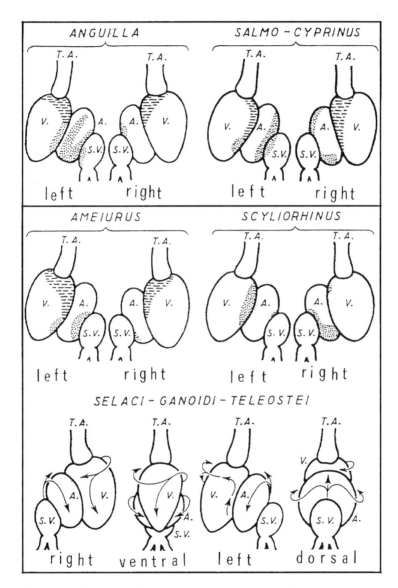

FIGURE 10-6 Comparative view of the epicardial activation in the fish heart. The upper and middle panels illustrate the regions initiating an action potential (*dotted*) and the areas of early activation (*hatched*). The lower panel shows that the activation of the epicardial areas in the sturgeon *Acipenser stuno* and in the dogfish *Scyliorhinus stellaris* follows nearly the same pattern as in some freshwater teleosts, starting in the sinoatrial region and then spreading toward the ventricle with a craniocaudal direction. Note that the left apical region of the ventricle is activated before the right. A. = atrium; S.V. = sinus venosus; T.A. = truncus arteriosus; V. = ventricle. (After Marchetti et al. 1962.)

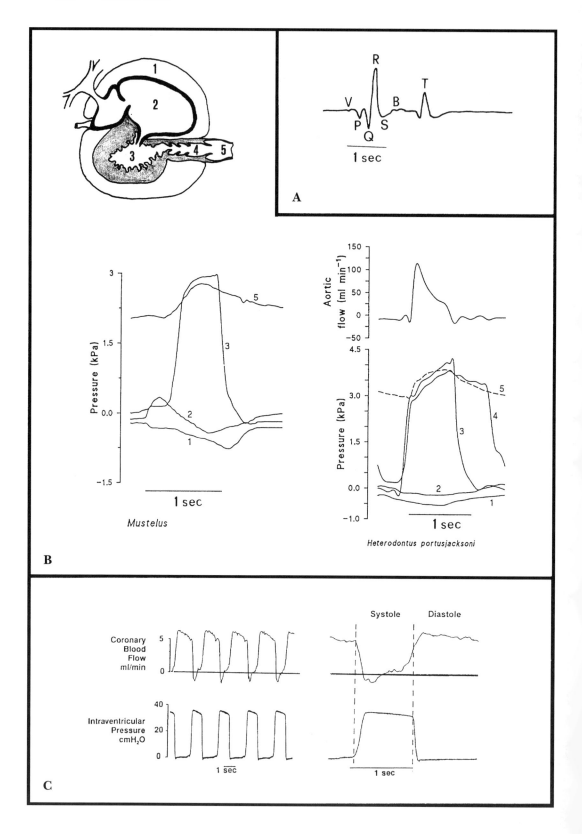

valve in the conus during which the volume of the ventricle remains unchanged while the pressure within it increases (Randall 1968; Satchell 1971). It would be of interest to examine this event in detail with more advanced techniques.

Electrical activation of the conus is signaled by a small deflection occurring during the Q–T interval, called the B wave. Records of the focal ECG in *Heterodontus* suggest the B wave passes up the conus rather slowly, with a velocity of 2–4 cm/sec (Tebecis 1967). This wave is soon followed by contraction of the conal myocardium, commencing from the base of the chamber at the ventriculoconal junction and before ventricular systole has ended. As evidenced in *Heterodontus* by pressure records taken above and below each of the three tiers of valves, during atrial and ventricular systole, the conal chamber is filled with blood and only the upper valves, large enough and perfectly competent, are closed. When conal pressure increases, these valves also open and outflow into the ventral aorta occurs. During conal systole, contraction of the myocardial fibers encircling the wall causes backflow, with consequent closure first of the lower and then of the middle tier of valves, as documented in *Triakis* by high-speed cinematographic analysis (Marchetti et al. 1962). At the end of conal systole, the upper valves close while the middle ones open, and the pressure in the upper conus falls to that in the ventricle. During conal contraction, blood flows into the ventral aorta as indicated by an upperward slope of pressure, which is followed by a rapid fall at the end of conal systole.

During the remainder of ventricular diastole, as blood flows through the gills to the peripheral vasculature, there is a further decline of pressure into the ventral aorta until the next increase with systole. The steepness of this downward slope of pressure depends on peripheral resistance and heart rate.

There is no consensus on the role of the conus (Satchell 1991; Bushnell et al. 1992). Satchell and Jones (1967) suggested that its major role is that of postponing backflow from the aorta and valve closure until later in the cardiac cycle, when the nadir of pericardial negative pressure related to ventricular emptying has passed. Johansen (1965) envisaged a depulsating action of the conus, in that it converts heart output to a smooth, extended outflow by prolonging both phases of accelerated and decelerated flow (Windkessel effect). Another, and perhaps more important, function can be that of prolonging aortic flow during diastole, as Hanson (1967) showed in *S. suckleyi*. Johansen (1965) and Johansen and Burggren (1980) considered the conus as an auxiliary cardiac pump, a view shared later by other workers on the basis of direct pressure and flow measurements in the ventral aorta and, more recently, by Pelster and Bemis (1991), who in their ontogenetic study on cardiac function in embryos of *R. erinacea* (at 44 and 88 days of age) detected conal pressure elevated during ventricular diastole.

Recently, improved technologies in cardiac imaging, including echo-Doppler cardiography and angiocardiography, have been used for reanalyzing the elasmobranch cardiac cycle in the leopard shark (*T. semifasciata*) at

FIGURE 10-7 The cardiac cycle of the shark. **A:** The electrocardiogram of *Heterodontus portusjacksoni*. **B:** Superimposed pressure and flow curves recorded in different regions of the heart of the dogfish *Mustelus canis* and the shark *Heterodontus portusjacksoni*. The numbers on the pressure profiles refer to the pericardium, cardiac chambers, and ventral aorta, as indicated in the schematic drawing (*upper left panel*). Note subambient pressures in pericardial cavity. **C:** Recordings of coronary blood flow and intraventricular pressure (with an expanded cycle on the right) from an anesthetized school shark *Galeorhinus australis*. Coronary flow reverses briefly at the very beginning of ventricular systole and then rises during the whole contraction phase, with a peak at the start of ventricular diastole. During the whole diastolic period coronary flow remains high, until it stops and reverses with the next protosystolic phase. (A, from Satchell 1971; B, *M. canis* after Sudak 1965, *H. portusjacksoni* after Satchell and Jones 1967 and Satchell 1970; C, after Davie and Franklin 1993.)

rest, during swimming, and in recovery (Lai et al. 1990). Important results obtained may be summarized as follows:

1. Throughout the cardiac cycle and in all phases of activity, the maximum and minimum pressures of the cardinal sinus are consistently higher than those of the pericardium. During swimming, increases in both mean ventricular diastolic transmural pressure and cardinal sinus pressure indicate that augmented cardiac filling by vis a tergo may be indeed the major mechanism for increasing cardiac output rather than by vis a fronte, as previously believed.
2. Atrial diastole is characterized by three filling periods, the first with atrial relaxation (occurring just after the P wave), the second with the vis a fronte effect of ventricular ejection (occurring soon after the QRS complex), and the third with sinus venosus contraction.
3. In contrast to classic descriptions of the elasmobranch cycle, ventricular filling is not entirely dependent on atrial contraction but clearly begins before it. In the absence of tachycardia (beats per minute equal to or more than 40), ventricular filling is biphasic, as in mammals, and is characterized by a period of early rapid filling (occurring after the T wave, which is early in diastole), a period of diastasis (absent at higher heart rates), and a late period of filling (occurring just after the P wave) caused by atrial systole.
4. Ventricular emptying is not complete, but the estimated 80% mean ejection fraction is higher than that reported for mammals; ventricular outflow into the conus arteriosus is monophasic and occurs just after the onset of the QRS.

On the whole, the picture of the shark heart emerging from this recent integrative approach shows more similarities with mammals in the patterns of ventricular filling and emptying than previously thought and seems to upset the long-held dogma of the major role played by the vis a fronte mechanism in selachians.

Coronary circulation

The major determinants of arterial coronary flow are perfusion (driving) pressure and vascular resistance. In reptiles, birds, and mammals the perfusion pressure is the aortic blood pressure just outside the aortic valves and hence is close to the intraventricular pressure. In fish, in which the coronary arteries arise postbranchially, the perfusion pressure is much lower than ventral aortic blood pressure and very close to the dorsal aortic postbranchial pressure. Thus, while in homeotherms coronary artery pressures and intraventricular pressures are almost equal, in fish coronary arterial pressures must be much lower than systolic intraventricular pressures. In elasmobranchs the only intraventricular pressure and ventral aortic flow known is that recently performed in anaesthetized school sharks, *G. australis*, by Davie and Franklin (1993) (see Figure 10-7). These authors recorded a coronary arterial pressure between 1.0 and 2.5 kPa, corresponding to values of 42–60% of peak intraventricular pressures, with maximum instantaneous flow (recorded in the ventral coronary artery) of 0.37 mL/min/kg body mass. On average, 86% of coronary flow occurred during diastole. Coronary arterial flow began during the last quarter of ventricular systole, peaked immediately after systole, when intraventricular pressure fell to just below zero, and briefly reversed during the initial period of systole (see Figure 10-7). This pattern contrasts with the continuous blood flow recorded in coronary arteries of unanesthetized trout (Axelsson and Farrell 1992). Such differences in coronary hemodynamics between elasmobranchs and trout, also found in isolated perfused heart preparations (Agnisola et al. 1993), may reflect different coronary distribution patterns of elasmobranchs and trout, with trout showing a type I ventricular myoangioarchitecture in contrast to the types III and IV that are characteristic of elasmobranchs. These different coronary patterns provide a further example of inherited organizational preconditions that constrain the functional aspects of the cardiac pump.

The working heart

CARDIAC OUTPUT

Blood flow through the body's tissues depends largely on metabolic demands. Since these demands change considerably under different physiological conditions, the total blood flow (cardiac output) needed by the body tissues at different times must vary also. The output of the heart is given by the equation: cardiac output = heart rate × stroke volume.

Stroke volume (usually abbreviated as SV) is equal to the difference between ventricular end diastolic volume and ventricular end-systolic volume. Cardiac output (CO) values range from 19.2 to 52.5 mL/min/kg in representative elasmobranchs (Farrell and Jones 1992).

CO undergoes changes in relation to changes in heart rate (chronotropic effect) and SV (inotropic effect), both controlled by aneural factors and by the autonomic innervation of the heart. Fish differ from homeotherm vertebrates in that they usually increase SV rather than heart rate. During exercise a number of elasmobranchs have been shown to maintain heart rate fairly constantly (Johansen et al. 1966; Lai et al. 1989a,b, 1990). Control of heart rate in elasmobranchs has been reviewed by Jensen (1970), Laurent et al. (1983), Butler and Metcalfe (1988), Farrell and Jones (1992), and Taylor (1992). As in amphibians and reptiles, Starling's law of the heart appears as the major mechanism controlling SV. Positive inotropism results from increased stretch of the cardiac chambers (atrium and ventricle) with corresponding lengthening of myocardial fibers that is induced by an increased filling pressure of the chamber (increased preload) when the return of venous blood to the heart augments. Hence, in response to a volume load, increased end-diastolic volume causes a greater force of contraction and a greater SV (Davie and Farrell 1991). Positive inotropism also results from an increased afterload. When a rise in the resistance of the peripheral circulation causes a rise in pressure, there will be greater blood volume in the chamber at the beginning of diastole, which will stretch the chamber wall, thus increasing systolic force. This mechanism, called homeometric regulation, adjusts SV to a pressure load (Satchell 1991; Farrell and Jones 1992).

The fact that atrial systole occurs before ventricular systole means that the corresponding degree of ventricular filling depends, among other factors, on the strength of atrial contraction. Since ventricular volume determines the strength of ventricular contraction, the atrium appears to be a primer pump if its contraction increases ventricular diastolic volume at the end diastole of the ventricular cycle, when diastolic volume determines ventricular performance. There is little doubt that atrial filling is the major determinant of ventricular filling in elasmobranch fish (Satchell 1970, 1971), although it is probably not the sole mechanism, as indicated by Lai et al. (1990).

STROKE WORK

The ejection of a volume of blood under pressure represents work (dynes/cm^2 × cm^3 = dynes cm). The work of the heart during each beat is calculated by the product of the SV times corresponding ejection pressure (P) as in the equation: stroke work = SV × P. Since pressure changes during the ejection phase, stroke work is better calculated as the integral of pressure and volume change.

Under most conditions pressure-volume work represents the major portion of useful cardiac work. Some pumps may be best arranged structurally to produce a high pressure difference from input to output; other pumps may develop the same power but invest in a high-volume flow, with only a modest increment in pressure. Clearly, efficiency is not considered here. In Figure 10-8 several species of elasmobranchs and teleosts have been ranked on the basis of the relative contribution of pressure and volume to the stroke work. For comparison, data for a typical avascular trabeculated heart, the frog heart, and for a typical compact, highly vascularized myocardium, the left ventricle of the rat, have been included. While the rat left ventricle moves relatively small volumes against considerable

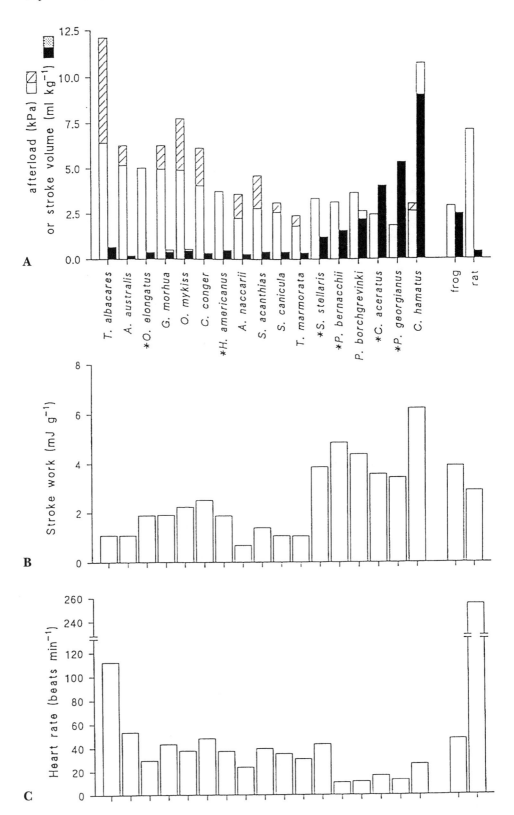

high pressures, the amphibian heart moves relatively large volumes against small pressures. On the left side of the bar graph are plotted hearts producing mainly pressure work (similar to the rat heart), while on the right side of the bar graph are placed hearts able to eject large volumes of blood against low pressures (similar to the frog heart). The figure also includes the values for afterloads and SV, measured under conditions of maximum stroke work. Clearly, in the species on the right side of the bar graph, the maximum stroke work is principally attained by an increase in volume, while in the species on the left side of the bar graph it is mainly attained by an increase in pressure. This comparative analysis shows that the first seven species on the left are teleosts possessing a mixed type of ventricle, while most of the species in which the volume component of the stroke work approaches or exceeds the pressure component possess a trabeculated spongy ventricle. This comparison indicates that ventricular myoarchitecture differs in accord with the functional specialization of operation primarily as a pressure pump (i.e., mixed-type ventricle with high ejection pressure) or as a volume pump (i.e., trabeculated spongy ventricle with low ejection pressure). It is of interest that the central part of the bar graph is occupied by four elasmobranchs; because of their lower ventral aorta pressure, the relative contribution of the volume movement to the cardiac work is correspondingly higher.

Aspects of cardiac metabolism and energetics

As in all muscles, the heart acts as a mechanochemical transducer, in that during contraction chemical energy is transformed into mechanical energy. The metabolic pathways by which adenosine triphosphate (ATP) is regenerated during cardiac activity are not identical in all fishes. They may use a variety of biochemical specializations and metabolic flexibilities in relation to different body composition and ecophysiological characteristics. Elasmobranch hearts, in contrast to teleost hearts, which are endowed with key enzymes necessary to utilize carbohydrates, fatty acids, and acetoacetate, seem poorly equipped for catabolizing fatty acids, as indicated by the very low levels of carnitine palmitoyltransferase (Zammit and Newsholme 1979). Moreover, under starvation, elasmobranch plasma, unlike teleost plasma, normally containing very low levels of nonesterified fatty acids, triacylglycerol, and fatty acid–binding proteins, shows notable elevation of ketone bodies (Zammit and Newsholme 1979). These differences may explain the different fuel preference and performance shown by the isolated working hearts of the skate *R. erinacea* and the teleost *Hemitripterus*. While the teleost heart functioned well when utilizing palmitate, this fatty acid was detrimental for the cardiac performance of the elasmobranch, which instead appeared more dependent on the catabolism of blood-borne ketone bodies such as acetoacetate (Driedzic and Hart 1984). In their study on mitochondrial and peroxisomal fatty acid–oxidation in the heart of *S. acanthias*, Moyes et al. (1990) found that the fatty-acid catabolizing enzyme carnitine palmitoyltransferase was below detectable limits; while mitochondria utilized oxygen when incubated with ketone bodies or pyruvate, they were unable to do it in the presence of long-chain fatty acids or carnitine derivatives, even under

FIGURE 10-8   Afterload, stroke volume, stroke work, and heart rate in fishes. **A:** Values for afterload (mean ventral aortic pressure minus sinus venosus pressure) and stroke volume. *Unshaded open columns* show in vitro basal or in vivo resting afterload. *Shaded columns* show in vitro basal or in vivo resting stroke volume. Increases of afterloads (*crosshatched columns*) and stroke volume (*shaded columns*) refer to conditions of maximal stroke work. Note in the species on the right side of the graph that the increase in stroke volume is attained without changing the afterload, while in most species on the left side there is an increase in afterload without changing the stroke volume. **B:** Corresponding values of heart stroke work. **C:** Corresponding values of stroke work. (From Agnisola and Tota 1994.)

osmotically stimulatory conditions, suggesting a lack of matrical coenzyme A synthase capable of activating long-chain free fatty acids. These authors also remarked that the inability of the elasmobranch heart and myotomal muscle mitochondria to burn acyl carnitines at high rates is unlike most vertebrate muscles, being more similar to certain invertebrate muscles, such as the heart mitochondria of squid.

In conclusion, these studies are all consistent with the inability of the elasmobranch hearts to utilize exogenous free fatty acids as metabolic fuels (Driedzic 1992). The hearts of the few elasmobranch species so far analyzed appear biochemically well equipped for sustaining periods of anaerobiosis (Sidell and Driedzic 1985) as indicated by the surprisingly low levels of cardiac ultrastructural damage observed in the dogfish *S. stellaris* subjected to periods of anoxia (Helle et al. 1983).

Cardiac efficiency can be estimated by dividing the external work of the heart by the energy equivalent of oxygen consumption. The resulting quotient, in which both the numerator and the denominator can be expressed either per beat or per unit of time, provides a satisfactory index of cardiac efficiency: cardiac efficiency = work/energy equivalent of oxygen consumption. Mechanical efficiency has been measured in an isolated perfused heart preparation of *S. acanthias* under controlled conditions of pressure loading and volume loading at two power outputs, 50% and 72% of maximum power outputs (Davie and Franklin 1992). It was found that oxygen consumption was linearly related to cardiac power output and that mechanical efficiency also increased with power output toward a maximum; maximum mechanical efficiency, around 21%, was well comparable with the corresponding values of mammalian hearts (ranging between 10% and 30%). Interestingly, while in mammalian hearts the cost of pressure generation is thought to be higher than that of flow generation, in the shark heart the work done at low SV and high pressure resulted in more efficiency than the work done at high SV and low pressure.

Neurohormonal control

Knowledge of the neurohumoral control of cardiac performance in elasmobranchs is rudimentary and mainly related to the nervous control of heart rate, particularly in relation to the central nervous generation of cardiorespiratory interactions (see Taylor 1992 for review). Present evidence indicates that the predominant mode of nervous cardioregulation is exerted by changes in the degree of cholinergic vagal tonus of the heart. An important reflexogenic area is represented by the branchial vasculature. For example, experiments by several workers have shown in *Mustelus* and in *S. acanthias* that the increase of blood pressure in the branchial vasculature is accompanied by bursts of action potentials in afferent nerve fibers from the gills, causing bradycardia. Elasmobranchs have also been among the first experimental animals on which cardiorespiratory synchrony has been explored in fish, since the pioneering studies of Schoenlein working on *T. marmorata* at the zoological station of Naples as long ago as 1895. The coordination between respiration and heart rate, showing 1:1 synchrony between heartbeat and ventilation, is evident in unrestrained, resting dogfish, is mediated by the vagus, and tends to be abolished when the animal is disturbed or active (with 2:1 relationships between ventilation and heart rate). Hypoxia in the great majority of fish causes a general circulatory response, analyzed in detail by Satchell (1991), which includes reflex bradycardia accompanied by an increase SV as a result of the augmented diastolic filling (Starling's law of the heart). In the dogfish, heart rate varied directly with Po_2; a normoxic vagal tone was induced by exposure to moderate hyperoxia, while extreme hyperoxia resulted in secondary reflex bradycardia, perhaps triggered by stimulation of venous receptors; hypoxia induced reflex bradycardia (Taylor 1992). Cardiac cholinergic inhibition increases with temperature. In fact, in the dogfish, little or no inhibitory influence on heart rate is detectable at 7°C, as documented by changes of heart rate from 20 to 22 beats/min following atropine treatment; an accompanying increase in cholinergic tone is recorded at progressively higher temperatures, as shown for example at 17°C by a rise of heart rate from 37 to 48 beats/min following atropine infusion.

In elasmobranchs, catecholamines are secreted into the blood from the chromaffin tissue even during spontaneous swimming, but secretion markedly increases under hypoxia, acidosis, and exercise (see Randall and Perry 1992 for review). Thus, it appears that despite the lack of cardiac sympathetic innervation in elasmobranchs, both under normal conditions and in emergency situations a humoral adrenergic influence may affect cardiac performance by consequent changes of heart rate and principally SV. How much a "basal" occurrence of catecholamines may influence the cardiac response to other neurotransmitters, autacoids, and endocrines remains unanswered.

Cardiac adaptation: future developments

It is not easy to predict the main lines of research on the cardiac biology of the elasmobranchs in the future. Some open questions have already been raised at different points in this review. Rather than deal with detailed aspects, it may be worthwhile to stress the need for an interdisciplinary approach that will help perceive how developmental processes, adaptation, and contingencies of phylogenetic history act together in the determination of the actual cardiac design. In elasmobranchs, as in all complex animals, the heart as a muscular pump is able to adjust its performance in relation to the changing demands of the organism. This ability is attained by three main mechanisms: (1) minute-to-minute regulation (Starling's law of the heart); (2) short-term regulation by biochemical changes within the myocardial cell (cardiac contractility, excitation-contraction coupling); and (3) long-term regulation by altered gene expression (molecular biology). These mechanisms, listed in Table 10-2, give a specific temporal dimension to cardiac adaptation. At the same time, when considered in the structural and ultrastructural context of the organ, they provide a suitable spatial illustration of upgrading performance incorporated into the cardiac muscle system. The paradigms of Table 10-2 show how the form-function relationship operates as an essential property of heart biology as we move from a whole-organ to a subcellular and molecular level and, at the same time, how through multiple interactions of these subsystems, entirely new properties may be added to the composite system. Clearly, a better understanding of elasmobranch cardiac

TABLE 10-2 CARDIAC PUMPING ADAPTATIONS

	Beat-to-beat response	*Short-term response*	*Long-term response*
Mechanism	Starling law of the heart	Excitation-contraction coupling, myocardial contractility	Activation of overall gene activity with consequent phenotype changes
	Length-dependent responses to changing preload and afterload with adjustment of cardiac output	Length-independent and length-dependent responses to neurotransmitters, hormones, and autacoids	Responses to chronic changes in workload, aging, endocrines
			Compensation for functional heterogeneity
			Putative asymmetric gene activation during cardiogenesis
Level	From whole organ to sarcomere and TnC	Cellular, subcellular (e.g., sarcolemma, T system)	Molecular biology

biology would benefit greatly from such a dynamic spatiotemporal perspective.

Multilevel dimensional constraints, the first response, are embodied in Starling's law of the heart. We have seen that in the elasmobranchs, as perhaps in all lower vertebrates possessing a trabeculated myocardium, the volume of the blood in a cardiac chamber (atrium or ventricle) at end-diastole is the key determinant of the work of the heart, more evidently than in the compact myocardium of the homeotherms. However, a shift in emphasis from earlier to later dimensional aspects of the Starling's law has signaled major changes in direction for myocardial physiologists. Considerable evidence from mammals now suggests that changes in myofilament overlap make only small contributions to the length-tension relationship, which instead is at least partly due to dependence on the calcium ion sensitivity of the cardiac myofilaments for regulation length. Troponin C, in addition to its function as the Ca^{2+} switch, appears the most likely candidate to act as the length transducer able to modify myocardial contraction (Lakatta 1992). To what extent is this true for elasmobranchs?

The second response (phasic control) reported in Table 10-2 regulates cardiac performance over a slower time period by changing myocardial contractility through mechanisms mostly related to changes in Ca^{2+} fluxes involved in excitation-contraction coupling. Some of these mechanisms are length-independent, while others might be length-dependent, like those related to stretch effect on the geometry and physical properties of membranes and organelles (e.g., sarcolemma, sarcoplasmic reticulum, caveoli) that in turn can influence transsarcolemmal ion flux. This response allows the myocardiocytes to respond to neurotransmitters such as catecholamines released during exercise, endocrines, and autacoids. It explains differences between energy requirements for "pressure work" and "volume work." The response opens the analysis for determinants of myocardial contractility such as muscle mechanics and biochemistry. Not only is knowledge of these aspects in fish modest, but it is particularly surprising that the great majority of the available data come from teleost species, as evidenced by the recent review of Tibbits et al. (1992). Pertinent literature on elasmobranchs is lacking.

The third response (tonic control) permits phenotypically reshaping the myocardiocyte (and other cardiac cellular types) in relation to long-term challenges and also compensation for regional heterogeneity. Although such heterogeneity is macroscopically evident in the elasmobranch heart ventricle as compared with that of the mammalian ventricle, nothing is known about its developmental and molecular determinism in elasmobranchs. This response is but one aspect of the universal homeostatic capability of the heart to regulate its growth, as marked in crucial developmental and phylogenetic steps.

It has been speculated (Katz 1990) that the time of the discovery of these mechanisms reflects inversely their appearance during the time of evolution. It is likely that regulation by altered gene expression, allowing adaptation to changes in the environment, appeared at a time very early in the evolution of life, before living organisms had achieved their present ability to control the milieu surrounding their DNA. Then, regulation by changes in myocardial contractility arose from biochemical processes related to the ability to control the internal cell environment, before the heart evolved as a complex multitissue organ pump.

REFERENCES

Aardal, S., and K. Helle. 1991. Comparative aspects of the endocrine myocardium. *Acta Physiol. Scand Suppl.* S559: 31–46.

Agnisola, C., and B. Tota. 1994. Structure and function of the fish cardiac ventricle: flexibility and limitations. *Cardioscience* 5: 145–53.

Agnisola, C., R. Venzi, D. F. Houlihan, and B. Tota. 1993. Coronary flow-pressure relationship in the working isolated fish heart: trout (*Oncorhynchus mykiss*) versus *torpedo* (*Torpedo marmorata*). *Philos. Trans. R. Soc. Lond.* 341B: 339–418.

Axelsson, M., and A. P. Farrell. 1992. Coronary blood flow in vivo in the Coho salmon (*Oncorhynchus kisutch*). *Am. J. Physiol.* 264: R963–71.

Berge, P. I. 1979. The cardiac ultrastructure of *Chimaera monstrosa* L. (Elasmobranchii: Holocephali). *Cell Tissue Res.* 201:181–95.

Bjenning, C., W. Driedzic, and S. Holmgren. 1989. Neuropeptide Y–like immunoreactivity in the cardiovascular nerve plexus of the elasmobranchs *Raja erinacea* and *Raja radiata*. *Cell Tissue Res.* 255: 481–86.

Bjenning, C., A. P. Farrell, and S. Holmgren. 1991. Bombesin-like immunoreactivity in skates, and in vitro effect of bombesin in the coronary vascular system of the longnose skate, *Raja rhina*. *Regul. Pept.* 35: 207–19.

Bjenning, C., N. Hazon, A. Balasubramanian, S. Holmgren, and J. M. Conlon. 1993a. Distribution and activity of dogfish NPY and peptide YY in the cardiovascular system of the common dogfish. *Am. J. Physiol.* 264: R1119–24.

Bjenning, C., S. Holmgren, and A. P. Farrell. 1993b. Neuropeptide Y potentiates contractile response to norepinephrine in skate coronary artery. *Am. J. Physiol.* 265: H661–65.

Block, B. A., and F. G. Carey. 1985. Warm brain and eye temperatures in sharks. *J. Comp. Physiol.* 156B: 229–36.

Bone, Q., N. B. Marshall, and J. H. S. Blaxter. 1995. *Biology of Fishes*. London: Blackie Academic and Professional.

Bride, M. 1977. Evolution de l'ultrastructure du coeur de xenope (*Xenopus laevis* Daud.) après l'établissement de l'innervation. *Biol. Cell.* 29: 167–76.

Brutsaert, D. L. 1989. The endocardium. *Annu. Rev. Physiol.* 51: 263–73.

Bushnell, P. J., D. R. Jones, and A. J. Farrell. 1992. The arterial system. In *Fish Physiology*, vol. 12, pt. A, ed. W. S. Hoar, D. J. Randall, and A. P. Farrell, 89–139. New York: Academic Press.

Butler, P. J., and J. D. Metcalfe. 1988. Cardiovascular and respiratory systems. In *Physiology of Elasmobranch Fishes*, ed. T. J. Shuttleworth, 1–47. New York: Springer-Verlag.

Butler, P. J., E. W. Taylor, M. F. Capra, and W. Davison. 1978. The effect of hypoxia on the level of circulating catecholamines in the dogfish *Scyliorhinus canicula*. *J. Comp. Physiol.* 127: 325–30.

Capra, M. F., and G. H. Satchell. 1977. The differential haemodynamic responses of the elasmobranch *Squalus acanthias* to naturally occurring catecholamines, adrenaline and noradrenaline. *Comp. Biochem. Physiol.* 58C: 41–47.

Clark, E. B., and N. Hu. 1982. Developmental hemodynamic changes in the chick embryo from stage 18 to 27. *Circ. Res.* 51: 810–15.

Davie, P. S., and A. P. Farrell. 1991. The coronary and luminal circulations of the myocardium of fishes. *Can. J. Zool.* 69: 1993–2001.

Davie, P. S., and C. E. Franklin. 1992. Myocardial oxygen consumption and mechanical efficiency of a perfused dogfish heart preparation. *J. Comp. Physiol.* 162A: 256–62.

———. 1993. Preliminary observation on blood flow in the coronary arteries of two school sharks (*Galeorhinus australis*). *Can. J. Zool.* 71: 1238–41.

De Andrés, V., R. Muñoz-Chápuli, and V. Sans Coma. 1993. Development of the coronary arteries and cardiac veins in the dogfish (*Scyliorhinus canicula*). *Anat. Rec.* 235: 436–42.

De Andrés, A. V., R. Muñoz-Chápuli, V. Sans Coma, and L. García Garrido. 1990. Anatomical studies of the coronary system in elasmobranchs: I. Coronary arteries in lamnoid sharks. *Am. J. Anat.* 187: 303–10.

———. 1992. Anatomical studies of the coronary system in elasmobranchs: II. Coronary arteries in hexanchoid, squaloid, and carcharhinoid sharks. *Anat. Rec.* 233: 429–39.

Donald, J. A., A. J. Vomachka, and D. H. Evans. 1992. Immunohistochemical localization of natriuretic peptides in the brains and hearts of the spiny dogfish, *Squalus acanthias*, and the Atlantic hagfish, *Myxine glutinosa*. *Cell Tissue Res.* 270: 535–45.

Driedzic, W. R. 1992. Cardiac energy metabolism. In *Fish Physiology*, vol. 12, pt. A, ed. W. S. Hoar, D. J. Randall, and A. P. Farrell, 219–66. New York: Academic Press.

Driedzic, W. R., and H. Gesser. 1988. Differences in force frequency relationships and calcium dependency between elasmobranch and teleost hearts. *J. Exp. Biol.* 140: 227–41.

Driedzic, W. R., and T. Hart. 1984. Relationship between exogenous fuel availability and performance by teleost and elasmobranch hearts. *J. Comp. Physiol.* 154B: 593–99.

Emery, S. H., C. Mangano, and V. Randazzo. 1985. Ventricle morphology in pelagic elasmobranch fishes. *Comp. Biochem. Physiol.* 82A: 635–43.

Farrell, A. P., and D. R. Jones. 1992. The heart. In *Fish Physiology*, vol. 12, pt. A, ed. W. S. Hoar, D. J. Randall, and A. P. Farrell, 1–88. New York: Academic Press.

Foxon, G. H. E. 1950. A description of the coronary arteries in dipnoan fishes and some remarks on their importance from the evolutionary standpoint. *J. Anat.* 84: 121–31.

Gegenbauer, C. 1866. Zur Vergleichenden Anatomie des Herzens: I. Über der Bulbus arteriosus der Fische. *Jena Z. Naturwiss.* 2: 365–83.

Grant, R. T., and M. Regnier. 1926. The comparative anatomy of the cardiac coronary vessels. *Heart* 14: 285–317.

Hamlett, W. C. 1987. Comparative morphology of the elasmobranch placental barrier. *Arch. Biol. (Brussels)* 98: 135–62.

Hamlett, W. C., J. A. Musick, A. M. Eulitt, R. L. Jarrell, and M. A. Kelly. 1996a. Ultrastructure of uterine trophonemata, accommodation for uterolactation, and gas exchange in the southern stingray, *Dasyatis americana. Can. J. Zool.* 74: 1417–30.

Hamlett, W. C., F. J. Schwartz, R. Schmeinda, and E. Cuevas. 1996b. Anatomy, histology, and development of the cardiac valvular system in elasmobranchs. *J. Exp. Zool.* 275: 83–94.

Hamlett, W. C., J. P. Wourms, and J. W. Smith. 1985. Stingray placental analogues: structure of trophonemata in *Rhinoptera bonasus. J. Submicrosc. Cytol.* 17: 541–50.

Hanson, D. 1967. Cardiovascular dynamics and aspects of gas exchange in Chondrichthyes. Ph.D. diss., University of Washington, Seattle.

Helle, K., A. Miralto, K. E. Pihl, and B. Tota. 1983. Structural organization of the normal and anoxic heart of *Scyllium stellare. Cell Tissue Res.* 231: 399–414.

Jensen, D. 1970. Intrinsic cardiac rate regulation in elasmobranchs: the horned shark, *Heterodontus francisci*, and the thornback ray, *Platyrhinoidis tdsedata. Comp. Biochem. Physiol.* 34: 289–96.

Johansen, K. 1965. Cardiovascular dynamics in fishes, amphibians, and reptiles. *Ann. N.Y. Acad. Sci.* 127: 414–42.

Johansen, K., and W. Burggren. 1980. Cardiovascular function in the lower vertebrates. In *Hearts and Heart-like Organs*, ed. G. H. Bourne. New York: Academic Press.

Johansen, K., D. E. Franklin, and R. L. Van Citters. 1966. Aortic blood flow in free-swimming elasmobranchs. *Comp. Biochem. Physiol.* 19: 151–60.

Johansen, K., and D. Hanson. 1967. Hepatic vein sphincters in elasmobranchs and their significance in controlling hepatic blood flow. *J. Exp. Biol.* 46: 195–203.

Katz, A. M. 1990. Molecular biology in cardiology: a paradigmatic shift. *J. Mol. Cell Cardiol.* 20: 355–66.

Lai, N. C., J. B. Graham, V. Bhargava, W. R. Lowell, and R. Shabetai. 1989a. Branchial blood flow distribution in the blue shark (*Prionace glauca*) and the leopard shark (*Triakis semifasciata*). *Exp. Biol.* 48: 273–78.

Lai, N. C., J. B. Graham, and W. R. Lowell. 1989b. Elevated pericardial pressure and cardiac output in the leopard shark, *Triakis semifasciata*, during exercise: the role of the pericardioperitoneal canal. *J. Exp. Biol.* 147: 263–77.

Lai, N. C., R. Shabetai, J. B. Graham, D. B. Holt, K. S. Sunnerhagen, and V. Bhargava. 1990. Cardiac function in the leopard shark, *Triakis semifasciata. J. Comp. Physiol.* 160A: 259–68.

Lakatta, E. G. 1992. Length modulation of muscle performance. Frank-Starling law of the heart. In *The Heart and Cardiovascular System*, ed. H. A. Fozzard, 1325–49. New York: Raven Press.

Laurent, P., S. Holmgren, and S. Nilsson. 1983. Nervous and humoral control of the fish heart: structure and function. *Comp. Biochem. Physiol.* 76A: 525–42.

Lewis, F. T. 1904. The question of sinusoids. *Anat. Anz.* 25: 261–79.

McWilliam, J. A. 1885. On the structure and rhythm of the heart in fishes, with special reference to the heart of the eel. *J. Physiol. (Lond.)* 6: 145–92.

Marchetti, R., V. Noseda, and F. Chiesa. 1962. Attivazione epicardica del cuore di selaci, condroganoidi e anfibi urodeli. *Riv. Biol.* 55: 187–213.

Maren, T. H. 1967. Special body fluids of the elasmobranch. In *Sharks, Skates, and Rays*, ed. P. W. Gilbert, R. F. Mathewson, and D. P. Rall, 287–92. Baltimore: Johns Hopkins University Press.

Martin, A. W. 1950. Some remarks on the blood volume of fish. In *Studies Honouring Trevor Kincaid*, ed. M. H. Hatch, 125–40. Seattle: University of Washington.

Maylie, J., and M. Morad. 1995. Evaluation of T- and L-type Ca++ currents in shark ventricular myocytes. *Am. J. Physiol.* 269: H1695–703.

Maylie, J., M. G. Nunzi, and M. Morad. 1980. Excitation-contraction coupling in ventricular muscle of dogfish (*Squalus acanthias*). *Bull. Mt. Desert Isl. Biol. Lab.* 19: 84–87.

Meghji, P., and G. Burnstock. 1984. The effect of adenyl compounds on the heart of the dogfish, *Scyliorhinus canicula. Comp. Biochem. Physiol.* 77C: 295–300.

Moyes, C. D., L. T. Buck, and P. W. Hochachka. 1990. Mitochondrial and peroxisomal fatty acid oxidation in elasmobranchs. *Am. J. Physiol.* 258: R756–62.

Muñoz-Chápuli, R., V. De Andrés, and C. Ramos. 1994. Tachykinin-like immunoreactivity in the sinus venosus of the dogfish (*Scyliorhinus canicula*). *Cell Tissue Res.* 278: 171–75.

Muñoz-Chápuli, R., A. Gallego, D. Maclas, C. Ramos, and V. De Andrés. 1995. Endocardial pores in the sinus venosus and atrium of the dogfish. *J. Fish Biol.* 47: 769–74.

Muñoz-Chápuli, R., and L. García Garrido. 1986. Cephalic blood vessels in elasmobranchs: anatomy and phylogenetic implications. In *Indo-Pacific Fish Biology: Proceedings of the 2nd International Conference on Indo-Pacific Fishes*, ed. T. Uyeno, R. Arai, T. Taniuchi, and K. Matsura, 164–72. Tokyo: Ichthyological Society of Japan.

Nabauer, M., and M. Morad. 1992. Modulation of contraction by intracellular Na+ via Na+-Ca++ exchange in single shark (*Squalus acanthias*) ventricular myocytes. *J. Physiol. (Lond.)* 457: 627–37.

Nair, M. G. K. 1970. The anatomy and embryology of the heart and its conducting system of the dogfish *Carcharias sorrah* Mull and Henle. *Zool. Anz.* 185: 265–74.

Nilsson, S. 1983. *Autonomic Nerve Function in the Vertebrates*. Berlin: Springer-Verlag.

Nilsson, S., and M. Axelsson. 1987. Cardiovascular control systems in fish. In *The Neurobiology of the Cardiorespiratory System*, ed. E. W. Taylor, 155–83. Manchester, England: Manchester University Press.

Nilsson, S., and S. Holmgren. 1992. Cardiovascular control by purines, 5-hydroxytryptamine, and neuropeptides. In *Fish Physiology*, vol. 12, pt. B, ed. W. S. Hoar, D. J. Randall, and A. P. Farrell, 301–41. New York: Academic Press.

Nilsson, S., S. Holmgren, and D. J. Grove. 1975. Effects of drugs and nerve stimulation on the spleen and arteries of two species of dogfish, *Scyliorhinus canicula* and *Squalus acanthias*. *Acta Physiol. Scand.* 95: 219–30.

O'Donoghue, C. H., and E. Abbott. 1928. The blood vascular system of the spiny dogfish *Squalus acanthias* and *Squalus suckleyi*. *Trans. R. Soc. (Edinb.)* 55: 823–90.

Olson, K. R. 1992. Blood and extracellular fluid volume regulation: role of the renin angiotensin system, kallikrein-kinin system, and atrial natriuretic peptides. In *Fish Physiology*, vol. 12, pt. B, ed. W. S. Hoar, D. J. Randall, and A. P. Farrell, 135–254. New York: Academic Press.

Ostadal, B., and T. H. Schiebler. 1970. Über die terminale Strombahn in Fischeherzen. *Z. Anat. Entwicklungsgesch.* 134: 101–10.

Parker, G. H. 1886. On the blood vessels of *Mustelus antarcticus*. *Philos. Trans. R. Soc. Lond.* 177B: 685–732.

Parker, G. H., and F. K. Davis. 1899. The blood vessels of the heart in Carcharias, *Raja*, and *Amia*. *Proc. Boston Soc. Nat. Hist.* 29: 163–78.

Pelster, B., and W. E. Bemis. 1991. Ontogeny of heart function in the little skate, *Raja erinacea*. *J. Exp. Biol.* 3: 87–97.

Poupa O., and L. Lindstrom. 1983. Comparative and scaling aspects of hearts and body weights with reference to blood supply of cardiac fibers. *Comp. Biochem. Physiol.* 76A: 413–21.

Raffaele, D. F. 1904. Ricerce sullo sviluppo del sistema vascolare nei Selacel. *Mitt. Zool. Sta. Naples* 10: 441–79.

Ramos, C., R. Muñoz-Chápuli, and P. Navarro. 1996. Ultrastructural study of the sinus venosus and sinoatrial valve in the dogfish (*Scyliorhinus canicula*). *J. Zool.* 238: 611–21.

Randall, D. J. 1968. Functional morphology of the heart in fishes. *Am. Zool.* 8: 179–89.

———. 1970. The circulatory system in fish physiology. In *Fish Physiology*, vol. IV, ed. W. S. Hoar and D. J. Randall, 132–72. New York: Academic Press.

Randall, D. J., and S. F. Perry. 1992. Catecholamines. In *Fish Physiology*, vol. 12, pt. B, ed. W. S. Hoar, D. J. Randall, and A. P. Farrell, 255–300. New York: Academic Press.

Reinecke, M., D. Betzler, W. G. Forssmann, M. Thorndyke, U. Askensten, and S. Falkmer. 1987. Electron microscopical, immunohistochemical, and biological evidence for the occurrence of cardiac hormones (ANPICDD) in chondrichthyes. *Histochemistry* 87: 531–38.

Saetersdal, T. S., E. Sorensen, R. Myklebust, and K. B. Helle. 1975. Granule containing cells and fibres in the sinus venosus of elasmobranchs. *Cell Tissue Res.* 163: 471–90.

Sanchez-Quintana, D., and J. M. Hurle. 1987. Ventricular myocardial architecture in marine fishes. *Anat. Rec.* 217: 263–73.

Santer, R. 1985. Morphology and innervation of the fish heart. *Adv. Anat. Embryol. Cell Biol.* 89: 1–102.

Satchell, G. H. 1970. A functional appraisal of the fish heart. *Fed. Proc.* 29: 1120–23.

———. 1971. *Circulation in Fishes*. Cambridge: Cambridge University Press.

———. 1991. *Physiology and Form of Fish Circulation*. Cambridge: Cambridge University Press.

Satchell, G. H., and M. P. Jones. 1967. The function of the conus arteriosus in the Port Jackson shark, *Heterodontus portusjacksoni*. *J. Exp. Biol.* 46: 373–82.

Shabetai, R., D. C. Abel, J. B. Graham, V. Bhargava, R. S. Keyes, and K. Witzum. 1985. Function of the pericardium and pericardioperitoneal canal in elasmobranch fishes. *Am. J. Physiol.* 248: H198–207.

Schoenlein, K. 1895. Beobachtungen ueber Blutkreislauf und Respiration bei einigen Fischen. *Z. Fisch Biol.* 32: 511–47.

Short, S., P. J. Butler, and E. W. Taylor. 1977. The relative importance of nervous, humoral, and intrinsic mechanisms in the regulation of heart rate and stroke volume in the dogfish *Scyliorhinus canicula*. *J. Exp. Biol.* 70: 72–92.

Sidell, B., and W. R. Driedzic. 1985. Relationship between cardiac energy metabolism and cardiac work demand in fishes. In *Circulation, Respiration, and Metabolism*, ed. R. Gilles, 386–401. Berlin: Springer-Verlag.

Smith, H. W. 1929. The composition of the body fluids of elasmobranchs. *J. Biol. Chem.* 81: 407–19.

Steffensen, J. F., and J. P. Lomholt. 1992. The secondary vascular system. In *Fish Physiology*, vol. 12, pt. A, ed. W. S. Hoar, D. J. Randall, and A. P. Farrell, 185–217. New York: Academic Press.

Sudak, F. N. 1965. Some factors contributing to the development of subatmospheric pressure in the heart chambers and pericardial cavity of *Mustelus canis* (Mitchill). *Comp. Biochem. Physiol.* 15: 199–215.

Suzuki, R., A. Takahashi, N. Hazon, and Y. Takei. 1991. Isolation of high-molecular-weight C-type natriuretic peptide from the heart of a cartilaginous fish (European dogfish, *Scyliorhinus canicula*). *FEBS Lett.* 282: 321–25.

Taylor, E. W. 1992. Nervous control of the heart and cardiorespiratory interactions. In *Fish Physiology*, vol. 12, pt. B, ed. W. S. Hoar, D. J. Randall, and A. P. Farrell, 343–87. New York: Academic Press.

Taylor, E. W., S. Short, and P. J. Butler. 1977. The role of the cardiac vagus in the response of the dogfish *Scyliorhinus canicula* to hypoxia. *J. Exp. Biol.* 70: 57–75.

Tebecis, A. K. 1967. A study of electrograms recorded from the conus arteriosus of an elasmobranch heart. *Aust. J. Biol. Sci.* 20: 843–46.

Tibbits, G. F., C. D. Moyes, and L. Hove-Madsen. 1992. Excitation-contraction coupling in the teleost heart. In *Fish Physiology*, vol. 12, pt. A, ed. W. S. Hoar, D. J. Randall, and A. P. Farrell, 267–304. New York: Academic Press.

Tota, B. 1983. Vascular and metabolic zonation in the ventricular myocardium of mammals and fishes. *Comp. Biochem. Physiol.* 76A: 423–37.

———. 1989. Myoarchitecture and vascularization of the elasmobranch heart ventricle. *J. Exp. Zool.* 2: 112–35.

Tota, B., V. Cimini, G. Salvatore, and G. Zummo. 1983. Comparative study of the arterial and lacunary systems of the ventricular myocardium of elasmobranch and teleost fishes. *Am. J. Anat.* 167: 15–32.

Tota, B., and A. Gattuso. 1996. Heart ventricle pumps in teleosts and elasmobranchs: a morphodynamic approach. *J. Exp. Zool.* 275: 162–71.

Uemura, H., M. Naruse, T. Hirohama, S. Nakamura, Y. Kasuya, and T. Aoto. 1990. Immunoreactive atrial natriuretic peptide in the fish heart and blood plasma examined by electron microscopy, immunohistochemistry, and radioimmunoassay. *Cell Tissue Res.* 260: 235–47.

Vogel, W. O. P. 1985. Systemic vascular anastomoses, primary and secondary vessels in fish, and the phylogeny of lymphatics. In *Cardiovascular Shunts*, ed. K. Johansen and W. Burggren, 143–59. Copenhagen: Munksgaard.

Wainwright, S. A., F. Vosburgh, and J. H. Hebrank. 1978. Shark skin: function in locomotion. *Science* 202: 747–49.

Yamauchi, A. 1980. Fine structure of the fish heart. In *Hearts and Heart-like Organs*, vol. 1, ed. G. H. Bourne. New York: Academic Press.

Zammit, V. A., and E. A. Newsholme. 1979. Activities of enzymes of fat and ketone body metabolism and effects of starvation on blood concentrations of glucose and fat fuels in teleost and elasmobranch fish. *Biochem. J.* 184: 313–22.

Zummo, G., and F. Farina. 1989. Ultrastructure of the conus arteriosus of *Scyliorhinus stellaris*. *J. Exp. Zool.* 2: 158–64.

MICHAEL H. HOFMANN

CHAPTER 11

Nervous System

Elasmobranchs separated some 400 million years ago from all other vertebrates and are of great importance for the investigation of the evolution of vertebrates. However, contemporary elasmobranchs are by no means just living fossils. They display a number of highly derived characteristics and cannot necessarily be considered to be more similar to ancestral vertebrates than are humans or any other animal. Only a closer look at individual characteristics and their cladistic analysis can distinguish between primitive and derived features.

This is especially important for an understanding of the evolution of the nervous system. In the older literature, many brain features of elasmobranchs were considered to be primitive, based on the misconception that living vertebrates can be arranged in a phylogenetic order from "lower" to "higher" forms, with the elasmobranchs on the "lower" end. Another reason why we intuitively tend to accept that elasmobranchs have simple brains is that we know very little about their life history and behavior. Unlike those of terrestrial animals, field observations of elasmobranchs over extended time are extremely difficult to achieve and few behavioral details are known. The movements of sharks and rays can be tracked by telemetry (Nelson 1990), but we still do not know what they are doing and what kind of interaction they have with conspecifics. Studies of elasmobranch behavior suggest relatively complex social interactions in several species (Klimley 1981, 1994), but progress is slow because of the technical problems.

Another reason why we suspect complex, as yet unknown, behavioral capabilities of elasmobranchs is the size and external form of their brains. Figure 11-1 shows dorsal and lateral views of the brains of six elasmobranch species. *Squalus acanthias*, the spiny dogfish, often used as a "typical" elasmobranch for dissection, has a relatively small brain. The same is true for *Squatina*, the angel shark. Both rays and galeomorph sharks indepen-

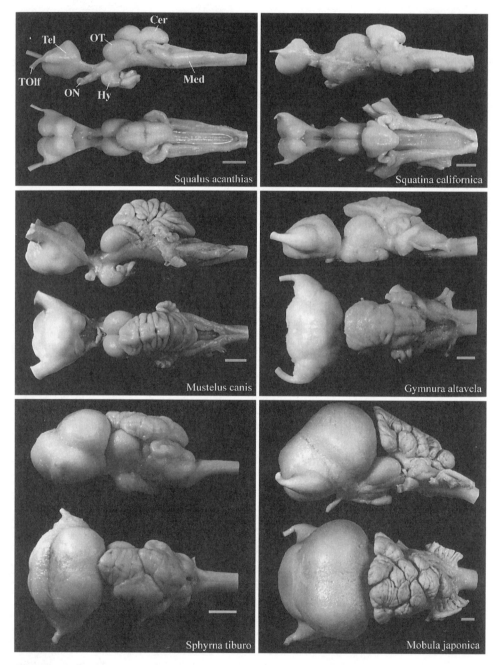

FIGURE 11-1 Sagittal and dorsal views of the brains of six elasmobranchs. The olfactory bulbs are omitted. Cer = cerebellum; Hy = hypothalamus; Med = medulla oblongata; ON = optic nerve; OT = optic tectum; TOlf = tractus olfactorius; Tel = telencephalon. Scale bars: 5 mm.

dently evolved large telencephalons and cerebellums with a relative brain weight comparable to birds and mammals (Northcutt 1978). The most complex brains are found in hammerhead sharks (e.g., *Sphyrna tiburo*, the bonnethead) and devil rays (e.g., *Mobula japonica*) (see Fig. 11-1). The functional significance of these enormous hypertrophies is unknown.

As in other vertebrates, the nervous system of elasmobranchs is divided into several categories. The spinal cord and spinal nerves are discussed in the next section, followed by the brainstem, cerebellum, and optic tectum. The forebrain consists of the diencephalon, the hypothalamus and preoptic region, and the telencephalon. They are discussed in the last three sections.

Spinal cord and spinal nerves

BASIC ORGANIZATION

The *spinal cord* consists of a series of segments extending from the caudal end of the rhombencephalon into the tail fin. Cell bodies and their dendrites are located mainly centrally, in the gray matter. The surrounding white matter contains mostly nerve fibers. The gray matter can be divided into the dorsal and ventral horns. Each trunk segment is supplied by a pair of spinal nerves. Before entering the spinal cord, the nerves divide into a dorsal and a ventral root. The dorsal root contains sensory fibers, which bifurcate and run in caudal and rostral directions in the dorsal white matter. Some fibers continue into the dorsal horn and, at least in *Pristiurus* and *Dasyatis*, there is evidence for a direct contact with motoneurons, suggesting the existence of a monosynaptic reflex arc (von Lenhossék 1894; Leonard et al. 1978). The dorsal root carries sensory information from mechanoreceptors in skin and muscles (excluding the lateral line). In the skin, free nerve endings provide tactile and temperature information (Roberts 1978). The presence of pain receptors is not clear. In mammals, small unmyelinated fibers (C fibers) are mainly responsible for the sensation of pain. Such fibers are apparently absent in elasmobranchs (Snow et al. 1993), but pain information could be transmitted to the brain by other fibers.

Two specialized mechanoreceptive organs innervated by the cord are known for elasmobranchs (see Roberts 1978). The Wunderer's corpuscles are located in the deep layers of the skin and respond to tactile stimulation as well as to fin movements. The endings of Poloumordwinoff are present only in rays. These intermuscular organs are found in the pectoral fins and are sensitive to stretch (Ridge 1977). In other vertebrates, specialized stretch receptors (muscle spindles) occur only in land vertebrates and are thought to be necessary for the control of posture. In contrast to other fishes, rays may have developed similar postural control mechanisms to coordinate the activity of the various pectoral fin muscles.

The ventral root of the spinal nerves consists predominantly of motor axons. However, some sensory fibers may be present in some of the roots (Grillner et al. 1976). The motor neurons are located in the ventral horn and supply the musculature of the corresponding myotome, and some of them innervate sympathetic and parasympathetic ganglia. The large motoneurons have extensive dendrites reaching into the surrounding white matter, the dorsal horn, and some cross to the contralateral side. Local interneurons are abundant but poorly studied.

The general bauplan of each spinal cord segment shows a number of variations at different levels. Some of the most rostral segments may lack the dorsal roots (Smeets et al. 1983). In those rays with an electric organ located in the tail region, the corresponding segments contain much larger motoneurons. Also in the tail region, there are large neurosecretory cells in a number of segments of the spinal cord (Dahlgren cells). They project onto diffuse neurohemal areas on the ventral surface of the spinal cord (Conlon et al. 1996). This caudal neurosecretory system may be homologous to the urophysis of teleosts. Liu and Demski (1993) found by means of tracer studies and electrical stimulation that the lumbar segments 60–64 contain motoneurons innervating the claspers, which are used by the male to introduce sperm into the female. This area is probably the target of descending gonadotropin-releasing hormone–immunoreactive fibers from a cell group in the mesencephalic tegmentum (Wright and Demski 1991).

FUNCTIONAL STUDIES

The spinal cord plays an important role in locomotion and body movements. It is not a simple relay station that merely transmits information between the brain and the trunk; instead, it is itself capable of coordinating locomotion and modifying it in response to sensory stimulation.

The "spinal dogfish" has been a valuable tool in the investigation of locomotor coordination and its sensory feedback control (Roberts 1988). If the spinal cord is isolated from the brain by a cut, dogfish are still able to swim and do so for extended periods. Oscillatory discharges can even be recorded from the ventral roots of the spinal nerves in immobilized (curarized) animals, although they are slower than normal. Sensory feedback is obviously not necessary for the occurrence of the rhythm. However, the swimming rhythm of the immobilized dogfish is much slower than required for sustained-level swimming. Only additional sensory input accelerates the rhythm enough to become fully functional. Further evidence for the importance of sensory input for functional swimming comes from experiments in which the body of a curarized dogfish was passively moved to simulate swimming. The recorded motor output of the ventral roots adjusts to the imposed rhythm in a certain frequency range. Furthermore, local cutaneous stimulation can modify the swimming rhythm, resembling an avoidance reaction directed against the stimulus.

CONNECTIONS WITH THE BRAIN

Ascending spinal projections can roughly be grouped into two fiber systems, the dorsal and lateral funiculi. The lateral funiculus is located dorsolaterally along the spinal cord. It contains axons supplying the brain with somatosensory and visceral afferent information. They terminate in various areas, including the cerebellum, reticular formation, mesencephalic tegmentum, and tectum, and some fibers may reach the thalamus (Hayle 1973; Ebbesson and Hodde 1981). The ascending fibers are associated with a cell group at the caudal brainstem, the nucleus funiculi lateralis (Smeets and Nieuwenhuys 1976).

The dorsal funiculus is less pronounced. Hayle (1973) even concluded that the dorsal column does not contain long ascending fibers, but Ebbesson and Hodde (1981) demonstrated a spinal projection to a dorsal column nucleus in the obex region. This nucleus, however, is rather inconspicuous and a corresponding medial lemniscus has not been reported for elasmobranchs (Smeets and Nieuwenhuys 1976).

Descending projections from the brain come from various areas. The ventral thalamus and periventricular hypothalamus are the most rostral ones. More caudal sources of descending fibers are the reticular formation, the optic tectum, the octaval nuclei, the nucleus descendens of the trigeminal nerve, the nucleus of the lateral funiculus, and the nucleus tractus solitarii. A rubrospinal projection has been reported for *Raja*, but not for *Scyliorhinus* (Smeets and Timerick 1981).

Brainstem

The *brainstem* is the rostral extension of the spinal cord and continues up to the border with the diencephalon. Two specializations of the brainstem, the tectum mesencephali and the cerebellum, are discussed in separate sections. The remaining brainstem consists of two parts, the dorsally located alar (or wing) plate, which is the extension of the dorsal (sensory) half of the spinal cord, and the basal plate, which is continuous with the ventral (motor) area of the spinal cord. Each plate can be further divided into visceral and somatic components. These areas form longitudinal columns in the brainstem. Additional columns are formed by the reticular formation and the octavolateralis nuclei. The brain stem is thus discussed under the headings: (1) somatic motor nuclei, (2) visceral motor nuclei, (3) reticular formation, (4) visceral sensory nuclei, (5) somatic sensory nuclei, (6) octavolateralis nuclei. A number of additional nuclei, especially in the mesencephalon, do not fit into these categories. They are discussed under the heading (7) additional components.

SOMATIC MOTOR NUCLEI

The *somatic motor nuclei* are located in the ventromedial brainstem. The most caudal one consists of motoneurons supplying the hypobranchial muscles. The others innervate the extrinsic eye muscles: abducens (VI), trochlear (IV), and oculomotor (III), in caudorostral order (Fig. 11-2). The abducens innervates the lateral rectus muscle, which is responsible for horizontal movements of the eye. This muscle is much larger in *Mustelus* than in *Raja* (Graf and Brunken 1984). In sharks, horizontal head movements are part of the normal swimming behavior and the eyes have to rotate continuously in the same plane to stabilize the visual field. This is not the case in rays, in which undulatory fin movements propel the animal without much lateral head deviation. The size and exact location of the abducens nucleus may also vary between sharks and rays as noted by Montgomery and Housley (1983). The oculomotor nucleus can be further subdivided into dorsal and ventral subnuclei (Northcutt 1978; Rosiles and Leonard 1980). The main function of all the oculomotor nuclei is to stabilize the image on the retina during head movements of the animal. For that purpose, they receive strong input from the vestibular system, as in other vertebrates (Puzdrowski and Leonard 1994).

VISCERAL MOTOR NUCLEI

The *visceral motor nuclei* are, in caudorostral order, the motor nuclei of the vagus (X), glossopharyngeus (IX), facialis (VII), and trigeminus (V) (see Fig. 11-2). The Edinger-Westphal nucleus at the rostral end of this column probably gives rise to preganglionic fibers innervating the dilatator iridis muscle of the eye via cranial nerve III. Although present in sharks, this nucleus was not found in *Raja* and *Hydrolagus* (Smeets et al. 1983). The iris of sharks, especially diurnal ones, respond relatively rapidly to changes in light intensity, in contrast to rays. The slow changes in rays might be mediated by the light sensitivity of the iris sphincter alone, whereas central control by the Edinger-Westphal nucleus is required for the faster responses of sharks (Young 1933; Kuchnow 1971).

The vagus innervates the heart and, together with the seventh and ninth motor nuclei, the respiratory apparatus (Withington-Wray et al. 1986; Barrett and Taylor 1985b). It is interesting to note that the respiratory rhythm can still be recorded from the motor nerves in paralyzed animals, although with different frequency (Barrett and Taylor 1985a). This is similar to the locomotor rhythm of "spinal dogfish" as described in the section on the spinal cord.

The motor nucleus of the trigeminal nerve is located just medial to the entrance of the nerve into the brainstem. It is involved in the jaw closure reflex, which is mediated through the sensory root of V. This is discussed in the section on somatic sensory nuclei.

RETICULAR FORMATION

In most studies, the *reticular formation* has been divided into three longitudinal zones. The median zone consists of an inferior and a superior raphe. In species with an electric organ (*Raja, Torpedo*), the inferior raphe is involved in the control of the electric discharge (Nakajima 1970; Fox and Richardson 1979). In other vertebrates the raphe nuclei are characterized by a high serotonin content. This is also true for elasmobranchs, but serotonin-containing neurons are also found more laterally in the brainstem (Ritchie et al. 1983). The medial zone consists of the rhombencephalic inferior, medial, and superior reticular nuclei and the mesencephalic nucleus of the fasciculus longitudinalis. The third component of the reticular formation, the lateral reticular zone, is a rather diffuse cell group,

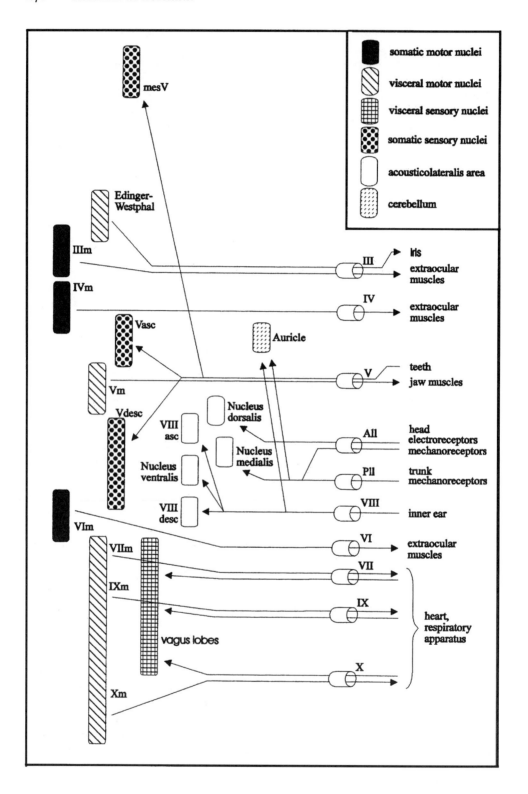

which for the most part cannot be clearly delimited in normal histological material (Smeets et al. 1983).

A detailed investigation of the reticular formation of several elasmobranchs has been performed by Stuesse et al. (1990, 1991a,b). These authors found that the complexity of this area is comparable to that of mammals. They could recognize most of the cell groups known for mammals in several elasmobranch species, with the exception of a dorsal raphe nucleus.

The reticular formation is an important relay station between brain and spinal cord. In particular, the medial rhombencephalic zone and the nucleus fasciculus longitudinalis project heavily into the spinal cord (Smeets and Timerick 1981). Other efferents reach cerebellum and tectum (Smeets et al. 1983). The reticular formation receives afferents from all major brain regions, including the cerebellum, tectum, and telencephalon.

VISCERAL SENSORY NUCLEI

The facialis (VII), glossopharyngeus (IX), and vagus (X) nerves all contribute afferent fibers to the area of the *visceral sensory nuclei*, which is located dorsolateral to the visceral motor column. The information carried by these fibers is poorly investigated in elasmobranchs, but in other vertebrates there is a prominent gustatory representation, besides temperature, pain, pressure, and respiratory chemoreceptor information (Northcutt 1978). The first relay station of these fibers is the vagus lobe, which was further subdivided by Barry (1987) into facialis, glossopharyngeus, and vagus sublobes. Stuesse et al. (1992) reinvestigated this area and found that the number of sublobes correspond to the number of gill arches (four in *Hydrolagus* and five in *Squalus*) rather than to the number of nerves supplying them. This would be in contrast to other vertebrates. They also found differences in the distribution of several neuropeptides in the sublobes.

The secondary projections of the visceral sensory nuclei are not experimentally investigated. A descending tractus solitarius can be seen in normal histological material. Ascending efferents can be followed for a short distance as well. In teleosts, this tract projects to a secondary gustatory nucleus in the rostral rhombencephalon, which in turn projects to the inferior lobes of the hypothalamus (Finger 1978; Morita et al. 1980).

SOMATIC SENSORY NUCLEI

The sensory components of the trigeminal nerve (V) constitute the *somatic sensory nuclei*. After entering the brain, they form a descending tract, terminating in the nucleus descendens nervi trigemini. This nucleus can be traced into the first few segments of the spinal cord. An ascending tract of the trigeminus courses toward the principal trigeminal nucleus. This nucleus is an important relay station to the forebrain in other vertebrates, but virtually nothing is known about its connections in elasmobranchs.

An interesting trigeminal cell group is located in the mesencephalon (mes V). It consists of large cells close to the ventricular surface of the optic tectum. They are primary sensory neurons and should be located outside the brain in the ganglion of the sensory root of the trigeminus. Their peculiar location has attracted further investigation. Many of these cells innervate mechanoreceptors in or around the teeth and send axons to trigeminal motor neurons. This constitutes a monosynaptic reflex arc controlling jaw closure (Roberts 1988). The peculiar location of these cells can be

FIGURE 11-2 Schematic drawing of the connections of cranial nerves III–X and the anterior and posterior lateral line nerves (All, Pll) to and from the brainstem. Only the major primary projections of one half of the brainstem are shown. Roman numerals refer to the corresponding cranial nerves. Suffixes after the numbers have the following meanings: m = motor nucleus; asc = nucleus ascendens; desc = nucleus descendens; mesV = mesencephalic trigeminal nucleus. This chart does not reflect the true topographical location and size of the cell groups. However, an attempt was made to maintain the approximate location from rostral to caudal (*top to bottom*) and from medioventral to dorsolateral (*left to right*).

understood from an ontogenetic point of view. Ganglion cells of the sensory roots are derivatives of the neural crest and migrate during development to a location lateral to the neural tube. The mes V cells, however, simply never migrate from their place of birth.

OCTAVOLATERALIS AREA

The dorsolateral part of the brainstem consists of a series of nuclei that receive lateral line, electrosensory, auditory, and vestibular input (Fig. 11-3). This column is called the *octavolateralis area*. The vestibular and auditory input is carried by the eighth nerve and the lateral line and electrosensory information by the lateral line nerves. It should be noted that in the earlier literature the lateral line nerves were erroneously considered part of the facialis and vagus system. Closer examination and embryologic studies revealed that the lateral line nerves derive from independent placodes and constitute additional cranial nerves that are not related to other nerves (Northcutt 1978). The exact number of lateral line nerves

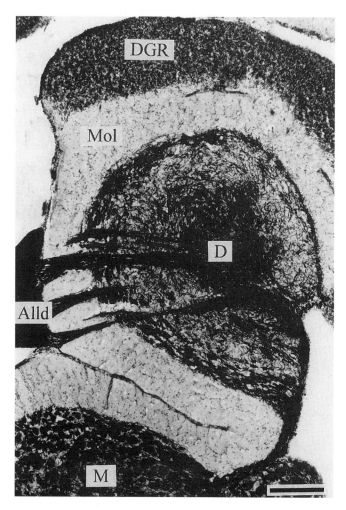

FIGURE 11-3 Cross section through the nucleus dorsalis of the octavolateralis area (*Urolophus halleri*). This nucleus is the primary target of electrosensory information. The section was stained according to a modification of the silver method of Gallyas (1979) and counterstained with cresyl violet. Alld = dorsal ramus of the anterior lateral line nerve; D = dorsal nucleus; DGR = dorsal granular ridge; M = medial nucleus; Mol = molecular layer. Scale bar: 500 μm.

is not clear. In elasmobranchs, there is evidence for up to six nerves, which are partly fused to form the anterior and posterior lateral line nerves.

The octavolateralis area can roughly be divided into a nucleus dorsalis, nucleus medialis, and nucleus ventralis. Five additional nuclei are associated with the eighth nerve, the nucleus octavus ascendens, nucleus octavus descendens, and nuclei C_1, C_2, and C_3.

The anterior lateral line nerve innervates mechanoreceptors and electroreceptors in the head. On entering the brain, the nerve fibers rearrange according to their modality. The dorsal root carries electrosensory fibers and projects to the nucleus dorsalis. The ventral root supplies the nucleus medialis with mechanosensory information from the head.

The posterior lateral line nerve supplies only mechanoreceptors on the trunk. It enters the brain as a separate root and projects to the nucleus medialis as well as to the auricles of the cerebellum. The dorsal and medial nuclei form separate lemniscal channels and innervate electrosensory and mechanosensory areas of the lateral mesencephalic complex, respectively.

The statoacoustic nerve carries vestibular and auditory information from the inner ear and enters the brain as a single root. It terminates in the nucleus ventralis and the nucleus octavus descendens and ascendens. The eighth nerve also projects to the auricles of the cerebellum and to the reticular formation (Barry 1987; Puzdrowski and Leonard 1993).

Nuclei C_1, C_2, and C_3 are rather distinct cell clusters that are the source of ascending pathways to the lateral mesencephalic complex. The extent of segregation of modalities in this channel is not completely clear. Corwin and Northcutt (1982) presented evidence that nuclei C_1 and C_2 are part of an ascending auditory pathway, whereas Bodznick and Schmidt (1984) found lateral line representation. Nucleus C_3 receives only sparse direct octaval input, if any at all (Puzdrowski and Leonard 1993).

More information about the octavolateralis area can be found in reviews (Bodznick and Boord 1986; Boord and Northcutt 1988; McCormick and Braford 1988; Boord and Montgomery 1989; McCormick 1992; Puzdrowski and Leonard 1993).

The physiology of medullary and higher octavolateralis areas is discussed in chapter 12, on special senses. However, one point should be mentioned here. The dorsal and medial nuclei are arranged in cerebellum-like fashion. They are capped by a molecular layer formed by dendrites of the secondary sensory neurons, stellate cells, and parallel fibers from the dorsal granular ridge. The molecular layer and the dorsal granular ridge are caudal extensions of the cerebellar auricles. The significance of this peculiar cerebellum-like arrangement is beginning to emerge. The electro- and mechanosensory systems are very sensitive to noise. The animal's own locomotor and respiratory movements are sources of reafferents, which have to be distinguished from meaningful environmental stimuli. The molecular layer of the lateral line lobes is thought to play an important role in several noise cancellation mechanisms at medullary levels (New and Bodznick 1990; Conley and Bodznick 1994; Montgomery and Bodznick 1994; Conley 1995).

ADDITIONAL COMPONENTS

The *mesencephalic tegmentum*, the most rostral portion of the brainstem, contains several cell groups that do not fall into any of the above categories. The lateral mesencephalic complex is the largest of them. It can be further subdivided into lateral, mediodorsal, and medioventral cell groups. They are targets of secondary electro-, mechano-, and auditory fibers from the brainstem, respectively (Corwin and Northcutt 1982; Boord and Northcutt 1982). The lateral mesencephalic complex may be homologous to the torus semicircularis of other fishes and non-mammals and hence to the inferior colliculus of mammals (Northcutt 1989).

A nucleus ruber has been identified on the basis of its cerebellar afferents (Ebbesson and Campbell 1973). It is located in the ventromedial tegmentum. A rubrospinal tract was not found in the shark *Scyliorhinus*, but is present in the ray *Raja* (Smeets and Timerick 1981). This is an interesting parallel to reptiles, in which such a tract is present in lizards but not in snakes. The presence or absence of distinct limblike structures (legs, large pectoral fins) may be correlated with a rubrospinal pathway (Smeets et al. 1983).

In amniotes, especially mammals, a substantia nigra can be identified by its high content of dopamine. In actinopterygian fishes, a mesencephalic dopamine system is apparently lacking, which led Parent et al. (1984) to suggest that it is a feature of the terrestrial way of life. However, Meredith and Smeets (1987) found large numbers of dopaminergic cells in the ventromedial tegmental area and more laterally in a cell group that they identified as the substantia nigra. Again, the animal used in their study was a ray and the presence of pectoral fins could be correlated with an extensive mesencephalic dopamine system.

Cerebellum

EXTERNAL FORM

The *cerebellum* of elasmobranchs can be divided into a pair of laterally situated auricles and an unpaired corpus cerebelli. The auricles consist of a dorsomedial upper leaf and a ventrolateral lower leaf, which is continuous with the octavolateralis area of the medulla. Whereas the auricles display relatively little variation, the corpus underwent remarkable modifications in several groups. The primitive condition is most likely the one found in squalomorphs, heterodontids, squatinomorphs, scyliorhinids, electric rays, and in the closest living relative of elasmobranchs, the holocephalimorphs (Northcutt 1978). In these species, the corpus consists of an anterior lobe, partly overlapping the optic tectum, and a posterior lobe, in part covering the fourth ventricle. A primary transverse fissure separates the otherwise smooth lobes (see Fig. 11-1, *Squalus, Squatina*; Fig. 11-4, *Platyrhinoidis*).

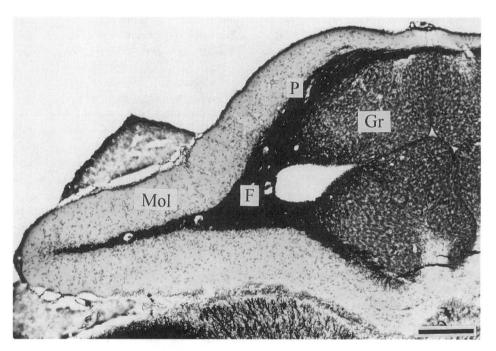

FIGURE 11-4 Cross section through the cerebellum of *Platyrhinoidis triseriata*. Myelin stain (modified after Gallyas 1979). F = fiber layer; Gr = granule cell layer; Mol = molecular layer; P = Purkinje cell layer. Scale bar: 500 μm.

In some rays and sharks, the corpus is greatly enlarged and both lobes are highly foliated and subdivided into sublobes. In some myliobatids, the cerebellum is markedly asymmetric. The species with the highest relative brain weight (the manta ray and the hammerhead) have also the most differentiated cerebellums (see Fig. 11-1). Ten to 15 lobes of different size can be recognized, but the size and number varies among individuals.

INTERNAL ORGANIZATION
The internal organization of the cerebellum is similar to that of other vertebrates. Four layers can be distinguished: an inner granular layer, a fiber zone, a Purkinje cell layer, and a molecular layer (see Fig. 11-4). In forms with multiple lobes, each lobe shows the typical cerebellar arrangement, suggesting that the increase in total volume is the result of multiplication of the basic structural plan rather than just an increase in size of one or two lobes.

As in other vertebrates, the small and densely packed granule cells give rise to axons forming the parallel fibers in the molecular layer. They synapse on Purkinje cell dendrites, organized in a plane perpendicular to the parallel fibers. Local interneurons, the stellate cells, are also present in the molecular layer, but basket cells have not been described. The Purkinje cell bodies form a thin, one-to-two-cell-thick layer around the granule cells layer, separated from the latter, especially laterally, by the fiber layer consisting of afferents and outgoing Purkinje cell axons. The granule cell layer contains also the larger Golgi cells, presumably receiving mossy fiber input. Climbing fibers have been described by Nicholson et al. (1969), but other authors have failed to recognize them in Golgi preparations. Fiebig (1988), however, could retrogradely label inferior olivary neurons after tracer injections into the molecular layer, which are the source of climbing fibers in amniotes. It is, however, not clear to what extent olivary fibers actually "climb" the molecular layer or whether they elicit complex spikes in Purkinje cells, a typical effect of climbing fiber activity in mammals (Paul 1982). Alvarez-Otero et al. (1993) provided evidence that climbing fibers are present but make direct synaptic contact with Purkinje cell bodies instead of with their dendrites.

AFFERENT CONNECTIONS
Besides the inferior olivary nucleus, many other cell groups project to the corpus cerebelli including the pretectal and accessory optic nuclei, interstitial nucleus of Cajal, nucleus ruber, oculomotor and possibly the trochlear nucleus, central periaqueductal gray, nucleus H, reticular formation, cerebellar nucleus, caudal part of nucleus F, octaval and trigeminal nuclei, part of the octavolateralis area, and spinal cord. Most of these cell groups seem to project only to the granular cell layer (Fiebig 1988). Injections into the molecular and Purkinje cell layers reveal only retrogradely filled cells in the inferior olive.

The pretectal and accessory optic nuclei receive direct retinal input and probably convey visual information to the cerebellum. Many of the other brainstem afferents are from secondary trigeminal or octavolateralis nuclei, and the spinal cord contributes somatosensory information. The cerebellum also receives efferent copies of motor commands; that is, some motoneurons send collaterals into the cerebellum. The oculomotor and trochlear nuclei are possible candidates for such an input, and below I report on physiological evidence of ascending projections from spinal locomotor rhythm generators.

The auricles are the only part of the cerebellum receiving direct cranial nerve input. The pars medialis (upper leaf) is innervated by the vestibular fibers, whereas the pars lateralis (lower leaf) predominantly receives lateral line input (Schmidt and Bodznick 1987). The auricles also receive input from a number of brainstem nuclei, summarized by Schmidt and Bodznick (1987).

EFFERENT CONNECTIONS
The only known output of the corpus cerebelli is through Purkinje cell axons. Tracing experiments confirmed that the majority of these

axons project to the nucleus cerebelli. This cell group lies at the base of the cerebellar peduncles. Developmental studies have shown that the group is derived from the cerebellar anlage and is thus part of the cerebellum (Rüdeberg 1961, cited by Smeets et al. 1983). Some of its cells project back into the cortex of the corpus (Fiebig 1988; Paul and Roberts 1984). The nucleus cerebelli projects to many of the nuclei giving rise to cerebellar afferents, including diencephalon, trochlear and oculomotor nuclei, nucleus ruber, reticular formation, and inferior olive (Smeets et al. 1983).

The efferent projections of the auricles are different from the corpus. The auricles do not project to the nucleus cerebelli but send parallel fibers into the molecular layer of the medullary electrosensory and lateral line lobes. Other efferents ascend as mossy fibers into the corpus cerebelli (Schmidt and Bodznick 1987).

FUNCTIONAL STUDIES

Behavioral changes after creating lesions in various parts of the cerebellum have provided some insights into its function. The early investigations have been summarized by Ten Cate (1935). Unilateral auricular lesions increase muscle tonus on the operated side, which lead to bending and circulating swimming. Removal of the corpus has no visible effect on swimming movements or equilibrium, but if the cerebellar peduncles are removed in addition, some behavioral changes similar to the ones after auricular removal are observed.

Paul and Roberts (1979) investigated in more detail the role of the cerebellum in modulating reflex movements of the fins and other movements in a dogfish by producing lesions and electrophysiological recordings of moving animals. The elevation reflex of a pectoral fin to tactile or electrical stimulation consists of a fast phasic and a longer latency tonic component. The whole reflex is organized in the spinal cord. The brainstem has a principally inhibitory effect on the spinal circuitry and the cerebellum inhibits the brainstem. After cerebellar lesions, the brainstem inhibition is unopposed and the reflex, especially its tonic component, is greatly reduced.

Isolating the spinal cord from the brainstem by a cut fully restores the reflex, and the tonic component is even more pronounced than in normal animals.

The cerebellum and brainstem probably do not act as diffuse inhibitors but may inhibit or disinhibit selected motor pools to achieve a spatially differentiated modulation (Paul and Roberts 1979). The habituation of the reflex is also influenced by the cerebellum. The phasic component, the only one remaining in decerebellated animals, shows no habituation to repetitive stimulation, and in the isolated spinal cord, neither component habituates even at a 1-Hz repetition rate.

Recordings from single cells (mainly Purkinje cells) in freely swimming animals show that many neurons are firing in phase with swimming movements even in the absence of sensory feedback, probably driven by corollary discharges from spinal locomotory pattern generators. They also respond to cutaneous stimulation, but only if the stimulation induces an actual movement. Passive tail bending was not effective (Paul 1982).

Responses to various kinds of sensory stimulation have been reported in the thornback ray, *Platyrhinoidis triseriata* (Tong and Bullock 1982; Fiebig 1988). Visual, electrosensory, and mechanosensory stimuli readily evoked single unit spikes or evoked potentials. An acoustic area has not been found in this species, but large differences in the importance of the auditory system within elasmobranchs might account for that. In fact, Bullock and Corwin (1979) found auditory responses in the posterior part of the corpus in a shark with a well-developed auditory system. Most sensory units in the cerebellum are unimodal. There is some evidence for somatotopy, but the maps are discontinuous and organized into patches, resembling the fractured somatotopy of the cerebellum in mammals (Fiebig 1988).

Tectum mesencephali

HISTOLOGY

The *mesencephalic tectum*, or roof, is well developed in elasmobranchs. In general, species with large eyes also have large tecta, but the lack of quantitative data does not allow recognition of possible evolutionary trends. Transverse sections through the tectum show a number of alternating fiber and cellular layers. The lamination is not as distinct as in teleosts or anurans, but up to six layers can be distinguished in some species. Probably as a result of a considerable variation within elasmobranchs, there is currently no agreement on the exact number of laminae. In all elasmobranchs, most neurons have migrated away from the periventricular zone and two basic patterns can be recognized (Northcutt 1989): in *Squalus* and *Squatina*, most neurons are located in the central zone, but in some galeomorph sharks, heterodontids, and some rays, the majority of neurons are in the outer half of the tectum (Fig. 11-5). There is not enough information on tectal layering within elasmobranchs to establish phylogenetic trends, but an outgroup analysis shows that the primitive condition for vertebrates is a third pattern in which most neurons are located in the periventricular zone (Northcutt 1989).

Although sometimes considered to be more primitive, the internal organization of the elasmobranch tectum is comparable to that of other vertebrates, including mammals. Several neuron types can be distinguished in Golgi material and specialized dendroaxonic and dendrodendritic synapses are present in addition to the normal axodendritic ones (Manso and Anadón 1991a,b).

AFFERENT CONNECTIONS
The majority of tectal afferents come from the retina. Ganglion cell axons cross in the optic chiasm and divide into a tractus opticus medialis and a tractus opticus lateralis, supplying the rostral and dorsomedial and lateral and caudal parts of the tectum, respectively. In all elasmobranchs, as in other vertebrates, retinal fibers terminate predominantly in the outer half of the tectum, but they seem to innervate different layers in different species (see Smeets et al. 1983). Retinotectal projections are completely crossed, except for a sparse ipsilateral component in *Platyrhinoidis* (Northcutt and Wathey 1980).

In addition to retinal fibers, the tectum receives input from all other major brain parts and the spinal cord. After tracer injections into the tectum of *Scyliorhinus* and *Raja*, Smeets (1982) labeled some neurons retrogradely in the dorsal half of the cervical spinal cord. Rhombencephalic cell groups projecting to the tectum include the reticular formation, nucleus tractus descendens nervi trigemini, and several octavolateralis nuclei. The nucleus cerebelli, the main output pathway of the corpus cerebelli, also projects strongly to the tectum.

Afferents at mesencephalic levels come from the contralateral tectum, nucleus tegmentalis lateralis, ventrolateral tegmental area, and nucleus ruber. From the diencephalon, the tectum receives afferents from the pretectal area, some dorsal and ventral thalamic nuclei, and from the nucleus medius of the hypothalamus. A projection from the telencephalon originates from a part of the dorsal pallium.

EFFERENT CONNECTIONS
Smeets et al. (1983) distinguish four main efferent pathways from the tectum. An ascending projection innervates the pretectum, several dorsal thalamic nuclei, and the inferior lobes of the hypothalamus. These projections are bilateral. A second pathway is crossed, innervating the contralateral tectum via the tectal commissure. Descending projections divide into a crossed and an uncrossed pathway distributing fibers to the isthmic area, the reticular formation, and several other rhombencephalic targets. Some descending fibers continue into the spinal cord. The fourth pathway is described as a rather diffuse ipsilateral tectotegmental system.

Luiten (1981), after tracer injections into the eye, retrogradely labeled cells in the optic tectum. Central neurons projecting to the eye have long been known to exist in birds, in which they originate in a cell group termed *isthmooptic nucleus*. Studies showed that projections to the retina are present in many vertebrates, but they originate from several nonhomologous cell groups. Some of these are

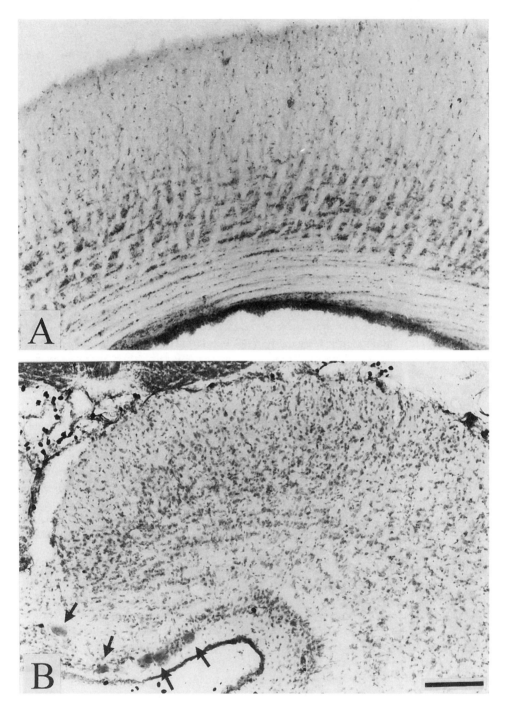

FIGURE 11-5 Cross sections through the optic tectum in *Squalus acanthias* (**A**) and *Platyrhinoidis triseriata* (**B**). Most cell bodies are located in the middle zone of the tectum in *Squalus*, but in the outer zone in *Platyrhinoidis*. *Arrows* in **B** mark the large cell bodies of the mesencephalic part of the trigeminus (mesV). Scale bar: 250 μm.

probably neuromodulatory and innervate a large number of other brain areas as well. The projection from the tectum to the retina may be different in that they provide a feedback loop to the retina, but even in birds, in which up to 12,000 neurons project to the retina, the function of the pathway is still unknown (Uchiyama 1989; Repérant et al. 1989).

PHYSIOLOGY
The physiology of the tectum is poorly known. Early lesion experiments gave conflicting results regarding the effect on locomotion and equilibrium (Ten Cate 1935). This might be explained by the fact that the extent of the lesion was not always well documented and tegmental areas were probably involved in some cases. Many authors reported no changes in locomotion and equilibrium after tectal lesions. As in other vertebrates, the tectum is probably an important multimodal integration center in which the location of objects is mapped relative to the animal. Another interesting function has been suggested by Wilson et al. (1974). They showed that small bilateral tectal lesions result in variable darkening of the skin and concluded that the tectum computes the light intensities from different directions to adjust the body color through hypothalamic centers.

The ability to discriminate visual targets in normal and tectally ablated sharks has been investigated by Graeber et al. (1973). After removal of the tectum, the animals were still able to discriminate between horizontal and vertical stripes. They were tested, however, about a month postoperatively and did show some signs of visual deficits during this time. The recovery of vision, at least some aspects of it, was obviously mediated by other areas.

Electrophysiological recordings have been made by some authors (Platt et al. 1974; Bullock and Corwin 1979; Northcutt and Bodznick 1983; Bullock 1984; Bleckmann et al. 1987, 1989; Bullock et al. 1990, 1991; Bodznick 1991; Hofmann and Bullock 1995). Mainly, compound field potentials in response to different sensory stimuli or to direct peripheral nerve shocks were recorded.

The behavior of single units has not been systematically studied. The tectum responds to visual, electrosensory, mechanosensory, and acoustic stimuli and probably also receives somatosensory input from the spinal cord. Response properties and latencies are quite different in different modalities. Best investigated is the response to photic stimulation. The latency of the first main tectal peak is, in general, longer than in teleosts but shorter than in hagfish and lampreys (Bullock et al. 1991). Within elasmobranchs, rays tend to have longer latencies than galeomorph sharks (60–150 msec versus 30–70 msec). In this respect, squalomorph sharks and heterodontids are more similar to rays. Evoked potentials in response to mechanical or electrosensory stimulation have much shorter latencies. Most of the latency of response to visual stimulation can be attributed to the retina. Simultaneous recordings from the optic nerve and tectum show that differences in peak latency are only in the millisecond range. Not only the first large potential, but also the later waves, which may peak at latencies of up to 1.5 sec, are already found in the retina (Bullock et al. 1991).

Investigations of temporal properties of the visual input to the tectum revealed some unexpected phenomena. If the eye is stimulated with a train of light flashes at frequencies from 2 Hz up to the flicker fusion frequency, the omission of a flash causes a distinct response with a fixed latency to the "expected" event (omitted stimulus potential) (Bullock et al. 1990). Gradations in light intensity cause large oscillatory potentials that can last for seconds (Hofmann and Bullock 1995). These temporal properties of the optic tectum are probably not limited to elasmobranchs; they have just not been looked for in other vertebrates.

Diencephalon

The forebrain, or prosencephalon, is commonly divided into telencephalon and diencephalon with the preoptic region included in the telencephalon and the hypothalamus in the diencephalon (Fig. 11-6). Studies on the segmental organization of the brain suggest a

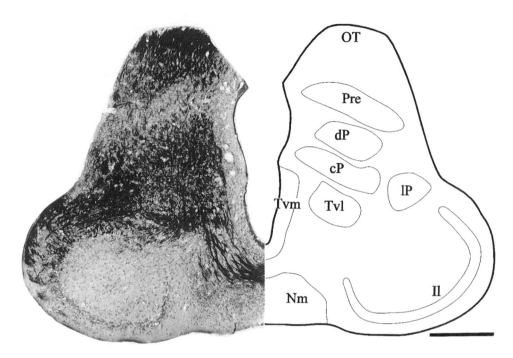

FIGURE 11-6 Cross section through the diencephalon of *Platyrhinoidis triseriata*. The *left half* of the figure shows a section stained for myelin (modified after Gallyas 1979) and counterstained with cresyl violet. The drawing on the *right side* shows the location of some of the major nuclei. cP = central posterior thalamic nucleus; dP = dorsal posterior thalamic nucleus; Il = inferior lobe of the hypothalamus; lP = lateral posterior thalamic nucleus; Nm = nucleus medius of the hypothalamus; OT = optic tract; Pre = pretectal area; Tvl = thalamus ventrolateralis; Tvm = thalamus ventromedialis. Scale bar: 1 mm.

different model (Rendahl 1924; Puelles et al. 1987; Northcutt 1995). The *diencephalon* consists of three divisions, the synencephalon and posterior and anterior parencephalon. The hypothalamus and preoptic region are part of the secondary prosencephalon, which also includes the telencephalon. The reasons for excluding the hypothalamus from the diencephalon cannot be discussed here in detail and the reader is referred to Northcutt (1995). In order to be compatible with this model, the hypothalamus and preoptic region are described in a separate section.

SYNENCEPHALON

The transition zone between midbrain and diencephalon is called the *synencephalon*. It consists of the pretectal area, the ventral optic nucleus, and the nucleus of the medial longitudinal fascicle. The pretectal area comprises four different nuclei, all of which receive direct retinal input. All but one also receive tectal afferents. The pretectal area projects to the tectum and the cerebellum.

A ventral optic nucleus has been identified by Northcutt (1990). It receives retinal and tectal input, and the only known efferents are to the cerebellum (Fiebig 1988). In anurans, it projects extensively to several brainstem nuclei, including the oculomotor and trochlear nuclei, and is thought to mediate reflex eye movements.

The nucleus of the medial longitudinal fascicle is known to project into the spinal cord (Smeets and Timerick 1981). Nothing is known about its afferents.

POSTERIOR PARENCEPHALON

The *posterior encephalon* includes the epithalamus, dorsal thalamus, and posterior tubercle.

The *epithalamus* consists of a long diverticulum, leading to the light sensitive pineal organ, and the habenula. Primitively, vertebrates possessed two diverticula, a left parietal eye and a right epiphysis (pineal). Elasmobranchs retained only the pineal, which has photoreceptors similar to the ones in the retina (Rüdeberg 1969) that are sensitive enough to detect light at intensities of the full moon (Hamasaki and Streck 1971).

The habenular nuclei are one of the most prominent and conspicuous cell groups in vertebrates. In all elasmobranchs, the left nucleus is larger than the right. They fuse dorsally and form a habenular commissure in which the stria medullaris also runs. A distinct fasciculus retroflexus is present and probably projects to the interpeduncular nucleus.

The *dorsal thalamus* consists of anterior, dorsal posterior, central posterior, and lateral posterior nuclei. All these nuclei project to the telencephalon (Fiebig and Bleckmann 1989). The exact targets are unknown, but at least the lateral posterior nucleus projects to the medial pallium (Smeets and Northcutt 1987). The anterior thalamic nucleus receives retinal as well as tectal input (Northcutt 1990). The dorsal posterior nucleus does not receive direct retinal input but is reciprocally connected with the tectum (Northcutt 1990). The central posterior nucleus receives afferents from the mechano- and electrosensory lateral mesencephalic complex, which is thought to be the homologue of the torus semicircularis. This projection has been electrophysiologically confirmed by Bleckmann et al. (1987). The lateral posterior nucleus receives strong electrosensory input from the midbrain (Schweitzer 1983). It projects to the telencephalon and to the inferior lobes of the hypothalamus (Smeets and Boord 1985).

The *posterior tubercle* is located ventrocaudal to the dorsal thalamus and dorsocaudal to the inferior lobes of the hypothalamus. The afferents are largely unknown, but it projects to the medial pallium of the telencephalon, the optic tectum, and spinal cord (Smeets and Timerick 1981; Smeets 1982; Smeets and Northcutt 1987).

ANTERIOR PARENCEPHALON

The anterior parencephalon contains the *ventral thalamus*, which can be further subdivided into three cell groups: a ventrolateral, a ventromedial, and an intermediate thalamic nucleus. All of them receive direct retinal and tectal input. The ventrolateral nucleus projects to the optic tectum (Smeets 1982) and the ventromedial nucleus to the spinal cord (Smeets and Timerick 1981).

Hypothalamus and preoptic region

The preoptic region and the hypothalamus are the interface between the brain and the hormonal system. Most of the neurosecretory cells are located here. This area is also an important center for the control of motivational behavior, like feeding, reproduction, aggression, and probably migration.

The largest cell group of the *preoptic area* is the preoptic nucleus. It is located periventricularly in the caudal part of the telencephalon between the septal area and the chiasmatic ridge (Fig. 11-7). In teleosts, it consists of distinct large-celled (magnocellular) and small-celled (parvocellular) parts, but such clear subdivisions have not been found in elasmobranchs (Smeets et al. 1983). The preoptic nucleus projects to the inferior lobes of the hypothalamus (Smeets and Boord 1985) and to the medial pallium (Smeets and Northcutt 1987). Immunocytochemical studies revealed that the preoptic nucleus is rich in neuroactive peptides like substance P (Rodriguez-Moldes et al. 1993), met- and leu-enkephalin (Vallarino et al. 1994), serotonin (Ritchie et al. 1983), phe-met-arg-phe (FMRF) amide (Vallarino et al. 1991), and neuropeptide Y (Vallarino et al. 1988). In other vertebrates, large numbers of gonadotropin-releasing hormone–immunoreactive cells are also present in the preoptic nucleus, but only a few are found in elasmobranchs in this area (Wright and Demski 1993).

Two other cell groups are found adjacent to the preoptic nucleus. The thalamic eminence has been identified dorsolateral to the preoptic nucleus (Smeets et al. 1983), but

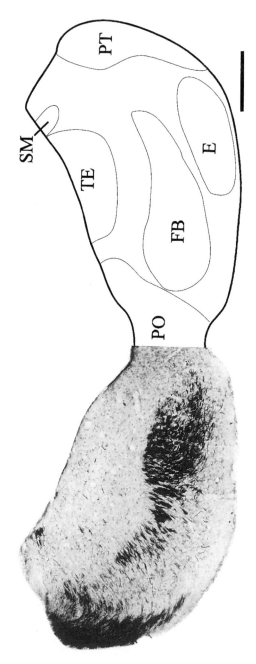

FIGURE 11-7 Cross section through the caudal end of the telencephalon impar, just rostral to the optic chiasm (*Platyrhinoidis triseriata*). E = entopeduncular nucleus; FB = forebrain bundle; PO = preoptic nucleus; PT = pallial tract; SM = stria medullaris; TE = thalamic eminence. Scale bar: 500 μm.

nothing is known about its connections in elasmobranchs. Caudal to the preoptic nucleus, at the level of the optic chiasm, the suprachiasmatic nucleus is located. It receives direct retinal input (Repérant et al. 1986) and is probably involved in the control of diurnal rhythms.

The *hypothalamus* is well developed in elasmobranchs. The nucleus medius is the most rostral of the hypothalamic nuclei and begins caudal to the optic chiasm. The large inferior lobes consist of the lateral lobular nucleus and a ventromedially located lateral tuberal nucleus. At the dorsal aspect of the inferior lobes, Smeets et al. (1983) identified a periventricular hypothalamic nucleus, but Northcutt (1995) suggested that this nucleus is actually part of the posterior tubercle.

The connections of the inferior lobes have been investigated by Smeets and Boord (1985) in the clearnose skate. Main brainstem afferents originate in the lateral mesencephalic complex and a nucleus termed F. The latter cell group has been suggested to constitute a nucleus coeruleus (Fiebig and Bleckmann 1989; Stuesse et al. 1990). The inferior lobes also receive strong input from the telencephalon, especially from the central nucleus of the dorsal pallium. Main efferent pathways reach the tegmentum, reticular formation, and the corpus cerebelli (Smeets and Boord 1985). However, the cerebellar projections have not been confirmed by Fiebig (1988) after tracer injections into the corpus cerebelli in the thornback ray, *Platyrhinoidis triseriata*.

Telencephalon

EXTERNAL FORM

The external form and the relative size of the telencephalon vary greatly among elasmobranchs. The smallest telencephalons are found in squalomorph sharks and, within batoids, in guitarfishes. In several elasmobranch radiations, the telencephalon has been enlarged independently. In batoids, the largest telencephalons are present in myliobatids, and here especially in the devil rays (*Mobula*, *Manta*). Within sharks, galeomorphs in general have larger telencephalons than squalomorphs. The largest are found in hammerhead sharks and, probably independently evolved, in *Ginglymostoma*. It is an interesting correlation that in forms with a large telencephalon, the cerebellum is also enlarged, but any evidence on functional correlations is missing.

Smaller telencephalons usually display a large ventricle surrounded by the telencephalic wall, a situation not unlike that of amphibians. In larger ones, the ventricle becomes completely obliterated and the walls are thickened. The lack of ventricular landmarks and the poor morphological differentiation of many areas makes comparison to other vertebrates and among elasmobranchs difficult.

ANATOMY

As in other vertebrates, the telencephalon can be divided into pallium and subpallium. The *pallium* consists of a lateral part, which receives olfactory input from the olfactory bulb, a dorsal pallial complex, and a medial pallium. The dorsal pallium became increasingly complex in some elasmobranch radiations. In *Squalus*, two subdivisions of the dorsal pallium have been recognized: a pars superficialis and a central nucleus (Northcutt 1978; Smeets 1990). In *Raja*, a skate with a moderately developed telencephalon, the pars superficialis consists of distinct lateral and medial parts (P3 and P2) (Smeets et al. 1983). These can also be recognized in *Platyrhinoidis* (PDSm and PDSl) (Fig. 11-8). The central nucleus is also subdivided into several parts, but there is currently no agreement on the exact number of subdivisions in different species. In the section shown in Figure 11-8, dorsal and ventral subdivisions can be distinguished in normal histological material. In *Raja* and *Ginglymostoma*, more subdivisions have been reported along the rostrocaudal extent of this area (Luiten 1981; Smeets 1990).

A distinct feature of the elasmobranch telencephalon is that the two hemispheres are fused in the midline. This is less pronounced in *Squalus*, but in some rays and galeomorph

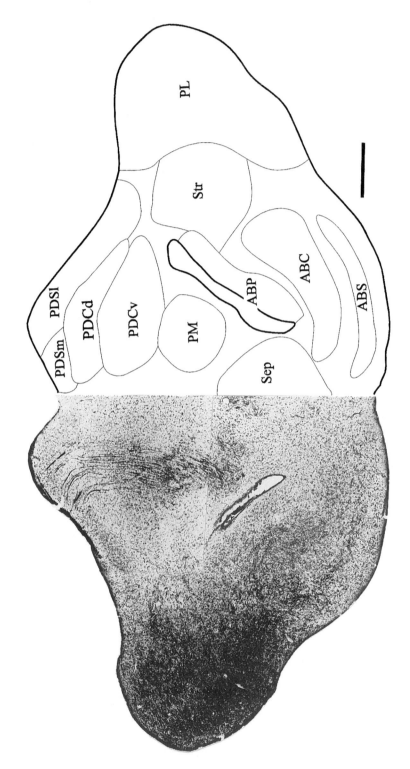

FIGURE 11-8 Cross section through the telencephalon of *Platyrhinoidis triseriata*. ABC = area basalis centralis; ABP = area basalis periventricularis; ABS = area basalis superficialis; PDCd = central nucleus of the dorsal pallium, pars dorsalis; PDCv = central nucleus of the dorsal pallium, pars ventralis; PDSl = dorsal pallium, pars superficialis lateralis; PDSm = dorsal pallium, pars superficialis medialis; PL = lateral pallium; PM = medial pallium; Sep = septal area; Str = striatum. Scale bar: 1 mm.

sharks the whole telencephalon constitutes a solid cell mass with both sides interconnected by the large pallial commissure. This extensive interhemispheric commissure is only present in elasmobranchs.

The *subpallium* can be divided into a medially situated septal area, the area basalis, and a laterally situated striatal area. The septal area consists of several subdivisions (e.g., four in *Scyliorhinus* and five in *Squalus* [Smeets 1990]). The area basalis can further be divided into the area basalis superficialis, the area basalis centralis, and the area basalis periventricularis. The area basalis superficialis has been termed tuberculum olfactorium in the older literature (Holmgren 1922; Bäckström 1924). It receives heavy input from the lateral pallium and also, to varying degrees, directly from the olfactory bulb. In this respect, it resembles the olfactory tubercle of frogs. The organization of the striatal area is quite different in different species. In *Squalus* and *Scyliorhinus*, it consists of clusters of granular cells and exhibits a laminar organization (Northcutt 1978). In *Raja*, several cell groups are found in comparable locations. Smeets et al. (1983) distinguished four different nuclei (SP 1–4).

It should be noted that there is currently no agreement concerning the homologies of the subpallial areas to other vertebrates, except for the septal nuclei. Some areas labeled here and by others as the striatum of elasmobranchs may in fact be homologous to pallial and subpallial parts of the amygdala of other vertebrates (Northcutt 1989). Furthermore, there is immunocytochemical evidence that at least part of the area basalis is comparable to the striatum or basal ganglia of terrestrial vertebrates (Northcutt et al. 1988). Therefore, the terminology used here does not imply any homologies to cell groups with the same name in other animals until further data are evaluated.

The intrinsic and extrinsic connections of the telencephalon are poorly known. Before the advent of modern tracer techniques, the telencephalon of elasmobranchs and other fishes was mainly thought to be involved in the processing of olfactory information and accordingly was sometimes called an olfactory lobe. Work during the last three decades, however, showed that the secondary olfactory projections are restricted to the lateral pallium and to a varying extent also to the area superficialis basalis (Bruckmoser and Dieringer 1973; Ebbesson and Heimer 1970; Smeets 1983; Dryer 1994). Tracer injections confirmed the existence of extensive ascending projections from a variety of diencephalic and brainstem nuclei (Luiten 1981; Smeets and Northcutt 1987; Schroeder and Ebbesson 1974). Visual information most likely is relayed in the dorsal thalamus and suprachiasmatic nucleus. A major source of electrosensory information is the posterior lateral thalamic nucleus. Other projections with unknown function arise from the preoptic region, the inferior lobes and medial nucleus of the hypothalamus, the posterior tubercle, ventral tegmental area, substantia nigra, reticular formation, raphe superior, and locus coeruleus. The exact telencephalic target of these projections is not known. At least some of the ascending projections reach the medial pallium (Smeets and Northcutt 1987). Other fibers innervate the area basalis (Ebbesson and Schroeder 1971).

The most prominent output pathway of the telencephalon is the pallial tract. It runs on the dorsal surface of the pallium caudally, turns ventrally (Fig. 11-7) to cross behind the optic chiasm, and mainly terminates in the inferior lobes of the hypothalamus (Smeets and Boord 1985) and in the lateral mesencephalic complex (Hofmann and Northcutt 1995). Cells contributing fibers to this pathway are located in the central nucleus of the dorsal pallium, the medial pallium, and a subpallial area termed SP1 by Smeets and Boord (1985). A number of other subpallial areas (area superficialis basalis, septum, and striatum) project through the forebrain bundles to the hypothalamus, ventral thalamus, and mesencephalic tegmentum (Ebbesson and Schroeder 1971). The majority of these fibers decussate in the anterior commissure. An additional descending fiber bundle is the stria medullaris (see

Fig. 11-7). It is believed to contain fibers from caudal parts of the telencephalon and projects to the habenula.

Medial to the olfactory tracts, which leave the telencephalon in a more lateral position in elasmobranchs, another tract was discovered by Fritsch (1878). It connects the olfactory organ with the septal-preoptic area and is known today as the nervus terminalis. It contains gonadotropin-releasing hormone–immunoreactive cells and fibers and is probably involved in the control of reproductive behavior (Demski 1989). Presumed homologues of this nerve have been found in all vertebrate classes. It appears to include both afferents and efferent components. Action potentials of the latter group have been recorded and are modulated by trigeminal stimulation (Bullock and Northcutt 1984). For discussions of the nervus terminalis, see Demski (1993) and Muske (1993).

PHYSIOLOGY

Several investigations have been made of the physiology of the telencephalon. They confirmed the representation of other modalities besides olfaction. Evoked potentials and sometimes single unit activity in response to visual, electrosensory, and lateral line stimuli have been obtained (Cohen et al. 1973; Vesselkin and Kovacevic 1973; Platt et al. 1974; Kuk'yanov and Nikonorov 1982; Bleckmann and Bullock 1989). The greatest responses were usually found in the medial pallium or central nucleus of the dorsal pallium. An exact determination of the recording site was not always possible and interspecies comparison is difficult because of the lack of detailed anatomical descriptions of the regions involved in many species.

Response properties have been investigated by some of the authors mentioned above. The amplitudes of the evoked potentials are usually small and subject to considerable variation among trials. Habituation is fast. Stimulus repetition rates higher than 2 Hz for vision and 0.2 Hz for mechanoreception lead to a suppression of the response. This is very different from the response properties in the primary sensory cortex of mammals (Bleckmann et al. 1989).

Behavioral studies of telencephalic function are rare. Initial experiments showed no deficits after telencephalic lesions (see Ten Cate 1935). In the 1970s, Graeber and collaborators performed experiments on visual discrimination in the nurse shark *Gingylmostoma* (Graeber et al. 1973; Graeber 1978, 1980). They found that sharks with tectal lesions could still discriminate horizontal from vertical bars to receive food rewards. Animals with lesions in the central nucleus of the pallium, a presumed target of ascending visual fibers, however, decreased in performance. The researchers concluded that the central nucleus is responsible for the discrimination of visual targets. Lesion studies, however, are difficult to interpret. Further studies are needed to determine whether the actual discrimination is performed in the central nucleus or whether a higher-order motivational control center is modulating feeding behavior at another level.

ACKNOWLEDGMENTS

I thank T. H. Bullock and R. G. Northcutt for teaching me so much about the brains of elasmobranchs and for their support during my studies. H. Bleckmann supported me during my writing, and all of them critically reviewed the manuscript. The material presented in the figures was kindly provided by R. G. Northcutt. The writing of this chapter and research in the author's laboratory were supported by a grant from the German Science Foundation (Ho 1602/1-1).

REFERENCES

Alvarez-Otero, R., S. D. Regueira, and R. Anadon. 1993. New structural aspects of the synaptic contacts on Purkinje cells in an elasmobranch cerebellum. *J. Anat.* 182: 13–21.

Bäckström, K. 1924. Contributions to the forebrain morphology in selachians. *Acta Zool.* 5: 123–240.

Barrett, D. J., and E. W. Taylor. 1985a. Spontaneous efferent activity in branches of the vagus nerve controlling heart rate and ventilation in the dogfish. *J. Exp. Biol.* 117: 433–48.

———. 1985b. The location of cardiac vagal preganglionic neurons in the brain stem of the dogfish *Scyliorhinus canicula. J. Exp. Biol.* 117: 449–58.

Barry, M. A. 1987. Central connections of the IXth, Xth, and XIth cranial nerves in the clearnose skate, *Raja eglanteria*. *Brain Res.* 425: 159–66.

Bleckmann, H., and T. H. Bullock. 1989. Central nervous physiology of the lateral line, with special reference to cartilaginous fishes. In *The Mechanosensory Lateral Line. Neurobiology and Evolution*, ed. S. Coombs, P. Görner, and H. Münz, 387–408. New York: Springer-Verlag.

Bleckmann, H., T. H. Bullock, and J. M. Jørgensen. 1987. The lateral line mechanoreceptive, mesencephalic, diencephalic, and telencephalic regions in the thornback ray, *Platyrhinoidis triseriata* (Elasmobranchii). *J. Comp. Physiol.* 161A: 67–84.

Bleckmann, H., O. Weiss, and T. H. Bullock. 1989. Physiology of lateral line mechanoreceptive regions in the elasmobranch brain. *J. Comp. Physiol.* 164A: 459–74.

Bodznick, D. 1991. Elasmobranch vision: multimodal integration in the brain. *J. Exp. Zool. Suppl.* 5: 108–16.

Bodznick, D., and R. L. Boord. 1986. Electroreception in chondrichthyes. In *Electroreception*, ed. T. H. Bullock and W. Heiligenberg, 225–56. New York: John Wiley & Sons.

Bodznick, D., and A. W. Schmidt. 1984. Somatotopy within the medullary electrosensory nucleus of the little skate, *Raja erinacea*. *J. Comp. Neurol.* 225: 581–90.

Boord, R. L., and J. C. Montgomery. 1989. Central mechanosensory lateral line centers and pathways among the elasmobranchs. In The *Mechanosensory Lateral Line. Neurobiology and Evolution*, ed. S. Coombs, P. Görner, and H. Münz, 323–40. New York: Springer-Verlag.

Boord, R. L., and R. G. Northcutt. 1982. Ascending lateral line pathways to the midbrain of the clearnose skate, *Raja eglanteria*. *J. Comp. Neurol.* 207: 274–82.

———. 1988. Medullary and mesencephalic pathways and connections of lateral line neurons of the spiny dogfish, *Squalus acanthias*. *Brain Behav. Evol.* 32: 76–88.

Bruckmoser, P., and N. Dieringer. 1973. Evoked potentials in the primary and secondary olfactory projection areas of the forebrain in elasmobranchia. *J. Comp. Physiol.* 87: 65–74.

Bullock, T. H. 1984. Physiology of the tectum mesencephali in elasmobranchs. In *Comparative Neurology of the Optic Tectum*, ed. H. Vanegas, 47–68. New York: Plenum Press.

Bullock, T. H., and J. T. Corwin. 1979. Acoustic evoked activity in the brain of sharks. *J. Comp. Physiol.* 129: 223–34.

Bullock, T. H., and R. G. Northcutt. 1984. Nervus terminalis in dogfish (*Squalus acanthias*, Elasmobranchii) carries tonic efferent impulses. *Neurosci. Lett.* 44: 155–60.

Bullock, T. H., M. H. Hofmann, F. K. Nahm, J. G. New, and J. C. Prechtl. 1990. Event-related potentials in the retina and optic tectum of fish. *J. Neurophysiol.* 64: 903–14.

Bullock, T. H., M. H. Hofmann, J. G. New, and F. K. Nahm. 1991. Dynamic properties of visual evoked potentials in the tectum of cartilaginous and bony fishes, with neuroethological implications. *J. Exp. Zool. Suppl.* 5: 142–55.

Cohen, D. H., T. A. Duff, and S. O. E. Ebbesson. 1973. Electrophysiological identification of a visual area in shark telencephalon. *Science* 182: 492–94.

Conley, R. A. 1995. Descending input from the vestibulolateral cerebellum suppresses electrosensory responses in the dorsal octavolateralis nucleus of the elasmobranch *Raja erinacea*. *J. Comp. Physiol.* 176A: 325–35.

Conley, R. A., and D. Bodznick. 1994. The cerebellar dorsal granular ridge in an elasmobranch has proprioreceptive and electroreceptive representations and projects homotopically to the medullary electrosensory nucleus. *J. Comp. Physiol.* 174A: 707–21.

Conlon, J. M., K. Yano, D. Waugh, and N. Hazon. 1996. Distribution and molecular forms of urotensin II and its role in cardiovascular regulation in vertebrates. *J. Exp. Zool.* 275: 226–38.

Corwin, J. T., and R. G. Northcutt. 1982. Auditory centers in the elasmobranch brain stem: deoxyglucose autoradiography and evoked potential recording. *Brain Res.* 236: 261–73.

Demski, L. S. 1989. Pathways for GnRH control of elasmobranch reproductive physiology and behavior. *J. Exp. Zool. Suppl.* 2: 4–11.

———. 1993. Terminal nerve complex. *Acta Anat.* 148: 81–95.

Dryer, L. 1994. Projections of the olfactory bulb in an elasmobranch fish, *Sphyrna tiburo*: segregation of inputs in the telencephalon. *Anat. Embryol.* 190: 563–72.

Ebbesson, S. O. E., and C. B. G. Campbell. 1973. On the organization of cerebellar efferent pathways in the nurse shark (*Ginglymostoma cirratum*). *J. Comp. Neurol.* 152: 233–54.

Ebbesson, S. O. E., and L. Heimer. 1970. Projections of the olfactory tract fibers in the nurse shark (*Ginglymostoma cirratum*). *Brain Res.* 17: 47–55.

Ebbesson, S. O. E., and K. C. Hodde. 1981. Ascending spinal systems in the nurse shark, *Ginglymostoma cirratum*. *Cell Tissue Res.* 216: 313–31.

Ebbesson, S. O. E., and D. Schroeder. 1971. Connections of the nurse shark's telencephalon. *Science* 173: 254–56.

Fiebig, E. 1988. Connections of the corpus cerebelli in the thornback guitarfish, *Platyrhinoidis triseriata* (Elasmobranchii): a study with WGA-HRP and extracellular granule cell recording. *J. Comp. Neurol.* 268: 567–83.

Fiebig, E., and H. Bleckmann. 1989. Cell groups afferent to the telencephalon in a cartilaginous fish (*Platyrhinoidis triseriata*). A WGA-HRP study. *Neurosci. Lett.* 105: 57–62.

Finger, T. E. 1978. Gustatory pathways in the bullhead catfish: II. Facial lobe connections. *J. Comp. Neurol.* 180: 691–705.

Fox, G. Q., and G. P. Richardson. 1979. The developmental morphology of *Torpedo marmorata*: electric organ-electrogenic phase. *J. Comp. Neurol.* 185: 293–316.

Fritsch, G. 1878. *Untersuchungen über den feineren Bau des Fischgehirns mit besonderer Berücksichtigung der Homologien bei anderen Wirbeltierklassen.* Berlin: Gutmannscher Verlag.

Gallyas, F. 1979. Silver staining of myelin by means of physical development. *Neurol. Res.* 1: 203–09.

Graeber, R. C. 1978. Behavioral studies correlated with central nervous system integration of vision in sharks. In *Sensory Biology of Sharks, Skates, and Rays*, ed. E. S. Hodgson and R. F. Mathewson, 195–225. Washington, D.C.: U.S. Government Printing Office.

———. 1980. Telencephalic function in elasmobranchs. A behavioral perspective. In *Comparative Neurology of the Telencephalon*, ed. S. O. E. Ebbesson, 17–39. New York: Plenum Press.

Graeber, R. C., S. O. E. Ebbesson, and J. A. Jane. 1973. Visual discrimination in sharks without optic tectum. *Science* 180: 413–15.

Graf, W., and W. J. Brunken. 1984. Elasmobranch oculomotor organization: anatomical and theoretical aspects of the phylogenetic development of vestibulo-oculomotor connectivity. *J. Comp. Neurol.* 227: 569–81.

Grillner, S., C. Perret, and P. Zangger. 1976. Central generation of locomotion in the spinal dogfish. *Brain Res.* 109: 255–69.

Hamasaki, D. I., and P. Streck. 1971. Properties of the epiphysis cerebri of the small-spotted dogfish shark *Scyliorhinus caniculus* L. *Vision Res.* 11: 189–98.

Hayle, T. H. 1973. A comparative study of spinal projections to the brain (except cerebellum) in three classes of poikilothermic vertebrates. *J. Comp. Neurol.* 149: 463–76.

Hofmann, M. H., and T. H. Bullock. 1995. Induced rhythms and apparent expectation in retina and optic tectum of an elasmobranch. *Proc. Joint Symp. Neural Comput.* 5: 115–25.

Hofmann, M. H., and R. G. Northcutt. 1995. Organization of the telencephalon of an elasmobranch fish, the thornback *Platyrhinoidis triseriata*. In *Learning and Memory*, ed. N. Elsner and R. Menzel, 913 (abstract). Stuttgart: Georg Thieme Verlag.

Holmgren, N. 1922. Points of view concerning forebrain morphology in lower vertebrates. *J. Comp. Neurol.* 34: 391–459.

Klimley, A. P. 1981. Grouping behavior in the scalloped hammerhead. *Oceanus* 24: 65–71.

———. 1994. The predatory behavior of the white shark. *Am. Sci.* 82: 122–33.

Kuchnow, K. P. 1971. The elasmobranch pupillary response. *Vision Res.* 11: 1395–406.

Kuk'yanov, A. S., and S. I. Nikonorov. 1982. Bilateral representation of visual afferentation in the forebrain pallium of the dogfish shark *Squalus acanthias*. *J. Evol. Biochem. Physiol.* 19: 486–92.

Lenhossék, M. von 1894. *Zur Kenntnis des Rückenmarkes der Rochen. Beiträge zur Histologie des Nervensystems und der Sinnesorgane.* Wiesbaden: Bergman.

Leonard, R. B., P. Rudomín, and W. D. Willis. 1978. Central effects of volleys in sensory and motor components of peripheral nerve in the stingray *Dasyatis sabina*. *J. Neurophysiol.* 41: 108–25.

Liu, Q., and L. S. Demski. 1993. Clasper control in the round stingray, *Urolophus halleri*: lower sensorimotor pathways. *Environ. Biol. Fish.* 38: 219–30.

Luiten, P. G. M. 1981. Two visual pathways to the telencephalon in the nurse shark (*Ginglymostoma cirratum*): II. Ascending thalamo-telencephalic connections. *J. Comp. Neurol.* 196: 539–48.

Manso, M. J., and R. Anadón. 1991a. The optic tectum of the dogfish *Scyliorhinus canicula* L. A Golgi study. *J. Comp. Neurol.* 307: 335–49.

———. 1991b. Specialized presynaptic dendrites in the stratum cellulare externum of the optic tectum of an elasmobranch *Scyliorhinus canicula* L. *Neurosci. Lett.* 129: 291–93.

McCormick, C. A. 1992. Evolution of central auditory pathways in anamniotes. In *The Evolutionary Biology of Hearing*, ed. D. B. Webster, R. R. Fay, and A. N. Popper, 323–50. New York: Springer-Verlag.

McCormick, C. A., and M. R. Braford, Jr. 1988. Central connections of the octavolateralis system: Evolutionary considerations. In *Sensory Biology of Aquatic Animals*, ed. J. Atema, R. R. Fay, A. N. Popper, and W. N. Tavolga, 733–56. New York: Springer-Verlag.

Meredith, G. E., and W. J. A. J. Smeets. 1987. Immunocytochemical analysis of the dopamine system in the forebrain and midbrain of *Raja radiata*: evidence for a substantia nigra and ventral tegmental area in cartilaginous fish. *J. Comp. Neurol.* 265: 530–48.

Montgomery, J. C., and G. D. Housley. 1983. The abducens nucleus in the carpet shark *Cephaloscyllium isabella*. *J. Comp. Neurol.* 221: 163–68.

Montgomery, J. C., and D. Bodznick. 1994. An adaptive filter that cancels self-induced noise in electrosensory and lateral line mechanosensory systems of fish. *Neurosci. Lett.* 174: 145–48.

Morita, Y., H. Ito, and H. Masai. 1980. Central gustatory paths in the crucian carp *Carassius carassius*. *J. Comp. Neurol.* 191: 119–32.

Muske, L. E. 1993. Evolution of gonadotropin-releasing hormone (GnRH) neuronal systems. *Brain Behav. Evol.* 42: 215–30.

Nakajima, Y. 1970. Fine structure of the medullary command nucleus of the electric organ of the skate. *Tissue Cell* 2: 47–59.

Nelson, D. R. 1990. Telemetry studies of sharks: a review, with applications in resource management. *NOAA Tech. Rep. Natl. Mar. Fish. Serv.* 90: 239–56.

New, J. G., and D. Bodznick. 1990. Medullary electrosensory processing in the little skate: II. Suppression of self-generated electrosensory interference during respiration. *J. Comp. Physiol.* 167A: 295–07.

Nicholson, C., Llinás, R., and Precht, W. 1969. Neural elements of the cerebellum in elasmobranch fishes; structural and functional characteristics. In *Neurobiology of Cerebellar Evolution and Development*, ed. R. Llinás, 215–43. Chicago: American Medical Association.

Northcutt, R. G. 1978. Brain organization in the cartilaginous fishes. In *Sensory Biology of Sharks, Skates, and Rays,* ed. E. S. Hodgson and R. F. Mathewson, 117–93. Washington, D.C.: U.S. Government Printing Office.

———. 1989. Brain variation and phylogenetic trends in elasmobranch fishes. *J. Exp. Zool. Suppl.* 2: 83–100.

———. 1990. Visual pathways in elasmobranchs: organization and phylogenetic implications. *J. Exp. Zool.* 5: 97–107.

———. 1995. The forebrain of gnathostomes: in search for a morphotype. *Brain Behav. Evol.* 46: 275–318.

Northcutt, R. G., and D. Bodznick. 1983. Areas of electrosensory activity in the mesencephalon of the spiny dogfish, *Squalus acanthias*. *Bull. Mt. Desert Isl. Biol. Lab.* 23: 33–36.

Northcutt, R. G., and J. C. Wathey. 1980. Guitarfish possess ipsilateral as well as contralateral retinofugal projections. *Neurosci. Lett.* 20: 237–42.

Northcutt, R. G., A. J. Reiner, and H. J. Karten. 1988. Immunohistochemical study of the telencephalon of the spiny dogfish, *Squalus acanthias*. *J. Comp. Neurol.* 277: 250–67.

Parent, A., D. Poitras, and L. Dubé. 1984. Comparative anatomy of central monoaminergic systems. In *Handbook of Chemical Neuroanatomy. Classical Neurotransmitters in the CNS*, vol. 2, pt. 1, ed. A. Björklund and T. Hökfeld, 409–39. Amsterdam: Elsevier.

Paul, D. H. 1982. The cerebellum of fishes: a comparative neurophysiological and neuroanatomical review. *Adv. Comp. Physiol. Biochem.* 8: 111–77.

Paul, D. H., and B. L. Roberts. 1979. The significance of cerebellar function for a reflex movement of the dogfish. *J. Comp. Physiol.* 134: 69–74.

———. 1984. Projections of cerebellar Purkinje cells in the dogfish, *Scyliorhinus*. *Neurosci. Lett.* 44: 43–6.

Platt, C. J., T. H. Bullock, G. Czéh, N. Kovačević, D. Konjević, and M. Gojković. 1974. Comparison of electroreceptor, mechanoreceptor, and optic evoked potentials in the brain of some rays and sharks. *J. Comp. Physiol.* 95: 323–55.

Puelles, L., J. A. Amad, and M. Martínez-de-la-Torre. 1987. Segment-related, mosaic neurogenetic pattern in the forebrain and mesencephalon of early chick embryos: I. Topography of AChE-positive neuroblasts up to stage HH18. *J. Comp. Neurol.* 266: 247–68.

Puzdrowski, R. L., and R. B. Leonard. 1993. The octavolateral system in the stingray, *Dasyatis sabina*: I. Primary projections of the octaval and lateral line nerves. *J. Comp. Neurol.* 332: 21–37.

———. 1994. Vestibulo-oculomotor connections in an elasmobranch fish, the Atlantic stingray, *Dasyatis sabina*. *J. Comp. Neurol.* 339: 587–97.

Rendahl, H. 1924. Embryologische und morphologische Studien über das Zwischenhirn beim Huhn. *Acta Zool.* 5: 241–344.

Repérant, J., D. Miceli, J. P. Rio, J. Pierre, J. Peyrichoux, and E. Kirpitchnikova. 1986. The anatomical organization of retinal projections in the shark *Scyliorhinus canicula* with special reference to the evolution of the selachian primary visual system. *Brain Res. Rev.* 11: 227–48.

Repérant, J., D. Miceli, N. P. Vesselkin, and S. Molotchnikoff. 1989. The centrifugal visual system of vertebrates: a century-old search reviewed. *Int. Rev. Cytol.* 118: 115–71.

Ridge, R. M. A. P. 1977. Physiological responses of stretch receptors in the pectoral fin of the ray *Raja clavata*. *J. Mar. Biol. Assoc. U.K.* 57: 535–41.

Ritchie, T. C., C. A. Livingston, M. G. Hughes, D. J. McAdoo, and R. B. Leonard. 1983. The distribution of serotonin in the CNS of an elasmobranch fish: immunocytochemical and biochemical studies in the Atlantic stingray, *Dasyatis sabina*. *J. Comp. Neurol.* 221: 429–43.

Roberts, B. L. 1978. Mechanoreceptors and the behavior of elasmobranch fishes with special reference to the acoustico-lateralis system. In *Sensory Biology of Sharks, Skates, and Rays,* ed. E. S. Hodgson and R. F. Mathewson, 331–90. Washington, D.C.: U.S. Government Printing Office.

———. 1988. The central nervous system. In *Physiology of Elasmobranch Fishes*, ed. T. J. Shuttleworth, 49–78. Berlin: Springer-Verlag.

Rodriguez-Moldes, I., M. J. Manso, M. Becerra, P. Molist, and R. Anadon. 1993. Distribution of substance P–like immunoreactivity in the brain of the elasmobranch *Scyliorhinus canicula*. *J. Comp. Neurol.* 335: 228–44.

Rosiles, J. R., and R. B. Leonard. 1980. The organization of the extraocular motor nuclei in the Atlantic stingray, *Dasyatis sabina. J. Comp. Neurol.* 193: 677–87.

Rüdeberg, C. 1969. Light and electron microscopic studies on the pineal organ of the dogfish *Scyliorhinus canicula* (L.). *Z. Zellforsch.* 96: 548–81.

Rüdeberg, S. I. 1961. Morphogenetic studies on the cerebellar nuclei and their homologization in different vertebrates including man. Ph.D. thesis, University of Lund, Sweden.

Schmidt, A., and D. Bodznick. 1987. Afferent and efferent connections of the vestibulolateral cerebellum of the little skate, *Raja erinacea. Brain Behav. Evol.* 30: 282–02.

Schroeder, D. M., and S. O. E. Ebbesson. 1974. Nonolfactory telencephalic afferents in the nurse shark (*Ginglymostoma cirratum*). *Brain Behav. Evol.* 9: 121–55.

Schweitzer, J. 1983. The physiological and anatomical localization of two electroreceptive diencephalic nuclei in the thornback ray, *Platyrhinoidis triseriata. J. Comp. Physiol.* 153A: 331–41.

Smeets, W. J. A. J. 1982. The afferent connections of the tectum mesencephali in two chondrichthyans, the shark *Scyliorhinus canicula* and the ray *Raja clavata. J. Comp. Neurol.* 205: 139–52.

———. 1983. The secondary olfactory connections in two chondrichthians, the shark *Scyliorhinus canicula* and the ray *Raja clavata. J. Comp. Neurol.* 218: 334–44.

———. 1990. The telencephalon of cartilaginous fishes. In *Cerebral Cortex*, vol. 8A: *Comparative Structure and Evolution of Cerebral Cortex*, pt. 1, ed. E. G. Jones and A. Peters, 3–30. New York: Plenum Press.

Smeets, W. J. A. J., and R. L. Boord. 1985. Connections of the lobus inferior hypothalami of the clearnose skate, *Raja eglanteria* (Chondrichthyes). *J. Comp. Neurol.* 234: 380–92.

Smeets, W. J. A. J., and R. Nieuwenhuys. 1976. Topological analysis of the brain stem of the sharks *Squalus acanthias* and *Scyliorhinus canicula. J. Comp. Neurol.* 165: 333–68.

Smeets, W. J. A. J., R. Nieuwenhuys, and B. L. Roberts. 1983. The *Central Nervous System of Cartilaginous Fishes. Structure and Functional Correlations.* Berlin: Springer-Verlag.

Smeets, W. J. A. J., and R. G. Northcutt. 1987. At least one thalamotelencephalic pathway in cartilaginous fishes projects to the medial pallium. *Neurosci. Lett.* 78: 277–82.

Smeets, W. J. A. J., and S. J. B. Timerick. 1981. Cells of origin of pathways descending to the spinal cord in two chondrichthyans, the shark *Scyliorhinus canicula* and the ray *Raja clavata. J. Comp. Neurol.* 202: 473–91.

Snow, P. J., M. B. Plenderleith, and L. L. Wright. 1993. Quantitative study of primary sensory neuron populations of three species of elasmobranch fish. *J. Comp. Neurol.* 334: 97–103.

Stuesse, S. L., W. L. R. Cruce, and R. G. Northcutt. 1990. Distribution of tyrosine hydroxylase- and serotonin-immunoreactive cells in the central nervous system of the thornback guitarfish, *Platyrhinoidis triseriata. J. Chem. Neuroanat.* 3: 45–58.

———. 1991a. Localization of serotonin, tyrosine hydroxylase, and leu-enkephalin immunoreactive cells in the brainstem of the horn shark, *Heterodontus francisci. J. Comp. Neurol.* 308: 277–92.

———. 1991b. Serotoninergic and enkephalinergic cell groups in the reticular formation of the bat ray and two skates. *Brain Behav. Evol.* 38: 39–52.

Stuesse, S. L., D. C. Stuesse, and W. L. R. Cruce. 1992. Immunohistochemical localization of serotonin, leu-enkephalin, tyrosine hydroxylase, and substance P within the visceral sensory area of cartilaginous fish. *Cell Tissue Res.* 268: 305–16.

Ten Cate, J. 1935. Physiologie des Zentralnervensystems der Fische. *Ergeb. Biol.* 11: 335–409.

Tong, S., and T. H. Bullock. 1982. The sensory functions of the cerebellum of the thornback ray, *Platyrhinoidis triseriata. J. Comp. Physiol.* 148: 399–410.

Uchiyama, H. 1989. Centrifugal pathways to the retina: influence of the optic tectum. *Vis. Neurosci.* 3: 183–206.

Vallarino, M., J. M. Danger, A. Fasolo, G. Pelletier, S. Saint-Pierre, and H. Vaudry. 1988. Distribution and characterization of neuropeptide Y in the brain of an elasmobranch fish. *Brain Res.* 448: 67–76.

Vallarino, M., M. T. Salsotto-Cattaneo, M. Feuilloley, and H. Vaudry. 1991. Distribution of FMRF amide–like immunoreactivity in the brain of the elasmobranch fish *Scyliorhinus canicula. Peptides* 12: 1321–328.

Vallarino, M., Bucharles, C., Facchinetti, F., and Vaudry, H. 1994. Immunocytochemical evidence for the presence of met-enkephalin and leu-enkephalin in distinct neurons in the brain of the elasmobranch fish *Scyliorhinus canicula. J. Comp. Neurol.* 347: 585–97.

Vesselkin, N. P., and N. Kovacevic. 1973. Nonolfactory afferent projections of the telencephalon of elasmobranchii. *J. Evol. Biochem. Physiol.* 9: 512–18.

Wilson, J. F., H. J. T. Goos, and J. M. Dodd. 1974. An investigation of the neural mechanisms controlling the colour range responses of the dogfish *Scyliorhinus canicula* L. by mesencephalic and diencephalic lesions. *Proc. R. Soc. Lond.* 187B: 171–90.

Withington-Wray, D. J., B. L. Roberts, and E. W. Taylor. 1986. The topographical organization of the vagal motor column in the elasmobranch fish,

Scyliorhinus canicula L. *J. Comp. Neurol.* 248: 95–104.

Wright, D. E., and L. S. Demski. 1991. Gonadotropin-releasing hormone (GnRH) immunoreactivity in the mesencephalon of sharks and rays. *J. Comp. Neurol.* 307: 49–56.

———. 1993. Gonadotropin-releasing hormone (GnRH) pathways and reproductive control in elasmobranchs. Environ. Biol. Fish. 38: 209–18.

Young, J. Z. 1933. Comparative studies on the physiology of the iris: I. Selachians. *Proc. R. Soc. Lond.* 112B: 228–41.

HORST BLECKMANN
MICHAEL H. HOFMANN

CHAPTER 12

Special Senses

Most species of sharks, skates, and rays recognized today are restricted to the marine habitat. However, some selachians enter freshwater and a few, like the stingray *Paratrygon motoro*, are even restricted to freshwater. Within their aquatic world cartilaginous fishes roam the deepest part of the oceans, edges of the continental slopes, and shallow coastal waters. Because selachians live in a variety of aquatic environments and have different feeding strategies, it is not possible to generalize about their sensory systems. However, while silently roaming the ocean these predatory animals constantly gather information about their aquatic environment through many sensory channels including the olfactory and gustatory systems, the eyes and ears. In addition to these major senses, which are also well known from other vertebrate species, cartilaginous fishes are equipped with special receptors that give them the ability to detect minute water displacements and very weak low-frequency electric fields. Our knowledge regarding the sensory, behavioral, and neural capabilities of elasmobranch fishes is still sparse, but the information we have indicates that sharks, skates, and rays are not small-brained animals that depend mostly on olfaction but powerful and effective predators equipped with sensory systems and brains that in many aspects equal those of higher vertebrates.

Olfactory and gustatory systems

Elasmobranchs have a particularly acute sense of olfaction and taste (Hodgson and Mathewson 1978; Kleerekoper 1978). Consequently, smell and taste play a major role in mediating their behavioral responses. Olfactory stimuli include biochemical products released by aquatic organisms, some of which may reveal the presence and location of food items, mates, or other conspecifics. Like bony fish, elasmobranchs detect olfactory stimuli through at least two different channels of chemoreception: olfaction (smell) and gustation (taste). Gustation, like olfaction, also plays an important part in enabling cartilaginous fishes to discriminate between food and nonfood objects.

ANATOMY
Olfactory system

The peripheral olfactory organs of selachians show considerable diversity in position, size, and form. However, in all selachians the nostrils are paired, encapsulated structures, which lie lateral on the ventral side of the head between the tip of the snout and the mouth (Fig. 12-1A). From the outside the olfactory organs are reached by the nasal aperture, which is separated by an overlapping median flap into an upper and a lower opening through which water flows. If the narial flaps

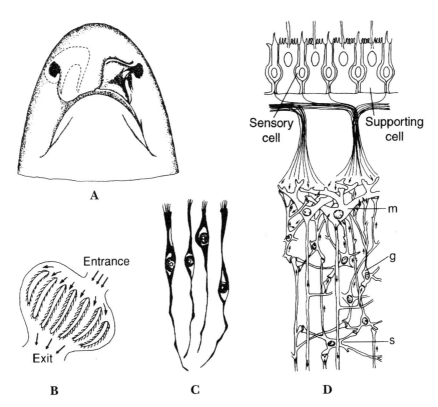

FIGURE 12-1   Anatomy of the olfactory system of selachians. **A:** Ventral view of the head of the shark *Scyliorhinus*. On the left side of the animal the nasal flap has been removed so that the entrance and exit of the olfactory groove becomes visible. **B:** The principal of water flow in the olfactory organ of the dogfish is shown. Leaves of tissue divide the olfactory chamber into a series of corridors on which the sensory cells are located. Water currents (*arrows*) are directed through the olfactory chamber with aid of cilia found on supporting cells. **C:** Examples of olfactory sensory cells found in the shark *Mustelus laevis*. **D:** Diagram of neural circuitry of the olfactory epithelium and bulb of the shark *Scyliorhinus*. The dendrites of the bipolar sensory cells (s) extend into the lumen of the olfactory chamber. Their axon links with other axons to synapse with the dendrites of the mitral cells (m) of the olfactory bulb. Also shown are granular cells (g) (which are local inhibitory cells) and supporting cells, which are interspersed between the receptor cells of the sensory epithelium. (A, D, redrawn from Kleerekoper 1978; C, redrawn from Asai 1913.)

are removed, an elliptical cup appears and platelike lamellae, which may be arranged in a series of folds that radiate outward from a central hub (see Fig. 12-1A,B). In most selachians the olfactory organs are entirely separated from the mouth; however, in many species they are either in close association with the mouth or connected to it by a groove, allowing the irrigation of the olfactory epithelium by means of the buccopharyngeal pump (Bell 1993).

Located in the convoluted lining of the olfactory membrane are the primary olfactory receptor cells, which have large somata (15–20 mm) and typically a peripheral dendritic process with cilia or microvilli projecting into the lumen of the olfactory cup (see Fig. 12-1C,D). These cells also send fibers posteriorly to the olfactory bulb, which is directly adjacent to the olfactory organ (see Fig. 12-1D). The projection from the olfactory epithelium to the olfactory bulb is regionally topographic (Dryer and Graziadei 1993), a condition that is found also in tetrapods (Ressler et al. 1994), but not in other fishes (Baier et al. 1994). The olfactory bulb is connected to the telencephalon by a long olfactory tract. The topographic projection from epithelium to the bulb is to some degree preserved in this tract (Dryer and Graziadei 1994).

Gustatory system

The receptors of the gustatory system are the taste buds or terminal buds (Fig. 12-2). In elasmobranchs taste buds are associated with papillae that occur in the epithelial lining of the mouth and pharynx. The structure of taste buds is similar to that described for other vertebrates. They occur upon small papillae that project slightly above the surrounding epithelium. At the apex of the papilla, epithelial cells are modified into sensory cells, which form barrel- or flask-shaped clusters (see Fig. 12-2). Each sensory cell terminates in a short, hairlike projection. The taste buds of cartilaginous fishes are innervated by fibers of the facial (VII), glossopharyngeal (IX), and vagus (X) nerves.

PERIPHERAL PHYSIOLOGY

The sensitivity and selectivity of the olfactory system of elasmobranchs has been investigated with the aid of compound electrical responses (electroolfactograms) and single unit recordings (Hodgson and Mathewson 1978; Silver 1979; Broun and Fesenko 1982; Nikonov et al. 1990). Single unit recordings show that olfactory fibers respond very sensitively to some amino acids (e.g., the threshold to serine is 10^{-14} to 10^{-13} mol/L). The physiological thresholds to other amino acids may be considerably higher (e.g., 10^{-10} mol/L for asparagine and 10^{-8} mol/L for lysine and valine). In general, different olfactory fibers may have different thresholds to various amino acids. This selectivity indicates that the receptors for different amino acids are probably located on different sensory cells and that the specificity of

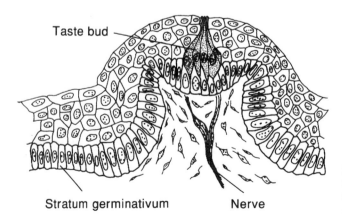

FIGURE 12-2 Section of a taste papilla and bud of spiny dogfish, *Squalus acanthias* (Squalidae). (Redrawn from Tester 1963.)

information provided by the receptors is not lost in the processing of information by the olfactory lobe.

BEHAVIOR

The keen sense of olfaction enables elasmobranchs to detect and locate food (see Kleerekoper 1978 for a review). In addition, olfaction may be implicated in reproductive behavior. Field studies show that sharks usually approach the source of olfactory stimuli from a downstream direction. Some sharks show a true gradient searching, or klinotaxis, in that they approach the source of an olfactory stimulus along an S-shaped track. Obviously the sharks turn in the direction of the nostril that receives the strongest olfactory stimulus. This can be deduced from experiments in which one nostril was occluded. In this case, the sharks turn persistently toward the open nostril when activated by an olfactory stimulus (Parker 1914; Kleerekoper 1978). Thus directional swimming results from the ability of the shark to detect small differences in the concentration of odorous materials in each nostril and from the ability to head upstream in the direction of equal stimulation.

Some sharks (e.g., the lemon shark) do not follow an olfactory gradient directly. On encountering a stimulus, they simply turn and swim upstream. Lemon sharks do not localize the source of an olfactory stimulus in open sea tests but continue to swim against the current past the stimulus source. Apparently they need additional visual, mechanosensory, or electrosensory cues for the final localization and identification of potential food items (Hodgson and Mathewson 1978). In the smooth dogfish, dilute quinine hydrochloride produces behavioral responses (jerks and other movements) when applied to the inside of the mouth and spiracle, but not when applied to other parts of the body (Sheldon 1909). Evidently, this bitter-tasting substance elicits a response from taste buds. Whether selachians can distinguish the four primary tastes—bitter, salty, sweet, and sour—is not known. Apart from Sheldon's experiments, little or no work has been done on gustation in selachians.

The nervus terminalis is distinct from, and morphologically rostral to, the olfactory nerve. In cartilaginous fishes this nerve may be part of a pheromone receptor system (Demski and Northcutt 1983), but thus far experimental evidence for this hypothesis is missing (Bullock and Northcutt 1984). For a recent review, see Demski (1993).

CENTRAL PATHWAYS
AND CENTRAL PHYSIOLOGY

Not much is known about the central processing of olfactory information in cartilaginous fishes. In the shark *Scyliorhinus canicula*, secondary olfactory potentials can be elicited after electrical stimulation of the nasal mucosa or the olfactory bulb. In addition, olfactory potentials of higher order can be evoked outside the primary olfactory area of the forebrain (Bruckmoser and Dieringer 1973). In the medial pallium of the shark *Squalus acanthias*, single units have been found that respond to both optic and olfactory nerve stimulation (Nikonorov 1983). Higher-order olfactory potentials show a strong fatigue with high-stimulus repetition rates and disappear completely at a repetition rate of 0.3 Hz. They are restricted to the caudal part of the ipsilateral forebrain. At low repetition rates late potentials with latencies up to 800 msec and regular electroencepghalogram (EEG) synchronous after-potentials may occur (Bruckmoser and Dieringer 1973).

Visual system

The visual sense of selachians has been consistently denigrated in favor of the olfactory modality. This is surprising since many cartilaginous fishes are littoral or epipelagic and clearly depend on vision as a source of sensory information. The various habitats of selachians may differ in the intensity of the light pouring down from the surface or in the spectral distribution of the light with depth (McFarland 1991). Most selachians have small but well-developed eyes (for reviews see Gilbert 1963; Cohen 1989), which enable them to see over a range of 10 log units of light intensities from very dim to very bright light (for a review see Gruber and Cohen 1978).

ANATOMY

The eyes of elasmobranchs are situated laterally in the head. They are usually so far separated from each other that binocular vision is probably not of importance. The exposed portion of the eyeballs projects outward only slightly beyond the skin and, at least in sharks, conforms to the streamlined surface of the head. Selachians possess well-developed upper and lower eyelids, which cannot completely cover the eyeball. For a full protection of the eye from mechanical stimuli, some sharks possess a third eyelid, or nictitating membrane, which can entirely cover the exposed portion of the eye (Gilbert 1963).

The anatomical organization of the elasmobranch eye is fairly typical for vertebrates, but some distinct features emerge (Fig. 12-3). The external shape of the eyeball is determined by its tough outer covering, the sclera. The fibrous portion of the sclera is continuous outwardly with the transparent cornea, which—at least in sharks—is usually oval-shaped with a long cephalo-caudal axis. The sclera is supported by a thick cartilaginous layer and the nutritive choroid, which contains a tapetum lucidum. In pelagic species the choroidal tapetum, which consists of a series of parallel, reflecting, platelike cells containing crystals of guanine, can be occluded in bright light by the migration of dark melanin granules over the reflecting plates (Heath 1991). The crystalline lens of the selachian eye is inelastic and spherical in shape. Its corneo-retinal diameter is usually slightly less than its vertical or horizontal diameter (Sivak 1991; Zigman 1991). The lens is supported by a dorsal suspensory ligament and a ventral pseudocampanule that, in some species, function as a protractor lentis muscle to produce accommodatory movements of the lens (Sivak 1991). The iris of the eye is part of the choroid layer. It bulges outward from the ciliary body over the lens as a low, truncated cone with an aperture at its apex, the pupil. Two groups of contractile fibers, one arranged radially and the other circularly, regulate the size of the pupil and thus the amount of light that reaches the retina. This is similar to higher vertebrates but in contrast to teleosts, which have generally immobile pupils (Sivak 1990). In

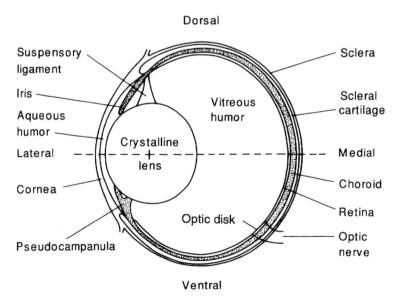

FIGURE 12-3 Diagramatic vertical section through the eye of a shark. (Redrawn from Montgomery 1988.)

selachians the pupillary aperture varies from circular to oval in shape when wide open. In bright light the pupil may only be a tiny circle or a mere slit, which may be vertical or horizontal. The reduction of a single large pupillary aperture to a series of tiny openings is carried to an extreme in many skates, in which the dorsal iris forms an operculum that has a scalloped border (Fig. 12-4) (Murphy and Howland 1991). The retina of elasmobranchs usually has rod and cone photoreceptors (Gruber and Cohen 1978, 1985; Gruber et al. 1991), but in the skate, only one type of photoreceptor was found (Szamier and Ripps 1983; Ripps and Dowling 1991). This photoreceptor, however, is remarkably adaptive. Under dim light conditions, it acts like a rod, while in bright light, its response resembles more the cone system of other animals (Dowling and Ripps 1991). Most elasmobranchs, however, have a duplex retina. In diurnal species the rod:cone ratio may be as low as 6:1, whereas in dim-light species a rod:cone ratio of 100:1 has been observed (Gruber et al. 1991). The duplex retina most likely serves to extend the dynamic range of vision from nocturnal to diurnal levels of ambient light, which can range over 11 log units. The selachian retina is organized into an outer and inner nuclear layer as well as a ganglion cell layer and into two plexiform or synaptic layers, also outer and inner (Cohen 1989).

The retina of some sharks contains a prominent visual streak, a horizontal band of higher cell density, in both cone and ganglion cell layers (Hueter 1991). Cone density may range from 6500 cones/mm^2 along the horizontal meridian to less than 500 cones/mm^2 in the dorsal and ventral periphery. Ganglion cell density ranges from nearly 1600 cells/mm^2 inside the streak to less than 500 cells/mm^2 in the periphery (Hueter 1991). This implies

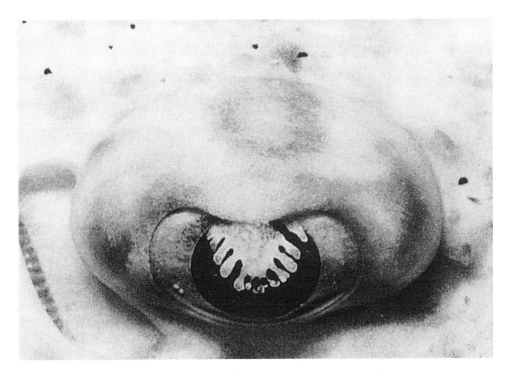

FIGURE 12-4 Crescent-shaped pupil of the clearnose skate (*Raja eglanteria*). The operculum of this skate has many fine lobes, which are associated with its axial margin. If maximally constricted, the serrated margin of the operculum overrides the ventral pupillary border, creating multiple pupillary apertures that are arranged in a crescent. (From Murphy and Howland 1991.)

that the shark's visual acuity is greater along the horizontal meridian than in the dorsal and ventral periphery. In the lemon shark the minimum visual angle is 4.5°. Spatial heterogeneity in the retina is not restricted to the lemon shark. In several other shark species there is also a higher cone concentration along the horizontal meridian (Franz 1931; Peterson and Rowe 1980) or in the central part of the retina (Hamasaki and Gruber 1965; Yew et al. 1984; Gruber and Cohen 1985).

PERIPHERAL PHYSIOLOGY

The outer segments of the visual receptor cells contain pigments that absorb light as the first step of the sensory transduction process. The visual pigments of selachians are based on either vitamin A_1 or vitamin A_2 (Cohen 1991). In species that are most active in dim light, the visual pigments within the rods are centered between 497 nm (e.g., in the shark *Mustelus californicus*) and 510 nm (e.g., in the ray *Raja clavata*); for a listing see Gruber and Cohen (1978) and Cohen (1991). In deepwater elasmobranchs the rod pigments may be more blue-sensitive, that is, these pigments have an absorbance maximum of about 480 nm (Crescitelli 1991; Cohen 1991).

The visual information processing by the retina in elasmobranchs is similar to that in other vertebrates (Ashmore and Falk 1980a,b, 1981, 1982). The horizontal cells respond to light with a graded hyperpolarization and a sensitivity about four times that of rod receptors. In contrast, bipolar cells show a graded depolarization to light with a sensitivity that is about 135 times that of the rod receptors. Ganglion cells represent the output of the retina. They respond to changes in illumination with action potentials. The receptive fields of horizontal cells cover up to 10 mm during to the lateral extent of their dendrites and electrotonic junctions with neighboring horizontal cells. Bipolar cells have a relatively small receptive field of about 0.15 mm. The receptive field of ganglion cells is organized into a central region and a concentric surround. Ganglion cells may show an increase in their firing rate when light is shone on the central region of their receptive field, whereas light shone on the surrounding area inhibits the background firing rate. Other ganglion cells may have a reverse response pattern.

BEHAVIOR

For predators such as the great white shark, *Carcharodon carcharias*, silhouetting prey against the daylight on the surface is a common mode of feeding (Gruber and Cohen 1985). Whether elasmobranchs are sensitive to UV light like some teleosts (e.g., see Hawryshyn 1991) or have the ability to sense the polarization of light (Waterman and Hashimoto 1974) is not known.

There are a number of studies in which the visual abilities of selachians have been investigated. Clark (1963) trained adult lemon sharks (*Negaprion brevirostris*) to press a visual target to obtain food. Similar studies have been done with bull sharks, *Carcharhinus leucas* (Wright and Jackson 1964), juvenile Pacific reef blacktip (*C. melanopterus*), and gray reef (*C. menisorrah = amblyrhynchos*) sharks, and with juvenile nurse sharks (*Ginglymostoma cirratum*) (Aronson et al. 1967). Graeber (1978) trained juvenile nurse sharks to discriminate between vertically and horizontally striped targets in a y-maze. Each target consisted of three black and three white 5-cm-wide stripes.

CENTRAL PATHWAYS
AND CENTRAL PHYSIOLOGY

Recent anatomical studies indicate that elasmobranchs possess up to ten primary retinofugal targets in addition to the optic tectum. Elasmobranchs have two visual thalamic nuclei, one of which receives tectal input and the other retinal and tectal input. The visual thalamic nuclei project to various regions of the telencephalon (Northcutt 1991). Thus, with respect to the visual system, the central organization of elasmobranch fishes appears to be similar to that of other gnathostomes.

Hueter (1991) mapped tectal visual receptive fields (RF) in the horn shark with aid of black disks. Receptive field size of units ranged between 5° and 10°. This agrees well with the 5.2°–8.8° receptive field size of ganglion cells of the lemon shark. The projection pattern in the lemon shark is from dorsal visual field to medial tectum, ventral visual field to lateral tectum, rostral visual field to rostral tectum, and caudal visual field to caudal tectum. This pattern is typical for vertebrates. The mapping experiments show that three times more tectal surface is devoted to vision inside the streak than to peripheral vision in the lemon shark. Receptive fields of peripheral units are large, as much as 10° or more in diameter, and irregular in shape. Streak unit receptive fields tend to be smaller, elliptical in shape, and horizontally aligned, with vertical diameters of 7°–10°. Similar elliptical fields projecting from the central visual field to the tectum of the little skate (*Raja erinacea*) have been described by Bodznick (1991). *R. erinacea* also has a retinotectally magnified visual streak.

Bullock et al. (1991) studied the properties of visual evoked potentials in the tectum opticum of 35 species of fish, including 12 species of elasmobranchs. The latency to the first peak is usually longer in rays than in sharks. Elasmobranchs, in general, have longer latencies than teleosts but are faster than agnathans.

Not much is known about visual information processing in diencephalic and telencephalic brain centers of cartilaginous fishes. Some units of the diencephalic posterior central thalamic nucleus of the thornback guitarfish, *Platyrhinoidis triseriata*, respond only if the animal receives a corresponding visual and hydrodynamic stimulus (Bleckmann et al. 1987). In the medial pallium of the telencephalon of the little skate, *R. erinacea*, and the guitarfish, *P. triseriata*, visual evoked potentials have been recorded (Bodznick and Northcutt 1984; Bleckmann et al. 1987). In *P. triseriata* a diffuse high-intensity light flash caused peak latencies of 68–72 msec, and the visual responses did not survive a repetition rate of 2 Hz. In the shark *Ginglymostoma cirratum*, optic nerve stimulation evokes short-latency telencephalic field potentials localized to the ipsilateral dorsal pallium pars centralis (or central nucleus). In this region, field potentials have latencies of 20–23 msec at stimulating currents twice threshold (Cohen et al. 1973).

In selachians, visual information also reaches the cerebellum (Tong and Bullock 1982). Receptive fields of visual units recorded from the cerebellum of *P. triseriata* are quite large, varying from 25° to 45°. The visual units respond to a light flash with up to six bursts of spikes. This pattern is reduced at repetition rates greater than 1/sec; above about 4/sec, units tend to fire irregularly (Tong and Bullock 1982). The threshold of cerebellar visual units to objects moving back and forth in a 90° arc with a radius of 50 mm around the eye in the horizontal plane was 0.2°–0.5°/sec. All units increased their response as the velocity of the object increased. Most cerebellar visual units were directionally sensitive (Tong and Bullock 1982).

Mechanosensory systems

Any moving, turning, or struggling animal generates both pressure waves and water displacements. Therefore, water displacements and pressure (sound) waves provide a potential source of biologically relevant information in the aquatic world.

Many biologically relevant sound sources have a strong dipole component. Sound pressure caused by a dipole falls off as one over the distance, that is to say, the pressure component of a dipole-caused sound wave covers a large area. For this reason, sound pressure detectors are well suited for informing an animal about distant acoustic events. In contrast to sound pressure, the local flow of a dipole sound source falls off as one over the distance cubed (Kalmijn 1988a). Therefore, water movements caused by a vibrating object can be detected by other aquatic organisms only if a wave source is close (centimeter to decimeter range).

Elasmobranch fish have at least three types of mechanosensory systems that enable them to sense acoustic (hydrodynamic) events. These are the inner ear, the lateral line, and the general cutaneous sense. Selachians, like bony fish, are nearly neutrally buoyant. For this reason they tend to move with the water mass. These absolute movements can be detected with otolith inner ear mechanoreceptors. In contrast, the lateral line detects only the spatial derivative of any imposed local flow, that is, the net movement between the animal and the surrounding water. If these net movements are sufficiently strong, the general cutaneous sense may also be stimulated.

The sensory cells of the inner ear and the lateral line are hair cells. At the top of each hair cell is a bundle of cilia, one of which is a single membrane–bound kinocilium with a true nine-plus-two ciliary pattern (nine peripheral doublets surrounding a central pair of simple microtubules). The rest of the cilia lack any internal structure and are termed stereovilli. The stereovilli of the hair cells of the vestibular system are graded in length, with the tallest border of them being closest to the kinocilium. In contrast, the stereovilli of the lateral line hair cells are all of a similar length. These conditions are unlike those found in the neuromasts of teleosts (for reviews see Bleckmann 1993, 1994). In any case, hair cells display a morphological polarization that has a physiological correlate. The adequate stimulus of a hair cell is a mechanical shearing force applied to the distal end of the hair bundle. Movements of the cilia alter the electrical potential within the hair cell, which in turn modulates the release of neurotransmitter at the afferent synapse. An individual hair cell is connected to the central nervous system by afferent nerve fibers. In addition, hair cells may be contacted by efferent nerve fibers whose cell bodies are within the brain. Hair cells are similar in structure and function throughout the various octavolateralis systems. It is the biomechanics of the auxiliary structure that determine whether the effective stimulus is head rotation, gravity, head vibration, or external water movements.

Vestibular and acoustic system

Like the olfactory and visual senses, the acoustic system of cartilaginous fishes acts as a teloreceptor to detect stimuli from some distance (for a review see Backus 1963). Many species of elasmobranchs are acutely sensitive to low-frequency underwater sounds. Evidence for this comes from behavioral experiments that show that sharks are attracted by sounds resembling those produced by struggling wounded fish (Kritzler and Wood 1961; Dijkgraaf 1963; Nelson and Gruber 1963; Nelson and Johnson 1972; Kelly and Nelson 1975; Myrberg 1978). Sharks are not only aroused by these sounds but are able to determine the direction of a sound source, even from distances as great as 250 m (Nelson 1967; Myrberg 1978). Since they lack any known pressure-to-displacement transducers, elasmobranch fish most likely sense only the displacement component of a sound wave (Banner 1967; Myrberg et al. 1972; Kelly and Nelson 1975; Corwin 1981a).

ANATOMY

The ears of elasmobranchs are located in cartilage walled otic capsules just lateral to the braincase in the posterior portion of the chondocranium. They are composed of three semicircular canals and the otolith organs: utriculus, sacculus, and lagena. A cochlea or cochlea rudiment is absent. The three semicircular canals, which are filled with labyrinthine fluid, are arranged at mutual right angles: two canals lie in a vertical plane and one in a horizontal one, provided the head is in a normal position. At one end, each canal widens into a spherical ampulla that harbors an epithelium made up of sensory hair cells and supporting cells (Fig. 12-5). The cilia of the hair cells project into the lumen of the ampulla where they are ensheathed in a jellylike cupula, which is displaced by movement of the endolymph within the canal. The three otolith organs are designed on the statocyst principle. Each organ is equipped with a hair cell macula, covered by a cupula in which are embedded calcium carbonate granules, or otoconia (Barber and Emerson 1980). These otoconia are specifically heavier than the surrounding endolymph, and the otolithic mass will therefore "seek" the lowest possible level, thus stimulating the underlying hair cells. In

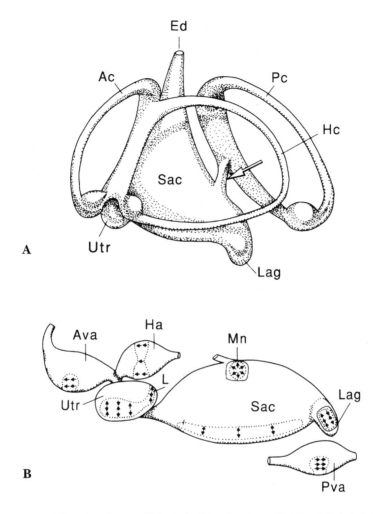

FIGURE 12-5 **A:** Drawing of exposed labyrinth of the electric ray, *Torpedo nobiliana*, in lateral view. *Arrow* points out the posterior canal duct. Ac = anterior semicircular canal; Ed = endolymphatic duct; Hc = horizontal semicircular canal; Lag = Lagena; Pc = posterior semicircular canal; Sac = sacculus; Utr = utriculus. **B:** Diagrammatic representation of the polarity of sensory hair bundles found in the cristae and maculae of the left labyrinth of the ray, *Raja clavata*. Part of the dorsal wall of the sacculus above the macula neglecta and of the posterior wall of the lagena has been cut away to show their two sensory areas. In the schematic drawing the orientation of the hair cells is symbolized by an *arrow*, the *arrowheads* indicating the position of the kinocilium. Ava = anterior vertical ampulla; Ha = horizontal ampulla; Lag = lagena; Mn = macula neglecta; Pva = posterior vertical ampulla; Sac = sacculus; Utr = utriculus. (A, redrawn from Corwin 1978; B, redrawn from Lowenstein et al. 1964.)

addition to the otolith organs, cartilaginous fishes possess a nonotolith organ, the macula neglecta (see Figs. 12-5B and 12-6). This macula is located dorsal to the sacculus in the posterior canal duct, which is positioned lateral to an opening in the dorsal chondocranium, the fenestra ovalis (see Fig. 12-5B) (Howes 1883).

The shape of the inner ear of selachians varies widely, and this variation probably has some functional significance. Selachians with poor sound detection abilities, for instance, have a small sacculus that is ventrally located far below the dorsal surface of the head. In this type of ear the endolymphatic duct is long. In sharks with improved sound detection abilities,

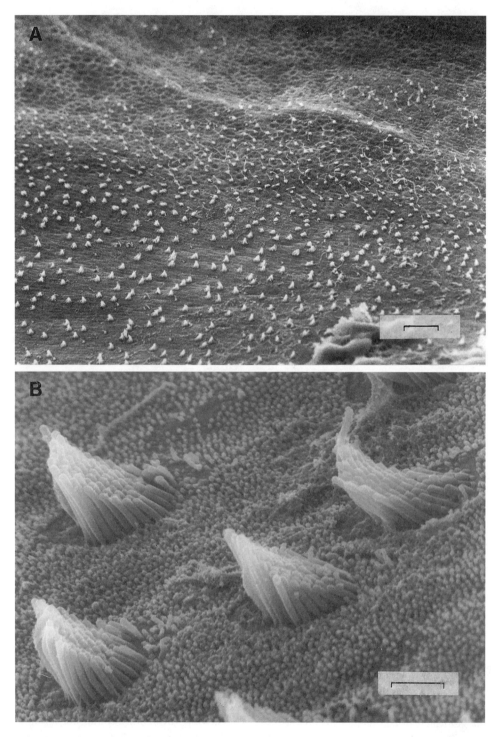

FIGURE 12-6 **A:** A scanning electron micrograph of a section of the macula neglecta of the shark *Mustelus*. Scale bar: 20 mm. **B:** At higher magnification the cilia complexes of macula neglecta hair cells become visible. Note their uniform size and form. Scale bar: 2 mm. (From Corwin 1981b.)

the membranous labyrinth is characterized by a triangular shape dominated by a large sacculus and a well-developed macula neglecta (Corwin 1991). Carcharhinid sharks typically have this type of ear (Tester et al. 1972). A large macula neglecta is characteristic of sharks that feed by raptorial predation. In contrast, the maculae neglectae in skates and rays or in sedentary sharks that feed on invertebrates are small (Corwin 1991). These comparative observations suggest that a large macula neglecta is an adaptation to the need to sense and localize a sound source.

PERIPHERAL PHYSIOLOGY

The semicircular canals of selachians are highly sensitive to angular accelerations (Lowenstein and Sand 1936, 1940a,b). The otolith bearing part of the macula utriculi, the posterior portion of the macula sacculi, and the adjoining macula lagenae contain gravity receptors (Lowenstein and Roberts 1950). Units that innervate these parts of the sensory epithelia show a resting discharge, the frequency of which increases or decreases with positional changes of the animal. The functional ranges of utriculus and sacculus overlap. Both contain sense endings responding to lateral and fore-and-aft tilting. Some fibers have their maximum discharge rate in side-up and nose-up and side-up and nose-down positions, respectively. The maculae also have out-of position receptors that respond to a change of position irrespective of the direction of the change. Gravity receptors of the lagena also respond to lateral and to for-and-aft tilts. Their maximum activity usually occurs in or near the normal position of the animal. The otolith organs, in addition, show clear reactions to linear translations in the three planes of space (Lowenstein and Roberts 1950).

Tester et al. (1972) were the first to propose that the macula neglecta is an important auditory (vibration) detector in sharks. Electrophysiological recordings show that potentials evoked by vibrational stimulation of the isolated ear from *Raja clavata* originate in both the macula neglecta and the otolithic sacculus (Lowenstein and Roberts 1951). At low-intensity stimulation, the responses consist of an increase in the discharge rate of spontaneously firing units. Higher stimulus intensities lead to the additional recruitment of previously quiescent fibers and to a phase coupling of the response. Adaptation to sustained vibrational stimuli occurs, and after cessation of prolonged stimuli a silent period is observed (Lowenstein and Roberts 1951). Fay et al. (1974) recorded microphonic potentials from the macula neglecta of a Carcharhinid shark while stimulating the animal with a vibrating probe placed against the surface of the head. Although this is not adequate to simulate an underwater acoustic stimulus (see Kalmijn 1988a), it at least shows that the macula neglecta is sensitive to vibration in the frequency range 100–600 Hz.

Threshold curves obtained by recording action potentials from the eighth nerve of the lemon shark, *Negaprion brevirostris*, have a low-frequency (<200 Hz) plateau of best sensitivity and gradually decreasing sensitivity between 200 and 600 Hz (Corwin 1981a). Click-evoked responses have a dynamic amplitude range of at least 30 dB with a latency shortening of 120–170 msec/dB and an amplitude increase of 4–11%/dB relative to a nearly saturated response (Corwin 1981a). Units in the eighth nerve fall into three classes: regularly spontaneous and nonacoustic, irregular spontaneous and acoustic, and nonspontaneous and acoustic. The best excitatory frequencies for the acoustic units range from 31 to 365 Hz, with the majority of units being most sensitive below 200 Hz. In the ear of the lemon shark there are two maculae that are capable of responding to sound. One, the nonotolithic macula neglecta, is composed of two large patches of sensory epithelium that line the walls of the posterior canal duct and extend cilia complexes toward a gelatinous cupula that fills the lumen of the duct. The other sound detector is the macula of the otolithic sacculus. In juvenile lemon sharks this epithelium contains up to 300,000 hair cells that extend their cilia toward a large mass of otoconia (Corwin 1981a).

BEHAVIOR

Parker (1909) observed that when he struck the side of a large wooden aquarium, the shark *Mustelus canis* responded with characteristic fin movements. The sensitivity to sound entirely disappeared when the eighth nerve was cut bilaterally. Struggling wounded teleosts generate pulsed low-frequency sounds (Nelson and Gruber 1963). Nelson et al. (1969) noticed that some sharks approach a vocalizing or a struggling wounded fish within seconds. Free living sharks may also respond to artificial sound with approach reactions (Kritzler and Wood 1961; Nelson and Gruber 1963; Banner 1972; Myrberg et al. 1969, 1972, 1976; Nelson 1967; Nelson and Johnson 1972, 1976). For instance, when Nelson and Gruber (1963) played preylike synthesized sounds back into the ocean, several species of free-ranging carcharhinid sharks were attracted, including the gray reef shark *Carcharhinus menisorrah*, the silvertip *C. albimarginatus*, the blacktip *C. melanopterus*, the reef whitetip *Triaenodon obesus*, and the lemon shark *Hemigaleops* (= *Negaprion*) *fosteri*. In one case, 35 gray reef sharks arrived within 3 minutes after the sound was turned on; many sharks were excited to the point of striking or biting the suspended transducer (Nelson and Johnson 1972). Using the natural responses of sharks, field observers were able to demonstrate that artificial sound sources could be localized at distances up to at least 250 m (Myrberg et al. 1972). Sudden or loud stimuli may produce withdrawal or escape responses in sharks (Myrberg 1978).

Measurements of threshold curves by use of conditioning methods demonstrate that sharks are especially sensitive to low frequency sound, that their best sensitivity is in the vicinity of 100 Hz, and that sensitivity decreases rapidly as the frequency of the stimulus increases toward 1000 Hz (Nelson 1967; Banner 1972; Kelly and Nelson 1975). Some sharks have the ability to discriminate among sounds on the basis of frequency alone. Lemon sharks, for instance, can distinguish between tones at 40 and 60 Hz that are presented at intensities randomly varied over a 20-dB range. The mechanism of this discrimination is unknown (Nelson 1967).

CENTRAL PATHWAYS AND CENTRAL PHYSIOLOGY

In selachians the eighth nerve terminates largely within the ipsilateral ventral octaval cell column, the reticular formation, and the cerebellum (e.g., Dunn and Koester 1984; Puzdrowski and Leonard 1993). The ventral octaval column is divided into as many as five nuclei that receive specific input from the various acoustic end organs. Several studies suggest that auditory information from medullary octaval nuclei is relayed to the lateral mesencephalic complex, a region that may be homologous to the torus semicircularis of tetrapods (for a review see McCormick 1992). Consequently, acoustic responses can be recorded from the medulla, cerebellum, and mesencephalon of cartilaginous fishes (Corwin and Northcutt 1982; Plassmann 1982), but also from circumscribed sites in the telencephalon (Bullock and Corwin 1979). In sharks, a good acoustic stimulus for generating evoked potentials is either a click with a resonance of a few hundred Hz or a rapidly rising tone burst of about 300 Hz. The lowest sound pressure threshold is −8 dB relative 1 μbar near the head, in response to a click delivered to the water surface (Bullock and Corwin 1979). In the thornback guitarfish, *P. triseriata*, there are three categories of medullary units that respond to vibratory (acoustic) stimuli (a rod that touched the fish's head immediately above the parietal fossa). Type 1 units are very sensitive to vibratory bursts (2–15 mm displacement in the frequency range 60–100 Hz), type 2 units are intermediately sensitive (50–150 mm, 60–100 Hz), and type 3 units are nearly insensitive (400–500 mm at frequencies below 60 Hz) (Plassmann 1983). Type 1 units can also be driven with an airborne acoustic stimulus; they are believed to get input from the macula neglecta since they are not affected by simultaneous presentation of static tilt or movement of pitch. However, type 1 units yield responses that are phase-locked to sinusoidal pitch, if simultaneously exposed to vibratory stimuli. Type 2 units, which possibly receive input from the macula sacculi or the lacinia utriculi, respond to both tilt and pitch, provided that continuous vibration is delivered at the same time. Type 3 units respond strongly to pitch even without simultaneous delivery of vibration. These units may be involved in the processing of vestibular information.

The responses of vibration sensitive neurons recorded from the midbrain of *Platyrhinoidis* resemble those of type 1 or type 2 units of the medulla (Plassmann 1983). In the midbrain and forebrain of sharks (*Negaprion acutidens* and *Carcharhinus melanopterus*) no interaction is seen when evoked potentials for visual, electric, and acoustic modalities are elicited at the same time or in succession (Bullock and Corwin 1979). Telencephalic evoked potentials show a slow recovery from tone bursts. In a typical telencephalic level, acoustically evoked potentials are distinct from 100 to 630 Hz and are best at about 300 Hz (Bullock and Corwin 1979).

Mechanosensory lateral line

Like the bony fish, sharks, skates, and rays have a special hydrodynamic sensory system, termed the mechanosensory lateral line (for a review see Boord and Campbell 1977). This sensory system was already present in the earliest heterostracans (Northcutt 1989) and provides all fishes with a sense of which humans have no direct experience.

ANATOMY
The mechanosensory lateral line is visible externally as rows of small pores found on the head and trunk. These pores connect the outside medium with the interior of a sophisticated subepidermal canal system. Typically, there are a posterior lateral line canal, which runs from the head to the tail; two cephalic lateral line canals, which run above or below the eye; a supratemporal canal, which crosses the head; and a mandibular canal on the side of the lower jaw (see Fig. 12-7C). There are variations in this pattern, especially within the skates and rays. Lateral line canals contain numerous sensory hillocks or neuromast organs (see Fig. 12-7D). Sharks, skates, and rays also have free-standing (superficial) neuromasts or pit organs, which lie between the bases of modified scales and are covered to a lesser or greater extent by the exposed scale crown. In rays and some sharks, pit organs are in slitlike grooves. The distribution of pit organs in sharks follows a generalized pattern: a rather distinct transverse line is found posterior to the mouth, another between the pectoral fins, two longitudinal, more or less distinct lines of pit organs are found on the sides, one along the lateral sensory canal and the other near the middorsal line. Other pit organs are scattered over the upper sides and may even be found on the tail (see Fig. 12-7A,B). In sharks, pit organs on the side of the body may be scattered in an apparently random fashion, extending well below the trunk lateral line canal. The number of pit organs on the side of the body can range from about 70 in *Squalus acanthias* to over 600 in *Sphyrna lewini* (Tester and Nelson 1967).

Both pit organs and canal neuromasts consist of sensory hair cells, supporting cells, and mantle cells that are situated above the basement membrane in the epidermis. The cilia of the hair cells extend into a fragile, tonguelike cupula. Pit organs, whose longitudinal diameter can cover 70–700 µm, may occupy a round or oval-shaped area between the bases of modified scales. The sensory hair cells occupy a spindle-shaped area extending transversely across the sense organ. In some sharks (e.g., in *Ginglymostoma cirratum*) pit organs are elevated on a platform or pedestal (Tester and Nelson 1967).

In cartilaginous fish two more kinds of end organs have been classified as mechanosensory lateral lines: the spiracular organs (also found in most nonteleost bony fishes), which are suspected to function as proprioceptors monitoring the position of the hyomandibula, and the vesicles of Savi, which may function in substrata vibration detection (Barry and Bennett 1989).

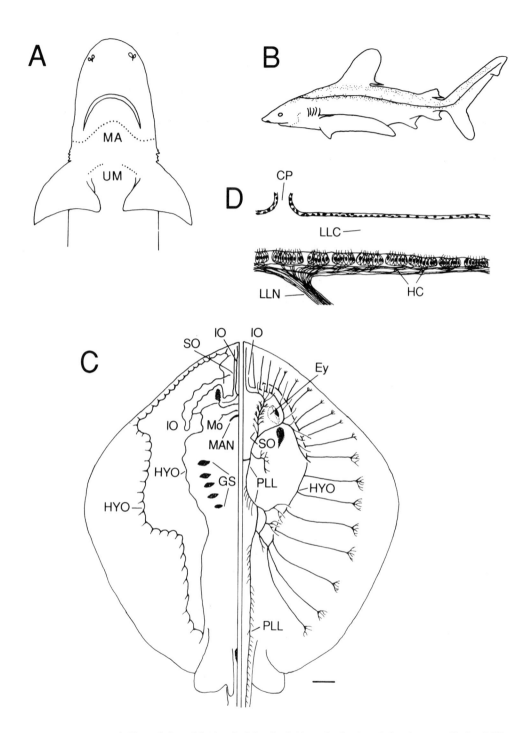

FIGURE 12-7 **A:** Ventral view of the head of the shark *Negaprion brevirostris* showing mandibular (MA) and umbilical (UM) lines of pit organs, each *dot* indicates the position of one pit organ. **B:** Lateral distribution of pit organs (*dots*) in *Carcharhinus longimanus*. **C:** Distribution of lateral line canals over the ventral (*left*) and dorsal (*right*) surface of the Atlantic stingray, *Dasyatis sabina*. Scale bar = 0.5 cm. Ey = eye; GS = gill slits; Hyo = hyomandibular canal; IO = infraorbital canal; MAN = mandibular canal; Mo = mouth; PLL = posterior (trunk) lateral line canal; SO = supraorbital canal. **D:** Longitudinal section of a lateral line canal of *Mustelus canis*. CP = canal pore; HC = hair cells; LLC = lateral line canal; LLN = lateral line nerve. Cupulae, which probably sit above the hair cell epithelia, are not shown. (A, B, redrawn from Tester and Nelson 1967; C, redrawn from Puzdrowski and Leonard 1993; D, redrawn from Johnson 1917.)

PERIPHERAL PHYSIOLOGY

Sand (1937) was the first to demonstrate that movement of the fluid in the lateral line canals of selachians causes lateral line nerve responses. Sand demonstrated this by inserting tubules into the ends of the hyomandibular loop of the lateral line canal in several species of *Raja* and by connecting these to a plumbing system that allowed the controlled flow of fluid through the canal. The slowest rate of perfusion was about 2 mm/sec; this flow was already well above threshold. Görner and Kalmijn (1989) recorded primary mechanosensory afferents from the posterior lateral line nerve of the thornback guitarfish, *P. triseriata*, while stimulating the animal with sinusoidal water movements. In this fish all lateral line fibers showed resting activity, which varied between 5 and 17 impulses per second. Within the test range of 1–128 Hz, all fibers responded to water movements. At frequencies below 10 Hz, acceleration thresholds of individual fibers were flat, at higher frequencies the thresholds increased (Fig. 12-8). At 128 Hz the threshold was 50 dB above the values found below 10 Hz. At 4 Hz the threshold was about 0.5 cm/sec^2 at the skin.

BEHAVIOR

Parker (1904) was the first to notice that a shark that has been deprived of hearing and sight still responds to water movements as long as the lateral line nerves are intact. There are no further behavioral experiments that offer conclusive evidence of the biological role of the lateral line system in selachians. However, since the lateral line of sharks, skates, and rays is sensitive to local water movements (Görner and Kalmijn 1989), there is little doubt that this sensory system provides cartilaginous fishes with a hydrodynamic sense that is used to detect nearby prey, predators, or conspecifics.

CENTRAL PATHWAYS AND CENTRAL PHYSIOLOGY

Our knowledge regarding central lateral line pathways is limited to only a few elasmobranchs, such as the batoids (thornback guitarfish, *P. triseriata*; clearnose skate, *Raja eglanteria*; and the little skate, *Raja erinacea*), the squalomorph sharks (spiny dogfish, *Squalus acanthias*), and the galeomorph sharks (carpet shark, *Cephaloscyllium isabella*). In these animals roots of the anterior and posterior lateral

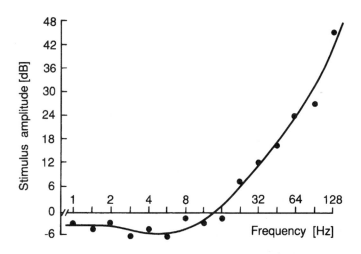

FIGURE 12-8 Threshold of a primary lateral line afferent to sinusoidal water movements. At 0 dB acceleration of a 1-Hz stimulus produced 0.5-cm displacement amplitude. (Redrawn from Görner and Kalmijn 1989.)

line nerves enter the ipsilateral brainstem and terminate in the medial octavolateralis nucleus. The axons of the secondary cells of the primary lateral line nuclei collectively course bilaterally, with contralateral dominance. The largest portion of this pathway terminates among the cells of the lateral mesencephalic nucleus or within the central zone of the optic tectum (e.g., Boord and Northcutt 1982, 1988). The final ascending pathway of the mechanosensory lateral line involves the relay of information from midbrain to diencephalon and, finally, to telencephalon (Boord and Montgomery 1989). In cartilaginous fish some lateral line information also reaches the cerebellum (Tong and Bullock 1982; Fiebig 1988).

The physiology of central lateral line areas was investigated to some degree in *P. triseriata* (Bleckmann et al. 1987, 1989). In this fish, lateral line responses can be recorded from the medulla, midbrain, diencephalon, and telencephalon. Within the test range of 6.5–200 Hz, all lateral line areas respond to minute water vibrations. In terms of displacement, best frequencies were between 75 and 200 Hz, with threshold values for evoked potentials being as low as 0.005 µm peak-to-peak water displacement. In general, diencephalic and telencephalic areas were less sensitive to water vibrations than hindbrain and midbrain responses (Bleckmann et al. 1989). In the medulla, lateral line responses are tonic or phasic-tonic and spikes usually phase-couple to a sinusoidal stimulus. In both the midbrain and forebrain, lateral line responses are more phasic; however, at higher stimulus intensities a tonic response component may appear. The dynamic amplitude range of lateral line units may cover 90 dB. Mesencephalic, diencephalic, and telencephalic lateral line receptive fields are usually contralateral (midbrain) or ipsi- and contralateral (forebrain). Forebrain lateral line cells may be bimodal in that an electrical or visual stimulus may alter the response to a lateral line stimulus (Bleckmann et al. 1989). In *P. triseriata*, some lateral line information reaches discrete areas in the cerebellum (Fiebig 1988).

Cutaneous mechanosensitivity

Mechanoreceptors other than the lateral line provide an animal with information about skin deformation. In the skin of sharks (*Scyliorhinus canicula*) three distinct classes of mechanosensitive units have been found (Nier 1976). With square-wave microvibration, in solid contact with placoid scales, one type of unit responds only to the pressure phase (indentation), the other only to the tension phase (relaxation) of a step stimulus. Both types of units yield PD-transfer functions with predominating D-components and slow adaptation. A third type of unit is spontaneously active and shows a mechanical and thermal (cold) sensitivity. The spontaneous activity of these units can be modulated by mechanical stimuli. As in the lateral line, spikes occur phase-locked according to the harmonic input. Absolute thresholds are around 17 µm (mean of 78 afferents). Suprathreshold intensity is encoded in spike rate and in the number of units activated. Synchronization thresholds of units that are sensitive to skin indentation depends on both the stimulus frequency and the superimposed skin indentation (Fig. 12-9). The receptive field of individual afferent fibers varies between 0.5 and 60 mm^2, the latter area covering about 150 placoid scales. For the majority of mechanosensory fibers an overlap of up to 100% of their receptive areas was found (Nier 1976).

Electrosensory system

Besides olfactory, visual, acoustic, hydrodynamic, and tactile cues, electric and magnetic fields present a wealth of information in natural waters. There are four main sources for biologically relevant electrical fields (for a review see Kalmijn 1988b):

1. Animate electric fields, produced incidentally by electrochemical disparities in the internal and external milieu of an animal. These DC bioelectric fields are usually weak (1–500 µV/cm) but can reach more than 1000 µV/cm in wounded crustaceans (Kalmijn 1974). Animate

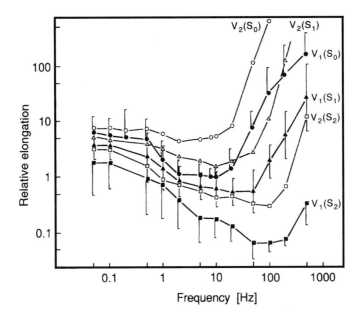

FIGURE 12-9 Amplitude-frequency responses of cutaneous afferents for different values of superimposed constant skin indentation. The first synchronization threshold (V_1) is the amplitude that causes a unit to respond phase-locked with one action potential per stimulus cycle. The second synchronization threshold (V_2) is the amplitude that causes a unit to respond phase-locked with two action potentials per stimulus cycle. Mean values and standard deviations of V_1 were averaged from corresponding values of 50 preparations, tested with sinusoidal vibration and additional skin indentations (S) $S_0 = 0$ mm; $S_1 = 150$ mm; $S_2 = 300$ mm). V_2 is indicated by mean values only. (Redrawn from Nier 1976.)

electric fields are low-frequency modulated by respiratory gill movements or other relative movements of body parts (Roth 1972).

2. Kinetic electric fields, induced whenever water (a conductor) is moving through the earth's magnetic field. Depending on flow speed, kinetic electric fields have a magnitude of up to 0.5 µV/cm.

3. Fields induced by a moving recipient animal. Every moving conductor (e.g., a shark, skate, or ray) induces electromotive forces whose gradients are perpendicular to both the magnetic field and the line of motion.

4. Bioelectric fields created by electric organs. Within elasmobranchs, skates are known to have a weakly electric organ in the tail region, which they use for communication (Bratton and Ayers 1987; Jacob et al. 1994). Strongly electric organs are present only in some torpenid rays and are used for prey capture (Belbenoit 1986).

Elasmobranchs have a keen electric sense, which they use to detect animate and inanimate electric fields (for reviews see Kalmijn 1974, 1988b, 1997; Bodznick and Boord 1986; Schweitzer 1986b). The receptors involved are the ampullae of Lorenzini (Dijkgraaf and Kalmijn 1963, 1966), which measure minute changes in potential difference between the water at skin openings and the basal surface of the receptor cells located in their interior (Murray 1960). Behavioral studies show that one important biological role of electroreception is in prey detection; in this context, the ampullary organs enable sharks, skates, and rays to detect the low-frequency bioelectric fields that emanate from prey animals. The ampullae of Lorenzini are also used for orientation and navigation, that is, for the measure of kinetic electric fields and the fields induced by a moving recipient animal (Kalmijn 1988b).

ANATOMY

The ampullae of Lorenzini are distributed dorsally and ventrally over the head in sharks, and head and pectoral fins in skates and rays, as shown in Figure 12-10A (Murray 1965). Each ampulla is composed of sensory and supporting cells and a tubelike canal that opens externally to pores in the epidermal surface (see Fig. 12-10B). The canal walls consist of tightly joined squamous epithelium (Waltmann 1966) whose specific resistance, at 6 million ohms per centimeter, is the highest of all animal tissues tested (Waltmann 1966). In contrast, the ampullary canals are filled with transparent mucopolysaccharides (Doyle 1963), whose ionic composition guarantees a low resistance (20–25 Ω/cm), that is, a resistance equaling that of seawater (Murray and Potts 1961). About five to six afferent fibers innervate the hundreds of receptor cells within an ampulla (Murray 1965; Waltman 1966); no receptor efferents are present (Bullock 1982). In marine elasmobranchs the ampullae of different receptor organs may be grouped together in clusters well beneath the skin with up to 400 ampullary tubes radiating out from a given cluster in many directions (see Fig. 12-10A) (Chu and Wen 1979). The number and locations of ampullary groups may vary among species (usually three to six are present per side) and each group is innervated by a separate ramus of the anterior lateral line nerve. In general, different species show a wide variation in receptor number and distribution, which probably reflects habitat and feeding differences rather than taxonomic positions. The ampullae of Lorenzini, which are particularly numerous in the head of hammerhead sharks (*Sphyrnidae*), appear to be present in all elasmobranch species, including the plankton feeding whale shark (*Rhincodon*

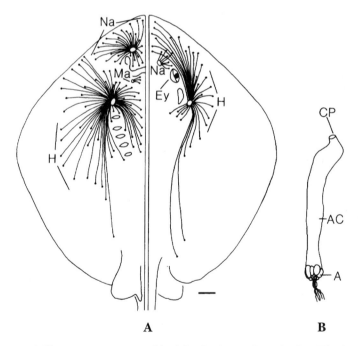

FIGURE 12-10 A: Electrosensory systems of the Atlantic stingray, *Dasyatis sabina*. The drawing shows the distribution of electroreceptor organs over the ventral (*left*) and dorsal (*right*) surfaces of the animal. *Dots* indicate openings of ampullary canals. Scale bar: 0.5 cm. Ey = eye; = Ma mandibular group; Na = nasal group; H = hyoid group. **B:** Ampullary organ of *Lorenzini*. A = ampulla; AC = ampullary canal; CP = canal pore. (A, redrawn from Puzdrowski and Leonard 1993; B, redrawn from Peabody 1897.)

typus), the basking shark (*Cetorhinus maximus*), the megamouth shark (*Megachasma pelagios*), and the devil fish (*Manta birostris*) (Kalmijn 1984).

PERIPHERAL PHYSIOLOGY

Afferents from the ampullae of Lorenzini form a relatively homogeneous population of units that exhibit a regular ongoing activity at average rates between 8 and 34 impulses/sec (Murray 1962; Montgomery 1984b; Bodznick and Schmidt 1984; New 1990; Tricas and New 1998). Neural activity increases by cathodal stimuli at the ampullary pore and adaptation is slow (Murray 1962; New 1990). Presentation of an anodal stimulus leads to a decrease of neural activity; however, a brief burst of spikes may follow stimulus offset (New 1990). The link between a cathodal stimulus and the increase in primary afferent activity involves a depolarizing, inward Ca^{2+} current, which is followed by a late outward K^+ current. Synaptic transmission is chemical, with L-glutamate or L-aspartate most likely acting as neurotransmitter. At concentrations between 10^{-7} and 10^{-3} M these substances increase resting and stimulus-evoked activity (Akoev et al. 1991). The electroreceptors of cartilaginous fishes are low-frequency sensors with highest sensitivity in the range 1–4 Hz and a rapid high-frequency (16–20 Hz) roll-off (Montgomery 1984b; New 1990; Tricas and New 1998). At a frequency of 1 Hz a stimulus amplitude of 40 nV/cm peak-to-peak may be sufficient to cause a neural response. The high sensitivity to low-frequency stimuli is probably facilitated by the regular ongoing discharge pattern of primary afferents.

The ampullae of Lorenzini encode the stimulus amplitude. An increase in electric field intensity causes an increase in firing rate with a high level of certainty because of a low endogenous variance in the resting interspike interval. Peak discharge slopes across the full stimulus intensity are different in different units (Tricas and New 1998). In some primary afferents the dynamic amplitude range extends to only 0.5 µV/cm; other primary afferents start to saturate at intensities above 25 µV/cm (Murray 1965; Montgomery 1984b; Tricas and New 1998). Primary afferents are most responsive to electric fields oriented parallel to the line between the respective ampulla and the corresponding skin opening (Bodznick and Boord 1986). The distribution of ampullary canals usually is such that the system provides complete directional information, including dorsal and ventral. Electrophysiological recordings have shown that the ongoing discharge rate of the ampullae is strongly modulated by an animal's own ventilatory activity: the discharge rate is depressed during expiration and accelerated during inspiration (Dijkgraaf and Kalmijn 1966; Montgomery 1984b; New and Bodznick 1990). The extreme sensitivity to small (1 µV) stimuli is preserved even in the presence of large (e.g., 1 mV) DC fields (Araneda and Bennett 1993). The receptor cells show small membrane potential oscillations near the spike-triggering threshold (Braun et al. 1994; Lu and Fishman 1994).

BEHAVIOR

The ability of elasmobranchs to detect weak, low-frequency electric fields is well documented (for reviews see Kalmijn 1974, 1984, 1988). When tested with uniform square-wave stimuli of 5 Hz, the dogfish *Scyliorhinus canicula* and the skate *Raja clavata* show behavioral reflexes at voltage gradients as low as 100 nV/cm. With monitoring of cardiac reflexes, thresholds of only 10 nV/cm were found (Dijkgraaf and Kalmijn 1962, 1966). In experiments at sea the dogfish *Mustelus canis* shows behavioral responses to voltage gradients as low as 5 nV/cm (Kalmijn 1982).

One biological relevance of electroreception became evident when Kalmijn (1966, 1971) proved that cartilaginous fishes respond to the bioelectric fields of their prey. When small flounder were presented to bottom-feeding sharks and skates, first under a thin layer of coarse sand or seawater agar permeable to electric fields, strong feeding responses were elicited even though the flounder were completely hidden from view. If, however, the agar was covered by a thin plastic film that screened the prey electrically, the flounder remained undetected (Kalmijn 1971). Apparent feeding responses occurred again when the prey fields were simulated by passing weak electric currents between two electrodes under agar or sand. In the final attack, the electric fields were nearly always more attractive than any lingering odor cues (Kalmijn 1971). In later experiments done at sea, Kalmijn (1982) proved that even free living sharks (*Mustelus canis* and *Prionace glauca*) execute feeding responses to dipole electric fields designed to mimic prey (Fig. 12-11A).

The electrosensory system of cartilaginous fish in principal is sensitive enough to provide information for orientation and navigation. For instance, sharks, skates, and rays may be able to estimate their passive drift with a flow of water from the electrical fields that ocean currents produce by interaction with the earth's magnetic field. Behavioral experiments clearly show that cartilaginous fishes can orientate with aid of uniform electric fields: in Figure 12-11B, the stingray *Urolophus halleri* enters an enclosure on the left and avoids a similar enclosure on the right relative to the direction of a uniform (DC) electric field if rewarded properly. A significant orientation even occurs if the voltage gradient is as low as 5 nV/cm (Kalmijn 1982). The applied electric fields were similar to those caused by ocean currents, which can have an amplitude of up to 500 nV/cm (Kalmijn 1984).

Although it seems clear that elasmobranchs can sense the earth's magnetic field and that their electrosensory system is sensitive enough to detect voltages induced by movements through the field, it is not clear how they actually do this. Both relative movements of the animal and movements of the water masses through which the fish swims determine the voltage gradients on the receptors. Two theories exist about possible mechanisms involved in the extraction of magnetic field direction by the electrosensory system (Kalmjin 1984; Paulin 1995).

That cartilaginous fishes can indeed use magnetic field information for orientation has been shown. In the absence of any imposed electric field the stingray *Urolophus* was conditioned to enter an enclosure in the magnetic east and to avoid a similar enclosure in

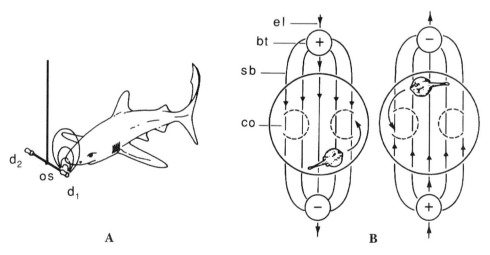

FIGURE 12-11 **A:** Feeding attacks of the blue shark *Prionace glauca* on electrically simulated prey. os = Odor source; d_1 = electrodes passing a current of 8 mA; d_2 = control electrodes. **B:** Orientation response of the stingray *Urolophus halleri* to a uniform electric field of 5 nV/cm. The ray was trained to enter a corral (co) on the left relative to the direction of the electric field. The electrodes (el) are contained in two plastic bottles (bt) that connect to the main body of seawater by two current dividers of 18 salt bridges (sb) each. (B, redrawn from Kalmjin 1982.)

the magnetic west. In the actual experiments, the magnetic fields were applied with the horizontal magnetic component randomly offered in either the normal or reversed direction (Kalmijn 1984). The behavioral experiments also revealed that *Urolophus* detects not only the direction, but also the polarity of the magnetic field. Some recent field experiments indicate that under natural conditions elasmobranchs may actually also use the geomagnetic field for navigation (Klimley 1993).

CENTRAL PATHWAYS
AND CENTRAL PHYSIOLOGY

The electrosensory fibers of elasmobranchs project via the dorsal root of the anterior lateral line nerve to the dorsal octavolateralis nucleus, which is the first-order medullary electrosensory nucleus (Bodznick and Northcutt 1980; Koester 1983). The principal ascending electrosensory pathway leaves the dorsal octavolateralis nucleus, crosses the midline within the medulla, and terminates within distinct areas of the midbrain (e.g., Boord and Northcutt 1988). From there, electrosensory information is relayed to the diencephalon and telencephalon (Platt et al. 1974; Bullock 1979; Bodznick and Northcutt 1984; Schweitzer and Lowe 1984; Bleckmann et al. 1987). Some electrosensory information also reaches the cerebellum (Tong and Bullock 1982; Fiebig 1988). Central pathology physiological studies are not extensive, but there several features emerge of electroreceptive activity in the medulla and midbrain and, to a lesser extent, at higher brain levels. Cells recorded from the dorsal octavolateralis nucleus have a lower rate of spontaneous activity than primary afferents and their responses are considerably more phasic than those of primary afferents (New 1990). Like primary afferents, these cells are excited by the cathodal pole and inhibited by the anodal pole of a stimulus source (New 1990); thresholds may be lower than 0.3 µV/cm (Schweitzer 1986b). With threshold stimuli, most medullary cells have small, ipsilateral excitatory receptive fields that result from convergent input from two to five adjacent receptor pores (Bodznick and Schmidt 1984). The convergence probably accounts for the increase in the sensitivity of second-order cells in comparison with primary afferents. Receptive fields usually progress smoothly across the body surface as neurons are recorded dorsoventrally (Bodznick and Schmidt 1984). No difference is found in the frequency response of secondary and primary neurons. Both are most sensitive in the range 0.5–10 Hz, followed by a rather abrupt cutoff frequency between 10 and 15 Hz (Andrianov et al. 1984; Montgomery 1984b; New 1990; Tricas and New 1998). There are no indications of a frequency filter mechanism in the medulla.

One problem inherent in electrosensory systems is that the animal must be able to detect extremely weak electric signals on top of its own large bioelectric fields. One clear physiological distinction between electrosensory afferents and secondary cells is reduced ventilatory modulation in secondary cells (Montgomery 1984a; New and Bodznick 1990). A mechanism by which this can be accomplished is common mode rejection. Electroreceptors on both sides of the body are modulated by ventilatory movements at the same time and this "common mode" is suppressed by medullary circuitry (Montgomery and Bodznick 1991; Bodznick et al. 1992). This common mode rejection system depends to a large degree on inhibitory commissural connections among the dorsal nuclei of the medulla (New and Bodznick 1990; Montgomery and Bodznick 1993; Duman and Bodznick 1997).

Another important noise cancellation mechanism was recently discovered in the electrosensory system of elasmobranchs. If an artificial stimulus is coupled with the ventilatory movements, the response of second-order neurons to this stimulus gradually decreases over a time period of several minutes (Bodznick 1993; Montgomery and Bodznick 1993). This response is mediated by an adaptive filter that suppresses unwanted reafference in the medulla. The neurological basis of this filter is the projection from a part of the cerebellum (auricles) into the molecular layer of the dorsal nucleus (Conley and Bodznick 1994; Conley 1995; Nelson and Paulin 1995; Hjelmstad et al. 1996). This adaptive filter mechanism, which cancels the sensory consequences of the animal's own movements, may play a role not only in the electrosensory system, but also in the mechanosensory lateral line and auditory system (Montgomery et al. 1995).

Information processing in the elasmobranch electrosensory system has also been studied in the midbrain of cartilaginous fishes (Schweitzer 1983, 1986b). Electrosensory midbrain units of the thornback guitarfish, *P. triseriata*, usually exhibit no spontaneous activity, and if stimulated they respond with only a few spikes per stimulus onset or turnoff or both. Unit thresholds range from less than 0.3 µV/cm (the weakest stimulus delivered) to 5 µV/cm. Midbrain responses in the shark *Carcharhinus*, measured with evoked potentials, exhibit even lower thresholds of 0.015 µV/cm (Platt et al. 1974). In *P. triseriata*, unit responses gradually increase with stimulus amplitude and reach a maximum at about 100 µV/cm. Higher stimulus intensities may cause a response depression. Best frequencies in response to sinusoidal stimuli range from 0.2 Hz (the lowest frequency delivered) to about 4 Hz. Interestingly, in the shark *Carcharhinus*, evoked activity is maximal at higher frequencies (20–30 Hz), and 10–15 Hz elicit maximal activity in the midbrain of *Potamotrygon* (Bullock 1979). Midbrain electrosensory units may have small, well-defined excitatory receptive fields that include 2–20 (*P. triseriata*) or 4–8 (*Raja radiata*) ampullary pores (Schweitzer 1986b; Andrianov et al. 1984). Receptive fields are somatotopically organized in that anterior, middle, and posterior body surfaces map on the rostral, middle, and caudal levels of the contralateral midbrain. Like primary afferents and secondary cells, units in the midbrain are sensitive to orientation of uniform electric fields: best orientation is, again, a function of the orientation of the ampullary canal, from which the unit receives its input. Units may be bimodal, responding vigorously to a hairbrush applied gently to the skin. The mechanosensitivity is not mediated by the ampullae of Lorenzini, since tactile stimuli adequate in the midbrain do not excite medullary electrosensory neurons. Measured near threshold, mechanosensory receptive fields are generally larger than the corresponding electrosensory receptive fields but overlap with them. Cross-modality interaction may be facilitatory or inhibitory (Schweitzer 1986a).

Ampullary input to the diencephalon or telencephalon of cartilaginous fishes has been demonstrated by Platt et al. (1974), Bullock (1979), Schweitzer (1983), Bodznick and Northcutt (1984), and Bleckmann et al. (1989). In rays (primarily *Torpedo ocellata*) and sharks (primarily *Scyliorhinus canicula*) telencephalic response latencies to DC pulses or sine waves are as short as 39 msec; with some labile responses as late as 330 msec, an "off" response is sometimes present (Platt et al. 1974). The response amplitudes show a decline above eight stimuli per second; no facilitation is observed. In the skate *Raja erinacea* and the guitarfish, *Platyrhinoidis*, the telencephalic electrosensory area corresponds to the medial pallium (Bodznick and Northcutt 1984; Bleckmann et al. 1987). Reversal of field polarity may change the latency and the shape of the evoked potentials. This indicates that some directional information is preserved up to the level of the telencephalon.

ACKNOWLEDGMENTS

We thank M. Hansel for drawing the figures. Drs. P. Görner, J. Mogdans, and W. C. Hamlett kindly commented on the manuscript. The writing of this chapter and research in the authors' laboratory were supported by a grant from the German Science Foundation (Bl 245).

REFERENCES

Akoev, G., G. N. Andrianov, T. Szabo, and B. Bromm. 1991. Neuropharmacological analysis of synaptic transmission in the Lorenzinian ampulla of the skate *Raja clavata*. *J. Comp. Physiol.* 168A: 639–46.

Andrianov, Y. N., G. R. Broun, O. B. Ilinskii, and V. M. Muraveiko. 1984. Frequency characteristics of skate electroreceptive central neurons responding to electrical and magnetic stimulation. *Neurophysiology* 16: 364–69.

Araneda, R. C., and M. V. L. Bennett. 1993. Electrical properties of electroreceptor cells isolated from skate ampulla of Lorenzini. *Biol. Bull.* 185: 310–11.

Aronson, L. R., F. R. Aronson, and E. Clark. 1967. Instrumental conditioning and light-dark discrimination in young nurse sharks. *Bull. Mar. Sci.* 17: 249–56.

Asai, T. 1913. Untersuchungen über die Struktur der Riechorgane bei *Mustelus laevis*. *Anat. Heft.* 49: 441–521.

Ashmore, J. F., and G. H. Falk. 1980a. The single-photon signal in rod bipolar cells of the dogfish retina. *J. Physiol. (Lond.)* 300: 151–66.

———. 1980b. Responses of rod bipolar cells in the dark-adapted retina of the dogfish, *Scyliorhinus canicula*. *J. Physiol. (Lond.)* 300: 115–50.

———. 1981. Photon-like signals following weak rhodopsin bleaches. *Nature* 289: 489–91.

———. 1982. An analysis of voltage noise in rod bipolar cells of dogfish retina. *J. Physiol. (Lond.)* 332: 273–97.

Backus, R. H. 1963. Hearing in elasmobranchs. In *Sharks and Survival*, ed. P. W. Gilbert, 243–54. Boston: D. C. Heath.

Baier, H., S. Rotter, and S. Korsching. 1994. Connectional topography in the zebrafish olfactory system: random positions but regular spacing of sensory neurons projecting to an individual glomerulus. *Proc. Natl. Acad. Sci. U.S.A.* 91: 11646–50.

Banner, A. 1967. Evidence of sensitivity to acoustic displacements in the lemon shark, *Negaprion brevirostris* (Poey). In *Lateral Line Detectors*, ed. P. H. Cahn, 265–73. Bloomington: Indiana University Press.

———. 1972. Use of sound in predation by young lemon sharks, *Negaprion brevirostris*. *Bull. Mar. Sci.* 22: 251–83.

Barber, V. C., and C. J. Emerson. 1980. Scanning electron microscopic observations on the inner ear of the skate, *Raja ocellata*. *Cell Tissue Res.* 205: 199–215.

Barry, M. A., and V. L. Bennett. 1989. Specialized lateral line receptor systems in elasmobranchs: the spiracular organs and vesicles of Savi. In *The Mechanosensory Lateral Line. Neurobiology and Evolution*, ed. S. Coombs, P. Görner, and H. Münz, 591–606. New York: Springer-Verlag.

Belbenoit, P. 1986. Fine analysis of predatory and defensive motor events in *Torpedo marmorata* (Pisces). *J. Exp. Biol.* 121: 197–226.

Bell, M.A. 1993. Convergent evolution of nasal structures in sedentary elasmobranchs. *Copeia* 1: 144–58.

Bleckmann, H. 1993. Role of the lateral line in fish behavior. In *Behaviour of Teleost Fishes*, ed. T. J. Pitcher, 201–46. London: Chapman & Hall.

———. 1994. *Reception of Hydrodynamic Stimuli by Aquatic and Semiaquatic Animals*. Stuttgart: Gustav Fischer Verlag.

Bleckmann, H., T. H. Bullock, and J. M. Jørgensen. 1987. The lateral line mechanoreceptive mesencephalic, diencephalic, and telencephalic regions in the thornback ray, *Platyrhinoidis triseriata* (Elasmobranchii). *J. Comp. Physiol.* 161A: 67–84.

Bleckmann, H., O. Weiss, and T. H. Bullock. 1989. Physiology of lateral line mechanoreceptive regions in the elasmobranch brain. *J. Comp. Physiol.* 164A: 459–74.

Bodznick, D. 1991. Elasmobranch vision: multimodal integration in the brain. *J. Exp. Zool. Suppl.* 5: 108–16.

———. 1993. The specificity of an adaptive filter that suppresses unwanted reafference in electrosensory neurons of the skate medulla. *Biol. Bull.* 185: 312–14.

Bodznick, D., and R. G. Northcutt. 1980. Segregation of electro- and mechanoreceptive inputs to the elasmobranch medulla. *Brain Res.* 195: 313–21.

———. 1984. An electrosensory area in the telencephalon of the little skate, *Raja erinacea*. *Brain Res.* 298: 117–24.

Bodznick, D., and A. W. Schmidt. 1984. Somatotopy within the medullary electrosensory nucleus of the little skate, *Raja erinacea*. *J. Comp. Neurol.* 225: 581–90.

Bodznick, D., and R. L. Boord. 1986. Electroreception in chondrichthyes. In *Electroreception*, ed. T. H. Bullock and W. Heiligenberg, 225–56. New York: John Wiley & Sons.

Bodznick, D., J. C. Montgomery, and D. J. Bradley. 1992. Suppression of common mode signals within the electrosensory system of the little skate, *Raja erinacea*. *J. Exp. Biol.* 171: 107–25.

Boord, R. L., and C. B. G. Campbell. 1977. Structural and functional organization of the lateral line system of sharks. *Am. Zool.* 17: 431–41.

Boord, R. L., and R. G. Northcutt. 1982. Ascending lateral line pathways to the midbrain of the clearnose skate, *Raja eglanteria*. *J. Comp. Neurobiol.* 207: 274–82.

———. 1988. Medullary and mesencephalic pathways and connections of lateral line neurons of the spiny dogfish, *Squalus acanthias*. *Brain Behav. Evol.* 32: 76–88.

Boord, R. L., and J. C. Montgomery. 1989. Central mechanosensory lateral line centers and pathways among the elasmobranchs. In *The Mechanosensory Lateral Line. Neurobiology and Evolution*, ed. S. Coombs, P. Görner, and H. Münz, 323–40. New York: Springer-Verlag.

Bratton, B. O., and J. L. Ayers. 1987. Observation on the electric organ discharge of two skate species (Chondrichthyes: Rajidae) and its relationship to behavior. *Environ. Biol. Fish* 20: 241–54.

Braun, H. A., H. Wissing, K. Schaefer, and M. C. Hirsch. 1994. Oscillation and noise determination and signal transduction in shark multimodal sensory cells. *Nature* 367: 270–73.

Broun, G. R., and E. E. Fesenko. 1982. Impulse response of single nerve fibers in the olfactory tract of black sea skates. *Dokl. Biol. Sci.* (Engl. transl. *Dokl. Akad. Nauk. SSSR*) 259: 405–7.

Bruckmoser, P., and N. Dieringer. 1973. Evoked potentials in primary and secondary olfactory projection areas of the forebrain in elasmobranchs. *J. Comp. Physiol.* 87A: 65–74.

Bullock, T. H. 1979. Processing of ampullary input in the brain: comparison of sensitivity and evoked responses among elasmobranchs and siluriform fishes. *J. Physiol. Paris* 75: 297–407.

———. 1982. Electroreception. *Annu. Rev. Neurosci.* 5: 121–71.

Bullock, T. H., and J. T. Corwin. 1979. Acoustic evoked activity in the brain of sharks. *J. Comp. Physiol.* 129A: 223–34.

Bullock, T. H., and R. G. Northcutt. 1984. Nervus terminalis in dogfish (*Squalus acanthias*, Elasmobranchii) carries tonic efferent impulses. *Neurosci. Lett.* 44: 155–60.

Bullock, T. H., M. H. Hofmann, J. G. New, and F. K. Nahm. 1991. Dynamic properties of visual evoked potentials in the tectum of cartilaginous and bony fishes, with neuroethological implications. *J. Exp. Zool. Suppl.* 5: 142–55.

Chu, Y. T., and Wen, M.C. 1979. A study of the lateral-line canals system and that of Lorenzini ampullae and tubules of elasmobranchiate fishes of China (in Chinese). Shanghai: Science and Technology Press.

Clark, E. 1963. Maintenance of sharks in captivity with a report on their instrumental conditioning. In *Sharks and Survival*, ed. P. W. Gilbert, 115–49. Boston: D. C. Heath.

Cohen, J. L. 1989. Functional pathways in the elasmobranch retina. *J. Exp. Zool. Suppl.* 2: 75–82.

———. 1991. Adaptations for scotopic vision in the lemon shark (*Negaprion brevirostris*). *J. Exp. Zool. Suppl.* 5: 76–84.

Cohen, D. H., T. A. Duff, and S. O. E. Ebbesson. 1973. Electrophysiological identification of a visual area in shark telencephalon. *Science* 182: 492–94.

Conley, R.A. 1995. Descending input from the vestibulolateral cerebellum suppresses electrosensory responses in the dorsal octavolateralis nucleus of the elasmobranch, *Raja erinacea*. *J. Comp. Physiol.* 176A: 325–35.

Conley, R. A., and D. Bodznick. 1994. The cerebellar dorsal granular ridge in an elasmobranch has proprioceptive and electroreceptive representations and projects homotopically to the medullary electrosensory nucleus. *J. Comp. Physiol.* 174A: 707–21.

Corwin, J. T. 1978. The relation of inner ear structure to the feeding behavior in sharks and rays. In *Scanning Electron Microscopy*, ed. O. Johari, 1105–12. Chicago: S.E.M.

———. 1981a. Peripheral auditory physiology in the lemon shark: evidence of parallel otolithic and non-otholitic sound detection. *J. Comp. Physiol.* 142A: 379–90.

———. 1981b. Postembryonic production and aging of inner ear hair cells in sharks. *J. Comp. Neurol.* 201: 541–53.

———. 1991. Functional anatomy of the auditory system in sharks and rays. *J. Exp. Zool. Suppl.* 2: 62–74.

Corwin, J. T., and R.G. Northcutt. 1982. Auditory centers in the elasmobranch brain stem: deoxyglucose autoradiography and evoked potential recording. *Brain Res.* 236: 261–73.

Crescitelli, F. 1991. Adaptations of visual pigments to the photic environment of the deep sea. *J. Exp. Zool. Suppl.* 5: 66–75.

Demski, L.S. 1993. Terminal nerve complex. *Acta Anat.* 148: 81–95.

Demski, L. S., and R. G. Northcutt. 1983. The terminal nerve: a new chemosensory system in vertebrates? *Science* 220: 435–37.

Dijkgraaf, S. 1963. Sound reception in the dogfish. *Nature* 197: 93–94.

Dijkgraaf, S., and A. J. Kalmijn. 1962. Verhaltensversuche zur Funktion der Lorenzinischen Ampullen. *Naturwissenschaften* 49: 400.

———. 1963. Untersuchungen über die Funktion der Lorenzinischen Ampullen an Haifischen. *Z. Vergl. Physiol.* 47: 438–56.

———. 1966. Versuche zur biologischen Bedeutung der Lorenzinischen Ampullen bei den Elasmobranchieren. *Z. Vergl. Physiol.* 53: 187–94.

Dowling, J. E., and H. Ripps. 1991. On the duplex nature of the skate retina. *J. Exp. Zool. Suppl.* 5: 55–65.

Doyle, J. 1963. The acid mucopolysaccharides in the glands of Lorenzini of elasmobranch fish. *Biochem. J.* 88: 7.

Dryer, L., and P. P. C. Graziadei. 1993. A pilot study on morphological compartmentalization and heterogeneity in the elasmobranch olfactory bulb. *Anat. Embryol.* 188: 41–51.

———. 1994. Projections of the olfactory bulb in an elasmobranch fish, Sphyrna tiburo: segregation of inputs in the telencephalon. *Anat. Embryol.* 190: 563–72.

Duman, C. H., and D. Bodznick. 1997. Distinct but overlapping populations of commissural and

GABAergic neurons in the dorsal nucleus of the little skate, *Raja erinacea*. *Brain Behav. Evol.* 49: 99–109.

Dunn, R. F., and D. M. Koester. 1984. Primary afferent projections of the octavus nerve to the inferior reticular formation and adjacent nuclei in the elasmobranch *Rhinobatos* spec. *Brain Res.* 323: 354–59.

Fay, R. R., J. I. Kendall, A. N. Popper, and A. L. Tester. 1974. Vibration detection by the macula neglecta of sharks. *Comp. Biochem. Physiol.* 47A: 1235–40.

Fiebig, E. 1988. Connections of the corpus cerebelli in the thornback guitarfish, *Platyrhinoidis triseriata* (Elasmobranchii): a study with WGA-HRP and extracellular granule cell recording. *J. Comp. Neurol.* 268: 567–83.

Franz, V. 1931. Die Akkomodation des Selachierauges und seine Abblendungsapparat, nebst Befunden an der Retina. *Zool. Jahrb. Abt. Zool. Physiol.* 49: 323–462.

Gilbert, P. W. 1963. The visual apparatus of sharks. In *Sharks and Survival*, ed. P. W. Gilbert, 283–26. Boston: D. C. Heath.

Görner, P., and A. J. Kalmijn. 1989. Frequency response of lateral line neuromasts in the thornback ray (*Platyrhinoidis triseriata*). In *Proceedings of the 2nd International Congress of Neuroethology*, ed. J. Erber, R. Menzel, H. J. Pflüger, and D. Todt, 82. Stuttgart: Georg Thieme.

Graeber, R. C. 1978. Behavioral studies correlated with central nervous system integration of vision in sharks. In *Sensory Biology of Sharks, Skates, and Rays*, ed. E. S. Hodgson and R. F. Mathewson, 195–225. Arlington, Va.: Office of Naval Research.

Gruber, S. H., and J. L. Cohen. 1978. Visual system of the elasmobranchs: state of the art 1960–1975. In *Sensory Biology of Sharks, Skates, and Rays*, ed. E. S. Hodgson and R. F. Mathewson, 11–116. Arlington, Va.: Office of Naval Research.

———. 1985. Visual system of the white shark, *Carcharodon carcharias*, with emphasis on retinal structure. *Mem. S. Calif. Acad. Sci.* 9: 61–72.

Gruber, S. H., E. R. Loew, and W. N. McFarland. 1991. Rod and cone pigments of the Atlantic guitarfish, *Rhinobatos lentiginosus*. *J. Exp. Zool. Suppl.* 5: 85–87.

Hamasaki, D. I., and S. H. Gruber. 1965. The photoreceptors of the nurse shark, *Ginglymostoma cirratum*. *Bull. Mar. Sci.* 15: 1051–59.

Hawryshyn, C. W. 1991. Light-adaptation properties of the ultraviolet-sensitive cone mechanism in comparison to other receptor mechanisms of goldfish. *Visual Neurosci.* 6: 293–301.

Heath, A. R. 1991. The ocular tapetum lucidum: a model system of elasmobranch sexual development and behavior. *J. Exp. Zool.* 5: 41–45.

Hjelmstad, G. O., G. Parks, and D. Bodznick. 1996. Motor corollary discharge activity and sensory responses related to ventilation in the skate vestibulolateral cerebellum: implications for electrosensory processing. *J. Exp. Biol.* 199: 673–81.

Hodgson, E. S., and R. F. Mathewson. 1978. Electrophysiological studies of chemoreception in elasmobranchs. In *Sensory Biology of Sharks, Skates, and Rays*, ed. E. S. Hodgson and R. F. Mathewson, 227–67. Arlington, Va.: Office of Naval Research.

Howes, G. B. 1883. The presence of a tympanum in the genus *Raia*. *J. Anat. Physiol.* 17: 188–90.

Hueter, R. E. 1991. Adaptations for spatial vision in sharks. *J. Exp. Zool. Suppl.* 5: 130–41.

Jacob, B. A., J. D. McEachran, and P. L. Lyons. 1994. Electric organs in skates: variation and phylogenetic significance (Chondrichthyes: Rajoidei). *J. Morphol.* 221: 45–63.

Johnson, S. E. 1917. Structure and development of the sense organs of the lateral line canal system of selachians (*Mustelus canis* and *Squalus acanthias*). *J. Comp. Neurol.* 28: 1–74.

Kalmijn, A. J. 1966. Electro-perception in sharks and rays. *Nature* 212: 1232–33.

———. 1971. The electric sense of sharks and rays. *J. Exp. Biol.* 55: 371–83.

———. 1974. The detection of electric fields from inanimate and animate sources other than electric organs. In *Handbook of Sensory Physiology*, vol. 3, pt. 3: *Electroreceptors and Other Specialized Receptors in Lower Vertebrates*, ed. A. Fessard, 147–200. New York: Springer-Verlag.

———. 1982. Electric and magnetic field detection in elasmobranch fishes. *Science* 218: 916–18.

———. 1984. Theory of electromagnetic orientation: a further analysis. In *Comparative Physiology of Sensory Systems*, ed. L. Bolis, R. D. Keynes, and S. H. P. Madrell, 525–60. Cambridge: Cambridge University Press.

———. 1988a. Hydrodynamic and acoustic field detection. In *Sensory Biology of Aquatic Animals*, ed. J. Atema, R. R. Fay, A. N. Popper, and W. N. Tavolga, 83–130. New York: Springer-Verlag.

———. 1988b. Detection of weak electric fields. In *Sensory Biology of Aquatic Animals*, ed. J. Atema, R. R. Fay, A. N. Popper, and W. N. Tavolga, 151–186. New York: Springer-Verlag.

———. 1997. Electric and near-field acoustic detection: a comparative study. *Acta Physiol. Scand.* 161: 25–38.

Kelly, J. C., and D. R. Nelson. 1975. Hearing thresholds of the horn shark, *Heterodontus francisci*. *J. Acoust. Soc. Am.* 58: 905–9.

Kleerekoper, H. 1978. Chemoreception and its interaction with flow and light perception in the locomotion and orientation of some elasmobranchs. In *Sensory Biology of Sharks, Skates,* and Rays, ed. E. S. Hodgson and R. F. Mathewson, 269–29. Arlington, Va.: Office of Naval Research.

Klimley, A. P. 1993. Highly directional swimming by scalloped hammerhead sharks, *Sphyrna lewini*, and subsurface irradiance, temperature, bathymetry, and geomagnetic field. *Mar. Biol.* 117: 1–22.

Koester, D. M. 1983. Central projections of the octavolateralis nerves of the clearnose skate, *Raja eglanteria*. *J. Comp. Neurol.* 221: 199–215.

Kritzler, H., and L. Wood. 1961. Provisional audiogram for the shark, *Carcharhinus leucas*. *Science* 133: 1480–82.

Lowenstein, O., and A. Sand. 1936. The activity of the horizontal semicircular canal of the dogfish, *Scyllium canicula*. *J. Exp. Biol.* 13: 416–28.

———. 1940a. The mechanism of the semicircular canal. A study of the responses of single-fibre preparations to angular accelerations and to rotation at constant speed. *Proc. R. Soc. Lond.* 129B: 256–75.

———. 1940b. The individual and integrated activity of the semicircular canals of the elasmobranch labyrinth. *J. Physiol.* 99: 89–01.

Lowenstein, O., and T. D. M. Roberts. 1950. The equilibrium function of the otolith organs of the thornback ray (*Raja clavata*). *J. Physiol.* 110: 392–415.

———. 1951. The localization and analysis of the responses to vibration from the isolated elasmobranch labyrinth. A contribution to the problem of the evolution of hearing in vertebrates. *J. Physiol.* 114: 471–89.

Lowenstein, O., M. P. Osborne, and J. Wersäll. 1964. Structure and innervation of the sensory epithelia of the labyrinth in the thornback ray (*Raja clavata*). *Proc. R. Soc. Lond.* 160B: 1–12.

Lu, J., and H. M. Fishman. 1994. Interaction of apical and basal membrane ion channels underlies electroreception in ampullary epithelia of skate. *Biophys. J.* 67: 1525–33.

McCormick, C. A. 1992. Evolution of central auditory pathways in anamniotes. In *The Evolutionary Biology of Hearing*, ed. R. R. Fay and A. N. Popper, 323–50. New York: Springer-Verlag.

McFarland, W. 1991. Light in the sea: the optical world of elasmobranchs. *J. Exp. Zool. Suppl.* 5: 3–12.

Montgomery, J. C. 1984a. Noise cancellation in the electrosensory system of the thornback ray; common mode rejection of input produced by the animals's own ventilatory movement. *J. Comp. Physiol.* 155A: 103–11.

———. 1984b. Frequency response characteristics of primary and secondary neurons in the electrosensory system of the thornback ray. *Comp. Biochem. Physiol.* 79A: 189–95.

———. 1988. Sensory physiology. In *Physiology of Elasmobranch Fishes*, ed. T. J. Shuttleworth, 79–98. New York: Springer-Verlag.

Montgomery, J. C., and D. Bodznick. 1991. Properties of medullary interneurons of the skate electrosense provide evidence for the neural circuitry mediating ventilatory noise suppression. *Biol. Bull.* 181: 326

———. 1993. Hindbrain circuitry mediating common mode suppression of ventilatory reafference in the electrosensory system of the little skate, *Raja erinacea*. *J. Exp. Biol.* 183: 203–15.

Montgomery, J. C., S. Coombs, R. A. Conley, and D. Bodznick. 1995. Hindbrain sensory processing in lateral line, electrosensory, and auditory systems: a comparative overview of anatomical and functional studies. *Audit. Neurosci.* 1: 207–31.

Murphy, C. J., and H. C. Howland. 1991. The functional significance of crescent-shaped pupils and multiple pupillary apertures. *J. Exp. Zool. Suppl.* 5: 22–28.

Murray, R. W. 1960. Electrical sensitivity of the ampullae of Lorenzini. *Nature* 187: 957.

———. 1962. The response of the ampullae of Lorenzini of elasmobranchs to electrical stimulation. *J. Exp. Biol.* 39: 119–28.

———. 1965. Receptor mechanisms in the ampullae of Lorenzini of elasmobranch fishes. *Cold Spring Harb. Symp. Quant. Biol.* 30: 235–62.

Murray, R. W., and Potts, W. T. W. 1961. The composition of the endolymph, perilymph and other body fluids of elasmobranchs. *Comp. Biochem. Physiol.* 2: 65–75.

Myrberg, A. A., Jr. 1978. Underwater sound—its effect on the behavior of sharks. In *Sensory Biology of Sharks, Skates, and Rays*, ed. E. S. Hodgson and R. F. Mathewson, 391–17. Arlington, Va.: Office of Naval Research.

Myrberg, A. A., Jr., A. Banner, and J. D. Richard. 1969. Shark attraction using a video-acoustic system. *Mar. Biol.* 2: 264–76.

Myrberg, A. A., Jr., C. R. Gordon, and A. P. Klimley. 1976. Attraction of free-ranging sharks by low-frequency sound, with comments on its biological significance. In *Sound Reception in Fish*, ed. A. Schuijf and A. D. Hawkins, 205–28. Amsterdam: Elsevier.

Myrberg, A. A., Jr., S. J. Ha, S. Walewski, and J. C. Banbury. 1972. Effectiveness of acoustic signals in attracting epipelagic sharks to an underwater sound source. *Bull. Mar. Sci.* 22: 926–49.

Nelson, D. R. 1967. Hearing thresholds, frequency discrimination, and acoustic orientation in the lemon shark, *Negaprion brevirostris* (Poey). *Bull. Mar. Sci.* 17: 741–68.

Nelson, D. R., and S. H. Gruber. 1963. Sharks: attraction by low-frequency sounds. *Science* 142: 975–77.

Nelson, D. R., and R. H. Johnson. 1972. Acoustic attraction of Pacific reef sharks: effect of pulse intermittency and variability. *Comp. Biochem. Physiol.* 42A: 85–95.

———. 1976. Some recent observations on acoustic attraction of Pacific reef sharks. In *Sound Reception in Fish*, ed. A. Schuijf and A. D. Hawkins, 229–40. Amsterdam: Elsevier.

Nelson, D. R., R. H. Johnson, and L. G. Waldorp. 1969. Responses in Bahamian sharks and groupers to low-frequency, pulsed sounds. *Bull. S. Calif. Acad. Sci.* 68: 131–37.

Nelson, M. E., and M. G. Paulin. 1995. Neural stimulations of adaptive reafference suppression in the elasmobranch electrosensory system. *J. Comp. Physiol.* 177A: 723–36.

New, J. G. 1990. Medullary electrosensory processing in the little skate: I. Response characteristics of neurons in the dorsal octavolateralis nucleus. *J. Comp. Physiol.* 167A: 285–94.

New, J. G., and D. Bodznick. 1990. Medullary electrosensory processing in the little skate: II. Suppression of self-generated electrosensory interference during respiration. *J. Comp. Physiol.* 167A: 295–307.

Nier, K. 1976. Cutaneous sensitivity to touch and low frequency vibration in selachians. *J. Comp. Physiol.* 109A: 345–55.

Nikonorov, S. I. 1983. Electrophysiological analysis of convergent relations between the olfactory and visual analyzers in the forebrain of *Squalus acanthias*. *J. Evol. Biochem. Physiol.* 18: 268–73.

Nikonov, A. A., Y. N. Ilyin, O. M. Zherelova, and E. E. Fesenko. 1990. Odour thresholds of the black sea skate (*Raja clavata*). Electrophysiological study. *Comp. Biochem. Physiol.* 95A: 325–28.

Northcutt, R. G. 1989. The phylogenetic distribution and innervation of craniate mechanoreceptive lateral lines. In *The Mechanosensory Lateral Line. Neurobiology and Evolution*, ed. S. Coombs, P. Görner, and H. Münz, 17–8. New York: Springer-Verlag.

———. 1991. Visual pathways in elasmobranchs: organization and phylogenetic implications. *J. Exp. Zool. Suppl.* 5: 97–107.

Parker, G. H. 1904. The function of the lateral-line organs in fishes. *Bull. U.S. Bur. Fish.* 24: 185–207.

———. 1909. The sense of hearing in the dogfish. *Science* 29: 428.

———. 1914. The directive influence of the sense of smell in the dogfish. *Bull. U.S. Bur. Fish.* 33: 61–68.

Paulin, M.G. 1995. Electroreception and the compass sense of sharks. *J. Theor. Biol.* 174: 325–39.

Peabody, J. E. 1897. The ampullae of Lorenzini of the Selachii. *Zool. Bull.* 1: 163–78.

Peterson, E. H., and M. H. Rowe. 1980. Different regional specialization of neurons in the ganglion cell layer and inner plexiform layer of the California horn shark, *Heterodontus francisci*. *Brain Res.* 201: 195–201.

Plassmann, W. 1982. Central projections of the octaval system in the thornback ray *Platyrhinoidis triseriata*. *Neurosci. Lett.* 32: 229–33.

———. 1983. Sensory modality interdependance in the octaval system of an elasmobranch. *Exp. Brain Res.* 50: 283–92.

Platt, C. J., T. H. Bullock, G. Czéh, N. Kovacevic, D. Konjevic, and M. Gojkovic. 1974. Comparison of electroreceptor, mechanoreceptor, and optic evoked potentials in the brain of some rays and sharks. *J. Comp. Physiol.* 95A: 323–55.

Puzdrowski, R. L., and R. B. Leonard. 1993. The octavolateral systems in the stingray, *Dasyatis sabina*: I. Primary projections of the octaval and lateral line nerves. *J. Comp. Neurol.* 332: 21–37.

Ressler, K. J., S. L. Sullivan, and L. B. Buck. 1994. A molecular dissection of spatial patterning in the olfactory system. *Curr. Opin. Neurobiol.* 4: 588–96.

Ripps, H., and J. E. Dowling. 1991 Structural features and adaptive properties of photoreceptors in the skate retina. *J. Exp. Zool. Suppl.* 5: 46–54.

Roth, A. 1972. Wozu dienen die Elektrorezeptoren der Welse? *J. Comp. Physiol.* 79A: 113–35.

Sand, A. 1937. The mechanism of the lateral line sense organs of fishes. *Proc. R. Soc. Lond.* 123B: 427–95.

Schweitzer, J. 1983. The physiological and anatomical localization of two electroreceptive diencephalic nuclei in the thornback ray, *Platyrhinoidis triseriata*. *J. Comp. Physiol.* 153A: 331–41.

———. 1986a. Functional organization of the electroreceptive midbrain in an elasmobranch (*Platyrhinoidis triseriata*): a single unit study. *J. Comp. Physiol.* 158A: 43–58.

———. 1986b. The neural basis of electroreception in elasmobranchs. In *Indo-Pacific Fish Biology: Proceedings of the Second International Conference on Indo-Pacific Fishes,* ed. T. Uyeno, T. Arai, T. Taniuchi, and K. Matsuura, 392–407. Tokyo: Ichthyological Society of Japan.

Schweitzer, J., and D. A. Lowe. 1984. Mesencephalic and diencephalic cobalt-lysine injections in an elasmobranch: evidence for two parallel electrosensory pathways. *Neurosci. Lett.* 44: 317–22.

Sheldon, R. E. 1909. The reactions of dogfish to chemical stimuli. *J. Comp. Neurol.* 19: 273–311.

Silver, W. L. 1979. Olfactory responses from a marine elasmobranch, the Atlantic stingray, *Dasyatis sabina*. *Mar. Behav. Physiol.* 6: 297–306.

Sivak, J. G. 1990. Optic variability of the fish lens. In *The Visual System of Fish*, ed. R. Douglas and J. Djamgoz, 63–80. London: Chapman & Hall.

———. 1991 Elasmobranch visual optics. *J. Exp. Zool. Suppl.* 5: 13–21.

Szamier, R. B., and H. Ripps. 1983. The visual cells of the skate retina: structure, histochemistry, and disc-shedding properties. *J. Comp. Neurol.* 215: 51–62.

Tester, A. L. 1963. Olfactation, gustation, and the common chemical sense in sharks. In *Sharks and Survival*, ed. P. W. Gilbert, 255–82. Boston: D. C. Heath.

Tester, A. L., J. I. Kendall, and W. B. Milisen. 1972. Morphology of the ear of the shark genus *Carcharhinus* with particular reference to the macula neglecta. *Pacif. Sci.* 26: 264–74.

Tester, A. L., and G. J. Nelson. 1967. Free neuromasts (pit organs) in sharks. In *Sharks, Skates, and Rays*, ed. P. W. Gilbert, R. F. Mathewson, and D. P. Rall, 503–31. Baltimore: Johns Hopkins University Press.

Tong, S. L., and T. H. Bullock. 1982. The sensory functions of the cerebellum of the thornback ray, *Platyrhinoidis triseriata*. *J. Comp. Physiol.* 148A: 399–410.

Tricas, T. C., and J. G. New. 1998. Sensitivity and response dynamics of elasmobranch electrosensory primary afferent neurons to near threshold fields. *J. Comp. Physiol.* 182A: 89–101.

Waltmann, B. 1966. Electrical properties and fine structure of the ampullary canals of Lorenzini. *Acta Physiol. Scand. Suppl.* 66: 1–60.

Waterman, T. H., and H. Hashimoto. 1974. E-vector discrimination by the goldfish optic tectum. *J. Comp. Physiol.* 95A: 1–12.

Wright, T., and R. Jackson. 1964. Instrumental conditioning of young sharks. *Copeia* 1964: 409–12.

Yew, D. T., Y. W. Chan, M. Lee, and S. Lam. 1984. A biophysical, morphological, and morphometrical survey of the eye of the small shark (*Hemiscyllium plagiosum*). *Anat. Anz.* 155: 355–63.

Zigman, S. 1991. Comparative biochemistry and biophysics of elasmobranch lenses. *J. Exp. Zool. Suppl.* 5: 29–40.

KENNETH R. OLSON

CHAPTER 13

Rectal Gland and Volume Homeostasis

Marine elasmobranchs are unusual among aquatic vertebrates in that their extracellular body fluids are nearly isosmotic to the environment, whereas the distribution of electrolytes between the hypersaline environment and body fluids is in considerable disequilibrium. Therefore, while the need to maintain water balance is almost inconsequential, regulation of salt balance is of paramount importance, and a unique salt-secreting organ, the rectal gland, has developed to assist in the salt elimination process. Because of its salt-secreting capacity and because of its applicability as a model for salt-secreting epithelia, the rectal gland has received considerable attention from physiologists and anatomists alike. The intent of this chapter is to describe the osmotic and volume regulatory characteristics of elasmobranchs, the role of the rectal gland in achieving these ends, and the anatomical features that are important to the secretory capacity of the gland. Detailed descriptions of other osmoregulatory organs such as kidney, gills, and gastrointestinal tract are provided elsewhere in this volume.

Fluid compartments

Vertebrate body fluids are broadly defined as belonging to either an intracellular or extracellular compartment. Most, if not all, cells are equipped with a number of transmembrane pumps and intracellular osmolyte regulating systems that enable them to regulate intracellular volume (Schmidt-Nielsen 1975; Ballantyne et al. 1987; Kleinzeller and Ziyadeh 1990;

McConnell and Goldstein 1990; Lang et al. 1990). However, individual cells appear to have little, if any, active role in directly regulating the volume of the fluid surrounding them, that is, the extracellular compartment (Olson 1992). Extracellular fluids are regulated by factors that govern water and osmolyte transfer across gill, skin, renal, and gastrointestinal (GI) epithelia. The extracellular compartment can be further divided into interstitial, intravascular, and transcellular fluids. Fluid movement between intravascular and interstitial compartments is affected by the net colloid osmotic (oncotic) and hydraulic transcapillary pressure difference between the compartments and by the permeability of the capillary endothelium to colloid, namely plasma protein. The lymphatic system, well characterized in higher vertebrates but poorly understood in fish, is also important in the return of interstitial fluid containing extravasated colloid back to the intravascular compartment. Transcellular fluids are enclosed within an epithelium and include fluid in the lumen of the GI tract; urinary and gall bladders; pericardial, peritoneal, and synovial spaces; renal tubules; and ocular and cerebrospinal fluid. Transcellular fluid volumes are also determined by osmotic and hydraulic pressure gradients and by epithelial permeability.

Water is not actively transported by biological membranes, and therefore fluid movement across both epithelial and endothelial surfaces is regulated through generation of osmotic and hydraulic pressure gradients and by adjusting the permeability of compartmental barriers. In elasmobranchs, two electrolytes, sodium (Na^+) and chloride (Cl^-), plus two nonelectrolytes, urea and trimethyl amine oxide (TMAO), account for the majority of the osmotically active particles of the extracellular fluids. Because these small molecules are freely permeable across the endothelial membrane, they have little effect on transcapillary fluid balance. However, they are the primary osmotic determinants of fluid movement between the elasmobranch and its environment. Thus regulation of the extracellular fluid volume is achieved by adjusting the total amount of extracellular osmolytes relative to the osmolarity of the environment.

Fluid compartment volumes of several elasmobranchs and a representative cyclostome and teleost are listed in Table 13-1. Although there is substantial variation in the values for

TABLE 13-1 FLUID COMPARTMENT VOLUMES IN CYCLOSTOMES, ELASMOBRANCHS, AND TELEOSTS

Species	Habitat	Blood volume	Extracellular volume	Total water volume
CYCLOSTOME				
Eptatretus stoutii	SW	206[a]	254[a]	746[a]
ELASMOBRANCH				
Squalus acanthias	SW	68[b]	121[b], 202[b], 122[c]	683[b], 712[d]
Raja binoculata	SW	80[b]	126[b]	787[b]
R. rhina	SW	72[b]	112[b]	781[b]
Potamotrygon hystrix	FW			793[e]
TELEOST				
Oncorhynchus mykiss	SW	47[f]	172[f]	736[g]
O. mykiss	FW	37[f]	222[f]	773[g]

SOURCES: [a]McCarthy and Conte (1966); [b]Thorson (1958); [c]Benyajati and Yokota (1989); [d]Robertson (1975); [e]Bittner and Lang (1980); [f]M. Kellogg and K. R. Olson (unpublished); [g]Eddy and Bath (1979).
Volume is expressed in mL/kg body weight.
FW = freshwater; SW = saltwater.

fluid volumes in fish that can be attributed to methodology (Olson 1992), there appears to be a phylogenetic trend for blood volume reduction in the higher forms of fish such as teleosts. Extracellular fluid volume, on the other hand, does not follow this trend, as elasmobranchs have the lowest volume of any of the fish measured. Elasmobranchs typically have larger blood volumes and smaller extracellular fluid spaces than teleosts, while total fluid volumes are comparable. Some of the discrepancy between elasmobranch and teleost blood volumes may be attributable to the current uncertainty of fluid compartment measurements in teleosts. Teleosts have been reported to have an additional intravascular compartment, the secondary circulation, which may be 1.5 times larger than the blood volume measured by most indicator dilution methods (Steffensen and Lomholt 1992). It has been suggested that this volume is inadvertently excluded as a result of inadequate mixing or poor distribution of intravascular indicators (Steffensen and Lomholt 1992; but see Olson 1996). If the secondary circulation is a true vascular volume, the blood volumes of teleosts and elasmobranchs are equivalent. Similarly, the relatively small extracellular volume of elasmobranchs determined by Thorson (1958) may also be an artifact of an inadequate mixing period (4 hours) of indicator. Nevertheless, it appears that extracellular and, especially, interstitial fluid volume in elasmobranchs may be among the lowest found in any vertebrate. Additional volume measurements are required to determine if this is indeed the case. If so, one must question why, especially since elasmobranchs live in a slightly hydrating environment.

Osmoregulation

Saltwater fish have developed a variety of adaptive strategies for coping in an environment with an osmolarity of around 1000 mOsm/L. Teleosts retain the extracellular ion concentrations established in their freshwater ancestors, around 150 mmol NaCl. While this offsets the intracellular osmolytes, potassium (K^+), Cl^-, and organic osmolytes, it is considerably hypoosmotic to the environment. Volume depletion in saltwater teleosts is countered by drinking seawater, while the ingested Na^+ and Cl^- (along with salt entering by diffusion) is cleared from the body by active extrusion across the gills. Elasmobranchs, on the other hand, are slightly hyperosmotic with their environment. They maintain plasma Na^+ and Cl^- around 250 mmol, while the remaining ~500 mOsm/L consists of organic osmolytes, namely urea and TMAO (Ballantyne et al. 1987). Cyclostomes such as the hagfish are also slightly hyperosmotic to their environment; however, Na^+ and Cl^- are the major constituents of their extracellular fluids (Holmes and Donaldson 1969).

Although earlier studies indicated that elasmobranchs were isosmotic with their environment, accurate comparisons of plasma and ambient osmolarity have shown that elasmobranch plasma is often slightly hyperosmotic to seawater (Table 13-2). Thus there is a slight but continuous influx of water into the fish. The excess water is excreted by the kidneys,

TABLE 13-2 OSMOTIC ACTIVITY AND PRINCIPAL OSMOLYTES IN FLUIDS OF *SQUALUS ACANTHIAS*

Fluid	Osmolarity (mOsm/L)	Sodium (mmol/L)	Chloride (mmol/L)	Urea (mmol/L)
Plasma	1018	286	246	351.0
Urine	780	337	203	~0
Rectal gland	1018	540	533	14.5
Seawater	930	440	495	~0

SOURCE: Burger and Hess (1960).

whose major accomplishment lies not only in the excretion of water but in the simultaneous retention of the organic osmolytes, urea, and TMAO. This process is described in detail in Chapter 14.

Table 13-2 shows that there is a substantial gradient for diffusive entry of both Na^+ and Cl^- into marine elasmobranchs. There is also an active accumulation of Na^+ and Cl^- by the gills as part of an acid-base regulatory system involved in hydrogen ion and bicarbonate excretion (Evans 1982, 1984). Although there is substantial efflux of salt across the gills, this is still less than the rate of influx, and thus there is a net salt accumulation by the gills (Shuttleworth 1988). Na^+ and Cl^- accumulation is further exacerbated when the fish ingests additional quantities of salt from both seawater and prey during feeding. As efficient as the elasmobranch kidney is in water, urea, and TMAO balance, it cannot increase salt excretion without a parallel increase in urine volume because it lacks the ability to produce urine with an electrolyte concentration greater than plasma. Thus the long-term capacity of the kidney to excrete salt is dependent on the commensurate osmotic influx of water. Only the rectal gland has the capability to secrete Na^+ and Cl^- in greater concentrations than that of the plasma. The significance of this gland, then, becomes apparent when water influx is reduced or, more important, when salt intake is increased.

The rectal gland

HISTORICAL PERSPECTIVE

The unique structure now known as the rectal gland was first described by Severini in 1645 (Loretz 1987) and was originally named the appendix digitiformis by Monroe in 1785 (in Hoskins 1917). In the early 1800s the rectal gland was described by a number of anatomists and given various names, such as bursa cloacae by Retzius (1819) (Hoskins 1917), Darmanhang by Carus (1834) (Hoskins 1917), caecal appendage by Owen (1846) (Hoskins 1917), superanal gland by Blanchard (1882a,b) (Hoskins 1917), and anal gland by Crawford (1899) (Hoskins 1917), as well as others (see Loretz 1987). The first histological description of the gland was provided by Blanchard (1878a,b) (Hoskins 1917), who, along with Leydig (1852) (Hoskins 1917) and Pillet (1885) (Hoskins 1917), compared the secretory tubules to the secretory cells in the mammalian intestinal Brunner's glands and crypts of Lieberkühn. Early hypotheses of the physiological attributes of the rectal gland were no less imaginative and, as summarized by Hoskins (1917), included similarities to the ink bag of cuttlefish and avian intestinal caeca (Home 1814), an accessory reproductive organ (Hyrtl 1858) (Hoskins 1917), the urinary bladder (Milne-Edwards 1862) (Hoskins 1917), and the appendix of mammals (Howes 1891) (Hoskins 1917). The rectal gland was also originally thought to have digestive capability by Blanchard (1882a,b); Crawford (1899); Pixell (1908) (Hoskins 1917), and to secrete a factor that contracted the intestine (Morgera 1910) (Hoskins 1917). It was not until the early 1960s that combined anatomical and physiological studies began to clearly identify the fine structure of the rectal gland and to clarify its role in salt metabolism.

RECTAL GLAND ANATOMY
General features

The rectal gland is a blind-end tube, suspended in the dorsal mesentery above the intestine and attached to the latter in the postvalvular region (Fig. 13-1). It weighs between 100 and 200 mg/kg body weight in most adult marine and brackish-water elasmobranchs (see Thorson et al. 1978; Oguri 1981), although in the spiny dogfish, *Squalus acanthias*, weights of 444 (Bonting 1966) and 600 (Burger 1972) mg/kg have been reported. Rectal glands are reduced (20–60 mg/kg) in euryhaline species and smaller still (15 mg/kg) in the freshwater stingray, *Potamotrygon circularis* (Thorson et al. 1978). In 255–780 g spotted dogfish, *Scyliorhinus canicula*, the gland ranges in length from 0.9 to 1.6 cm; in a 273-kg tiger shark, *Galeocerdo cuvieri*, the gland was 15.8 cm long and weighed 30.8 g (Oguri 1981). In sharks such as *Squalus*, the gland is long and slender, whereas in the stingray *Urolophus jamaicensis*, it is S-shaped and lobulated.

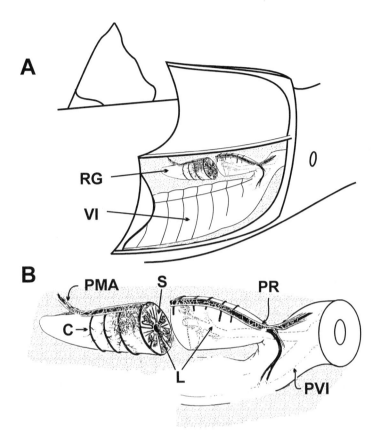

FIGURE 13-1 General features of the rectal gland. **A:** General appearance of the gland (RG) in *Squalus acanthias* (side view, anterior to left) and its relationship to the valvular intestine (VI). **B:** the three vascular pathways in the gland. The posterior mesenteric artery (PMA) travels along the dorsum of the capsule and gives rise to numerous circumferential arteries (C) that encircle the gland. Circumferential arteries form progressively smaller arterioles that ultimately feed sinusoidal capillaries (S) radiating inward, along with the tubules, toward the lumen. Luminal veins (not shown) drain the capillaries into the postvalvular intestine (PVI), thus forming the first pathway. The second pathway is derived from arteriovenous anastomoses in the capsular arteries that connect directly to a capsular venous plexus; it either flows retrograde along the PMA or follows the posterior ramus (PR) of the PMA into the PVI. The PR also continues directly into the VI, forming a third, shunt pathway.

The adult rectal gland has three concentric tissue layers that surround the hollow lumen of the central canal: an outer capsule, a middle secretory parenchyma, and an inner margin of transitional epithelium. The capsule is covered with a visceral peritoneum and is invested with blood vessels, connective tissue, smooth muscle, and a network of nerves. The secretory parenchyma consists of radially oriented tubules and an extratubular matrix of connective tissue interspersed with capillaries and nerve fibers. A single tubule may traverse the entire parenchyma, or and more commonly, it may branch as it moves outward from the central canal into three to five generations of daughter tubules. In the peripheral portion of the secretory parenchyma the tubules are tightly packed, lined with a single type of columnar cell, and have a narrow lumen (Fig. 13-2). Here the tubules are highly ordered, radiating away from the central canal of the gland, although an occasional tubule may turn parallel to the axis of the gland. The extratubular matrix is compact and capillaries are closely

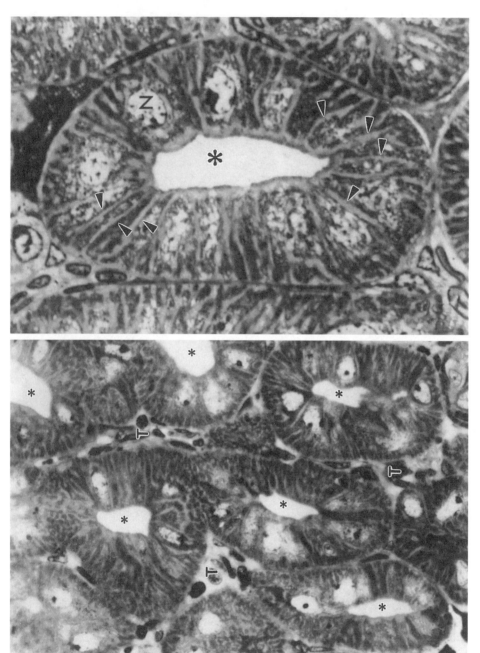

FIGURE 13-2 Cross sections of *Squalus acanthias* rectal gland tubules. **A:** Six tubules with clear lumens (*asterisks*) and surrounded by intertubular connective tissue (T) (light micrograph, ×450). **B:** Cross section of a single tubule showing tubular columnar cells with large central nucleus (N) and clear lumen (*asterisk*). Light bands between cells (*arrowheads*) are extensively interdigitated lateral membranes (light micrograph, ×1250). Micrographs courtesy of Dr. K. J. Karnaky, Jr. (From Eveloff et al. 1979, with permission.)

associated with, and run parallel to, the tubules. In the central portion of the secretory parenchyma the tubules become more randomly oriented and the tubular lumen is large. Capillaries are often replaced by venous sinuses, and the extratubular matrix is less compact. Nerve fibers with vasoactive intestinal peptide immunoreactivity are closely associated with the tubular cells (Stoff et al. 1988). These fibers, like the tubules they are associated with, are well ordered in the outer secretory parenchyma and ramify extensively in the venous sinusoids of the inner parenchyma. The inner layer of transitional epithelium is four to eight cells thick. At the caudal end of the gland, where it becomes embedded in the postvalvular intestine, the secretory parenchyma is reduced and the ductal epithelium predominates.

Embryological development

Hoskins (1917) provided a detailed description of the development of the rectal gland in *Squalus acanthias*. The anlage of the rectal gland first appears in 15-mm-long embryos as a slight bulge in the right dorsolateral surface of the intestine midway between the spiral valve and the cloaca. A definite bud appears in 19-mm embryos, and as it develops it turns anteriorly and attains its tubular features by the time the embryos are 22 mm long. Primary tubules appear in 28-mm embryos, and when the embryo is 33 mm long, secondary tubules have begun to form. Further elaboration of the tubules in length and number of branches (up to six generations) continues until the embryo is around 200 mm long, at which time the gland has attained the gross characteristics of the adult.

As the gland initially develops in the embryo its vasculature is derived from capillaries of the intestine. The circulation is random, sinusoidal, and large peripheral sinuses are connected to central sinuses by wide channels. As the embryo grows beyond 100 mm, the sinusoidal circulation is lost and the adult circulation, where a peripheral, circumferential arterial system feeds capillaries that drain into a central venous system, becomes prominent.

FINE STRUCTURE

Secretory cells

Initial electron microscopic studies by Doyle (1962), Bulger (1963), Komnick and Wohlfarth-Bottermann (1966), and van Lennep (1968) clearly showed that the tubular secretory cells of all elasmobranch rectal glands were similar and share features common to other salt-secreting epithelia. The secretory tubules generally consist of a single type and a single layer, of columnar epithelium (see Fig. 13-2). Bulger (1963) noted that cells could be categorized as "light" or "dark" based on the density of the cytoplasmic matrix; however, it is not clear if these are different cells or if they represent variations in cell activity. A rare mucus cell may also be present in the more medial ducts. Tubular cells in all areas of the secretory parenchyma are similar (Bulger 1963) but may vary in height from 8 to 18 µm, depending on the degree of distention of the lumen (Doyle 1962). From six to nine cells surround the lumen in *Urolophus* (Doyle 1962). The two most notable features of tubular secretory cells are the numerous mitochondria and the extensive elaboration of the basolateral membrane.

The columnar secretory cells are structurally bipolar and are arranged radially around the tubule (see Fig. 13-2). Nuclei are centrally located and the Golgi apparatus is nearby, along with various 350-Å-diameter vacuoles, the latter being especially numerous in the supranuclear region (Doyle 1962). Other cellular organelles common to all cells are also present. The most striking intracellular feature of the secretory cells is the densely packed mitochondria that can be found in all but the apical (luminal) region of the cell (Figs. 13-3 and 13-4). In much of the basal and lateral area the mitochondria are so numerous that there appears to be little room for other cellular organelles. Mitochondria in *Squalus* have a somewhat larger diameter and the cristae are closer together than in *Urolophus*, but the physiological significance of this is unclear (Doyle 1962).

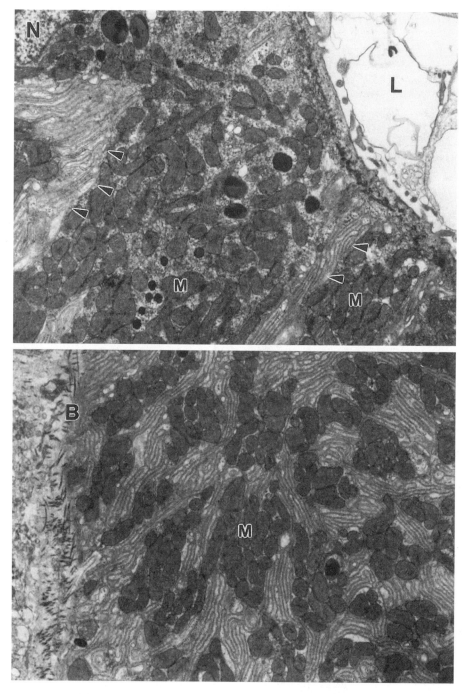

FIGURE 13-3 Electron micrographs of tubular cells. **Top:** Longitudinal section of tubular cell. Apical membrane facing tubular lumen (L) has only several blunt microvilli and few mitochondria (M) are nearby, whereas mitochondria become densely packed along infoldings of lateral membranes (*arrowheads*). Cell nucleus (N) is visible in upper left (×9000). **Bottom:** Oblique section through basal membrane (B) showing infoldings of basolateral membranes and their relationship to the densely packed mitochondria (M) (×9000). Micrographs courtesy of Dr. K. J. Karnaky, Jr. (From Eveloff et al. 1979, with permission.)

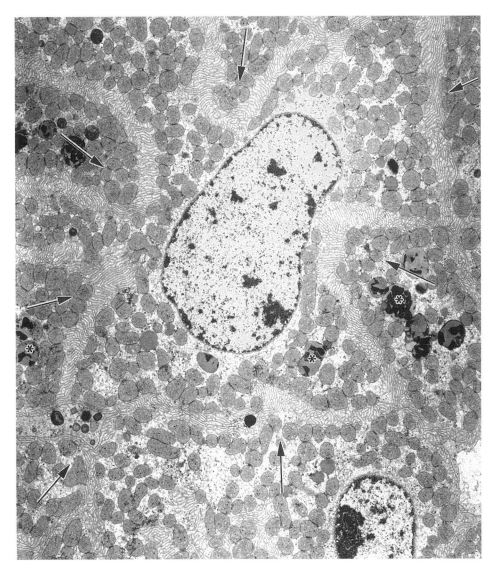

FIGURE 13-4 Electron micrograph of cross section through tubular cell at the level of the nucleus, showing extensive primary cytoplasmic projections, secondary infolding of lateral membranes between adjacent cells, and densely packed mitochondria. *Arrows* indicate primary projections from adjacent cells; clusters of lysosomes (*asterisks*) are also evident (×10,500). Micrograph courtesy of Dr. S. A. Ernst. (From Ernst et al. 1981, with permission.)

Few mitochondria are present in the apical cytoplasm, where membrane-enclosed vesicles predominate (see Fig. 13-3). In *Urolophus* these vesicles are larger (550 Å diameter) than the supranuclear vesicles and have thinner walls (Doyle 1962). They also appear to fuse with the apical membrane.

The apical membrane is studded with stubby microvilli (0.2 µm diameter and 0.4 µm high) and "naplike" filaments project outward from the apical membrane between and on the microvilli (Figs. 13-3 and 13-5; see also Bulger 1963). Freeze-fracture replicas of the apical membrane (Fig. 13-6) show that the microvilli are randomly distributed, and occasionally two or three are fused together to form a short ridge (Ernst et al. 1981). With exception of microvilli, the apical membranes are relatively flat.

Unlike the apical membrane, the basal and lateral membranes are extensively elaborated to increase their surface area (Bulger 1963; Komnick and Wohlfarth-Bottermann 1966; van Lennep 1968; Ernst et al. 1981). Amplification of the lateral membrane is achieved through primary cytoplasmic projections that cause neighboring cells to bulge into each other, thus forming a jigsaw puzzle–like system of interlocking cells (see Fig. 13-4). This extensive cell-cell interdigitation is continuous all the way to the apical membrane, and it is strikingly evident in cross sections (see Fig. 13-5) and in freeze-fracture replicas (see Fig. 13-6) of the apical surface, where it appears even more prevalent (Ernst et al. 1981; Forrest et al. 1982). Superimposed on the primary projections are secondary labyrinthine infoldings of the plasma membrane that form tortuous intercellular channels and, in so doing, greatly augment the cell surface (see Figs. 13-3 and 13-4). The secondary system is found on both lateral and basal membranes, whereas the primary jigsaw puzzle–like projections are not observed in the basal membrane because of its close abutment to the planar basal lamina. The secondary infoldings are deeper and more numerous in *Squalus* than they are in *Urolophus* (Doyle 1962), and they are greatly

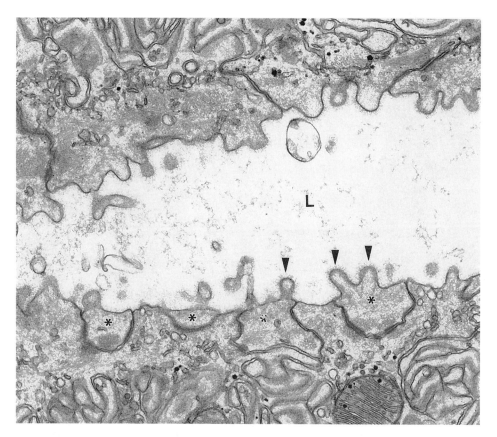

FIGURE 13-5 Transmission electron micrograph of secretory cell apical membranes lining the lumen (L) of the secretory duct. Microvillar surface is covered with glycoprotein-rich coat (*arrowheads*). Interdigitating projections from neighboring cells (*asterisks*) increase the length of the tight (occluding) junctional complex and provide an extensive paracellular pathway for diffusion of select molecules (×19,500). Micrograph courtesy of Dr. S. A. Ernst. (From Ernst et al. 1981, with permission.)

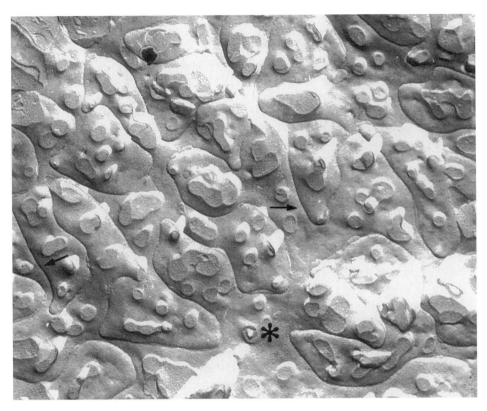

FIGURE 13-6 Freeze-fracture replica of apical membrane. Much of the surface belongs to a single cell (*asterisk*) whose surface extends throughout the figure and is bounded by the tight junctional borders between cells (two areas shown by *arrows*). Surface microvilli are cross-fractured in this micrograph (×12,100). Micrograph courtesy of Dr. S. A. Ernst. (From Ernst et al. 1981, with permission.)

reduced in the basal membrane of the Port Jackson shark, *Heterodontus portusjacksoni* (van Lennep 1968).

Cell junctions

A key link in rectal gland form and function is the junctional complex along the apical margin of tubular cells (Ernst et al. 1981; Forrest et al. 1982). Here one to five parallel strands of tight (occluding) junctions, typical of "leaky" epithelia, separate adjacent cells. Single junctional strands may even be discontinuous. These junctions are relatively shallow (56 nm in *Squalus* and 18 nm in the stingray *Dasyatis sabina*). However, their most striking feature is the length density (length of junctional complex per unit of apical surface area), which is 67 m/cm^2 in *Squalus* and 75 m/cm^2 in *Dasyatis* (Ernst et al. 1981). This provides an extensive (see Fig. 13-6) and selective paracellular diffusional pathway that is important in Na$^+$ secretion, as described subsequently, while simultaneously restricting the diffusion of other ions into the lumen and presumably preventing loss of plasma urea as well (Forrest et al. 1982). The length density in *Squalus* is greater in the inner region of the secretory parenchyma than it is in the outer region, namely, 102 versus 80 m/cm^2 (Forrest et al. 1982), which may indicate regional differences in secretory activity. The anatomical characteristics of *Squalus* tight junctions do not change when the fish is adapted to dilute (68%) seawater or when secretion rate of perfused glands is maximally stimulated (Forrest et al. 1982). Thus a change in the rate of rectal gland secretion is independent of alterations in junctional morphology.

Excretory canal

A stratified epithelium, four to eight cells thick, lines the central canal and continues into the intestinal lumen as the excretory canal (Bulger 1965; Ernst et al. 1981). Basal cells abut the basement membrane and lie beneath an intermediate cell layer, the latter surrounded by large extracellular spaces (Bulger 1965). Four types of surface cells, granular, mucus, flask-shaped, and occasional cells with large mitochondria, line the lumen. Granular surface cells are the most abundant and are cuboidal, with a central nucleus and relatively few, but unusually large mitochondria. There is relatively little interdigitation between cells, which reduces the length density of the tight junctions on surface epithelial cells to only one-tenth that of the tubular secretory cells (Ernst et al. 1981). Microvilli are prominent on the apical membrane, whereas the basolateral membranes lack the extensive infolding common in the secretory cells. Tight junctions in cells lining the lumen are formed from 10 to 18 interlocking strands, with a junctional depth of 0.6–0.8 µm, imparting an impermeable barrier to the epithelial surface. Round mucus cells are more prevalent in the ductal epithelium (Bulger 1965).

VASCULATURE

The rectal gland vasculature has been examined following vinyl acetate (Bulger 1963) or latex (Hayslett et al. 1974) infusion and also by scanning electron microscope (SEM) analysis of methyl methacrylate corrosion replicas (Kent and Olson 1982). A single artery, originating from the dorsal aorta, supplies the rectal gland (see Fig. 13-1). This vessel has been variously described as the rectal artery (Bulger 1963), posterior mesenteric artery (Kent and Olson 1982; Holmgren and Nilsson 1983), and digitiform artery (Hoskins 1917). This artery enters the distal (anterior) third of the gland and bifurcates into anterior and posterior rami that travel the length of the dorsal aspect of the gland. Paired circumferential arteries arise at 3 mm intervals from the rami (Figs. 13-1 and 13-7) and as they encircle the gland they give rise to consecutively smaller branches, thereby forming an arteriolar plexus in the outer capsule. The large posterior ramus continues along the dorsal surface of the gland and enters the postvalvular intestine.

Capsular arterioles perfuse two distinct circulations in the gland. They either supply capillaries in the secretory parenchyma or they connect directly to capsular venules through arteriovenous anastomoses (AVAs) (Fig. 13-8).

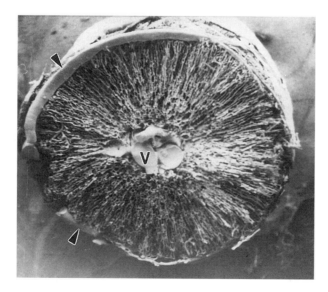

FIGURE 13-7 Corrosion replica of rectal gland vasculature viewed perpendicular to long axis of gland. Circumferential arteries (*arrowheads*) feed parenchymal capillaries that in turn drain into one of several central veins (V) (SEM, ×15). (From Kent and Olson 1982, with permission.)

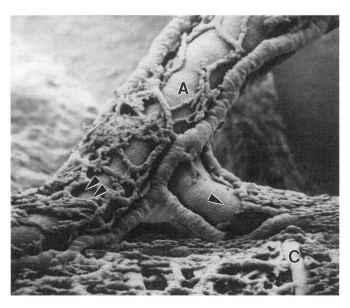

FIGURE 13-8 Vascular corrosion replica of posterior mesenteric artery (A) as it bifurcates into the anterior (*double arrowheads*) and posterior (*single arrowhead*) rami. A circumferential artery (C, lower right) supplies the parenchymal capillaries beneath the surface (*not visible*) and the capsular venous plexus on the surface of the gland and surrounding the posterior mesenteric artery (SEM, ×40). (From Kent and Olson 1982, with permission.)

Pronounced constrictions in methacrylate replicas of the AVAs (Fig. 13-9) suggest that flow through the AVAs is tightly regulated (Kent and Olson 1982). Capsular venules are commonly paired and lie on either side of the corresponding arteries and arterioles. Numerous small vessels crisscross between the paired venules and form a vascular web over the arterial vasculature. As the venules anastomose into progressively larger veins they form a dense venous plexus in the capsule that lies directly beneath the large circumferential arteries. This plexus is drained by large veins that either travel along the posterior mesenteric artery to join the posterior cardinal veins (see Fig. 13-8) or travel with the posterior arterial ramus into the postvalvular intestine.

The vasculature of the secretory parenchyma consists almost exclusively of capillaries or postcapillary venules. These vessels originate perpendicularly from capsular arterioles and travel radially, in the connective tissue between tubules, toward the central lumen (Fig. 13-10). Capillaries in the outer two-thirds of the secretory parenchyma are relatively straight; however, adjacent capillaries are frequently interconnected by anastomotic branches. Capillaries in the inner secretory parenchyma have extensive anastomoses and the vasculature becomes more sinusoidal. Secretory capillaries have a fenestrated endothelium (Ernst et al. 1981) and lie in close juxtaposition to the basal membranes of the secretory cells (Fig. 13-11). The innermost venous sinusoids coalesce into one of several central veins that line the rectal gland duct, the latter ultimately forming a single central vein that exits the posterior of the gland in the tissue forming the excretory duct. Because the flow of blood in the tubules is concurrent with tubular secretory flow, the secretory product cannot be amplified through countercurrent exchange.

It is evident that blood in the posterior mesenteric artery may follow one of three possible pathways (Kent and Olson 1982): (1) It may proceed directly to the postvalvular intestine via the posterior ramus without entering the rectal gland microcirculation; (2) it

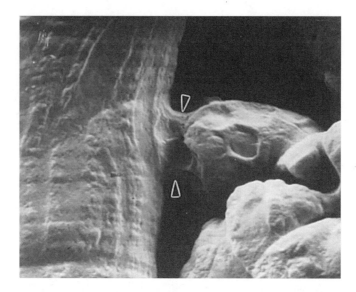

FIGURE 13-9 Corrosion replica of arteriovenous anastomoses from capsular artery (*on left*) to capsular vein. Note constriction at *arrowheads*. Indentations in replica are from endothelial cells (SEM, ×770). (From Kent and Olson 1982, with permission.)

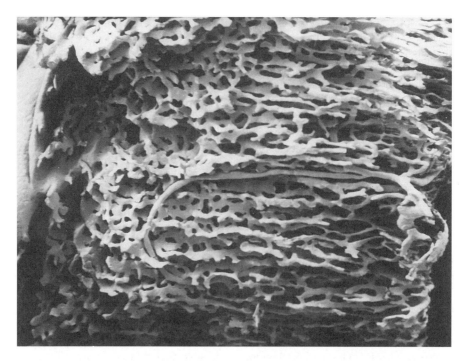

FIGURE 13-10 Corrosion replica of secretory parenchyma vascular sinusoids. Blood flows directly from outer parenchyma (*far right*) to veins in the central canal (*far left*). Outer sinusoids are linearly arranged and become irregular in the inner third of the gland (SEM, ×70). (From Kent and Olson 1982, with permission.)

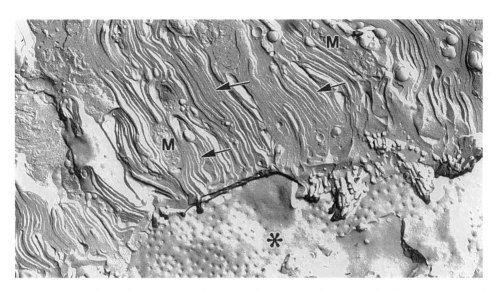

FIGURE 13-11 Freeze-fracture replica through basal membrane of secretory cell (*above*) and fenestrated capillary (*asterisk*). Infoldings of basolateral membrane (*arrows*) and mitochondria (M) of secretory cells are evident (×13,300). Micrograph courtesy of Dr. S. A. Ernst. (From Ernst et al. 1981, with permission.)

may enter the capsular sinusoids via the AVAs; or (3) it may perfuse the secretory tubules. Results from in vivo measurements of blood flow with radioactive microspheres suggest that flow to the gland is intermittent, varying from less than 0.07% to as much as 7% of the cardiac output in different fish (Kent and Olson 1982). Burger (1962) also observed that the intermittent secretory rate in Squalus in vivo correlated with varying Cl⁻ concentrations in arteriovenous blood and he suggested this was a reflection of variations in rectal gland perfusion. He (1962) also estimated rectal gland blood flow to be between 3 and 23 mL/hr/kg body weight. Flow in perfused rectal glands may be correlated with secretion rate (Hayslett et al. 1974), although Shuttleworth and Thompson (1986) observed that flow through perfused *Scyliorhinus* glands did not change even after nearly a 20-fold increase in Na⁺ secretion. A similar dissociation between perfusion flow and secretion has been observed by Solomon et al. (1984a). However, neither the in vivo nor in vitro studies have been able to resolve changes in perfusion distribution within the gland; in fact, in most in vitro cannulated preparations, only the tubular sinusoid pathway remains patent. The specific function of the dense capsular sinus and the influence of vasoactive factors on rectal gland activity remain to be determined.

MECHANISMS OF SECRETION

The pioneering studies by Burger and colleagues (Burger and Hess 1960; Burger 1962, 1965, 1972) clearly established the sodium chloride secretory activity of the rectal gland and laid the foundation for extensive anatomical and physiological studies that have continued up to the present time. These studies showed that rectal gland secretions were isosmotic with plasma; however, the osmolyte was nearly all NaCl (Table 13-2). Thus neither was the secretory product an ultrafiltrate of plasma nor did its formation involve generation of an appreciable osmotic gradient. Rectal gland secretion could be stimulated in intact *Squalus*, after a lag-time of 30–75 minutes, by subdermal, intramuscular, or intravascular injection of saline, sucrose, or distilled water. Often secretory rates were correlated with the tonicity or volume of the injected fluid and secretion was unaffected by blockade of nerves to the gland. Infusion of the saline load directly into the rectal gland artery did not shorten the lag-time or affect the duration of increased secretion. Injection of the aforementioned solutions into the extradural or ventricular spaces of the brain did not stimulate secretion. Rectal gland secretion was also maintained when the fish were volume loaded by placing them in dilute (~75%) seawater for up to 9 days. Collectively these studies showed that rectal gland secretion could be stimulated by osmotic (especially NaCl) and volume loading and that a hormonal signal was probably involved. More recent studies suggest that intravascular volume expansion may be the primary stimulus for rectal gland secretion.

Ligation of the rectal gland impairs an elasmobranch's ability to excrete an exogenous salt load but it is not a life-threatening situation. Following rectal gland ligation, plasma Na$^+$ and Cl$^-$ may increase over several days, but then they usually return toward normal, and may even reach it (Burger 1962, 1965; Haywood 1975). Burger (1965) has shown that under these conditions, the excess salt accumulated by the fish is excreted by the kidney (Cl$^-$ excretion increases approximately threefold). However, because the kidneys cannot concentrate Na$^+$ and Cl$^-$ in the urine above the concentrations of the plasma, there is a concomitant doubling of the volume of urine excreted. The long-term consequences of renal diuresis on the volume of various extracellular fluid compartments in elasmobranchs without functioning rectal glands has not been examined.

Secretion rate

The methods for examining rectal gland secretory processes and their control inevitably made the transition from in vivo studies to in vitro studies of perfused organs, perfused tubules, and more recently, to secretory cells in culture. This has ultimately led to molecular analysis of the transporters and membrane channels. These studies have not only contributed greatly to our understanding of the secretion process in rectal glands, but they have led to an appreciation of the basic mechanisms common in many transporting epithelia.

Rectal gland secretion rate in *Squlaus*, in vivo, averages 470 µL/hr/kg body weight and the secreted fluid has a Cl$^-$ concentration of 492 mmol (Burger 1962). The Cl$^-$ excretion rate is 231 µmol/hr/kg. Assuming that the gland weighs 500 mg/kg body weight, the secretory flow rate per gram gland weight is 940 µL/hr/g and the Cl$^-$ excretion rate per gram gland weight is 462 µmol/hr/g (Table 13-3). Secretion rate oscillates over a 24-hour period but it does not appear to stop completely at any time

TABLE 13-3 SECRETORY FLOW RATE AND CHLORIDE EXCRETION BY *SQUALUS* RECTAL GLAND IN VIVO AND BY THE PERFUSED GLAND IN VITRO

Method	Treatment	Flow (µL/hr/g)		Cl$^-$ Secretion (µmol/h/g)	
		CONTROL	TREATED	CONTROL	TREATED
In vivo	None[a]	940	—	462	—
	10 µg/kg atriopeptin[b]	670	2440	344	1241
	Volume expanded[c]	270	1110	82	422
In vitro	2.5 × 10^{-4} mol/L theophylline + 5 × 10^{-5} M dibutyryl cAMP[d]	130	3330	61	1596
	10^{-6} mol/L VIP[e]	648	4398	297	1872
	0.3 mg/mL veratrine[f]	460	1270	128	478

SOURCES: [a]Burger (1962); [b]Solomon et al. (1985a); [c]cross-perfused, Solomon et al. (1985b); [d]Forrest et al. (1982); [e]Stoff et al. (1979); [f]Stoff et al. (1988).
cAMP = cyclic adenosine monophosphate; VIP = vasoactive intestinal peptide.

(Burger 1962). Secretion rates in perfused glands are comparable to those observed in vivo and maximal secretory activity can be invoked in intact fish and in perfused glands with similar results (Table 13-3). The perfused preparation has been used extensively to examine the secretory process and the regulatory factors involved.

Ion transport processes

Evolution of the model of Na^+ and Cl^- transport by the rectal gland has proceeded through several steps that were based on a secondary cotransport of Cl^- coupled to an active Na^+,K^+-adenosinetriphosphatase (ATPase) (Epstein 1979; Shuttleworth 1988; Riordan et al. 1994). A detailed account of the early studies leading to the currently accepted model can be found in the excellent review by Shuttleworth (1988).

A key step in unraveling the mechanism of rectal gland secretion was the localization of Na^+,K^+-ATPase on the basolateral membrane. Initial evidence for this was obtained by Goertemiller and Ellis (1976), who used cytochemical techniques to identify Na^+,K^+-ATPase–like activity on the cytoplasmic side of the basolateral membrane and to show that there was little evidence for the enzyme in the apical membrane. This was confirmed in an autoradiographic study by Eveloff et al. (1979) in which they found that [^{3}H]-ouabain binding was high in the basolateral membranes and especially concentrated in the membranous area closest to the highest density of mitochondria (Fig. 13-12). Although this seemingly put the transport process on the wrong side of the cell, it soon became apparent that the Na^+,K^+-ATPase transporter was an indirect but integral component of the secretory process. More recently, Dubinsky and Monti (1986) were able to isolate apical and basolateral membranes and confirm that Na^+,K^+-ATPase activity was exclusively restricted to the latter.

A second key finding in resolving the secretory process were observations by Hayslett et al. (1974), Siegel et al. (1976), and Silva et

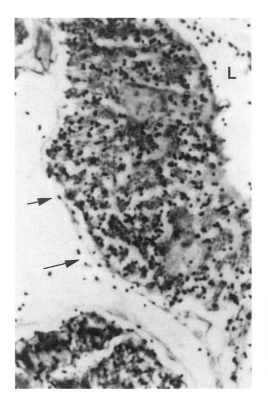

FIGURE 13-12 Autoradiograph of [3H]-ouabain binding to a freeze-dried section of tubular secretory cells. Highest grain density is in parenchymal cells with sparse labeling in lumen (L) or capsule (*arrows*) (×1660). Micrograph courtesy of Dr. K. J. Karnaky, Jr. (From Eveloff et al. 1979, with permission.)

al. (1977) that the electrochemical gradient across the rectal gland favored the downhill secretion of Na^+, whereas Cl^- must be translocated against an electrochemical gradient. More recent studies with microelectrode and patch-clamp techniques have localized the carriers and channels and further defined these processes (Gögelein et al. 1987a,b; Greger et al. 1985, 1987a,b, 1988; Devor et al. 1995).

The current model of rectal gland secretion, summarized by Riordan et al. (1994), is shown schematically in Figure 13-13. In this model, Cl^- is secreted against its concentration gradient, but only by secondary transport; that is to stay, the movement of Cl^- ions is not directly coupled to an energy consuming process. Na^+ on the other hand, is actively transported out of the secretory cell toward the blood rather than into the lumen. Secretion

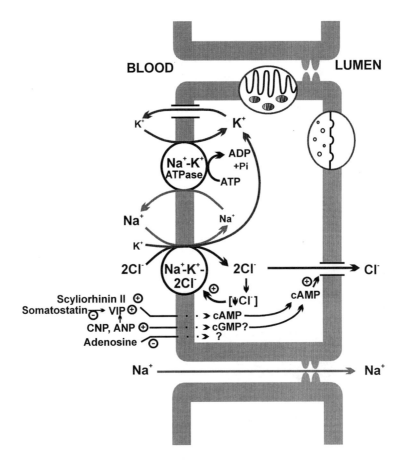

FIGURE 13-13 Mechanism of chloride secretion by the rectal gland and its control. A cyclic andenric monophosphate (cAMP)–sensitive Cl^- channel regulates the diffusional efflux of Cl^- across the apical membrane, down its electrochemical gradient. $Na^+,-K^+$-ATPase in the convoluted basolateral membranes lowers intracellular Na^+, thereby creating a favorable gradient for the downhill entry of Na^+, which is coupled to the uphill entry of K^+ and Cl^- via the $Na^+,-K^+$-$2Cl^-$ symporter. The resultant transepithelial potential moves Na^+ via a paracellular path through the tight junctions into the lumen. Secretion is regulated by hormonal, neuronal, or paracrine signals that raise intracellular cAMP or affect other factors. The $Na^+,-K^+$-$2Cl^-$ symporter is stimulated by a fall in intracellular $[Cl^-]$ pursuant to increased Cl^- efflux across the apical membrane. ADP = adenosine diphosphate; ANP = atrial natriuretic peptide; ATP = adenosine triphosphate; cAMP = cyclic adenosine monophosphate; cGMP = cyclic guanosine monophosphate; Cl^- = chloride ion; K^+ = potassium ion; Na^+ = sodium ion; Pi = phosphatidylinositol; VIP = vasoactive intestinal peptide.

of Na^+ into the tubular fluid is a passive, paracellular process, driven by the negative potential in the lumen that was established through Cl^- secretion.

The two transmembrane carrier systems and multiple ion channels important in the secretory process are shown in Figure 13-13. Na^+,K^+-ATPase, common in all animal cells and located in the basolateral membrane of rectal gland secretory cells, actively transports K^+ into the cell in exchange for Na^+. Na^+ is transported out of the cell into the extracellular space on the blood side of the epithelium. This produces a steep electrochemical gradient for the passive reentry of Na^+ into the tubular cell and, in doing so, provides the energy to simultaneously drive the coupled inward, uphill translocation of K^+ and two Cl^- by a non-electrogenic Na^+-K^+-$2Cl^-$ symporter. Na^+ entering the cell is again pumped out of the cell by the Na^+,K^+-ATPase, and excess intracellular K^+ is prevented by passive diffusion out of the cell via a basolateral potassium-specific (K^+) channel (Gögelein et al. 1987a; Greger et al. 1988). The Na^+-K^+-$2Cl^-$ symporter raises intracellular Cl^- above its electrochemical equilibrium, and Cl^- passively diffuses out of the cell into the lumen via a chloride-selective channel located only on the apical membrane. This net Cl^- secretion creates a negative potential in the lumen (with respect to the interstitial, i.e., blood, side) that becomes the driving force for passive Na^+ diffusion from the basolateral interstitium through the sodium selective tight junctions and into the lumen. Thus neither Cl^- nor Na^+ movements into the tubular fluid are directly coupled to an active transport process, and while Cl^- secretion follows a transcellular route, Na^+ movements are strictly paracellular. Under certain circumstances, nitrate may substitute for Cl^- on the low-affinity binding site of the Na^+-K^+-$2Cl^-$ transporter, resulting in a net nitrate secretion of up to nearly 30% of the maximal Cl^- secretory rate (Silva and Epstein 1993). Thus rectal gland function may encompass more than NaCl excretion.

Control of rectal gland secretion is achieved through regulating the permeability of the chloride channel in the apical membrane (Greger et al. 1988; Riordan et al. 1994). This channel has high sequence homology with another chloride-regulatory channel, the cystic fibrosis transmembrane conductance regulator (CFTR), and the two are believed to be similar (Riordan et al. 1994; Devor et al. 1995). The CFTR has 12 transmembrane domains, several of which constitute the chloride channel. The channel is activated by phosphorylation of one of the internal (cytoplasmic) loops, the R domain, by protein kinase A, a process initiated by an increase in cyclic adenosine monophosphate (cAMP). Thus hormonal stimulation of tubular cell cAMP appears to be a key regulatory step in rectal gland secretion.

Activity of the Na^+-K^+-$2Cl^-$ symporter is also regulated, but this process is secondary to an increased Cl^- conductance by the apical Chloride channel. The Na^+-K^+-$2Cl^-$ symporter has 12 transmembrane domains and appears to be activated by a fall in intracellular Cl^- concentration (Riordan et al. 1994). Although the energy for net NaCl secretion by the gland is utilized by the basolateral Na^+,K^+-ATPase, the activity of this enzyme does not appear to be regulated by events associated with changes in glandular secretion. This is consistent with the studies of Eveloff et al. (1979), who showed that neither [^{3}H]-ouabain binding nor Na^+,K^+-ATPase activity were affected when the glandular secretion was maximally stimulated by dibutyryl cAMP plus theophylline.

Control of secretion

The initial studies of Burger (1962, 1965) showed that, in vivo, rectal gland secretion in *Squalus* was stimulated by increases in plasma volume or tonicity. The stimulatory effect of volume expansion, but not tonicity, of the expanding fluid on secretion has been confirmed in anesthetized *Squalus* (Erlij and Rubio 1986) and in in situ cross-perfused glands of pithed sharks (Solomon et al. 1984a,b, 1985b). Erlij and Rubio (1986) also showed that central venous volume or pressure was perhaps the most potent secretory stimulus and that the receptors were near the heart or in the head region. The increased secretion observed when denervated glands were perfused with blood from a volume expanded donor fish (Solomon et al. 1984a,b, 1985b) strongly suggests that a humoral signal is responsible for at least part of the efferent loop of the secretory stimulus.

Some of the most plausible candidates for the humoral limb regulating rectal gland secretion are hormones of the natriuretic peptide (NP) family. Atrial natriuretic peptides (ANP) are secreted by the mammalian atria in response to an increase in central venous volume or pressure, and they increase renal excretion of water and electrolytes (Needleman et al. 1989; Brenner et al. 1990). The information on fish, especially elasmobranchs, is still limited; however, a number of NPs belonging several NP subfamilies have been identified and their physiological effects have been investigated (Olson 1992). The secretory abilities of mammalian ANP and CNP, the latter a peptide found in both mammals and elasmobranchs, have been examined in detail in *Squalus* using both perfused rectal glands and monolayers of cultured tubular cells (Solomon et al. 1985a, 1992; Valentich et al. 1995). Initially, it was thought that the NP-stimulated secretion was indirect and was mediated through ANP stimulation of vasoactive intestinal peptide (VIP). VIP has been identified in the nerve bodies and fibers in *Squalus* rectal glands (Holmgren and Nilsson 1983; Stoff et al. 1988); its release is stimulated by ANP (Silva et al. 1987), CNP (Solomon et al. 1992), and other factors that may stimulate intramural nerves in the gland (Stoff et al. 1988). VIP is a potent stimulant of rectal gland secretion through a direct increase in intracellular cAMP (Stoff et al. 1979). However, Karnaky et al. (1991) showed that ANP also stimulated Cl^- secretion by cultured rectal gland cells in the absence of VIP. This study raises the possibility of a direct ANP effect on tubular cells, mediated by cyclic granosine monophosphate (cGMP) or perhaps by another intracellular signaling system.

Unlike the response observed in *Squalus*, secretion rate in rectal glands from the lesser spotted dogfish, *Scyliorhinus canicula*, is unaffected by VIP but is stimulated by an intestinal factor, initially named rectin (Anderson et al. 1995). Rectin has recently been isolated and sequenced from the intestine of *Scyliorhinus* and found to be an 18-amino-acid peptide identical to the tachykinin scyliorhinin II (Anderson et al. 1995). The intestinal origin of scyliorhinin II suggests that one of the key stimuli for rectal gland secretion in *Scyliorhinus* is feeding. Because *Scyliorhinus* have largely an invertebrate diet and gorge food (Anderson et al. 1995), their problems with salt loading may be more immediate than those encountered by *Squalus* and require a more responsive salt-excretory system. ANP, angiotensin II, bombesin, VIP, glucagon, and somatostatin immunoreactivities have also been localized, along with reninlike, and angiotensin converting enzyme–like activity, in distinct areas of the rectal gland of *S. canicula*, suggesting multiple or complex regulatory processes (Masini et al. 1994).

Information regarding inhibitors of rectal gland secretion is scanty. Kelley et al. (1990, 1991) have shown that adenosine can be released by the perfused rectal gland of *Squalus* and can inhibit Cl^- secretion. In this preparation adenosine stimulates high-affinity A_1 adenosine receptors and inhibits secretion by a mechanism that is independent of the cAMP system. Somatostatin, present in rectal gland nerves (Holmgren and Nilsson 1983), also inhibits secretion (Silva et al. 1985). However, somatostatin acts by inhibiting VIP-stimulated secretion; it neither directly inhibits secretion, nor does it block secretion stimulated by cAMP analogs (Stoff et al. 1979, 1988; Silva et al. 1982). Somatostatin can be released from glandular neurons by another peptide also found in glandular neurons, bombesin (Silva et al. 1990). Neuropeptide Y (NPY) has recently been found to inhibit secretion by perfused glands and by monolayers of cultured secretory cells (Silva et al. 1993). The action of NPY appears to be distal to cAMP production (Silva et al. 1993), and both NPY and adenosine may be primary inhibitors of rectal gland activity.

Conclusion

Clearly, the rectal gland has proven to be an ideal model with which to examine Cl^- secretion, owing to the seeming anatomical simplicity of the secretory tubules and to the ease with which glands can be perfused and homogenous populations of cells isolated. However, it has also become evident that the overall regulation of rectal gland secretion, in vivo, is a highly complex process, integrated at multiple levels. There is no current evidence to suggest that the secretory stimuli are necessarily the same in all elasmobranchs or that the secretagogues have a common origin or chemical identity. As pointed out by Shuttleworth (1988), it is unfortunate that so few species have been examined, and perhaps our myopic view of the physiological significance of rectal gland function will change when other elasmobranch models are pursued. As molecular techniques develop there will undoubtedly be further advances in understanding the membrane channels involved in ion movements, and perhaps additional insight can be gained on the role of the extraparenchymal vasculature. Rectal gland function should also be examined in the context of the physiological consequences of environmental pollution. Solomon et al. (1989) have shown that organotin compounds, common pollutants from antifouling paints, have toxic effects on rectal gland secretion, and Silva et al. (1992) found that gland function is affected by inorganic mercury. The long-term effects of these and other toxic chemicals on rectal gland function and elasmobranch homeostasis need to be evaluated.

REFERENCES

Anderson, W. G., J. M. Conlon, and N. Hazon. 1995. Characterization of the endogenous intestinal peptide that stimulates the rectal gland of *Scyliorhinus canicula*. *Am. J. Physiol.* 268: R1359–64.

Ballantyne, J. S., C. D. Moyes, and T. W. Moon. 1987. Compatible and counteracting solutes and the evolution of ion and osmoregulation in fishes. *Can. J. Zool.* 65: 1883–88.

Benyajati, S., and S.D. Yokota. 1989. Hormonal regulation of renal function during environmental dilution and volume loading in the spiny dogfish. In *Proceedings of the 11th International Symposium on Comparative Anatomy and Endocrinology*, Malaga, Spain (abstract).

Bittner, A., and S. Lang. 1980. Some aspects of the osmoregulation of Amazonian freshwater stingrays (*Potamotrygon hystrix*): I. Serum osmolality, Na^+ and Cl^- content, water content, hematocrit and urea level. *Comp. Biochem. Physiol.* 67A: 9–13.

Bonting, S. L. 1966. Studies on soidum-potassium–activated adenosinetriphosphatase: XV. The rectal gland of the elasmobranchs. *Comp. Biochem. Physiol.* 17: 953–66.

Brenner, B. M., B. J. Ballermann, M. E. Gunning, and M. L. Zeidel. 1990. Diverse biological actions of atrial natriuretic peptide. *Pharmacol. Rev.* 70: 665–99.

Bulger, R. E. 1963. Fine structure of the rectal (salt-secreting) gland of the spiny dogfish, *Squalus acanthias*. *Anat. Rec.* 147: 95–127.

———. 1965. Electron microscopy of the stratified epithelium lining the excretory canal of the dogfish rectal gland. *Anat. Rec.* 151: 589–608.

Burger, J. W. 1962. Further studies on the function of the rectal gland in the spiny dogfish. *Physiol. Zool.* 35: 205–17.

———. 1965. Roles of the rectal gland and the kidneys in salt and water excretion in the spiny dogfish. *Physiol. Zool.* 38: 191–96.

———. 1972. Rectal gland secretion in the stingray *Dasyatis sabina*. *Comp. Biochem. Physiol.* 42: 31–32.

Burger, J. W., and W. N. Hess. 1960. Function of the rectal gland in the spiny dogfish. *Science* 131: 670–71.

Crawford, J. 1899. The rectal gland in elasmobranchs. *Proc. R. Soc. (Edinb.)* 23.

Devor, D. C., J. N. Forrest, Jr., W. K. Suggs, and R. A. Frizzell. 1995. cAMP-activated Cl^- channels in primary cultures of spiny dogfish (*Squalus acanthias*) rectal gland. *Am. J. Physiol. Cell Physiol.* 268: C70–79.

Doyle, W. L. 1962. Tubule cells of the rectal salt-gland of *Urolophus*. *Am. J. Anat.* 111: 223–37.

Dubinsky, W .P., and L. B. Monti. 1986. Resolution of apical from basolateral membrane of shark rectal gland. *Am. J. Physiol.* 251: C721–26.

Eddy, F. B., and R. N. Bath, 1979. Ionic regulation in rainbow trout (*Salmo gairdneri*) adapted to fresh water and dilute sea water. *J. Exp. Biol.* 83: 181–92.

Epstein, F. H. 1979. The shark rectal gland: a model for the active transport of Cl^-. *Yale J. Biol. Med.* 52: 517–23.

Erlij, D., and R. Rubio. 1986. Control of rectal gland secretion in the dogfish (*Squalus acanthias*): steps in the sequence of activation. *J. Exp. Biol.* 122: 99–112.

Ernst, S. A., S. R. Hootman, J. H. Schreiber, and C. V. Riddle. 1981. Freeze-fracture and morphometric analysis of occluding junctions in rectal glands of elasmobranch fish. *J. Membr. Biol.* 58: 101–14.

Evans, D. H. 1982. Mechanisms of acid extrusion by two marine fishes: the teleost *Opsanus beta* and the elasmobranch *Squalus acanthias*. *J. Exp. Biol.* 97: 289–99.

———. 1984. The roles of gill permeability and transport mechanisms in euryhalinity. In *Fish Physiology*, vol. X, ed. W. S. Hoar and D. J. Randall, 239–83. New York: Academic Press.

Eveloff, J., K. J. Karnaky, Jr., P. Silva, F. H. Epstein, and W. B. Kinter. 1979. Elasmobranch rectal gland cell. Autoradiographic localization of [^{3}H]ouabain-sensitive Na$^+$,K$^+$-ATPase in rectal gland of dogfish *Squalus acanthias*. *J. Cell Biol.* 83: 16–32.

Forrest, J. N., Jr., J. L. Boyer, T. A. Ardito, H. V. Murdaugh, Jr., and J. B. Wade. 1982. Structure of tight junctions during Cl$^-$ secretion in the perfused rectal gland of the dogfish shark. *Am. J. Physiol.* 242: C388–92.

Goertemiller, C. C., Jr., and R. A. Ellis. 1976. Localization of ouabain-sensitive, potassium-dependent nitrophenyl phosphatase in the rectal gland of the spiny dogfish, *Squalus acanthias*. *Cell Tissue Res.* 175: 101–12.

Gögelein, H., R. Greger, and E. Schlatter, E. 1987a. Potassium channels in the basolateral membrane of the rectal gland of *Squalus acanthias*. Regulation and inhibitors. *Pflugers Arch.* 409: 107–13.

Gögelein, H., E. Schlatter, and R. Greger. 1987b. The "small" conductance Cl$^-$ channel in the luminal membrane of the rectal gland of the dogfish (*Squalus acanthias*). *Pflugers Arch.* 409: 122–25.

Greger, R., H. Gögelein, and E. Schlatter. 1987a. Potassium channels in the basolateral membrane of the rectal gland of the dogfish (*Squalus acanthias*). *Pflugers Arch.* 409: 100–6.

———. 1988. Stimulation of NaCl secretion in the rectal gland of the dogfish *Squalus acanthias*. *Comp. Biochem. Physiol.* 90A: 733–37.

Greger, R., E. Schlatter, and H. Gögelein. 1985. Cl$^-$ channels in the apical cell membrane of the rectal gland "induced" by cAMP. *Pflugers Arch.* 403: 446–48.

———. 1987b. Cl- channels in the luminal membrane of the rectal gland of the dogfish (*Squalus acanthias*). Properties of the "larger" conductance channel. *Pflugers Arch.* 409: 114–21.

Hayslett, J. P., D. A. Schon, M. Epstein, and C. A. M. Hogben. 1974. In vitro perfusion of the dogfish rectal gland. *Am J. Physiol.* 226: 1188–92.

Haywood, G. P. 1975. A preliminary investigation into the roles played by the rectal gland and kidneys in the osmoregulation of the striped dogfish, *Poroderma africanum*. *J. Exp. Zool.* 193: 167–76.

Holmes, W. N., and E. M. Donaldson. 1969. The body compartments and the distribution of electrolytes. In *Fish Physiology*, vol. 1, W. S. Hoar and D. J. Randall, 1–89. New York: Academic Press.

Holmgren, S., and S. Nilsson. 1983. Bombesin-, gastrin/CCK–, 5-hydroxytryptamine–, neurotensin-, somatostatin-, and VIP-like immunoreactivity and catecholamine fluorescence in the gut of the elasmobranch, *Squalus acanthias*. *Cell Tissue Res.* 234: 595–618.

Hoskins, E. R. 1917. On the development of the digitiform gland and the post-valvular segment of the intestine in *Squalus acanthias*. *J. Morphol.* 28: 329–67.

Karnaky, K. J., J. D. Valentich, M. G. Currie, W. F. Oehlenschlager, and M. P. Kennedy. 1991. Atriopeptin stimulates chloride secretion in cultured shark rectal gland cells. *Am. J. Physiol.* 260: C1125–30.

Kelley, G. G., O. S. Aassar, and J. N. Forrest, Jr. 1991. Endogenous adenosine is an autacoid feedback inhibitor of Cl$^-$ transport in the shark rectal gland. *J. Clin. Invest.* 88: 1933–39.

Kelley, G. G., E. M. Poeschla, H. V. Barron, and J. N. Forrest, Jr. 1990. A1 adenosine receptors inhibit Cl$^-$ transport in the shark rectal gland. Dissociation of inhibition and cyclic AMP. *J. Clin. Invest.* 85: 1629–36.

Kent, B., and K. R. Olson. 1982. Blood flow in the rectal gland of *Squalus acanthias*. *Am. J. Physiol.* 243: R296–303.

Kleinzeller, A., and F. N. Ziyadeh. 1990. Cell volume regulation in epithelia—with emphasis on the role of osmolytes and the cytoskeleton. In *Cell Volume Regulation*, 4th ed., ed. K. W. Beyenbach, 59–86. Farmington, Conn.: Karger.

Komnick, H., and K. E. Wohlfarth-Bottermann. 1966. Zur Cytologie der Rectaldrüsen von Knorpelfischen: I. Die Feinstruktur der Tubulusepithelzellen. *Z. Zellforsch.* 74: 123–44.

Lang, F., H. Volkl, and D. Haussinger. 1990. General principles in cell volume regulation. In *Cell Volume Regulation*, 4th ed., ed. K. W. Beyenbach, 1–25. Farmington, Conn.: Karger.

Loretz, C. A. 1987. Rectal gland and crypts of Lieberkühn: is there a phylogenetic basis for functional similarity? *Zool. Sci.* 4: 933–44.

McCarthy, J. E., and F. P. Conte. 1966. Determination of the volume of vascular and extravascular fluids in the Pacific hagfish (*Polistotrema stouti*). *Am. Zool.* 6: 605.

McConnell, F. M., and L. Goldstein. 1990. Volume regulation in elasmobranch red blood cells. In *Cell Volume Regulation*, 4th ed., ed. K. W. Beyenbach, 114–31. Farmington, Conn.: Karger.

Masini, M. A., B. Uva, M. Devecchi, and L. Napoli. 1994. Renin-like activity, angiotensin-I converting enzyme–like activity, and osmoregulatory peptides in the dogfish rectal gland. *Gen. Comp. Endocrinol.* 93: 246–54.

Needleman, P., E. H. Blaine, J. E. Greenwald, M. L. Michener, C. B. Saper, P. T. Stockmann, and H. E. Tolunay. 1989. The biochemical pharmacology of atrial peptides. *Annu. Rev. Pharmacol. Toxicol.* 29: 23–54.

Oguri, M. 1981. On the rectal gland weight of some elasmobranchs. *Bull. Jpn. Soc. Sci. Fish.* 47: 1019–21.

Olson, K. R. 1992. Blood and extracellular fluid volume regulation: role of the renin-angiotensin system, kallikrein-kinin system, and atrial natriuretic peptides. In *Fish Physiology*, vol. 12, pt. B, ed. W. S. Hoar, D. J. Randall, and A. P. Farrell, 136–232. New York: Academic Press.

———. 1996. The secondary circulation in fish: anatomical organization and physiological significance. *J. Exp. Zool.* 275: 172–85.

Riordan, J. R., B. Forbush, III, and J. W. Hanrahan. 1994. The molecular basis of Cl^- transport in shark rectal gland. *J. Exp. Biol.* 196: 405–18.

Robertson, J. D. 1975. Osmotic constituents of the blood plasma and parietal muscle of *Squalus acanthias* L. *Biol. Bull.* 148: 303–19.

Schmidt-Nielsen, B. 1975. Comparative physiology of cellular ion and volume regulation. *J. Exp. Biol.* 194: 207–20.

Shuttleworth, T. J. 1988. Salt and water balance—extrarenal mechanisms. In *Physiology of Elasmobranch Fishes*, ed. T. J. Shuttleworth, 171–99. New York: Springer-Verlag.

Shuttleworth, T. J., and J. L. Thompson. 1986. Perfusion-secretion relationships in the isolated elasmobranch rectal gland. *J. Exp. Biol.* 125: 373–84.

Siegel, N. J., D. A. Schon, and J. P. Hayslett. 1976. Evidence for active chloride transport in dogfish rectal gland. *Am. J. Physiol.* 230: 1250–54.

Silva, P., and Epstein, F. H. 1993. Secretion of nitrate by rectal gland of *Squalus acanthias. Comp. Biochem. Physiol.* 104A: 255–59.

Silva, P., F. H. Epstein, K. J. Karnaky, Jr., S. Reichlin, and J. N. Forrest, Jr. 1993. Neuropeptide Y inhibits Cl^- secretion in the shark rectal gland. *Am. J. Physiol.* 265: R439–46.

Silva, P., F. H. Epstein, and R. J. Solomon. 1992. The effect of mercury on Cl^- secretion in the shark (*Squalus acanthias*) rectal gland. *Comp. Biochem. Physiol.* 103C: 569–75.

Silva, P., S. Lear, S. Reichlin, and F. H. Epstein. 1990. Somatostatin mediates bombesin inhibition of Cl^- secretion by rectal gland. *Am. J. Physiol.* 258: R1459–63.

Silva, P., J. S. Stoff, and F. H. Epstein. 1982. Hormonal control of Cl^- secretion in the rectal gland of *Squalus acanthias*. In *Chloride Transport in Biological Membranes*, ed. J. A. Zadunaisky, 277–93. New York: Academic Press.

Silva, P., J. Stoff, M. Field, L. Fine, J. N. Forrest, and F. H. Epstein. 1977. Mechanism of active Cl^- secretion by shark rectal gland: role of Na^+,K^+-ATPase in Cl^- transport. *Am. J. Physiol.* 233: F298–306.

Silva, P., J. S. Stoff, D. R. Leone, and F. H. Epstein. 1985. Mode of action of somatostatin to inhibit secretion by shark rectal gland. *Am. J. Physiol.* 249: R329–34.

Silva, P., J. S. Stoff, R. J. Solomon, S. Lear, D. Kniaz, R. Greger, and F. H. Epstein. 1987. Atrial natriuretic peptide stimulates salt secretion by shark rectal gland by releasing VIP. *Am. J. Physiol.* 252: F99–103.

Solomon, R., S. Lear, R. Cohen, K. Spokes, P. Silva, Jr., M. Silva, H. Solomon, and P. Silva. 1989. The effect of organotin compounds on chloride secretion by the in vitro perfused rectal gland of *Squalus acanthias. Toxicol. Appl. Pharmacol.* 100: 307–14.

Solomon, R., A. Protter, F. H. Epstein, and P. Silva. 1992. C-type natriuretic peptides stimulate Cl^- secretion in the rectal gland of *Squalus acanthias. Am. J. Physiol.* 262: 707–11.

Solomon, R., M. Taylor, D. Dorsey, P. Silva, and F. H. Epstein. 1985a. Atriopeptin stimulation of rectal gland function in *Squalus acanthias. Am. J. Physiol.* 18: R348–54.

Solomon, R., M. Taylor, J. S. Stoff, P. Silva, and F. H. Epstein. 1984a. In vivo effect of volume expansion on rectal gland function: I. Humoral factors. *Am. J. Physiol.* 246: R63–66.

Solomon, R. J., M. Taylor, R. Rosa, P. Silva, and F. H. Epstein. 1984b. In vivo effect of volume expansion on rectal gland function: II. Hemodynamic changes. *Am. J. Physiol.* 246: R67–71.

Solomon, R. J., M. Taylor, S. Sheth, P. Silva, and F. H. Epstein. 1985b. Primary role of volume expansion in stimulation of rectal gland function. *Am. J. Physiol.* 248: R638–40.

Steffensen, J. F., and J. P. Lomholt. 1992. The secondary vascular system. In *Fish Physiology*, vol. 12, pt. A, ed. W. S. Hoar, D. J. Randall, and A. P. Farrell, 185–213. New York: Academic Press.

Stoff, J. S., R. Rosa, R. Hallac, P. Silva, and F. H. Epstein. 1979. Hormonal regulation of active Cl^- transport in the dogfish rectal gland. *Am. J. Physiol.* 6: F138–44.

Stoff, J. S., P. Silva, R. Lechan, R. Solomon, and F. H. Epstein. 1988. Neural control of shark rectal gland. *Am. J. Physiol.* 255: R212–16.

Thorson, T. 1958. Measurement of the fluid compartments of four species of marine Chondrichthyes. *Physiol. Zool.* 31: 16–23.

Thorson, T. B., R. M. Wotton, and T. A. Georgi. 1978. Rectal gland of freshwater stingrays,

Potamotrygon spp. (Chondrichthyes: Potamotrygonidae). *Biol. Bull.* 154: 508–16.

Valentich, J. D., K. J. Karnaky, Jr., and W. M. Moran. 1995. Phenotypic expression and natriuretic peptide–activated Cl⁻ secretion in cultured shark (*Squalus acanthias*) rectal gland epithelial cells. In *Cellular and Molecular Approaches to Fish Ionic Regulation*, ed. C. M. Wood and T. J. Shuttleworth, 173–205. New York: Academic Press.

van Lennep, E. W. 1968. Electron microscopic histochemical studies on salt-excreting glands in elasmobranchs and marine catfish. *J. Ultrastruct. Res.* 25: 94–108.

ERIC R. LACY
ENRICO REALE

CHAPTER 14

Urinary System

Elasmobranchs are a highly successful group of cartilaginous fish that are largely marine. Some species are euryhaline, being able to tolerate and even thrive in both fresh and salt water, and a few species are restricted to freshwater. Each of these environments demands different strategies for maintaining osmotic and ionic homeostasis. Elasmobranchs that adapt to or are restricted to a marine environment have concentrations of tissue and body fluids slightly greater than the osmolality of the surrounding seawater (Bottazzi 1897; Rodier 1900). This internal hypertonicity results in a net influx of water as well as diffusional loss of ions into the seawater, a situation different from that of teleost fish, in which the internal milieu is markedly hypotonic to the seawater (Smith 1931). This osmoregulatory strategy of elasmobranchs is achieved in large part by maintaining high concentrations of urea and trimethyl amine oxide, which account for about one-third of tissue and plasma osmolytes (Rodier 1899; Hoppe-Seyler 1930; Kempton 1953; review in De Vlaming and Sage 1973; Somero 1986). Although the gills, rectal gland, and kidney all play significant roles in marine and euryhaline elasmobranch osmoregulation, the kidney is the primary site where plasma urea regulation takes place (Boylan 1967; Forster 1967). The observation that nearly all urea filtered through the glomeruli was reabsorbed in the kidney (Clarke and Smith 1932) led investigators (Marshall 1934; Smith 1936) to speculate that a specific tubule segment must be responsible for the transepithelial transport of urea. According to Haller (1902), a "special segment" was interposed between the neck and proximal tubule in the spiny dogfish nephron. Numerous microscopical studies failed to elucidate this "special segment" in any elasmobranch but did show a level of nephron complexity in both configuration and epithelial diversity that rivaled that of mammals (Kempton 1940, 1943, 1956, 1962). Indeed, the complexity was so great that classical methods of anatomical reconstruction could not be used for the elasmobranch nephron. It was not until the advent of mainframe computers that the nephron configuration was reconstructed from serial sections of the kidney of the little skate, *Raja erinacea* (Smith et al. 1983; Lacy et al. 1984, 1985). The results showed a renal countercurrent system formed by each nephron, verifying earlier reports made from tubular micropuncture studies (Thurau and Acquisto 1969; Lacy et al. 1975; Boylan 1972). The fact that stenohaline freshwater elasmobranchs do not use urea as an osmolyte (Griffith et al. 1973; Goldstein and Forster 1971) and do not possess a renal countercurrent system (Hentschel et al. 1986; Lacy et al. 1985) gave strong support to the notion that it was this unique arrangement of tubule segments and not an anatomically different type of segment (epithelium) that was responsible for the efficient renal reabsorption of urea (Lacy et al. 1985). Stenohaline freshwater elasmobranchs (Potamotrygonidae) have secondarily invaded this environment from a marine habitat and are confined to the rivers of South America. Their internal milieu and osmoregulatory strategy appears to be similar to that of freshwater teleosts, although far less is known about this extremely interesting group of stingrays (Smith 1936; Thorson 1967, 1970; Thorson et al. 1967). Their renal tubular architecture has recently been elucidated and shows a complexity greater than that of teleosts but strikingly different than that of their marine counterparts (Lacy and Reale 1995; G. M. Grabowski, E. Reale, and E. R. Lacy, unpublished observations).

Taken together, it is clear that in elasmobranch fish, the structure of the kidney reflects its function. This chapter synthesizes a disparate literature on elasmobranch renal architecture. Even less is known about nephron function in marine or freshwater species. Table 14-1 gives an extensive list of elasmobranch species in which renal structure has been morphologically investigated. The distinction between marine, freshwater, or euryhaline made by these authors is often vague, not available, or questionable (see, e.g., Table 1 of the work of Smith 1936). Likewise, common names vary in different geographical regions.

Renal lobules and zones

The kidneys are paired organs flanking the dorsal aorta, situated in the most caudal part of the abdominal cavity. The most anterior portion of the renal tissue is often described as "rudimentary" and often intermingles with the gonads in males. The kidneys are semilunar in skates such as *Raja erinacea* (Leydig 1852; Borcea 1906; Lacy and Reale 1985a), extensively elongated in marine sharks such as *Squalus acanthias* (Leydig 1852) or freshwater rays, *Potamotrygon humerosa* (G. M. Grabowski, E. Reale, and E. R. Lacy, unpublished observations) but always dorsoventrally flattened, with remarkable lobations, especially on the ventral surface (Fig. 14-1).

One or two orange-brown interrenal glands (depending upon the species) lie along the medial border of the kidneys. Each lobe of the kidney consists of small lobules, which have lighter and darker regions reflective of differences in the concentrations of tubular and vascular components and characterizing two renal zones. Two major zones are present in all marine and euryhaline species thus far examined: (1) one, usually thin, in which tubules are tightly packed and the vasculature is reduced, is the bundle zone; (2) another, larger one, in which tubules are segregated from one another by wide vascular spaces resembling lacunae, is the sinus zone. Generally the bundle zone lies dorsally and the sinus zone ventrally, but these relationships are not always consistent, particularly at the extreme margins of the kidney and the lobules (Lacy et al. 1984, 1985). In the freshwater stingray, there

TABLE 14-1 MARINE AND FRESHWATER ELASMOBRANCHS USED FOR MICROSCOPICAL STUDIES ON THE KIDNEY

Species	Synonyms	Common name	Authors
Acanthias vulgaris	Squalus acanthias; Spinax acanthias	Piked dogfish; Spilled dogfish; Spiny dogfish	Bargmann (1937); Borcea (1904, 1906); Bohle and Walvig (1964); Boyd and DeVries (1986); Gambaryan (1987); Ghouse et al. (1968); Haller (1902); Hentschel (1988); Kempton (1940, 1943, 1957); Lacy and Reale (1985a,b, 1986, 1989, 1990); Lacy et al. (1985, 1987, 1989); Reale and Lacy (1987); Rhodin (1972)
Carcharias glaucus	?	?	Bargmann (1937); Borcea (1904, 1906)
Carcharias leucas		Bull shark	Crockett et al. (1973); Oguri et al. (1970)
Carcharias limbatus		Blacktip shark	Crockett et al. (1973); Oguri et al. (1970)
Carcharias obscurus		Dusky shark	Kempton (1956)
Cephaloscyllium uter		California swell shark	Kempton (1956, 1962)
Dasyatis akajei		Akaei stingray	Nishimura et al. (1970); Ogawa and Hirano (1982); Oguri et al. (1970); Sokabe et al. (1969)
Dasyatis americana		Southern stingray	Kempton (1956, 1962)
Dasyatis guttata		?	Crockett et al. (1973)
Dasyatis marinus		?	Bargmann (1954)
Dasyatis pastinaca		Stingray	Gambaryan (1987)
Dasyatis sabina		Atlantic stingray	Lacy and Reale (1991a); Oguri et al. (1970)
Dasyatis say	Dasyatis sayi	Bluntnose stingray	Kempton (1956, 1962)
Ginglymostoma cirratum		Common nurse shark	Oguri et al. (1970)
Heptanchus maculatus		?	Daniel (1934)
Heterodontus francisci		California horned shark; Horn shark	Kempton (1956, 1962)
Heterodontus japonicus		?	Nishimura et al. (1970); Oguri et al. (1970); Sokabe et al. (1969)
Holorhinus tobijei		Marine stingray	Ogawa and Hirano (1982)
Mustelus canis	Galeus canis	Smooth dogfish	Borcea (1904, 1906); Kempton (1953, 1956); Lacy and Reale (1989, 1990); Lacy et al. (1989); Nash (1931); Reale and Lacy (1987)
Mustelus mustelus	Mustelus laevis; Mustelus vulgaris	?	Bargmann (1937); Borcea (1904, 1906)
Myliobatis aquila	Noctula		Meyer (1888)
Myliobatis californicus		Bat ray	Kempton (1956, 1957, 1962)
Narcine brasiliensis		Lesser electric ray	Kempton (1956, 1957, 1962)

(table continues)

TABLE 14-1 *(continued)*

Species	Synonyms	Common name	Authors
Orectolobus japonicus		?	Nishimura et al. (1970); Oguri et al. (1970); Sokabe et al. (1969)
Potamotrygon circularis		River ray	Nishimura et al. (1970); Oguri et al. (1970)
Potamotrygon humerosa		River ray	Grabowski et al. (1992); Lacy et al. (1989a,b, 1991a,b)
Potamotrygon magdelanae		River ray	Ogawa and Hirano (1982)
Potamotrygon motoro		Motoro ray	Crockett et al. (1973)
Prionace glauca		Blue shark	Kempton (1956, 1962)
Pristis pectinatus		Common sawfish	Crockett et al. (1973)
Pristis perotteti		Largetooth sawfish	Crockett et al. (1973)
Pristiurus melanostomus		Sawfish	Borcea (1904, 1906)
Raja asterias		?	Bargmann (1937); Borcea (1906)
Raja batis	*Raja undulata*	Gray skate; Common skate	Borcea (1904, 1906); Leydig (1852)
Raja clavata		Thresher; Thornback skate	Bargmann (1937); Bargmann and von Hehn (1971); Borcea (1904, 1906); Gambaryan (1987); Kölliker (1845)
Raja diaphanes		?	Nash (1931)
Raja erinacea		Little skate	Boyd and DeVries (1986); Capréol and Sutherland (1968); Decker and Reale (1991); Deetjen and Antkowiak (1970); Elger et al. (1984); Hentschel (1988); Hentschel et al. (1986); Kempton (1956); Lacy and Reale (1981, 1985a,b, 1986, 1990, 1991a,b); Lacy et al. (1975, 1984, 1985, 1987, 1989); Nash (1931); Reale and Lacy (1987); Stolte et al. (1977)
Raja laevis		Barndoor skate	Capréol and Sutherland (1968)
Raja macrorhynchus		?	Borcea (1906)
Raja marginata			Meyer (1888)
Raja miraletus			Meyer (1888)
Raja mosaica		?	Borcea (1904, 1906)
Raja naevus		Cuckoo skate	Borcea (1906)
Raja nasuta			Parker (1881)
Raja ocellata		Winter skate	Hentschel (1988); Kempton (1962)
Raja oxyrhynchus		Sharpnose skate	Bargmann (1937); Meyer (1888)
Raja punctata			Meyer (1888)
Raja radula		?	Borcea (1904)
Raja stabuliforis		?	Nash (1931)
Rhinobatos productus		Shovelnosed guitarfish	Kempton (1956, 1962)

(table continues)

TABLE 14-1 *(continued)*

Species	Synonyms	Common name	Authors
Rhinoptera bonasus		Atlantic cow-nosed ray	Kempton (1956, 1962); Lacy and Reale (1989, 1990); Lacy et al. (1989)
Rhizoprionodon terraenovae		Atlantic sharpnose shark	Lacy and Reale (1989); Lacy et al. (1989)
Scyliorhinus caniculus	*Scyliorhinus catulus*; *Scyllium canicula*	Lesser spotted dogfish; Mediterranean dogfish	Bargmann (1937, 1954); Bargmann and von Hehn (1971); Borcea (1904, 1906); Elger and Hentschel (1983); Hentschel (1987, 1988, 1991); Hentschel and Elger (1982); Hentschel and Walter (1993); Kozlik (1939)
Scyllium stellare	*Scyliorhinus stellaris*	Stellar cat shark	Borghese (1966); Hentschel (1988)
Sphyrna lewini		Scalloped hammerhead shark	Lacy et al. (1989, 1991b)
Sphyrna tiburo		Bonnethead shark	Oguri et al. (1970)
Sphyrna zygaena		Smooth hammerhead shark	Oguri et al. (1970); Sokabe et al. (1969)
Squatina squatina	*Squatina angelus*; *Rhina squatina*	European angel shark	Borcea (1904, 1906); Marples (1936); Meyer (1888)
Torpedo marmorata	*Torpedo galvani*	Marbled torpedo ray	Borcea (1906); Hentschel (1988); Leydig 1852)
Torpedo ocellata	*Torpedo torpedo*	Ocellated torpedo ray	Bargmann (1937, 1954); Krause (1923); Meyer (1888)
Triakis semifasciata		Leopard shark	Kempton (1956, 1962)
Triakis scyllia		Banded dogfish	Nishimura et al. (1970); Ogawa and Hirano (1982); Oguri et al. (1970); Sokabe et al. (1969)
Trygon pastinaca		?	Bargmann (1937); Borcea (1906); Leydig (1852)
Trygon violaceus		?	Bargmann (1937)

are also two zones, termed *simple zone* and *complex zone*, which occur in the peripheral and central portions of the organ, respectively. The renal corpuscles separate bundle zone from sinus zone in marine and euryhaline species (Fig. 14-2A); they bisect the complex zone in the river rays (Fig. 14-2B).

Configuration of the renal tubules

Whereas in freshwater elasmobranchs the renal tubules have a relatively simple configuration, their arrangement in marine and euryhaline species is extremely complex and more highly organized than any of the "lower" and most of the "higher" vertebrates. The tubule is extremely long, measuring up to 9 cm in *Raja stabuliforis* (Nash 1931), about 3.3 cm in *Squalus acanthias* (Ghouse et al. 1968), about 5 cm (Deetjen and Antkowiak 1970) or 8 cm (Stolte et al. 1977) in *Raja erinacea*, and 2.8 cm in *Dasyatis pastinaca* (Gambaryan 1987).

In the kidneys of all marine and euryhaline elasmobranchs thus far examined, the tubule leaves the renal corpuscle and is arranged in four hairpin loops (Figs. 14-3 and 14-4), two directed generally dorsally and tightly packed in a bundle and two wandering loopingly in

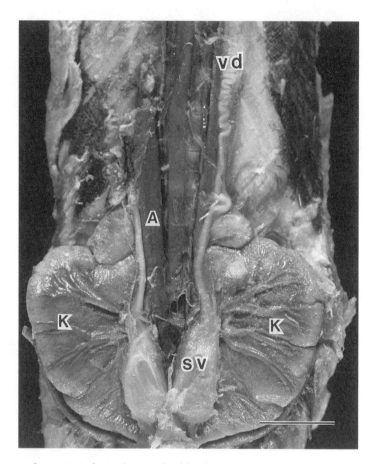

FIGURE 14-1 Low-power photomicrograph of fixed and methylmethacrylate-injected dorsal body wall of a little skate with *in situ* kidneys (K) and aorta (A). sv = Seminal vesicle; vd = vas deferens (×2.3; scale bar: 1 cm).

the ventral part of the kidney. All bundles are found together in the bundle zone (see Figs. 14-2A, 14-3, and 14-4), while the other less organized segments are found in the sinus zone (see Figs. 14-2A and 14-4). The large renal corpuscles are found along the border of the two zones (see Fig. 14-2A). The tubule leaves the renal corpuscle and first moves toward the dorsal kidney surface into the bundle zone, then turns back on itself to the parent corpuscle, forming the first loop (see Fig. 14-3). From this location it descends into the ventral region, where it meanders among large blood sinuses, returning to the same renal corpuscle and completing its second loop. It passes again into the bundle zone along with the first loop, where it twists tightly on itself as it turns back to touch its parent renal corpuscle (see Fig. 14-3). It then goes ventrally, generating the final, fourth loop. Upon this last return to the renal corpuscle, the tubule is interposed between the afferent and efferent glomerular arterioles adjacent to the hilar region of the glomerulus. From this point the tubule joins the two other nephron loops in the bundle zone, traveling rather straight, and then at the end of the bundle (loops), the tubule continues to join other tubules to form larger collecting ducts. These latter conduits become progressively larger, often dive between adjacent lobules, and emerge on the ventral renal surface, where

Urinary System 359

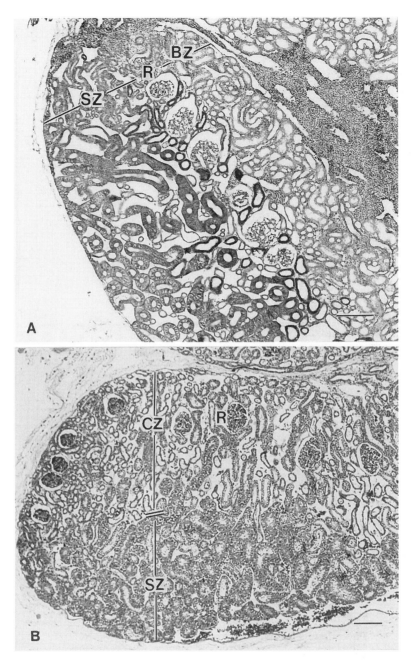

FIGURE 14-2 Kidney of little skate (**A**) and river ray (**B**), paraffin sections, hematoxylin-eosin. In **A** the bundle zone (BZ) and sinus zone (SZ) are seen separated by the renal corpuscles (R). The urinary pole of renal corpuscles is directed toward the bundle zone (×50; scale bar: 20 µm). In **B** the complex zone (CZ) and simple zone (SZ) can be seen. The renal corpuscles (R) bisect the complex zone (×40; scale bar: 20 µm). (From Lacy and Reale 1985a.)

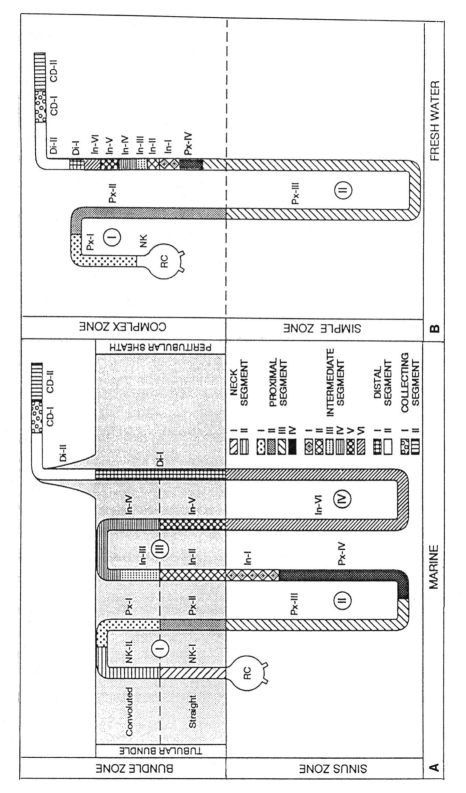

FIGURE 14-3 Models of marine and freshwater elasmobranch renal tubule showing the sequence and position but not the length of tubular segments and their subdivisions. RC = renal corpusles. (Modified from Lacy and Reale 1991a.)

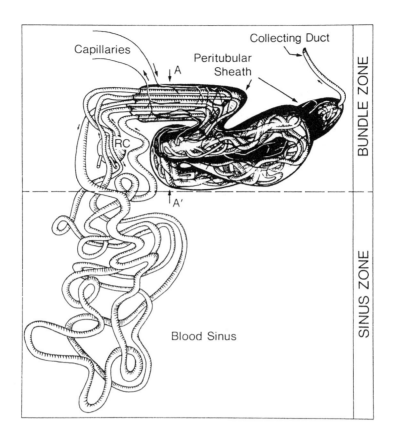

FIGURE 14-4 Elasmobranch kidney, showing bundle and sinus zones.

they coalesce along the medial renal margin into a ureter. The different arrangements of the renal tubule can be observed in the dissecting microscope (Fig. 14-5).

The two loops and final distal segment of each nephron lie tightly adjacent to one another, forming a bundle of tubular segments in which there are interspersed capillaries but no segments from other tubules. This tubular bundle is tightly wrapped together by overlapping squamous cells, which form a casing around the bundle, termed the *peritubular* or *peribundular sheath* (Lacy et al., 1985) (Figs. 14-4 and 14-6A). The cells of the sheath are connected by tight junctions, suggesting that they act as a barrier separating the environment inside the sheath (tubules, capillaries, scanty connective tissue, interstitial fluid) from that outside the sheath (Lacy and Reale 1986). As is discussed in the section on circulation, capillaries enter the proximal sheath (nearest the renal corpuscle), anastomose around and between the tubules, and exit also at the proximal end of the sheath (Figs. 14-4 and 14-7). Thus there exists the potential for a countercurrent multiplier system as found in other vertebrate organs, including the mammalian kidney. Taken together, it appears that each nephron of the marine and euryhaline elasmobranch has its own countercurrent multiplier system. This is distinctly different than that found in the mammalian kidney, in which the renal countercurrent multiplier system is formed by nearly the entire population of nephrons, which have segments in the inner and outer medulla. In the mammalian kidney, interstitial fluid within the medulla is shared by all the nephrons. Although there is compelling anatomic evidence for the presence of a renal countercurrent multiplier system in

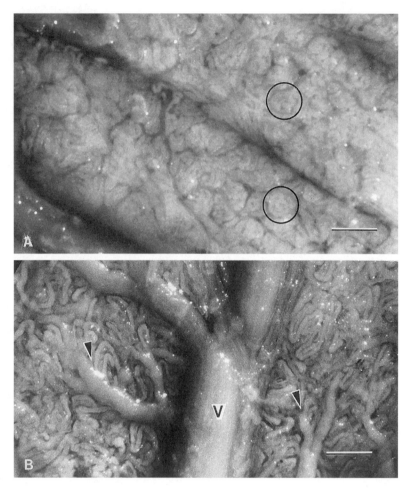

FIGURE 14-5 Little skate kidney, photomicrographs of the dorsal surface (**A**) and of the ventral surface (**B**) by incident light. In **A** two of the numerous tubular bundles are encircled; in **B** large vessels (V) and numerous tubules of the sinus zone can be seen. The *arrowheads* indicate collecting ducts confluent into larger tubules (×12; scale bar: 1 mm). (From Lacy et al. 1985.)

elasmobranch fish (Lacy et al. 1985), there is no physiological data yet to either support or refute this contention.

In the freshwater river ray (*Potamotrygon humerosa*) the nephron forms a single loop in the complex zone and a second loop in the simple zone (see Fig. 14-2B). Tubular bundles and surrounding peritubular sheaths are absent. The tubule lies adjacent to the parent renal corpuscle at the end of both loops. However, at the end of the second loop the tubule lies adherent to the corpuscle in a long groove formed by multiple efferent and afferent arterioles. After exiting the vascular groove the tubule extends to the urinary pole of the glomerulus. After a short path, it finally reaches the interlobular septa, where it merges into a collecting duct (see Figs. 14-2 and 14-3).

The renal corpuscle

The formation of urine in both marine and freshwater elasmobranchs begins, as in almost all other vertebrates, by ultrafiltration of the blood across a highly selective series of barriers. The entire filtration unit is called the renal corpuscle, which consists of Bowman's capsule and glomerulus, both separately shown

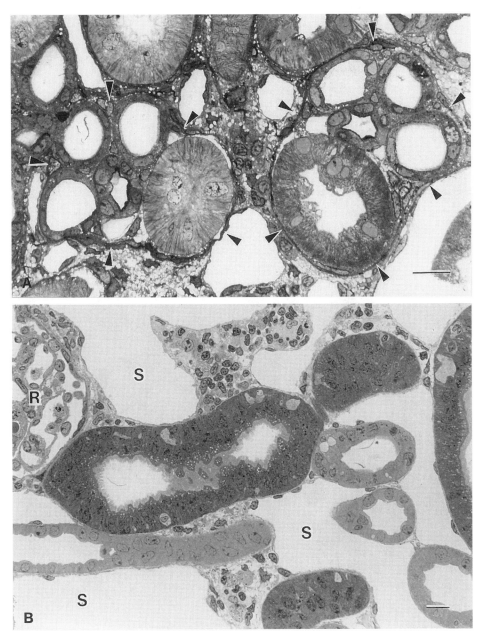

FIGURE 14-6 Spiny dogfish kidney, semithin sections. In **A** cross sections of two bundles are seen, each composed of five tubules and blood vessels and each surrounded by the peribundular sheath (*arrowheads*) (×500; scale bar: 20 μm). In **B** tubules are seen meandering in the large blood sinuses (S), with a renal corpuscle (R) on the left side (×320; scale bar: 20 μm).

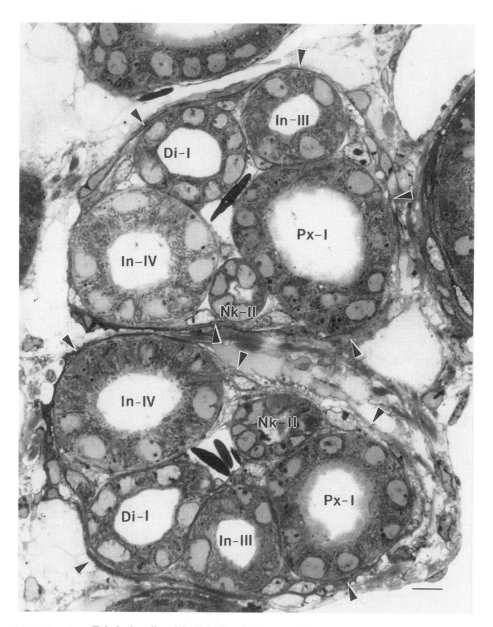

FIGURE 14-7 Tubular bundles of the little skate kidney, semithin section, showing cross sections of the five tubules (Nk-II, Px-I, In-III, In-IV, and Di-I), of blood capillaries (with erythrocytes in their lumens), and of the peribundular sheath (*arrowheads*) (×800; scale bar: 10 μm).

in Figure 14-8. The size of the renal corpuscles in marine and freshwater elasmobranchs is extremely variable (see Fig. 14-2B), as noted by Nash (1931) and Bargmann (1937), who cautioned against giving "average" glomerular sizes for elasmobranchs. The large size ranges could reflect different developmental stages of the organ. In the mammalian kidney, the juxtamedullary renal corpuscles are larger than the cortical (Peter 1909; Bankir and Farman 1973). A dorsomedial size gradient of the renal corpuscles has been described by Borcea (1906) in the kidney of the developing spiny dogfish. A mediolateral gradient with large

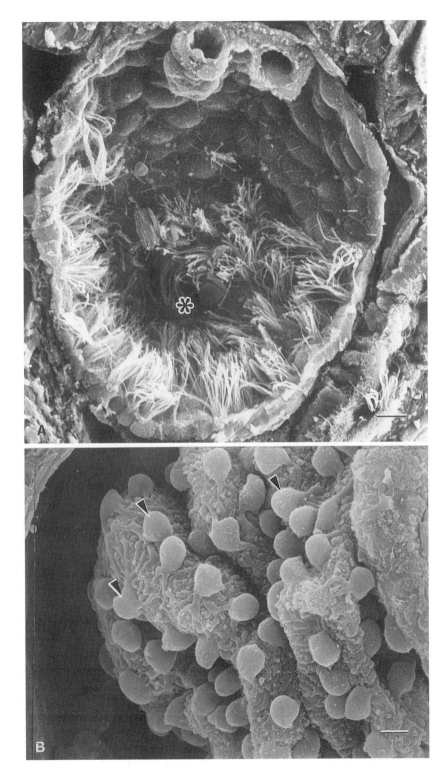

FIGURE 14-8 Little skate kidney; scanning electron micrographs of the components of a renal corpuscle, Bowman's capsule (**A**) and glomerulus (**B**). The *asterisk* in **A** lies at the beginning of the neck segment at the urinary pole, which is marked by numerous flagellar cells. The *arrowheads* in **B** indicate the bodies of some podocytes (×750; scale bar: 10 μm). (From Lacy et al. 1987.)

glomeruli close to the middle line (toward the dorsal aorta) and gradually decreasing in size in the direction of the kidney lateral border can be clearly seen in the Microfil casts of the renal arterial system of *Raja erinacea* (Hentschel 1988). Near the small glomeruli, developing nephrons have been observed (Hentschel 1988). All these data suggest that the larger renal corpuscles are older than the smaller ones, the former lying closer to renal and interenal arteries from which the afferent arterioles originate. Therefore, the size distribution of the renal corpuscles is not random.

In marine elasmobranchs, Bowman's capsule nearly surrounds the glomerulus, forming the outer boundary of the urinary space. It is formed by a single layer of cells that have different morphologies depending on the region in which they are found. From the vascular pole, where the afferent and efferent vessels enter and exit the glomerulus, toward the urinary pole, the cells are squamous and provided with a single long (about 10 µm) cilium (Figs. 14-8 and 14-9A). Near the beginning of the neck segment these cells are gradually replaced by low cuboidal cells bearing unique, extremely long (~50 µm) flagellar ribbons (see Figs. 14-8A and 14-9A). These unusual cells continue into the initial segments of the renal tubule, where they intermingle with the tubular (brush border) epithelial cells. Flagellar cells are consistently found in the neck and proximal segments of the tubule but are in the intermediate segment of selected species only, as in the skate (Lacy and Reale 1985b). They are consistently absent in the distal and collecting duct segments of all species. The microscopic anatomy of these cells is described with the neck segment.

As in marine elasmobranchs, Bowman's capsules in freshwater species display renal corpuscles with squamous ciliated cells toward the vascular pole and low cuboidal flagellar cells toward the urinary pole where they meet the neck segment epithelium.

The general structure of the glomerulus is similar in marine and freshwater species. The glomerulus is formed when the afferent vessel enters Bowman's capsule, splits into multiple capillary loops, and then exits the capsule as a single or as multiple efferent vessels. Specialized epithelial cells, podocytes, cover these capillary loops, forming the inner boundary of the urinary space (Figs. 14-8B and 14-10). Near the vascular pole of the glomerulus, the podocytes lie adjacent to specialized epithelial cells, peripolar cells, which form a collar around the afferent arteriole and also border the squamous Bowman's capsule cells. Peripolar cells are described in the section dealing with the juxtaglomerular apparatus.

The podocytes have large, nearly circular cell bodies that bulge into the urinary space (Figs. 14-8B, 14-10, and 14-11). The foot processes (pedicels) of these cells surround the capillary wall and are joined by scattered occluding and gap junctions (Lacy et al. 1987). The pedicels sit on a thick, continuous basement membrane, under which is usually found a substantial layer of mesangium that separates a thinner and highly discontinuous endothelial basement membrane (see Fig. 14-11). A glomerular filtration barrier like that found in mammals, in which there is a single basement membrane shared by the epithelium on the outside and the endothelium on the inside, can occasionally be observed. Thus at least some of the factors regulating glomerular ultrafiltration in elasmobranchs are different from those in "higher" vertebrates. The thick epithelial basement membrane is composed of a lamina densa and lamina rara containing glycoproteins and proteoglycans (Boyd and De Vries 1986; Decker and Reale 1991). The intervening mesangium consists of long processes of mesangial cells, collagen fibrils, anchoring fibrils, and microfibrils, and in the river ray kidney, elaunin and elasticlike fibrils (Lacy et al. 1991b). The thin endothelial cells possess fenestrations (40–70 µm in diameter), which are closed by a diaphragm (see Fig. 14-11B).

The juxtaglomerular apparatus

The juxtaglomerular apparatus (JGA) consists of vascular, tubular, neural, extracellular matrix, and epithelial components anatomically arranged so that they directly communicate

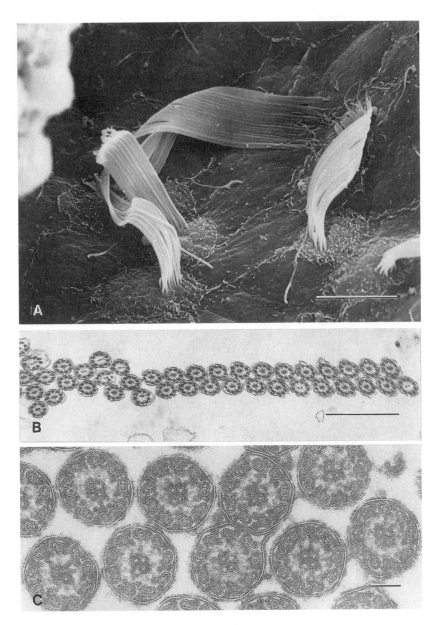

FIGURE 14-9 Flagellar ribbons in scanning (**A**) and transmission (**B, C**) electron microscopy. In **A** flagellar cells of Bowman's capsule, little skate (×5000; scale bar: 10 μm). In **B** and **C** low and high magnifications showing the alignment of the flagella in compact rows and the fine structure of single flagellum (B, ×20,000; C, ×80,000; scale bars: B, 1 μm; C, 0.1 μm).

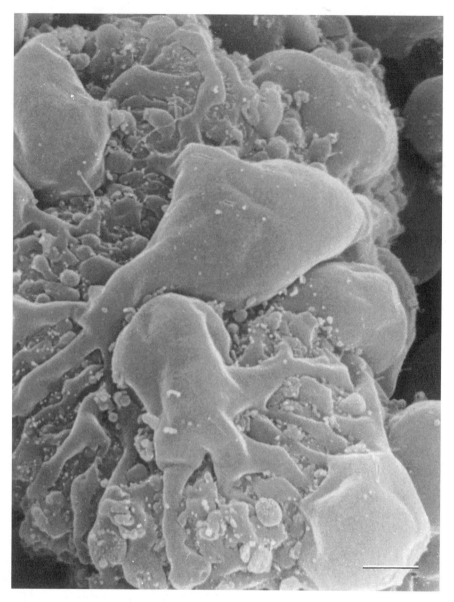

FIGURE 14-10 Little skate kidney, scanning electron micrograph of some podocytes. The bodies of the cells and their interdigitating processes are wrapped around a glomerular capillary (×3000; scale bar: 5 μm).

and regulate on another's function and thus the physiology of both the kidney and other more peripheral organs (Taugner and Hackenthal 1989). Although the JGA and renin-angiotensin system were once believed to have evolved after the elasmobranchs, there is now compelling anatomical, physiological, and biochemical evidence for the presence of them in these cartilaginous fish (Henderson et al. 1981, 1988).

In marine elasmobranchs the JGA consists of one or occasionally two afferent arterioles; one or frequently several efferent arterioles; the macula densa, a specialized portion of the distal tubule between these arterioles; an extraglomerular mesangium; and peripolar cells,

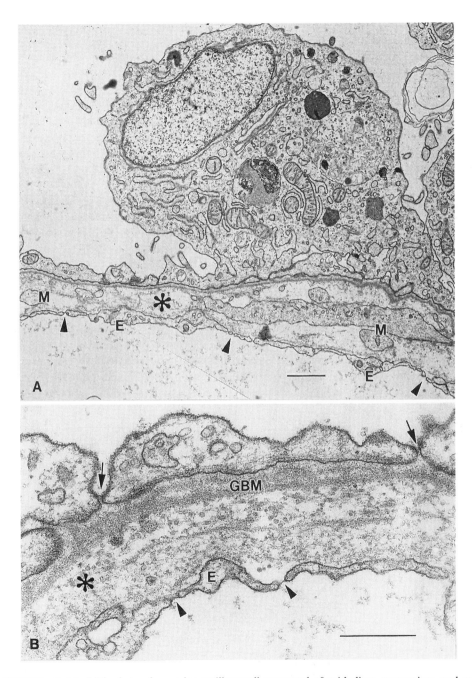

FIGURE 14-11 Little skate, glomerular capillary wall composed of epithelium, mesangium, and endothelium. **A:** On the right side can be seen the cell body of a podocyte with nucleus and, in the cytoplasm, some cisternae of the endoplasmic reticulum, a Golgi complex, dense bodies, and a few mitochondria. The cell body and some pedicels lie on a continuous, distinct basement membrane. This separates the epithelium from the mesangial matrix (*asterisk*) and from the processes of mesangial cells (M). The endothelium (E) is fenestrated (arrowheads); its basement membrane is discontinuous (×10,000; scale bar: 0.5 μm).
B: On higher magnification, the glomerular capillary wall shows the filtration slits between adjacent pedicels (*arrows*), the basement membrane (GBM), and fibrillar material in the mesangial matrix (*asterisk*). The *arrowheads* indicate fenestrations of the endothelium (E) (×40,000; scale bar: 0.5 μm).

which form a collar around the hilar vessels. Some of these components can be seen in Figure 14-12.

The afferent arteriole in both marine and freshwater species consists of a thin wall composed of endothelial cells surrounded by a single complete layer of flattened smooth muscle cells, which most frequently contain granules, close to the glomerulus (Fig. 14-13). However, these cells can also be found scattered in the wall of the afferent arteriole at considerable distances from the renal corpuscle. Ultrastructurally, the granules have a crystalline substructure similar to that reported for mammals. The granules, in addition to extensive glycogen particles and most other cell organelles, are positioned in the adluminal side of the cell (Figs. 14-14 and 14-15). Prominent bundles of intracellular microfilaments are found adjacent to the endothelium.

The efferent vessels are very short, thin-walled, and devoid of typical smooth muscle cells. However, pericytes—themselves surrounded by an intermittent basement membrane—lie adjacent to the fenestrated endothelium.

Between the afferent and efferent arterioles of marine species runs the distal tubule, which, in addition to contacting both vessels, has one surface facing the glomerulus. It is along this surface that the tubular epithelium is specialized into a macula densa, characterized by densely packed high columnar cells with convex luminal surfaces (see Fig. 14-12). The basolateral surfaces of these cells are extensively dilated, compared to the closely apposed and uniformly wide intercellular spaces of epithelial cells on the other side of the distal tubule (see Fig. 14-12). The Golgi apparatus is basal or lateral to the nucleus and there are few mitochondria.

In freshwater stingrays, *Potamotrygon humerosa*, there are some glomeruli that possess a single afferent and one or two efferent arterioles with an intervening distal tubule and

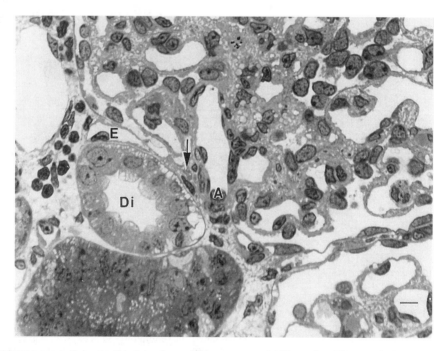

FIGURE 14-12 Spiny dogfish, juxtaglomerular apparatus, semithin section. The macula densa of the distal tubule (Di), afferent (A) and efferent (E) arteriole, and a few cells of the extraglomerular mesangium (*arrow*) are recognizable (×500; scale bar: 10 μm).

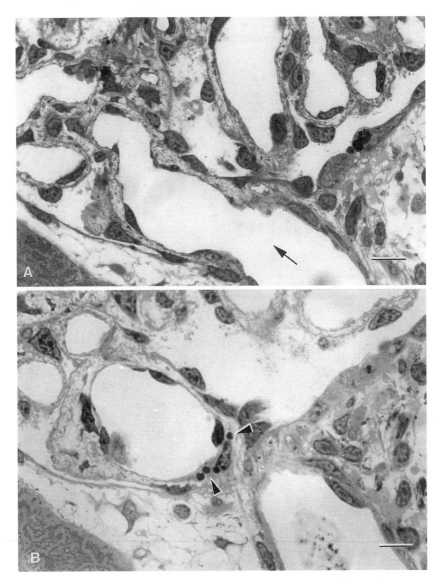

FIGURE 14-13 Freshwater ray, consecutive sections of a series. **A:** The afferent arteriole (*arrow*) enters the renal corpuscle. **B:** Some granules (*arrowheads*) are seen inside smooth muscle cells of the tunica media (×900; scale bar: 10 μm).

macula densa as described previously (Y. Kon, M. Sugimura, K. Murakami, E. R. Lacy, and E. Reale, unpublished observations). Probably these are characteristic aspects of the smaller, ventral, younger renal corpuscles. The large, probably old, renal corpuscles have multiple (up to 11) efferent arterioles exiting the renal corpuscle linearly aligned, and within a lateral indentation or vascular groove of Bowman's capsule. The usually flattened smooth muscle cells of the afferent vessel have membrane-bound granules on their abluminal region (see Fig. 14-15) like those described in the same location in the marine species or in higher vertebrates. Granular cells can be found in the tunica media of the arterioles at their entry site into the glomerulus (see Fig. 14-13B) as well as at considerable

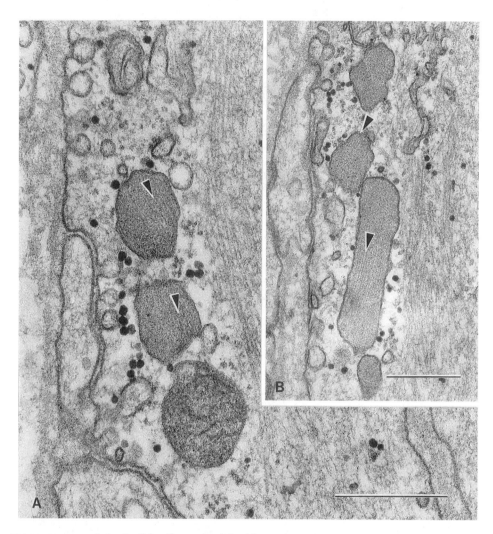

FIGURE 14-14 Spiny dogfish, afferent arteriole with membrane-bound granules inside smooth muscle cells of the tunica media. The granule contents have a paracrystalline structure (*arrowheads*) (A, ×60,000; B, ×40,000; scale bars: 0.5 µm). (From Lacy and Reale 1990.)

distance before this. Granular cells in the extremely short efferent arterioles touch the macula densa and the extraglomerular mesangium. The macula densa portion of the distal tubule, characterized by prominent intercellular spaces, is found adjacent to the extraglomerular mesangium (Fig. 14-16). Dilated intercellular spaces are characteristic of cells in which major movement of solutes and water occurs (DiBona et al. 1969; Grantham et al. 1969; Kaissling and Kriz 1982; Kirk et al. 1985).

The glomerular mesangium extends beyond the hilar region and occupies the space between the afferent and efferent vessels and the macula densa portion of the distal tubule at the vascular pole or between efferent vessels and macula densa along the lateral groove. In marine euryhaline and freshwater species there are but a few cells in this site (Fig. 14-12). Studies in mammals have shown that mesangial cells form a functional syncytium, since they are all connected by communicating-type gap junctions (review in Taugner and Hackenthal 1989). These cells have contractile properties in vitro (Mahieu et al. 1980; Ausiello et al. 1980). In vivo, however, and in the rat glomerulus they do not contain immunohistochemically demonstrable actin (Elger et al. 1993). Thus, the hypothesis that mesangial cells have the capacity to regulate glomerular function through a mechanical action is questionable. Gap junctions have been described between the pedicels in the glomerulus of the skate (Lacy et al. 1987) but not between mesangial cells.

Peripolar cells may be a component of the JGA (Ryan et al. 1979; Gardiner and Lindop 1985). These epithelial cells are juxtaposed between podocytes and Bowman's capsule cells, are low cuboidal to squamous, and possess distinctive vacuoles and granules, both of which are membrane-bound, extremely variable in size, and filled with homogeneous material of different light microscopic staining (Fig. 14-17A) or electron density (Fig. 14-17B). The contents of these granules is currently unknown, as is the function of these cells, which are found in both marine and freshwater elasmobranchs as well as in higher vertebrates (Lacy and Reale 1989; Grabowski et al. 1994; Thumwood et al. 1993; Kon et al. 1994). The position of peripolar cells and their reported response to

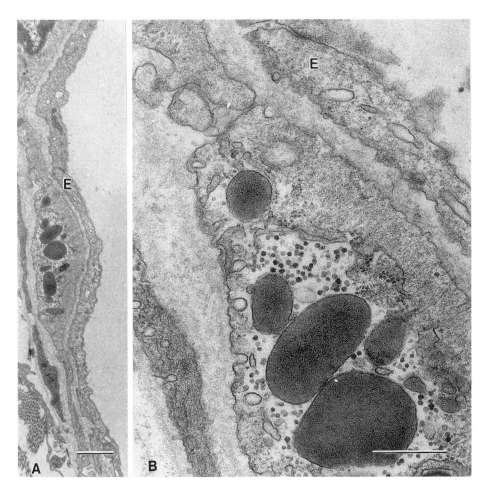

FIGURE 14-15 Freshwater ray, membrane-bound granules in a smooth muscle cell of an afferent arteriole. The granules lie with glycogen particles in the abluminal side of the cell. E = endothelium (A, ×8000; scale bar: 1 μm. B, ×40,000; scale bar: 0.5 μm).

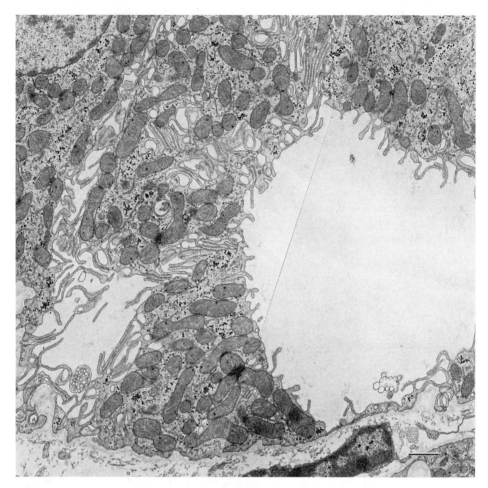

FIGURE 14-16 Freshwater ray, macula densa of the distal tubule. The intercellular spaces are broad, partially filled with microvilli-microplicae–like cytoplasmic projections of the bordering cells (×8000; scale bar: 1 μm).

changes in dietary salt suggests that they may be a part of the complex renin-angiotensin and JGA system, but definitive evidence is currently lacking.

The presence of renin granules as part of the JGA would suggest that vasoactive components are produced. Indeed, angiotensin II–like immunoreactivity has been reported in elasmobranch tissues (Galli and Kiang 1990). Partially purified renal tissue extracts from sharks increases systemic blood pressure in this species and in laboratory rats as well as generating angiotensin from porcine and synthetic angiotensinogen (Hazon et al. 1989; Henderson et al. 1981; Uva et al. 1992).

Vasculature

Blood supply to the kidneys of marine elasmobranchs originates from two major sources: arterial and a renal portal system (Fig. 14-18). Renal arteries derive largely from the adjacent dorsal aorta and, to a lesser extent, from intercostal and iliac vessels. Renal portal vessels derive from the caudal vein and segmental veins of the posterior body wall (Parker 1881; Rand 1905; Daniel 1934).

Figure 14-19 gives a composite overview of the blood circulation in elasmobranchs synthesized from numerous sources (Daniel 1934; Marples 1936; Ghouse et al. 1968; Hentschel 1988; Lacy et al. 1985; Lacy and Reale 1994;

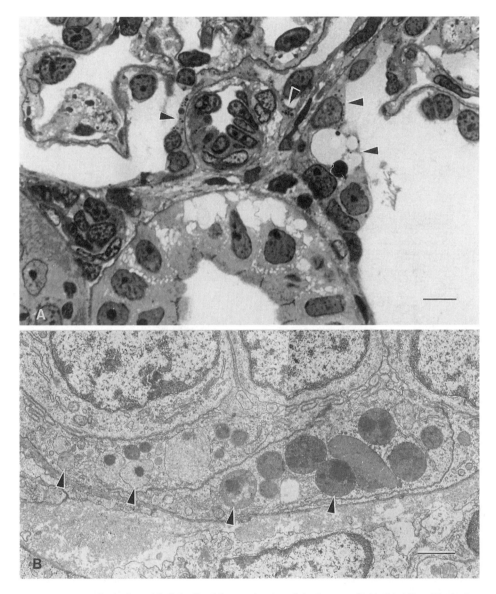

FIGURE 14-17 Peripolar epithelial cells of the renal corpuscle in the smooth dogfish (**A**) and in the hammerhead (**B**). In both species the granules (*arrowheads*) have different sizes and contents (A, ×900; scale bar: 10 μm. B, ×8000; scale bar: 1 μm).

Parker 1881; Romer and Parsons 1977; Satchel 1971). It is uncertain if this schema is correct for all species or phylogenetic groups of elasmobranchs.

Renal arteries enter the dorsal surface of the kidney and provide major branches to each of the lobes. Subsequently, these can provide branches directly to the glomeruli (afferent arterioles), which in some cases can be extremely long, or intrarenal branches, which may also give rise to afferent arterioles. Interrenal arteries also give rise to branches, interbundular capillaries, which supply the interstitial tissue between the tubular bundles (countercurrent multiplier system) in the bundle zone.

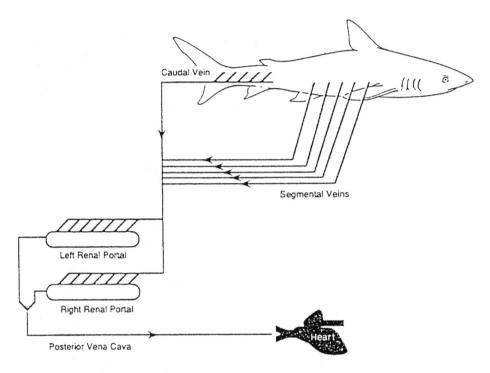

FIGURE 14-18 Schematic drawing of the portal system in the shark. (From Stoskopf 1993.)

The sinus zone receives blood from at least three sources: capillaries, generated by the extremely short, thin-walled efferent arterioles from postglomerular capillaries; interbundular capillaries, which empty directly into the sinus zone; and the renal portal veins from the tail and body wall. Taken together, current evidence shows that in marine elasmobranchs arterial blood goes into the bundle zone and venous blood (postcapillary) goes into the sinus zone. The arterial origin and venous path of the capillaries within the peribundular sheath of marine elasmobranchs is currently not known.

Blood from the sinus zone leaves the kidney directly via renal veins. The small interstitial capillaries and veins are sometimes surrounded by a few large smooth muscle cells arranged around the vessel wall like a sphincter (Fig. 14-20). These interesting structures are termed turban-organs (*Turbanorgane*, Leydig 1852; Meyer 1888; Bargmann 1937, 1954) and present the possibility for regulation of blood flow at the level of the veins, which is not the regulatory site characteristically found in mammals. Similar venous sphincters have been described around small veins of the interlobular septa in the kidney of the elasmobranch *Trygon violaceus* (Bargmann 1937) as well as of the renal interstitial tissue of the teleost *Rachycentron canadum* (Howse et al. 1992).

Little has been published on the ultrastructure of the elasmobranch renal vasculature. The arteries and arterioles in the little skate, *Raja erinacea*, and in the river ray, *Potamotrygon humerosa*, have a thin tunica media composed of flattened smooth muscle cells; an endothelium composed of elongated cells parallel to the vessel major axis and characterized by some mitochondria; few cisternae of the rough endoplasmic reticulum, Golgi apparatus, dense bodies and, in small arteries, elongated bodies that have a substructure of longitudinal striations corresponding to the specific endothelial granules (Weibel-Palade granules) of higher vertebrates (Weibel and Palade 1964). These granules have been shown

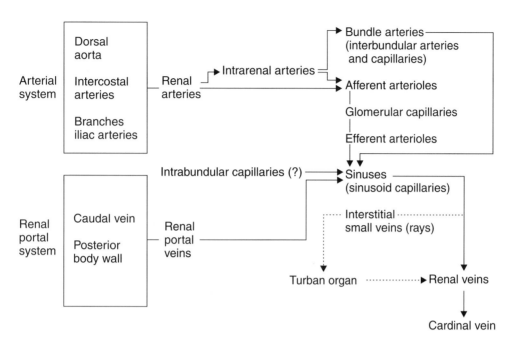

FIGURE 14-19 Schematic overview of the renal vessels in elasmobranchs.

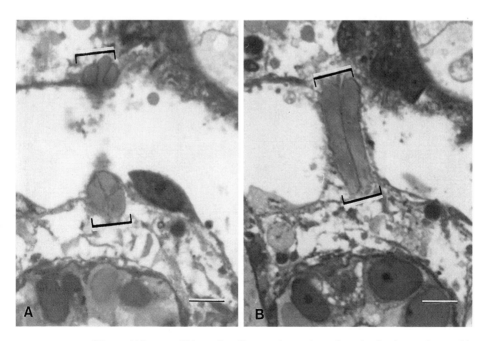

FIGURE 14-20 Stingray kidney, semithin section. Consecutive sections of a series showing a minute sphincter ("turban organ") around a small vein (*framed area*). The bundle of smooth muscle cells circularly arranged around the vessel has been cut along the diameter in **A** and tangentially in **B** (×960; scale bar: 10 μm).

TABLE 14-2 RENAL TUBULE SEGMENTS AND SEGMENT SUBDIVISIONS: COMPARATIVE SYNOPSIS OF THE TERMINOLOGY

Haller (1902)	Kempton (1940, 1943, 1957, 1962)	Borghese (1966)	Ghouse et al. (1968)	Hickman and Trump (1969)	Deetjen and Antkowiak (1970)	Hentschel (1988)	Lacy et al. (1985)
1. Abschnitt (1st segment) with ciliary cells	Nk with flagellar and non-flagellar cells; inside a transparent streak	Nk (in a bundle of 5 tubules all associated with RC); flagellar and non-flagellar cells	Segment I: thin; bb and ciliated cells; it returns to RC as large Segment II: without bb and flagellar cells	Nk, thin, with ciliated cells	Segment I: thin walled Segment II: close to RC	Nk	Nk I Nk II
2. Abschnitt (2nd segment) "Streifenstück"; special segment; without ciliary cells; without bb cells	"Streifenstück" does not exist in this location; it is in continuity with the distal segment (see below)						
3. Abschnitt (3rd segment) long, large; with bb cells	Proximal segment with bb and flagellar cells; large, distally thin; it returns to RC to form continuation around Nk-loop; in continuity with Di	Large dorsal canal, thick (thin at its end); bb cells; flagellar cells; in continuity with the next segment but with interposition of a ventral l;loop (2 of the 5 tubules); with Falk	Segment III: bb cells, dense bodies Segment IV: bb and ciliated cells Segment V: largest tubule (thin at its end); bb and ciliated cells	Proximal tube with bb cells PI A: thin, numerous mitochondria PI B: large, numerous dense bodies PII: still larger, longest part; striation due to numerous mitochondria	Segment III: large, with brush border Segment IV: as above, most of the tubule Segment V: returns to RC	PI A PI B: as Hickman and Trump (1969) PII	Px-I: thin Px-II: large Px-III: still larger, matrix rich mitochondria; endosomes Px-IV: numerous flagellar cells

Urinary System 379

					IS: intermediate segment	In-I, In-II, In-III, In-IV, In-V, In-VI: all with extensive, uniform intercellular spaces
4. Abschnitt (4th segment) same structure as the 2nd segment; with part of this it forms a "Knäuel" surrounded by a "Kapsel"	Distal segment, striated ("Streifenstück" of Haller?); mesial lateral (inside transparent streak)	Small dorsal canal; thin, without brush border (not Falk); returns to MC; MD absent	Segment VI: without bb cells; MD present	Distal tubule, wrapped around Nk; without bb cells	Segment VI: loose coil around segment II; returns to RC Segment VII: convolutions Segment VIII: thin, returns to RC, then flanks segment II forming a countercurrent system	EDT: early distal tubule LDT-A: late distal tubule LDT-B: late distal tubule B	Di-I: large intercellular spaces; with MD Di-II
5. Abschnitt (5th segment) secreting	Collecting duct	Collecting duct	Collecting duct	Collecting duct	Collecting duct	Small collecting duct Large collecting duct	CD-I CD-II

RC = renal corpuscle; Nk = neck segment; P = Px = proximal segment; In = intermediate segment; Di = distal segment; CD = collecting duct; MD = macula densa; FAlk = alkaline phosphatase; bb = brush border.

Stolte et al. (1977) in their model of the little skate nephron indicate three proximal tubules (equivalent to Px-III, Px-IV, In-I, and In-VI, according to the nomenclature of Lacy et al. 1985); three distal coilings (equivalent to the convoluted part of the tubular bundle formed by Nk-II, Px-I, In-III, In-IV, and Di-I); and a countercurrent system connecting proximal tubules with distal coilings (equivalent to the straight part of the tubular bundle formed by Nk-I, Px-II, In-II, In-V, and Di-I). The collecting duct system is subdivided into three parts: initial, middle, and terminal. According to Hentschel (1988), the subdivisions In-I to -III of Lacy et al. (1985) form, all together, a short intermediate segment. In-IV of Lacy et al. belongs to the early distal segment (DSE); In-V and -VI belong to the late distal segment (DSL); Di-I and -II represent the collecting tubule (CT) and collecting duct (CD), respectively; and CT-I and -II constitute the large collecting duct. In a previous paper (Hentschel 1987), DSE was named the distal segment (DS) and DSL, CT, and CD formed the collecting tubule (CT). Lacy et al. (1985) define all nephron segments having similar structural (and presumably functional) characteristics with the same terminology. The transition from a segment or segment subdivision to the successive subdivision is usually very sharp.

to be the storage sites of von Willebrand's factor (Simionescu and Simionescu 1988). Most of the organelles occur in the cytoplasm close to the nucleus. A thin basement membrane separates the endothelium from the tunica media. In this, the usually flattened smooth muscle cells, surrounded by a noticeable basement membrane, have the usual organelles. Some smooth muscle cells of the arteries and glomerular arterioles display membrane-bound granules (see JGA description). Processes of these modified smooth muscle cells extend through the basement membrane to lie in close apposition to the endothelial cells. Otherwise, the space between endothelial and smooth muscle cells contains microfibrils (elastic fiber microfibrils), tiny collagen fibrils, and elastic fibers (Lacy et al. 1989).

The endothelium of efferent arterioles is fenestrated as soon as they exit the glomerulus; the tunica media is composed of a single discontinuous layer of pericytes. The large sinuses of the sinus zone in the skate and shark and of the intertubular space in the river ray are lined with large, irregularly polygonal epithelial cells with scattered assemblies of fenestrations; the endothelial basement membrane is thin, intermittent, and externally reinforced by collagen fibrils and cytoplasmic cell processes, which probably belong to pericytes.

Renal tubular epithelium

Until the configuration of the renal tubules was elucidated (Lacy et al. 1984, 1985), which then allowed a definitive sequence of the epithelium to be defined (Lacy and Reale 1985b), terminology for various segments of the nephron was confusing. A synopsis of the various nomenclatures applied to the marine elasmobranch renal tubule is shown in Table 14-2.

Despite a highly complex nephron, the overall pattern remains consistent with that of all other vertebrates. The renal tubule has five distinctive segments, beginning at the renal corpuscle: neck, proximal, intermediate, distal, and collecting duct, each of which has subdivisions (see Fig. 14-3). These subdivisions are often defined by subtle changes in ultrastructure and may not be exactly equivalent among species. The description given here is based on the extensive data available for the little skate, *Raja erinacea* (Lacy and Reale, 1981, 1985a,b, 1991a,b).

NECK SEGMENT

Despite major differences in renal and tubular architecture between marine and freshwater species, the neck segment (Nk) appears to be the same in both groups of fish but is significantly shorter in the latter groups (*P. humerosa*). The epithelium consists of two cell types (Fig. 14-21): the nonflagellar cells, which bear few if any microvilli or microplicae on their apical surface, and flagellar cells, each with multiple flagella arranged in rows. The length of the Nk in marine samples allows the identification of two subdivisions, Nk-I and Nk-II. The first subdivision has numerous columnar flagellar cells and a narrow diameter. In the second subdivision, especially toward its end, the number of flagellar cells progressively decreases, the cells become cuboidal to squamous, and therefore the tubule diameter is large. A similar distinction for the Nk, which is extremely short, is not evident in the freshwater elasmobranch.

The flagellar cells are continuous with those lining the urinary pole of Bowman's capsule and are identical to those found in other parts of the nephron. Multiple (10–30), long (~50 µm) flagella, aligned in one to several parallel rows, project from the apical surface of these cells (see Fig. 14-9). The rows are perpendicular to the long axis of the tubule and, because of their shape and length, are always bent in the direction of urine flow. The axonemes of adjacent flagella are bound together by a prominent surface coat and transmembrane structures (Lacy et al. 1991a), thus forming a single entity analogous to a flexible paddle that beats in unison (Bargmann, 1937; Kempton 1943, 1962, 1964; Lacy et al. 1989, 1991a). Each flagellum begins deep within the cell with a basal body and conspicuous rootlets. Intermembranous particles are

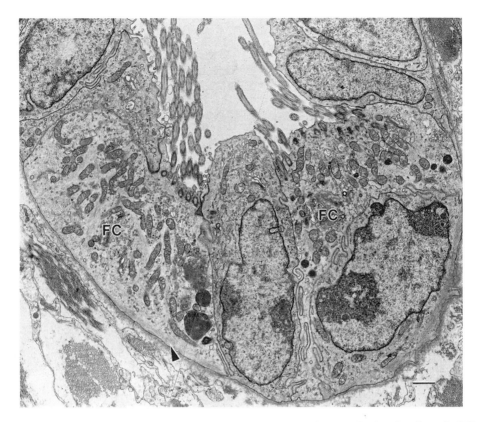

FIGURE 14-21 Little skate, neck segment (Nk-I) close to the renal corpuscle. Two flagellar cells (FC) are interposed between smooth surfaced nonflagellar cells. Bundles of filament are seen in the basal region of the flagellar cells (*arrowhead*) (×6000; scale bar: 1 μm). (From Lacy and Reale 1991a.)

arranged in a typical "flagellar" necklace. Each axoneme possesses the usual 9 + 2 microtubular pattern, with the central pair of microtubules orthogonal to the main tubular axis, indicative of the beat direction (see Fig. 14-9C). Direct observations of renal tubules, time lapse cinematography, and video microscopy have shown an extremely rapid, powerful beating of these flagellar ribbons (Koelliker 1845; Leydig 1852; Bargmann 1937; Kempton 1943, 1962, 1964; Lacy et al. 1989; Lacy et al. 1991a). These cells probably function to propel urine down the tubule and could generate a negative pressure in Bowman's space to assist glomerulation filtration, since blood pressure (glomerular filtration) is low in these animals. Observations of tubular fluid transit times would support this proposal since linear velocity in early tubular segments where flagel-

lar cells are abundant is about 3.5 times greater than in more distal tubular segments where flagellar cells are absent (Deetjen and Antkowiak 1970).

In marine species, the nonflagellar cells have a rather smooth surface, which proximally (toward the renal corpuscle) slightly bulges into the tubular lumen (see Fig. 14-21) and distally progressively flattens, as the whole epithelium does in Nk-II. The nucleus is large and is surrounded by scanty cytoplasm with few organelles but consistent basal bundles of intermediate filaments (about 10–12 nm thick) and microfilaments (about 5–7 nm thick) and, in Nk-II, numerous autophagosomes (Lacy and Reale 1985b, 1991a).

In freshwater species, Nk is usually short and lined by squamous flagellar and nonflagellar cells, whose nucleus and perinuclear cytoplasm irregularly bulge toward the interstitium, the luminal surface (outside the flagellar ribbons) remaining rather flat. In both marine and freshwater elasmobranchs, the transition from the Nk segment to the proximal segment is abrupt, marked by a remarkable increase of the cell size and by the sudden appearance of the brush border (G. M. Grabowski, E. Reale, and E. R. Lacy, unpublished observations).

PROXIMAL SEGMENT

The proximal segment (Px) is easily distinguished and is long, with portions in both the bundle and sinus zones of marine species (see Fig. 14-6B). The tubular epithelium consists of flagellar cells morphologically indistinguishable from those in the Nk segment and cells bearing a distinct brush border (Figs. 14-22 and 14-23). In the little skate, *R. erinacea*, there are four subdivisions, defined

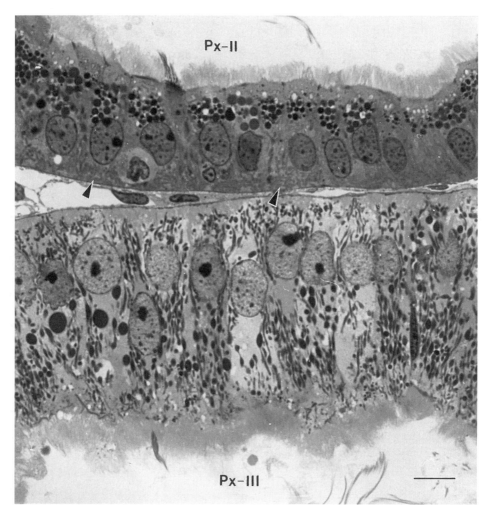

FIGURE 14-22 Smooth dogfish, semithin section, showing the second (Px-II) and third (Px-III) subdivisions of the proximal tubule. In the Px-II epithelial brush border, cells with deeply stained granules are evident, the mitochondria are barely visible (*arrowheads*), and the brush border is remarkable. In Px-III, the mitochondria are very evident, and the brush border massive. In both tubule subdivisions flagellar cells are numerous (×1080; scale bar: 10 μm).

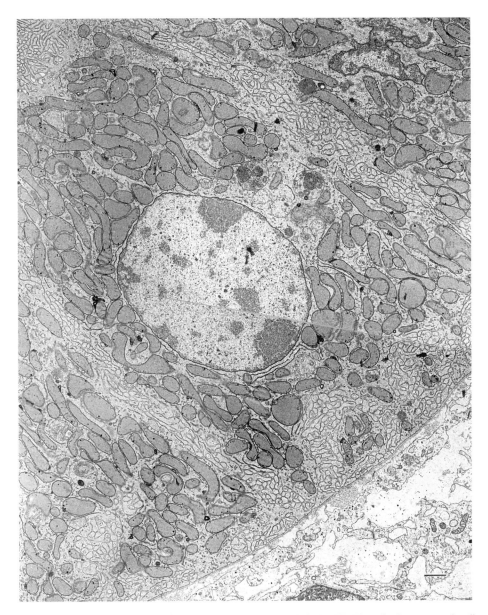

FIGURE 14-23 Little skate, third subdivision of the proximal tubule (Px-III). The tall columnar nonflagellar cell bears a brush border visible partly on the upper left corner of the micrograph. The basolateral surface is covered with tightly packed microvillilike projections. The mitochondria are numerous, with rare cristae but a large extensive matrix (see also Fig. 14-21) (×6000; scale bar: 1 μm). (From Lacy and Reale 1991a.)

by morphological differences in these brush border cells. The transition between the Nk and Px occurs in the countercurrent bundle at the distal end of the first loop, far away from the renal corpuscle (see Fig. 14-3B). There is also a transition between the first (Px-I) and second (Px-II) subdivisions of the proximal segment within the tubular bundle, close to the renal corpuscle. After exiting the bundle by piercing the peritubular sheath, Px goes to the same renal corpuscle and lies along Bowman's capsule. It then extends into the sinus zone as Px-III, and deep in this zone it makes a transition to the final subdivision, Px-IV.

The following characteristics increase along the first three subdivisions (Px-I, -II, -III) but decrease in the final subdivision (Px-IV): height of the epithelial cells and the apical brush border, apical staining (periodic acid–Schiff reaction) for glycoproteins, both the number and extension of basolateral plasma membrane invaginations, and the outer diameter of the tubule. The mitochondria have markedly different morphologies along the length of this segment and are a key distinguishing feature of the various subdivisions. Their cristae have the "classical appearance" in Px-I, -II, and -IV; they are rare and, in particular, extremely short in Px-III (Fig. 14-24). Freeze-fracture shows that zonulae occludentes of all Px subdivisions are composed of several superimposed strands (Lacy and Reale 1991a) (Fig. 14-25A). Therefore, the tight junction in the proximal tubule of the marine species differs profoundly from what has been described in higher vertebrates (review in Orci et al. 1981). In all higher vertebrates Px zonulae occludentes consist of one strand, which correlates with their being a typical example of a "leaky" epithelium.

There are some marked elasmobranch species differences in the Px segment. For example, in the little skate the transition region from Px-II to -III is lined by epithelial cells possessing a series of structures such as coated pits and vesicles, apical tubules connecting with deeper vesicles, and lysosomelike dense bodies, suggesting that a significant amount of receptor-mediated endocytosis occurs here (Lacy and Reale 1991a). In addition, the last subdivision, which has numerous flagellar cells, has a final transition region to the intermediate segment lined by cells of markedly different heights, which alter the luminal diameter (Lacy and Reale 1985b). In both the spiny dogfish (*S. acanthias*) and smooth dogfish (*Mustelus canis*), the Px-II segment is characterized by nonflagellar cells with numerous apical granules of markedly different size and shape as well as their staining patterns (see Fig. 14-22). It is not unusual to find large inclusions in the proximal tubule of vertebrates, as was noted by Gérard and Cordier (1934), who speculated that they represented the remnants of material from the tubular fluid that these cells had reabsorbed.

Px-III segments from the spiny dogfish have been isolated and perfused in vitro (Sawyer and Beyenbach 1985). This subdivision exhibits a net secretion of NaCl and fluid, which confirmed earlier work of Beyenbach and Fromter (1985) showing active Cl^- secretion in shark proximal tubule. Stolte et al. (1977) reported that in the little skate, in vivo micropuncture of proximal tubular segments (Px-III) showed this segment was the principal site for Mg^{2+}, phosphate, and sulfate secretion as well as reabsorption of Na^+ and Cl^- in excess of water.

The proximal segment of the freshwater ray, *P. humerosa*, has a general structure similar to that described previously in marine species. The apical region of the brush border cells contains part of the smooth endoplasmic reticulum that, at the transition from Px-II to Px-III, is arranged in stacks of flattened cisternae, which are straight or wrapped around a lipid droplet (Fig. 14-26B). These structures measure up to 9 µm in diameter. Numerous flagellar cells, dense bodies, and lipid droplets characterize the tiny Px-IV subdivision (see Fig. 14-26B). There are no reports on the function of the proximal segment in freshwater species.

INTERMEDIATE SEGMENT

The intermdiate (In) segment is positioned between the proximal and distal segments and is analogous to that of Henle's loop in mammals, but structurally and most assuredly functionally it is quite different. It is the longest segment in all marine species and forms two of the four loops (see Fig. 14-4). In marine elasmobranchs it begins as a sharp transition from Px-IV (Fig. 14-27A) in the sinus zone and extends into the bundle zone, where it is tightly wrapped in the countercurrent arrangement of tubules. After ex-

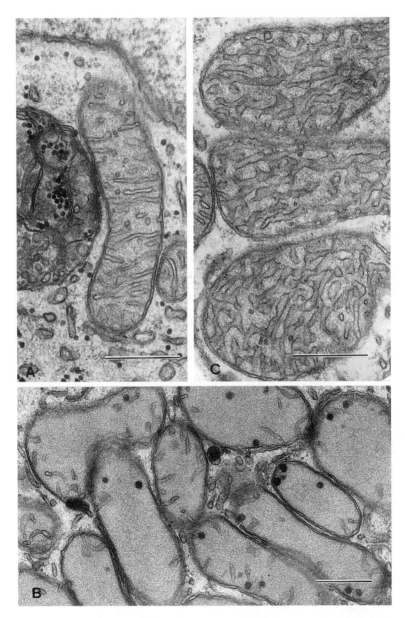

FIGURE 14-24 Little skate, mitochondria heterogeneity in the different segments of the renal tubule. **A:** Usual crista type is found in most of the tubular cells, including flagellar cells; **B:** scanty cristae and abundant matrix in Px-III brush cells; **C:** tubular type seen in the intermediate section (In-IV) subdivision (A, C, ×40,000; B, ×30,000; scale bars: 0.5 μm).

tending to the end of the bundle and looping back along the bundle (loop III), it pierces the peritubular sheath again as it exits and goes to Bowman's capsule. It then reaches the sinus zone, where it meanders among the other tubules from its own as well as other nephrons, and finally goes back to the same renal corpuscle, thus forming loop IV (see Fig. 14-27B). At the vascular pole of the renal corpuscle, it becomes the distal segment. In freshwater species it is confined to the complex zone (see Fig. 14-3).

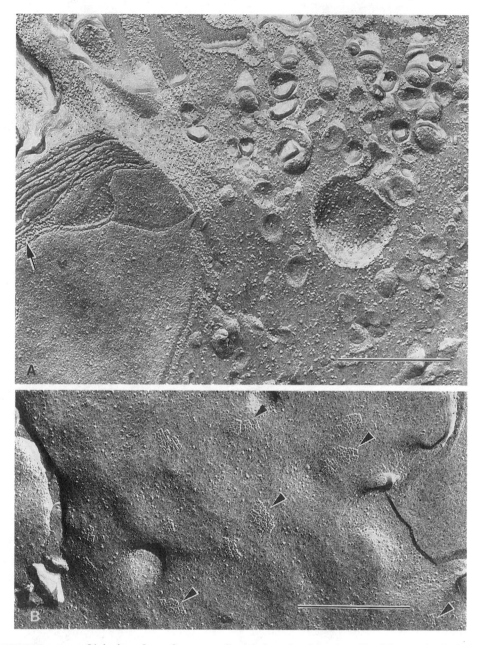

FIGURE 14-25 Little skate, freeze-fracture replicas. **A** shows brush border cells of the proximal tubule (Px-III). The zonula occludens is composed of numerous superimposed strands, occasionally surrounding particles of a gap junction (*arrow*). In **B**, the replica reveals assemblies of orthogonally arranged particles (*arrowheads*) on the face P of a cleaved pedicel plasma membrane (A, B, ×60,000; scale bars: 0.5 μm). (A, from Lacy and Reale 1991a; B, from Lacy et al. 1987.)

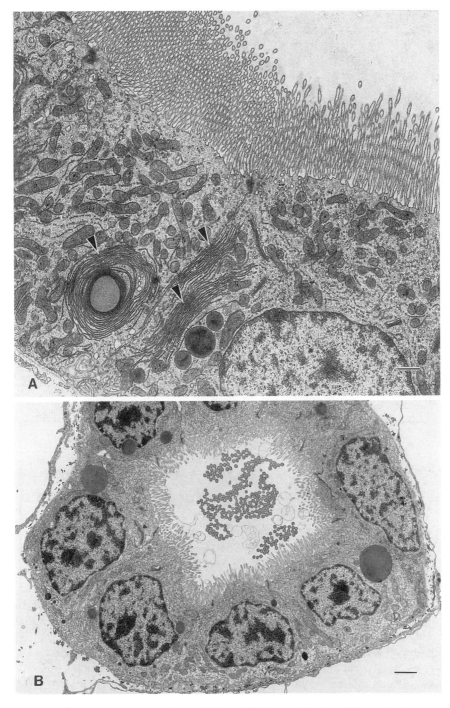

FIGURE 14-26 Freshwater ray, proximal tubule subdivisions. **A:** Stacks of flattened cisternae of the smooth endoplasmic reticulum surrounding a lipid droplet (*arrowheads*) characterize the second subdivision of the tubule (×6000; scale bar: 1 μm). **B:** Fourth subdivisions, which, as in marine species, are marked by brush border cells containing lipid droplets and dense bodies as well as by numerous flagellar cells (×2600; scale bar: 2 μm).

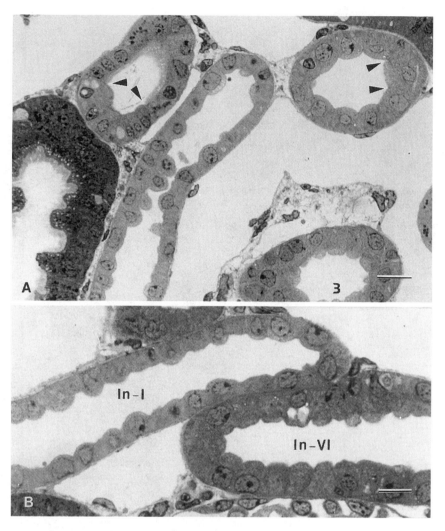

FIGURE 14-27 Spiny dogfish, sinus zone. A shows transitions from Px-IV (cells with brush border, *arrowhead*) to In-I (cells without brush border). In **B** the first (In-I) and sixth (In-VI) subdivisions of In can be seen. In = intermediate segment; Px = proximal segment (×450; scale bar: 20 μm).

Although in marine species there are six subdivisions defined in the little skate by ultrastructural criteria, the most distinguishing characteristic is the absence of a brush border; prominent basolateral membrane interdigitations (with the exception of In-VI), which usually are arranged roughly parallel to one another; and numerous, elongated mitochondria, which are not oriented in any particular way. Haller (1902) described what is now known as the intermediate segment, or at least some subdivisions of it, as the *Streifenabschnitt*, or striated segment (Table 14-2), probably because of the basolateral infoldings apparent in light microscopy.

Figure 14-28 shows some subdivisions of In in marine species. As with the Px, there appear to be some species-specific characteristics of In. For example, in the dogfish *Squalus acanthias*, the first subdivision (In-I) is thin, with a consistently wide lumen lined by cuboidal cells, which have no intracellular granules and whose apical border bulges into the lumen. This is in contrast to In-I of the little skate,

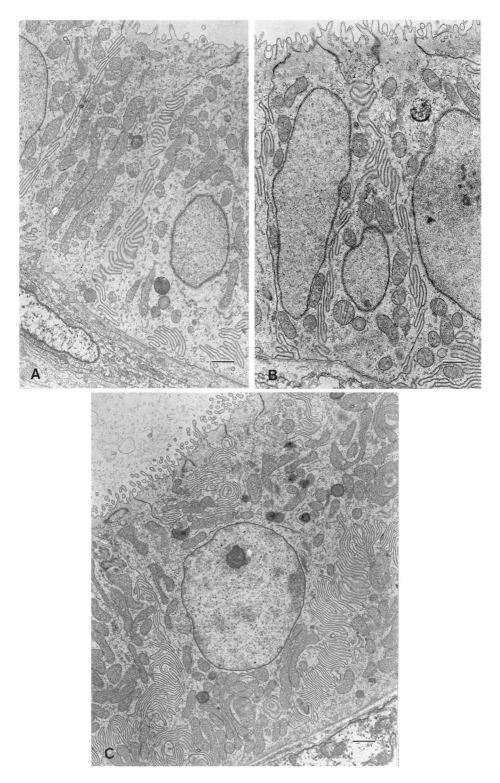

FIGURE 14-28 Little skate, aspects of some subdivisions of the intermediate (In) segment; **A:** In-II; **B:** In-III; **C:** In-IV. The intercellular space between adjacent epithelial cells is gradually longer from In-II to In-IV. In these last subdivisions the lateral cell borders are highly and tightly interdigitating; the interdigitations are superimposed in stacks, which are visible by light microscopy. The mitochondria are of tubular type, without a particular orientation (×6000; scale bar: 1 μm). (From Lacy and Reale 1991b.)

which is lined by irregular epithelium varying from columnar to squamous, all of which contain numerous granules up to 3 µm in diameter and whose contents stain heterogenously. These cells also contain multivesicular bodies, granules with lipid layers, and caveolae, all structures suggestive of an extensive endocytotic machinery (Lacy and Reale 1985b, 1991b).

In-II cells are cuboidal to columnar, with a large centrally placed and lightly stained nucleus and long mitochondria (see Fig. 14-28A). In the basal plasma membrane of In-II cells, as well as in that of the In-V, freeze-fracture replicas demonstrate orthogonal arrays of particles (Lacy and Reale 1991b). In-III has round mitochondria but also a pale nucleus in cuboidal cells (see Fig. 14-28B). In-IV is characteristic in that it is the largest tubule in the countercurrent bundle. In addition, it has highly interdigitated and complex lateral borders and a great density of mostly elongated mitochondria (see Fig. 14-28C). These accommodate an extensive inner membrane arranged in a tight network of anastomosing tubules. Thus, in the renal tubules of marine species, the inner mitochondrial membrane can (1) possess cristae with the "classical arrangement" (see Fig. 14-24A); (2) form only a few stubby cristae, which project bluntly into an extensive matrix (see Fig. 14-24B); or (3) be ordered in a tubular network (see Fig. 14-24C). It is interesting that In-V is morphologically indistinguishable from In-II, suggesting similar functions. In-VI is bordered by columnar epithelial cells, which have characteristic dome-shaped apical regions in which there are numerous mitochondria, and basal but not many lateral interdigitations; unlike the previous subdivisions, the basal interdigitations are numerous, the lateral ones scanty. In the Atlantic sharpnose shark, In-VI possesses apical secretory granules that stain brightly for glycoproteins.

There are few functional data for the In other than the work of Friedman and Hebert (1990) and Hebert and Friedman (1990) showing that In-IV, which is in the tubular bundle, is a "diluting segment" in the spiny dogfish.

In freshwater species (*P. humerosa*), all of the In, which is comparatively short, is confined to the complex zone. Along the entire length of this segment the epithelial cells have a Golgi apparatus that is para- or supranuclear, little rough endoplasmic reticulum but abundant smooth endoplasmic reticulum, numerous glycogen particles, and few dense bodies. From the beginning to the end of this segment there is a progressive increase in cell height (up to 50% from In-I to In-VI), in the amount of mitochondria, and in the interdigitations of the lateral plasma membrane, which become clearly visible by light microscopy, although the width of the intercellular compartment remains remarkably constant. There is no granular subdivision in the freshwater elasmobranch intermediate segment. The function of the intermediate segment in freshwater species is not known.

DISTAL SEGMENT

The distal (Di) tubule begins where the nephron lies in close apposition to the vascular pole of the renal corpuscle, as is the case for all vertebrates.

In marine species there are two morphologically different subdivisions of the (Di) that correspond to their positions in the kidney (see Fig. 14-3). Di-I extends from the vascular pole of the renal corpuscle into the countercurrent bundle. Di-I pierces the peritubular sheath at its proximal end, traverses the bundle in a nearly straight course to the distal end of the tubular bundle, pierces the end of the peritubular sheath, and, still surrounded by some of the sheath cells, continues as a single tubule, Di-II. This subdivision now makes a direct path in the interbundular and subcapsular connective tissue for its anastomoses with other distal segments to form the collecting duct.

In marine species the epithelium lining Di-I, except for the macula densa (see JGA description previously), consists of only nonflagellar cells, which are darkly staining, low cuboidal, with convex apical border and lateral borders sharply defined when sections are observed with the light microscope. Ultrastructurally (Fig. 14-29A) these cells have

Urinary System 391

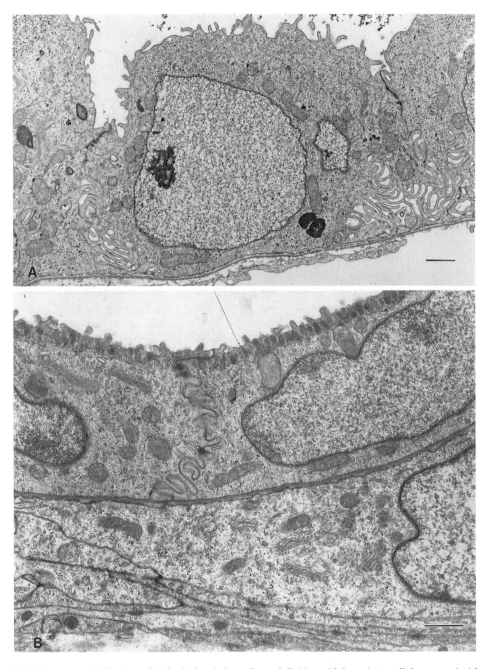

FIGURE 14-29 Little skate, distal tubules. **A** shows first subdivision with large intercellular spaces inside the tubular bundle (×8000); **B** shows squamous epithelium with apical, glycoprotein-containing granules enwrapped by the end of the peribundular sheath (×10,000; scale bar: 1 μm). (From Lacy and Reale 1991b.)

irregularly shaped microvilli or microplicae, a paucity of mitochondria, reduced cisternae of the rough endoplasmic reticulum but abundant smooth endoplasmic reticulum, and a supranuclear Golgi complex. Lateral cell borders are elaborated into interdigitating microplicae, or folds, which are more elaborate basally than apically. In the scalloped hammerhead shark and the little skate there are apical secretory granules that stain positively for sulfated glycoconjugates.

Di-II cells in *R. erinacea* are squamous to low cuboidal and have an irregularly shaped, flattened nucleus. The lateral cell borders are moderately interdigitating, never open. These cells also have sulfated glycoconjugates in apical secretory granules, which change from round to oblong in shape as the Di-II nears the collecting duct (see Fig. 14-29B).

In freshwater rays (see Fig. 14-3) the distal tubule either traverses a short distance between the afferent and efferent arteriole, when there are one or two of these vessels, or lies in the lateral groove adjacent to multiple efferent arterioles (see JGA description previously). Beyond the vascular pole this segment continues in the complex zone with epithelial cells that are distinctly high columnar. The apical border bulges into the lumen, comparable to that of their marine counterpart. These cells have numerous mitochondria, some cisternae of the rough endoplasmic reticulum, a few dense bodies, and an intercellular space filled with numerous long, flattened, interdigitating folds. The transitions are abrupt from In (with long, uniformly narrow intercellular spaces) to Di (with open intercellular spaces filled with microvilli-microplicae–like projections) and from Di to the collecting duct.

COLLECTING DUCT

There are two subdivisions to the collecting duct (CD) in marine species, both of which are found in the connective tissue septa between lobes and in the renal capsule. CD-I is lined by a simple epithelium of cuboidal to columnar cells (Fig. 14-30). CD-II is distinctly different, being lined by a pseudostratified epithelium (Lacy and Reale 1985b). Common to both subdivisions are cells lining the lumen that possess pleomorphic, deeply indented nuclei, and apical secretory granules, which are slightly larger in CD-II than in CD-I. Ultrastructurally, these granules have an irregularly banded substructure. In CD-II the basal epithelial cell layer is devoid of secretory granules but contains filament bundles directed to numerous desmosomes. Fibrocytes alternating with bundles of collagen fibrils surround CD-I and CD-II.

In freshwater rays, the structure of the CD subdivisions is similar to that of the marine species: the transition from Di to CD is abrupt; the simple cuboidal-columnar epithelium becomes rapidly pseudostratified and then stratified; the apical granules contain glycoconjugates. Confluent CD subdivisions generate large ducts located on the ventral surface of the kidney. The largest ducts are surrounded by layers of smooth muscle cells, indicating a propulsion of their contents as suggested in renal tubules of teleosts (Townsley and Scott 1963).

Overview

Elasmobranchs evolved a unique strategy of maintaining tissue hypertonicity to achieve osmotic homeostasis in marine environments. The complex structure of their kidneys reflects the important role that this organ plays in the process. The ability of some marine species to move to and from freshwater suggests a highly controlled series of regulatory mechanisms for both salt and water balance, of which currently little is known. Observations of kidney structure in strictly stenohaline freshwater elasmobranchs suggest the necessity of a countercurrent system for adaptation to a marine environment; on the other hand, this structure does not prevent adaptation of marine species to a freshwater environment. The recent discovery of a juxtaglomerular apparatus and components of the renin-angiotensin system in these fish of both environments offers the opportunity to not only elucidate the early evolutionary history of these control mechanisms

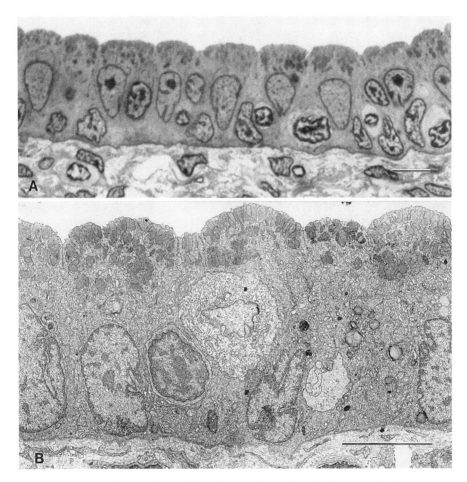

FIGURE 14-30 Second subdivisions of the collecting duct with tall, columnar cells bearing elongated secretory granules in their apical cytoplasm are shown in (**A**) the Atlantic sharpnose shark and (**B**) the spiny dogfish (A, ×1050; B, ×2400; scale bar: 10 µm).

but surely will provide insight into mechanisms important in human renal function.

ACKNOWLEDGMENTS

Eric R. Lacy received support from the Alexander von Humboldt-Stiftung and the National Science Foundation (DCB 8903369). Enrico Reale received support from the Deutsche Forschungsgemeinschaft (SF13146 RE 257/7-1)

REFERENCES

Ausiello, D. A., J. I. Kreisberg, C. Roy, and M. J. Karnovsky. 1980. Contraction of cultured rat glomerular mesangial cells after stimulation with angiotensin II and arginine vasopressin. *J. Clin. Invest.* 65: 754–60.

Bankir, L., and N. Farman. 1973. Hétérogénéité des glomerules chez le lapin. *Arch. Anat. Microsc.* 62: 281–91.

Bargmann, W. 1937. Untersuchungen über Histologie und Histophysiologie der Fischniere. *Z. Zellforsch.* 26: 765–88.

———. 1954. Über die Endomeninx der Fische (zugleich ein Beitrag zur Kenntnis der Turbanorgane). *Z. Zellforsch.* 40: 88–100.

Bargmann, W., and G. von Hehn. 1971. Über das Nephron des Elasmobrancher. *Z. Zellforsch.* 114: 1–21.

Beyenbach, K. W., and E. Fromter. 1985). Electrophysiological evidence for Cl⁻ secretion in shark renal proximal tubules. *Am. J. Physiol.* 248: F282–95.

Bohle, A., and Walvig, F. 1964. Beitrag zur vergleichenden Morphologie der epitheloiden

Zellen der Nierenarteriolen unter besonderer Berücksichtigung der epitheloiden Zellen in den Nieren von Seewasserfischen. *Klin. Wochenschr.* 42: 415–421.

Borcea, J. 1904. Quelques considérations sur l'appareil urinaire des Elasmobranches. *Bull. Soc. Zool. Fr.* 29: 143–48.

———. 1906. Recherches sur le système urogénital des Elasmobranches. *Arch. Zool. Exp. Gen.* 4: 199–484.

Borghese, E. 1966. Studies on the nephrons of an elasmobranch fish, *Scyliorhinus stellaris* (L.). *Z. Zellforsch.* 72: 88–92.

Bottazzi, F. 1897. La pression osmotique du sang des animaux marins. *Arch. Ital. Biol.* 28: 61–72.

Boyd, R. B., and A. L. DeVries. 1986. A comparison of anionic sites in the glomerular basement membranes from different classes of fishes. *Cell Tissue Res.* 245: 513–17.

Boylan, J. W. 1967. Gill permeability in *Squalus acanthias*. In *Sharks, Skates, and Rays*, ed. P. W. Gilbert, R. F. Mathewson, and D. P. Rall, 197–206. Baltimore: Johns Hopkins University Press.

———. 1972. A model for passive urea reabsorption in the elasmobranch kidney. *Comp. Biochem. Physiol.* 42A: 27–30.

Capréol, S. V., and L. E. Sutherland. 1968. Comparative morphology of juxtaglomerular cells: I. Juxtaglomerular cells in fish. *Can. J. Zool.* 46: 249–256.

Clarke, R. W., and H. W. Smith. 1932. Absorption and excretion of water and salts by the elasmobranch fishes: III. The use of xylose as a measure of the glomerular filtrate in *Squalus acanthias*. *J. Cell. Comp. Physiol.* 1: 131–43.

Crockett, D. R., J. W. Gerst, and S. Blankenship. 1973. Absence of juxtaglomerular cells in the kidneys of elasmobranch fishes. *Comp. Biochem. Physiol.* 44A: 673–675.

Daniel, J. F. 1934. *The Elasmobranch Fishes*. Berkeley: University of California Press.

Decker, B., and E. Reale. 1991. The glomerular filtration barrier of the kidney in seven vertebrate classes. Comparative morphological and histochemical observations. *Basic Appl. Histochem.* 35: 15–36.

Deetjen, P., and D. Antkowiak. 1970. The nephron of the skate *Raja erinacea*. *Bull. Mt. Desert Isl. Biol. Lab.* 10: 5–10.

De Vlaming, V. L., and M. Sage. 1973. Osmorgulation in the euryhaline elasmobranch *Dasyatis sabina*. *Comp. Biochem. Physiol.* 45A: 31–44.

DiBona, D. R., M. M. Civan, and A. Leaf. 1969. The cellular specificity of the effect of vasopressin on toad urinary bladder. *J. Membr. Biol.* 1: 79–91.

Elger, M., and H. Hentschel. 1983. Ultrastrukturelle Befunde am Opisthonephros des Katzenhaies, *Scyliorhinus caniculus* L. *Verh. Anat. Res.* 77: 589–90.

Elger, M., D. Drenckhahn, R. Nobiling, P. Mundel, and W. Kriz. 1993. Cultured rat mesangial cells contain smooth muscle alpha-actin not found in vivo. *Am. J. Pathol.* 142: 497–509.

Elger, M., H. Hentschel, H. McDuffey, B. Graves, and B. Schmidt-Nielsen. 1984. Characterization of two zones in the renal tissue of thre little skate, *Raja erinacea* Mitch. *Bull. Mt. Desert Isl. Biol. Lab.* 23: 62–63.

Forster, R. P. 1967. Osmoregulatory role of the kidney in cartilaginous fishes (Chondrichthyes). In *Sharks, Skates, and Rays*, ed. P. W. Gilbert, R. F. Mathewson, and D. P. Rall, 187–95. Baltimore: Johns Hopkins University Press.

Friedman, P. A., and S. C. Hebert. 1990. Diluting segment in the kidney of dogfish shark: I. Localization and characterization of chloride absorption. *Am. J. Physiol.* 258: R398–408.

Galli, S., and K. Kiang. 1990. Angiotensin II–like immunoreactivity in sharks: response to hemorrhage and sodium depletion. *FASEB J.* 4: 4000.

Gambaryan, S. P. 1987. Structural characteristics of the nephron in marine cartilagenous fish. *J. Evol. Biochem. Physiol.* 23: 142–45.

Gardiner, D. S., and G. B. M. Lindop. 1985. The granular peripolar cell of the human glomerulus: a new component of the juxtaglomerular apparatus? *Histopathology* 9: 675–85.

Gérard, M. P., and M. R. Cordier. 1934. Comment interpréter les formations granulaires recontrées dans le tube contourné du rein, chez les Vertébrés? *Bull. Acad. R. Med. Belg.* 14: 160–85.

Ghouse, A. M., B. Parsa, J. W. Boylan, and J. C. Brennan. 1968. The anatomy, micro-anatomy, and ultrastructure of the kidney of the dogfish *Squalus acanthias*. *Bull. Mt. Desert Isl. Biol. Lab.* 8: 22–29.

Goldstein, L., and R. P. Forster. 1971. Osmoregulation and urea metabolism in the little skate *Raja erinacea*. *Am. J. Physiol.* 220: 742–46.

Grabowski, G. M., E. Reale, and E. R. Lacy. 1992. Absence of a countercurrent system in the freshwater elasmobranch (*Potamotrygon humerosa*). Paper presented at the 8th annual meeting of the American Elasmobranch Society, University of Illinois at Urbana-Champaign, June.

———. 1994. Renal morphology of the freshwater elasmobranch, *Potamotrygon humerosa* (submitted).

Grantham, J. J., C. E. Ganote, M. B. Burg, and J. Orloff. 1969. Paths of transtubular flow in isolated renal collecting tubules. *J. Cell Biol.* 41: 562–76.

Griffith, R. W., P. K. T. Pang, A. K. Srivastava, and G. E. Pickford. 1973. Serum composition of freshwater stingrays (*Potamotrygonidae*) adapted to fresh and dilute sea water. *Biol. Bull.* 144: 304–20.

Haller, B. 1902. Über die Urniere von *Acanthias vulgaris*. *Gegenbaurs Morphol. Jahrb.* 29: 283–316.

Hazon, N., R. J. Balment, M. Perrott, and L. B. O'Toole, 1989. The renin-angiotensin system and vascular and dipsogenic regulation in elasmobranchs. *Gen. Comp. Endocrinol.* 74: 230–36.

Hebert, S. C., and P. A. Friedman. 1990. Diluting segment in kidney of dogfish shark: II. Electrophysiology of apical membranes and cellular resistances. *Am. J. Physiol.* 258: R409–17.

Henderson, I. W., J. A. Oliver, A. McKeever, and N. Hazon. 1981. Phylogenetic aspects of the renin-angiotensin system. In *Advances in Animal and Comparative Physiology*, ed. G. Pethes and V. L. Frenyo, 355–63. Oxford: Pergamon Press.

Henderson, I. W., L. B. O'Toole, and N. Hazon. 1988. In *Physiology of Elasmobranch Fishes*, ed. T. Shuttleworth. New York: Springer-Verlag.

Hentschel, H. 1987. Renal architecture of the dogfish, *Scyliorhinus caniculus* (Chondrichthyes, Elasmobranchii). *Zoomorphology* 107: 115–25.

———. 1988. Renal blood vascular system in the elasmobranch *Raja erinacea* Mitch, in relation to kidney zones. *Am. J. Anat.* 183: 130–47.

———. 1991. Developing nephrons in adolescent dogfish, *Scyliorhinus caniculus* (L.), with reference to ultrastructure of early stages, histogenesis of the renal countercurrent system and nephron segmentation in marine elasmobranchs. *Am. J. Anat.* 190: 309–33.

Hentschel, H., and M. Elger. 1982. Structural organization of the elasmobranch kidney. *Verh. Dtsch. Zool. Ges.* 75: 263.

Hentschel, H., and P. Walter. 1993. Heterogeneous distribution of glycoconjugates in the kidney of dogfish, *Scyliorhinus caniculus* (L.), with reference to changes in the glycosylation pattern during ontogenetic development of the nephron. *Anat. Rec.* 235: 21–32.

Hentschel, H., M. Elger, and B. Schmidt-Nielsen. 1986. Chemical and morphological differences in the kidney zones of the elasmobranch *Raja erinacea* Mitch. *Comp. Biochem. Physiol.* 84A: 553–57.

Hentschel, H., P. Herter, S. Mähler, and M. Elger. 1992. Morphological and immunohistochemical evidence for acid secretion in the renal tubule of elasmobranch fish. *Pflügers Arch. Eur. J. Physiol.* 420 (suppl. 1): R72.

Hentschel, H., S. Mähler, P. Herter, and M. Elger. 1993. Renal tubule of dogfish, *Scyliorhinus caniculus*: a comprehensive study of structure with emphasis on intramembrane particles and immunoreactivity for H^+-K^+-adenosine triphosphatase. *Anat. Rec.* 235: 511–532.

Hickman, C. P., and B. F. Trump. 1969. The kidney. In *Fish Physiology*, ed. W. S. Hoar and D. J. Randall. New York: Academic Press.

Hoppe-Seyler, F. A. 1930. Die Bedingungen und die Bedeutung biologischer Methylierungsprozesse. *Z. Biol.* 90: 433–66.

Howse, H. D., R. M. Overstreet, W. E. Hawkins, and J. S. Franks. 1992. Ubiquitous perivenous smooth muscle cords in viscera of the teleost, with special emphasis on liver. *J. Morphol.* 212: 175–89.

Kaissling, B., and W. Kriz. 1982. Variability of intercellular spaces between macula densa cells: a transmission electron microscopic study in rabbits and rats. *Kidney Int.* 22 (suppl. 12): 9–17.

Kempton, R. T. 1940. The morphology of the dogfish renal tubule. *Bull. Mt. Desert Isl. Biol. Lab.* 42: 28–34.

———. 1943. Studies on the elasmobranch kidney: I. The structure of the renal tubule of the spiny dogfish (*Squalus acanthias*). *J. Morphol.* 73: 247–63.

———. 1953. Studies on the elasmobranch kidney: II. Reabsorption of urea by the smooth dogfish, *Mustelus canis*. *Biol. Bull.* 104: 45–56.

———. 1956. The problem of the "special segment" of the elasmobranch kidney tubule. *YB Am. Philos. Soc.* 210–12.

———. 1957. The elasmobranch kidney tubule. *Anat. Rec.* 128: 575.

———. 1962. Studies on the elasmobranch kidney: III. The kidney of the lesser electric ray, *Narcine brasiliensis*. *J. Morphol.* 111: 217–25.

———. 1964. Some anatomical features of the elasmobranch kidney. *Biol. Bull.* 127: 377.

Kirk, K. L., P. D. Bell, D. W. Barfuss, M. Ribadeneira. 1985. Direct visualization of the isolated and perfused macula densa. *Am. J. Physiol.* 248: F890–94.

Kölliker, A. 1845. Über Flimmerbewegungen in den Primordialnieren. *Arch. Anat. Physiol. Wiss. Med.* 518–23.

Kon, Y., D. Alcorn, K. Murakami, M. Sugimura, and G. B. Ryan. 1994. Immunohistochemical studies of renin-containing cells in the developing sheep kidney. *Anat. Rec.* 239: 191–97.

Kozlik, F. 1939. Über den Bau des Nierenkanälchens. Vergleichend-anatomische Untersuchungen. *Z. Anat. Entwicklungsgesch.* 109: 624–48.

Krause, R. 1923. *Mikroskopische Anatomie der Wirbeltiere in Einzeldarstellung:* IV. *Teleostier, Plagiostomen, Zyklostomen und Leptokardier.* Berlin: Walter De Gruyter.

Lacy, E. R., M. Castellucci, and E. Reale. 1987. The elasmobranch renal corpuscle: fine structure of Bowman's capsule and the glomerular capillary wall. *Anat. Rec.* 218: 294–305.

Lacy, E. R., L. Luciano, and E. Reale. 1989a. Flagellar cells and ciliary cells in the renal tubule of elasmobranchs. *J. Exp. Zool. Suppl.* 2: 186–92.

———. 1991a. Membrane specializations in flagellar ribbons of elasmobranch fish. *Tissue Cell* 23: 223–34.

———. 1991b. Elastic-like tissue, a new component of the renal glomerular capillary wall in a cartilaginous fish. *Anat. Embryol.* 183: 475–81.
Lacy, E. R., and E. Reale. 1981. The brush border segment of the nephron in an elasmobranch. Thin section and freeze fracture observations. *Verh. Anat. Ges.* 75: 611–12.
———. 1985a. The elasmobranch kidney: I. Gross anatomy and general distribution of the nephrons. *Anat. Embryol. (Berl.)* 173: 23–34.
———. 1985b. The elasmobranch kidney: II. Sequence and structure of the nephrons. *Anat. Embryol. (Berl.)* 173: 163–86.
———. 1986. The elasmobranch kidney: III. Fine structure of the peritubular sheath. *Anat. Embryol. (Berl.)* 173: 299–305.
———. 1989. Granulated peripolar epithelial cells in the renal corpuscle of marine elasmobranch fish. *Cell Tissue Res.* 257: 61–67.
———. 1990. The presence of a juxtaglomerular apparatus in elasmobranch fish. *Anat. Embryol. (Berl.)* 182: 249–62.
———. 1991a. Fine structure of the elasmobranch renal tubule: neck and proximal segments of the little skate. *Am. J. Anat.* 190: 118–32.
———. 1991b. The fine structure of the elasmobranch renal tubule: intermediate, distal, and collecting duct segments of the little skate. *Am. J. Anat.* 192: 478–97.
———. 1994. Functional morphology of the elasmobranch nephron and retention of urea. In *Cellular and Molecular Approaches to Fish Ionic Regulation*, ed. C. M. Wood and T. J. Shuttleworth, pp. 107–147. New York: Academic Press.
Lacy, E. R., E. Reale, D. S. Schlusselberg, W. K. Smith, and D. S. Woodward. 1984. Computer-assisted reconstruction of the renal countercurrent system in a marine elasmobranch fish. *Anat. Rec.* 208: 97A.
Lacy, E. R., B. Schmidt-Nielsen, R. G. Galaske, and H. Stolte. 1975. Configuration of the skate (*Raja erinacea*) nephron and ultrastructure of two segments of the proximal tubule. *Bull. Mt. Desert Isl. Biol. Lab.* 15: 54–56.
Lacy, E. R., E. Reale, D. S. Schlusselberg, W. K. Smith, and D. S. Woodward. 1985. A renal countercurrent system in marine elasmobranch fish: a computer-assisted reconstruction. *Science* 227: 1351–54.
Lacy, E. R., J. C. Williams, R. Rivers, K. J. Karnaky, Jr., L. A. Mackanos, G. M. Grabowski, L. Luciano, and E. Reale. 1989b. Motion analysis of flagellar ribbon and ciliary beat cycyes in teleost and elasmobranch fish nephrons. *Eur. J. Cell Biol.* 49 (suppl. 27): 54.
Leydig, F. 1852. Beiträge zur mikroskopischen Anatomie und Entwicklungsgeschichte der Rochen und Haie. Leipzig: Engelmann.

Mahieu, R. P., J. B. Foidart, C. H. Dubois, C. A. Dechenne, and J. Deheneffe. 1980. Tissue culture of normal rat glomeruli: contractile activity of the cultured mesangial cells. *Invest. Cell Pathol.* 3: 121–28.
Marples, B. J. 1936. The blood vascular system of the elasmobranch fish *Squatina squatina* (Linné). *Trans. R. Soc. (Edinb.)* 58: 817–40.
Marshall, E. K. 1934. The comparative physiology of the kidney in relation to theories of renal secretion. *Physiol. Rev.* 14: 133–59.
Meyer, P. 1888. Über Eigenthuemlichkeiten in den Kreislaufsorganen der Selachier. *Mitt. Zool. St. Neapel* 8: 307–73.
Nash, J. 1931. The number and size of glomeruli in the kidneys of fishes, with observations on the morphology of the renal tubules of fishes. *Am. J. Anat.* 47: 425–45.
Nishimura, H., M. Oguri, M. Ogawa, H. Sokabe, and M. Imai. 1970. Absence of renin in kidneys of elasmobranchs and cyclostomes. *Am. J. Physiol.* 218: 911–15.
Ogawa, M., and T. Hirano. 1982. Studies on the nephron of a freshwater stingray, *Potamotrygon magdalenae*. *Zool. Mag.* 91: 101–5.
Oguri, M., M. Ogawa, and H. Sokabe. 1970. Absence of juxtaglomerular cells in the kidneys of Chondrichthyes and Cyclostomi. *Bull. Jpn. Soc. Sci. Fish.* 36: 881–84.
Orci, L., F. Humbert, D. Brown, and A. Perrelet. 1981. Membrane ultrastructure in urinary tubules. *Int. Rev. Cytol.* 73: 183–242.
Parker, T. J. 1881. On the venous system of the skate (*Raja nausata*). *Trans. N.Z. Inst.* 13: 413–18.
Peter, K. 1909. Untersuchungen über Bau und Entwicklung der Niere. Jena: Gustav Fischer Verlag.
Rand, H. W. 1905. The skate as a subject for classes in comparative anatomy; injection methods. *Am. Nat.* 39: 364–79.
Reale, E., and E. R. Lacy. 1987. Some observations on the elasmobranch kidney. *Arch. Biol. (Brussels)* 98: 231–41.
Rhodin, J. A. G. 1972. Fine structure of elasmobranch arteries, capillaries and veins in the spiny dogfish, *Squalus acanthias*. *Comp. Biochem. Physiol.* 42A: 59–64.
Rodier, E. 1899. Observations et expériences comparatives sur l'eau de mer, le sang et les liquides internes des animaux marins. *Trav. Lab. Soc. Sci. Stat. Zool. Arachon* 103–23.
———. 1900. Sur la pression osmotique du sang et des liquides internes chez les poissons sélaciens. *C. R. Seances. Acad. Sci.* 131: 1008–10.
Romer, A. S., and T. S. Parsons. 1977. *The Vertebrate Body*, 5th ed. Philadelphia: W. B. Saunders.
Ryan, G. B., J. P. Coghlan, and B. A. Scoggins. 1979. The granulated peripolar epithelial cell:

a potential secretory component of the renal juxtaglomerular complex. *Nature* 277: 655–56.

Satchell, G. H. 1971. *Circulation in Fishes*. London: Cambridge University Press.

Sawyer, D. B., and K. W. Beyenbach. 1985. Mechanism of fluid secretion in isolated shark renal proximal tubules. *Am. J. Physiol.* 249: F884–90.

Simionescu, N., and M. Simionescu. 1988. The cardiovascular system. In *Cell and Tissue Biology, A Textbook of Histology*, ed. L. Weiss. Baltimore: Urban and Schwarzenberg.

Smith, H. W. 1931. The absorption and excretion of water and salts by the elasmobranch fishes: II. Marine elasmobranchs. *Am. J. Physiol.* 98: 296–310.

———. 1936. The retention and physiological role of urea in Elasmobranchii. *Biol. Rev.* 11: 49–82.

Smith, W. K., D. S. Schlusselberg, B. G. Culter, D. J. Woodward, and E. R. Lacy. 1983. Hierarchical database design for biological modelling. In *National Computer Graphics Association Conference Proceedings*, 106–16.

Sokabe, H., M. Ogawa, M. Oguri, and H. Nishimura. 1969. Evolution of the juxtaglomerular apparatus in the vertebrate kidneys. *Texas Rep. Biol. Med.* 27: 867–85.

Somero, G. N. 1986. From dogfish to dogs: trimethylamines protect proteins from urea. *News Physiol. Sci.* 1: 9–19.

Stolte, H., R. G. Galaske, G. M. Eisenbach, C. Lechene, B. Schmidt-Nielsen, and J. W. Boylan. 1977. Renal tubule ion transport and collecting duct function in the elasmobranch little skate, *Raja erinacea*. *J. Exp. Zool.* 199: 403–10.

Stoskopf, M. K. 1993. Clinical pathology of sharks, skates, and rays. In *Fish Medicine*, ed. M. K. Stoskopf, 754–57. Philadelphia: W. B. Saunders.

Taugner, R., and E. Hackenthal. 1989. *The Juxtaglomerular Apparatus: Structure and Function*. Berlin: Springer-Verlag.

Thorson, T. B. 1967. Osmoregulation in freshwater elasmobranchs. In *Sharks, Skates, and Rays*, ed. P. W Gilbert, R. F. Mathewson, and D. P. Rall, 265–70. Baltimore: Johns Hopkins University Press.

———. 1970. Freshwater stingrays, *Potamotrygon* spp.: failure to concentrate urea when exposed to saline medium. *Life Sci.* 9: 893–900.

Thorson, T. B., C. M. Cowan, and D. E. Watson. 1967. *Potamotrygon* spp.: elasmobranchs with low urea content. *Science* 158: 375–77.

Thumwood, C. M., J. McCausland, D. Alcorn, and G. B. Ryan. 1993. Scanning and transmission electron-microscopic study of peripolar cells in the newborn lamb kidney. *Cell Tissue Res.* 274: 597–604.

Thurau, K., and P. Acquisto. 1969. Localization of the diluting segment in the dogfish nephron: a micropuncture study. *Bull. Mt. Desert Isl. Biol. Lab.* 9: 60–63.

Townsley, P. M., and M. A. Scott. 1963. Systolic muscular action of the kidney tubules of flounder. *J. Fish Res. Board Can.* 20: 243–44.

Uva, B., M. A. Masini, N. Hazon, L. B. O'Toole, I. W. Henderson, and P. Ghiani. 1992. Renin and angiotensin converting enzyme in elasmobranchs. *Gen. Comp. Endocrinol.* 86: 407–12.

Weibel, E. R., and G. Palade. 1964. New cytoplasmic components in arterial endothelia. *J. Cell Biol.* 23: 101–02.

WILLIAM C. HAMLETT
THOMAS J. KOOB

CHAPTER 15

Female Reproductive System

Elasmobranch fishes successfully utilize a remarkably broad and complex array of reproductive modes, all of which result in the production of a small number of relatively large offspring. Internal fertilization and an adaptable reproductive tract, coupled with neural and stromal mechanisms for intricate endocrine regulation, facilitated evolution of both a specialized form of oviparity and diverse viviparous strategies. However, since elasmobranchs are anamniotes, the morphological and physiological mechanisms recruited for specific reproductive modes evolved innovations unique among vertebrates. Evolution of these reproductive modes has relied not only on adaptations of the maternal reproductive system but also on concomitant embryonic adaptations, many of which are also unique among vertebrates. These successful innovations are intriguing aspects of elasmobranch reproductive biology that capture the biologist's imagination today as they have for hundreds of years.

Internal fertilization is common to all elasmobranchs and, as such, requires a suite of biological activities, including behavioral, morphological, and physiological mechanisms, to ensure successful copulation and fertilization. Mating behavior and copulation have been observed in only a few elasmobranch species (Dempster and Herald 1961; Johnson and Nelson 1978; Kimley 1980; Luer and Gilbert 1985; Myrberg and Gruber 1974; Tricas and Le Feuvre 1985; Carrier and Pratt 1994). As in other vertebrates with similarly complex endocrine control, female receptivity likely depends on hormonal status, and signaling of this receptivity to the male may involve behavioral and chemical cues. That females participate in precopulatory mating behavior is clear from observations on some sharks. Females allow the typical grasping and biting behavior of the male and generally remain passive during copulation. Subsequent to insemination, sperm are stored, presumably in the oviducal gland (Pratt 1993), in some cases for months or longer, before they are released to fertilize the ovulated eggs.

Reproductive cycles

Of all the areas relating to elasmobranch reproductive biology, reproductive cycles are perhaps the best understood. For oviparous species, the general trend is for year-round egg production with seasonal periods when a greater proportion of the adult females are laying eggs. This appears true for some species of sharks and skates in northern latitudes (*Scyliorhinus canicula* [Sumpter and Dodd 1979]; *Raja erinacea* [Richards et al. 1963; Fitz and Daiber 1963]; *Raja naevus* [DuBuit 1976]), and cat sharks in temperate latitudes (*Apristurus bruneus* and *Parmaturus xaniurus* [Cross 1988]). However, far too few observations have been reported to assume that this pattern is true for all oviparous species. There are reports suggesting that some oviparous species utilize well-circumscribed seasonal cycles, for example, species in warm waters (*Heterodontus portusjacksoni* [McLaughlin and O'Gower 1971]; *Raja eglanteria* [Luer and Gilbert 1985]). More work is clearly needed here before a complete story will emerge. Moreover, it must be emphasized that the reproductive cycles of individual females are not known for any species. Given the fact that within populations of oviparous elasmobranchs a large proportion of the adult females are not laying eggs during the year, cycles for individual females must include a protracted period when they are not reproductively active.

Reproductive periods of viviparous species are generally well-circumscribed annual cycles, with the entire population undergoing nearly synchronous mating, gestation, and parturition. Utilized by most aplacental and placental sharks is a relatively long gestation, in most cases approaching a year, followed immediately by mating and the ensuing pregnancy. Species that typify this pattern include *Mustelus canis* (Dodd 1983), *Sphyrna lewini* and *Sphyrna tudes* (Chen et al. 1988), *Squatina californica* (Natanson and Cailliet 1986), and *Rhizoprionodon terraenovae* (Parsons 1981). A few species, like *Sphyrna tiburo*, depart from this pattern in having shorter pregnancies but retaining the yearly cycle (Manire et al. 1995). Parturition may be timed to specific ecological factors like temperature or relative abundance of prey for the neonates. Females return each year to specific nursery areas to give birth (Castro 1993), possibly because these areas afford a more advantageous niche for the offspring.

The reproductive cycles of several viviparous shark species differ from this typical yearly pattern. Pregnancy in *Squalus acanthias* lasts nearly 2 years, with mating in the spring of one year and parturition in the fall of the following year (Templeman 1944; Hisaw and Albert 1947). In *Prionace glauca* (Pratt 1979), and *Carcharinus milberti* (Springer 1960; Wass 1973) females are pregnant for a full year but deliver young every 2 years. During the intervening year females are not pregnant. Pregnancy in the frilled shark, *Chlamydoselachus anguineus*, is thought to span 3.5 years, if not longer (Tanaka et al. 1990). In at least one viviparous shark, *Gollum attenuatus*, the population appears asynchronous, since at any one time females can be carrying newly ovulated eggs or developing or near term embryos (Yano 1993).

In contrast to viviparous sharks, pregnancy in most rays lasts several months only but nonetheless occurs within a yearly cycle. Pregnancy takes place usually in the spring-summer-fall months. Ray species that typify this cycle include *Pteromylaeus bovina* (Ranzi 1934), *Dasyatis violacea* (Ranzi 1934), *Dasyatis sayi* (Snelson et al. 1989), *Dasyatis sabina* (Snelson et al. 1988), *Dasyatis centroura* (Capape 1993), and *Urolophus halleri* (Babel 1967). Only a few species of ray are known to depart from this pattern, including *Myliobatis californicus* (Martin and Cailliet 1988), with a nearly year-long gestation, and *Torpedo marmorata* (Mellinger 1974; Capape 1979) with a year-long pregnancy followed by 2 nonpregnant years.

Unfortunately, little is known about the ecological factors that regulate reproductive cycles in female elasmobranchs. The synchrony within populations of viviparous species suggests that specific cues trigger these yearly events, cues perhaps directly related to their immediate environment. We can only speculate whether the elasmobranch reproductive system responds to external factors such as light, temperature, and prey abundance or, alternatively, is regulated by some as yet unidentified internal biological rhythm.

Gross and microscopic anatomy

The female reproductive system consists of paired or single ovaries and oviducts (Fig. 15-1). The oviducts are differentiated into an ostium; anterior oviduct; oviducal gland; isthmus in some species; the dilated terminal region, which can be modified as a uterus; the cervix; and a common urogenital sinus.

OVARY

The elasmobranch ovary performs three principal functions: generation of germ cells, acquisition and accumulation of yolk, and biosynthesis and secretion of hormones, most notably steroids. Germ cell production presumably follows pathways identical to those in other vertebrates and is not discussed further here. What little is known about sequestration and chemical composition of yolk will be presented in order to emphasize the importance of yolk for nearly all elasmobranchs, despite specific differences in

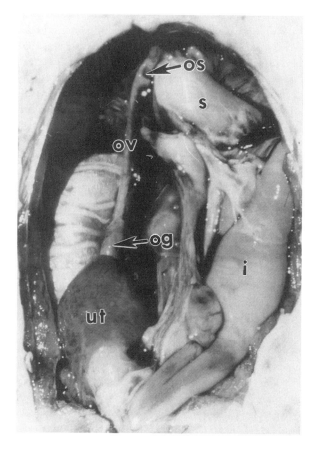

FIGURE 15-1 Gravid female yellow spotted stingray, *Urolophus jamaicensis*. os = Ostium; s = stomach; ov = oviduct; og = oviducal gland; ut = uterus; i = intestine.

reproductive modes. Ovarian steroidogenesis has received considerable attention recently and is discussed later in relation to reproductive cycles and possible functions of circulating steroids.

Female elasmobranchs have either paired ovaries, both of which contribute to egg production, or a single ovary solely responsible for producing eggs. No correlation appears to exist between reproductive mode and ovarian symmetry, although in most viviparous species one ovary predominates. In oviparous species, ovaries are either paired, as in most skates, or asymmetrical as in cat sharks. Among viviparous sharks, the right ovary is generally the principal generator of ovulated eggs. In these species, the left ovary is either absent or rudimentary as in *Carcharhinus* (Jensen 1976), *Prionace* (Pratt 1979), *Pristophorus* (Daniel 1928), *Galeus* (Daniel 1928), *Mustelus* (Daniel 1928), *Sphyrna* (Chen et al. 1988; Schlernitzauer and Gilbert 1966). However, several viviparous shark species have two functioning ovaries, for example, *Notorynchus maculatus* (Daniel 1928), *C. anguineus* (Gudger 1940), *Pristis cuspidatus*, and *Rhyncobatus djiddensis* (Setna and Sarangdhar 1948), *S. acanthias* (Hisaw and Albert 1947), and *Squalus brevirostris* (Kudo 1956). In most rays, the left ovary develops into the sole producer of mature ova.

The elasmobranch ovary in adult females consists of oocytes, which differentiate in extraovarian sites and migrate to the ovary early in ontogeny (Beard 1903–4; Woods 1902), developing follicles of various sizes, preovulatory follicles undergoing atresia, and corpora lutea, all of which are embedded in a loose connective tissue stroma. The relative number and size of developing follicles depends on the stage in the reproductive cycle and reproductive mode. In reproductively active oviparous species, such as most skates (Koob et al. 1986) and cat sharks (Dodd 1972; Castro et al. 1988), a graded series of paired follicles is present at all times during egg production. Each pair requires at least several weeks to attain ovulatory size. The ovaries in subadult and reproductively inactive females contain only small follicles with little or no yolk.

In viviparous species, follicle development is more diverse. However, in most species, the exception being the oophagous sharks, one set of small follicles is selected for subsequent development to ovulation. These follicles develop in concert, proliferating thecal and granulosa cells and accumulating yolk at the same rate. Follicle development can take several months, as in most rays and some sharks, or nearly a year and, in a few species, 2 years or more. The diversity in follicle development originates in the timing with respect to the previous ovulation and pregnancy. Follicle development for the next pregnancy can occur concomitantly with the present pregnancy, primarily during the latter half of pregnancy, or mostly between pregnancies. Follicle growth during pregnancy is typical of sharks with year-long gestations. Follicle growth between pregnancies is found in both shark and ray species with comparatively short pregnancies. However, in few species of both sharks and rays, follicle growth requires 2 years or longer. Regardless of the timing involved, all of these means for follicle development ensure that the next pregnancy will ensue soon after parturition, thereby optimizing lifetime fecundity.

Follicles consist of an oocyte surrounded by granulosa cells and delimited by a basal lamina (Fig. 15-2). The overall histology of follicles has been examined in a variety of ovoviviparous and viviparous species (see Koob and Callard 1991). The cells immediately surrounding this basal lamina form thecal layers. Granulosa cells in small follicles form a single layer having squamous or cuboidal morphology. As follicles grow and differentiate and the oocyte accumulates yolk, granulosa cell morphology changes and lipid-rich inclusions begin to appear and enlarge. These inclusions may be the source of precursors for steroid biosynthesis or they may indicate that granulosa cells produce, secrete, and supply the oocyte with yolk granules (Tsang and Callard 1983). The theca undergoes coordinated changes with follicle growth. Increased

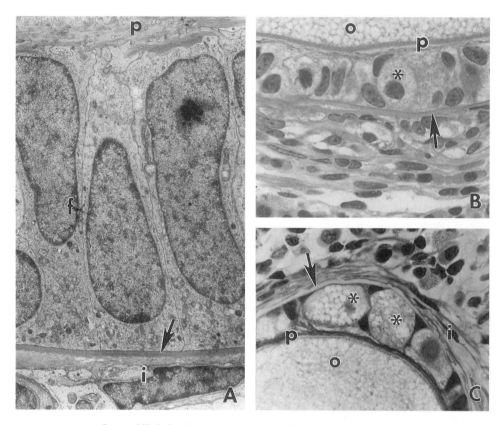

FIGURE 15-2 Ovary of *Urolophus jamaicensis*, with *arrows* showing basement membrane (**A–C**), and the *asterisk* showing lipid-laden follicle cells (**B, C**). **A** is a transmission electron micrograph (×4000); **B** and **C** are light micrographs (×600); all reproduced at 73%. f = Follicle cells; o = oocyte, p = zona pellucida.

vascularization, layering of distinct cell types, elaboration of lipid-rich inclusions, and transformation of cell morphology accompany follicle development.

Atretic follicles can be found in the ovaries of most reproductively active elasmobranchs (Dodd 1983). They form by degeneration of preovulatory follicles and thus contain the same cellular elements, including an outer layer of thecal cells surrounding a collapsed layer of lipid-rich granulosa cells. A rich vascular supply invades the central lumen. Atresia can apparently be triggered in any size of follicle. Depending on when during follicle development atresia commences and the duration of degeneration, atretic follicles can contain more or less yolk in the oocyte. The function of atresia and the follicles it produces is not known, and no correlation is apparent between the formation of these structures and specific reproductive events.

Ultrastructure of the ovarian follicle of the yellow spotted ray, *Urolophus jamaicensis*, has been studied using light and electron microscopy (see Fig. 15-2) (Jezior and Hamlett 1995; Hamlett et al. 1999b). The ovary of *U. jamaicensis* is embedded in the epigonal gland, a lymphomyeloid organ. The covering of the ovary is composed of a germinal epithelium that is cuboidal and dome-shaped with microvilli. Adjacent cells have elaborate intercellular folds that create dilated intercellular spaces. In previtellogenic follicles, the follicle cells are simple cuboidal and contain modest amounts of synthetic or transport organelles. As vitellogenesis proceeds, the epithelium becomes multilaminar. Follicle cells are columnar as yolk precursors are transported from the maternal circulation through the follicle cell cytoplasm to the oocyte. Large, round cells that contain lipidlike substances occur in the follicle wall. These cells decrease in size and number as folliculogenesis proceeds and eventually disappear prior to ovulation. Columnar follicular cells and the oocyte have cellular extensions that impinge upon the zona pellucida. Transosomes are follicle cell extensions that indent the oocyte membrane. The tips of transosomes become enclosed by a layer of oocyte plasmalemma. The tips pinch off and become resident in the ooplasm. Dense staining material occurs on the inner surface of the transosome membrane derived from the follicle cell. In other animals, this material has been described as ribosomelike. This study is the first to document the presence of transosomes in a group other than Aves or reptiles. Follicle cells are supported by an extremely thick basal lamina. Subjacent to the lamina is the vascularized theca with fibroblasts embedded in a collagenous network. There is no differentiation into definitive theca interna and externa. In vitellogenic eggs, extensive inward foldings of the follicular epithelium occur, thereby generating more surface area for the transport of yolk precursors to the oocyte. Atretic follicles are common.

Following ovulation, the follicle wall collapses, forming a corpus luteum, which consists of lipid-filled cells derived from the granulosa, initially surrounding a cavity but later filling the entire structure (Hisaw and Hisaw 1959). Thecal cells do not contribute to corpora lutea other than being structural elements and providing a conduit for the extensive vascularization to the granulosa-filled central lobes. The functional life of the corpora lutea correlates with the reproductive mode. In oviparous species, their size decreases relatively quickly. In viviparous species, they persist throughout gestation, although continually decreasing in size.

By far the predominant type of ovum produced by female elasmobranchs is one of relatively large size containing abundant yolk. Large eggs are characteristic of oviparous sharks and skates, ovoviviparous sharks, and most viviparous sharks and rays. In only a few viviparous species in which embryonic development relies primarily on uterine derived nutrients does the ovulated egg contain little or no yolk. Yolk provides the sole nutrient source in oviparous and some aplacental species, and in most viviparous species it forms the bulk of the nutrients for early embryonic development. Yolk acquisition by the ovary and the composition of yolk are thus of fundamental importance for elasmobranch reproduction.

Accumulation of yolk in developing follicles is a slow process requiring several months to over a year in most species. The chronology of yolk deposition is best illustrated in viviparous species with well-circumscribed breeding cycles. In *S. acanthias*, follicles begin to accumulate yolk approximately one-quarter of the way into the 2-year pregnancy. Follicles reach one third of ovulatory size halfway through pregnancy. The remainder of yolk accumulation, that is, two-thirds of follicle growth, requires the latter half, or 1 year, of pregnancy. Even in species in which the duration of pregnancy is comparatively short, yolk accumulation in follicles is protracted. Babel (1967) estimated that follicle growth to ovulation in *U. halleri* requires 2 years, even though gestation lasts only 3 months. In other stingrays, follicle growth is a seasonal process requiring 2–3 months before ovulation.

The chemical composition of yolk has been poorly studied with contemporary analytical methods. However, significant efforts were made in the late 19th and early 20th centuries to identify and characterize the principal components of yolk, as well as albumen, especially as they relate to embryonic development in both oviparous and viviparous elasmobranchs. Needham (1931, 1942) provided thorough reviews of the results and conclusions of these pioneering studies. Only a brief summary of conclusions can be given here. As might be expected, urea is present in the yolk of all elasmobranch eggs examined. Protein accounts for a relatively large proportion of total organic contents, and several distinct proteins constitute this fraction, among which are glycoproteins. Lipids account for the remainder of the organic constituents. Finally, minerals are present in the yolk of all elasmobranch eggs, but the relative amount differs according to reproductive mode.

In the only contemporary compositional study of elasmobranch yolk, a variety of free oligosaccharides were isolated from ova of freshly laid *S. canicula* eggs and their structure determined by nuclear magnetic resonance analyses (Plancke et al. 1996). These oligosaccharides apparently derive from yolk glycoproteins by enzymatic processing. Although the exact nature and timing of this processing was not determined, these observations raise important physiological questions about the generation of yolk during oogenesis and the mechanisms for its utilization during embryogenesis.

One principal component of yolk is a form of vitellogenin, a lipophosphoprotein yolk-granule precursor produced by the liver. Measurement of vitellogenin in plasma of reproductively active female *S. canicula* and *R. erinacea* have shown a chronically elevated level during egg production (Craik 1978a; Perez and Callard 1989). Follicles grow continuously during ovulatory cycles in oviparous species, consequently requiring a constant supply of vitellogenin. Isotopic labeling studies demonstrated that plasma vitellogenin is sequestered by the ovary and subsequently deposited in yolk granules of growing follicles (Craik 1978b). The mechanisms and control of vitellogenesis are only poorly known, and for oviparous species only. While ovary-derived estrogen is undoubtedly involved in inducing vitellogenin biosynthesis and secretion by the liver, the mechanisms by which follicles sequester yolk precursors are entirely unexplored. Whether endocrine factors regulate vitellogenin uptake by follicles is not known.

OSTIUM

In all species thus far examined, the ostium is a funnel shaped opening populated by simple columnar cells with cilia. In some species both oviducts open to a single common ostium, and in others each oviduct has its own ostium.

ANTERIOR OVIDUCT

In all female elasmobranchs, ova released into the peritoneal cavity are collected by the ostium and transported by the oviduct to the oviducal gland. These structures have received only scant attention, probably because their function is evident. The only published study that describes the manner by which ovulated eggs migrate to and through the oviduct is that of Metten (1939) in the oviparous dogfish *S. canicula*. Cilia in adult females located on the peritoneal wall as well as the mesovarium and associated structures move the ovulated egg toward the ostium. In the dogfish, cilia appear to be a sexually dimorphic character, since they are not present in male peritoneum or subadult female. Cilia on the luminal wall of the ostium and oviduct then move the egg toward the oviducal gland. That ciliary action is the sole mode of egg transport is supported by the apparent lack of smooth muscle in the oviduct. Whether the cilia activity is under hormonal control is not clear.

It seems likely that ciliary transport of ovulated eggs into and through the oviduct is universal in elasmobranchs. Based on the lack of specialized secretory cells in the oviduct, it would appear that the sole function of the oviduct is egg transport, accomplished by ciliary action alone.

Diversity of oviducal gland morphology and function

Encapsulation of fertilized eggs with egg jelly occurs in all oviparous and nearly all viviparous elasmobranchs. All oviparous species produce structurally complex capsules with site-specific morphological and biochemical attributes (Fig. 15-3). Egg capsules produced by species employing yolk sac viviparity vary, from substantial structures, as in the whale shark and frilled shark, to thin, diaphanous membranes, or "candles," as in the spiny dogfish *S. acanthias*. In these species, the capsule either functions throughout much of pregnancy (*Rhincodon typus*) or remains intact only through the early

FIGURE 15-3 Representative egg capsules of oviparous sharks and skates. *Upper panel: Cephaloscyllium ventriosum; middle panel: Heterodontus mexicanus; lower panel: Bathyraja interrupta.*

phases of embryonic development (*S. acanthias*). Thin capsules are also typical of those viviparous species utilizing uterine secretions, placentas, or oophagy. But even here, capsule morphology varies. In some, the fertilized eggs plus jelly fill the entire capsule and the developing embryos break out to complete development free of capsular materials, for example, some sharks and rays. In other viviparous species, the egg and jelly occupy only a small proportion of the capsule lumen but subsequently fill the capsule as the embryo grows from nutrients supplied by the uterus. In addition, in oophagous species, the ovulated eggs that are ingested by the developing embryo are packaged in egg capsules.

Egg encapsulation is evidently a fundamental process that elasmobranchs have not abandoned during the evolutionary divergence in reproductive modes, with the exception, perhaps, of some rays. An understanding of the adaptation of specific morphological and physiological mechanisms for the various reproductive modes employed by elasmobranchs is impossible without a full appreciation of egg encapsulation. Regardless of which type of egg capsule is produced, all derive from a specialized gland unique to the elasmobranchs, the oviducal gland, also known as the shell, or nidamental, gland (Figs. 15-4 and 15-5) (Prasad 1948).

Capsule formation is a rapid process, requiring 12–24 hours in oviparous species (Koob et al. 1986), and probably approaching a similar duration in viviparous species. In order to accomplish this rapid assembly, oviducal glands synthesize capsule precursors over protracted periods prior to ovulation and store them in specialized cytoplasmic compartments. Oviducal gland size correlates

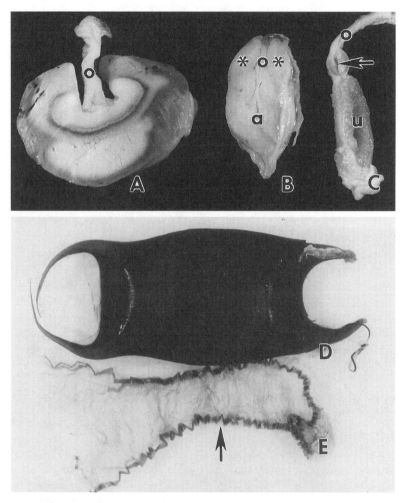

FIGURE 15-4 Diversity of oviducal gland morphology and egg covering morphology is demonstrated by (**A**) *Raja erinacea* (oviparous), (**B**) *Rhizoprionodon terraenovae* (placental viviparous), and (**C**) *Urolophus jamaicensis* (aplacental viviparous with trophonemata). *Asterisks* indicate lateral coils (**B**) and *arrows* show the oviductal gland (**C**). o = Oviduct; a = antrum; u = uterus. Egg covering diversity is exemplified by thick, rigid eggcase of *R. erinacea* (**D**) and thin, pliable and plaited (*arrow*) egg envelope of *R. terraenovae* (**E**).

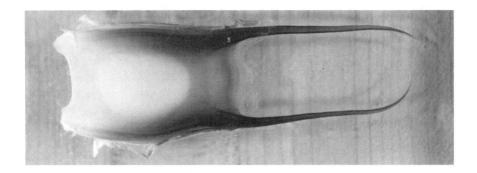

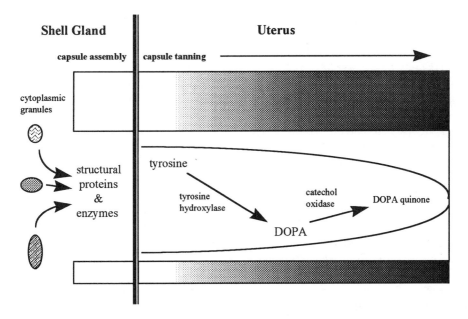

FIGURE 15-5 Illustration of egg capsule formation in little skate, *Raja erinacea*.

with the reproductive cycle, reaching maximum size just before ovulation. In oviparous species, gland size increases coincident with ovarian recrudescence, then fluctuates in size only slightly during the ovulatory cycle. The glands in viviparous species grow in concert with follicle development, attain full size just before ovulation, then diminish in size after discharging capsule precursors for encapsulation. Upon the appropriate stimulus, which is clearly not passage of the egg, since eggless capsules have been observed in both oviparous and viviparous species, cells secrete and assemble these stored precursors.

Oviducal glands are present in the anterior region of the reproductive tract in all female elasmobranchs. However, their overall development and structural complexity correlate with reproductive mode. In oviparous species, the oviducal gland dominates the reproductive tract, and its overall morphology reflects the shape of the capsule. In viviparous species, the size and shape of the gland correspond to the size and complexity of the capsules produced.

Oviducal gland morphology and function have recently been surveyed (Hamlett et al. 1998). Oviducal glands consistently display four fundamental zones regardless of the type of reproduction of the particular species. The zones correspond to lamellae that extend full width across the gland lumen. Formerly zones of the oviducal gland were designated as albumen secreting and shell secreting. However, this oversimplified terminology does not accurately apply to all species. We have adopted the terminology recently introduced (Hamlett et al. 1999a) that refers to the four basic zones based on morphology rather than a purported function that may not be applicable across species lines. This allows comparisons to be accurately made between species.

Oviducal glands have a proximal club zone, papillary zone, baffle zone, and terminal zone. Variations in the makeup of each zone may show species variability, but the fundamental organization is maintained. The club and papillary zones replace the zone formerly designated the albumen zone. The club zone is so named because of its shape when viewed in transverse section with the light microscope. Similarly the papillary zone is characterized by a papillary or conical profile when viewed in section. The club and papillary zones are responsible for producing the various jelly coats that surround the egg. Recently Koob and Straus (1998) have investigated the role of jelly in *Raja erinacea*. They report that the jellies from various regions of the egg capsule have differing carbohydrate compositions. They also conclude that the jelly seems to function as a structural device to support hydrodynamically the egg and developing embryo. The baffle zone produces the various types of egg investments seen in elasmobranchs. In oviparous species such as the sharks *S. canicula* and *S. stellaris* and the skates *R. erinacea*, *R. eglanteria*, and *R. clavata* tubular glands produce secretory components that pass to secretory ducts. These ducts are confluent with a spinneret that has paired baffle plates that manipulate the secretory material as it emerges from the secretory duct. Secretory materials from adjacent secretory ducts blend in transverse grooves that extend across the full width of the gland; thus one transverse groove is responsible for one secreted layer.

Baffle plates are likewise present in the aplacental spiny dogfish, *S. acanthias*, and the placental dogfish shark, *M. canis*. The yellow spotted stingray, *U. jamaicensis*, is unusual in that it does not produce an egg investiture other than jelly coats and therefore lacks baffle plates. Despite variations among species, the capsule-producing oviducal glands studied appear to use the same basic assembly process to produce tough, flexible, and selectively permeable egg capsules, candles, or egg envelopes. This process involves the extrusion of a capsule material through dies, each of which opens between flattened baffle plates and extrudes a flattened ribbon containing precisely and complexly oriented molecules. The dies discharge into transverse grooves in the main lumen of the gland, each groove thus secreting a single lamella of the egg covering.

ISTHMUS

In at least two viviparous species a short segment of the reproductive tract joins the nidamental gland with the uterus. In *S. acanthias*, this segment appears to function as a closing device or sphincter, thus isolating the contents of the uterus, possibly preventing refluxing of the uterine contents back into the abdominal cavity (Widakowich 1907). At ovulation, however, it must expand through alterations in its mechanical properties to allow passage of the large, encapsulated eggs, a change that is likely regulated by ovary-derived steroids (Koob et al. 1983). Within the isthmus of *Torpedo*, a leaflike, membranous structure may function in isolating the lumen of the uterus to prevent refluxing of the copious amounts of histotroph produced in this species (Widakowich 1907). Whether similar structures operate, or even exist, in other viviparous species is not known but certainly seems likely.

POSTERIOR OVIDUCT OR UTERUS

In oviparous species, the uterus harbors the egg capsule during capsule sclerotization and thereafter until oviposition, which in skates can be several days (Koob et al. 1986). What contribution, if any, the uterus makes to egg encapsulation is not known. Certain regions of the uterine endometrium (Otake 1990) appear specialized for synthesis and secretion of materials that may contribute to capsule surface structure or chemistry or facilitate biochemical events associated with capsule polymerization.

The diversity of elasmobranch viviparous modes signals the complexity of uterine function. The elasmobranch uterus is able to accommodate the conceptus, regulate the intrauterine environment, provide oxygen for anabolic respiratory demands, provide nutrients, and mediate embryonic waste disposal. The degree to which any of these activities develop is commensurate with the mode of viviparity. However, comparatively little is known about the physiological and biochemical mechanisms employed for any of these functions. Nonetheless, a relatively large body of gross morphological observations over the past 100 years, together with substantial light and electron microscopic studies in recent years, are largely responsible for our current understanding of the role of the uterus in the various viviparous modes. The structural specializations of the uterus particular to each form of viviparity are described in detail under Reproductive Modes.

Two aspects of intrauterine development have captivated much of the contemporary empirical work on viviparous sharks and rays: first, the mechanisms for the production, transfer, and utilization of the nutrients required for embryonic development after the yolk provided in the ovulated egg has been consumed, especially in the placental and oophagous sharks and aplacental rays; and second, regulation of the intrauterine milieu, a physiological necessity shared by all viviparous forms. The uterus is central to each of these reproductive mechanisms.

CERVIX

The juncture between the uterus and the urogenital sinus is bounded by a constriction, the wall of which is composed of dense connective tissue and which is relatively inextensible. The location and function of this structure is analogous to the cervix in other vertebrates. During pregnancy it remains closed, but at parturition it must admit the passage of the term fetuses. Experimental evidence obtained in *S. acanthias* suggests that this change in properties is mediated by a relaxin like molecule originating in the ovary (Koob et al. 1984). The morphology, properties, and function of the cervix in other elasmobranchs remain largely unexplored.

UROGENITAL SINUS

The uterus independently joins a common urogenital sinus. This structure has received little attention from any standpoint. However, two observations suggest an important role during reproductive events. Steven (1934) observed that the urogenital sinus in female skates enlarges concomitant with sexual maturation, suggesting it to be a secondary sexual characteristic. The reason for this enlargement may derive from the relative size of the egg capsule, as well as the fact that female skates harbor the egg capsule in the urogenital sinus for several hours before oviposition.

Even more intriguing is the recent observation that in *Squatina guggenheim* and *S. occulata*, the developing embryos spend the last half of gestation in an enlarged chamber consisting of both the uterus and the urogenital sinus (Sunye and Vooren 1997). The urogenital sinus becomes greatly enlarged in order to undergo this transition and accommodate the relatively large embryos. While the sinus does not develop vascularization comparable to the uterine portion of the chamber, the flaccid wall of the sinus displays spontaneous contractions and distentions similar to those of the uterus.

Reproductive modes

OVIPARITY

Elasmobranch oviparity is characterized by the production of relatively large eggs, encapsulation of fertilized eggs in structurally complex and remarkably durable shells, and incubation periods lasting months to over a year. Following insemination and storage of sperm, egg capsules are laid at daily to weekly intervals for periods of several months or longer (Mellinger 1983; Luer and Gilbert 1985; Koob et al. 1986; Castro et al. 1988). Enough evidence exists to suggest that females select oviposition sites and attach eggs to suitable objects or place them in advantageous locations, perhaps returning to these places year after year. After oviposition, embryonic development proceeds without further maternal contributions. Essentially all of the nutrients necessary for development to hatching are contained in the oviposited egg, although water, minerals, and possibly other solutes may be sequestered from the environment. In later stages, yolk platelets are transferred by ciliary action from the external yolk sac to the internal yolk sac and thence to the intestine of the embryo via the yolk stalk (Fig. 15-6). No larval stage appears in any species, and following the completion of development, the hatchling emerges looking much like a miniature adult. The only distinguishing feature is that the internal yolk sac serves to nourish the embryo during the first few weeks after hatching (Figs. 15-7 and 15-8). Approximately 40% of all shark species and all skates employ oviparity of this type; no rays are oviparous.

Oviparous elasmobranchs encapsulate eggs in morphologically complex shells produced by a specialized oviducal gland located in the proximal reproductive tract (see Figs. 15-3 and 15-4). Capsule formation begins before ovulation, is one third to one half completed when the fertilized egg enters the capsule lumen, and then the remainder of the capsule is formed (Hobson 1930). The formed capsule enters the uterine portion of the reproductive tract, where it is held for several days prior to oviposition.

Egg capsules at oviposition are equipped with structures that serve as anchoring devices. Cat shark (Scyliorhinidae) egg capsules have remarkably long, coiled tendrils arising from each of the four corners. The female winds these tendrils around a suitable object during the oviposition process. Horn sharks (Heterodontidae) produce auger shaped capsules with two spiral flanges running the length of the capsule. It has been hypothesized, but never observed, that after oviposition the

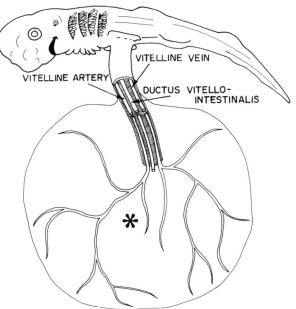

FIGURE 15-6 The yolk-dependent oviparous shark embryo is connected to the external yolk sac (*asterisk*) via a yolk stalk that contains vitelline artery, vein, and ciliated ductus vitellointestinalis. (From Hamlett 1986.)

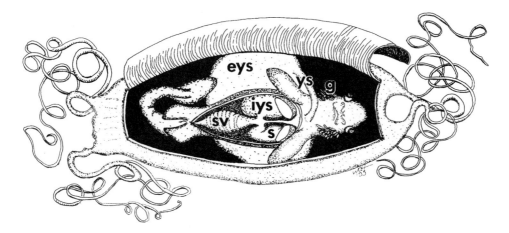

FIGURE 15-7 Oviparous shark embryos contain yolk in its external yolk sac (eys) and transmit it to internal yolk sac (iys), stomach (s), and spiral intestine containing a spiral valve (sv) for digestion and absorption. External gill filaments (g) aid in respiration prior to development of gill lamellae. (From Hamlett et al. 1993a.)

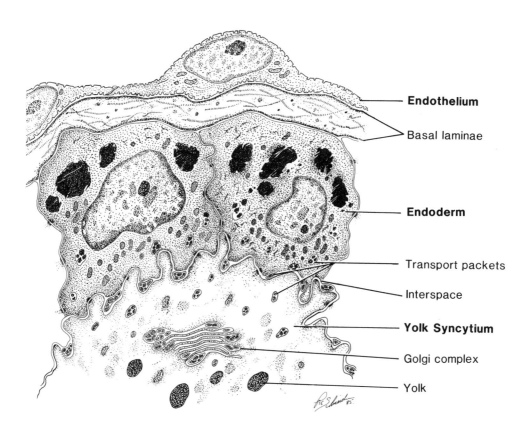

FIGURE 15-8 Yolk platelets are partially digested in the yolk syncytium of the external yolk sac and products transported to the endoderm in membrane-limited packets. Metabolites are then passed to yolk sac endothelium. (From Hamlett 1986.)

female picks up the capsule with her mouth and inserts the pointed end in a rock or coral crevice. This hypothesis derives from observations of capsules in situ where capsules have been found in such situations. Skate (Rajadae) egg capsules are equipped with masses of loose, sticky fibers, which presumably either stick to objects or accumulate bound particles. However, naturally oviposited skate egg capsules have never been observed in situ, leaving it unclear to date how anchoring is effected, if indeed it is.

Encapsulated along with the ovum is a gelatinous egg jelly, often referred to as albumen, which fills the lumen of the capsule not occupied by the egg itself. Neither the composition nor function of this material is currently known. It functions only during the initial one third of the incubation period, after which it disappears. The mechanism underlying liquefaction may involve enzymes produced by specialized glands on the head of the developing embryo (Ouang 1931). One consequence of dissolution is eclosion of the capsule, or the opening of four slits (in the corners of shark capsules; in the distal horns of most skate capsules) initially plugged with a dense egg jelly, thereby allowing the ingress of seawater. For the remaining two-thirds of incubation, the embryo develops in seawater.

The oviducal gland in oviparous species produces a multilaminate, proteinaceous matrix that is chemically stabilized following assembly in the gland lumen. Both the morphology and biochemistry of capsule formation are complex, as reflected in the structure and properties of the formed capsule. The ultrastructure and composition of the capsules produced by these glands have been studied adequately in only two species, *S. canicula* and *R. erinacea*. It seems likely that capsule formation in these two divergent species typifies that in most oviparous elasmobranchs, since certain structural and biochemical characteristics are held in common. First, capsules of both species are laminates made up of distinct layers organized in an alternating pattern with respect to orientation of the constituent matrix. Second, sclerotization of the assembled precursors by a quinone tanning mechanism occurs after the forming capsule passes out of the gland and enters the uterine portion of the reproductive tract.

Dogfish (*S. canicula*) capsules are constructed largely from an unusual collagenlike molecule, with possible affinities to type IV collagen, which forms the fibrils of the inner layers of the body wall as well as the tendrils and seams (Gathercole et al. 1993; Feng and Knight 1992, 1994a,b; Knight and Hunt 1974; Knight and Feng, 1992, 1994a,b). Following secretion from the shell gland, this collagen undergoes changes in macromolecular organization, resulting in formation of a liquid crystalline ultrastructure (Knight et al. 1993). Mixed together with these fibrils are hydrophobic granules containing protein rich in tyrosine (Knight and Feng 1994b).

Based on histochemical tests on both the oviducal gland and the capsule, stabilization of the dogfish capsule following assembly in the gland appears to involve in part a form of quinone tanning (Threadgold 1957; Rusaouën 1976; Feng and Knight 1992). Phenoloxidase activity, tyrosine-rich protein, and a peroxidase were identified in specific compartments within the oviducal gland and capsule wall. In addition, free 3,4-dihydroxyphenylalanine was identified in extracts of the gland; however, whether this or another catechol is incorporated in the capsule matrix is still an open question. Regardless of what the exact mechanism for stabilization is, the resulting material has extraordinary tensile properties (Hepworth et al. 1994).

In the skate *R. erinacea*, six major structural proteins ranging in molecular weight from 20 to 95 kDa form the bulk of the capsule matrix (Koob and Cox 1993). All of these proteins are relatively rich in glycine, serine, and tyrosine. The two small proteins with molecular weights of 20 and 22 kDa are exceptional in that they contain 50% glycine and 20–25% tyrosine. While none of the proteins exhibit amino acid compositions typical of interstitial collagens, they may contain collagenous domains analogous to one or more of the minor collagens found in other vertebrate connective tissues, or to the dogfish capsule collagen described previously.

Following secretion, fusion, and assembly of the capsule precursors in the oviducal gland, the newly formed white material is only chemically and mechanically metastable. Disulfide bonding of the 27- and 95-kDa proteins is partly responsible for the initial stability of the pretanned material (Koob and Cox 1990, 1993). Once the formed capsule leaves the oviducal gland, it begins to develop pigment, or tan. Chemical and mechanical stability develop coincident with progression of tanning. The tanning process requires at least 12 hours in utero.

Stabilization of the capsule matrix in this skate and likely most other oviparous species is attained in part through a quinone tanning mechanism in which approximately 50% of the tyrosine residues present in the newly assembled matrix are converted to 3,4-dihydroxyphenylalanine (DOPA) by a tyrosine hydroxylating activity (Koob 1992) (see Fig. 15-5). Some of these DOPA residues are then oxidized by a catechol oxidase forming dopachrome, the quinone of DOPA (Koob and Cox 1988). While the ensuing events are currently unclear, it seems likely that the dopachrome subsequently interacts with other capsule constituents, possibly amino acid side chains of neighboring proteins, forming covalent cross-links that polymerize the matrix into a cohesive unit. In addition to the structural proteins and enzymes, minerals are incorporated in the assembled material during capsule formation in the skate oviducal gland (Koob 1992). Calcium and, to a lesser extent, magnesium are present in the newly assembled capsule; however, they are not bound within the matrix. As the tanning process proceeds, these minerals become chemically bound, indicating that products of the tanning process generate an affinity for these divalent cations. Following exposure to seawater, significantly more magnesium and somewhat more calcium are bound. The function of these bound minerals is currently unknown.

The structure of the oviducal gland in *S. canicula* has been the subject of detailed morphological and microscopic studies, as it represents the typical condition in reproductively active oviparous species (Borcea 1905; Metten 1939; Rusaouën 1976; Threadgold 1957; Widakowich 1906; for a review see Knight et al. 1996a). The gland consists of numerous long, tubular structures lined with columnar secretory cells interspersed with ciliated cells. These tubules empty via a short ciliated duct into lamellae bordering the lumen of the gland. Lamellae span the entire width of the gland lumen and are responsible for producing the laminated structure of the body of the capsule. Specialized regions of the gland located antero-laterally form the anterior and posterior tendrils (Feng and Knight 1994b).

The secretory epithelial cells lining the gland tubules are packed with cytoplasmic granules containing capsule precursors. The amount and type of granules vary with the exact location within the gland. Morphological evidence, coupled with histochemical reactions, has demonstrated at least six distinct zones from anterior to posterior. Histochemical tests on the various morphologically distinguishable regions of the gland have delineated several discrete types of granules (Threadgold 1957; Rusaouën 1976; Feng and Knight 1992). Cells in the zone closest to the upper oviduct contain granules that react primarily with stains directed toward carbohydrates, indicating the presence of neutral oligosaccharides and possibly sulfated polysaccharides. Cells in the next zone exhibit elements of an oxidative phenolic tanning system, including a phenoloxidase, peroxidase, and tyrosine-rich protein. In the following zone, which is responsible for secreting the bulk of the capsule wall, cells store abundant numbers of granules containing the major fibrillar component of the capsule. Cytoplasmic granules in the cells of the posterior region are filled primarily with phenolic protein, phenoloxidase, and peroxidase, although carbohydrates account for a portion of the stored precursors in these zones.

Capsule formation begins with secretion of cytoplasmic granules containing capsule precursors into the lumen of the tubules. The granules coalesce within the tubule, forming a homogeneous material that is then transported by ciliary action to the lamellae. Extruded material from a row of tubule spinnerets in one lamella fuses to form a single layer. In the case of the fibrous layers of the capsule wall, the fibrils are organized within each layer in a dense parallel array. This sheet then emerges from the lamella and is joined to the neighboring sheets from the adjacent lamellae. The varying composition of the capsule layers (Knight and Hunt 1974; Knight and Feng 1992) results from the segregation of precursors within the discrete zones of the gland (see Knight et al. 1996b).

The morphology of the uterus in the oviparous little skate has recently been examined (Hamlett and Koob 1996; Koob and Hamlett 1998). The uterus in oviparous sharks and skates houses the eggs during the crucial chemical events resulting in the polymerization of the assembled egg capsule materials. Capsule sclerotization involves a quinone tanning mechanism in which catechols are introduced in utero and subsequently oxidized to quinones prior to oviposition. The uterus in *R. erinacea* is morphologically specialized for well-defined functional characteristics associated with egg capsule formation: (1) regionally distinct structural modifications, in both the uterine wall and the epithelial lining, for active movement of the capsule through the uterine lumen; (2) biosynthesis and secretion of materials into the lumen; (3) vascular facilitation of oxygen transfer in order to fuel the oxidation process involved in capsule tanning; and (4) intercellular spaces for removal of water from the uterine lumen. The first three characteristics are qualitatively similar to those operating in viviparous species. The uterus throughout its length has longitudinal folds punctuated by secretory crypts. Proceeding from the anterior to the posterior end of the uterus, there is a progressive decrease in the number of cilia and the depth of the lamina propria and an increase in vascularity and the width of the muscularis.

The capacity of the elasmobranch uterus to develop the physiological means to satisfy the demands peculiar to each reproductive mode relies on a few, shared basic principles: structural accommodation of the eggs and embryos, supply of oxygen to the uterine lumen, and biosynthesis and secretion of structural or nutritional materials. Regardless of the reproductive mode, similar design characteristics are shared by all elasmobranch uteri (Hamlett and Hysell 1998). It is evident from the study of the little skate that the function of the uterus in oviparous species is more complex than serving as a simple and primitive conduit to transport eggs from the oviducal gland to the external environment. The skate uterus employs certain mechanisms qualitatively similar to those operating in viviparous species. These mechanisms in viviparous species have been touted as the hallmark of viviparity, the development of which was crucial for the evolution of viviparous reproductive strategies. We believe that, when considering the evolution of reproductive modes in elasmobranchs, the similarities between physiological mechanisms in the uteri of oviparous and viviparous species must be incorporated into phylogenetic schemes.

Egg capsules serve to protect the embryo throughout development. To accomplish this, they must withstand prolonged exposure to the corrosive action of seawater. In addition, they must be able to prevent attacks by predatory organisms. That egg predation threatens many oviparous species is apparent by the fact that empty shark and skate egg capsules have been found with evidence of predatory activity (Cox et al. 1997). Egg capsules may function in ways other than mechanical protection, including regulation of solute permeation, antimicrobial activity, and antifouling properties, though none of these activities have been unequivocally proven to operate in situ. Capsule permeability has been the most studied of these properties. Low-molecular-weight solutes, including cations, molecular oxygen, and urea, are freely permeable; thus the capsule does not appear to regulate osmolality in the oviposited egg or during early development

(Hornsey 1978; Foulley and Mellinger 1980; Evans 1981; Kormanik 1993). Regulation of the intracapsular milieu occurs only during the initial third of development; thereafter, seawater fills the cavity.

VIVIPARITY

Diverse viviparous modes operate in extant sharks and rays. Viviparous sharks are known especially for their placental and oophagous modes; however, other equally effective strategies combining distinct intrauterine mechanisms are widespread throughout the taxon. In contrast, rays exclusively utilize an aplacental mode based principally on uterus-derived histotroph. Despite this diversity in uterine mechanisms for facilitating embryonic development, some key features of viviparity are shared, and of these many are identical in kind, if not degree, to that in oviparous species. Most viviparous species ovulate relatively large eggs. In only a few rays and sharks is the ovulated egg of small size. Nearly all viviparous species encapsulate eggs along with egg jelly in tertiary egg envelopes produced by the oviducal gland. The initial phase of embryogenesis occurs inside these capsules. Thereafter, the capsules function in various ways depending on the reproductive mode. In some aplacental species, the partially developed embryos break out of the capsule early in gestation, while in others embryos remain inside the capsule, which eventually becomes an integral part of the placenta. In oophagous species, the capsule is used to deliver ovulated eggs to the embryos in utero. In all viviparous species, the uterus undergoes morphological and physiological modifications to accommodate the conceptus, to regulate the intrauterine milieu, and in most viviparous species, to nourish the developing embryos. The exact nature of these modifications is specific to the reproductive mode.

Egg capsules of viviparous species have received little scrutiny from either a biochemical or structural perspective. Nothing is known about the protein composition or the chemical mechanisms for stabilizing the assembled matrix in these species. Beyond descriptions of gross morphology, little is known about the organization of these capsules, except for *Mustelus canis* and *Carcharhinus plumbeus*. The capsule in *M. canis*, a placental species, is a thin laminate consisting of at least four orthogonally arranged fibrous layers (Lombardi and Files 1993). Knowledge of the functional properties of egg capsules in viviparous species is also acutely lacking. While it is well recognized that these structures must have properties important for embryonic development, especially in placental species in which the capsule is integrated into the placenta (Kormanik 1992; Hamlett 1993; Lombardi and Files 1993), permeation studies have been accomplished in only two species (Kormanik 1993; Lombardi and Files 1993). In contrast, the structure and organization of oviducal glands responsible for producing egg capsules in viviparous species have been examined in more detail (Borcea 1905; Nalini 1940; Prasad 1945a,b). Although considerably smaller than in oviparous species, the oviducal glands in viviparous species exhibit similar morphology (see Fig. 15-4).

In the majority of viviparous species, particularly the matrotrophic viviparous species, nutrients are supplied to the developing embryos in utero. Specialized structures are elaborated for the transfer of these nutrients from the mother and for their uptake by the embryo. Implicit in the abundant descriptions of these morphological specializations is that the uterus mediates nutrient transfer. However, the origin and nature of the organic nutrients remains little explored. The early work, especially that of Ranzi (1932, 1934) and Needham (1942), focused on the overall composition of the fluids surrounding the embryo, that is, overall organic and inorganic content. These studies were important because they formed the foundation for establishing that the organic content of term embryos surpassed that in the ovulated eggs, thereby proving that the uterus must supply nutrients. Unfortunately, there has been little progress relating to the composition of uterine fluids since these pioneering efforts, despite the obvious importance of nutritional constraints on viviparous reproduction. Moreover, little attention has been directed toward determining the source of the organic constituents in the uterine fluids, that is, a maternal source (liver, gut?) from which they are transported to the embryo through the circulation and uterine vascularity, or the uterus itself. Both are likely involved (Hamlett et al. 1985a, 1993; Hamlett 1993), but the degree to which each contributes may be species-specific.

Files and Lombardi (1993) and Lombardi et al. (1993) have approached these questions with contemporary analytical techniques. They identified and measured free amino acids and other small metabolites in the uterine fluids of four viviparous sharks, *S. acanthias*, *C. plumbeus*, *M. canis*, and *R. terraenovae*. The relative amounts of these compounds correlated with the type of viviparity, with *S. acanthias* having the least and *R. terraenovae* the highest concentrations. They also demonstrated that proteins separated by polyacrylamide gel electrophoresis were most concentrated in *R. terraenovae* and virtually absent in *S. acanthias*. Some of the proteins in the uterine fluids of these sharks appeared to correspond, at least with respect to apparent molecular weights, with proteins found in the maternal serum.

Aplacental yolk sac viviparity

Three forms of aplacental viviparity operate in sharks and rays. What has traditionally been termed *ovoviviparity*, now called *aplacental yolk sac viviparity*, is utilized by approximately one quarter of contemporary shark species. In this mode, embryos rely primarily on the substantial yolk in the ovulated egg for the organic material required by development. The uterus in these species specializes in regulating the intrauterine milieu. Elaborate vascularization and differentiation of the uterine endometrium during pregnancy mediates transfer between the maternal circulation and the uterine lumen. Embryonic development in these species is similar to that in oviparous species. Yolk is mobilized from the external yolk sac and transported via the yolk stalk to the embryonic intestine as well as being digested in the yolk syncytial-endoderm complex (Fig. 15-8).

Pregnancy in *S. acanthias* has been the subject of considerable study with respect to regulation of the intrauterine milieu. Ovulated eggs are encapsulated together in a thin, diaphanous capsule along with albumen. The initial part of embryonic development occurs within the egg capsule, which is closely apposed to the walls of the uterus. After approximately half of gestation is completed, the capsule breaks open and the embryos complete development to term free in uterine fluids. Early studies established that the composition of intrauterine fluids is not like maternal blood; nor is it similar to seawater, even though the uterine lumen is routinely flushed with seawater during the later developmental stages (Burger 1967). More recent studies have confirmed and greatly extended these early observations (Evans et al. 1982; Kormanik 1988, 1992, 1993). In conjunction with experiments on the physiology of embryos when challenged by various solutes, these authors have elucidated fundamental relationships between maternal regulation of the intrauterine environment and the developing ability of the embryo to osmoregulate. During early pregnancy, the sodium and chloride concentrations in uterine fluids are intermediate between maternal plasma and seawater, with urea levels approaching those in maternal blood. Later in pregnancy, uterine fluids are essentially identical to seawater and urea concentrations are low. These data indicate that the mother actively regulates the uterine milieu early in pregnancy but contributes little to the fluids later in pregnancy, except to acidify the fluid, which detoxifies the ammonia that accumulates (Kormanik 1993). The development of embryonic osmoregulatory mechanisms is concomitant with the transition to seawater in utero.

During gestation in *S. acanthias*, the uterine mucosa is thrown into a series of longitudinally oriented folds. According to Jollie and Jollie (1967b), the gravid uterus consists of (1) two layers of mucosal epithelial cells and their basal lamina; (2) an extensively ramified system of juxtaepithelial capillaries; (3) a sparse connective tissue region with reticular fibers; and (4) a well-developed muscular layer. Although both cellular layers between the capillary bed and the uterine lumen are epithelial, they are dissimilar. The juxtaluminal layer alone possesses numerous apical invaginations and smooth-surfaced apical "microvesicuations." Abundant, closely packed, smooth-surfaced vacuoles are restricted to the apical cytoplasm. The juxtacapillary epithelial cells are categorized by evidence of micropinocytotic activity, particularly abundant at the basal surface. Capillaries of the vascular bed are closely apposed to the juxtacapillary epithelium. Micropinocytotic pits and vesicles occur at both the luminal and basal surfaces of the endothelial plasmalemma (Jollie and Jollie 1967b). The increased vascularity in late pregnancy might be related to the oxygen demands of the large embryos or the acidification of the uterine fluids.

The study by Jollie and Jollie (1967a) of the yolk sac of the spiny dogfish, *S. acanthias*, is the only ultrastructural study of a nonplacental viviparous shark. Gestation in this species is prolonged (ca. 18–20 months). During early development, the yolk sac functions in both nutritive and respiratory exchange; during midgestation its function is solely nutritive; and neither function is served during late development. The yolk sac in early-stage embryos (50–90 mm) is composed of five distinguishable regions: (1) an ectodermal epithelium one to two cells thick, whose cytoplasm contains a prominent Golgi complex, many smooth-surfaced "microvesicles," rough endoplasmic reticulum, but no keratin; (2) a wide zone of "reticular" fibers; (3) vitelline capillaries embedded in a connective tissue matrix containing fibroblastlike cells; (4) pyramidal endodermal cells, whose basal surfaces interdigitate with invaginations of the apical surface of the yolk cytoplasm; and (5) peripheral yolk cytoplasm. By the time *Squalus* embryos have attained a length of 150–190 mm, at midgestation, a bilaminar mesodermal layer has made its appearance between the ectoderm and the vitelline capillary layer. Mesodermal cells possess a prominent Golgi apparatus, many endocytotic pits, and abundant intracytoplasmic filaments. Jollie and Jollie (1967a) consider these cells to be smooth muscle. The yolk cytoplasmic layer, endoderm, capillary endothelium, and intervening spaces contain 100 Å granules that originate in the Golgi of the yolk cytoplasmic layer and are believed to represent yolk degradation products. Term (295 mm) embryos possess a pendulous yolk sac that is not used for respiration or nutrition. The yolk sac retains its regional organization, except that the yolk cytoplasmic layer is absent and the bilaminar mesodermal layer has lost its tissuelike organization. The ectodermal epithelium has increased its thickness to four or five cell layers.

Recently, the whale shark, *Rhinocodon typus*, has been shown to be aplacental yolk sac viviparous (Joung et al. 1996). A 10.6-m-total-length (TL) pregnant female was found to harbor about 300 embryos. Many embryos still had attached yolk sacs and were enclosed in egg cases. Empty egg cases were also found. Thousands of ova less than 1 cm in diameter were found in the ovary. This report is the record for the largest number of embryos carried by any shark.

Aplacental viviparity with uterine villi or trophonemata
In rays, aplacental viviparity involves production and secretion into the uterine lumen of an organically rich histotroph, which is then ingested by the embryo or absorbed by external gill filaments. This strategy necessitates development of uterine vascularization and endometrium to effect histotroph supply, in addition to its function in regulating the intrauterine milieu. The most highly developed of these strategies occurs in some rays in which the uterine epithelium forms elongate villi, termed *trophonemata* (Wood-Mason and Alcock 1891), thereby significantly increasing the surface area for histotroph secretion and respiratory exchange

(Figs. 15-9 and 15-10). Developing embryos consume histotroph either through specialized structures or by ingestion. The result of this form of nutrient supply is that the offspring grow to very large sizes. In general, rays produce very few, relatively large offspring.

Gestation in stingrays is shorter than in most sharks. Ray embryos develop generally from 2 to 4 months, while viviparous sharks produce young in 10–12 months. Rays also have smaller litters, ranging from 1 to 15. In contrast to most sharks, stingrays produce much smaller eggs, the yolk content of which is insufficient for the embryos to grow to term.

The trophonemal uterine lining closely envelops the embryo, and this may form the basis of earlier reports that trophonemata enter the spiracles, where they secrete directly into the embryonic gut. These observations have not been confirmed by contemporary observations (Hamlett et al. 1985d, 1996a).

Needham (1942) reported 13.3% organic substance in the uterine milk of *D. violacea* and 1.2% for *Torpedo ocellata*. Needham (1942) and Amoroso (1960) reported that there is a 21% loss in organic substances from egg to embryo for *T. ocellata* and a 1628% increase for *D. violacea*. In *Rhinoptera bonasus*, histotroph production continues until parturition, at which time it ceases (Smith 1980). At three-quarter term, the yolk sac and stalk are almost completely absorbed, with only 3 mm of the yolk stalk remnant protruding. At full term the umbilicus is totally absorbed, leaving only a small scar. The embryos ingest the histotroph in utero and may reach 405 mm disk width. Bearden (1959) reported that the stomachs and intestines of fetal *Myliobatis freminvilli* are filled with yellowish material considered to be histotroph; this is also the case with *R. bonasus*.

In *R. bonasus*, the trophonemata provide a continuous supply of metabolites and account for a 3000-times embryo weight increase from the egg (Hamlett et al. 1985d). This far exceeds the efficiency of the shark yolk sac placenta. The increase in embryonic organic material during development is 840% for the placental blue shark, *P. glauca*, and 1050% for the dogfish, *Mustelus laevis* (Ranzi 1934; Needham 1942). This may be accounted for in several ways. The cow-nose ray has only one functional uterus and a single enclosed embryo as opposed to the placental blue shark, *P. glauca*, which may carry up to 130 fetuses (Compagno 1984). On an individual basis, there is a larger energetic investment in the single ray fetus. At term its disk width is 405 mm, which is almost half that of the adult. Placental sharks are considerably smaller at

FIGURE 15-9 Stingray embryos, with yolk stalk (st) and yolk sac (ys), reside in a uterus (ut) adorned with trophonemata (t). (From Hamlett et al. 1993a.)

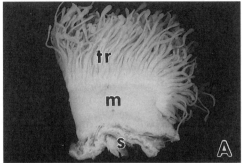
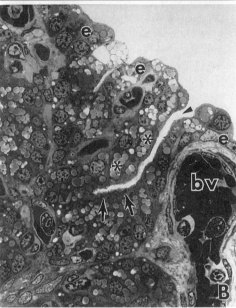

FIGURE 15-10 A: Fresh freehand section of uterus of *Dasyatis americana* containing full-term fetuses reveals villiform trophonemata (t), smooth muscle myometrium (m), and outer serosa (s) or perimetrium. **B:** Light micrograph of trophonematum of *D. americana.* Crypt cells are filled with dark secretion granules (*arrows*) and large lipid deposits (*asterisks*). Surface epithelial cells (e) are cuboidal and overlie subsurface peripheral blood vessels (bv) (×3000, reproduced at 73%). (A, From Hamlett et al. 1996a.)

birth relative to the adult. The entire surface area of the uterus of rays undergoes hypertrophy to form trophonemata that greatly increase the surface area available for production of histotroph. Trophonemata are not restricted to any area of the uterus, as is the placental attachment site in sharks. The entire surface of the uterus is available for secretion.

According to Ranzi (1934), trophonemata epithelial secretion in *Dasyatis* and *Myliobatis* commences when a leucocyte enters the cell and disintegrates. Shortly thereafter, lipid and protein granules appear and are secreted. No evidence of this process was found in *R. bonasus* (Hamlett et al. 1985d) and *Dasyatis americana* (Hamlett et al. 1996a). Ranzi's conclusions may have been based on fixation artifacts, or what he interpreted as a leucocyte within a cell may have been a capillary containing a leucocyte between ridges.

The Southern stingray, *D. americana*, displays aplacental viviparity with trophonemata where embryos are retained in the maternal uterus throughout gestation and initially nourished by yolk sac contents. There is a 3750% increase in wet weight from the egg to the term fetus. Trophonemata are 1.5 cm long, narrower at the base and spatulate at the tip. Surface epithelial cells form a pattern of surface cables, each with a small blood vessel at its core. In females containing fertilized eggs, the epithelium is simple cuboidal (Fig. 15-10). In contrast, in uteri containing late-term fetuses, the epithelium is squamous. Epithelial cells with PAS-positive cytoplasmic vesicles form invaginated crypts. Epithelial cells produce proteinaceous, mucous, and lipid secretions; thus the term *uterolactation* has been coined to describe this phenomenon (Hamlett et al. 1996a).

In the fetal Southern stingray, both the stomach and spiral intestine function early in development to digest and absorb nutrient histotroph (Hamlett et al. 1996b). The gastric mucosa consists of a surface columnar mucous epithelium that is confluent with gastric pits, or foveolae. Gastric glands are populated by oxynticopeptic and enteroendocrine cells. Surface mucous cells are pyramidal with apical microvilli. Oxynticopeptic cells are low columnar, with a distinct and elaborate tubulovesicular system in the apical cytoplasm. Oxynticopeptic cells produce both hydrochloric acid and pepsinogen. Microvilli line the lumen of the gastric glands and cells have elaborate interdigitating lateral folds. Enteroendocrine cells are characterized by basal granules and a prominent rough endoplasmic reticulum. The fetal intestine is filled with green-tinged viscous fluid, lipid-rich histotroph in the process of being emulsified by bile. A core of submucosa supports spiral intestinal plicae that form the spiral valve, from which villi project. The most prominent characteristic of the cells are enormous supranuclear vesicles that are formed by coalescence of smaller endocytotic vesicles. The apical cytoplasm has a profusion of smooth tubules, endoplasmic reticulum, and lysosomes. The large vesicles are interpreted as storage depots for continually ingested histotroph. Small vesicles may then bud off to be digested via the lysosomal system.

Aplacental viviparity with oophagy and intrauterine cannibalism
The last form of aplacental viviparity, oophagy, occurs in only a few species of the lamnoid sharks. Early development in these sharks is essentially identical to that in aplacental yolk sac species. Embryos develop primarily from material in the yolk sac. Thereafter, development relies on the continued supply of yolk in the form of ovulated eggs. The embryos actively feed on these eggs and store the yolk for processing in the cardiac portion of the stomach. The sand tiger, *Carcharias taurus*, also practices intrauterine cannibalism.

Sharks in the order Lamniformes consist of 16 species in seven families. The families are the sand tigers, goblin sharks, crocodile sharks, megamouth sharks, threshers, basking sharks, and the mackerel sharks. Only nine species from four families have been collected during pregnancy, and all species thus far examined have embryos that ingest uterine eggs. The sand tiger has been the most thoroughly studied (Gilmore et al. 1983; Gilmore 1991; Hamlett 1983; Stribling et al. 1980; Hamlett et al. 1993a) and is the only documented intrauterine cannibal. In this species and the thresher, *Alopias* (Gruber and Compagno 1981), only one embryo survives to term in each oviduct, while multiple embryos do so in all other species examined.

Sand tigers congregate in specific mating grounds on the East Coast from North Carolina to Florida from March to May, and females give birth 1–2 months prior to mating again. Sand tigers appear to mate every year and females spend most of their adult life pregnant. Following insemination, sperm is stored within the female genital tract. Initial ovulations release a pair of eggs, each one destined for transport to one of the two oviducts. Subsequent ovulations release two, then three, and then multiple (up to 18) ova per oviduct. Single or multiple ova are fertilized and are subsequently encapsulated in the oviducal gland and transported to the uterus. Fertilization of ova occurs over a 1- to 2-week period. After this, multiple ovulations continue but no further ova are fertilized (Gilmore et al. 1983; Gilmore 1991).

Scanning electron microscopy of intrauterine embryos of the sand tiger (Fig. 15-11) (Hamlett et al. 1993a) have revealed that a distinctive lateral line system and jaws are formed by the time the embryo is 27 mm TL. Tooth buds appear on each jaw by 30 mm TL and teeth have erupted by 35 mm TL. By 40–45 mm TL, functional dentition is established and multiple rows of recurved teeth appear.

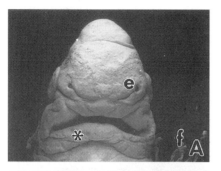

FIGURE 15-11 Scanning electron micrographs of intrauterine sand tiger embryos, *Carcharias taurus*. **A:** A 27-mm embryo with external nares (e), branchial filaments (f), and formed lower jaw (*asterisk*) (×17.6) **B:** A 33-mm embryo with tooth buds (*arrows*) and lateral line system (*asterisk*) (×10.4). **C:** A 45-mm embryo with recurved teeth (×13). **D:** A 60-mm embryo with dentition resembling that of adult (×38.5). All reproduced at 73%. (A–D from Hamlett et al. 1993a.)

Embryos 60 mm TL are present in the uterine lumen. Predation on uterine siblings enclosed in egg capsules commences when the dominant embryo is 80–100 mm in length. All siblings have been ingested by the time the surviving embryo reaches 350 mm. The embryos, one in each uterus, continue to feed on egg capsules containing multiple unfertilized ova. With continued ingestion of ova the stomachs of the surviving embryos become markedly distended. About 1 month prior to birth, ovulation ceases and the stomach size then decreases. The fetuses are born at a length of 100 cm (Gilmore 1991).

Placental viviparity

Placental viviparity occurs only in sharks and accounts for only about 10% of extant species. In all cases, the first phase of embryonic development is similar to the aplacental strategy in which eggs are encapsulated in some form of capsule and embryos develop from yolk in the ovulated egg (Hamlett and Wourms 1984). During the middle of gestation, uterine secretions augment the declining yolk stores and only during the latter phases of pregnancy does the placenta form, and in all cases the placenta is a yolk sac, or epitheliochorial placenta (Hamlett et al. 1985a; Hamlett 1986, 1987, 1989, 1990, 1993). Uterine specializations are concomitant with placentation. In some species the umbilical cord develops villous extensions, appendiculae, that may serve as a paraplacental nutrient absorptive site (Hamlett 1993; Hamlett et al. 1993c).

In viviparous shark species with a placenta, the tertiary egg envelope is greatly reduced in thickness, reflecting altered function of the oviducal gland. Similar reduction or loss of oviducal glands and the shell membrane is also observed in viviparous reptiles (Bauchot 1965; Guillette et al. 1981; Hoffman 1970; Weekes 1935). The gross structure of the shark egg envelope is that of a diaphanous, filmy membrane rather than an egg "case." In all placental species thus far examined, the egg envelope is retained throughout gestation, with the exception of the blue shark, *P. glauca* (Otake and Mizue 1985) and the Telok Anson shark, *Scoliodon laticaudus* (Teshima et al. 1978), in which the egg envelope is not incorporated into the uteroplacental complex. In most placental sharks, all metabolic exchange between the uterus and fetus must, therefore, be effected through or across the egg envelope. The only study of the physiology of transport across the egg envelope in placental species is the recent work of Graham et al. (1995). They showed that the egg envelope in *M. canis* allows transfer of low-molecular-weight metabolites (less than 100 Da) from the maternal circulation through the envelope to the intracapsular fluids. Permeability experiments on isolated egg envelopes established that the molecular weight cutoff for permeation is between 1000 and 5000 Da (Lombardi and Files 1993). Morphological information based on ultrastructural observations in *C. plumbeus* (Hamlett et al. 1985a–c) revealed that the envelope is acellular and relatively homogeneous on the placental surface but has delaminations on the uterine side. This situation also occurs in the Atlantic sharpnose shark (Hamlett 1993; Hamlett et al. 1993b). It has not been determined if the egg envelope of placental species is composed of collagen.

What exact role the egg envelope in placental sharks plays in the transport dynamics of the placenta awaits further systematic analysis. Aside from the permeability studies on *M. canis*, the only other available information on permeability deals with the egg case in oviparous species. In *S. canicula*, Foulley and Mellinger (1980) demonstrated the permeability of the egg case to water and some organic molecules and showed that it is more permeable to sodium ions than urea. In the skate, *R. erinacea*, Evans (1981) found that the egg case can establish and maintain both osmotic and ionic gradients between the fluids contained within the egg case and external seawater; however, Kormanik (1993) established that the little skate egg case is freely permeable to low-molecular-weight solutes. Foulley et al. (1981) showed asymmetric permeability characteristics in the egg case of *S. canicula*. Passive permeability was favored toward the embryo. If similar passive or facilitated mechanisms are active in placental species, this might provide for the asymmetric transport of nutrient metabolites toward the fetus and might reflect the organization of the layering of the egg envelope. In the Atlantic sharpnose shark, no electron dense bodies are observed traversing the envelope, but many such bodies populate the interspace between the placental surface of the envelope and the placental epithelium. Material may be transported in an insoluble form and reassemble prior to being incorporated into the placenta (Hamlett 1993).

During ontogeny, the yolk sac of some viviparous sharks differentiates into a yolk sac placenta that persists to term (Figs. 15-12 and 15-13). The placenta is noninvasive and nondeciduate. Hematrophic transport is the major route of nutrient transfer from mother to fetus. The placental unit of the sandbar

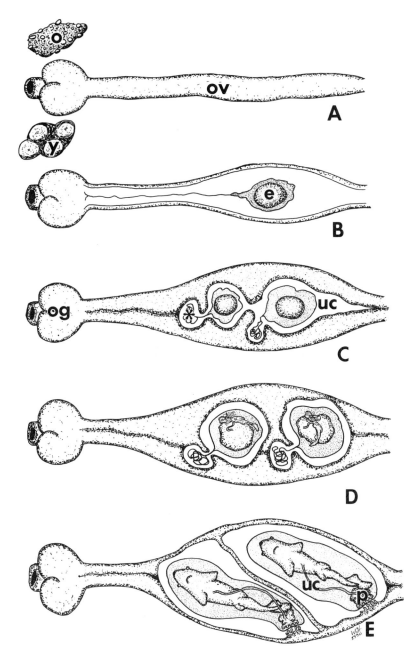

FIGURE 15-12 Ontogenetic development of uterine compartments in placental sharks. **A–E:** Immature placental sharks possess an ovary (o) with developing oocytes and a small oviduct (ov), not differentiated into a uterus. At sexual maturity, yolked eggs (y) are shed from the ovary and enter the oviducal gland, where an egg envelope is formed around each egg. Fertilized eggs (e) reside in uterine compartments (uc) and the excess egg envelope is sequestered in an egg envelope reservoir (*asterisk*). The yolk stalk differentiates into an umbilical cord (uc) and the yolk sac becomes a part of functional placenta (p). (From Hamlett et al. 1993a.)

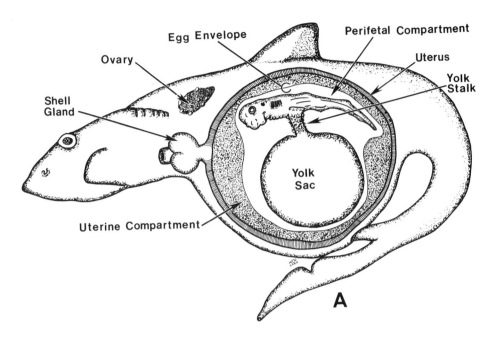

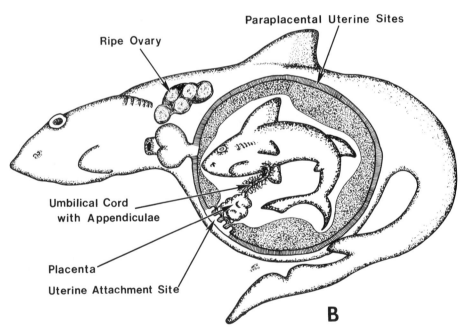

FIGURE 15-13 **A:** Placental shark embryo in the preimplantation stage. During early stages of gestation, the ovary contains small, developing oocytes, while the uterus harbors yolk-dependent early-term embryos. **B:** At term in placental sharks, the yolk stalk has been transformed into an umbilical cord, which may contain appendiculae. The yolk sac contributes to the functional placenta. Specialized attachment sites for distal portion of placenta mediate metabolic exchange. (A, B, from Hamlett 1993.)

shark, *C. plumbeus*, consists of (1) an umbilical stalk; (2) the smooth proximal portion of the placenta; (3) the distal rugose portion; (4) the egg envelope; and (5) the maternal uterine tissues (Fig. 15-14) (Hamlett et al. 1985a–c). Exchange of metabolites must be effected through the intervening egg envelope (Figs. 15-15 and 15-16). The distal, rugose portion of the placenta is the fetal attachment site. It consists of (1) surface epithelial cells; (2) a collagenous stroma with vitelline capillaries; and (3) an innermost boundary cell layer. In vitro exposure of full-term placentas to solutions of trypan blue and horseradish peroxidase (HRP) reveals little uptake by the smooth portion of the placenta but rapid absorption by the surface epithelial cells of the distal, rugose portion. HRP enters these cells by an extensive apical system of smooth-walled membranous anastomosing canaliculi and tubules. Prominent whorl-like inclusions that occupy the basal cytoplasm of the surface cells adjacent to the pinocytotically active endothelium of the vitelline capillaries are proposed to be yolk proteins that are transferred from the mother to the embryo throughout gestation.

During mid- and late gestation, the uterus of sandbar sharks possesses specialized sites for exchange of metabolites between the mother and embryo. Uterine attachment sites are highly vascular rugose elevations of the maternal uterine lining that interdigitate with the fetal placenta. The maternal epithelium remains intact and there is no erosion. Comparison with the uterine epithelium of nonplacental sharks, mammalian epitheliochorial placentas, and selected transporting epithelia reveals that the structure of the maternal shark placenta is consistent with its putative multiple functions, namely: (1) nutrient transfer; (2) transport of macromolecules, such as immunoglobulins; (3) respiration; and (4) osmotic and ionic regulation.

The Atlantic sharpnose shark develops a yolk sac placenta composed of (1) uterine mucosa, (2) egg envelope, and (3) fetal yolk sac mucosa (Hamlett 1993; Hamlett et al. 1993a–c). The transporting uterine mucosa is a squamous epithelial bilayer with prominent lateral and basal infoldings between contiguous cells. The surface cells have prominent secretion vesicles that empty their contents to the exterior. Immediately beneath the epithelium is a basal lamina and a profuse vascular supply with a continuous endothelium. The epithelium of paraplacental uterine sites is mucous. The tertiary egg envelope is retained throughout gestation and separates the distal part of the yolk sac from the maternal uterine mucosa. The egg envelope is compact on the yolk sac surface but displays delaminations on the uterine surface.

The fetal yolk sac is composed of two portions, namely, a proximal saccular region and a heavily vascularized, rugose distal portion. The proximal portion has ultrastructural characteristics of a steroid hormone–producing tissue, including massive smooth endoplasmic reticulum frequently forming whorled arrays. However, definitive evidence that the yolk sac is an endocrine organ is lacking. The distal portion of the fetal yolk sac is composed of a squamous epithelial bilayer that is separated from the underlying vascular network by a continuous basal lamina. The endothelium of the vessels is fenestrated. Cytoplasmic characteristics of these cells include an extensive Golgi complex, smooth-walled caveolae, vesicles with electron-dense contents that are presumably endocytotic in nature, and dense bodies that are suggested to be lysosomes involved in the digestion of material, perhaps yolk metabolites (Hamlett 1993; Hamlett et al. 1993a,b).

The ultrastructural characteristics of the smooth proximal portion of the shark placenta suggests that it could be secreting steroid hormones (Hamlett 1990). Although the human placenta produces steroid hormones, it lacks the complete repertoire of enzymes necessary for synthesis of all the hormones it produces. An exchange between the fetus and placenta by the so-called fetoplacental unit is responsible for the final hormone production (Van Tienhoven 1983). It is feasible that similar mechanisms might be operating in the shark placenta. Whether

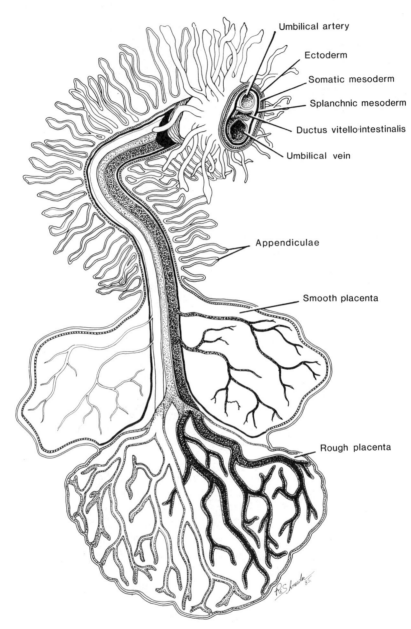

FIGURE 15-14 Stereo diagram of germ layer contributions to shark yolk sac placenta. Appendiculae are vascularized outpocketings of ectoderm that are filled with connective tissue. Smooth placenta has contributions from ectoderm and somatic mesoderm. Rough placenta is richly supplied with mesodermally derived vasculature. Endodermal cells coat blood vessels and the inner aspect of the ectoderm. (From Hamlett 1986.)

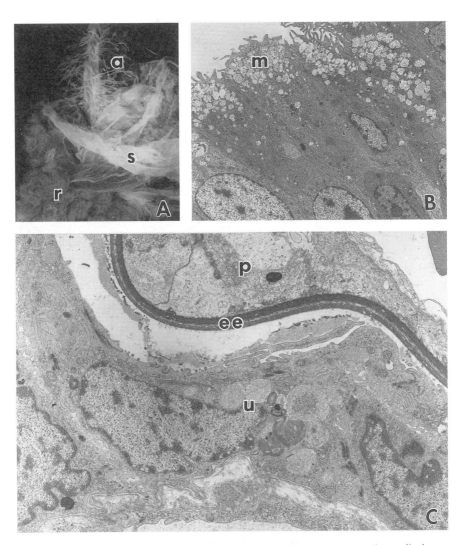

FIGURE 15-15 **A:** Gross photograph of placenta of *Rhizoprionodon terraenovae*. a = Appendiculae emerging from umbilical cord; r = distal, rugose portion; s = proximal, smooth portion. **B:** Columnar epithelium of term paraplacental uterus contains prominent mucus droplets (m). **C:** Egg envelope (ee) separates maternal uterus (u) from distal portion of fetal placenta (p). (A, from Hamlett 1987; B, from Hamlett et al. 1993a.)

the shark placenta functions as an endocrine organ awaits investigation; however, preliminary studies employing radioimmunoassay procedures have identified measurable levels of some steroids in the placenta of the blacknose shark, *Carcharhinus acronotus* (Hamlett 1990). Determining the site of production of the various steroid hormones and their levels during development will help to understand the hormonal control of pregnancy in placental sharks.

The presence of crystalline inclusions in the ovary, uterus, preimplantation embryo, and fetal membranes of mammals has been considered as a means of storage of material that is to be used by the developing embryo (Hoffman and Olson 1984). Hamlett et al. (1985a) noted the similarity in ultrastructural appearance of non-membrane-bound inclusions in the distal portion of the term placenta of the sandbar shark, *C. plumbeus*, with teleost (Walzer and Schonenberger

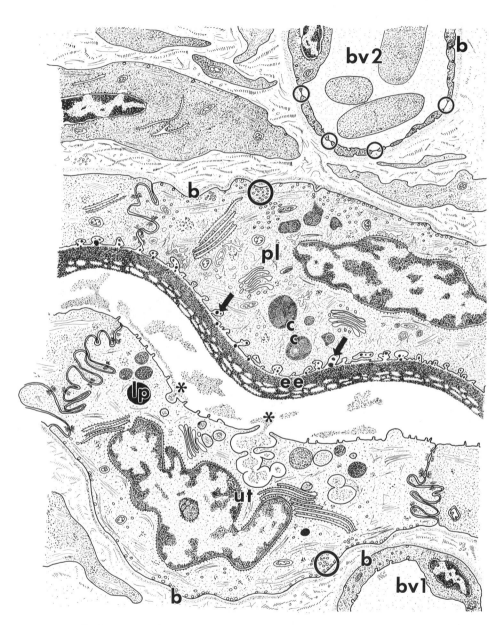

FIGURE 15-16 At term, the placenta is characterized by squamous uterine epithelium (ut) that has basal pinocytotic vesicles (*large circles*), lipid inclusions (lp) and secretory vesicles that release their contents to uterine lumen (*asterisks*). Subjacent capillaries (bv1) are continuous. The egg envelope (ee) separates fetal and maternal portions of placenta. Electron-dense vesicles (*arrows*), of uterine origin, are endocytosed by the distal portion of fetal placenta (pl). Epithelium has basal pinocytotic vesicles (*large circles*) and crystalline inclusions that are interpreted as yolk precursors. Subjacent capillaries (bv2) are fenestrated (*small circles*). B = basal lamina. (From Hamlett 1993.)

1979a,b) and amphibian (Ward 1980) yolk protein precursors. The inclusions may conceivably represent residual bodies derived from partial lysosomal digestion of yolk. Residual bodies are frequently characterized by membrane fragments, myelin figures, or whorled inclusions (De Duve and Wattiaux 1966). Based on ultrastructural analyses of the shark yolk sac (Hamlett et al. 1987) and placenta, it has been suggested (Hamlett 1989) that yolk protein precursors of hepatic origin continue to be produced during pregnancy in placental sharks. Initial yolk stores in the egg mass are depleted and used for embryogenesis and subsequent growth. As these stores are diminishing, the maternal liver produces yolk protein precursors that are secreted into the blood, where they are taken up by the new crop of developing oocytes for the next reproductive cycle. The same yolk precursors could then be transported across the uterus and fetal placenta to nourish the fetuses during the latter stages of gestation.

Paraplacental uterus

Ultrastructural characteristics of the paraplacental portion of the uterus at term indicate that it functions as a lubricating structure by elaborating mucus from prominent columnar cells (see Fig. 15-15). Wandering phagocytic cells also populate this region. Light microscopic studies of paraplacental regions of uteri containing midterm fetuses suggest that secretory products other than mucus may be produced during midgestation, namely nutrient histotroph, or "uterine milk."

Umbilical cord and appendiculae

In most carcharhinid species the umbilical cord is a smooth, glistening structure (Budker 1953) that contains the umbilical artery, umbilical vein, extraembryonic coelom, and the ductus vitellointestinalis, all of which persist from the yolk stalk.

The term *appendiculae* was introduced by Alcock (1890) when he described villiform extensions of the umbilical cord in some placental sharks (Fig. 15-17). Genera that possess appendiculae include *Rhizoprionodon, Zygaena, Scoliodon, Hemigaleus, Sphyrna,* and *Paragaleus* (Southwell and Prashad 1919; Setna and Sarangdhar 1948; Budker 1949, 1953; Schlernitzauer and Gilbert 1966).

Considerable morphological diversity in appendiculae exists among and within genera. Southwell and Prashad (1919) described four types of appendiculae and three types of placentas in sharks from the Indian Ocean. They suggested that the presence or absence of appendiculae and the complexity of their branching are associated with the type of placenta. Their hypothesis is that species with a less efficient placenta develop appendiculae to augment the metabolic activities of the placenta. Their descriptions of both appendiculae and the placenta are based on gross observations and a limited amount of histological data, represented as line drawings. Others have presented descriptions of histology of appendiculae in other species (Budker 1949, 1953, 1958; Southwell and Prasad 1919; Setna and Sarangdhar 1948; Schlernitzauer and Gilbert 1966; Teshima et al. 1978).

In early stages of development in the Atlantic sharpnose shark, the yolk stalk is a smooth, cylindrical structure. By the time the embryo is 25 mm TL, modest longitudinal folds form on the surface of the yolk stalk alternating with punctate depressions. These are first indications of appendiculae (Hamlett 1993; Hamlett et al. 1993c). By the time the embryo has grown to 4 cm TL, appendiculae are present as rounded, longitudinal protuberances. Coincident with the development of appendiculae, external branchial filaments make their appearance. By the time the embryo attains 6 cm TL, yolk stalk is adorned with a bushy covering of cylindrical and modestly branched appendiculae. Scanning electron microscopy reveals the surface epithelium to be columnar with a profusion of small blood vessels populating the connective tissue core in close proximity to the surface cells. The surface contour is smooth and modestly undulating. Epithelium

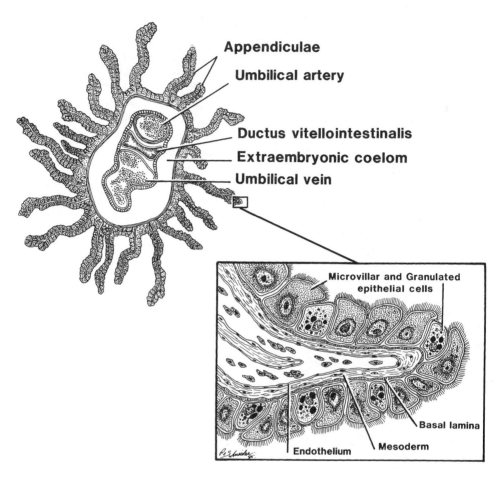

FIGURE 15-17 A transversely sectioned umbilical cord illustrates umbilical artery and vein, ciliated ductus vitellointestinalis, extraembryonic coelom, and appendiculae. An enlargement demonstrates histological features of an appendicula: epithelium with microvillar and granulated cells, basal laminae, mesodermal connective tissue elements, and endothelium. (From Hamlett 1986.)

frequently displays surface bulges and foliose cell apices. Closer examination reveals two distinct cell types: one with elongate microvilli and one with low relief microvilli. Adjacent cells are joined by continuous apical tight junctions. The yolk stalk of 10-cm TL embryos is covered with a dense mat of branched appendiculae. The bases of the appendiculae are rounded but flatten distally as they branch. The yolk stalk of an 18-cm TL embryo shows appendiculae basically characteristic of a term fetus.

In appendiculae of the Atlantic sharpnose shark, epithelial cells are of two types: granulated and microvillar (Hamlett 1983, 1986, 1987, 1993; Hamlett et al. 1993b). Microvillar cells form pyriform processes insinuated between adjacent cells. Their cytoplasm contains lipid and apical micropinocytotic vesicles. Uptake experiments suggest that these cells are capable of limited endocytosis in vitro. Dilated intercellular spaces frequently occur between adjacent cells. These may represent functional transport channels or may be produced as a stress response associated with capture of animals. Granulated cells undergo synthetic and secretory cycles. In any particular section, several cells in different functional stages are observed. Cells seemingly devoid of most cytoplasmic organelles have, in reality, just undergone exocytosis. A large aggregation of mitochondria occupies the base of the cells, having been displaced by secretion granules. The rough endoplasmic reticulum is active and the supranuclear Golgi complex is prominent. As the synthesis of material proceeds, the apical cytoplasm becomes engorged with secretion granules. Following exocytosis, the synthesis-secretion cycle repeats. The chemical nature of the secretion vesicles at present is unknown and the function of the secretory cells also is unresolved. What has emerged is a concept of what functions appendiculae might perform. The microvillar cells are able to absorb exogenous material the size of protein; thus they may serve as a paraplacental nutrient-absorptive fetal membrane. They could absorb material in the periembryonic fluid that is produced by the maternal uterus. Granulated cells may be the source of material that is absorbed by the microvillar cells or their secretion may serve other, as yet unelucidated, functions. Their elaborations may be a lubricant, an antibacterial factor, an immunological component, or any of several other possibilities (Hamlett 1993).

External gill filaments

External gill or branchial filaments have been considered as respiratory structures in early embryos of sharks, skates, and rays. Ranzi (1932, 1934) suspected these transitory structures might also function as sites of nutrient absorption, particularly in viviparous forms. In viviparous sharks, the embryo initially floats free in the uterus and is nourished via yolk stores sequestered during oogenesis. During this time, external gill filaments are present and may function to absorb uterine secretions. Hamlett et al. (1985e) have demonstrated that external gill filaments in 4.5-cm TL embryos of the Atlantic sharpnose shark can produce endocytosis in macromolecular protein tracer HRP, after 10 minutes of in vitro incubation. The yolk sac develops into an epitheliochorial yolk sac placenta after yolk stores are depleted. External gills may thus serve as a nutrient absorptive membrane before the establishment of the yolk sac placenta, as well as performing their respiratory function. It is likely that external gills in the Southern stingray may also perform a nutrient absorptive function by uptake of histotroph prior to the retraction of the yolk sac (Hamlett et al. 1996a). Smith (1980) noted that the quantity and viscosity of histotroph in the cownose ray increased during development. External gills may, therefore, function to absorb uterine milk in fetal rays.

Reproductive endocrinology

An integrated suite of endocrine tissues and the factors they produce control reproduction in female elasmobranchs, in a way similar to that in other vertebrates (Callard et al. 1989). Neural tissues, particularly the ventral lobe of the pituitary, are involved in regulating ovarian function. The ovaries are clearly responsible for producing steroids and possibly peptide hormones. Ovary-derived steroids are crucial in the regulation, of reproductive tract function and possibly in the control of reproductive behavior. These conclusions, however, derive from fragmentary studies on a few species only. Moreover, the exact nature of endocrine regulation, specifically, which factors control specific reproductive events, is poorly known and not well studied. Nonetheless, enough information exists to derive certain patterns that illustrate, if not elucidate, the principal factors in endocrine regulation of elasmobranch reproduction.

Increasing levels of estradiol and testosterone in the circulation correlate with ovarian recrudescence in two oviparous species, *S. canicula* (Sumpter and Dodd 1979) and *R. erinacea* (Koob et al. 1986). In the lesser spotted dogfish, circulating estradiol and testosterone levels rise in synchrony and correlate with the increase in ovary size when follicles differentiate and grow before egg laying commences. They reach their highest levels during active egg-laying season. In the little skate, serum estradiol levels similarly rise coincident with ovarian recrudescence and follicle development. These observations suggest that as follicles grow to maturity, they differentiate for biosynthesis and secretion of both androgens and estrogens and that both testosterone and estradiol are important endocrine regulators in female elasmobranchs.

CYCLES

For only three species is there enough information to adequately define the changes in circulating titers of principal steroids during reproductive cycles. Fortunately, these species utilize three distinct reproductive modes, ovoviviparity (*R. erinacea*), aplacental yolk sac viviparity (*Squalus acanthias*), and placental viviparity (*Sphyrna tiburo*) (Fig. 15-18). In the little skate, *R. erinacea*, changes in plasma levels of estradiol, testosterone, and progesterone were correlated with ovulation, egg encapsulation, egg retention, and oviposition in individual females in captivity (Koob et al. 1986). Estradiol and testosterone predominate during the preovulatory phase of each cycle, with testosterone being the major circulating preovulatory steroid of the two. Testosterone titers peak 5–6 days before ovulation, then fall gradually to low levels at about the time of ovulation. Estradiol levels remain relatively constant throughout the ovulatory cycle, although a small, gradual decrease occurs before ovulation. Progesterone is absent until 1–2 days prior to ovulation, then titers rise dramatically, remain elevated for 1 day only, and subsequently quickly decline to baseline levels at oviposition. This pattern, which may be typical for oviparous species, is illustrated in Figure 15-18.

During early pregnancy in *S. acanthias*, plasma levels of estradiol and testosterone are relatively low (Tsang and Callard 1987a,b). Levels of both steroids increase as pregnancy progresses and reach their maximum plasma titers near term. In contrast, circulating progesterone titers are high in early and midpregnancy, then fall to low levels in late pregnancy. In the bonnethead shark, *S. tiburo*, estradiol and testosterone are elevated during the period between pregnancies when follicles grow to attain their preovulatory size (Manire et al. 1995). After ovulation, circulating levels of both steroids decline rapidly. Estradiol titers reach lowest levels during early pregnancy, then slowly rise at the end of gestation. Testosterone levels remain relatively elevated during most of pregnancy but then drop precipitously near parturition. Progesterone titers increase during the preovulatory phase, peak around the time of ovulation, and remain elevated only during the first part of pregnancy. Thereafter they slowly decline, reaching lowest levels around the time of implantation, then remain low for the remainder of pregnancy. Fluctuations in steroid titers during reproductive cycles in these viviparous sharks are depicted in Figure 15-18 and compared to those in the little skate.

SOURCE

The source of circulating titers of estradiol, testosterone, and progesterone is clearly the ovary, but which compartment in the ovary synthesizes and secretes these steroids is not clear. The correlation between increasing follicle size and rising estradiol and testosterone titers described previously indicates that developing follicles secrete these steroids. The prevalence of estradiol and testosterone in the preovulatory phase, when follicles are most dynamic, also points to a follicular compartment as the principal source. Even in *S. acanthias*, where follicle development occurs through the latter two thirds of pregnancy, high levels of estradiol and testosterone correlate with follicular development (Tsang and Callard 1987a). In vitro experiments on isolated follicular compartments from both oviparous (*S. canicula* and *R. erinacea*) and viviparous species (*S. acanthias*) have clearly delineated the biosynthetic capacity of follicles and corpora lutea. In oviparous species, overall steroidogenic potential increases with advancing size of preovulatory follicles (Dodd et al. 1983; Tsang and Callard 1983). The predominant product of small follicles is estradiol, while that of large follicles is testosterone. In considering the source of circulating titers of these two steroids it must be recalled that in egg-laying females follicles of all sizes are continually present. While their differential

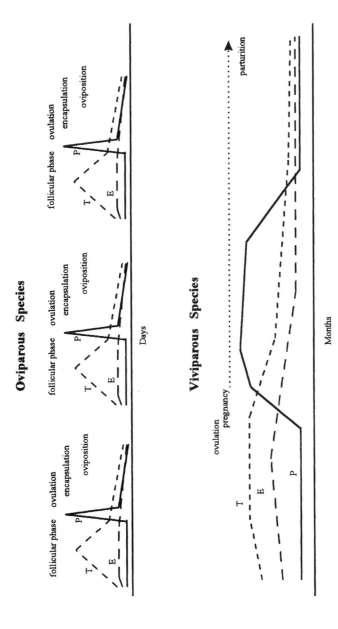

FIGURE 15-18 Schematic representation of circulating plasma levels of estradiol (E), testosterone (T), and progesterone (P) in oviparous and viviparous species during reproductive cycles. Testosterone is the predominant circulating steroid during the preovulatory phase of both reproductive modes. In oviparous species progesterone peaks before ovulation and is elevated for only a brief period before egg encapsulation, while in viviparous species progesterone titers rise after ovulation and remain elevated during the initial phase of pregnancy.

ability to produce estradiol and testosterone may explain the overall output of these steroids, it does not necessarily account for the fluctuations in their titers during the ovulatory cycle. Granulosa cells from *S. acanthias* synthesize both estradiol and testosterone. Thecal tissue is also able to synthesize estradiol, indicating that the whole follicle may coordinate biosynthesis and secretion of the major preovulatory steroids (Tsang and Callard 1983). Since preovulatory follicles grow synchronously in viviparous species, the total follicular output of estradiol and testosterone likely derives from and is determined by these follicles (Fig. 15-19).

In viviparous species the principal source of circulating progesterone appears to be the corpus luteum. Elevated levels of progesterone correspond to the histologically functional life of these structures (Tsang and Callard 1987a,b). Moreover, in vitro studies established that corpora lutea from spiny dogfish synthesize progesterone, and the amount produced in culture correlated with circulating levels (Tsang and Callard 1987a,b). In oviparous species, since progesterone peaks before ovulation and formation of corpora lutea, it seems likely that it originates in corpora lutea from previous ovulatory cycles. Luteal tissue from little skate ovaries produces progesterone in culture, and this amount of basal progesterone synthesis increases with time after corpora lutea formation.

Several interesting aspects of this endocrine system should be emphasized here. First, given the very high levels of testosterone in female elasmobranchs during preovulatory phases of reproductive cycles as described previously, as well as at isolated times in other species (Rasmussen and Murru 1992; Rasmussen and

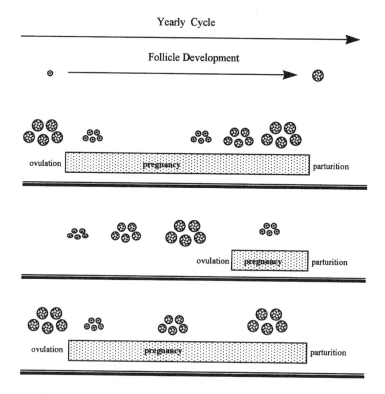

FIGURE 15-19 Timing of follicular development in viviparous species. Follicle development occurs during the latter half of pregnancy in *Squalus acanthias* (*upper panel*); between pregnancies in *Sphyrna tiburo, Rhizoprionodon terraenovae*, and *Dasyatis* spp. (*middle panel*); or throughout pregnancy in *Sphyrna lewini* (*lower panel*).

Gruber 1993), circulating androgens must play an important role within the reproductive system. Unfortunately, nothing is known about the function of androgens in the female; however, one credible suggestion is that testosterone performs a regulatory function related to sperm storage (Manire et al. 1995). Second, fluctuations in circulating progesterone differ between oviparous and viviparous species. In *R. erinacea*, progesterone peaks prior to ovulation and remains elevated for only a short period, dropping to lower levels well before oviposition. In contrast, progesterone levels in the spiny dogfish and bonnethead shark increase only after ovulation and they remain elevated through a good part of pregnancy, only dropping during the latter half. Progesterone may well have different functions in these two reproductive modes. Last, and perhaps most intriguing, is the complexity of this system. In oviparous species, the ovaries contain nonvitellogenic follicles, vitellogenic follicles, preovulatory follicles, and corpora lutea continuously throughout the egg-laying period. Moreover, there is a steady progression through these phases of follicle development and eventual degeneration. How the ovary regulates biosynthesis and secretion by these distinct and dynamic follicular compartments to effect the observed fluctuations in estradiol, testosterone, and progesterone is a mystery. In viviparous species in which follicular development is concomitant with pregnancy, the means for regulating hormone production by the various ovarian compartments is equally enigmatic.

REGULATION

Regulation of ovarian function is in part directed by the hypothalamus and ventral lobe of the pituitary (Sherwood and Lovejoy 1993; Wright and Demski 1993). An authentic gonadotropin-releasing hormone originates in the hypothalamus and travels to the ventral lobe of the pituitary via a systemic blood route. The ventral lobe contains essentially all of the gondodotrophic activity, whose secretion is stimulated by gonadotropin-releasing hormone (Sumpter et al. 1978; Dodd et al. 1983).

Gonadotropins then travel to the ovary to direct steroidogenesis. While this scheme is the product of summarized conclusions from a variety of studies, the exact nature of the interactions between hypothalamus and pituitary, pituitary and ovary, and feedback among all three endocrine organs is unfortunately not known. In vitro studies have shown that ventral lobe extracts stimulate steroid synthesis in both theca and granulosa from *S. acanthias* (Klosterman and Callard 1986). These extracts were also capable of stimulating progesterone synthesis by corpora luteal tissue and cells. Accordingly, it seems clear that gonadotrophic activity that stimulates steroidogenesis is present in the ventral lobe.

FUNCTION

At present, very little is known about the exact reproductive processes regulated by ovary-derived hormones. However, on the basis of correlations between reproductive events and peaks in circulating steroid titers, as well as experimental studies, several general hypotheses about the functions of these steroids seem reasonable. These are schematized in Figure 15-20. Estradiol and testosterone function primarily during the preovulatory period when follicles develop, grow, and accumulate yolk. As mentioned previously, the function of testosterone is entirely unknown. For estradiol, however, several actions have been proposed. It seems likely that estradiol regulates the production of vitellogenin by the liver. Injections of estradiol increase circulating vitellogenin levels in both oviparous species, such as *S. canicula* (Craik 1978c,d) and *R. erinacea* (Perez and Callard 1989), and viviparous ones, such as *S. acanthias* (Ho et al. 1980). Moreover, in the absence of circulating estradiol resulting from ablation, vitellogenin disappears from the plasma but can be replaced if estradiol is administered exogenously (Perez and Callard 1989).

Estradiol is also likely to be involved in regulating reproductive tract function. Increasing estradiol levels correlate with growth of the oviducal gland in *R. erinacea* (Koob et al. 1986), indicating that it may stimulate biosynthesis

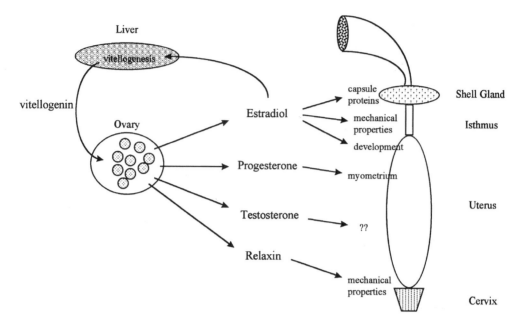

FIGURE 15-20 Known and putative functions of estradiol and progesterone during reproductive cycles. The role of elevated testosterone levels during reproductive cycles in females is not known.

and storage of egg capsule precursors. Long-term treatment of immature *S. canicula* with estradiol resulted in secretion of capsular material, indicating that chronically elevated estradiol levels can regulate oviducal gland function either directly or through associated factors (Dodd and Goddard 1961). Estradiol may also regulate the size and properties of certain regions of the reproductive tract at critical times during ovulatory cycles and pregnancy. Exogenous treatment with estradiol has been shown to cause an increase in extensibility of the isthmus in *S. acanthias* (Koob et al. 1983). Moreover, estradiol potentiates the effects of another ovary-derived hormone, relaxin, on the cervix in the little skate and spiny dogfish (Koob and Callard 1991; Koob et al. 1984). Identification of specific steroid receptors in reproductive tract tissues lends further support for a regulatory role for the steroid whose receptor is identified. An estradiol receptor was identified in the oviduct of *R. erinacea* (Reese and Callard 1987), clearly establishing that estradiol regulates reproductive tract function in this species.

Even less is known about the function of progesterone in elasmobranchs. Since progesterone levels are elevated during the first phase of pregnancy in viviparous species, it must have some role in the maintenance of pregnancy, differentiation of the uterine endometrium, and perhaps the formation of placental structures (Callard et al. 1992). In addition, it may have a role in feedback regulation to the ovary, since follicle development is slowed while circulating progesterone levels are elevated (Callard et al. 1992) In *R. erinacea*, progesterone peaks for 1 day only just prior to ovulation, indicating that it may have a part in egg encapsulation or oviposition (Fig. 15-20). Progesterone treatment of females carrying newly encapsulated eggs caused early oviposition (Koob and Callard 1985). Whether this response was a direct effect of progesterone on the reproductive tract or a result of altering levels of other circulating endocrines was not determined. But the result clearly suggests that progesterone is involved in egg encapsulation and oviposition in this species. The fact that progesterone receptors have also been identified in the skate reproductive tract (Callard et al. 1993) indicates that it plays an important role in oviparous as well as viviparous species.

It should be apparent from the paucity of experimental work on ovary-derived steroids and their functions that much needs to be accomplished before we have a clear understanding of the role these steroids play during reproductive cycles in elasmobranchs. It would be a mistake to assume that they have functions identical to those in other vertebrates, especially given the diversity of reproductive modes within this taxon.

Conclusions

Elasmobranchs evolved reproductive strategies that have allowed survival for over 400 hundred million years. Given the diversity of reproductive strategies, their disparate ecomorphologies, the wide range in ecological niches utilized by contemporary elasmobranchs, and a rich fossil record, it should be possible to devise a hypothetical scheme for the factors instrumental in the evolution of these reproductive modes. However, no such synthesis has been entirely successful to date. Most analyses have focused on morphological and physiological adaptations in the female and developing embryo, with particular emphasis on trophic relationships between mother and offspring as the dominant theme (Wourms 1977, 1981; Wourms et al. 1988). While essential for understanding the nature of transitions that have occurred, trophic relationships alone do not adequately explain selective factors responsible for the evolution of elasmobranch reproductive modes.

Several factors have been posited for the evolution of elasmobranch reproductive modes, particularly viviparity, which has been the principal focus of contemporary analysis. Development within the maternal environment is thought to offer protection from predators and other hazards. Unfortunately, data for predation on pregnant females, offspring, and eggs are critically lacking, putting this hypothesis beyond analysis at present. Development in utero clearly provides physiological regulation of the extraembryonic environment. How important regulating the intrauterine environment might be is not known, since most studies have been concerned primarily with metabolic needs of developing embryos and mechanisms through which females effect their transfer. Viviparity results in increased size of full-term embryos, and large offspring would gain advantages related to predation and feeding. Again, no data on predation and feeding for neonates of either oviparous or viviparous species exist to test this hypothesis.

Compagno (1990), in the most recent attempt, concluded that reproductive modes are not strongly correlated with either ecomorphotypes or phylogenetic relationships among extant elasmobranchs. A cladistic comparison of neoselachian interrelationships with reproductive modes, while confusing, suggested parallel and convergent evolution of viviparity among the major groups. Moreover, reproductive modes apparently arose independently of shifts in the ecomorphotype.

All speculation in the literature about the evolution of elasmobranch reproduction has centered on viviparity per se. There have been no attempts to explain how or why the astonishing diversity in viviparous modes arose. Moreover, there have been no speculations as to why oviparity has been retained in such a large proportion of elasmobranch species. Perhaps we have not yet gained sufficiently detailed information on reproductive physiology of elasmobranchs, their evolutionary history, or their past and present relationship with the marine ecosystem to comprehend fully the origin and diversification of their remarkable reproductive strategies.

REFERENCES

Alcock, A. 1890. Observations on the gestation of some sharks and rays. *J. Asiat. Soc. Bengal* 59(2): 51–56.

Amoroso, E. C. 1960. Viviparity in fishes. *Symp. Zool. Soc. (Lond.)* 1: 153–81.

Babel, J. S. 1967. Reproduction, life history, and ecology of the round stingray, *Urolophus halleri* Cooper. *Calif. Fish Game Bull.* 137: 1–104.

Bauchot, R. 1965. La placentation chez les reptiles. *Ann. Biol.* 4: 547–75.

Beard, J. 1903–4. The germ cells. *J. Anat. Physiol.* 38: 81–102, 205–32, 341–59.

Bearden, C. 1959. A life history study of the eagle ray, *Myliobatis freminvilli* Leseur 1824, in Delaware Bay. M.S. thesis, University of Delaware, Dover.

Borcea, I. 1905. Recherches sur le système urogenital des Elasmobranches. *Arch. Zool. Exp.* 4: 199–484.

Budker, P. 1949. Note preliminaire sur le placenta et le cordon ombilical de trois selaciens vivipares de la côte occidentale d'Afrique. In *C. R. 13th Congress Int. Zool., Paris*, 337–338. Paris: Masson.

———. 1953. Sur le cordon ombilical des squales vivipares. *Bull. Mus. Nat. Hist. Paris* 25: 541–45.

———. 1958. La viviparité chez les selaciens. In *Traité de Zoologie*, ed. P. P. Grassé, vol. 13, pt. 2, 1755–90. Paris: Masson.

Burger, J. W. 1967. Problems in electrolyte economy of the spiny dogfish, *Squalus acanthias*. In *Sharks, Skates, and Rays*, ed. P. W. Gilbert, R. F. Mathewson, and D. P. Rall, 177–85. Baltimore: Johns Hopkins University Press.

Callard, I. P., L. A. Fileti, L. E. Perez, L. A. Sorbera, G. Gianmoukos, L. L. Klosterman, P. Tsang, and J. A. McCracken. 1992. Role of the corpus luteum and progesterone in the evolution of vertebrate viviparity. *Am. Zool.* 32: 264–75.

Callard, I. P., L. L. Klosterman, L. A. Sorbera, L. A. Fileti, and J. C. Reese. 1989. Endocrine regulation in elasmobranchs: archetype for terrestrial vertebrates. *J. Exp. Zool. Suppl.* 2: 12–22.

Callard, I. P., L. A. Fileti, and T. J. Koob. 1993. Ovarian steroid synthesis and the hormonal control of the elasmobranch reproductive tract. *Environ. Biol. Fish.* 38: 175–85.

Capape, C. 1979. La torpille marbrée, *Torpedo marmorata* Risso, 1810 (Pisces, Rajiformes) des côtes tunisiennes: nouvelles données sur l'écologie et la biologie de la reproduction de l'espèce, avec une comparaison entre les populations méditerranéennes et atlantiques. *Ann. Sci. Nat. Zool. Biol. Anim.* 1: 79–97.

———. 1993. New data on the reproductive biology of the thorny stingray, *Dasyatis centroura* (Pisces: Dasyatidae) from off the Tunisian coasts. *Environ. Biol. Fish.* 38: 73–80.

Carrier, J. C., and H. L. Pratt. 1994. Group reproductive behaviors in free-living nurse sharks, *Ginglymostoma cirratum*. *Copeia* 1994: 646–56.

Castro, J. I. 1993. The shark nursery of Bulls Bay, South Carolina, with a review of the shark nurseries of the southeastern coast of the United States. *Environ. Biol. Fish.* 38: 37–48.

Castro, J. I., P. M. Bubucis, and N. A. Overstrom. 1988. The reproductive biology of the chain dogfish, *Scyliorhinus retifer*. *Copeia* 1988: 740–46.

Chen, C. T., T. C. Leu, and S. J. Joung. 1988. Notes on reproduction in the scalloped hammerhead, *Sphyrna lewini*, in northeastern Taiwan waters. *Fish. Bull.* 86: 389–93.

Compagno, L. J. V. 1984. *Sharks of the World*. FAO Species Catalogue, vol. 4, pts. 1 and 2. Rome: United Nations Food and Agriculture Organization.

———. 1990. Alternative life-history styles of cartilaginous fishes in time and space. *Environ. Biol. Fish.* 28: 33–75.

Cox, D. L., P. Walker, and T. J. Koob. 1997. Predation on eggs of the starry ray. *Trans Am. Fish. Soc.* In press.

Craik, J. C. A. 1978a. Plasma levels of vitellogenin in the elasmobranch *Scyliorhinus canicula* L. *Comp. Biochem. Physiol.* 60B: 9–18.

———. 1978b. Kinetic studies of vitellogenin metabolism in the elasmobranch *Scyliorhinus canicula* L. *Comp. Biochem. Physiol.* 61A: 355–61.

———. 1978c. An annual cycle of vitellogenesis in the elasmobranch *Scyliorhinus canicula* L. *J. Mar. Biol. Assoc. U.K.* 58: 719–26.

———. 1978d. The effects of oestrogen treatment on certain plasma constituents associated with vitellogenesis in the elasmobranch *Scyliorhinus canicula* L. *Gen. Comp. Endocrinol.* 35: 455–64.

Cross. J.N. 1988. Aspects of biology of two Scyliorhinid sharks, *Apristurus brunneus* and *Parmaturus xaniurus*, from the upper continental slope off southern California. *Fish. Bull.* 86: 691–702.

Daniel, J. F. 1928. *The Elasmobranch Fishes*. Berkeley: University of California Press.

De Duve, C., and R. Wattiaux. 1966. Functions of lysosomes. *Ann. Rev. Physiol.* 28: 435–92.

Dempster, R. P., and E. S. Herald. 1961. Notes on the horn shark, Heterodontus francisci, with observations on mating activity. *Occas. Pap. Calif. Acad. Sci.* 33: 1–7.

Dodd, J.M. 1972. Ovarian control in cyclostomes and elasmobranchs. *Am. Zool.* 12: 325–39.

———. 1983. Reproduction in cartilaginous fishes (Chondrichthyes). In *Fish Physiology*, vol. 9, pt. A, 31–95. San Diego: Academic Press.

Dodd, J. M., M. H. I. Dodd, and R. T. Duggan. 1983. Control of reproduction in elasmobranch fishes. In *Control Processes in Fish Physiology*, ed. J. C. Rankin, T. J. Pitcher, and R. T. Duggan. London: Croom-Helm.

Dodd, J. M., and C. K. Goddard. 1961. Some effects of oestradiol benzoate on the reproductive ducts of the female dogfish *Scyliorhinus canicula*. *Proc. Zool. Soc. Lond.* 137: 325–31.

DuBuit, M. H. 1976. The ovarian cycle of the cuckoo ray, *Raja naevus* (Müller and Henle) in Celtic Sea. *J. Fish Biol.* 8: 199–207.

Evans, D. H. 1981. The egg case of the oviparous elasmobranch, *Raja erinacea*, does osmoregulate. *J. Exp. Biol.* 92:337–40.

Evans, D. H., A. Oikari, G. A. Kormanik, and L. Mansberger. 1982. Osmoregulation by the prenatal spiny dogfish, *Squalus acanthias. J. Exp. Biol.* 101: 295–305.

Feng, D., and D. P. Knight. 1992. Secretion and stabilization of layers of the egg capsule of the dogfish *Scyliorhinus canicula. Tissue Cell* 24: 773–90.

———. 1994a. The effect of pH on fibrillogenesis of collagen in the egg capsule of dogfish, *Scyliorhinus canicula. Tissue Cell* 26: 649–59.

———. 1994b. Structure and formation of the egg capsule tendrils in the dogfish *Scyliorhinus canicula. Philos. Trans. R. Soc. Lond.* 343B: 285–302.

Files, T., and J. Lombardi. 1993. Free amino acids in the uterine fluids of four species of viviparous sharks (*Squalus acanthias, Carcharhinus plumbeus, Mustelus canis* and *Rhizoprionodon terraenovae*). *Comp. Biochem. Physiol.* 104B: 583–88.

Fitz, E. S., and F. C. Daiber. 1963. An introduction to biology of *Raja eglanteria* Bosc 1802 and *Raja erinacea* Mitchill 1825 as they occur in Delaware Bay. *Bull. Bingh. Oceanogr. Coll.* 18: 69–97.

Foulley, M. M., and J. Mellinger. 1980. La diffusion de l'eau tritiée, de l'urée ^{14}C et d'autres substances à travers la coque de l'oeuf de Roussette, *Scyliorhinus canicula. C. R. Acad. Sci. (Paris)* 290D: 427–30.

Foulley, M. M., F. Wrisez, and J. Mellinger. 1981. Observations sur la perméabilité asymetrique de la coque de l'oeuf Rousette (*Scyliorhinus canicula*). *C. R. Acad. Sci. III* 293: 389.

Gathercole, L. J., E. D. Atkins, T. Goldbeck-Wood, and K. Barnard. 1993. Molecular bending and networks in a basement membrane–like collagen: packing in dogfish egg capsule collagen. *Int. J. Biol. Macromol.* 15: 81–88.

Gilmore, R. G. 1991. The reproductive biology of lamnoid sharks. *Underw. Nat.* 19: 64–67.

Gilmore, R. G., J. W. Dodrill, and P. A. Linley. 1983. Reproduction and embryonic development of the sand tiger shark, *Odontaspis taurus* (Rafinesque). *Fish. Bull.* 81: 201–25.

Graham, C. R., Jr., C. Bond, V. P. Chacko, and J. Lombardi. 1995. NMR studies of glucose and alanine utilization and maternal-embryonic nutrient transfer in the smooth dogfish, *Mustelus canis. Comp. Biochem. Physiol.* 111A: 199–207.

Gruber, S. H., and L. J. V. Compagno. 1981. Taxonomic status and biology of the bigeye thresher, *Alopias superciliosus. Fish. Bull.* 79: 617–40.

Guillette, L. J., S. Spielvogel, and F. L. Moore. 1981. Luteal development, placentation, and plasma progesterone concentration in the viviparous lizard *Sceloporus jarrovi. Gen. Comp. Endocrinol.* 43: 20–29.

Gudger, E. W. 1940. The breeding habits, reproductive organs, and external embryonic development of *Chlamydoselachus*, based on notes and drawings by Bashford Dean. In *The Bashford Dean Memorial Volume, Archaic Fishes*, 525–633. New York: American Museum of Natural History.

Hamlett, W. C. 1983. Maternal-fetal relations in elasmobranch fishes. Ph.D. diss., Clemson University, South Carolina.

———. 1986. Prenatal nutrient absorptive structures in selachians. In *Indo-Pacific Fish Biology*, ed. T. Uyeno, R. Arai, T. Taniuchi, and K. Matsuura, 333–43, Tokyo: Ichthyological Society of Japan.

———. 1987. Comparative morphology of the elasmobranch placental barrier. *Arch. Biol. (Brussels)* 98:135–62.

———. 1989. Evolution and morphogenesis of the placenta in sharks. *J. Exp. Zool. Suppl.* 2: 35–52.

———. 1990. Elasmobranch species as models for studies of placental viviparity and its endocrine regulation. *J. Exp. Zool. Suppl.* 4: 129–31.

———. 1993. Ontogeny of the umbilical cord and placenta in the Atlantic sharpnose shark, *Rhizoprionodon terraenovae. Environ. Biol. Fish.* 38: 253–67.

Hamlett, W. C., and M. K. Hysell. 1998. Uterine specializations in elasmobranchs. *J. Exp. Zool.* 282: 438–59.

Hamlett, W. C., and T. J. Koob. 1996. Microscopic structure of the gravid uterus in *Raja erinacea*. Paper presented at the 12th Annual Meeting of the American Society of Ichthyologists and Herpetologists and the American Elasmobranch Society, June, New Orleans, Louisiana.

Hamlett, W. C., and J. P. Wourms. 1984. Ultrastructure of the preimplantation shark yolk sac placenta. *Tissue Cell* 16:613–25.

Hamlett, W. C., D. J. Allen, M. D. Stribling, F. J. Schwartz, and L. J. A. DiDio. 1985e. Permeability of embryonic shark external gill filaments. Electron microscopic observations using horseradish peroxidase as a macromolecular tracer. *J. Submicrosc. Cytol.* 17:31–40.

Hamlett, W. C., A. M. Eulitt, R. L. Jarrell, and M. A. Kelly. 1993a. Uterogestation and placentation in elasmobranchs. *J. Exp. Zool.* 266: 347–67.

Hamlett, W. C., M. Hysell, M. Jezior, T. Rozycki, N. Brunette, and K. Tumilty. 1999a. Fundamental zonation in elasmobranch oviducal glands. In *Proceedings of the 5th Indo-Pacific Fish Conference, Nouméa*, ed. B. Séret and J.-Y. Sire. Paris: French Ichthyological Society. and ORSTOM. In press.

Hamlett, W. C., M. Jezior, and R. Spieler. 1999b. Ultrastructural analysis of folliculogenesis in the ovary of the yellow spotted stingray, *Urolophus jamaicensis*. *Ann. Anat.* In press.

Hamlett, W. C., D. P. Knight, T. Koob, M. Jezior, T. Luong, T. Rozycki, N. Brunette, and M. Hysell. 1998c. Survey of oviducal gland structures and function in elasmobranchs. *J. Exp. Zool.* 282: 399–420.

Hamlett, W. C., M. A. Miglino, and L. J. A. DiDio. 1993b. Subcellular organization of the placenta in the Atlantic sharpnose shark, *Rhizoprionodon terraenovae*. *J. Submicrosc. Cytol. Pathol.* 25: 535–45.

Hamlett, W. C., M. A. Miglino, D. J. Federman, P. Schafer, and L. J. A. DiDio. 1993c. Fine structure of the term umbilical cord in the Atlantic sharpnose shark, *Rhizoprionodon terraenovae*. *J. Submicrosc. Cytol. Pathol.* 25: 547–57.

Hamlett, W. C., J. A. Musick, A. M. Eulitt, R. L. Jarrell, and M. A. Kelly. 1996a. Ultrastructure of uterine trophonemata, accommodation for uterolactation and gas exchange in the Southern stingray, *Dasyatis americana*. *Can. J. Zool.* 74: 1417–30.

———. 1996b. Ultrastructure of fetal alimentary organs: stomach and spiral intestine in the Southern stingray, *Dasyatis americana*. *Can. J. Zool.* 74: 1431–43.

Hamlett, W. C., F. J., Schwartz, and L. J. A. DiDio. 1987. Subcellular organization of the yolk syncytial-endoderm complex in the preimplantation yolk sac of the shark *Rhizoprionodon terraenovae*. *Cell Tissue Res.* 247:275–85.

Hamlett, W. C., J. P. Wourms, and J. S. Hudson. 1985a. Ultrastructure of the full term shark yolk sac placenta: I. Morphology and cellular transport at the fetal attachment site. *J. Ultrastruct. Res.* 91:192–206.

———. 1985b. Ultrastructure of the full term shark yolk sac placenta: II. The smooth, proximal segment. *J. Ultrastruct. Res.* 91:207–20.

———. 1985c. Ultrastructure of the full term shark yolk sac placenta: III. The maternal attachment site. *J. Ultrastruct. Res.* 91:221–31.

Hamlett, W. C., J. P. Wourms, and J. W. Smith. 1985d. Stingray placental analogues: structure of trophonemata in *Rhinoptera bonasus*. *J. Submicrosc. Cytol.* 17:541–50.

Hepworth, D. G., L. J. Gathercole, D. P. Knight, D. Feng, and J. F. V. Vincent. 1994. Correlation of ultrastructure and tensile properties of a collagenous composite material, the egg capsule of the dogfish, *Scyliorhinus* spp., a sophisticated collagenous material. *J. Struct. Biol.* 112: 231–40.

Hisaw, F. L., and A. Albert. 1947. Observations on the reproduction of the spiny dogfish, *Squalus acanthias*. *Biol. Bull.* 92: 187–99.

Hisaw, F. L., Jr., and F. L. Hisaw. 1959. Corpora lutea of elasmobranch fishes. *Anat. Rec.* 135: 269–77.

Ho, S. M., G. Wulczyn, and I. P. Callard. 1980. Induction of vitellogenin synthesis in the spiny dogfish, *Squalus acanthias*. *Biol. Bull.* 19: 37–38.

Hobson, A. D. 1930. A note on the formation of the egg case of the skate. *J. Mar. Biol. Assoc. U. K.* 16: 577–81.

Hoffman, L. H. 1970. Placentation in the garter snake *Thamnophis sirtalis*. *J. Morphol.* 131: 57–88.

Hoffman, L. H., and G. E. Olson. 1984. Crystalline inclusions in embryonic and maternal cells. In *Ultrastructure of Reproduction*, ed. J. Van Blerkom and P. M. Motta, 235–46. The Hague: Martinus Nijhoff.

Hornsey, D. J. 1978. Permeability coefficients of the egg-case membrane of *Scyliorhinus canicula* L. *Experientia* 34: 1596.

Jensen, N. H. 1976. Reproduction of the bull shark, *Carcharhinus leucas*, in the Lake Nicaraguan–Rio San Juan system. In: *Investigations of the Ichtyofauna of Nicaraguan Lakes*, ed. T. Thornson. Lincoln: University of Nebraska Press.

Jezior, M., and W. C. Hamlett. 1995. Ultrastructural organization of the ovarian follicle in the yellow spotted ray, *Urolophus jamaicensis*. *Notre Dame Sci. Q.* 34(5): 9.

Johnson, R. H., and D. R. Nelson. 1978. Copulation and possible olfaction-mediated formation in two species of carcharhinid sharks. *Copeia* 1978: 539–42.

Jollie, W. P., and L. G. Jollie. 1967a. Electron microscopic observations on the yolk sac of the spiny dogfish, *Squalus acanthias*. *J. Ultrastruct. Res.* 18: 102–26.

———. 1967b. Electron microscopic observations on accommodations to pregnancy in the uterus of the spiny dogfish, *Squalus acanthias*. *J. Ultrastruct. Res.* 20: 161–78.

Joung, S. J., C. T., Chen, E. Clark, S. Uchida, and W. Y. P. Huang. 1996. The whale shark, *Rhincodon typus*, is a livebearer: 300 embryos found in a "megamamma" supreme. *Environ. Biol. Fish.* 46: 219–23.

Kimley, A. P. 1980. Observations of courtship and copulation in the nurse shark, *Ginglymostoma cirratum*. *Copeia* 1980: 878–82.

Klosterman, L. L., and I. P. Callard. 1986. Progesterone production by enzymatically dispersed cells from copora lutea of the spiny dogfish, *Squalus acanthias*. *Biol. Bull.* 26: 119–21.

Knight, D. P., and D. Feng. 1992. Formation of the dogfish egg capsule, a coextruded, multilayer laminate. *Biomimetics* 1: 151–75.

———. 1994a. Interaction of collagen with hydrophobic protein granules in egg capsule of dogfish *Scyliorhinus canicula*. *Tissue Cell* 155–67.

———. 1994b. Some observations on the collagen fibrils of the egg capsule of the dogfish *Scyliorhinus canicula*. *Tissue Cell* 26: 385–401.

Knight, D. P., D. Feng, M. Stewart, and E. King. 1993. Changes in macromolecular organization in collagen assemblies during secretion in the nidamental gland and formation of the egg capsule wall in the dogfish *Scyliorhinus canicula*. *Philos. Trans. R. Soc. Lond.* 341B: 419–36.

Knight, D. P., D. Feng, and M. Stewart. 1996a. Structure and function of the selachian egg case. *Biol. Rev.* 71: 81–111.

Knight, D. P., X. W. Hu, L. J. Gathercole, M. Rusaouën-Innocent, M. W. Ho, and R. Newton. 1996b. Molecular orientations in an extruded collagenous composite, the marginal rib of the egg capsule of the dogfish *Scyliorhinus canicula*; a novel lyotropic liquid crystalline arrangement and its origin in the spinnerets. *Philos. Trans. R. Soc. Lond.* 351B: 1205–22.

Knight, D. P., and S. Hunt. 1974. Fibril structure of collagen in egg capsule of dogfish. *Nature* 249: 379–80.

Koob, T. J. 1992. Tyrosine hydroxylation during egg capsule tanning in little skate, *Raja erinacea*. *Biol. Bull.* 31: 29–31.

Koob, T. J., and I. P. Callard. 1985. Progesterone causes early oviposition in *Raja erinacea*. *Biol. Bull.* 25: 138–39.

———. 1991. Reproduction in female elasmobranchs. In *Comparative Physiology*, vol. 10, ed. R. K. H. Kinne, 155–209. Basel: Karger.

Koob, T. J., and D. L. Cox. 1988. Egg capsule catechol oxidase from little skate, *Raja erinacea* Mitchill. *Biol. Bull.* 175: 202–11.

———. 1990. Introduction and oxidation of catechols during formation of skate (*Raja erinacea* Mitchill 1825) egg capsule. *J. Mar. Biol. Assoc. U.K.* 70: 395–411.

———. 1993. Stabilization and sclerotization of *Raja erinacea* egg capsule proteins. *Environ. Biol. Fish.* 38: 151–57.

Koob, T. J., and W. C. Hamlett. 1998. Microscopic structure of the gravid uterus in the little skate, *Raja erinacea*. *J. Exp. Zool.* 282: 421–37.

Koob, T. J., and J. W. Straus. 1998. On the role of egg jelly in *Raja erinacea*. *Bull. Mt. Desert Island Biol. Lab.* 37: 117–19.

Koob, T. J., J. L. Laffan, B. Elger, and I. P. Callard. 1983. Effects of estradiol on Verschlussvorrichtung of *Squalus acanthias*. *Biol. Bull.* 23: 67–68.

Koob, T. J., J. L. Laffan, and I. P. Callard. 1984. Effects of relaxin and insulin on reproductive tract size and early fetal loss in *Squalus acanthias*. *Biol. Reprod.* 31: 231–38.

Koob, T. J., P. Tsang, and I. P. Callard. 1986. Plasma estradiol, testosterone, and progesterone levels during ovulatory cycle of little skate, *Raja erinacea*. *Biol. Reprod.* 35: 267–75.

Kormanik, G. A. 1988. Time course of the establishment of uterine sea water conditions in late-term pregnant spiny dogfish (*Squalus acanthias*). *J. Exp. Biol.* 137: 443–511.

———. 1992. Ion and osmoregulation in prenatal elasmobranchs: evolutionary implications. *Am. Zool.* 32: 294–302.

———. 1993. Ionic and osmotic environment of developing elasmobranch embryos. *Environ. Biol. Fish.* 38: 233–40.

Kudo, S. 1956. On *Squalus brevirostris*, in Hyuga-Nada. *Rep. Nankai Fish. Res. Lab.* 3: 66–72.

Lombardi, J., and T. Files. 1993. Egg capsule structure and permeability in the viviparous shark, *Mustelus canis*. *J. Exp. Zool.* 267: 76–85.

Lombardi, J., K. B. Jones, C. A. Garrity, and T. Files. 1993. Chemical composition of uterine fluid in four species of viviparous sharks (*Squalus acanthias*, *Carcharhinus plumbeus*, *Mustelus canis*, and *Rhizoprionodon terraenovae*). *Comp. Biochem. Physiol.* 105A: 91–102.

Luer, C. A., and P. W. Gilbert. 1985. Mating behavior, egg deposition, incubation period, and hatching in clearnose skate, *Raja eglanteria*. *Environ. Biol. Fish.* 13: 161–71.

Manire, C. A., L. E. L. Rasmussen, D. L. Hess, and R. E. Hueter. 1995. Serum steroid hormones and reproductive cycle of female bonnethead shark, *Sphyrna tiburo*. *Gen. Comp. Endocrinol.* 97: 366–76.

Martin, L. K., and G. M. Cailliet. 1988. Aspects of reproduction of bat ray, *Myliobatis californica*. *Copeia* 1988: 754–62.

Mellinger, J. 1974. Croissance et reproduction de la torpille (*Torpedo marmorata*): III. L'appareil genital femelle. *Bull. Biol. Fr. Belg.* 108: 107–50.

———. 1983. Egg-case diversity among dogfish, *Scyliorhinus canicula* L.: a study of egg laying rate and nidamental gland secretory activity. *J. Fish. Biol.* 22: 83–90.

Metten, H. 1939. Studies on the reproduction of the dogfish. *Philos. Trans. R. Soc. Lond.* 230B: 217–38.

McLaughlin, R. H., and A. K. O'Gower. 1971. Life history and underwater studies of heterodontid sharks. *Ecol. Monogr.* 41: 271–89.

Myrberg, A. A. Jr., and S. H. Gruber. 1974. The behavior of the bonnet head shark, *Sphyrna tiburo*. *Copeia* 1974: 358–74.

Nalini, K. P. 1940. Structure and function of the nidamental gland of *Chiloscyllium griseum* (Mull. and Henle). *Proc. Indian Acad. Sci.* 12B: 189–214.

Natanson, L. J., and G. M. Cailliet. 1986. Reproduction and development of the Pacific angel shark, *Squatina californica*, off Santa Barbara, California. *Copeia* 1986: 987–94.

Needham, J. 1931. *Chemical Embryology*. Cambridge: Cambridge University Press.

———. 1942. *Biochemistry and Morphogenesis*. Cambridge: Cambridge University Press.

Otake, T. 1990. Classification of reproductive modes in sharks with comments on female reproductive tissues and structures. In *Elasmobranchs as Living Resources: Advances in the Biology, Ecology, Systematics, and the Status of the Fisheries*, ed. H. L. Pratt, S. H. Gruber, and T. Taniuchi, 518. National Oceanic and Atmospheric Administration Technical Report, National Marine Fisheries Service 90.

Otake, T., and K. Mizue. 1985. The fine structure of the placenta of the blue shark, *Prionace glauca*. *Jpn. J. Ichthyol.* 32: 52–59.

Ouang, T. Y. 1931. La glande de l'éclosion chez les Plagiostomes. *Ann. Inst. Oceanogr.* 10: 281–370.

Parsons, G. R. 1981. The reproductive biology of the Atlantic sharpnose shark, *Rhizoprionodon terraenovae* (Richardson). *Fish. Bull.* 81: 61–73.

Perez, L. E., and I. P. Callard. 1989. Evidence for progesterone inhibition of vitellogenesis in the skate. *Am. Zool.* 27: 357A.

Plancke, Y., F. Delplace, J. M. Wieruszeski, E. Maes, and G. Strecker. 1996. Isolation and structures of glycoprotein-derived free oligosaccharides from the unfertilized eggs of *Scyliorhinus caniculus*. *Eur. J. Biochem.* 235: 199–206.

Prasad, R. R. 1945a. The structure, phylogenetic significance, and function of the nidamental glands of some elasmobranchs of the Madras Coast. *Proc. Nat. Inst. Sci. India* 11: 282–302.

———. 1945b. Further observations on the structure and function of the nidamental glands of a few elasmobranchs of the Madras Coast. *Proc. Indian Acad. Sci.* 22B: 368–73.

———. 1948. Observations on the nidamental glands of *Hydrolagus colliei*, *Raja rhina*, and *Platyrhinoidis*. *Copeia* 1948: 54–57.

Pratt, H. L. 1979. Reproduction in the blue shark, *Prionace glauca*. *Fish. Bull.* 77: 445–70.

———. 1993. The storage of spermatozoa in the oviducal glands of western North Atlantic sharks. *Environ. Biol. Fish.* 38: 139–49.

Ranzi, S. 1932. Le basi fisio-morfologiche dello sviluppo embrionale dei Selaci: pt. I. *Pubbl. Sta. Zool. (Naples)* 13:209–90.

———. 1934. Le basi fisio-morfologiche dello sviluppo embrionale dei Selaci: pts. 2, 3. *Pubbl. Sta. Zool. (Naples)* 13: 331–437.

Reese, J. C., and I. P. Callard. 1987. Receptors for estradiol-17B in the oviduct of skate *Raja erinacea*. *Biol. Bull.* 27: 28–29.

Richards, S. W., D. Merriman, and L. H. Calhoun. 1963. Studies on the marine resources of Southern New England: IX. The biology of the little skate, *Raja erinacea*, Mitchill. *Bull. Bingh. Oceanogr. Coll.* 18: 5–65.

Rasmussen, L. E. L., and S. H. Gruber. 1993. Serum concentrations of reproductively related circulating steroid hormones in the free-ranging lemon shark, *Negaprion brevirostris*. *Environ. Biol. Fish.* 38: 167–74.

Rasmussen, L. E. L., and F. L. Murru. 1992. Long-term studies of serum concentrations of reproductively related steroids in individual captive carcharhinids. *Aust. J. Mar. Freshw. Res.* 43: 273–81.

Rusaouën, M. 1976. The dogfish shell gland, a histochemical study. *J. Mar. Biol. Assoc. U.K.* 23: 267–83.

Schlernitzauer, D. A., and P. W. Gilbert. 1966. Placentation and associated aspects of gestation in the bonnethead shark, *Sphyrna tiburo*. *J. Morphol.* 120: 219–32.

Setna, S. B., and P. N. Sarangdhar. 1948. Observations on the development of *Chiloscyllium griseum* M. & H., Pristis cuspidatus Lath., and *Rhyncobatus djiddensis* (Forsk.). *Rec. Indian Mus.* 46: 1–24.

Sherwood, N. M., and D. A. Lovejoy. 1993. Gonadotropin-releasing hormone in cartilaginous fishes: structure, location, and transport. *Environ. Biol. Fish.* 38: 197–208.

Smith, J. W. 1980. The life history of the cownose ray, *Rhinoptera bonasus* (Mitchill 1815), in lower Chesapeake Bay, with notes on management of the species. M.S. thesis, College of William and Mary.

Snelson, F. F., Jr., S. E. Williams-Hooper, and T. H. Schmid. 1988. Reproduction and ecology of the Atlantic stingray, *Dasyatis sabina*, in Florida coastal lagoons. *Copeia* 1988: 729–39

———. 1989. Biology of the bluntnose stingray, *Dasyatis sayi*, in Florida coastal lagoons. *Bull. Mar. Sci.* 45: 15–25.

Southwell, T., and B. Prasad. 1919. Embryological and developmental status of Indian fishes. *Rec. Indian. Mus.* 16: 216–40.

Springer, S. 1960. Natural history of the sandbar shark, *Eulamia milberti*. *Fish. Bull.* 61: 1–38.

Steven, G. A. 1934. Observations on the growth of the claspers and cloaca in *Raia clavata* Linnaeus. *J. Mar. Biol. Assoc. U.K.* 19: 887–99.

Stribling, M. D., W. C. Hamlett, and J. P. Wourms. 1980. Developmental efficiency of oophagy, a method of embryonic nutrition displayed by the sand tiger shark (*Eugomphodus taurus*). *Bull. S. C. Acad. Sci.* 42: 111.

Sumpter, J. P., and J. M. Dodd. 1979. The annual reproductive cycle of the female lesser spotted dogfish, *Scyliorhinus canicula* L., and its endocrine control. *J. Fish. Biol.* 15: 687–95.

Sumpter, J. P., N. Jenkins, and J. M. Dodd. 1978. Gonadotrophic hormones in the pituitary of the dogfish (*Scyliorhinus canicula* L.): distribution and physiological significance. *Gen. Comp. Endocrinol.* 36: 275–85.

Sunye, P. S., and C. M. Vooren. 1997. On cloacal gestation in angel sharks from southern Brazil. *J. Fish Biol.* 50: 86–94.

Tanaka, S., Y. Shiobara, S. Hioki, H. Abe, G. Nishi, K. Yano, and K. Suzuki. 1990. The reproductive biology of the frilled shark, *Chlamydoselachus anguineus*, from Suruga Bay, Japan. *Jpn. J. Ichthyol.* 37: 273–91.

Templeman, W. 1944. The life history of the spiny dogfish (*Squalus acanthias*) and vitamin A values of dogfish liver oil. *Dept. Nat. Resource Newf. Res. Bull.* 15: 1–102.

Teshima, K., M. Ahmad, and K. Mizue. 1978. Studies on sharks: XIV. reproduction in the Telok Anson shark collected from Perak River, Malaysia. *Jpn. J. Icthyol.* 25:181–89.

Threadgold, L. T. 1957. A histochemical study of the shell gland of *Scyliorhinus canicula*. *J. Histochem. Cytochem.* 5: 159–66.

Tricas, T. C., and E. M. Le Feuvre. 1985. Mating of the white tip shark, *Triaenodon odesus*. *Mar. Biol.* 84: 233–37.

Tsang, P., and I. P. Callard. 1983. In vitro steroid production by ovarian granulosa cells of *Squalus acanthias*. *Biol. Bull.* 23: 78–79.

———. 1987a. Morphological and endocrine correlates of the reproductive cycle of the aplacental viviparous dogfish *Squalus acanthias*. *Gen. Comp. Endocrinol.* 66: 182–89.

———. 1987b. Luteal progesterone production and regulation in the viviparous dogfish *Squalus acanthias*. *J. Exp. Biol.* 241: 377–82.

Van Tienhoven, A. 1983. *Reproductive Physiology of Vertebrates*, 2d ed. Ithaca, N.Y.: Cornell University Press.

Walzer, C., and N. Schonenberger. 1979a. Ultrastructure and cytochemistry of the yolk syncytial layer in the alevin of trout (*Salmo fario trutta* L.) after hatching: I. The vitellolysis zone. *Cell Tissue Res.* 196: 59–73.

———. 1979b. Ultrastructure and cytochemistry of the yolk syncytial layer in the alevin of trout (*Salmo fario trutta* L. and *Salmo gairdneri* R.) after hatching: II. The cytoplasmic zone. *Cell Tissue Res.* 196:75–93.

Ward, R. T. 1980. The origin of protein and fatty yolk in *Rana pipiens*: V. Unusual paracrystalline configurations within the yolk precursor complex. *J. Morphol.* 165: 255–60.

Wass, R. C. 1973. Size, growth, and reproduction of the sandbar shark, *Carcharhinus milberti*, in Hawaii. *Pacif. Sci.* 27: 305–18.

Weekes, H. C. 1935. A review of placentation among reptiles with particular regard to the function and evolution of the placenta. *Proc. Zool. Soc. Lond.* 1935: 625–45.

Widakowich, V. 1906. Über Bau und Funktion des Nidamentalorgans von *Scyllium canicula*. *Z. Wiss. Zool.* 80: 1–21.

———. 1907. Über eine Verschlussvorrichtung im Eileiter von *Squalus acanthias*. *Zool. Anz.* 31: 636–43.

Woods, F.A. 1902. Origin and migration of the germ cells in *Acanthias*. *Am. J. Anat.* 1: 307–20.

Wood-Mason, J., and A. Alcock. 1891. On the uterine villiform papillae of *Pteroplatea micrura*, and their relation to the embryo. *Proc. R. Soc. Lond.* 49: 359–67.

Wourms, J. P. 1977. Reproduction and development in Chondrichthyan fishes. *Am. Zool.* 17: 379–410.

———. 1981. Viviparity: maternal-fetal relationships in fishes. *Am. Zool.* 21: 473–515.

Wourms, J. P., B. D. Grove, and J. Lombardi. 1988. The maternal-embryonic relationship in viviparous fishes. In *Fish Physiology*, vol. 11, ed. W. S. Hoar and D. J. Randall, 1–134. San Diego: Academic Press.

Wright, D. E., and L. S. Demski. 1993. Gonadotropin-releasing hormone (GnRH) pathways and reproductive control in elasmobranchs. *Environ. Biol. Fish.* 38: 209–18.

Yano, K. 1993. Reproductive biology of the slender smoothhound, *Gollum attenuatus*, collected from New Zealand waters. *Environ. Biol. Fish.* 38: 59–71.

WILLIAM C. HAMLETT

CHAPTER 16

Male Reproductive System

Because of their key phylogenetic position, elasmobranchs offer a unique opportunity to study the anatomy, physiology, and biochemistry of reproduction in an ancient group. As elasmobranchs are the oldest extant gnathostomes, having branched from the main evolutionary line some 400 million years ago, we can use extant species to elucidate highly conserved and successful reproductive mechanisms and make inferences about evolution.

The cystic nature of the testis and spermatogenesis in elasmobranchs offers technical advantages for the study of germ cell development and regulation. Only a relatively few species have been well studied, namely, *Squalus acanthias*, *Scyliorhinus canicula*, and *Raja erinacea*. Even in these cases our knowledge is often restricted by animal migration, season abundance, and availability of mature animals. Generalizing from such a restricted data base is a problem. Analysis of this group with contemporary biological methods and expansion of number of species used as models are of critical importance.

Gross and microscopic organization

The internal organs of the male include the testes, genital ducts (including the efferent ductules, epididymis, ductus deferens, and seminal vesicle), Leydig gland, and the alkaline gland. Peritoneum covers the genital ducts and the elongate kidneys.

The testes in all elasmobranchs are paired, elongate, dorsoventrally flattened organs that are supported by a mesorchium. The testes are relatively more flattened in skates and rays and more cylindrical in sharks. Testis weight constitutes 1–5% of body weight. The epigonal organ, a lymphomyeloid organ, envelops the testis and generally extends beyond the testis tissue in some species. Gross observation of the testis reveals a pregerminal fold running the length of the testis, which is the site of origin of the spermatogenic sequence. The testis performs the dual functions of germ cell generation (spermatogenesis) and the synthesis and secretion of steroid hormones (steroidogenesis).

During embryogenesis, primordial germ cells from the yolk sac endoderm migrate to the gonadal ridge, where they proliferate by mitosis. Stem cells are retained to give rise to future germ cells and others enter meiosis to transform diploid spermatogonia to spermatids. The maturation of spermatids into motile spermatozoa occurs during the process of spermiogenesis, in which morphological modifications to the spermatid occur, including elongation of the flagellum, condensation of the nucleus, formation of a definitive acrosome, reduction in the volume of the cytoplasm, and formation of the mitochondrial spiral in the midpiece.

Testes in elasmobranchs have been described as consisting of three structural types: radial, diametric, and compound (Fig. 16-1) (Pratt 1988). The Lamniformes have radial testes, where the follicle development radiates from the germinal zone (Fig. 16-1A). Squalomorph, galeomorph, and carcharhinid shark species have diametric testes, where the germinal zone consists of a strip along the distolateral surface and follicle development proceeds diametrically, that is, along the width of the testis toward the efferent ductules located medially (Fig. 16-1B). The skates combine radial and diametric forms into a compound testis. The germinal zone is located on the ventral surface. Lobes develop radially and migrate diametrically (Fig. 16-1C).

Spermatocyst development

In the elasmobranch testis, supporting Sertoli cells mature in concert with synchronously developing isogenetic clones of germ cells. The term *spermatoblast* refers to a single Sertoli cell and its associated germ cells. Many spermatoblasts together form a closed spherical entity bounded by a basement membrane to form the structural and functional unit of the elasmobranch testis, the spermatocyst. Callard (1991) redefined and clarified terminology that previously referred to ampullae follicles and spermatocysts.

Spermatocysts are considered to be the unit of evolution in the chordate testis (Grier 1992). In all elasmobranchs thus far examined, the fundamental details of spermatocyst development and spermatogenesis in general are similar. All germ cells pass through spermatogenesis as a synchronously developing clone of cells. Incomplete cytokinesis allows members of the clone to remain transiently connected via cytoplasmic bridges; hence all germ cells in a given spermatocyst are in the same stage of spermatogenesis. At spermiation the bridges are broken.

Spermatocyst formation begins with the association of a single germ cell and a single Sertoli cell. During spermatocyst development, each germ cell is embraced by extensive cytoplasmic extensions of the Sertoli cell (Holstein 1969; Roosen-Runge 1977). Sertoli cells in vertebrates generally play a vital role in physical support and clustering of germ cells, in controlling the microenvironment for germ cell development, and through formation of the blood-testis barrier (Fawcett 1975; Ritzen et al. 1989). Sertoli cells are joined to one another by extensive lateral junctional complexes located at the base of the cell that effectively isolates adluminal from abluminal compartments.

Good descriptions of spermatogenesis in elasmobranchs have been presented (Mellinger 1965; Stanley 1966; Holstein 1969) and Parsons and Grier (1992) have discussed seasonal changes in shark spermatogenesis.

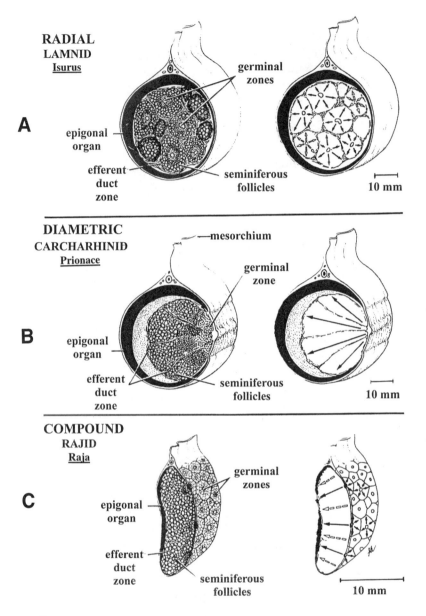

FIGURE 16-1 Testes types in mature male elasmobranchs (cross section, anterior view). *Left side:* Sizes of follicles are exaggerated to illustrate development. *Right side:* Arrows indicate path of development of seminiferous follicles. (From Pratt 1988.)

Spermatocyst development commences with the association of a single germ cell and a single Sertoli cell and terminates at sperm release when the Sertoli cells fragment to release the spermatozoa along with Sertoli cell remnants into the lumen of the efferent ductules. Initially spermatogonia undergo repeated mitotic divisions to yield 16 germ cells per Sertoli cell. The results of subsequent meiosis are 64 spermatozoa per Sertoli cell. Stanley (1966) has estimated that mature spermatocysts contain 500 Sertoli cells in *S. canicula* and 250 in *Torpedo marmorata*, each with 64 spermatids or 32,000 and 16,000 spermatozoa respectively. Parsons and Grier (1992) estimate 460 Sertoli cells per mature spermatocyst in *Sphyrna tiburo*. Since Sertoli cells are shed along with germ cells, both cell types must undergo cycles of proliferation and loss in sexually mature adults.

Spermatocyst genesis in mature adults originates at fixed germinal sites on the lateral aspect of the testis. Displacement of more developmentally advanced spermatocysts by successively less mature stages and to loss by spermiation are the reasons given for the progression of spermatocysts across the testis (Callard 1991).

Mellinger (1965) described 18 different stages in spermatocyst development in *S. canicula*, but Parsons and Grier (1992) condensed their description of the process in *S. tiburo* to seven stages. Both papers may be consulted for an in-depth discussion of the process. For brevity, the scheme of Parsons and Grier is used here. In *Urolophus jamaicensis*, spermatocysts with four or fewer spermatogonia in a peripheral position and Sertoli cells lining the lumen were assigned to stage 1 (Fig. 16-2A). Stage 2 was characterized by the presence of up to eight spermatogonia and the migration of Sertoli cell nuclei to a peripheral position in the spermatocyst just internal to the basement membrane (Fig. 16-2B). Meiotic primary spermatocytes containing large nuclei with prominent chromosomes and the peripheral position of Sertoli cells characterize stage 3 (Fig. 16-2C). Secondary spermatocytes with small, round nuclei containing completely heterochromatic condensed chromosomes are characteristic of stage 4 (Fig. 16-2D). Stage 5 includes spermatids with elliptical nuclei and emerging flagella (Fig. 16-2E). In stage 6, spermatozoa begin to aggregate loosely around the periphery of the spermatocyst (Fig. 16-2F). Prior to spermiation in stage 7, the linearly arrayed spermatozoa (Fig. 16-2G) become even more tightly aggregated (Fig. 16-2H). In *S. tiburo*, spermatocyst diameter increased until approximately stage 4, after which the diameter decreased. Diameter ranged from 20 to a maximum of 400 μm. In *S. canicula*, spermatocyst diameter was maximal at 350 μm and decreased to 240 μm at sperm release (Mellinger 1965; Stanley 1966).

Sperm bundle formation in the testis

Stanley and Lambert (1985) described the role of actin and myosin in sperm bundle formation in the ratfish, *Hydrolagus collei*, and these cytoskeletal proteins are most likely responsible for the similar phenomenon in *R. erinacea*. During spermiogenesis, the acrosome becomes apposed to the Sertoli cell plasma membrane lining the spermatocyst. As spermatids elongate they are gathered into an increasingly compact bundle by cytoskeletal elements in the Sertoli cell oriented parallel to the long axis of the spermatids. Simultaneously, the Sertoli cell membrane is endocytosed between spermatid attachment sites, thereby removing intervening Sertoli cell membrane and thus allowing the spermatids to approach each other to form a tightly packed bundle.

Retzius (1902) was the first to describe the elasmobranch spermatozoan with its spiral head. Stanley subsequently described the ultrastructure of spermiogenesis in *Squalus suckleyi* (1971a,b) and the mature spermatozoan of *H. collei* (1983). Prior to final acrosome formation, a segment of the nuclear envelope near the Golgi complex adheres to the acrosome. After nuclear-acrosomal adhesion, the nucleus rotates 180°, taking the acrosome to the pole opposite the flagellum. A fibrous layer, the fibrous nuclear sheath, spreads over the nuclear surface from the site of acrosomal adherence during nuclear elongation. Intranuclear fibrils assume longitudinal orientation from an initially random arrangement and become helical before final condensation. The axial components of the midpiece arise from two filamentous bundles associated with the centrioles. The nucleus assumes its helical arrangement beginning at the posterior end and progressing anteriorly. The fibrous sheath dissolves coincident with nuclear spiralization. The two fibrous elements of the midpiece fuse. Elongate mitochondria fragment

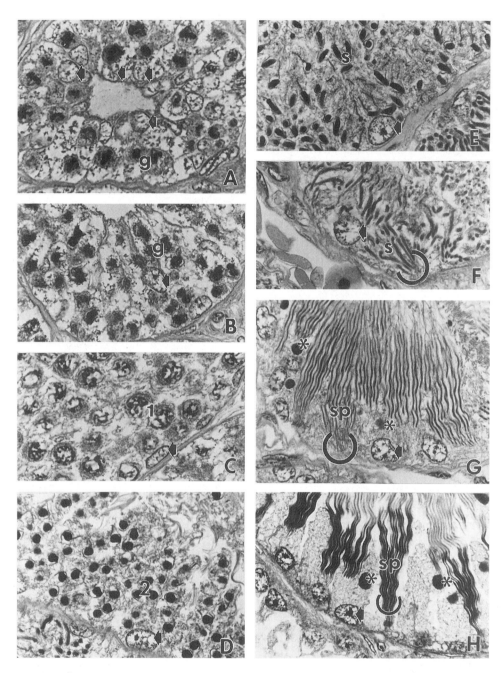

FIGURE 16-2 Spermatogenesis in *Urolophus jamaicensis*. **A:** Germ cells (g) are in active mitosis and Sertoli cells (*arrows*) occur at the lumen. **B:** Sertoli cell nuclei (*arrow*) migrate toward the periphery and germ cells (g) line the lumen. **C:** Nuclei of primary spermatocytes (1) are large and contain prominently condensed chromosomes. **D:** Secondary spermatocytes (2) contain completely heterochromatic chromatin. **E:** Spermatids (s) are characterized by ellipsoidal nuclei. **F:** Spermatids (s) are bundled together while being embedded on the Sertoli cells' apex (*arrow*). **G:** By late in spermatogenesis, spermatozoa (sp) associate in linear arrays in the Sertoli cells. **H:** In the terminal stages of spermatogenesis, spermatozoa (sp) form tight bundles. Problematic bodies (*asterisks*) also appear in the Sertoli cell cytoplasm. (All ×600, reproduced at 73%.)

into smaller spheres that then align along the midaxial piece. The morphology of the mature spermatozoan correlates with its spiral motility pattern. Sperm rotate along their long axis with little lateral tail bending and rapid rotation of the head to result in essentially straight line movement.

Gross and microscopic structure of the male duct system

There is considerable variation regarding terminology of genital ducts (cf. Borcea 1906; Matthews 1950; Botte et al. 1963; Stanley 1963). From the testis, spermatozoa and Sertoli cell fragments move through the efferent ductules located in the anterior end of the mesorchium, the mesentery that attaches the testis to the body wall. The efferent ductules are single in skates and number four in *S. acanthias*.

Most authors refer to the small bore, highly coiled initial segment of the extratesticular ducts as the *epididymis*. This segment is continuous with the broader in diameter, sinuous ductus deferens, also called the *vas deferens* or *Wolffian duct*. The ductus deferens is continuous as the very large diameter seminal vesicle, also referred to as the *ampulla of the ductus deferens*. Jones and Jones (1982), Jones et al. (1984), and Jones and Lin (1993) refer to the derivative of the mesonephric duct as the *ductus epididymis* rather than the *ductus deferens*, the term used by most authors.

I have chosen to use the most widely held terminology, namely *epididymis*, *ductus deferens*, and *seminal vesicle*. In *R. erinacea*, the epididymis may be further subdivided into (1) a head that projects craniad from the entrance of the efferent ductules, (2) a yellow pigmented body, and (3) an unpigmented tail (Fig. 16-3) (Hamlett et al. 1999). Both the epididymis and ductus deferens receive a viscous fluid produced by the adjacent Leydig gland. In sexually immature specimens the ductus is a thin, straight tube, but in sexually mature

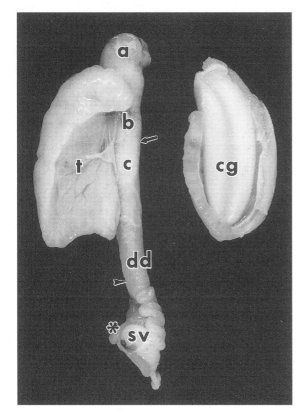

FIGURE 16-3 Testis, genital ducts, and clasper gland of mature *Raja erinacea*, showing the contracted empty alkaline gland (*asterisk*) and clasper gland (cg), with the Leydig gland (*arrow*), and kidney below (*arrowhead*). a = Head of epididymis; b = body of epididymis; c = tail of epididymis; cg = clasper gland; dd = ductus deferens; sv = seminal vesicle; t = testis.

males it is coiled and covers most of the kidney. The two broad seminal vesicles receive no ducts from Leydig gland. Each seminal vesicle has a lateral dilation termed the *sperm sac*. This is a misnomer because these structures do not store sperm but are actually Marshall's alkaline gland. The seminal vesicles unite to form the single urogenital sinus that ends as the urogenital papilla emptying into the cloaca. The cloaca is common to both sexes and is located between the pelvic fins; it receives material from the digestive, urinary, and reproductive tracts.

LUMINAL CONTENTS: SERTOLI CELL BODIES, SERTOLI CELL CYTOPLASTS, AND LEYDIG GLAND BODIES

Three main types of particulate material have been identified as occupying the male genital ducts in *Heterodontus portusjacksoni*, namely Sertoli cell bodies, Sertoli cell cytoplasts, and Leydig gland bodies. Sertoli cell bodies originate in the supranuclear region of Sertoli cells just prior to spermiation (see Figs. 16-2G,H). They have previously been termed *problematic bodies* (Holstein 1969; Collenot and Damas 1980) and *Sertoli bodies* (Simpson and Wardle 1967). They are large, oval membrane bound, eosinophilic bodies. In *Squalus*, they have been recovered from the epididymis, and cytochemical techniques reveal that their proteins are rich in lysine, cysteine, and tryptophan (Pudney and Callard 1986; Collenot and Damas 1975). Sertoli cell cytoplasts (Pudney and Callard 1986) have also been referred to as "small eosinophilic particles" (Jones and Jones 1982) and "cytoplasmic bodies" (Jones et al. 1984). Sertoli cell cytoplasts are remnants of Sertoli cells exclusive of the Sertoli cell bodies (Callard et al. 1989). They are membrane bound structures containing mitochondria, ribosomes, lipids, and endoplasmic reticulum. Since Sertoli cells may be a source of testicular steroids in elasmobranchs (Pudney and Callard 1986; Callard et al. 1989) and Sertoli cell cytoplasts contain organelles appropriate for steroid synthesis, it is possible they are responsible for the steroidogenic activity of shark semen (Simpson et al. 1963, 1964a,b). This may provide the mechanism for local control of extragonadal sperm duct activity. Leydig gland bodies are large, eosinophilic, non-membrane-bound secretions of Leydig glands.

EFFERENT DUCTULES

In *H. portusjacksoni*, the efferent ductules are lined by ciliated columnar epithelium and numerous intraepithelial leukocytes. These cells may be removing particulate cellular debris from the lumen (Jones and Lin 1993). Micropuncture studies also indicate that there is little or no net fluid resorption in the efferent ductules (Jones et al. 1984).

EPIDIDYMIS

In *R. erinacea*, the head of the ductus epididymis is narrow, convoluted (see Fig. 16-3) and unpigmented. It has a simple ciliated columnar epithelium, and individual spermatozoa are randomly scattered in the lumen (Fig. 16-4A). Present in the lumen is material that may represent Sertoli cell bodies and Sertoli cell cytoplasts. The body of the epididymis has a simple columnar epithelium (Fig. 16-4B). Some of the cells are ciliated and others are characterized by apical blebbing or apocrine secretion. A similar process has been confirmed in the epididymis of *U. jamaicensis* by transmission electron microscopy (Hamlett, unpublished). Luminal contents may include epididymal cell components, Sertoli cell bodies, Sertoli cell cytoplasts, and Leydig gland bodies. The biochemical composition, ultrastructure, and the contribution of these components is currently being investigated (Koob and Hamlett, unpublished). The role these components may play in sperm aggregation or maintenance is unknown. Sperm are still present as individuals in a nonaggregated state.

In *Heterodontus*, ultrastructural characteristics of epididymal cells suggest that their main function is protein secretion; however, micropuncture studies did not demonstrate an increase in luminal protein concentration. It has been suggested that protein associates with individual spermatozoa in the lumen im-

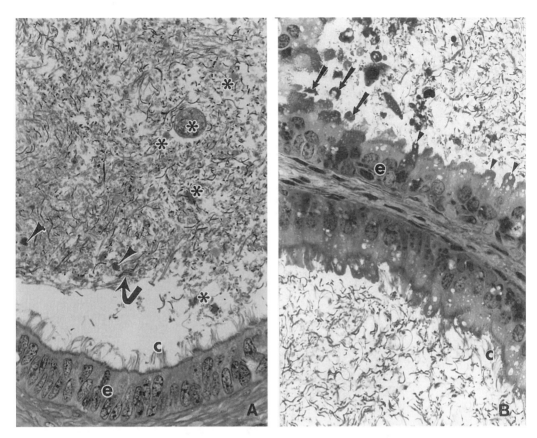

FIGURE 16-4 *Raja erinacea.* **A:** Head of epididymis. The epithelium (e) is simple columnar with cilia (c). For the most part, sperm occur as individuals, but occasionally several are found together in array (*curved arrow*). Eosinophilic spheres that are interpreted as Sertoli cell bodies (*arrowheads*) can be seen. Other material with heterogeneous profiles (*asterisks*) occurs in the lumen and may represent Sertoli cell cytoplasts or Leydig gland bodies. **B:** Epithelial cells (e) show apical blebs (*arrowheads*) that are detached as cytoplasmic fragments (*arrows*). (All ×600, reproduced at 73%.)

mediately upon secretion (Jones et al. 1984). It is possible that the sperm associated protein may play a role in aggregation of sperm in the tail of the epididymis and into the ductus deferens and seminal vesicle. Net fluid transport was not detected in *Heterodontus*, yet a net resorption of sodium occurred in this region (Jones et al. 1984).

DUCTUS DEFERENS

In *R. erinacea*, the epithelium of the ductus deferens is uniformly simple ciliated columnar (Fig. 16-5A). Sperm are present as individuals and associate laterally joined at their heads to form bundles (Fig. 16-5B). Matrical material makes its appearance and provides the nidus for sperm clumps to assemble into spermatozeugmata. The sources of the matrical material are as yet undetermined but may include residual Sertoli cell bodies, Sertoli cell cytoplasts, Leydig gland bodies (Fig. 16-5C), and as yet unidentified components.

The pigmented terminal segment of the ductus deferens is wider than the initial segment and contains columnar secretory epithelium with cilia. In *H. portusjacksoni*, the ciliated cells seem to be involved in heterophagic digestion, since the apical cytoplasm contains numerous vacuoles and dense bodies (Jones and Lin 1993). Sertoli cell bodies

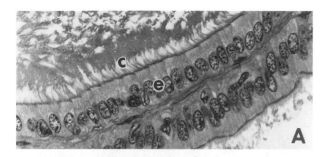
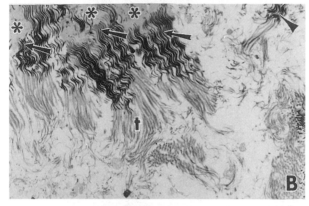
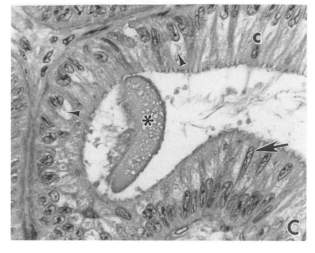

FIGURE 16-5 *Raja erinacea.* **A:** Ductus deferens has a nonsecretory simple columnar epithelium (e) with cilia (c). **B:** Some sperm aggregate with heads aligned in lateral register (*arrowhead*) and tails loosely aligned (t) without being associated with matrix, while others previously aligned (*arrows*) associate with foci of homogeneous matrix components (*asterisks*) in the ductus lumen. **C:** The Leydig gland has simple columnar epithelium with two types of cells: (1) pyriform ciliated cells (c) that are truncated basally while the nucleus (*arrow*) is near the cell apex; predominating are secretory cells (2), which are characterized by a light supranuclear zone (*arrowheads*) corresponding to the location of the Golgi complex, which is involved in the synthesis and secretion of homogeneous material (*asterisk*) that accumulates in the gland lumen. (All ×600, reproduced at 73%.)

are not present, suggesting they either disintegrate or are resorbed in the epididymis. Micropuncture studies reveal that 60% of testicular fluid is resorbed in the ductus deferens and there appears to be no net resorption of sodium from the lumen (Jones et al. 1984).

LEYDIG GLANDS

Leydig glands are a series of branched tubular glands that secrete into the epididymis and ductus deferens (see Fig. 16-5C). The Leydig gland is the modified anterior section of the mesonephros that produces and transmits a milky secretion that congeals into refractile bodies following fixation in formalin (see Fig. 16-5C). In immature males, urine is formed by these tubules. In *R. erinacea*, the epithelium is simple columnar. Some of the cells are ciliated while others are secretory and produce eosinophilic secretions (see

Fig. 16-5C) that aggregate into irregular masses in the lumen. Jones and Lin (1993) conclude that Leydig gland secretions are the main source of the increase in protein concentration of the luminal fluid in the ductus deferens in *Heterodontus*.

SEMINAL VESICLE

In *R. erinacea*, the seminal vesicle is characterized by low, simple columnar to cuboidal epithelium with cilia (Fig. 16-6A). Grossly discernable thin-walled partitions occur in the seminal vesicle. Some end blindly in the lumen while others connect to the vesicle wall to produce blind-ending bays. Spermatozoa continue to form definitive spherical bundles (Fig. 16-6B). Masses of matrical material forms in the lumen (Fig. 16-6B) and associates with individual and previously bundled sperm to produce spermatozeugmata (Fig. 16-6C). Spermatozeugmata enlarge as matrical masses and sperm clumps continue to associate.

The source of the matrical material is likely from Leydig gland secretions from the epididymis and ductus deferens, but other contributions cannot be ruled out such as from seminal vesicle epithelium, transudate from subjacent vascular beds, or other as yet undetermined sources. Changes in luminal ionic composition may also occur.

The ureter becomes entwined with the terminal portion of the seminal vesicle and the two terminate in the anterior wall of the urogenital sinus. The urogenital sinus vents into the common cloaca by means of a single large papilla.

Based on the micropuncture studies of Jones et al. (1984), it is concluded that the luminal fluids of efferent ducts and the epididymis, when compared to blood plasma, showed higher levels of sodium, potassium, and protein and had a greater osmolarity. Spermatocrit samples from the epididymis and ductus deferens reveal that virtually all fluid is resorbed. When testicular (9%) samples were compared with epididymal samples (96%), the percentage of sperm motility increased, despite any ultrastructural change. These data suggest that the epididymis and Leydig gland secretions play important roles in ion and water transport, protein secretion, and maturation of spermatozoa. The nature of the sperm matrices from various segments of the male genital tract and their relation to sperm disaggregation and reaggregation is currently being analyzed (Koob and Hamlett, unpublished). In *Heterodontus*, only the seminal vesicle has a muscular tunic; all other segments of the duct system convey sperm via ciliary activity (Jones and Lin 1993).

Alkaline gland

Marshall discovered that the so-called *sperm sacs* or *urinary bladders* in various species of *Raja* were actually specialized structures that contained a fluid of high electrolyte content and alkalinity. Smith (1929) subsequently analyzed the chemical composition of the fluid and Maren et al. (1963) further characterized the secretions and termed the gland by its primary chemical attribute; giving the current name, the *alkaline* (Marshall's) *gland* of the skate. The paired glands are situated at the base of the kidney and the seminal vesicles (see Fig. 16-3). The glands are almost always distended and filled with a clear fluid. The luminal epithelium is simple columnar and thrown into prominent folds (Fig. 16-7) with subjacent vascularization. The epithelium maintains a 100-fold concentration gradient of hydroxyl ions and a 50-fold gradient of carbon dioxide from plasma to gland lumen and maintains a pH of 9.2. It appears to secrete hydroxyl ions buffered by carbon dioxide (Maren et al. 1963; Smith 1985). It has been suggested that the gland secretions neutralize the acid urine, pH 5.8, and may be involved in sperm protection (Smith 1929) or the formation of copulatory plugs of skates (Callard 1988).

Claspers

The male copulatory appendages of elasmobranchs, *myxopterygia* or *claspers*, are paired, grooved extensions of the posterior bases of the pelvic fins (Fig. 16-8). General accounts of

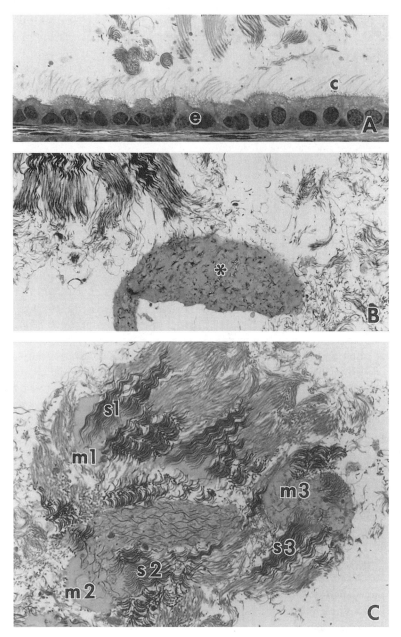

FIGURE 16-6 *Raja erinacea.* **A:** The seminal vesicle has simple low columnar to cuboidal epithelium (e) with cilia (c). **B:** Masses of homogeneous matrical material accumulate in the lumen (*asterisk*). **C:** It appears that individual foci of laterally bound sperm (s1, s2, s3) previously associated with matrical masses (m1, m2, m3) combine to form more complex spermatozeugmata. (All ×600, reproduced at 73%.)

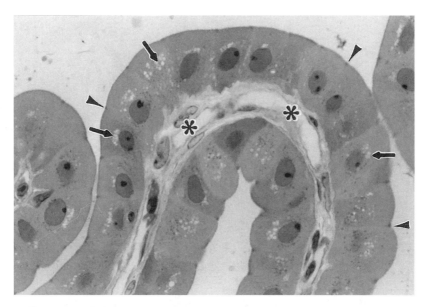

FIGURE 16-7 The alkaline gland of *Raja erinacea* has simple columnar epithelium, the cells of which are joined by terminal bars (*arrowheads*), have a distinct assemblage of supranuclear secretory vesicles (*arrows*), and rest atop a connective tissue layer containing blood vessels (*asterisks*). (×600.)

clasper structure have appeared (Daniel 1928; Breder and Rosen 1966), and Gilbert and Heath (1972) considered their function in *Mustelus canis* and *S. acanthias*. Leigh-Sharpe described the external morphology of 87 species (1920, 1921, 1922, 1924, 1926) and Compagno (1988) published a comprehensive study of claspers in the carcharhinids; a representative sample may be seen in Figure 16-8. Although varying in detail (cf. Compagno 1988), each clasper has a dorsal, longitudinal groove through which semen passes to the female. Details of the skeletal structure of claspers can be found in chapter 3 of this volume.

SIPHON SAC IN SHARKS

In sharks the clasper groove communicates anteriorly with a muscular sac, the siphon sac, or siphon, which is situated ventrally under the skin (Gilbert and Heath 1955). Gilbert and Heath (1972) presented the definitive explanation of the structure and function of the clasper siphon sac in *S. acanthias* and *M. canis* and can be consulted for details. In summary, each siphon sac opens through the apopyle to the clasper groove. In *S. acanthias* the sacs are small, 12% of body length, whereas in *M. canis* they measure 30% of body length. In mature animals, the sacs contain a quantity of sticky fluid that is secreted by goblet cells of the surface epithelium. The secretion has a pH of 5.8. Mann (1960) demonstrated that siphon sac secretion of mature *S. acanthias* contains a high concentration of serotonin (5-hydroxytryptamine), 437.5 μg/mL or 6.5% of the dry material. In sexually immature males, serotonin was absent or present only in trace amounts in semen removed directly from the seminal vesicles. The siphons in these immature specimens contained some 200 times less serotonin (0.017–0.048%) than found in sexually mature males. In other vertebrates, serotonin is a powerful stimulator of smooth muscle contraction and is implicated in various vital functions including regulation of blood pressure, pulmonary and renal circulation, and cardiac output. Serotonin stimulates isolated rat uterus and when administered to dogs intravenously, elicits strong uterine contractions followed by inhibition (Mann 1960). Serotonin may play a

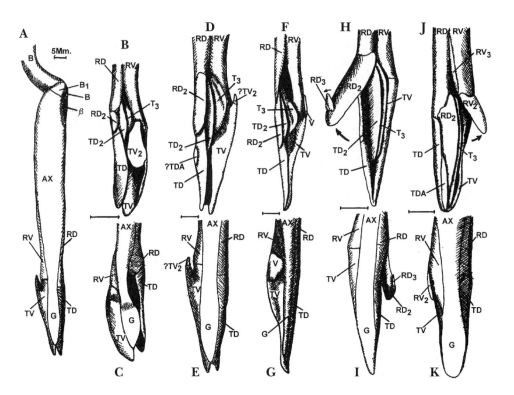

FIGURE 16-8 Clasper morphology from carcharhinids. **A, D, E:** *Trakis scyllium*; **B, C:** *Proscyllium habereri*; **F, G:** *Triakis acutipinna*, **H, I:** *Rhizoprionodon acutus*; **J, K:** *Sphyrna tiburo*. AX = axial cartilage; B = basipterygium; B_1 = intermediate segment; β = beta cartilage; G = end-style, RD = dorsal marginal cartilage; RD_2 = accessory dorsal marginal cartilage; RD_3 = accessory dorsal marginal cartilage 2; RV = ventral marginal cartilage; RV_2 = accessory ventral marginal cartilage; RV_3 = accessory ventral marginal cartilage 2; T_3 = accessory terminal cartilage; TD_2 = dorsal terminal 2 cartilage; TDA = accessory dorsal terminal cartilage; TV = ventral terminal cartilage; TV_2 = ventral terminal 2 cartilage. (From Compagno 1988.)

role in copulation and ejaculation in male elasmobranchs. Mann and Prosser (1963) demonstrated that siphon sac secretion of 5-hydroxytryptamine in *Squalus* uterus in vitro caused uterine contractions. They suggested that by stimulating uterine contractions during copulation, 5-hydroxytryptamine influenced sperm transport and fertilization.

Prior to copulation seawater is pumped into the sacs by repeated flexion of the claspers. An alternative way for water to enter the siphon is when the clasper is held in a fixed position as the shark swims. During copulation, contraction of the compressor muscle ejects seawater from the siphon, which serves to wash sperm from the urogenital papilla into the female. Mating is brief in sharks, on the order of 15–20 minutes.

CLASPER GLAND IN BATOIDS

Batoids do not possess siphons but do have paired, ventrally situated, solid, subdermal clasper glands that project into a clasper sac near the pelvic fins (Figs. 16-3 and 16-9). Garman (1912) diagrammed the clasper skeleton and Leigh-Sharpe (1926) described the external anatomy of the clasper and gland. LaMarca (1964) presented a comprehensive study of the clasper and clasper gland in *U. jamaicensis* and should be consulted for full details. Briefly summarized, the clasper gland of *Urolophus* is a solid structure that is ensheathed with striated muscle. The gland proper consists of a series of tubules lined by columnar epithelium that drain into a series of surface papillae. Upon contraction of the striated muscle of the gland, secretions are

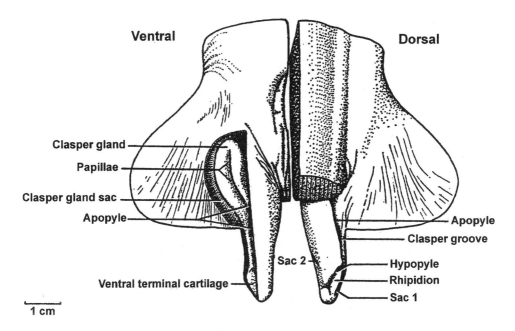

FIGURE 16-9 Clasper and clasper gland anatomy in *Urolophus jamaicensis*. (From LaMarca 1964.)

forced from the papillae into the surrounding clasper gland sac. Contraction of the clasper gland sac forces secretion down the clasper groove, where it is mixed with spermatozoa from the cloaca. Fresh clasper gland secretion is rich in muco- and glycoprotein and phospholipid and has a slightly acid pH. It is concluded that the gland secretion in *Urolophus* seals the clasper groove into a closed tube. This is supported by the fact that clasper gland secretions coagulate on contact with seawater. This protects the semen from dilution in seawater and prevents escape before reaching the hypopyle (LaMarca 1964). Other functions of the clasper gland secretions include provision of a transport medium for the sperm and lubrication to facilitate clasper insertion. Mating in batoids is prolonged compared to sharks. Copulation in *R. eglanteria* may last from 1–4 hours (Luer and Gilbert 1985).

COPULATION

Mating behavior in elasmobranchs has not frequently been observed (Gudger 1912; Matthews 1950; Dempster and Herald 1961; Myerberg and Gruber 1974; Johnson and Nelson 1978; Klimley 1980; Luer and Gilbert 1985; Tricas and Le Feuvre 1985; Uchida et al. 1990; Rouse 1992). Gordon described precopulatory behavior in captive sand tiger sharks (1993) and recently Nordell (1994) has reported mating in *Urolophus halleri*. Carrier et al. (1994) have presented a comprehensive account of mating behaviors in nurse sharks, *Ginglymostoma cirratum*. Their paper can be consulted for full details, but briefly, in observations of 50 mating events in nurse sharks, 8% included copulation. Based on field observations, the following mating stages have been established for the nurse shark. Stage 1 is precoupling, in which a male approaches a stationary female or one that is swimming. If the female is swimming, the male exhibits a "following" behavior such that he assumes a position alongside and slightly behind the female with his head close to her pectoral fin. Stage 2 is coupling, in which the male grasps the pectoral fin of the female. Positioning and alignment characterize stage 3. The male rolls the female to invert her to a ventral-side-up position. Stage 4 involves insertion of the clasper and copulation. If the right fin is grasped, the right clasper is inserted. Likewise, if the left fin is grasped, the left clasper is inserted. During copulation, the male thrusts and undulates his body. Postcopulation involves withdrawal of the clasper and release of the pectoral fin. The pairs either separates immediately or displays quiescent activity in a dorsal-side-up position.

Clasper innervation and central nervous system control has been studied in *Urolophus halleri* (Liu and Demski 1993). Several large myelinated nerves innervate the clasper muscles and skin. Electrical stimulation at low levels elicits various clasper movements, including elevation, medial and lateral extension, rotation, and opening. Stimulation of nerve 60 caused the clasper to move laterally and stimulation of nerves 61 and 62 produced elevation, rotation, and opening of the clasper. Retrograde labeling with either horseradish peroxidase or cobalt-lysine confirmed that neurons were located in a well-developed motor horn in a discrete segment of the spinal cord. Cobalt-labeled sensory cells were ipsilateral in the dorsal root ganglia of nerves 61–63.

Extensive pathways have been described in elasmobranchs, including stingrays, that are related to immunoreactive (ir) gonadotropin-releasing hormone (GnRH) (Demski 1989; Wright and Demski 1991, 1993). Beaded GnRH-ir fibers occur throughout the spinal cord and in areas where clasper neurons are located. These fibers are thought to arise from a prominent GnRH-ir nucleus in the midbrain (Wright and Demski 1991, 1993). It is reasonable to think that GnRH may be involved in the control of clasper movements.

Reproductive endocrinology

The morphological zonation of the elasmobranch testis has made it possible to isolate individual spermatocysts on the basis of size and differentiation via transillumination (Sourdaine and Jegou 1988). All spermatocysts within a single ampulla have germ cells that are at the same relative stage of development. Different zones of differentiation can be discerned moving from the peripheral wall of the testis to the medial hilus, namely, (1) spermatogonia in several layers and Sertoli cells, (2) primary spermatocytes, (3) secondary spermatocytes, (4) spermatids, and (5) spermatozoa in association with Sertoli cells and Sertoli cell fragments at spermiation (Callard et al. 1989; Fasano et al. 1989a,b). Chieffi and Lupo di Prisco (1961) demonstrated the presence of steroids in elasmobranch testis; however, the actual site of synthesis of steroids remains controversial. Interstitial cells with features of steroid-secreting cells have been reported for several species (Fasano et al. 1989a,b). Alternatively, Sertoli cells have been proposed as the source of steroids in *S. acanthias* (Callard et al. 1989) and *S. canicula* (Sourdaine et al. 1990). Callard et al. (1989) demonstrated that enzymes of the steroidogenic pathway from cholesterol to androgens are present in cultured spermatocysts. Sourdaine et al. (1990) concluded that germ cells probably influence steroidogenesis and responsiveness of Sertoli cells.

In elasmobranchs, typical Leydig cells have been reported to be absent and Sertoli cells are said to the primary steroidogenic source (Callard 1991). Early reports based on light microscopy provided various statements as to the presence of absence of Leydig cells in elasmobranchs. Some reported that they are absent in *S. stellaris*, *S. canicula* (Stephan 1902), and *S. acanthias* (Simpson and Wardle 1967); rare in *Cetorhinus maximus* (Matthews 1950) and *S. canicula* (Collenot 1969); or present in *S. stellaris* and *S. canicula* (Chieffi et al. 1961; Della Corte et al. 1961). Many of these studies relied on the histochemical reaction for $\Delta 5$-3β-hydroxysteroid dehydrogenase activity, which in mammals is in Leydig cells but in elasmobranchs is in Sertoli cells (Simpson and Wardle 1967; Collenot and Damas 1980). Using electron microscopy Pudney and Callard (1984) described stromal cells in *S. acanthias* that have cytological characteristics associated with steroidogenesis, including agranular endoplasmic reticulum, mitochondria with tubular cristae, and lipid droplets. The organelles were sparse, the cells infrequently encountered, and they resembled undifferentiated mammalian Leydig cells rather than definitive fully functioning ones. Partially because these cells showed no fine structural change during spermatogenesis but Sertoli cells did, Pudney and Callard concluded that the Leydig-like cells are unlikely to contribute significantly to the total steroid output of the elasmobranch testis. This is not in general agreement with Fasano et al. (1989a,b); hence the question of definitive Leydig cells remains unresolved.

Using electron microscopy, Holstein (1969) described agranular endoplasmic reticulum and lipid droplets in Sertoli cells of *S. acanthias* and concluded the cells had an endocrine function. This observation was subsequently confirmed by Pudney and Callard (1984). Simpson and Wardle (1967) reported $\Delta 5$-3β-hydroxysteroid dehydrogenase activity in Sertoli cell cytoplasm, semen, and the lumens of mature spermatocysts in *S. acanthias*. The seminal fluid of the same species contains remnants of Sertoli cells (Pudney and Callard 1986), steroidogenic activity (Simpson et al. 1964a,b), and endogenous steroids (Simpson et al. 1963). Most recently, evidence of steroid production by Sertoli cells was demonstrated by culturing Sertoli cell monolayers or intact spermatocysts in the presence of radiolabeled precursors (Dubois et al. 1989). The complete array of steroidogenic activity as seen in whole testicular microsomes is present in the cultures.

Analysis of organic extracts of testes of *S. stellaris* revealed the presence of progesterone, androstenedione, testosterone, and estradiol (Chieffi et al. 1961). The same study showed that animals caught during May at the onset of spermatogenic activity had greater 17 α-hydroxylase and $C17,20$-lyase activity than those collected in September, prior to a period of spermatogenic inactivity. The apical portion of Sertoli cells shed at spermiation may be the source of steroids and steroidogenic activity seen in the semen of *S. acanthias* (Pudney and Callard 1984) and other sharks (Jones et al. 1984). Little information is available regarding the plasma steroid levels in male elasmobranchs. Jenkins and Dodd (1980) reported 10–28 ng/mL of plasma testosterone in *S. canicula*. Idler and Truscott (1966) reported free testosterone and glucuronide in the peripheral plasma of the skate. Rasmussen and Gruber (1990) measured immunoreactive levels of testosterone levels from 0.8 to 358 ng/mL in several species of carcharhinid sharks. Whether the testis is the sole source of plasma steroids is not known with certainty. Presumably androgens and other testicular hormones diffuse into the blood from Sertoli cells and reach the general circulation after passing into testicular capillaries. An alternative route may be entry into the seminal fluids and resorption in the excurrent ducts.

A functional pituitary-testicular axis appears to exist in elasmobranchs. The anatomical presence of a differentiated median eminence is in question, and only a primitive portal system has been described (Schreibman 1986; Honma and Chiba 1987). The adenohypophysis forms a structure not seen in other vertebrates, the ventral lobe. This lobe is not connected to the hypothalamus by either the portal system or nerve endings (Dodd 1983; Gorbman et al. 1983) but has hormone activity (Sumpter et al. 1978a,b). The hypothalamus of elasmobranchs contains GnRH activity (King and Millar 1980; Powell et al. 1986; Sherwood and Lovejoy 1993). Only a single gonadotropin has been identified from the elasmobranch ventral lobe (Sumpter et al. 1978a). A strong decrease in peripheral androgen levels is effected by total hypophysectomy (Sumpter et al. 1978b), while injections of ventral lobe homogenate restore normal androgen levels. It is surmised that GnRH acting at the level of the ventral lobe elicits gonadotropin release, which in turn acts at the gonadal level, inducing steroid production.

Biochemical, immunoreactive, and bioassay methods have detected GnRH-like activity in elasmobranchs (Lovejoy and Sherwood 1989; Powell et al. 1986) and immunocytochemical studies have localized GnRH-like material in the preoptic area and hypothalamus (Demski 1989). Intratesticular injections of mammalian GnRH long-acting analogue induced an increase in plasma levels of androgen in hypophysectomized *T. marmorata* (Fasano et al. 1989a,b). Callard (1990), working on preovulatory follicles and corpora lutea, has confirmed this direct GnRH action on gonadal steroidogenesis. Since GnRH activity is absent in the testes, a reasonable notion is that circulating GnRH reaches the gonad (Pierantonio et al. 1993). Intragonadal steroids may exert a paracrine or autocrine role in regulation of steroidogenesis in the testis and in specific germ cell stages. Gonadal steroids may also feed back to affect the pituitary (Pierantonio et al. 1993). Alternate transport routes may include neurons in the spinal cord and cerebrospinal fluid (Demski 1989; Wright and Demski 1993).

Estrogen receptors (Callard and Mak 1985; Callard et al. 1985), androgen receptors (Cuevas and Callard 1990), and chromatin-binding sites for estrogen-receptor complexes have been identified in the dogfish testis. Estrogen and androgen receptors are typically similar to their mammalian counterparts but have several unique features. *Squalus* testicular receptors are similar to mammalian receptors in their high affinity and steroid specificity. Estrogen receptors in dogfish testes are able to form complexes with chromatin receptor sites from dogfish testes but not from rat testes (Ruh et al. 1986). Rat estrogen receptors, however, effectively bind to homologous chromatin but not to dogfish testicular chromatin. Estrogen receptors in *Squalus* are most numerous in areas with stem cells and spermatogonia, where they may exert paracrine regulation (Callard and Mak 1985; Callard et al. 1985). Androgen receptors are most numerous in premeiotic stages (Cuevas and Callard 1990), and androgen and estrogen may cooperate in regulating gene expression prior to meiosis (Callard 1991).

Several polypeptide growth factors originate in Sertoli cells and include recognized mitogenic principles. Estrogen is a known mitogen (Callard 1990), and estrogen stimulates the expression of the protooncogene coding for a nuclear transcriptional regulator, increases production of growth factors, and synergizes with growth factors. Steroids and growth factors may thus interact in regulating spermatogenesis.

Gonadal cycles

Elasmobranchs are considered seasonal breeders, and a seasonal cycle in spermatogenesis has been described in *S. acanthias* (Simpson and Wardle 1967), *S. canicula* (Dobson 1974), *Mustelus manazo* and *M. griseus* (Teshima 1981), *S. tiburo* (Parsons and Grier 1992), and *Dasyatis sabina* (Maruska et al. 1996). Dodd and Sumpter (1984) reviewed reproductive cycles in elasmobranchs and Dobson and Dodd (1977a–c) described environmental control of reproductive cycles. The most commonly used method for quantifying seasonal testicular changes is the gonadosomatic index (GSI). It is determined by expressing a given dimension of the testis, for example weight, as a percent of the same dimension of the entire animal. The GSI provides information on regression and recrudescence and is often used as an indicator of mating season. Some species show marked seasonal changes in testicular development, including *S. tiburo*, *M. manazo*, *M. griseus*, *Carcharhinus limbatus*, and *Rhizoprionodon terraenovae*, while others such as *Prionace glauca* (Parsons and Grier 1992) show little seasonal change in sperm production.

In *D. sabina* (Maruska et al. 1996), GSI showed three phases correlated with structural changes in the testis. An inactive phase characterized by low GSI lasted from March through July. The enlargement phase begins in mid-August and is followed by a peak in testicular growth by October. The diminution phase occurs from October through April, with a decrease in GSI. These data indicate that continuous spermatogenesis occurs through the fall and winter and that the peak of sperm production lags maximum GSI by 3–4 months. The diameter of seminal vesicles, which peaks in February, also lags maximum GSI by 4 months. The conclusion is that males store sperm in the seminal vesicles for use at the end of spermatogenesis when females are ready to ovulate. The authors suggest that the condition of the epididymis and seminal vesicles may be a more reliable indicator of impending fertilization than GSI, particularly in species that lack a distinct, brief mating period.

An annual cycle of brain GnRH has been described showing seasonality related to reproductive activity (Powell et al. 1986). GnRH immunoreactivity has also been found in plasma (King and Millar 1980; Powell et al. 1986; Sherwood and Lovejoy 1993). Since there are no definitive connections between the hypothalamus and the ventral lobe, the general circulation may be the most practical route.

Sperm aggregation and acquisition of motility

Testicular spermatozoa are immotile but acquire motility after traversing the ductus deferens in species with internal fertilization, such as reptiles, birds, mammals, and elasmobranchs (Bedford 1979). In sharks, at copulation and ejaculation, sperm are transferred from the seminal vesicle through the urogenital papilla and out the clasper groove by seawater expelled by the siphon sac (Matthews 1950). Spermatozoa acquire the potential for modest motility while within the terminal regions of the genital ducts but acquire active, robust motility at ejaculation. In *Triakis scyllia*, Minamikawa and Morisawa (1996) measured organic and inorganic constituents, pH and osmotic pressure in uterine fluid, artificial uterine fluid (300 mM NaCl, 8 mM KCl, 5.5 mM $CaCl_2$, 2 mM $MgCl_2$, 350 mM urea, 20 mM HEPES at pH 7.8), blood plasma, and seminal plasma and correlated the effects on sperm motility. *Triakis* has aplacental viviparity of the yolk sac type. They conclude that ionic levels and hexose have important roles in initiation of sperm motility and may be involved in the maintenance of sperm in the female reproductive tract.

Elasmobranchs maintain high blood plasma levels of urea, which accounts for a considerable proportion of the plasma osmotic pressure (Holmes and Donaldson 1969).

Triakis blood plasma also had high urea concentrations. Inorganic and organic components of uterine fluid was similar to blood plasma (Table 16-1). Na^+ levels in seminal plasma were lower than in blood, and levels of seminal K^+ were much higher than in blood. Sperm showed little motility in nonelectrolyte solutions at all concentrations but moved vigorously in NaCl or KCl solutions at a concentration of 500 mM and osmolarity 1000 mOsm/kg. Sperm mixed with seawater from the siphon sac and the uterine fluid showed osmolarity of 1000 mOsm/kg. It is likely that proper osmolarity and ion concentrations are a necessary prerequisite for initiation of sperm motility, although other factors such as siphon sac secretions may help to trigger the changes.

Concentration of glucose was 13-fold higher in uterine fluid than seminal fluid (Table 16-2) and sperm motility was high in both uterine fluid and blood plasma containing high concentrations of glucose. Sperm motility decreased markedly in media lacking glucose but with the same ionic composition as uterus fluid. It is reasonable to conclude that hexose, especially glucose, has a role in maintenance of sperm motility in the female reproductive tract.

Sperm are maintained in the seminal vesicle of elasmobranchs in various configurations, either as laterally aligned masses of sperm with no extracellular matrix, as linearly aligned clumps of sperm completely surrounded by a matrix (spermatophores), or as clumps of sperm that are surrounded by matrix but with the tails of the peripherally located sperm projecting out from the matrix into the luminal fluid of the seminal vesicle (spermatozeugmata) (Pratt and Tanaka 1994). Edwards (1842) used the term *spermatophore* to describe masses of sperm surrounded by a "capsule" in cephalopods and Ballowitz (1890) suggested the term *spermatozeugmata* for the sperm structures in insects, in which groups of sperm are held together without being enclosed by a "capsule." Ballowitz (1895) also used the abbreviated term *spermozeugma*. Spermatozoa are said to be adherent to each other

TABLE 16-1 INORGANIC CONSTITUENTS, OSMOLALITY, AND PH OF BLOOD PLASMA, SEMINAL PLASMA, AND UTERUS FLUID IN THE SHARK *TRIAKIS SCYLLIA*

	Na^+ (mM)	K^+ (mM)	Ca^{2+} (mM)	Mg^{2+} (mM)	Cl^- (mM)	Osmolality (mOsm/kg)	pH
Blood plasma	280 ± 6.1 (12)	9.0 ± 0.78 (11)	4.7 ± 0.18 (12)	2.2 ± 0.21 (12)	244 ± 22.8 (12)	996 ± 15.7 (11)	— —
Uterus fluid	299 ± 33.4 (4)	7.8 ± 0.65 (4)	5.4 ± 0.47 (4)	1.9 ± 0.19 (3)	273 ± 38.1 (3)	1027 ± 147.8[b] (3)	7.8 ± 0.33 (6)
Seminal plasma	198 ± 9.0[a] (4)	25.2 ± 4.16[a] (3)	6.2 ± 1.55 (3)	5.5 ± 1.20[a] (3)	201 ± 17.6 (3)	865 ± 18.0 (3)	6.6 ± 0.16[c] (5)

SOURCE: From Minamikawa and Morisawa (1996).
Experiment numbers are indicated in parentheses.
Values are means ± S.E.M.
$p < 0.05$.
[a]Differs significantly from corresponding blood plasma and uterus fluid values.
[b]Differs significantly from corresponding blood value.
[c]Differs significantly from uterus fluid value.

TABLE 16-2 ORGANIC CONSTITUENTS OF BLOOD PLASMA, SEMINAL PLASMA, AND UTERUS FLUID IN THE SHARK *TRIAKIS SCYLLIA*

	Urea (mM)	Protein (mg/mL)	Glucose (mM)	Cholesterol (mg/100mL)
Blood plasma	176 ± 1.93 (10)	20.7 ± 1.62 (10)	174 ± 13.1 (10)	60.7 ± 6.03 (10)
Uterus fluid	175 ± 2.98 (8)	11.8 ± 1.97 (8)	157 ± 9.5 (8)	47.3 ± 3.59 (8)
Seminal plasma	165 ± 5.34 (3)	38.4 ± 4.20[a] (3)	12 ± 7.6[a] (3)	38.3 ± 3.21 (3)

SOURCE: From Minamikawa and Morisawa (1996).
Experiment numbers are indicated in parentheses.
Values are means ± S.E.M.
$p < 0.05$.
[a]Significantly different from the corresponding blood plasma and uterus fluid values.

by a sticky substance that permeates the spermatozeugmata, namely, the matrix. Nielsen et al. (1968) point out that many authors have confused spermatophores and spermatozeugmata and that most spermatophores described in fish are actually spermatozeugmata. The use of such terms as *capsule* is unfortunate. The word *capsule* (Latin for "little box") is defined as a surrounding sheath or a structure in which something is enclosed, and it implies a separate entity such as a membrane or connective tissue investiture. The term *mass* has also been used. *Mass* is defined as a quantity of matter that forms a body of indefinite shape or size, usually a relatively large size; a lump or body made up of cohesive particles. The term *aggregate* has been used and is defined as a massing of materials to form a "clump." A more appropriate term for the investing material of elasmobranch sperm is a *matrix*. *Matrix* is defined in anatomy as any nonliving, intercellular substance in which living cells are embedded. Hence, extracellular matrices of various cellular origins constitute the material that is responsible for the various sperm structures seen in elasmobranchs, namely, spermatophores and spermatozeugmata. The term *capsule* should not be used when referring to a peripheral layer of matrix.

Pratt and Tanaka (1994) surveyed and described the various types of sperm associations in elasmobranchs. A *naked sperm* is an individual sperm with no surrounding matrix. A sperm *aggregate* is a generic term that can be applied to any cohesive mass composed of matrix and sperm, regardless of the way in which the sperm are arrayed in the matrix. *Spermatophores* are clumps of linearly aligned sperm surrounded by an eosinophilic matrix (Fig. 16-10). The important point is that no part of the sperm projects past the perimeter of the matrix and that there is no definite capsule. *Spermatozeugmata* are fundamentally cohesive masses of matrix and sperm in which the most peripheral sperm are arrayed with their tails projecting past the perimeter of the matrix and thus present a fuzzy appearance (see Fig. 16-10).

Spermatozeugmata have been further categorized according to the way in which sperm are arrayed and the size of the spermatozeugmata (Pratt and Tanaka 1994): (1) clump type, (2) single-layered, and (3) compound or multilayered. Clump-type spermatozeugmata occur in *S. acanthias*, where the sperm are aligned laterally head to head with no matrix. In *H. collei* the aligned sperm combine with matrix to form spermatozeugmata, in which aligned sperm occupy the center of the matrix

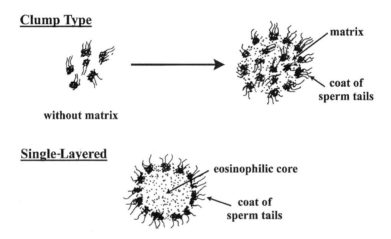

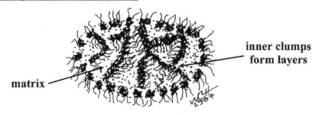

FIGURE 16-10 Diagrammatic representation of spermatophore and the different types of spermatozeugmata in elasmobranchs. *Spermatophores:* Structures of matrix components with sperm aligned at random, no part of the sperm projecting past the matrix perimeter. *Clump-type spermatozeugmata:* Clumps of laterally aligned sperm combine with matrix; resulting structure has sperm clumps randomly arrayed in the matrix interior, but the peripheral sperm have heads embedded in matrix and tails projecting radially, forming a fuzzy covering. *Single-layered spermatozeugmata:* Have an eosinophilic central zone devoid of sperm and peripheral sperm with their tails projecting outward to form a fuzzy coat. *Compound or multilayered spermatozeugmata:* Have standard peripheral sperm component while inner sperm clumps form internal layers with sperm tails pointing to center of matrix. (Information compiled from Pratt and Tanaka 1994.)

and peripherally located sperm have their tails projecting outward from the matrix. It is possible that the array seen in *Squalus* is a transitional stage prior to or during the elaboration of matrix but before definitive spermatozeugmata as such have formed. *Hydrolagus* may demonstrate the terminal situation, in which matrix has been elaborated and has associated with previously aligned sperm.

Single-layered spermatozeugmata occur in *Carcharhinus falciformis, C. limbatus, C. obscurus, C. plumbeus, C. porosus,* and *R. eglanteria* (Pratt and Tanaka 1994). In this configuration sperm are located only at the periphery of the matrix, where their tails project from the matrix.

Compound, or multilayered, spermatozeugmata are characteristic of *P. glauca, R. terraenovae,* and *Sphyrna lewini* (Pratt and Tanaka 1994). Peripheral clumps of sperm have their tails protruding from the matrix, while inner sperm clumps form irregular projections or layers (see Fig. 16-10). The inner sperm have their tails pointing toward the center of the matrix.

Sperm aggregation in the male genital ducts in *R. eglanteria* has recently been described (Hamlett et al. 1999). Sperm are released from Sertoli cells as aligned bundles. In the proximal epididymis, sperm dissociate as individuals. During transit through the epididymis, sperm begin to associate and re-aggregate with their heads in lateral register. By the time they arrive in the ductus deferens, masses of laterally aligned sperm are present as sperm aggregates. In the seminal vesicle, definitive spermatozeugmata are formed.

The factors that mediate lateral sperm alignment without the involvement of matrix are unknown. Potential candidates include nonjunctional contacts. Cells migrating in vertebrate embryos generally do not involve organized intercellular junctional complexes. Yet the interacting plasmalemmae often come within 10–20 nm. As several known transmembrane proteins extend above the plasmalemma by 10–20 nm, two cell surface proteins could interact directly to mediate adhesion.

Substances that bind sperm clumps to matrix may include integrins. Integrins are transmembrane heterodimers. Some integrins bind to only one matrix molecule while others bind to more than one. Binding of integrins to ligands depends on extracellular divalent cations Ca^{2+} or Mg^{2+}, depending on the integrin. The relationship between sperm-matrix binding and sperm motility has yet to be investigated.

Spermatozoa form spherical bundles as they pass through the seminal vesicle in the Port Jackson shark (Jones and Jones 1982; Jones et al. 1984). Bundle formation has also been described in other species including *Spinax niger, Squalus acanthias, S. stellaris, Chimaera monstrosa, T. marmorata* and *T. torpedo* (Redenz and Belonoschkin 1929; Botte et al. 1963).

Various suggestions have been offered as to why sperm aggregation in male elasmobranchs may be advantageous. Matthews (1950) suggested that spermatophores in the basking shark helped to prevent loss of semen during copulation when the semen must traverse the clasper to the female tract. A turgid spermatophore may be more hydraulically suited than a fluid semen to passage to the female without undue loss. Pratt and Tanaka (1994) mention that spermatozeugmata of carcharhinids and sphyrnids are porous, which may provide a mechanism to ensure adequate nourishment reaching the sperm yet still package them for transport at copulation.

The folds found in the seminal vesicles may increase surface area for nutrient exchange and for physical support (Pratt and Tanaka 1994). Matthews (1950) thought that the septa might play a role in spermatophore formation. He believed that epithelial secretions of the epididymis, ductus deferens, and the Leydig gland mix with spermatozoa and that in the upper portion of the seminal vesicle, secretions form globules by action of cilia in the septal bays. He suggested that the "tumbling mill" action increases the size of spermatophores. Pratt and Tanaka (1994) point out that sperm aggregates occur throughout the seminal vesicle in the shortfin mako, and they doubt that the "tumbling mill" theory is active in this species.

Summary

The body of knowledge of the workings of the male reproductive system in elasmobranchs is paltry when compared to that of the female. The testis system, however, offers a model to study the control of spermatogenesis in a stage-by-stage system, namely, the spermatocyst. The male reproductive system of elasmobranchs demonstrates fundamental differences from amniotes, in which germinal tissue is arranged in tubules that consist of a permanent population of Sertoli cells associated with successive stages of germ cell development (Loir et al. 1995). The spermatocyst is the functional unit of the testis in elasmobranchs and corresponds to a clone of isogenetic germ cells with associated Sertoli cells. Sertoli cells and primary spermatogonia initially are located in the tissue interstitium but become associated with a basement membrane. The newly formed structure is a spermatocyst, with several Sertoli cells being associated with a clone of germ cells. Leydig cells are either absent, as in *S. stellaris* and *S. acanthias* (Holstein 1969; Simpson and Wardle 1967; Stephan 1902); rare, as in *C. maximus* and *S. canicula* (Collenot 1969); or undifferentiated, as in *S. acanthias* (Pudney and Callard 1984), and their role in steroid production is at best minimal. Collenot and Damas (1980) demonstrated tight junctions and desmosomelike structures between Sertoli cells in *S. canicula*. They also reasoned, based on the fact that mitotic divisions propagate from spermatoblast to spermatoblast in *S. canicula* and *T. marmorata* (Stanley 1966), that gap junctions may exist between Sertoli cells. The existence of cytoplasmic bridges between germ cells in the same spermatoblast has been shown by Stanley (1966) and Sourdaine and Jegou (1988). These structures may also allow germ cells to synchronize their activities.

Of considerable interest is the role played by the matrix in which sperm are carried when the sperm is delivered to the female. Analysis of the sperm matrix is needed to ascertain its role beyond that of simply serving as a transport medium. How do the matrix and the sperm interact with the female tract and its secretions? What is the role of the male matrix in, for example, sperm activation, motility, and nutrition? These and other questions remain to be answered. Male elasmobranchs provide a unique opportunity for the study of cell and molecular mechanisms of sperm-female interactions.

REFERENCES

Ballowitz, E. 1890. Untersuchungen über die Struktur der Bau der kontraktilen Elemente die Spermatozoen der Insekten. *Z. Wiss. Zool.* 50: 317–407.

———. 1895. Die Doppelspermatozoen der Dyticiden. *Z. Wiss. Zool.* 60: 458–99.

Bedford, J. M. 1979. *The Spermatozoon.* Baltimore: Urban & Schwarzenberg.

Borcea, J. 1906. Systeme uro-genital des elasmobranches. *Arch. Zool. Exp. Genet.* 4: 119–484.

Botte, V., G. Chieffi, and H. P. Stanley. 1963. Histological and histochemical observations on the male reproductive tract of *Scyliorhinus stellaris, Torpedo marmorata*, and *T. torpedo. Pubbl. Sta. Zool. (Naples)* 33: 224–42.

Breder, C. M., and D. E. Rosen. 1966. *Modes of Reproduction in Fish.* Garden City, N.Y.: Natural History.

Callard, G. V. 1988. Reproductive physiology: B The male. In *Physiology of Elasmobranch Fishes*, ed. T. Shuttleworth, 292–317. New York: Springer-Verlag.

———. 1990. Spermatogenesis. In *Vertebrate Endocrinology: Fundamentals and Medical Implication*, vol. 4, pt. A, ed. P. K. T. Pang and M. P. Schreibman. New York: Academic Press.

———. 1991. Reproduction in male elasmobranchs. In *Comparative Physiology*, vol. 10, ed. R. K. H. Kinne, 104–54. Basel: Karger.

Callard, G. V., and P. Mak. 1985. Exclusive nuclear location of estrogen receptors in *Squalus testis. Proc. Natl. Acad. Sci. U.S.A.* 82: 1336–40.

Callard, G. V., J. A. Pudney, P. Mak, and J. Canick. 1985. Stage-dependent changes in steroidogenic enzymes and estrogen receptors during spermatogenesis in the testis of the dogfish *Squalus acanthias. Endocrinology* 117: 1328–35.

Callard, G. V., P. Mak, W. DuBois, and M. Cuevas. 1989. Regulation of spermatogenesis: the shark testis model. *J. Exp. Zool. Suppl.* 2: 353–64.

Carrier, J. C., H. L. Pratt, and L. K. Martin. 1994. Group reproductive behaviors in free-living nurse sharks, *Ginglymostoma cirratum. Copeia* 1994: 646–56.

Chieffi, G., and C. Lupo di Prisco. 1961. Identification of estradiol-17β, testosterone, and its precursors from *Scyliorhinus stellaris* testes. *Nature* 190: 169–70.

Chieffi, G., F. Della Corte, and V. Botte. 1961. Osservazioni sui tessuto interstiziale del testiculo dei Selachi. *Boll. Zool.* 28: 211–17.

Collenot, G. 1969. Structure et activité endocrine du testicle de *Scyliorhinus canicula* L. depuis l'éclosion jusqu'à l'état adulte. *Ann. Embryol. Morphol.* 2: 461–77.

Collenot, G., and D. Damas. 1975. Mise en evidence de la nature protéique des corps énigmatiques présents dans le testicule de *Scyliorhinus canicula* L. (Elasmobranche). *Cah. Biol. Mar.* 16: 39–64.

———. 1980. Etude ultrastructurale de la cellule de Sertoli au cours de la spermiogénèse chez *Scyliorhinus canicula* L. *Cah. Biol. Mar.* 21: 209–19.

Compagno, L. J. V. 1988. *Sharks of the Order Carcharhiniformes*. Princeton, N.J.: Princeton University Press.

Cuevas, M. E., and G. V. Callard. 1990. Characteristics and stage-dependent distribution of a putative androgen receptor in nuclear extracts of dogfish testis. In *Proceedings of the 72nd Annual Meeting of the Endocrine Society*, 125 (abstract).

Daniel, J. F. 1928. *The Elasmobranch Fishes*. Berkeley: University of California Press.

Della Corte, F., V. Botte, and G. Chieffi. 1961. Ricerca istochimica dell'attivita della steroide 3β-olo-deihydrogenase nel testiculo *Torpedo marmorata* Risso e di *Scyliorhinus stellaris* (L.). *Atti. Soc. Peloritana Sci. Fis. Mat. Nat.* 7: 393–97.

Demski, L. 1989. Pathways for GnRH control of elasmobranch reproductive physiology and behavior. *J. Exp. Zool. Suppl.* 2: 4–11.

Dempster, R. P., and E.S. Herald. 1961. Notes on the horn shark, *Heterodontus francisci*, with observations on mating activities. *Occas. Pap. Calif. Acad. Sci.* 33: 1–7.

Dobson, S. 1974. Endocrine control of reproduction in the male *Scyliorhinus canicula*. Ph.D. thesis, University of Wales.

Dobson, S., and J. M. Dodd. 1977a. Endocrine control of the testis in the dogfish *Scyliorhinus canicula* L.: I. Effects of partial hypophysectomy on gravimetric, hormonal, and biochemical aspects of testis function. *Gen. Comp. Endocrinol.* 32: 41–52.

———. 1977b. Endocrine control of the testis in the dogfish *Scyliorhinus canicula* L: II. Histological and ultrastructural changes in the testis after partial hypophysectomy (ventral lobectomy). *Gen. Comp. Endocrinol.* 32: 52–71.

———. 1977c. The roles of temperature and photoperiod in the response of the testis of the dogfish, *Scyliorhinus canicula* L. to partial hypophysectomy (ventral lobectomy). *Gen. Comp. Endocrinol.* 32: 114–15.

Dodd, J. M. 1983. Reproduction in cartilaginous fishes (Chondrichthyes). In *Fish Physiology*, vol. 9, ed. W. S. Hoar and D. J. Randall, 31–95. New York: Academic Press.

Dodd, J. M., and J. P. Sumpter. 1984. Fishes. In *Marshall's Physiology of Reproduction*, vol. 1: *Reproductive Cycles of Vertebrates*, ed. G. E. Lamming, 21–126. New York: Churchill Livingstone.

DuBois, W., P. Mak, and G. V. Callard. 1989. Sertoli cell functions during spermatogenesis: the shark testis mode. *Fish Physiol. Biochem.* 7: 221–27.

Edwards, H. M. 1842. Sur la structure et les fonctions de quelques zoophytes, mollusques et crustacés des côtes de la France. *Ann. Sci. Nat.* 18(3): 321–50.

Fasano, S., R. Pierantonio, and G. Chieffi. 1989b. Reproductive biology of elasmobranch with emphasis on endocrines. *J. Exp. Zool. Suppl.* 2: 53–61.

Fasano, S., R. Pierantonio, S. Minucci, L. DiMatteo, M. D'Antonia, and G. Chieffi. 1989a. Effects of intratesticular injections of estradiol and GnRH (GnRHA, HOE766) on plasma androgen levels in intact and hypophysectomized *Torpedo marmorata* and *Torpedo ocellata*. *Gen. Comp. Endocrinol.* 75: 349–54.

Fawcett, D. W. 1975 Ultrastructure and function of the Sertoli cell. In *Male Reproductive System, Handbook of Physiology*, ed. D. W. Hamilton and R. O. Greep, 21–56. Washington, D.C.: American Physiological Society.

Garman, S. 1912. The plagiostoma. *Mem. Harvard Mus. Comp. Zool.* 36: 409–91.

Gilbert, P. W., and G. W. Heath. 1955. The functional anatomy of the claspers and siphon sacs in the spiny dogfish (*Squalus acanthias*) and smooth dogfish (*Mustelus canis*). *Anat. Rec.* 121: 433 (abstract).

———. 1972. The clasper-siphon sac mechanism in *Squalus acanthias* and *Mustelus canis*. *Comp. Biochem. Physiol.* 42A: 97–119.

Gorbman, A., W. W. Dickhoff, S. R. Vigna, N. Clark, and C. L. Ralph. 1983. *Comparative Endocrinology*. New York: John Wiley & Sons.

Gordon, I. 1993. Pre-copulatory behavior of captive sandtiger sharks, *Carcharias taurus*. *Environ. Biol. Fish.* 38: 159–64.

Grier, H. J. 1992. Chordate testis: the extracellular matrix hypothesis. *J. Exp. Zool.* 261: 151–60.

Gudger, E. W. 1912. Summary of work done on the fishes of Tortugas. *Carnegie Inst. Wash. Yearb.* 11: 148–50.

Hamlett, W. C., M. Hysell, T. Rozycki, N. Brunette, K. Tumilty, A. Henderson, and J. Dunne. 1999. Sperm aggregation and spermatozeugmata formation in the male genital ducts in the clearnose skate, *Raja eglanteria*. In *Proceedings of the 5th Indo-Pacific Fish Conference, Nouméa*, ed. B. Séret and J.-Y. Sire. Paris: French Ichthyological Society and ORSTOM. In press.

Holmes, W. N., and E. M. Donaldson. 1969. *Fish Physiology*, vol. 1, 1–89. New York: Academic Press.

Holstein, A. F. 1969. Zur Frage der lokalen Steuerung der Spermatogenese beim Dorhai (*Squalus acanthias* L). *Z. Zellforsch.* 93: 265–81.

Honma, Y., and A. Chiba. 1987. Fine structure and vascularization of the hypothalamo-hypophyseal complex in the Japanese elasmobranchs. In *Proceedings of the First Congress of the Asia Oceania Society for Comparative Endocrinology*, Nagoya, Japan.

Idler, D. R., and B. Truscott. 1966. Identification and quantification of testosterone in peripheral plasma of skate. *Gen. Comp. Endocrinol.* 7: 375–83.

Jenkins, N., and J. M. Dodd. 1980. Effects of synthetic mammalian GnRH and dogfish hypothalamic extracts on levels of androgen and estradiol in the circulation of the dogfish (*Scyliorhinus canicula* L.). *J. Endocrinol.* 86: 171–77.

Jones, N., and R. C. Jones. 1982. The structure of the male genital system of the Port Jackson shark, *Heterodontus portusjacksoni*, with particular reference to the genital ducts. *Aust. J. Zool.* 30: 523–41

Jones, R. C., N. Jones, and D. Dakiew. 1984. Luminal composition and maturation of spermatozoa in the male genital ducts of the Port Jackson shark, *Heterodontus portusjacksoni*. *J. Exp. Zool.* 230: 417–26.

Jones, R. C., and M. Lin. 1993. Structure and functions of the genital ducts of the male Port Jackson shark, *Heterodontus portusjacksoni*. *Environ. Biol. Fish.* 38: 127–38.

Johnson, R. H., and D. R. Nelson. 1978. Copulation and possible olfaction-mediated pair formation in two species of carcharhinid sharks. *Copeia* 1978: 539–42.

King, J. A., and R. P. Millar. 1980. Comparative aspects of lutenizing hormone–releasing hormone structure and function in vertebrate phylogeny. *Endocrinology* 106: 707–17.

Klimley, A. P. 1980. Observations of courtship and copulation in the nurse shark. *Copeia* 1980: 878–82.

LaMarca, M. J. 1964. The functional anatomy of the clasper and clasper gland of the yellow stingray, *Urolophus jamaicensis* (Cuvier). *J. Morphol.* 114: 303–24.

Leigh-Sharpe, W. H. 1920. The comparative morphology of the secondary sexual characters of elasmobranch fishes: I. *J. Morphol.* 34: 245–65.

———. 1921. The comparative morphology of the secondary sexual characters of elasmobranch fishes: II. *J. Morphol.* 35: 359–80.

———. 1922. The comparative morphology of the secondary sexual characters of elasmobranch fishes: III, IV, and V. *J. Morphol.* 36: 191–243.

———. 1924. The comparative morphology of the secondary sexual characters of elasmobranch fishes: VI and VII. *J. Morphol.* 39: 553–77.

———. 1926. The comparative morphology of the secondary sexual characters of elasmobranch fishes: VIII, IX, X, and XI. *J. Morphol.* 42: 307–48.

Liu, Q., and L. S. Demski. 1993. Clasper control in the round stingray, *Urolophus halleri*: lower sensorimotor pathways. *Environ. Biol. Fish.* 38: 219–30.

Loir, M., P. Sourdaine, S. M. L. C. Mendis-Handagama, and B. Jegou. 1995. Cell-cell interactions in the testis of teleosts and elasmobranchs. *Microsc. Res. Tech.* 32: 533–52.

Lovejoy, D. A., and N. M. Sherwood. 1989. Gonadotropin-releasing hormone in ratfish, *Hydrolagus colliei*: distribution between the sexes, and possible relationships with chicken II and salmon II forms. *Comp. Biochem. Physiol.* 92B: 111–18.

Luer, C. A., and P. W. Gilbert. 1985. Mating behavior, egg deposition, incubation period, and hatching in the clearnose skate, *Raja eglanteria*. *Environ. Biol. Fish.* 13: 161–71.

Mann, T. 1960. Serotonin (5-hydroxytryptamine) in the male reproductive tract of the spiny dogfish. *Nature* 188: 941–42.

Mann, T. and C. L. Prosser. 1963. Uterine response to 5-hydroxytryptamine in the clasper siphon secretion of the spiny dogfish, *Squalus acanthias*. *Biol. Bull.* 125: 384–85.

Maren, T. H., J. A. Rawls, J. W. Burger, and A. C. Myers. 1963. The alkaline Marshall's gland of the skate. *Comp. Biochem. Physiol.* 10: 1–16.

Maruska, K. P., E. G. Cowie, and T. C. Tricas. 1996. Periodic gonadal activity and protracted mating in elasmobranch fishes. *J. Exp. Zool.* 276: 219–32.

Matthews, H. L. 1950. Reproduction in the basking shark, *Cetorhinus maximus*. *Philos. Trans. R. Soc. Lond.* 234B: 247–316.

Mellinger, J. 1965. Stades de la spermatogénèse chez *Scyliorhinus caniculus* L.: description, données histochimiques, variations normales et experimentales. *Z. Zellforsch.* 67: 653–73.

Minamikawa, S., and M. Morisawa. 1996. Acquisition, initiation, and maintenance of sperm motility in the shark, *Triakis scyllia*. *Comp. Biochem. Physiol.* 113A: 387–92.

Myerberg, A. A., and S. H. Gruber. 1974. The behavior of the bonnethead shark, *Sphyrna tiburo*. *Copeia* 1974: 358–74.

Nielsen, J. G., A. Jespersen, and O. Munk. 1968. Spermatophores in *Ophidioidea* (Pisces, Percomorphi). In *Galathea Report*, vol. 9, ed. T. Wolf, 239–54. Copenhagen: Danish Science Press.

Nordell, S. E. 1994. Observations of the mating behavior and dentition of the round stingray, *Urolophus halleri*. *Environ. Biol. Fish.* 39: 219–29.

Parsons, G. R., and H. J. Grier. 1992. Seasonal changes in shark testicular structure and spermatogenesis. *J. Exp. Zool.* 261: 173–84.

Pierantonio, R., M. D'Antonio, and S. Fasano. 1993. Morpho-functional aspects of the hypothalamus-pituitary-gonadal axis of elasmobranch fishes. *Environ. Biol. Fish.* 38: 187–96.

Powell, R. C., R. P. Millar, and J. A. King. 1986. Diverse molecular forms of GnRH in an elasmobranch and a teleost fish. *Gen. Comp. Endocrinol.* 63: 77–85.

Pratt, H. L. 1988. Elasmobranch gonad structure: a descriptive survey. *Copeia* 1988: 719–29.

Pratt, H. L., and S. Tanaka. 1994. Sperm storage in male elasmobranchs: a description and survey. *J. Morphol.* 219: 297–308.

Pudney, J., and G. V. Callard. 1984. Development of the agranular endoplasmic reticulum in Sertoli cell of the shark *Squalus acanthias* during spermatogenesis. *Anat. Rec.* 209: 311–21.

———. 1986. Sertoli cell cytoplasts in the semen of the spiny dogfish (*Squalus acanthias*). *Tissue Cell* 18: 375–82.

Rasmussen, L. E. L., and S. H. Gruber. 1990. Serum levels of circulating hormones in free-ranging carcharinoid sharks. In *Elasmobranchs as Living Resources,* ed. H. L. Pratt, T. Taniuchi, and S. H. Gruber, 143–55. Seattle: National Oceanic and Atmospheric Administration.

Redenz, E., and B. Belonoschkin. 1929. Form und Funktion des nebendens einiger Elasmobranchier. *Zellforsch. Mikrosk. Anat.* 9: 663–82.

Retzius, G. 1902. Über einen Spiralfaserapparat am Kopfe der Spermien der Selachien. *Biol. Untersuch.* NF 10: 61–64.

Ritzen, E. M., V. Hansson, and F. S. French. 1989. The Sertoli cell. In *The Testis,* 2d ed., ed. W. Burger and G. de Kretser, 269–302. New York: Raven Press.

Roosen-Runge, E. C. 1977. *The Process of Spermatogenesis in Animals.* London: Cambridge University Press.

Rouse, N. 1992. Life styles of sharks. In *Discovering Sharks,* ed. S. Gruber. Sandy Hook, N.J.: American Littoral Society.

Ruh, M. F., R. H. Singh, P. Mak, and G. V. Callard. 1986. Tissue and species specificity of nuclear acceptor sites for the estrogen receptor in *Squalus* testis. *Endocrinology* 118: 811–18.

Schreibman, M. P. 1986. Pituitary gland. In *Vertebrate Endocrinology: Fundamentals and Biomedical Implications,* vol. 1., ed. P. K. T. Pang and M. P. Schreibman, 11–55. New York: Academic Press.

Sherwood, N. M., and D. A. Lovejoy. 1993. Gonadotropin-releasing hormone in cartilaginous fishes: structure, location, and transport. *Environ. Biol. Fish.* 38: 197–208.

Simpson, T. H., and C. S. Wardle. 1967. A seasonal cycle in the testis of the spurdog, *Squalus acanthias,* and the sites of 3ß-hydroxysteroid dehydrogenase activity. *J. Mar. Biol. Assoc. U.K.* 47: 699–708.

Simpson, T. H., R. S. Wright, and H. Gottfried. 1963. Steroids in the semen of dogfish *Squalus acanthias. J. Endocrinol.* 26: 489–98.

Simpson, T. H., R. S. Wright, and S. V. Hunt. 1964a. Steroid biosynthesis in the testis of the dogfish *Squalus acanthias. J. Endocrinol.* 31: 29–38.

Simpson, T. H., R. S. Wright, and J. Renfrew. 1964b. Steroid biosynthesis in the semen of dogfish *Squalus acanthias. J. Endocrinol.* 31: 11–20.

Smith, H. W. 1929. The composition of the body fluids of elasmobranchs. *J. Biol. Chem.* 81: 407–19.

Smith, P. L. 1985. Electrolyte transport by alkaline gland of little skate, *Raja erinacea. Am. J. Physiol.* 248: R346–52.

Sourdaine, P., and B. Jegou. 1988. Dissociation and identification of intact seminiferous lobules from the testes of the dogfish (*Scyliorhinus canicula* L.). *Cell Tissue Res.* 225: 199–207.

Sourdaine, P., D. H. Garnier, and B. Jegou. 1990. The adult dogfish (*Scyliorhinus canicula* L.) testis: a model to study stage-dependent changes in steroid levels during spermatogenesis. *J. Endocrinol.* 127: 451–60.

Stanley, H. P. 1963. Urogenital morphology in the chimaeroid fish, *Hydrolagus collei* (Lay and Bennett). *J. Morphol.* 112: 99–127.

———. 1966. The structure and development of the seminiferous follicle in *Scyliorhinus caniculus* and *Torpedo marmorata* Elasmobranchii. *Z. Zellforsch.* 75: 453–68.

———. 1971a. Fine structure of spermiogenesis in the elasmobranch fish *Squalus suckleyi:* I. Acrosome formation, nuclear elongation, and differentiation of the midpiece axis. *J. Ultrastruct. Res.* 36: 86–102.

———. 1971b. Fine structure of spermiogenesis in the elasmobranch fish *Squalus suckleyi:* II. Late stages of differentiation and structure of the mature spermatozoon. *J. Ultrastruct. Res.* 36. 103–18.

———. 1983. The fine structure of spermatozoa of *Hydrolagus collei* (Chondrichthyes, Holocephali). *J. Ultrastruct. Res.* 83: 184–94.

Stanley, H. P., and C. C. Lambert. 1985. The role of a Sertoli cell actin-myosin system in sperm bundle formation in the ratfish, *Hydrolagus collei* (Chondrichthyes, Holocephali). *J. Morphol.* 186: 223–36.

Stephan, P. 1902. L'évolution de la cellule de Sertoli des sélaciens après la spermatogénèse. *C. R. Soc. Biol. Fil.* 54: 775–76.

Sumpter, J. P., B. K. Follett, N. Jenkins, and J. M. Dodd. 1978a. Studies on the purification and properties of gonadotropin from ventral lobes of the pituitary gland of the dogfish (*Scyliorhinus canicula* L.). *Gen. Comp. Endocrinol.* 36: 264–74.

———. 1978b. Gonadotrophic hormone in the pituitary gland of the dogfish *Scyliorhinus canicula* L.: distribution and physiological significance. *Gen. Comp. Endocrinol.* 36: 275–85.

Teshima, K. 1981. Studies on the reproduction of Japanese smooth dogfishes, *Mustelus manazo* and *M. griseus*. *J. Shimonoseki Univ. Fish.* 29: 113–99.

Tricas, T. C., and E. M. Le Feuvre. 1985. Mating of the white tip shark, *Triaenodon obesus*. *Mar. Biol.* 84: 233–37.

Uchida, S., M. Toda, and Y. Kamei. 1990. Reproduction of elasmobranchs in captivity. *NOAA Technical Report NMFS* 90: 211–37.

Wright, D. E., and L. S. Demski. 1991. Gonadotropin hormone-releasing hormone of sharks and rays. *J. Comp. Neurol.* 307: 49–56.

———. 1993. Gonadotropin-releasing hormone (GnRH) pathways and reproductive control in elasmobranchs. *Environ. Biol. Fish.* 38: 209–18.

LEONARD J. V. COMPAGNO

APPENDIX

Checklist of Living Elasmobranchs

This is an annotated checklist of living elasmobranchs that includes only described valid species and omits most synonyms and also known, undescribed species. A generalized listing of geographic distribution and habitat is included for each genus. Comments on classification are offered where appropriate.

SHARKS

ORDER HEXANCHIFORMES: COW AND FRILLED SHARKS

  FAMILY CHLAMYDOSELACHIDAE GARMAN, 1884. Frilled Sharks

   Genus *Chlamydoselachus* Garman, 1884. Wide-ranging but spotty in the western North Atlantic, eastern Atlantic, southwestern Indian Ocean, western Pacific, and eastern Pacific continental and insular slopes, occasionally on the shelves. Possibly more than one species.
     Chlamydoselachus anguineus Garman, 1884. Frilled shark.

  FAMILY HEXANCHIDAE GRAY, 1851. Cow Sharks

   Genus *Heptranchias* Rafinesque, 1810. Sometimes placed in a separate family Heptranchiidae. Western North Atlantic, eastern Atlantic and Mediterranean, Indo-West Pacific, and western South Pacific, upper slopes and lower shelves.
     Heptranchias perlo (Bonnaterre, 1788). Sharpnose sevengill shark.

Genus *Hexanchus* Rafinesque, 1810. Sixgill sharks. Circumglobal in temperate and tropical seas, continental and insular shelves and slopes, sea mounts.
Hexanchus griseus (Bonnaterre, 1788). Bluntnose sixgill shark.
Hexanchus nakamurai Teng, 1962. Bigeye sixgill shark (including *H. vitulus* Springer & Waller, 1969 as a synonym).

Genus *Notorynchus* Ayres, 1855. Sometimes placed in a separate family Notorynchidae. Temperate seas in the South Atlantic, Western Indian Ocean, and Pacific Ocean, continental shelves.
Notorynchus cepedianus (Peron, 1807). Broadnose sevengill shark.

FAMILY HEPTRANCHIIDAE. Western North Atlantic, eastern Atlantic, and Mediterranean, Indo-West Pacific, and western South Pacific upper slopes and lower shelves.

FAMILY NOTORYNCHIDAE. Temperate seas in the South Atlantic, western Indian Ocean, and Pacific Ocean continental shelves.
Notorynchus cepedianus (Peron, 1807). Broadnose seven-gill shark.

ORDER SQUALIFORMES: Dogfish Sharks

FAMILY ECHINORHINIDAE GILL, 1862. Bramble Sharks

Genus *Echinorhinus* Blainville, 1816. Bramble sharks. Temperate seas in the Atlantic Ocean, western Indian Ocean, and Pacific Ocean, continental and insular slopes and shelves, sea mounts.
Echinorhinus brucus (Bonnaterre, 1788). Bramble shark.
Echinorhinus cookei Pietschmann, 1928. Prickly shark.

FAMILY SQUALIDAE BLAINVILLE, 1816. Dogfish Sharks

Genus *Cirrhigaleus* Tanaka, 1912. Roughskin dogfish. Wide-ranging in the southwestern Indian; west-central Pacific, and western North Atlantic Oceans.
Cirrhigaleus asper (Merrett, 1973). Roughskin spurdog.
Cirrhigaleus barbifer Tanaka, 1912. Mandarin dogfish.

Genus *Squalus* Linnaeus, 1758. Spurdogs or spiny dogfish. Circumglobal in temperate and tropical seas, continental and insular shelves and slopes, sea mounts. At least six undescribed species in addition to the following:
Squalus acanthias Linnaeus, 1758. Piked dogfish.
?*Squalus blainvillei* (Risso, 1826). Longnose spurdog.
?*Squalus brevirostris* Tanaka, 1912. Japanese shortnose spurdog.
Squalus cubensis Howell-Rivero, 1936. Cuban dogfish.
Squalus japonicus Ishikawa, 1908. Japanese spurdog.
Squalus megalops (Macleay, 1881). Shortnose spurdog.
Squalus melanurus Fourmanoir, 1979. Blacktail spurdog.
Squalus mitsukurii Jordan & Snyder, in Jordan & Fowler, 1903. Shortspine spurdog.
?*Squalus probatovi* Myagkov & Kondyurin, 1986. Cunene dogfish.
Squalus rancureli Fourmanoir, 1978. Cyrano spurdog.

FAMILY CENTROPHORIDAE BLEEKER, 1859. Gulper Sharks

Genus *Centrophorus* Müller & Henle, 1837. Gulper sharks. Virtually circumglobal, except for the eastern Pacific and possibly the western South Atlantic continental and insular slopes and outer shelves.
Centrophorus acus Garman, 1906. Needle dogfish.
Centrophorus atromarginatus Garman, 1913. Dwarf gulper shark.
Centrophorus granulosus (Bloch & Schneider, 1801). Gulper shark.
Centrophorus harrissoni McCulloch, 1915. Longnose gulper shark.
Centrophorus isodon (Chu, Meng, & Liu, 1981). Blackfin gulper shark.
Centrophorus lusitanicus Bocage & Capello, 1864. Lowfin gulper shark.
Centrophorus moluccensis Bleeker, 1860. Smallfin gulper shark.
Centrophorus niaukang Teng, 1959. Taiwan gulper shark.
Centrophorus squamosus (Bonnaterre, 1788). Leaf-scale gulper shark.
Centrophorus tesselatus Garman, 1906. Mosaic gulper shark.

Genus *Deania* Jordan & Snyder, 1902. Birdbeak dogfishes. Western North Atlantic, eastern Atlantic, western Indian, western Pacific Oceans, continental and insular slopes.

Deania calcea (Lowe, 1839). Birdbeak dogfish.
Deania hystricosum (Garman, 1906). Rough longnose dogfish.
Deania profundorum (Smith & Radcliffe, 1912). Arrowhead dogfish.
Deania quadrispinosum (McCulloch, 1915). Longsnout dogfish.

FAMILY Etmopteridae Fowler, 1934. Lantern Sharks

Genus *Aculeola* de Buen, 1959. Eastern South Pacific continental slopes.
Aculeola nigra de Buen, 1959. Hooktooth dogfish.

Genus *Centroscyllium* Müller & Henle, 1841. Combtooth dogfishes. Virtually circumglobal on continental and insular slopes.
Centroscyllium excelsum Shirai & Nakaya, 1990. Highfin dogfish.
Centroscyllium fabricii (Reinhardt, 1825). Black dogfish.
Centroscyllium granulatum Günther, 1887. Granular dogfish.
Centroscyllium kamoharai Abe, 1966. Bareskin dogfish.
Centroscyllium nigrum Garman, 1899. Combtooth dogfish.
Centroscyllium ornatum (Alcock, 1889). Ornate dogfish.
Centroscyllium ritteri Jordan & Fowler, 1903. Whitefin dogfish.

Genus *Etmopterus* Rafinesque, 1810. Lantern sharks. Circumglobal on the outer continental and insular slopes, with a few species oceanic. Approximately 11 undescribed species in addition to the following:
Etmopterus baxteri Garrick, 1957. New Zealand lantern shark.
Etmopterus bigelowi Shirai & Tachikawa, 1993. Blurred smooth lantern shark.
Etmopterus brachyurus Smith & Radcliffe, 1912. Shorttail lantern shark.
Etmopteru bullisi Bigelow & Schroeder, 1957. Lined lantern shark.
Etmopterus carteri Springer & Burgess, 1985. Cylindrical lantern shark.
Etmopterus compagnoi Fricke & Koch, 1990. Brown lantern shark.
Etmopterus decacuspidatus Chan, 1966. Combtooth lantern shark.
Etmopterus gracilispinis Krefft, 1968. Broadband lantern shark.
Etmopterus granulosus (Günther, 1880). Southern lantern shark.
Etmopterus hillianus (Poey, 1861). Caribbean lantern shark.
Etmopterus litvinovi Parin & Kotlyar, in Kotlyar, 1990. Smalleye lantern shark.
Etmopterus lucifer Jordan & Snyder, 1902. Blackbelly lantern shark.
Etmopterus molleri (Whitley, 1939). Slendertail lantern shark.
Etmopterus perryi Springer & Burgess, 1985. Dwarf lantern shark.
Etmopterus polli Bigelow, Schroeder, & Springer, 1953. African lantern shark.
Etmopterus princeps Collett, 1904. Great lantern shark.
Etmopterus pusillus (Lowe, 1839). Smooth lantern shark.
Etmopterus pycnolepis Kotlyar, 1990. Dense-scale lantern shark.
Etmopterus robinsi Schofield & Burgess, 1997.
?*Etmopterus schmidti* Dolganov, 1986. Darkbelly lantern shark.
Etmopterus schultzi Bigelow, Schroeder, & Springer, 1953. Fringefin lantern shark.
Etmopterus sentosus Bass, D'Aubrey, & Kistnasamy, 1976. Thorny lantern shark.
Etmopterus spinax (Linnaeus, 1758). Velvet belly lantern shark.
Etmopterus splendidus Yano, 1988. Splendid lantern shark.
?*Etmopterus tasmaniensis* Myagkov & Pavlov, in Gubanov, Kondyurin, & Myagkov, 1986. Tasmanian lantern shark.
Etmopterus unicolor (Engelhardt, 1912). Brown lantern shark.
Etmopterus villosus Gilbert, 1905. Hawaiian lantern shark.
Etmopterus virens Bigelow, Schroeder, & Springer, 1953. Green lantern shark.

Genus *Miroscyllium* Shirai & Nakaya, 1990. Western North Pacific continental slopes.
Miroscyllium sheikoi (Dolganov, 1986). Rasptooth dogfish.

Genus *Trigonognathus* Mochizuki & Ohe, 1990. Western North Pacific Ocean, Japan, continental slopes.
Trigonognathus kabeyai Mochizuki & Ohe, 1990. Viper dogfish.

FAMILY SOMNOSIDAE JORDAN, 1888. Sleeper Sharks

Genus *Centroscymnus* Bocage & Capello, 1864. Velvet dogfishes. Western North Atlantic, eastern Atlantic, southwestern Indian, west-central Pacific, eastern South Pacific Oceans, continental and insular slopes.

Centroscymnus coelolepis Bocage & Capello, 1864. Portugese dogfish.
Centroscymnus crepidater (Bocage & Capello, 1864). Longnose velvet dogfish.
Centroscymnus cryptacanthus Regan, 1906. Shortnose velvet dogfish.
Centroscymnus macracanthus Regan, 1906. Largespine velvet dogfish.
Centroscymnus owstoni Garman, 1906. Roughskin dogfish.
Centroscymnus plunketi (Waite, 1909). Plunket shark.

Genus *Scymnodalatias* Garrick, 1956. Spineless velvet dogfishes. North-central Atlantic, southeastern Indian, South Pacific Oceans, continental and insular slopes, semi-oceanic.

Scymnodalatia albicauda Taniuchi & Garrick, 1986. Whitetail dogfish.
Scymnodalatias garricki Kukuyev & Konovalenko, 1988. Azores dogfish.
Scymnodalatias oligodon Kukuyev & Konovalenko, 1988. Sparsetooth dogfish.
Scymnodalatias sherwoodi (Archey, 1921). Sherwood dogfish.

Genus *Scymnodon* Bocage & Capello, 1864. Largetooth velvet dogfishes. Eastern North Atlantic, South Atlantic, Indian, western Pacific Oceans, slopes and oceanic.

Scymnodon ichiharai Yano & Tanaka, 1984. Japanese velvet dogfish.
Scymnodon ringens Bocage & Capello, 1864. Knifetooth dogfish.
Scymnodon squamulosus (Günther, 1877). Velvet dogfish. Recently placed in Genus *Zameus* Jordan & Fowler, 1903.

Genus *Somniosus* Le Sueur, 1818. Sleeper sharks (three or four species). Wide-ranging in Atlantic, Pacific, and Arctic Oceans, also off sub-Antarctic islands, continental and insular shelves (high latitudes) and slopes, also on sea mounts.

Subgenus *Rhinoscymnus* Gill, 1862:
Somniosus (Rhinoscymnus) rostratus (Risso, 1810). Little sleeper shark.
Subgenus *Somniosus* Le Sueur, 1818:
Somniosus (Somniosus) microcephalus (Bloch & Schneider, 1801). Greenland shark.
Somniosus (Somniosus) pacificus Bigelow & Schroeder, 1944. Pacific sleeper shark.

FAMILY OXYNOTIDAE GILL, 1872. Rough Sharks

Genus *Oxynotus* Rafinesque, 1810. Western North and eastern Atlantic and Mediterranean, western Pacific Oceans, continental and insular shelves and slopes.

Oxynotus bruniensis (Ogilby, 1893). Prickly dogfish.
Oxynotus caribbaeus Cervigon, 1961. Caribbean roughshark.
Oxynotus centrina (Linnaeus, 1758). Angular roughshark.
Oxynotus japonicus Yano & Murofushi, 1985. Japanese roughshark.
Oxynotus paradoxus Frade, 1929. Sailfin roughshark.
?*Oxynotus shubnikovi* Myagkov, in Gubanov, Kondyurin, & Myagkov, 1986. Flatiron shark.

FAMILY DALATIIDAE GRAY, 1851. Kitefin Sharks

Genus *Dalatias* Rafinesque, 1810. Wide-ranging in most warm-temperate and tropical seas, absent in the eastern Pacific Ocean, continental and insular slopes.

Dalatias licha (Bonnaterre, 1788). Kitefin shark.

Genus *Euprotomicroides* Hulley & Penrith, 1966. South Atlantic Ocean, continental slopes and oceanic.

Euprotomicroides zantedeschia Hulley & Penrith, 1966. Taillight shark.

Genus *Euprotomicrus* Gill, 1865. Circumglobal in temperate seas, oceanic.

Euprotomicrus bispinatus (Quoy & Gaimard, 1824). Pygmy shark.

Genus *Heteroscymnoides* Fowler, 1934. Southwestern Indian, western South Atlantic Oceans, oceanic.

Heteroscymnoides marleyi Fowler, 1934. Longnose pygmy shark.

Genus *Isistius* Gill, 1865. Cookie-cutter sharks. Circumglobal in temperate and tropical seas, oceanic.
   Isistius brasiliensis (Quoy & Gaimard, 1824). Cookie-cutter shark.
   ?*Isistius labialis* Meng, Chu, & Li, 1985. South China cookie-cutter shark.
   Isistius plutodus Garrick & Springer, 1964. Largetooth cookie-cutter shark.

Genus *Mollisquama* Dolganov, 1984. Eastern South Pacific Ocean on submarine ridges west of northern Chile, oceanic or slopes.
   Mollisquama parini Dolganov, 1984. Pocket shark.

Genus *Squaliolus* Smith & Radcliffe, in Smith, 1912. Spined pygmy sharks. Western South Atlantic, eastern North pacific, western Pacific Oceans, oceanic and on continental and insular shelves and slopes.
   Squaliolus aliae Teng, 1959. Smalleye pigmy shark.
   Squaliolus laticaudus Smith & Radcliffe, 1912. Spined pygmy shark.

ORDER PRISTIOPHORIFORMES

FAMILY PRISTIOPHORIDAE BLEEKER, 1859. Saw Sharks

Genus *Pliotrema* Regan, 1906. Southern Africa and Madagascar, continental shelves and slopes.
   Pliotrema warreni Regan, 1906. Sixgill sawshark.

Genus *Pristiophorus* Müller & Henle, 1837. Fivegill saw sharks. Western Indian, western Pacific, western North Atlantic Oceans, continental and insular shelves and slopes. At least four undescribed Indo-Pacific species in addition to the following:
   Pristiophorus cirratus (Latham, 1794). Longnose saw shark.
   Pristiophorus japonicus Günther, 1870. Japanese saw shark.
   Pristiophorus nudipinnis Günther, 1870. Shortnose saw shark.
   Pristiophorus schroederi Springer & Bullis, 1960. Bahamas saw shark.

ORDER SQUATINIFORMES

FAMILY SQUATINIDAE BONAPARTE, 1838. Angel Sharks

Genus *Squatina* Dumeril, 1806. Wide ranging on the continental shelves and upper slopes of most seas, mostly in temperate waters. At least three or four undescribed species in addition to the following:
   Squatina aculeata Dumeril, in Cuvier, 1817. Sawback angel shark.
   Squatina africana Regan, 1908. African angel shark.
   Squatina argentina Marini, 1930. Argentine angel shark.
   Squatina australis Regan, 1906. Australian angel shark.
   Squatina californica Ayres, 1859. Pacific angel shark.
   Squatina dumeril Lesueur, 1818. Sand devil.
   Squatina formosa Shen & Ting, 1972. Taiwan angel shark.
   Squatina guggenheim Marini, 1936. Angular angel shark.
   Squatina japonica Bleeker, 1858. Japanese angel shark.
   Squatina nebulosa Regan, 1906. Clouded angel shark.
   Squatina occulta Vooren & da Silva, 1991. Hidden angel shark.
   Squatina oculata Bonaparte, 1840. Smoothback angel shark.
   Squatina squatina (Linnaeus, 1758). Angel shark.
   Squatina tergocellata McCulloch, 1914. Ornate angel shark.
   Squatina tergocellatoides Chen, 1963. Ocellated angel shark.

ORDER HETERODONTIFORMES

FAMILY HETERODONTIDAE GRAY, 1851. Bullhead Sharks

Genus *Heterodontus* Blainville, 1816. Indo-Pacific, continental and insular shelves and uppermost slopes. One undescribed species in addition to the following:
   Heterodontus francisci (Girard, 1854). Horn shark.
   Heterodontus galeatus (Günther, 1870). Crested bullhead shark.

Heterodontus japonicus (Maclay & Macleay, 1884). Japanese bullhead shark.
Heterodontus mexicanus Taylor & Castro-Aguirre, 1972. Mexican horn shark.
Heterodontus portusjacksoni (Meyer, 1793). Port Jackson shark.
Heterodontus quoyi (Freminville, 1840). Galapagos bullhead shark.
Heterodontus ramalheira (Smith, 1949). White-spotted bullhead shark.
Heterodontus zebra (Gray, 1831). Zebra bullhead shark.

ORDER ORECTOLOBIFORMES: Carpet Sharks

FAMILY Parascylliidae Gill, 1862. Collared Carpet Sharks

Genus *Cirrhoscyllium* Smith & Radcliffe, 1913. Barbeled carpet sharks. Western North Pacific Ocean, continental and insular shelves.
Cirrhoscyllium expolitum Smith & Radcliffe, 1913. Barbelthroat carpet shark.
Cirrhoscyllium formosanum Teng, 1959. Taiwan saddled carpet shark.
Cirrhoscyllium japonicum Kamohara, 1943. Saddled carpet shark.

Genus *Parascyllium* Gill, 1862. Collared carpet sharks. Australia, continental shelves and upper slopes. One undescribed species in addition to the following:
Parascyllium collare Ramsay & Ogilby, 1888. Collared carpet shark.
Parascyllium ferrugineum McCulloch, 1911. Rusty carpet shark.
Parascyllium variolatum (Dumeril, 1853). Necklace carpet shark.

FAMILY Brachaeluridae Applegate, 1974. Blind Sharks

Genus *Brachaelurus* Ogilby, 1907. Continental shelf of eastern Australia.
Brachaelurus waddi (Bloch & Schneider, 1801). Blind shark.

Genus *Heteroscyllium* Regan, 1908. Continental shelf of Queensland, Australia.
Heteroscyllium colcloughi (Ogilby, 1907). Blue-gray carpet shark.

FAMILY Orectolobidae Gill, 1896. Wobbegongs

Genus *Eucrossorhinus* Regan, 1908. Australia and Indo-Australian archipelago, continental shelves.
Eucrossorhinus dasypogon (Bleeker, 1867). Tasseled wobbegong.

Genus *Orectolobus* Bonaparte, 1834. Western Pacific Ocean, continental shelves. At least one undescribed species in addition to the following:
Orectolobus japonicus Regan, 1906. Japanese wobbegong.
Orectolobus maculatus (Bonnaterre, 1788). Spotted wobbegong.
Orectolobus ornatus (de Vis, 1883). Ornate wobbegong.
Orectolobus wardi Whitley, 1939. Northern wobbegong.

Genus *Sutorectus* Whitley, 1939. Continental shelves of Australia.
Sutorectus tentaculatus (Peters, 1864). Cobbler wobbegong.

FAMILY Hemiscylliidae Gill, 1862. Long-tailed Carpet Sharks

Genus *Chiloscyllium* Müller & Henle, 1837. Bamboo sharks. Indo-West Pacific Ocean, continental shelves.
Chiloscyllium arabicum Gubanov, in Gubanov & Schleib, 1980. Arabian carpet shark.
Chiloscyllium burmensis Dingerkus & DeFino, 1983. Burmese bamboo shark.
Chiloscyllium griseum Müller & Henle, 1838. Gray bamboo shark.
Chiloscyllium hasselti Bleeker, 1852. Indonesian bamboo shark.
Chiloscyllium indicum (Gmelin, 1789). Slender bamboo shark.
Chiloscyllium plagiosum (Bennett, 1830). White-spotted bamboo shark.
Chiloscyllium punctatum Müller & Henle, 1838. Brown-banded bamboo shark.

Genus *Hemiscyllium* Müller & Henle, 1838. Epaulette sharks. Australia and Indo-Australian Archipelago, possibly Western Indian Ocean, continental shelves.
Hemiscyllium freycineti (Quoy & Gaimard, 1824). Indonesian speckled carpet shark.
Hemiscyllium hallstromi Whitley, 1967. Papuan epaulette shark.
Hemiscyllium ocellatum (Bonnaterre, 1788). Epaulette shark.
Hemiscyllium strahani Whitley, 1967. Hooded carpet shark.
Hemiscyllium trispeculare Richardson, 1843. Speckled carpet shark.

FAMILY GINGLYMOSTOMATIDAE GILL, 1862. Nurse Sharks

Genus *Ginglymostoma* Müller & Henle, 1837. Eastern Pacific, Atlantic Oceans, continental shelves.
  Ginglymostoma cirratum (Bonnaterre, 1788). Nurse shark.

Genus *Pseudoginglymostoma* Dingerkus, 1986. Western Indian Ocean, tropical East Africa, continental shelves. Familial placement provisional.
  Pseudoginglymostoma brevicaudatum (Günther, in Playfair & Günther, 1866). Shorttail nurse shark.

Genus *Nebrius* Rüppell, 1837. Indo-West Pacific Ocean, continental shelves.
  Nebrius ferrugineus (Lesson, 1830). Tawny nurse shark. Commonly referred to *N. concolor* Rüppell, 1837.

FAMILY STEGOSTOMATIDAE GILL, 1862

Genus *Stegostoma* Müller & Henle, 1837. Indo-West Pacific Ocean, continental shelves.
  Stegostoma fasciatum (Hermann, 1783). Zebra shark.

FAMILY RHINCODONTIDAE MÜLLER & HENLE, 1839

Genus *Rhincodon* Smith, 1829. Circumglobal in all warm-temperate and tropical seas, shelves and oceanic.
  Rhincodon typus Smith, 1828. Whale shark.

ORDER LAMNIFORMES: MACKEREL SHARKS

FAMILY MITSUKURINIDAE JORDAN, 1898. Goblin Sharks

Genus *Mitsukurina* Jordan, 1898. Atlantic, Indian and western Pacific Oceans, continental slopes.
  Mitsukurina owstoni Jordan, 1898. Goblin shark.

FAMILY ODONTASPIDIDAE MÜLLER & HENLE, 1839. Sand Tiger Sharks

Genus *Carcharias* Rafinesque, 1810. Largetooth sand tigers (formerly placed in *Eugomphodus* Gill, 1862). Atlantic, Indo-West Pacific Oceans, shelves.
  Carcharias taurus Rafinesque, 1810. Sand tiger, spotted raggedtooth, or gray nurse shark.
  ?*Carcharias tricuspidatus* Day, 1878. Indian sand tiger (possible synonym of *C. taurus*).

Genus *Odontaspis* Agassiz, 1838. Smalltooth sand tigers. Virtually circumglobal in warm-temperate and tropical seas, outer shelves and slopes, seamounts.
  Odontaspis ferox (Risso, 1810). Smalltooth sand tiger or bumpytail raggedtooth.
  Odontaspis noronhai (Maul, 1955). Bigeye sand tiger.

FAMILY PSEUDOCARCHARIIDAE COMPAGNO, 1973. Crocodile Sharks

Genus *Pseudocarcharias* Cadenat, 1963. Virtually circumtropical, oceanic.
  Pseudocarcharias kamoharai (Matsubara, 1936). Crocodile shark.

FAMILY MEGACHASMIDAE TAYLOR, COMPAGNO, & STRUHSAKER, 1983. Megamouth Sharks

Genus *Megachasma* Taylor, Compagno, & Struhsaker, 1983. Pacific, eastern Indian, and South Atlantic Oceans, oceanic and on continental shelves.
  Megachasma pelagios Taylor, Compagno, & Struhsaker, 1983. Megamouth shark.

FAMILY ALOPIIDAE BONAPARTE, 1838. Thresher Sharks

Genus *Alopias* Rafinesque, 1810. Circumglobal in warm-temperate and tropical seas, oceanic, continental shelves and upper slopes. Possibly a fourth species in the eastern Pacific Ocean.
  Alopias pelagicus Nakamura, 1935. Pelagic thresher.
  Alopias superciliosus (Lowe, 1839). Bigeye thresher.
  Alopias vulpinus (Bonnaterre, 1788). Thresher shark.

FAMILY CETORHINIDAE GILL, 1862. Basking Sharks

Genus *Cetorhinus* Blainville, 1816. Circumglobal in temperate seas, off continental and insular shelves.

Cetorhinus maximus (Gunnerus, 1765). Basking shark.

FAMILY LAMNIDAE MÜLLER & HENLE, 1838. Mackerel Sharks

Genus *Carcharodon* Smith, in Müller & Henle, 1838. Circumglobal in all seas but most commonly recorded in temperate seas, continental and insular shelves and slopes, oceanic.

Carcharodon carcharias (Linnaeus, 1758). Great white shark.

Genus *Isurus* Rafinesque, 1810. Makos. Circumglobal in all warm seas, continental and insular shelves, oceanic.

Isurus oxyrinchus Rafinesque, 1810. Shortfin mako.
Isurus paucus Guitart Manday, 1966. Longfin mako.

Genus *Lamna* Cuvier, 1816. Mackerel sharks. Wide-ranging in temperate to cold seas, continental and insular shelves, oceanic.

Lamna ditropis Hubbs & Follett, 1947. Salmon shark.
Lamna nasus (Bonnaterre, 1788). Porbeagle shark.

ORDER CARCHARHINIFORMES: GROUND SHARKS

FAMILY SCYLIORHINIDAE GILL, 1862. Cat Sharks

Genus *Apristurus* Garman, 1913. Ghost or demon cat sharks. Circumglobal, on continental and insular slopes and occasionally shelves.

Apristurus acanutus Chu, Meng, & Li, in Meng, Chu & Li, 1985. Flatnose cat shark.
Apristurus aphyodes Nakaya & Stehmann, 1998
Apristurus atlanticus (Koefoed, 1932). Atlantic ghost cat shark.
Apristurus brunneus (Gilbert, 1892). Brown cat shark.
Apristurus canutus Springer & Heemstra, in Springer, 1979. Hoary cat shark.
?*Apristurus fedorovi* Dolganov, 1985. Stout cat shark.
Apristurus gibbosus Meng, Chu & Li, 1985. Humpback cat shark.
Apristurus herklotsi (Fowler, 1934). Longfin cat shark.
Apristurus indicus (Brauer, 1906). Smallbelly cat shark.
Apristurus investigatoris (Misra, 1962). Broadnose cat shark.
Apristurus japonicus Nakaya, 1975. Japanese cat shark.
Apristurus kampae Taylor, 1972. Longnose cat shark.
Apristurus laurussoni (Saemundsson, 1922). Iceland cat shark (including
  A. maderensis Cadenat & Maul, 1966 as a synonym).
Apristurus longicephalus Nakaya, 1975. Longhead cat shark.
Apristurus macrorhynchus (Tanaka, 1909). Flathead cat shark.
Apristurus macrostomus Meng, Chu, & Li, 1985. Broadmouth cat shark.
Apristurus manis (Springer, 1979). Ghost cat shark.
Apristurus microps (Gilchrist, 1922). Smalleye cat shark.
Apristurus micropterygeus Meng, Chu & Li, in Chu, Meng, & Li, 1986. Smalldorsal cat shark.
Apristurus nasutus de Buen, 1959. Largenose cat shark.
Apristurus parvipinnis Springer & Heemstra, in Springer, 1979. Smallfin cat shark.
?*Apristurus pinguis* Deng, Xiong, & Zhan, 1983. Fat cat shark.
Apristurus platyrhynchus (Tanaka, 1909). Spatulasnout cat shark.
Apristurus profundorum (Goode & Bean, 1896). Deepwater cat shark.
Apristurus riveri Bigelow & Schroeder, 1944. Broadgill cat shark.
Apristurus saldanha (Barnard, 1925). Saldanha cat shark.
Apristurus sibogae (Weber, 1913). Pale cat shark.
Apristurus sinensis Chu & Hu, in Chu, Meng, Hu, & Li, 1981. South China cat shark.
Apristurus spongiceps (Gilbert, 1895). Spongehead cat shark.
Apristurus stenseni (Springer, 1979). Panama ghost cat shark.
Apristurus verweyi (Fowler, 1934). Borneo cat shark.

Genus *Asymbolus* Whitley, 1939. Australian spotted cat sharks. Australia and New Caledonia, continental shelves and uppermost slopes. At least seven undescribed species in addition to the following:
   Asymbolus analis (Ogilby, 1895). Grey spotted cat shark.
   Asymbolus vincenti (Zietz, 1908). Gulf cat shark.

Genus *Atelomycterus* Garman, 1913. Coral cat sharks. Northern Indian and Western Pacific Oceans, continental shelves.
   Atelomycterus fasciatus Compagno & Stevens, 1993. Banded sand cat shark.
   Atelomycterus macleayi Whitley, 1939. Australian marbled cat shark.
   Atelomycterus marmoratus (Bennett, 1830). Coral cat shark.

Genus *Aulohalaelurus* Fowler, 1934. Blackspotted cat sharks. Australia and New Caledonia, continental and insular shelves.
   Aulohalaelurus kanakorum Seret, 1990. New Caledonia cat shark.
   Aulohalaelurus labiosus (Waite, 1905). Black-spotted cat shark.

Genus *Cephaloscyllium* Gill, 1862. Swell sharks. Indo-Pacific Ocean, continental and insular shelves and upper slopes. Possibly nine undescribed species in addition to the following:
   Cephaloscyllium fasciatum Chan, 1966. Reticulated swell shark.
   Cephaloscyllium isabellum (Bonnaterre, 1788). Draughtsboard shark.
   Cephaloscyllium laticeps (Dumeril, 1853). Australian swell shark.
   Cephaloscyllium silasi (Talwar, 1974). Indian swell shark.
   Cephaloscyllium sufflans (Regan, 1921). Balloon shark.
   Cephaloscyllium umbratile Jordan & Fowler, 1903. Japanese swell shark.
   Cephaloscyllium ventriosum (Garman, 1880). Swell shark.

Genus *Cephalurus* Bigelow & Schroeder, 1941. Eastern Pacific Ocean, continental slopes. Possibly one additional species:
   Cephalurus cephalus (Gilbert, 1892). Lollipop cat shark.

Genus *Galeus* Rafinesque, 1810. Sawtail cat sharks. Virtually circumglobal, continental shelves and slopes. Possibly one additional species.
   Galeus arae (Nichols, 1927). Roughtail cat shark.
   Galeus atlanticus (Vaillant, 1888). Atlantic sawtail cat shark.
   Galeus boardmani (Whitley, 1928). Australian sawtail cat shark.
   Galeus cadenati Springer, 1966. Longfin sawtail cat shark.
   Galeus eastmani (Jordan & Snyder, 1904). Gecko cat shark.
   Galeus gracilis Compagno & Stevens, 1993. Slender sawtail cat shark.
   Galeus longirostris Tachikawa & Taniuchi, 1987. Longnose sawtail cat shark.
   Galeus melastomus Rafinesque, 1810. Blackmouth cat shark.
   Galeus murinus (Collett, 1904). Mouse cat shark.
   Galeus nipponensis Nakaya, 1975. Broadfin sawtail cat shark.
   Galeus piperatus Springer & Wagner, 1966. Peppered cat shark.
   Galeus polli Cadenat, 1959. African sawtail cat shark.
   Galeus sauteri (Jordan & Richardson, 1909). Blacktip sawtail cat shark.
   Galeus schultzi Springer, 1979. Dwarf sawtail cat shark.
   Galeus springeri Konstantinou & Cozzi, 1998. Springer's sawtail cat shark.

Genus *Halaelurus* Gill, 1862. Tiger cat sharks. Indian, western Pacific, southeastern Pacific Oceans, continental and insular shelves and slopes. At least one undescribed species.

   Subgenus *Bythaelurus* Compagno, 1988:
   ?*Halaelurus (Bythaelurus?) alcocki* Garman, 1913?. Arabian cat shark.
   Halaelurus (Bythaelurus) canescens (Günther, 1878). Dusky cat shark.
   Halaelurus (Bythaelurus) clevai Seret, 1987. Broadhead cat shark.
   Halaelurus (Bythaelurus) dawsoni Springer, 1971. New Zealand cat shark.
   Halaelurus (Bythaelurus) hispidus (Alcock, 1891). Bristly cat shark.
   Halaelurus (Bythaelurus) immaculatus Chu & Meng, in Chu, Meng, Hu, & Li, 1982. Spotless cat shark.
   Halaelurus (Bythaelurus) lutarius Springer & D'Aubrey, 1972. Mud cat shark.

Subgenus *Halaelurus* Gill, 1862:
Halaelurus (Halaelurus) boesemani Springer & D'Aubrey, 1972. Speckled cat shark.
Halaelurus (Halaelurus) buergeri (Müller & Henle, 1838). Darkspot, black-spotted, or Nagasaki cat shark.
Halaelurus (Halaelurus) lineatus Bass, D'Aubrey, & Kistnasamy, 1975. Lined cat shark.
Halaelurus (Halaelurus) natalensis (Regan, 1904). Tiger cat shark.
Halaelurus (Halaelurus) quagga (Alcock, 1899). Quagga cat shark.

Genus *Haploblepharus* Garman, 1913. Shy sharks. Southern Africa, continental shelves. At least one undescribed species.
Haploblepharus edwardsii (Voigt, in Cuvier, 1832). Puff-adder shy shark.
Haploblepharus fuscus Smith, 1950. Brown shy shark.
Haploblepharus pictus (Müller & Henle, 1838). Dark shy shark.

Genus *Holohalaelurus* Fowler, 1934. Izak cat sharks. Southern and East Africa, upper continental slopes and lower shelves. Possibly two undescribed species.
Holohalaelurus punctatus (Gilchrist & Thompson, 1914). African spotted cat shark.
Holohalaelurus regani (Gilchrist, 1922). Izak cat shark.

Genus *Parmaturus* Garman, 1906. Filetail cat sharks. Western Pacific, eastern North Pacific, western North Atlantic Oceans, continental and insular slopes. Possibly two undescribed species.
Parmaturus campechiensis Springer, 1979. Campeche cat shark.
Parmaturus macmillani Hardy, 1985. New Zealand filetail.
Parmaturus melanobranchius (Chan, 1966). Blackgill cat shark.
Parmaturus pilosus Garman, 1906. Salamander shark.
Parmaturus xaniurus (Gilbert, 1892). Filetail cat shark.

Genus *Pentanchus* Smith & Radcliffe, in Smith, 1912. Philippines, slopes.
Pentanchus profundicolus (Smith & Radcliffe, 1912). Onefin cat shark.

Genus *Poroderma* Smith, 1837. Barbeled cat sharks. Temperate southern Africa, continental shelves.
Poroderma africanum (Gmelin, 1789). Striped cat shark or pyjama shark.
Poroderma pantherinum (Smith, in Müller & Henle, 1838). Leopard cat shark.

Genus *Schroederichthys* Springer, 1966. Narrowtail cat sharks. Eastern South Pacific and western South Atlantic Oceans, South America, continental shelves and upper slopes.
Schroederichthys bivius (Smith, in Müller & Henle, 1838). Narrowmouth cat shark.
Schroederichthys chilensis (Guichenot, in Gay, 1848). Red-spotted cat shark.
Schroederichthys maculatus Springer, 1966. Narrowtail cat shark.
Schroederichthys tenuis Springer, 1966. Slender cat shark.

Genus *Scyliorhinus* Blainville, 1816. Spotted cat sharks. Atlantic, southwestern Indian, western Pacific Oceans, outer shelves and upper slopes.
Scyliorhinus besnardi Springer & Sadowsky, 1970. Polka-dot cat shark.
Scyliorhinus boa Goode & Bean, 1896. Boa cat shark.
Scyliorhinus canicula (Linnaeus, 1758). Small-spotted cat shark.
Scyliorhinus capensis (Smith, in Müller & Henle, 1838). Yellow-spotted cat shark.
Scyliorhinus cervigoni Maurin & Bonnet, 1970. West African cat shark.
Scyliorhinus comoroensis Compagno, 1989. Comoro cat shark.
Scyliorhinus garmani (Fowler, 1934). Brown-spotted cat shark.
Scyliorhinus haeckelii (Ribeiro, 1907). Freckled cat shark.
Scyliorhinus hesperius Springer, 1966. White-saddled cat shark.
Scyliorhinus meadi Springer, 1966. Blotched cat shark.
Scyliorhinus retifer (Garman, 1881). Chain cat shark.
Scyliorhinus stellaris (Linnaeus, 1758). Nursehound.
Scyliorhinus tokubee Shirai, Hagiwara, & Nakaya, 1992. Izu cat shark.
Scyliorhinus torazame (Tanaka, 1908). Cloudy cat shark.
Scyliorhinus torrei Howell-Rivero, 1936. Dwarf cat shark.

FAMILY PROSCYLLIIDAE FOWLER, 1941. Finback Cat Sharks

Genus *Ctenacis* Compagno, 1973. East Africa, northwestern Indian Ocean, continental slopes.
Ctenacis fehlmanni (Springer, 1968). Harlequin cat shark.

Genus *Eridacnis* Smith, 1913. Ribbontail cat sharks. Western North Atlantic, Indo-West Pacific Oceans, continental slopes.
Eridacnis barbouri (Bigelow & Schroeder, 1944). Cuban ribbontail cat shark.
Eridacnis radcliffei Smith, 1913. Pygmy ribbontail cat shark.
Eridacnis sinuans (Smith, 1957). African ribbontail cat shark.

Genus *Proscyllium* Hilgendorf, 1904. Graceful cat sharks. Western Pacific, continental shelves and uppermost slopes. Possibly including *P. venustum* (Tanaka, 1913)
Proscyllium habereri Hilgendorf, 1904. Graceful cat shark.

FAMILY PSEUDOTRIAKIDAE GILL, 1893. False Cat Sharks. Also, an undescribed genus and species from the continental slopes of the northern Indian Ocean.

Genus *Gollum* Compagno, 1973. Vicinity of New Zealand, upper slopes.
Gollum attenuatus (Garrick, 1954). Slender smooth hound.

Genus *Pseudotriakis* Capello, 1868. Wide-ranging but spotty in the North Atlantic, southwestern Indian, and West-Central Pacific Oceans, continental and insular slopes. including *P. acrales* Jordan & Snyder, 1904.
Pseudotriakis microdon Capello, 1868. False cat shark.

FAMILY LEPTOCHARIIDAE GRAY, 1851. Barbeled Hound Sharks

Genus *Leptocharias* Smith, in Müller & Henle, 1838. Eastern Atlantic Ocean, continental shelves.
Leptocharias smithii (Müller & Henle, 1839). Barbeled hound shark.

FAMILY TRIAKIDAE GRAY, 1851. Hound Sharks

Genus *Furgaleus* Whitley, 1951. Australia, continental shelves.
Furgaleus macki (Whitley, 1943). Whiskery shark.

Genus *Galeorhinus* Blainville, 1816. Eastern Atlantic and southwestern Indian, western South Pacific, eastern Pacific, western South Atlantic Oceans, continental and insular shelves and upper slopes.
Galeorhinus galeus (Linnaeus, 1758). Tope shark.

Genus *Gogolia* Compagno, 1973. Western South Pacific Ocean, Papua-New Guinea, insular shelf.
Gogolia filewoodi Compagno, 1973. Sailback hound shark.

Genus *Hemitriakis* Herre, 1923. Whitefin tope sharks. At least one undescribed species.
Hemitriakis abdita Compagno & Stevens, 1993. Deepwater sicklefin hound shark.
Hemitriakis japanica (Müller & Henle, 1839). Japanese tope shark.
Hemitriakis falcata Compagno & Stevens, 1993. Sicklefin hound shark.
Hemitriakis leucoperiptera Herre, 1923. Whitefin tope shark.

Genus *Hypogaleus* Smith, 1957. Western Indian, western Pacific Oceans, continental and insular shelves and uppermost slopes.
Hypogaleus hyugaensis (Miyosi, 1939). Blacktip tope shark.

Genus *Iago* Compagno & Springer, 1971. Bigeye hound sharks. Indian and western Pacific Oceans, continental and insular shelves and upper slopes. Possibly two undescribed species.
Iago garricki Fourmanoir, 1979. Longnose hound shark.
Iago omanensis (Norman, 1939). Bigeye hound shark.

Genus *Mustelus* Linck, 1790. Smooth hounds. Circumglobal in all temperate and tropical seas, continental and insular shelves and uppermost slopes. At least six undescribed species.
Mustelus antarcticus Günther, 1870. Gummy shark.
Mustelus asterias Cloquet, 1821. Starry smooth hound.

Mustelus californicus Gill, 1864. Gray smooth hound.
Mustelus canis (Mitchell, 1815). Dusky smooth hound.
Mustelus dorsalis Gill, 1864. Sharpnose smooth hound.
Mustelus fasciatus (Garman, 1913). Striped smooth hound.
Mustelus griseus Pitschmann, 1908. Spotless smooth hound.
Mustelus henlei (Gill, 1863). Brown smooth hound.
Mustelus higmani Springer & Lowe, 1963. Smalleye smooth hound.
Mustelus lenticulatus Phillipps, 1932. Spotted estuary smooth hound or rig.
Mustelus lunulatus Jordan & Gilbert, 1883. Sicklefin smooth hound.
Mustelus manazo Bleeker, 1854. Star-spotted smooth hound.
Mustelus mento Cope, 1877. Speckled smooth hound.
Mustelus minicanis Heemstra, 1997. Common name?
Mustelus mosis Hemprich & Ehrenberg, 1899. Arabian, hardnose, or Moses smooth hound.
Mustelus mustelus (Linnaeus, 1758). Smooth hound.
Mustelus norrisi Springer, 1940. Narrowfin or Florida smooth hound.
Mustelus palumbes Smith, 1957. Whitespot smooth hound.
Mustelus punctulatus Risso, 1826. Blackspot smooth hound.
Mustelus schmitti Springer, 1940. Narrownose smooth hound.
Mustelus sinusmexicanus Heemstra, 1997. Common name?
Mustelus whitneyi Chirichigno, 1973. Humpback smooth hound.

Genus *Scylliogaleus* Boulenger, 1902. Southwestern Indian Ocean, South Africa, continental shelf.
Scylliogaleus quecketti Boulenger, 1902. Flapnose houndshark.

Genus *Triakis* Müller & Henle, 1838. Leopard sharks. Western North Pacific, eastern South Atlantic and southwestern Indian Oceans (southern Africa), eastern Pacific Ocean, continental and insular shelves.

Subgenus *Cazon* de Buen, 1959:
Triakis (Cazon) acutipinna Kato, 1968. Sharpfin houndshark.
Triakis (Cazon) maculata Kner & Steindachner, 1866. Spotted houndshark.
Triakis (Cazon) megalopterus (Smith, 1849). Spotted gully shark.

Subgenus *Triakis* Müller & Henle, 1838:
Triakis (Triakis) scyllium Müller & Henle, 1839. Banded houndshark.
Triakis (Triakis) semifasciata Girard, 1854. Leopard shark.

FAMILY HEMIGALEIDAE HASSE, 1879. Weasel Sharks

Genus *Chaenogaleus* Gill, 1862. Northern Indian and western Pacific Oceans, continental shelves.
Chaenogaleus macrostoma (Bleeker, 1852). Hooktooth shark.

Genus *Hemigaleus* Bleeker, 1852. Sicklefin weasel sharks. Northern Indian and Western Pacific Oceans, continental shelves. Possibly one undescribed species.
Hemigaleus microstoma Bleeker, 1852. Sicklefin weasel shark.

Genus *Hemipristis* Agassiz, 1843. Indo-West Pacific Ocean from South Africa to Australia, continental shelves.
Hemipristis elongatus (Klunzinger, 1871). Snaggletooth shark.

Genus *Paragaleus* Budker, 1935. Sharpnose weasel sharks. Eastern tropical Atlantic, Indian, and West-Central Pacific Oceans, continental shelves.
Paragaleus leucolomatus Compagno & Smale, 1985. Whitetip weasel shark.
Paragaleus pectoralis (Garman, 1906). Atlantic weasel shark.
Paragaleus randalli Compagno, Krupp, & Carpenter, 1996. Slender weasel shark.
Paragaleus tengi (Chen, 1963). Straighttooth weasel shark.

FAMILY CARCHARHINIDAE JORDAN & EVERMANN, 1896. Requiem Sharks

Genus *Carcharhinus* Blainville, 1816. Gray sharks. Circumglobal in all warm-temperate and tropical seas, continental and insular shelves and uppermost slopes, oceanic or semioceanic, freshwater tropical and warm-temperate rivers and lakes. One undescribed species.

Carcharhinus acronotus (Poey, 1860). Blacknose shark.
Carcharhinus albimarginatus (Rüppell, 1837). Silvertip shark.
Carcharhinus altimus (Springer, 1950). Bignose shark.
Carcharhinus amblyrhynchoides (Whitley, 1934). Graceful shark.
Carcharhinus amblyrhynchos (Bleeker, 1856). Gray reef shark.
Carcharhinus amboinensis (Müller & Henle, 1839). Pigeye or Java shark.
Carcharhinus borneensis (Bleeker, 1859). Borneo shark.
Carcharhinus brachyurus (Günther, 1870). Bronze whaler.
Carcharhinus brevipinna (Müller & Henle, 1839). Spinner shark.
Carcharhinus cautus (Whitley, 1945). Nervous shark.
Carcharhinus dussumieri (Valenciennes, in Müller & Henle, 1839). Whitecheek shark.
Carcharhinus falciformis (Bibron, in Müller & Henle, 1839). Silky shark.
Carcharhinus fitzroyensis (Whitley, 1943). Creek whaler.
Carcharhinus galapagensis (Snodgrass & Heller, 1905). Galapagos shark.
Carcharhinus hemiodon (Valenciennes, in Müller & Henle, 1839). Pondicherry shark.
Carcharhinus isodon (Valenciennes, in Müller & Henle, 1839). Finetooth shark.
Carcharhinus leiodon Garrick, 1985. Smooth tooth blacktip shark.
Carcharhinus leucas (Valenciennes, in Müller & Henle, 1839). Bull shark.
Carcharhinus limbatus (Valenciennes, in Müller & Henle, 1839). Blacktip shark.
Carcharhinus longimanus (Poey, 1861). Oceanic whitetip shark.
Carcharhinus macloti (Müller & Henle, 1839). Hardnose shark.
Carcharhinus melanopterus (Quoy & Gaimard, 1824). Blacktip reef shark.
Carcharhinus obscurus (Lesueur, 1818). Dusky shark.
Carcharhinus perezi (Poey, 1876). Caribbean reef shark.
Carcharhinus plumbeus (Nardo, 1827). Sandbar shark.
Carcharhinus porosus (Ranzani, 1839). Smalltail shark.
Carcharhinus sealei (Pietschmann, 1916). Blackspot shark.
Carcharhinus signatus (Poey, 1868). Night shark.
Carcharhinus sorrah (Valenciennes, in Müller & Henle, 1839). Spottail shark.
Carcharhinus tilsoni (Whitley, 1950). Australian blacktip shark.

Genus *Galeocerdo* Müller & Henle, 1837. Circumglobal in warm-temperate and tropical seas, continental and insular shelves, semioceanic.

Galeocerdo cuvier (Peron & Lesueur, in Lesueur, 1822). Tiger shark.

Genus *Glyphis* Agassiz, 1843. River sharks. Northern Indian and western South Pacific Oceans, continental shelves and tropical rivers. Possibly three undescribed species.

Glyphis gangeticus (Müller & Henle, 1839). Ganges shark.
Glyphis glyphis (Müller & Henle, 1839). Speartooth shark.
Glyphis siamensis (Steindachner, 1896). Irrawaddy river shark.

Genus *Isogomphodon* Gill, 1862. Tropical Atlantic coast of South America, continental shelves.

Isogomphodon oxyrhynchus (Müller & Henle, 1839). Daggernose shark.

Genus *Lamiopsis* Gill, 1862. Northern Indian and western Pacific Oceans, continental shelves.

Lamiopsis temmincki (Müller & Henle, 1839). Broadfin shark.

Genus *Loxodon* Müller & Henle, 1839. Indo-West Pacific Ocean, continental shelves.

Loxodon macrorhinus Müller & Henle, 1839. Sliteye shark.

Genus *Nasolamia* Compagno & Garrick, 1983. Eastern Pacific Ocean, continental shelves.

Nasolamia velox (Gilbert, in Jordan & Evermann, 1898). Whitenose shark.

Genus *Negaprion* Whitley, 1940. Lemon sharks. Circumtropical, continental and insular shelves.
    Negaprion acutidens (Rüppell, 1837). Sharptooth lemon shark.
    Negaprion brevirostris (Poey, 1868). Lemon shark (possibly including *N. fronto* [Jordan & Gilbert, 1882]).

Genus *Prionace* Cantor, 1849. Circumglobal in all temperate and tropical seas, oceanic but penetrating narrow continental shelves.
    Prionace glauca (Linnaeus, 1758). Blue shark.

Genus *Rhizoprionodon* Whitley, 1929. Sharpnose sharks. Circumtropical on continental shelves, may penetrate mouths of rivers.
    Rhizoprionodon acutus (Rüppell, 1837). Milk shark.
    Rhizoprionodon lalandei (Valenciennes, in Müller & Henle, 1839). Brazilian sharpnose shark.
    Rhizoprionodon longurio (Jordan & Gilbert, 1882). Pacific sharpnose shark.
    Rhizoprionodon oligolinx Springer, 1964. Gray sharpnose shark.
    Rhizoprionodon porosus (Poey, 1861)?. Caribbean sharpnose shark.
    Rhizoprionodon taylori (Ogilby, 1915). Australian sharpnose shark.
    Rhizoprionodon terraenovae (Richardson, 1836). Atlantic sharpnose shark.

Genus *Scoliodon* Müller & Henle, 1837. Spadenose sharks. Indo-West Pacific Ocean, continental shelves.
    Scoliodon laticaudus Müller & Henle, 1838. Spadenose shark.

Genus *Triaenodon* Müller & Henle, 1837. Indo-Pacific Ocean, continental and insular shelves.
    Triaenodon obesus (Rüppell, 1837). Whitetip reef shark.

FAMILY SPHYRNIDAE GILL, 1872. Hammerhead Sharks

Genus *Eusphyra* Gill, 1862. Northern Indian and western Pacific Oceans, continental shelves.
    Eusphyra blochii (Cuvier, 1817). Winghead shark.

Genus *Sphyrna* Rafinesque, 1810. Hammerhead and bonnethead sharks. Circumglobal in warm-temperate and tropical seas, shelves and adjacent epipelagic zone. Possibly one or two additional species.
    Subgenus *Mesozygaena* Compagno, 1988:
    Sphyrna (Mesozygaena) corona Springer, 1940. Mallethead shark.
    Sphyrna (Mesozygaena) media Springer, 1940. Scoophead shark.
    Sphyrna (Mesozygaena) tudes (Valenciennes, 1822). Smalleye hammerhead.
    Subgenus *Platysqualus* Swainson, 1839:
    Sphyrna (Platysqualus) tiburo (Linnaeus, 1758). Bonnethead shark.
    Subgenus *Sphyrna* Rafinesque, 1810:
    Sphyrna (Sphyrna) lewini (Griffith & Smith, in Cuvier, Griffith, & Smith, 1834). Scalloped hammerhead.
    Sphyrna (Sphyrna) mokarran (Rüppell, 1837). Great hammerhead.
    Sphyrna (Sphyrna) zygaena (Linnaeus, 1758). Smooth hammerhead.

RAYS (BATOIDS)

ORDER PRISTIFORMES

FAMILY PRISTIDAE BONAPARTE, 1838. Sawfishes

Genus *Anoxypristis* White & Moy-Thomas, 1941. Northern Indian and western Pacific Oceans, continental shelves and possibly freshwater.
    Anoxypristis cuspidata (Latham, 1794). Knifetooth, pointed, or narrow sawfish.

Genus *Pristis* Linck, 1790. Narrowtooth sawfishes. Circumtropical, continental shelves and freshwater.
    Pristis clavata Garman, 1906. Dwarf or Queensland sawfish.
    Pristis microdon Latham, 1794. Greattooth or freshwater sawfish.
    Pristis pectinata Latham, 1794. Smalltooth or wide sawfish.

Pristis perotteti Valenciennes, in Müller & Henle, 1841. Largetooth sawfish.
Pristis pristis (Linnaeus, 1758). Common sawfish.
Pristis zijsron Bleeker, 1851. Green sawfish.

ORDER RHINIFORMES

FAMILY RHINIDAE MÜLLER & HENLE, 1841. Wedgefishes

Genus *Rhina* Bloch & Schneider, 1801. Indo-West Pacific Ocean, continental shelves.
Rhina ancylostoma Bloch & Schneider, 1801. Bowmouth guitarfish or shark ray.

Genus *Rhynchobatus* Müller & Henle, 1837. Wedgefishes. Eastern Atlantic and Indo-West Pacific Oceans, continental shelves. Two undescribed species.
Rhynchobatus australiae Whitley, 1939. White-spotted shovelnose ray.
Rhynchobatus djiddensis (Forsskael, 1775). White-spotted wedgefish or giant guitarfish.
Rhynchobatus laevis (Bloch & Schneider, 1801). Smooth nose wedgefish.
Rhynchobatus luebberti Ehrenbaum, 1914. African or spikenose wedgefish.

ORDER RHINOBATIFORMES: GUITARFISHES

FAMILY RHINOBATIDAE MÜLLER & HENLE, 1837. Guitarfishes

Genus *Aptychotrema* Norman, 1926. Australasian shovelnose rays. West-Central and western South Pacific Ocean, continental shelves. Two undescribed species.
Aptychotrema rostrata (Shaw & Nodder, 1794). Eastern shovelnose ray.
Aptychotrema vincentiana (Haake, 1885). Southern shovelnose ray.

Genus *Rhinobatos* Linck, 1790. Guitarfishes. Circumglobal in warm-temperate and tropical seas, continental shelves, and uppermost slopes. Subgeneric arrangement provisional. Possibly three undescribed species.

Subgenus *Acroteriobatus* Giltay, 1928:
Rhinobatos (Acroteriobatus) annulatus Smith, in Müller & Henle, 1841. Lesser guitarfish.
Rhinobatos (Acroteriobatus) blochii Müller & Henle, 1841. Bluntnose guitarfish or fiddlefish.
Rhinobatos (Acroteriobatus) leucospilus Norman, 1926. Grayspot guitarfish.
Rhinobatos (Acroteriobatus) ocellatus Norman, 1926. Speckled guitarfish.
Rhinobatos (Acroteriobatus) salalah Randall & Compagno, 1995. Salalah guitarfish.
Rhinobatos (Acroteriobatus) variegatus Nair & Lal Mohan, 1973. Stripenose guitarfish.
Rhinobatos (Acroteriobatus) zanzibarensis Norman, 1926. Zanzibar guitarfish.

Subgenus *Glaucostegus* Bonaparte, 1846:
Rhinobatos (Glaucostegus) cemiculus St. Hilaire, 1817. Blackchin guitarfish.
Rhinobatos (Glaucostegus) glaucostigmus Jordan & Gilbert, 1884. Slaty-spotted guitarfish.
Rhinobatos (Glaucostegus) halavi (Forsskael, 1775). Halavi guitarfish.
Rhinobatos (Glaucostegus) horkelii Müller & Henle, 1841. Brazilian guitarfish.
Rhinobatos (Glaucostegus) lentiginosus Garman, 1880. Freckled or Atlantic guitarfish.
Rhinobatos (Glaucostegus) leucorhynchus Günther, 1866. Whitenose guitarfish.
Rhinobatos (Glaucostegus) microphthalmus Teng, 1959. Smalleyed guitarfish.
Rhinobatos (Glaucostegus) percellens (Walbaum, 1792). Southern guitarfish.
Rhinobatos (Glaucostegus) petiti Chabanaud, 1929. Madagascar guitarfish.
Rhinobatos (Glaucostegus) planiceps Garman, 1880. Flathead guitarfish.
Rhinobatos (Glaucostegus) productus Girard, 1854. Shovelnose guitarfish.
Rhinobatos (Glaucostegus) typus Bennett, 1830. Giant shovelnose ray.

Subgenus *Platypornax* Whitley, 1939:
Rhinobatos (Platypornax) thouin (Anonymous, 1798). Clubnose guitarfish.

Subgenus *Rhinobatos* Linck, 1790:
Rhinobatos (Rhinobatos) albomaculatus Norman, 1930. White-spotted guitarfish.
Rhinobatos (Rhinobatos) annandalei Norman, 1926. Bengal guitarfish.
Rhinobatos (Rhinobatos) formosensis Norman, 1926. Taiwan guitarfish.
Rhinobatos (Rhinobatos) holcorhynchus Norman, 1922. Slender guitarfish.

Rhinobatos (Rhinobatos) hynnicephalus Richardson, 1846. Ringstraked guitarfish.
Rhinobatos (Rhinobatos) irvinei Norman, 1931. Spineback guitarfish.
Rhinobatos (Rhinobatos) lionotus Norman, 1926. Smoothback guitarfish.
Rhinobatos (Rhinobatos) punctifer Compagno & Randall, 1987. Spotted guitarfish.
Rhinobatos (Rhinobatos) rhinobatos (Linnaeus, 1758). Common guitarfish or violinfish.
Rhinobatos (Rhinobatos) schlegelii Müller & Henle, 1841. Brown guitarfish.
Subgenus *Scobatus* Whitley, 1939:
Rhinobatos (Scobatus) granulatus Cuvier, 1829. Sharpnose guitarfish.
Rhinobatos (Scobatus) obtusus Müller & Henle, 1841. Widenose guitarfish.
Not assigned to subgenus:
Rhinobatos prahli Acero & Franke, 1995. Gorgona guitarfish.
?*Rhinobatos spinosus* Günther, 1870?. Spiny guitarfish (possibly young of some other species).

Genus *Trygonorrhina* Müller & Henle, 1838. Fiddler rays. Australia, continental shelves. One undescribed species.
Trygonorrhina fasciata Müller & Henle, 1841. Southern fiddler ray (including *T. guanerius* Whitley, 1932).
Trygonorrhina melaleuca Scott, 1954. Magpie fiddler ray.

Genus *Zapteryx* Jordan & Gilbert, 1880. Banded guitarfishes. Western South Atlantic (South America) and eastern North Pacific Oceans, continental shelves.
Zapteryx brevirostris (Müller & Henle, 1841). Shortnose guitarfish.
Zapteryx exasperata (Jordan & Gilbert, 1880). Banded guitarfish.

FAMILY PLATYRHINIDAE JORDAN, 1923. Thornbacks

Genus *Platyrhina* Müller & Henle, 1838. Fanrays. Western North Pacific Ocean, continental shelves.
Platyrhina limboonkengi Tang, 1933. Amoy fanray.
Platyrhina sinensis (Bloch & Schneider, 1801). Fanray.

Genus *Platyrhinoidis* Garman, 1881. Eastern North Pacific Ocean, continental shelves.
Platyrhinoidis triseriata (Jordan & Gilbert, 1881). Thornback.

FAMILY ZANOBATIDAE FOWLER, 1928. Panrays

Genus *Zanobatus* Garman, 1913. Panrays. Tropical West Africa, Eastern Atlantic Ocean; Indian Ocean, India. Status of Indian representative uncertain.
Zanobatus schoenleinii (Müller & Henle, 1841). Striped panray
?*Zanobatus atlantica* (Chabanaud, 1928). African panray

ORDER TORPEDINIFORMES: Electric Rays

FAMILY NARCINIDAE GILL, 1862. Numbfishes

Genus *Benthobatis* Alcock, 1898. Blind rays. Western Atlantic, northern Indian, western North Pacific, eastern South Pacific Oceans, continental and insular slopes. At least one undescribed species.
Benthobatis marcida Bean & Weed, 1909. Pale or deep-sea blind ray.
Benthobatis moresbyi Alcock, 1898. Dark blind ray.

Genus *Diplobatis* Bigelow & Schroeder, 1948. Painted electric rays. Eastern Pacific, western Atlantic Oceans, continental shelves.
Diplobatis ommata (Jordan & Gilbert, in Jordan & Bollman, 1889). Target ray.
Diplobatis pictus Palmer, 1950. Painted electric ray.

Genus *Discopyge* Heckel, in Tschudii, 1846. Eastern South Pacific and western South Atlantic Oceans (South America), continental shelves.
Discopyge tschudii Heckel, in Tschudi, 1844. Apron ray.

Genus *Narcine* Henle, 1834. Numbfishes. Circumglobal in warm-temperate and tropical seas, continental and insular shelves, and uppermost slopes. Five to seven undescribed species.
?*Narcine bicolor* (Shaw, 1804). Bicolored electric ray.
Narcine brasiliensis (Olfers, 1831). Lesser electric ray.

Narcine brevilabiata Bessednov, 1966. Shortlip electric ray.
Narcine brunnea Annandale, 1909. Brown electric ray.
Narcine entemedor Jordan & Starks, 1895. Cortez electric ray.
?*Narcine firma* Garman, 1913. Stout electric ray.
Narcine indica Henle, 1834. Indian electric ray.
Narcine lingula Richardson, 1840. Rough electric ray.
?*Narcine maculata* (Shaw, 1804). Dark-spotted electric ray.
?*Narcine nigra* Dumeril, 1852. Black electric ray.
Narcine prodorsalis Bessednov, 1966. Tonkin electric ray.
Narcine rierai (Lloris & Rucabado, 1991). Mozambique electric ray (formerly in Heteronarce).
Narcine tasmaniensis Richardson, 1840. Tasmanian numbfish.
Narcine timlei ((Bloch & Schneider, 1801)). Black-spotted electric ray.
Narcine vermiculatus Breder, 1926. Vermiculated electric ray.
Narcine westralensis McKay, 1966. Banded numbfish.

FAMILY NARKIDAE FOWLER, 1934. Sleeper Rays

Genus *Crassinarke* Takagi, 1951. Western North Pacific Ocean, Sea of Japan and South China Sea, continental shelf.
Crassinarke dormitor Takagi, 1951. Sleeper torpedo.

Genus *Heteronarce* Regan, 1921. Soft sleeper rays. Western and northern Indian Ocean, continental and insular shelves and upper slopes. Possibly at least one undescribed species.
Heteronarce bentuvai (Baranes & Randall, 1989). Elat electric ray.
Heteronarce garmani Regan, 1921. Natal electric ray.
Heteronarce mollis (Lloyd, 1907). Soft electric ray.
Heteronarce prabhui Talwar, 1981. Quilon electric ray.

Genus *Narke* Kaup, 1826. Onefin sleeper rays. Indo-West Pacific Ocean, continental and insular shelves and upper slopes. Possibly two undescribed species.
Narke capensis (Gmelin, 1789). Cape numbfish or onefin electric ray.
Narke dipterygia (Bloch & Schneider, 1801). Spottail electric ray.
Narke japonica (Temminck & Schlegel, 1850). Japanese spotted torpedo.

Genus *Temera* Gray, 1831. Southeast Asia, continental shelves.
Temera hardwickii Gray, 1831. Finless sleeper ray.

Genus *Typhlonarke* Waite, 1909. Legged torpedos. New Zealand, insular shelves and uppermost slopes.
Typhlonarke aysoni (Hamilton, 1902). Blind legged torpedo.
Typhlonarke tarakea Phillipps, 1929. Slender legged torpedo.

FAMILY HYPNIDAE GILL, 1862. Coffin Rays

Genus *Hypnos* Dumeril, 1852. Australia, continental shelf and uppermost slope.
Hypnos monopterygius (Shaw & Nodder, 1795). Coffin ray.

FAMILY TORPEDINIDAE BONAPARTE, 1838. Torpedo Rays

Genus *Torpedo* Houttuyn, 1764. Circumglobal, continental and insular shelves and slopes.
Subgenus *Tetronarce* Gill, 1862:
Torpedo (Tetronarce) californica Ayres, 1855. Pacific torpedo.
Torpedo (Tetronarce) fairchildi Hutton, 1872. New Zealand torpedo.
Torpedo (Tetronarce) macneilli (Whitley, 1932). Australian torpedo.
Torpedo (Tetronarce) microdiscus Parin & Kotlyar, 1985. Smalldisk torpedo.
Torpedo (Tetronarce) nobiliana Bonaparte, 1835. Great, Atlantic, or black torpedo.
Torpedo (Tetronarce) peruana Chirichigno, 1963. Peruvian torpedo.
Torpedo (Tetronarce) puelcha Lahille, 1928. Argentine torpedo.
Torpedo (Tetronarce) semipelagica Parin & Kotlyar, 1985. Semipelagic torpedo.
Torpedo (Tetronarce) tokionis (Tanaka, 1908?). Trapezoid torpedo.
Torpedo (Tetronarce) tremens de Buen, 1959?. Chilean torpedo.

Subgenus *Torpedo* Houttuyn, 1764:
Torpedo (Torpedo) alexandrinsis Mazhar, 1987. Alexandrine torpedo.
Torpedo (Torpedo) andersoni Bullis, 1962. Florida torpedo.
Torpedo (Torpedo) bauchotae Cadenat, Capape & Desoutter, 1978. Rosette torpedo.
Torpedo (Torpedo) fuscomaculata Peters, 1855. Black-spotted torpedo.
Torpedo (Torpedo) mackayana Metzelaar, 1919. Ringed torpedo.
Torpedo (Torpedo) marmorata Risso, 1810. Spotted or marbled torpedo.
Torpedo (Torpedo) panthera Olfers, 1831. Leopard torpedo.
?*Torpedo (Torpedo) polleni* Bleeker, 1866. Reunion torpedo.
Torpedo (Torpedo) sinuspersici Olfers, 1831. Gulf torpedo.
?*Torpedo (Torpedo) suissi* Steindachner, 1898. Red Sea torpedo.
Torpedo (Torpedo) torpedo (Linnaeus, 1758). Ocellate or common torpedo.
?*Torpedo (Torpedo) zugmayeri* Engelhardt, 1912. Baluchistan torpedo.

ORDER RAJIFORMES: Skates

FAMILY Arhynchobatidae Fowler, 1934. Softnose Skates

Genus *Arhynchobatis* Waite, 1909. New Zealand, insular shelves.
Arhynchobatis asperrimus Waite, 1909. Longtailed skate.

Genus *Bathyraja* Ishiyama, 1958. Softnose skates. Circumglobal, most diverse in high latitudes, continental and insular shelves and slopes. At least two undescribed species.
Bathyraja abyssicola (Gilbert, 1896). Deep-sea skate.
Bathyraja aguja (Kendall & Radcliffe, 1912). Aguja skate.
Bathyraja aleutica (Gilbert, 1895). Aleutian skate.
Bathyraja andriashevi Dolganov, 1985. Little-eyed skate.
Bathyraja bergi Dolganov, 1985. Bottom skate.
Bathyraja brachyurops (Fowler, 1910). Broadnose skate.
Bathyraja caeluronigricans Ishiyama & Ishihara, 1977. Purple-black skate.
Bathyraja diplotaenia (Ishiyama, 1950). Dusky-pink skate.
Bathyraja eatonii (Günther, 1876). Eaton's skate.
Bathyraja fedorovi Dolganov, 1985. Cinnamon skate.
Bathyraja griseocauda (Norman, 1937). Graytail skate.
Bathyraja hesperafricana Stehmann, 1995. West African skate.
Bathyraja irrasa Hureau & Ozouf-Costaz, 1980. Kerguelen sandpaper skate.
Bathyraja isotrachys (Günther, 1877). Raspback skate.
Bathyraja kincaidi (Garman, 1908). Sandpaper skate.
Bathyraja lindbergi Ishiyama & Ishihara, 1977. Commander skate.
Bathyraja longicauda (de Buen, 1959). Slimtail skate.
Bathyraja maccaini Springer, 1972. McCain's skate.
Bathyraja maculata Ishiyama & Ishihara, 1977. White-blotched skate.
Bathyraja matsubarai (Ishiyama, 1952). Dusky-purple skate.
Bathyraja meridionalis Stehmann, 1987. Dark-belly skate.
Bathyraja microtrachys (Osburn & Nichols, 1917). Finespined skate.
Bathyraja minispinosa Ishiyama & Ishihara, 1977. Smallthorn skate.
Bathyraja notoroensis Ishiyama & Ishihara, 1977. Notoro skate.
Bathyraja pallida (Forster, 1967). Pallid skate.
Bathyraja papilonifera Stehmann, 1985. Butterfly skate.
Bathyraja parmifera (Bean, 1881). Alaska skate.
Bathyraja peruana McEachran & Miyake, 1984. Peruvian skate.
Bathyraja richardsoni (Garrick, 1961). Richardson's skate.
Bathyraja scaphiops (Norman, 1937). Cuphead skate.
Bathyraja schroederi (Krefft, 1968). Whitemouth skate.
Bathyraja shuntovi Dolganov, 1985. Narrownose skate.
Bathyraja simoterus (Ishiyama, 1967). Hokkaido skate.
Bathyraja smirnovi (Soldatov & Lindberg, 1913). Golden skate.
Bathyraja smithii (Müller & Henle, 1841). African softnose skate.
Bathyraja spinicauda (Jensen, 1914). Spinetail or spinytail skate.
Bathyraja spinosissima (Beebe & Tee-Van, 1941). Pacific white skate.

Bathyraja trachouros (Ishiyama, 1958). Eremo skate.
Bathyraja trachura (Gilbert, 1892). Roughtail skate
Bathyraja tzinovskii Dolganov, 1985. Creamback skate.
Bathyraja violacea (Suvorov, 1935). Okhotsk skate.

Genus *Irolita* Whitley, 1931. Round skates. Australia, shelves and uppermost slopes. One undescribed species.
Irolita waitei (McCulloch, 1911). Southern round skate.

Genus *Notoraja* Ishiyama, 1958. Velvet skates. Western Pacific Ocean, slopes. At least four undescribed species.
Notoraja asperula (Garrick & Paul, 1974). Prickly deepsea skate.
Notoraja laxipella (Yearsley & Last, 1992). Eastern looseskin skate.
Notoraja ochroderma McEachran & Last, 1994. Pale skate.
Notoraja spinifera (Garrick & Paul, 1974). Spiny deepsea skate.
Notoraja subtilispinosa Stehmann, 1985. Velvet skate.
Notoraja tobitukai (Hiyama, 1940). Leadhued skate.

Genus *Pavoraja* Whitley, 1939. Peacock skates. Australia and New Zealand, shelves and slopes. At least five undescribed species.
Pavoraja alleni McEachran & Fechhelm, 1982. Allens skate.
Pavoraja nitida (Günther, 1880). Peacock skate.

Genus *Psammobatis* Günther, 1870. Sand skates. Eastern South Pacific and western South Atlantic coasts of South America, shelves.
Psammobatis bergi Marini, 1932. Blotched sand skate.
Psammobatis extenta Garman, 1913. Zipper sand skate.
Psammobatis lentiginosa McEachran, 1983. Freckled sand skate.
Psammobatis parvacauda McEachran, 1983. Smalltail sand skate.
Psammobatis normani McEachran, 1983. Shortfin sand skate.
Psammobatis rudis Günther, 1870. Smallthorn sand skate.
Psammobatis rutrum Jordan, 1890. Spade sand skate.
Psammobatis scobina (Philippi, 1857). Raspthorn sand skate.

Genus *Pseudoraja* Bigelow & Schroeder, 1954. Western North Atlantic Ocean, continental slopes.
Pseudoraja fischeri Bigelow & Schroeder, 1954. Fanfin skate.

Genus *Rhinoraja* Ishiyama, 1952. Jointnose skates. North Pacific Ocean, western South Atlantic coast of South America, Antarctic, shelves and slopes. This may not be a monophyletic genus, as it differs from *Bathyraja* primarily in having a basal joint in the rostral cartilage.
Rhinoraja albomaculata (Norman, 1937). White-dotted skate.
Rhinoraja interrupta (Gill & Townsend, 1897). Bering skate.
Rhinoraja kujiensis (Tanaka, 1916). Dapple-bellied softnose skate.
Rhinoraja longi Raschi & McEachran, 1991. Aleutian dotted skate
Rhinoraja longicauda Ishiyama, 1952. White-bellied softnose skate.
Rhinoraja macloviana (Norman, 1937). Patagonian skate.
Rhinoraja magellanica (Philippi, 1902, or Steindachner, 1903). Magellan skate.
Rhinoraja multispinis (Norman, 1937). Multispine skate.
Rhinoraja murrayi (Günther, 1880). Murray's skate.
Rhinoraja obtusa (Gill & Townsend, 1897). Blunt skate.
Rhinoraja odai Ishiyama, 1952. Oda's skate.
Rhinoraja rosispinis (Gill & Townsend, 1897). Flathead skate.
Rhinoraja taranetzi Dolganov, 1985. Mud skate.

Genus *Rioraja* Whitley, 1939. Rio skates. Western South Atlantic coast of South America, continental shelves.
Rioraja agassizi (Müller & Henle, 1841). Rio skate.

Genus *Atlantoraja* Menni, 1972. La Plata skates. Western South Atlantic coast of South America, continental shelves. Possibly one undescribed species.
Atlantoraja castelnaui (Ribeiro, 1907). Spotback skate.
Atlantoraja cyclophora (Regan, 1903). Eyespot skate.
Atlantoraja platana (Günther, 1880). La Plata skate.

Genus *Sympterygia* Müller & Henle, 1837. Fanskates. Eastern South Pacific and western South Atlantic coasts of South America, continental shelves.
  Sympterygia acuta Garman, 1877. Bignose fanskate.
  Sympterygia bonapartei Müller & Henle, 1841. Smallnose fanskate.
  Sympterygia brevicaudata Cope, 1877. Shorttail fanskate.
  Sympterygia lima (Poeppig, 1835). Filetail fanskate.

FAMILY RAJIDAE BLAINVILLE, 1816. Skates

Genus *Amblyraja* Malm, 1877. Stout skates. Circumglobal, most records in higher latitudes and the Northern Hemisphere but also in deep water in the tropics and in the Southern Hemisphere, shelves and slopes. One undescribed species.
  Amblyraja badia (Garman, 1899). Broad skate.
  Amblyraja doellojuradoi (Pozzi, 1935). Southern thorny skate.
  Amblyraja frerichsi (Krefft, 1968). Thickbody skate.
  Amblyraja georgiana (Norman, 1938). Antarctic starry skate.
  Amblyraja hyperborea (Collette, 1879). Arctic skate.
  Amblyraja jenseni (Bigelow & Schroeder, 1950). Jensen's skate.
  Amblyraja radiata (Donovan, 1808). Thorny skate.
  Amblyraja reversa (Lloyd, 1906). Reversed skate.
  Amblyraja robertsi (Hulley, 1970). Bigmouth skate.
  Amblyraja taaf (Meisner, 1987). Whiteleg skate.

Genus *Breviraja* Bigelow & Schroeder, 1948. Lightnose skates. Western Atlantic Ocean, continental slopes. One undescribed species in Eastern Atlantic.
  Breviraja claramaculata McEachran & Matheson, 1985. Brightspot skate.
  Breviraja colesi Bigelow & Schroeder, 1948. Lightnose skate.
  Breviraja marklei McEachran & Miyake, 1987. Nova Scotia skate.
  Breviraja mouldi McEachran & Matheson, 1995. Blacknose skate.
  Breviraja nigriventralis McEachran & Matheson, 1985. Blackbelly skate.
  Breviraja spinosa Bigelow & Schroeder, 1950. Spinose skate.

Genus *Dactylobatus* Bean & Weed, 1909. Skilletskates. Western North Atlantic Ocean, continental slopes.
  Dactylobatus armatus Bean & Weed, 1909. Skilletskate.
  Dactylobatus clarki (Bigelow & Schroeder, 1958). Hookskate.

Genus *Dipturus* Rafinesque, 1810. Longnosed skates. Virtually circumglobal in cool-temperate to tropical seas, shelves, and slopes except possibly the Eastern North Pacific Ocean. Possibly 15 undescribed species.
  Dipturus batis (Linnaeus, 1758). Gray skate.
  Dipturus bullisi (Bigelow & Schroeder, 1962). Tortugas skate.
  Dipturus campbelli (Wallace, 1967). Blackspot skate.
  Dipturus chilensis (Guichenot, 1848). Yellownose skate.
  Dipturus crosnieri (Seret, 1989). Madagascar skate.
  Dipturus doutrei (Cadenat, 1960). Javalin skate.
  Dipturus ecuadoriensis (Beebe & Tee-Van, 1941). Ecuador skate.
  Dipturus garricki (Bigelow & Schroeder, 1958). San Blas skate.
  Dipturus gigas (Ishiyama, 1958). Giant skate.
  Dipturus gudgeri (Whitley, 1940). Bight skate.
  Dipturus innominatus (Garrick & Paul, 1974). New Zealand smooth skate.
  Dipturus johannisdavesi (Alcock, 1899). Travancore skate.
  Dipturus kwangtungensis (Chu, 1960). Kwangtung skate.
  Dipturus lanceorostratus (Wallace, 1967). Rattail skate.
  Dipturus laevis (Mitchill, 1817). Barndoor skate.
  Dipturus leptocaudus (Krefft & Stehmann, 1974). Thintail skate.
  Dipturus linteus (Fries, 1838). Sailskate or sailray.
  Dipturus macrocaudus (Ishiyama, 1955). Bigtail skate.
  Dipturus nasutus (Banks in Müller & Henle, 1841). New Zealand rough skate.
  Dipturus nidarosiensis (Collett, 1880). Norwegian skate.
  Dipturus olseni (Bigelow & Schroeder, 1951). Spreadfin skate.

Dipturus oregoni (Bigelow & Schroeder, 1958). Hooktail skate.
Dipturus oxyrhynchus (Linnaeus, 1758). Sharpnose skate.
Dipturus pullopunctatus (Smith, 1964). Slime skate.
Dipturus springeri (Wallace, 1967). Roughbelly skate.
Dipturus stenorhynchus (Wallace, 1967). Prownose skate.
Dipturus teevani (Bigelow & Schroeder, 1951. Caribbean skate.
Dipturus tengu (Jordan & Fowler, 1903). Acutenose or tengu skate.
Dipturus trachydermus (Krefft & Stehmann, 1974). Roughskin skate.

Genus *Fenestraja* McEachran & Compagno, 1982. Pygmy skates. Western Atlantic Ocean, Western Indian Ocean (Madagascar, India), Western Pacific (Celebes), continental and insular slopes.
Fenestraja atripinna (Bigelow & Schroeder, 1950). Blackfin pygmy skate.
Fenestraja cubensis (Bigelow & Schroeder, 1950). Cuban pygmy skate.
Fenestraja ishiyamai (Bigelow & Schroeder, 1962). Plain pygmy skate.
Fenestraja maceachrani (Seret, 1989). Madagascar pygmy skate.
Fenestraja mamillidens (Alcock, 1889). Prickly skate.
Fenestraja plutonia (Garman, 1881). Pluto skate.
Fenestraja sibogae (Weber, 1913). Siboga pygmy skate.
Fenestraja sinusmexicanus (Bigelow & Schroeder, 1950). Gulf of Mexico pygmy skate.

Genus *Gurgesiella* de Buen, 1959. Finless pygmy skates. Western Atlantic, Eastern South Pacific Oceans, continental and insular slopes.
Gurgesiella atlantica (Bigelow & Schroeder, 1962). Atlantic pygmy skate.
Gurgesiella dorsalifera McEachran & Compagno, 1980. Onefin skate.
Gurgesiella furvescens de Buen, 1959. Dusky finless skate.

Genus *Leucoraja* Malm, 1877. Rough skates. Western North Atlantic, Eastern Atlantic, Mediterranean and southwestern Indian Ocean, Australia, shelves and slopes. Two undescribed species.
Leucoraja circularis (Couch, 1838). Sandy skate or ray.
Leucoraja compagnoi (Stehmann, 1995). Tigertail skate.
Leucoraja erinacea (Mitchill, 1825). Little skate.
Leucoraja fullonica (Linnaeus, 1758). Shagreen skate or ray.
Leucoraja garmani (Whitley, 1939). Rosette skate.
Leucoraja lentiginosa (Bigelow & Schroeder, 1951). Freckled skate.
Leucoraja leucosticta (Stehmann, 1971). Whitedappled skate.
Leucoraja melitensis (Clark, 1926). Maltese skate or ray.
Leucoraja naevus (Müller & Henle, 1841). Cuckoo skate or ray.
Leucoraja ocellata (Mitchill, 1815). Winter skate.
Leucoraja wallacei (Hulley, 1970). Yellowspot or blancmange skate.
Leucoraja yucatanensis (Bigelow & Schroeder, 1950). Yucatan skate.

Genus *Malacoraja* Stehmann, 1970. Soft skates. Western North Atlantic Ocean, Eastern Atlantic, slopes.
Malacoraja kreffti (Stehmann, 1978). Krefft's skate or ray.
Malacoraja senta (Garman, 1885). Smooth skate.
Malacoraja spinacidermis (Barnard, 1923). Prickled skate or ray, roughskin skate (provisionally including *Raja mollis* Bigelow & Schroeder, 1950).

Genus *Neoraja* McEachran & Compagno, 1982. Pygmy skates. Western North Atlantic Ocean, Eastern Atlantic, continental slopes.
Neoraja africana Stehmann & Seret, 1983. West African pygmy skate
Neoraja caerulea (Stehmann, 1976). Blue pygmy skate.
Neoraja carolinensis McEachran & Stehmann, 1984. Carolina pygmy skate.
Neoraja stehmanni (Hulley, 1972). South African pygmy skate.

Genus *Okamejei* Ishiyama, 1958. Spiny rasp skates. Indian Ocean and Western Pacific, shelves and slopes. At least two undescribed species.
Okamejei acutispina (Ishiyama, 1958). Sharpspine skate.
Okamejei australis (Macleay, 1884). Sydney skate.
Okamejei boesemani (Ishihara, 1987). Black sand skate.

Okamejei cerva (Whitley, 1939). White-spotted skate.
Okamejei heemstrai (McEachran & Fechhelm, 1982). East African skate.
Okamejei hollandi (Jordan & Richardson, 1909). Yellow-spotted skate.
Okamejei kenojei (Müller & Henle, 1841). Spiny rasp, swarthy, or ocellate spot skate.
Okamejei lemprieri (Richardson, 1846). Australian thornback skate.
Okamejei meerdervoorti (Bleeker, 1860). Bigeye skate.
Okamejei philipi? (Lloyd, 1906). Aden ringed skate.
Okamejei pita (Fricke & Al-Hussar, 1995). Pita skate.
Okamejei powelli (Alcock, 1898). Indian ringed skate.
Okamejei schmidti (Ishiyama, 1958). Browneye skate.

Genus *Raja* Linnaeus, 1758. Ocellate skates. Eastern Atlantic Ocean and Southwestern Indian Ocean. One undescribed species.
?*Raja africana* Capape, 1977. African skate or ray.
Raja asterias Delaroche, 1809. Atlantic starry skate.
Raja brachyura Lafont, 1873. Blonde skate or ray.
Raja clavata Linnaeus, 1758. Thornback skate or ray.
Raja herwigi Krefft, 1965. Cape Verde skate.
Raja maderensis Lowe, 1841. Madeira skate or ray.
Raja microocellata Montagu, 1818. Smalleyed skate or ray, painted skate.
Raja miraletus Linnaeus, 1758. Brown or twineye skate or ray.
Raja montagui Fowler, 1910. Spotted skate or ray.
Raja polystigma Regan, 1923. Speckled skate or ray.
Raja radula Delaroche, 1809. Rough skate or ray.
Raja rondeleti Bougis, 1959. Rondelet's skate or ray.
Raja straeleni Poll, 1951. Biscuit skate.
Raja undulata Lacepede, 1802. Undulate skate or ray.

Genus *Rajella* Stehmann, 1970. Gray skates. Eastern South Pacific Ocean, western North Atlantic, eastern Atlantic, southwestern Indian Ocean, western South Pacific (Indonesia, Australia), outer shelves and slopes. Four undescribed species.
?*Rajella alia* (Garman, 1899). Blake skate.
Rajella annandalei (Weber, 1913). Indonesian round skate.
Rajella barnardi (Norman, 1935). Bigthorn skate.
Rajella bathyphila (Holt & Byrne, 1908). Deepwater skate or ray.
Rajella bigelowi (Stehmann, 1978). Bigelow's skate or ray.
Rajella caudaspinosa (von Bonde & Swart, 1923). Munchkin skate.
Rajella dissimilis (Hulley, 1970). Ghost skate.
Rajella fuliginea (Bigelow & Schroeder, 1954). Sooty skate.
Rajella fyllae (Lütken, 1888). Round skate or ray.
Rajella kukujevi (Dolganov, 1985). Mid-Atlantic skate.
Rajella leopardus (von Bonde & Swart, 1923). Leopard skate.
Rajella nigerrima (de Buen, 1960). Blackish skate.
Rajella purpuriventralis (Bigelow & Schroeder, 1962). Purplebelly skate.
Rajella ravidula (Hulley, 1970). Smoothback skate.
Rajella sadowskyii (Krefft & Stehmann, 1974). Brazilian skate.

Genus *Rostroraja* Hulley, 1972. Eastern Atlantic and southwestern Indian Ocean, shelves and slopes.
Rostroraja alba (Lacepede, 1803). White, bottlenose, or spearnose skate.

Species formerly included in the genus *Raja* and probably requiring new genera in some cases, which remain to be described. These are arranged in three groups:

Undescribed **"Genus A"** for the **"North Pacific Assemblage"** of McEachran & Dunn (1998), including *Dipturus*-like giant species:
Raja binoculata Girard, 1854. Big skate.
Raja cortezensis McEachran & Miyake, 1988. Cortez skate.
Raja inornata Jordan & Gilbert, 1880. California skate.

Raja pulchra Liu, 1932. Mottled skate.
Raja rhina Jordan & Gilbert, 1880. Longnose skate.
Raja stellulata Jordan & Gilbert, 1880. Pacific starry skate.

Undescribed **"Genus B"** for the **"Amphi-American Assemblage"** of McEachran & Dunn (1998), including mostly *Raja*-like species from the Western Atlantic and Eastern Pacific:

Raja ackleyi Garman, 1881. Ocellate skate.
Raja bahamensis Bigelow & Schroeder, 1965. Bahama skate.
Raja cervigoni Bigelow & Schroeder, 1964. Venezuela skate.
Raja eglanteria Bosc, 1802. Clearnose skate.
Raja equatorialis Jordan & Bollman, 1890. Equatorial skate.
Raja texana Chandler, 1921. Roundel skate.
Raja velezi Chirichigno, 1973. Rasptail skate.

Western Pacific species, including two named species and at least two undescribed taxa from Australia that were placed by Last & Stevens (1994) in *Raja* without assigning them to subgenera; and a recently described skate from Korea:

Raja koreana Jeong & Nakabo, 1997. Korean skate.
Raja polyommata Ogilby, 1910. Argus skate.
Raja whitleyi Iredale, 1938. Melbourne skate.

FAMILY ANACANTHOBATIDAE VON BONDE & SWART, 1924. Legskates

Genus *Anacanthobatis* von Bonde & Swart, 1924. Smooth leg skates. Western North Atlantic, southwestern Indian Ocean, western Pacific continental slopes. Possibly three undescribed species.

Anacanthobatis americanus Bigelow & Schroeder, 1962. American legskate.
Anacanthobatis borneensis Chan, 1965. Borneo leg skate.
Anacanthobatis donghaiensis (Deng, Xiong, & Zhan, 1983). East China leg skate.
Anacanthobatis folirostris (Bigelow & Schroeder, 1951). Leaf-nose leg skate.
Anacanthobatis longirostris Bigelow & Schroeder, 1962. Longnose leg skate.
Anacanthobatis marmoratus (von Bonde & Swart, 1924). Spotted leg skate.
Anacanthobatis melanosoma (Chan, 1965). Blackbodied leg skate.
Anacanthobatis nanhaiensis (Meng & Li, 1981). South China leg skate.
Anacanthobatis ori (Wallace, 1967). Black leg skate.
Anacanthobatis stenosoma (Li & Hu, 1982). Narrow leg skate.

Genus *Cruriraja* Bigelow & Schroeder, 1948. Rough leg skates. Western North Atlantic, eastern South Atlantic and western Indian Ocean continental and insular shelves and slopes.

Cruriraja andamanica (Lloyd, 1909). Andaman leg skate.
Cruriraja atlantis Bigelow & Schroeder, 1948. Atlantic leg skate.
Cruriraja cadenati Bigelow & Schroeder, 1962. Broadfoot leg skate.
Cruriraja durbanensis (von Bonde & Swart, 1924). Smooth nose leg skate.
Cruriraja parcomaculata (von Bonde & Swart, 1924). Roughnose leg skate.
Cruriraja poeyi Bigelow & Schroeder, 1948. Cuban leg skate.
Cruriraja rugosa Bigelow & Schroeder, 1958. Rough leg skate.
Cruriraja triangularis Smith, 1964. Triangular leg skate.

ORDER MYLIOBATIFORMES: STINGRAYS

FAMILY PLESIOBATIDAE NISHIDA, 1990. Giant Stingarees

Genus *Plesiobatis* Nishida, 1990. Wide-ranging in Indo-West Pacific Ocean from South Africa to Hawaii, continental and insular slopes. Formerly included in the genus *Urotrygon* and the Family Urolophidae but placed in its own genus and Family by Nishida (1990).

Plesiobatis daviesi (Wallace, 1967). Giant stingaree.

FAMILY HEXATRYGONIDAE HEEMSTRA & SMITH, 1980. Sixgill Stingrays

Genus *Hexatrygon* Heemstra & Smith, 1980. Wide-ranging in Indo-West Pacific Ocean from South Africa to Hawaii, continental and insular slopes. Several species described but possibly only one species.

Hexatrygon bickelli Heemstra & Smith, 1980. Sixgill stingray. Provisionally including *H. longirostrum* Chu & Meng, 1981; *H. yangi* Shen & Liu, 1984; *H. taiwanensis* Shen, 1986; and *H. brevirostra* Shen, 1986.

FAMILY UROLOPHIDAE MÜLLER & HENLE, 1841. Stingarees. Recently McEachran et al. (1996) have separated the American members of this family, comprising the genera *Urobatis* and *Urotrygon*, into their own family Urotrygonidae.

Genus *Trygonoptera* Müller & Henle, 1841. Shovelnose stingarees. Australia, continental shelves to slope edge. Two undescribed species.

Trygonoptera mucosa (Whitley, 1939). Western shovelnose stingaree.
Trygonoptera ovalis Last & Gomon, 1987. Striped stingaree.
Trygonoptera personalis Last & Gomon, 1987. Masked stingaree.
Trygonoptera testacea Banks, in Müller & Henle, 1841. Common stingaree.

Genus *Urobatis* Garman, 1913. Shorttail round stingrays. Western North Atlantic and eastern Pacific continental shelves.

Urobatis concentricus Osburn & Nichols, 1916. Bull's-eye stingray.
Urobatis halleri (Cooper, 1863). Round stingray.
Urobatis jamaicensis (Cuvier, 1817). Yellow stingray.
Urobatis maculatus Garman, 1913. Cortez round stingray.
Urobatis marmoratus (Philippi, 1893). Chilean round stingray.
Urobatis tumbesensis (Chirichigno & McEachran, 1979). Tumbes round stingray.

Genus *Urolophus* Müller & Henle, 1837. Stingarees. Western Pacific continental shelves and upper slopes. Four undescribed species.

Urolophus armatus Valencienes, in Müller & Henle, 1841. New Ireland stingaree.
Urolophus aurantiacus Müller & Henle, 1841. Sepia stingray.
Urolophus bucculentus Macleay, l884. Sandyback stingaree.
Urolophus circularis McKay, 1966. Circular stingaree.
Urolophus cruciatus (Lacepede, 1804). Banded or crossback stingaree.
Urolophus expansus McCulloch, 1916. Wide stingaree.
Urolophus flavomosaicus Last & Gomon, 1987. Patchwork stingaree.
Urolophus gigas Scott, 1954. Spotted or Sinclair's stingaree.
Urolophus javanicus (Martens, 1864). Java stingaree.
Urolophus kaianus Günther, 1880. Kai stingaree.
Urolophus lobatus McKay, 1966. Lobed stingaree.
Urolophus mitosis Last & Gomon, 1987. Mitotic or blotched stingaree.
Urolophus orarius Last & Gomon, 1987. Coastal stingaree.
Urolophus paucimaculatus Dixon, 1969. Sparsely-spotted, Dixon's, or white-spotted stingaree.
Urolophus sufflavus Whitley, 1929. Yellowback stingaree.
Urolophus viridis McCulloch, 1916. Greenback stingaree.
Urolophus westraliensis Last & Gomon, 1987. Brown stingaree.

Genus *Urotrygon* Gill, 1864. Longtail round stingrays. Eastern Pacific and western Atlantic continental shelves.

Urotrygon aspidura (Jordan & Gilbert, 1882). Roughtail round stingray.
Urotrygon chilensis (Günther, 1871). Thorny round stingray.
Urotrygon microphthalmum Delsman, 1941. Smalleyed round stingray.
Urotrygon munda Gill, 1863. Shortfin round stingray.
Urotrygon nana Miyake & McEachran, 1988. Dwarf round stingray.
Urotrygon reticulata Miyake & McEachran, 1988. Reticulate round stingray.
Urotrygon rogersi (Jordan & Starks, 1895). Lined round stingray.
Urotrygon simulatrix Miyake & McEachran, 1988. Stellate round stingray.
Urotrygon venezuelae Schultz, 1949. Venezuela round stingray.

FAMILY POTAMOTRYGONIDAE GARMAN, 1877. River Stingrays. Recently McEachran et al. (1996), following work by Lovejoy (1996), have included the Eastern Hemisphere marine dasyatid genus *Taeniura* and the two American marine species in the dasyatid genus *Himantura* (*H. pacifica* and *H. schmardae*) in this family, but without generic allocation of the latter. These taxa are tentatively retained in Dasyatidae pending further clarification of the taxonomic limits of the Dasyatidae. At least one undescribed genus and species.

Genus *Paratrygon* Dumeril, 1865. Rivers of northern Bolivia, eastern Peru, and northern Brazil.

Paratrygon aireba Müller & Henle, 1841. Discus ray.

Genus *Plesiotrygon* Rosa, Castello, & Thorson, 1987. Upper and mid Amazon River and tributaries in Ecuador and Brazil.

Plesiotrygon iwamae Rosa, Castello, & Thorson, 1987. Long-tailed river stingray.

Genus *Potamotrygon* Garman, 1877. Short-tailed river stingrays. Rivers of Colombia, Venezuela, Bolivia, Guyana, French Guiana, Surinam, Peru, Brazil, Argentina, Uruguay, and Paraguay. Two or more undescribed species.

Potamotrygon brachyura (Günther, 1880). Short-tailed river stingray.
Potamotrygon castexi Castello & Yagolkowski, 1969. Vermiculate river stingray.
Potamotrygon constellata (Vaillant, 1880). Thorny river stingray.
Potamotrygon dumerilii (Castelnau, 1855). Anglespot river stingray.
Potamotrygon falkneri Castex & Maciel, 1963. Largespot river stingray.
Potamotrygon henlei (Castelnau, 1855). Bigtooth river stingray.
Potamotrygon histrix (Müller & Henle, in Orbigny, 1834). Porcupine river stingray.
Potamotrygon humerosa Garman, 1913. Roughback river stingray.
Potamotrygon leopoldi Castex & Castello, 1970. White-blotched river stingray.
Potamotrygon magdalenae (Valenciennes, in Dumeril, 1865). Magdalena river stingray.
Potamotrygon motoro (Natterer, in Müller & Henle, 1841). Ocellate river stingray.
Potamotrygon ocellata (Engelhardt, 1912). Red-blotched river stingray.
Potamotrygon orbignyi (Castelnau, 1855). Smooth back river stingray.
Potamotrygon schroederi Fernandez Yepez, 1957. Rosette river stingray.
Potamotrygon schuemacheri Castex, 1964. Parana river stingray.
Potamotrygon scobina Garman, 1913. Raspy river stingray.
Potamotrygon signata Garman, 1913. Parnaiba river stingray.
Potamotrygon yepezi Castex & Castello, 1970. Maracaibo river stingray.

FAMILY DASYATIDAE JORDAN, 1888. Whiptail Stingrays

Genus *Dasyatis* Rafinesque, 1810. Fintail stingrays. Circumglobal in all temperate and tropical seas, continental and insular shelves and uppermost slopes, also in tropical and warm-temperate rivers and lakes. At least five undescribed species.

Dasyatis acutirostra Nishida & Nakaya, 1988. Sharpnose stingray.
Dasyatis akajei (Müller & Henle, 1841). Red stingray.
Dasyatis americana Hildebrand & Schroeder, 1928. Southern stingray.
Dasyatis annotata Last, 1987. Plain maskray.
Dasyatis bennetti (Müller & Henle, 1841). Bennett's cowtail or frilltailed stingray.
Dasyatis brevicaudata (Hutton, 1875). Shorttail or smooth stingray.
Dasyatis brevis (Garman, 1880). Diamond stingray.
Dasyatis centroura (Mitchill, 1815). Roughtail stingray.
Dasyatis chrysonota (Smith, 1828). Blue or marbled stingray.
Dasyatis fluviorum Ogilby, 1908. Estuary stingray.
Dasyatis garouaensis (Stauch & Blanc, 1962). Smooth freshwater stingray, Niger stingray.
Dasyatis geijskesi Boeseman, 1948. Wingfin stingray.
Dasyatis gigantea (Lindberg, 1930). Giant stumptail stingray.
Dasyatis guttata (Bloch & Schneider, 1801). Longnose stingray.
Dasyatis izuensis Nishida & Nakaya, 1988. Izu stingray.
Dasyatis kuhlii (Müller & Henle, 1841). Blue-spotted stingray or mask ray.
Dasyatis laevigata Chu, 1960. Yantai stingray.
Dasyatis laosensis Roberts & Karnasuta, 1987. Mekong freshwater stingray.

Dasyatis lata (Garman, 1880). Brown stingray.
Dasyatis leylandi Last, 1987. Painted maskray.
Dasyatis longa (Garman, 1880). Longtail stingray.
Dasyatis margarita (Günther, 1870). Daisy stingray.
Dasyatis margaritella Compagno & Roberts, 1984. Pearl stingray.
Dasyatis matsubarai Miyosi, 1939. Pitted stingray.
Dasyatis microps (Annandale, 1908). Thickspine giant stingray.
Dasyatis multispinosa (Tokarev, 1959). Multispine giant stingray.
Dasyatis navarrae (Steindachner, 1892). Blackish stingray.
Dasyatis pastinaca (Linnaeus, 1758). Common stingray.
Dasyatis rudis (Günther, 1870). Smalltooth stingray.
Dasyatis sabina (Lesueur, 1824). Atlantic stingray.
Dasyatis sayi (Lesueur, 1817). Bluntnose stingray.
Dasyatis sinensis (Steindachner, 1892). Chinese stingray.
Dasyatis thetidis Ogilby, in Waite, 1899. Thorntail or black stingray.
?*Dasyatis tortonesei* Capape, 1977. Tortonese's stingray.
Dasyatis ushiei Jordan & Hubbs, 1925. Cow stingray.
Dasyatis zugei (Müller & Henle, 1841). Pale-edged stingray.

Genus *Himantura* Müller & Henle, 1837. Whip rays. Tropical Indo-West Pacific Ocean, with a few species in the tropical western Atlantic and eastern Pacific Oceans, on the continental and insular shelves and in tropical rivers and lakes. At least four undescribed species.

Himantura alcocki (Annandale, 1909). Pale-spot whip ray.
Himantura bleekeri (Blyth, 1860). Whiptail stingray.
Himantura chaophraya Monkolprasit & Roberts, 1990. Giant freshwater stingray or whip ray.
Himantura fai Jordan & Seale, 1906. Pink whip ray.
?*Himantura fluviatilis* (Hamilton-Buchanan, 1822/Annandale, 1910). Ganges stingray.
Himantura gerrardi (Gray, 1851). Sharpnose stingray, bluntnose whiptail ray or whip ray, banded whiptail ray (possibly a species complex).
Himantura granulata (Macleay, 1883). Mangrove whip ray.
Himantura imbricata (Bloch & Schneider, 1801). Scaly stingray or whip ray.
Himantura jenkinsii (Annandale, 1909). Pointed-nose stingray or golden whip ray (? = *Himantura draco* [Compagno & Heemstra, 1984]).
Himantura krempfi (Chabanaud, 1923). Marbled freshwater whip ray.
Himantura marginata (Blyth, 1860). Black-edge whip ray.
Himantura microphthalma (Chen, 1948). Smalleye whip ray.
Himantura oxyrhyncha (Sauvage, 1878). Longnose marbled whip ray.
?*Himantura pacifica* (Beebe & Tee-Van, 1941). Pacific whip ray.
Himantura pareh (Bleeker, 1852). Pareh whip ray.
Himantura pastinacoides (Bleeker, 1852). Round whip ray.
?*Himantura schmardae* (Werner, 1904). Chupare stingray.
Himantura signifer Compagno & Roberts, 1982. White-edge freshwater whip ray.
Himantura toshi Whitley, 1939. Black-spotted whip ray or coachwhip ray.
Himantura uarnacoides (Bleeker, 1852). Whitenose whip ray.
Himantura uarnak (Forsskael, 1775). Honeycomb or leopard stingray or reticulate whip ray (probably a species complex).
Himantura undulata (Bleeker, 1852). Leopard whip ray (? = *H. fava* [Annandale, 1909]).
Himantura walga (Müller & Henle, 1841). Dwarf whip ray.

Genus *Pastinachus* Rüppell, 1829. Feathertail stingrays. Tropical Indo-West Pacific on the continental shelves, also tropical rivers and lakes. Often placed in genus *Hypolophus* Müller & Henle, 1837 or in *Dasyatis*. At least two additional species.

Pastinachus sephen (Forsskael, 1775). Feathertail or cowtail stingray.

Genus *Pteroplatytrygon* Fowler, 1910. Circumglobal in tropical and temparate seas, oceanic. Often placed in genus *Dasyatis*.

Pteroplatytrygon violacea (Bonaparte, 1832). Pelagic stingray.

Genus *Taeniura* Müller & Henle, 1837. Ribbontail stingrays. Tropical and warm-temperate eastern Atlantic, Mediterranean, and Indo-West Pacific continental and insular shelves.
- *Taeniura grabata* (Geoffroy St. Hilaire, 1817). Round fantail stingray.
- *Taeniura lymma* (Forsskael, 1775). Ribbon-tailed stingray, Blue-spotted ribbontail or fantail ray.
- *Taeniura meyeni* Müller & Henle, 1841. Fantail stingray, round ribbontail ray, speckled stingray (including *T. melanospilos* Bleeker, 1853).

Genus *Urogymnus* Müller & Henle, 1837. Porcupine rays. Tropical eastern Atlantic and Indo-West Pacific continental shelves and tropical rivers and lakes of West Africa.
- *Urogymnus asperrimus* (Bloch & Schneider, 1801). Porcupine ray. Possibly including *U. africanus* (Bloch & Schneider, 1801).
- *Urogymnus ukpam* (Smith, 1863). Pincushion ray or thorny freshwater stingray.

FAMILY GYMNURIDAE FOWLER, 1934. Butterfly Rays

Genus *Aetoplatea* Valenciennes, in Müller & Henle, 1841. Fintail butterfly rays. Northern Indian Ocean and western Pacific continental shelves. Genus possibly not distinct from *Gymnura*.
- *Aetoplatea tentaculata* Valenciennes, in Müller & Henle, 1841. Tentacled butterfly ray.
- *Aetoplatea zonura* Bleeker, 1852. Zonetail butterfly ray.

Genus *Gymnura* Kuhl in van Hasselt, 1823. Butterfly rays. Circumglobal in temperate and tropical seas, continental shelves.
- *Gymnura afuerae* (Hildebrand, 1946). Peruvian butterfly ray.
- *Gymnura altavela* (Linnaeus, 1758). Spiny butterfly ray.
- *Gymnura australis* (Ramsay & Ogilby, 1885). Australian butterfly ray.
- *Gymnura bimaculata* (Norman, 1925). Twin-spot butterfly ray.
- *Gymnura crebripunctata* (Peters, 1869). Mazatlan butterfly ray.
- ?*Gymnura hirundo* (Lowe, 1843). Madeira butterfly ray.
- *Gymnura japonica* (Schlegel, 1850). Japanese butterfly ray.
- *Gymnura marmorata* (Cooper, 1863). California butterfly ray.
- *Gymnura micrura* (Bloch & Schneider, 1801). Smooth butterfly ray.
- *Gymnura natalensis* (Gilchrist & Thompson, 1911). Diamond ray or backwater butterfly ray.
- *Gymnura poecilura* (Shaw, 1804). Longtail butterfly ray.

FAMILY MYLIOBATIDAE BONAPARTE, 1838. Eagle Rays

Genus *Aetobatus* Blainville, 1816. Bonnet rays. Circumglobal in all warm-temperate and tropical seas, continental and insular shelves and possibly semioceanic.
- *Aetobatus flagellum* (Bloch & Schneider, 1801). Longheaded eagle ray.
- *Aetobatus narinari* (Euphrasen, 1790). Spotted eagle ray or bonnetray. Possibly including *Myliobatis punctatus* Maclay & Macleay, 1885, formerly placed in *Pteromylaeus*.
- ?*Aetobatus guttatus* (Shaw, 1804). Indian eagle ray.

Genus *Aetomylaeus* Garman, 1908. Smooth tail eagle rays. Indo-West Pacific continental and insular shelves and possibly semioceanic.
- *Aetomylaeus maculatus* (Gray, 1832). Mottled eagle ray.
- ?*Aetomylaeus milvus* (Valenciennes, in Müller & Henle, 1841). Ocellate eagle ray or vulturine ray.
- *Aetomylaeus nichofii* (Bloch & Schneider, 1801). Banded or Nieuhof's eagle ray.
- *Aetomylaeus vespertilio* (Bleeker, 1852). Ornate or reticulate eagle ray.

Genus *Myliobatis* Cuvier, 1817. Eagle rays. Circumglobal in temperate and tropical seas, with most diversity in temperate waters, continental shelves.
- *Myliobatis aquila* (Linnaeus, 1758). Common eagle ray or bull ray.
- *Myliobatis australis* Macleay, 1881. Southern eagle ray.
- *Myliobatis californicus* Gill, 1865. Bat ray.
- *Myliobatis chilensis* Philippi, 1892?. Chilean eagle ray.
- *Myliobatis freminvillii* Lesueur, 1824. Bullnose ray.
- *Myliobatis goodei* Garman, 1885. Southern eagle ray.

Myliobatis hamlyni Ogilby, 1911. Purple eagle ray.
Myliobatis longirostris Applegate & Fitch, 1964. Longnose eagle ray.
Myliobatis peruanus Garman, 1913. Peruvian eagle ray.
?*Myliobatis rhombus* Basilewsky, 1855. Rhombic eagle ray.
Myliobatis tenuicaudatus Hector, 1877. New Zealand eagle ray.
Myliobatis tobijei Bleeker, 1854. Kite ray.

Genus *Pteromylaeus* Garman, 1913. Bull rays. Eastern North Pacific, eastern Atlantic and Mediterranean, southwestern Indian Ocean continental shelves.
Pteromylaeus asperrimus (Jordan & Evermann, 1898). Roughskin bull ray.
Pteromylaeus bovinus (Geoffroy St. Hilaire, 1817). Bull ray or duckbill ray.

FAMILY RHINOPTERIDAE JORDAN & EVERMANN, 1896. Cow-Nose Rays

Genus *Rhinoptera* Kuhl in Cuvier, 1829. Circumglobal in tropical and warm temperate seas, continental shelves. Of the eleven or more nominal species, only five are valid according to Schwartz (1990).
?*Rhinoptera adspersa* Valenciennes, in Müller & Henle, 1841. Rough cow-nose ray.
Rhinoptera bonasus (Mitchill, 1815). Cow-nosed ray.
?*Rhinoptera brasiliensis* Müller & Henle, 1841. Brazilian cow-nose ray.
?*Rhinoptera hainanensis* Chu, 1960? Hainan cow-nose ray.
Rhinoptera javanica Müller & Henle, 1841. Javanese cow-nose ray or flapnose ray.
?*Rhinoptera jayakari* Boulenger, 1895. Oman cow-nose ray.
Rhinoptera marginata (Geoffroy St. Hilaire, 1817). Lusitanian cow-nose ray.
Rhinoptera neglecta Ogilby, 1912. Australian cow-nose ray.
?*Rhinoptera peli* Bleeker, 1863. African cow-nose ray.
?*Rhinoptera sewelli* Misra, 1947. Indian cow-nose ray.
Rhinoptera steindachneri Evermann & Jenkins, 1891. Hawkray or Pacific cow-nose ray.

FAMILY MOBULIDAE GILL, 1893. Devil Rays

Genus *Manta* Bancroft, 1828. Mantas. Virtually circumglobal in all warm seas. Possibly one species, but this is uncertain and the genus is in need of a world-wide review. Genus *Indomanta* Whitley, 1939 synonymized with this genus.
Manta birostris (Donndorff, 1798). Manta. The most commonly cited regional species are *M. hamiltoni* (Newman, 1849) from the eastern Pacific and *M. alfredi* (Krefft, 1868) from the western South Pacific Ocean.

Genus *Mobula* (Rafinesque, 1810). Devil rays. Wide-ranging in all warm seas, mostly in inshore waters but some species oceanic or semioceanic.
Mobula eregoodootenkee Garman, 1913. Longfin devil ray or ox ray. Often cited as "*Mobula eregoodootenkee* (Cuvier, 1829)," but Cuvier did not propose a binomial for this species; possibly a synonym of *M. diabolus* (Shaw, 1804).
Mobula hypostoma (Bancroft, 1831). Atlantic devil ray. Including *Ceratobatis robertsi* Boulenger, 1897
Mobula japonica (Müller & Henle, 1841). Spinetail devil ray. Possibly = *M. mobular*
Mobula kuhlii (Valenciennes, in Müller & Henle, 1841). Shortfin devil ray.
Mobula mobular (Bonnaterre, 1788). Giant devil ray or devil ray.
Mobula munkiana Di Sciara, 1988. Pygmy devil ray.
Mobula rochebrunei (Vaillant, 1879). Lesser Guinean devil ray.
Mobula tarapacana (Philippi, 1892). Sicklefin devil ray.
Mobula thurstoni (Lloyd, 1908). Bentfin or smooth tail devil ray. Possibly = *M. eregoodoo* (Cantor, 1849).

CONTRIBUTORS

HORST BLECKMANN, Professor, Zoological Institute, University of Bonn, 53115 Bonn, Germany. E-mail unb306@ibm.rhrz.uni-bonn.de

QUENTIN BONE, Professor Emeritus, Marine Biological Association of the United Kingdom, The Laboratory, Plymouth Pl1 2PB, United Kingdom. E-mail Q.bone@pml.ac.uk

PATRICK J. BUTLER, Professor of Zoology and Comparative Physiology, University of Birmingham, Birmingham B15 2TT, United Kingdom. E-mail p.j.butler@bham.ac.uk

LEONARD J. V. COMPAGNO, Curator of Fishes and Director, Shark Research Centre, South African Museum, Queen Victoria Street, Cape Town, 8000 Cape Town, South Africa. E-mail srcsam@uctvms.uct.ac.za

WILLIAM C. HAMLETT, Associate Professor of Anatomy, Indiana University School of Medicine, South Bend Center for Medical Education, University of Notre Dame, Notre Dame, Indiana 46556. E-mail hamlett.1@nd.edu

MICHAEL H. HOFMANN, Assistant, Zoological Institute, University of Bonn, 53115 Bonn, Germany. E-mail unb316@ibm.rhrz.uni-bonn.de

SUSANNE HOLMGREN, Professor, Department of Zoophysiology, Zoological Institute, University of Goteborg, 41390 Goteborg, Sweden. E-mail S.Holmgren@zool.gu.se

NORMAN E. KEMP, Professor of Biological Sciences, University of Michigan, Ann Arbor, Michigan 48109.

THOMAS J. KOOB, Section Chief, Skeletal Biology, Shriners Hospital, Tampa Florida 33612. E-mail thomas.koob@resnet.fmhi.usf.edu

ERIC R. LACY, Professor of Anatomy, Medical University of South Carolina, Charleston, South Carolina 29425. E-mail lacyer@musc.edu

KAREL F. LIEM, Henry Bryant Bigelow Professor and Curator of Ichthyology, Museum of Comparative Zoology, Harvard University, Cambridge, Massachusetts 02134. E-mail csouza@oeb.harvard.edu

RAMÓN MUÑOZ-CHÁPULI, Associate Professor, Department of Animal Biology, Faculty of Sciences, University of Malaga, 29071 Malaga, Spain. E-mail chapuli@ccuma.sci.uma.es

STEFAN NILSSON, Professor, Department of Zoophysiology, Zoological Institute, University of Goteborg, 41390 Goteborg, Sweden. E-mail: S.Nilsson@zool.gu.se

KENNETH R. OLSON, Professor of Physiology, Indiana University School of Medicine, South Bend Center for Medical Education, University of Notre Dame, Notre Dame, Indiana 46556. E-mail olson.1@nd.edu

ENRICO REALE, Professor, Cell Biology and Electron Microscopy Section, Medizinische Hochschule Hannover, 3000 Hannover, Germany.

GEOFFREY H. SATCHELL, Professor Emeritus, Department of Physiology, Otago Medical School, Dunedin, New Zealand.

ADAM P. SUMMERS, Department of Organismic and Evolutionary Biology, University of Massachusetts, Amherst, Massachusetts 01003. E-mail summers@bio.umass.edu

BRUNO TOTA, Professor and Chairman, Department of Cellular Biology, University of Calabria, 87030 Cosenza, Italy. E-mail tota@pobox.unical.it

INDEX

Page numbers for entries occurring in figures are followed by an *f*; those for entries in tables are followed by a *t*.

Abdominal pores, 7
Abdominal veins, 215
Abducens (VI) nerve, 95, 277
Acanthias vulgaris. See *Squalus acanthias*
Acanthodii, 45
Acetylcholine, 123, 151, 164, 225
Acousticolateralis placodes, 50
Acoustic system, 308–13
Acrodus, 47f
Actin, 447
Action potentials, 126–27, 138–39, 252, 311
Adductor branchialis muscle, 100
Adductor mandibulae, 95–96, 109, 110, 176, 178
Adelphophagy, 19
Adenohypophysis, 459
Adenosine, 257
Adenosine triphosphatase (ATPase), 118, 134, 345, 346f, 347
Adenosine triphosphate (ATP), 128, 129–30, 132, 165, 265
Adrenaline. See Epinephrine
Aetobatus, 5
Afferent branchial arteries, 199–200, 220f
Afferent connections
    of cerebellum, 283
    of tectum mesencephali, 285
Afferent pseudobranchial artery, 221
Alar plate, 276
Albumen (egg jelly), 403, 404, 405, 408, 412, 415
Alkaline gland (sperm sac), 444, 450, 453, 455f
Alopias, 243f, 420
Alopias superciliosus, 78
Alopiidae, 7, 20f, 21

Ameloblasts, 60, 63, 64
Ampullae of Lorenzini, 6, 71, 317–19, 322
Anacanthobatidae, 35–37
Anacanthobatis, 35
Anal fins, 1, 7, 88–91
Anaspida, 45
Anatolepis, 45
Androgen receptors, 460
Androgens, 432, 435, 458, 459, 460
Androstenedione, 459
Angel sharks, 8, 9, 18, 83, 84, 85
Angiotensin II, 348
Angiotensin-converting enzyme (ACE), 257
Animate electric fields, 316–17
Anoxypristis, 8
Anterior cardinal veins, 213–15
Anterior cerebral arteries, 208
Anterior cerebral vein, 214
Anterior dorsolateral arteries, 212
Anterior gastropancreaticosplenic artery, 212
Anterior intestinal artery, 212
Anterior intestinal vein, 216
Anterior mesenteric artery, 212–13
Anterior oviduct, 404–8
Anterior parencephalon, 288, 289
Apical membrane, 337–38, 339f
Aplacental viviparity
    with oophagy and intrauterine cannibalism, 420–21
    with uterine villi or trophonemata, 417–20
Aplacental yolk sac viviparity (ovoviparity), 403, 416–17, 432
Appendiculae, 422, 429–31

501

Appendicular muscles, 94t
Appendicular skeleton, 85–91
Apristurus, 25
Apristurus burneus, 399
Aptychotrema, 139
Arhynchobatidae, 35, 36f
Arteriovenous anastomoses (AVAs), 340–41, 342f
Asterocanthus, 47f
Atlantic sharpnose shark. See *Rhizoprionodon terraenovae*
Atlantic shortfin mako. See *Isurus oxyrhynchus*
Atretic follicles, 402
Atrial diastole, 262
Atrial natriuretic peptides (ANP), 348
Atrial systole, 263
Atrioventricular orifice, 247–48
Atrium, 247–48
Atropine, 164, 257, 266–67
Axial skeleton, 71–82. See also Cranium

Baffle zone, 408
Barbeled houndsharks, 26f, 27
Basal membrane, 338–39
Basal plate, 71, 76, 276
Basibranchial copula, 82
Basibranchials, 82
Basidorsals, 83
Basiventrals, 83
Basking sharks, 20f, 21, 465
Bathyraja interrupta, 405f
Batoidea, 9
Batoids
  circulatory system of, 218–19, 221, 232
  clasper gland of, 456–57
  endoskeleton of, 90f
  external body form of, 29–41
  fins of, 88, 91
  girdles of, 85
  locomotion in, 113
  muscle fibers of, 115–16, 132–41
  muscular system of, 102–3, 104
  nervous system of, 291
  neurocranium of, 74
  placoid scales of, 52
  respiration in, 111
  skin coloration of, 51
  splanchnocranium of, 79, 80, 82
  systematics of, 10
  vertebral column of, 83, 84
Behavior
  acoustic and vestibular systems and, 312
  electrosensory system and, 319–21
  gustatory system and, 303
  mechanosensory lateral line and, 315
  olfactory system and, 303
  visual system and, 306
Benthobatis, 3
Betaine, 134
Big-eyed thresher, 78
Bile, 159
Bioelectric fields, 317
Blackmouth cat shark. See *Galeus melanostomus*
Blind sharks, 22, 23f
Blood plasma, 461, 462t, 463t
Blood pressure, 167f, 188–89, 219t, 240, 374
Blood-water barrier, 179, 182f
Blue shark. See *Prionace glauca*
B lymphocytes, 231, 232
Bombesin, 154t, 155, 158, 164, 165f, 166, 348
Bonnethead shark. See *Sphyrna tiburo*
Bowman's capsule, 362–64, 365f, 366, 367f, 371, 383, 385

Brachaeluridae, 22, 23f
Bradyodonti, 48
Brain, 1, 276. See also *specific structures*
Brainstem, 276–82, 284
Bramble sharks, 12, 13f, 82, 85
Branchial arches. See Gill arches
Branchial cardiac branch, 257
Branchial veins, 214–15
Branchiomeric muscles, 94t, 95–100
Breviraja, 35
Brown trout, 231–32
Buccopharyngeal vein, 214
Bullhead sharks, 7, 9, 18–19, 74. See also Heterodontidae; Heterodontiformes
Bull shark. See *Carcharhinus leucas*
Bundle zone, 354, 358, 361f, 375, 376, 382
Buoyancy, 225, 242, 307–8
Butterfly rays, 38f, 40
B wave, 261

Caerulein, 164
Calcium currents, 252
Calcitonin, 155
Callorhynchus milii, 58
Canalis auricularis, 248
Cannibalism, 19, 420–21
Carbachol, 156f
Carcharhinidae, 26f, 28
Carcharhiniformes
  circulatory system of, 204
  external body form of, 25–29
  systematics of, 9, 10
Carcharhinoids, 3, 79, 88, 201f, 445
Carcharhinus, 47f, 322, 401
Carcharhinus acronotus, 427
Carcharhinus albimarginatus, 312
Carcharhinus amblyrhynchos. See *Carcharhinus menisorrah*
Carcharhinus falciformis, 226, 465
Carcharhinus glaucus, 355t
Carcharhinus leucas, 52, 306, 355t
Carcharhinus limbatus, 355t, 460, 465
Carcharhinus longimanus, 314f
Carcharhinus melanopterus, 306, 312, 313
Carcharhinus menisorrah, 59f, 60, 61f, 63, 65f, 306, 312
Carcharhinus milberti, 399
Carcharhinus obscurus, 132, 226, 355t, 465
Carcharhinus plumbeus, 160, 230
  female reproductive system of, 415, 416, 422–25, 427
  male reproductive system of, 465
Carcharhinus porosus, 465
Carcharhinus signatus, 226
Carcharhinus sorrah, 245
Carcharhinus taurus, 19, 46, 420–21
Carcharias. See *Carcharhinus*
Carcharinoids, 91
Carcharodon, 5, 47f
Carcharodon carcharias
  circulatory system of, 226
  digestive system of, 162
  heart of, 251
  skin coloration of, 51
  teeth of, 48
  visual system of, 306
Carcharodon megalodon, 5
Cardiac adaptation, 267–68
Cardiac cycle, 258–62
Cardiac output, 263
Cardinal sinuses, 224, 240
Cardinal sinus vein, 223
Carnitine palmitoyltransferase, 265

Carotid arteries, 204, 206, 208–9, 229
Carotid rete, 229
Carpet sharks, 9, 22–25
Cartilaginous fishes. *See* Chondrichthyans
Catecholamines, 151, 153f, 166, 224, 225, 233, 257, 267
Cat sharks, 5, 25–26, 79, 410
Caudal aorta, 211
Caudal fin, 2, 3, 7–8, 88, 91
Caudal heart, 242
Caudal sinus, 222–23
Caudal vein, 215, 223, 241f, 242
Cavernous bodies (CB), 230–32
CB. *See* Cavernous bodies
CCK. *See* Cholecystokinin
Celiac artery, 150
Celiac trunk, 212
Cenozoic era, 3
Central venous sinus, 188, 189f
Centrophoridae, 13f, 14
Centrophoriformes, 9
Centrophorus granulosus, 53
Centrophorus lusitanicus, 53
Centrophorus scalpratus (Endeavour dogfish), 178, 183f, 186, 187, 193
Centroscymnus crepidater, 15
Centrum, 83
Cephalic arteries, 204–9
Cephaloscyllium, 222
Cephaloscyllium isabella, 223f, 242, 315
Cephaloscyllium uter, 355t
Cephaloscyllium ventriosum, 405f
Ceratobranchials, 82
Ceratotrichia, 56–57, 85
Cerebellum, 274f, 275, 282–84, 307, 321
Cervix, 409
Cetorhinidae, 20f, 21
Cetorhinus, 198
Cetorhinus maximus, 54
  digestive system of, 145, 148, 159
  electrosensory system of, 319
  epigonal organ of, 232
  external body form of, 5, 6
  male reproductive system of, 458, 466
Chimaera, 245
Chimaera colliei, 51
Chimaera monstrosa, 148, 154t
  digestive system of, 157, 158
  heart of, 247, 248, 251
  male reproductive system of, 465
Chimaeroids, 1, 2
  digestive system of, 145
  placoid scales of, 52
  systematics of, 9
  teeth of, 48, 58
Chlamydoselachidae, 10
Chlamydoselachus
  circulatory system of, 200, 202, 203
  external body form of, 5
  locomotion in, 112
  systematics of, 9
Chlamydoselachus anguineus, 399, 401
Chloride, 330, 331, 332, 343, 344–45, 346–47, 348, 349, 416
Chloride channels, 347
Cholecystokinin (CCK), 154t, 155, 157, 158, 159, 164
Chondrichthyans, 1
  fins of, 91
  muscular system of, 105
  placoid scales of, 45
  systematics of, 8–10
  teeth of, 58, 60

Choroidal tapetum, 304
Chromatin-binding sites, 460
Chromatophores, 49, 50, 51
Cilia, 308, 311, 313
Circulation. *See also* Vasculature
  coronary, 262
  in the gills, 182–89
  in the gut, 165–67
Circulatory system, 198–216, 218–34. *See also* Afferent branchial arteries; Cephalic arteries; Dorsal aorta; Efferent branchial arteries; Hepatic vein sphincters; Hypobranchial arteries; Lateral abdominal vein; Mononuclear phagocytic system; Secondary blood system; Spiracle; Venous sinuses; Venous system; Ventral aorta
Cirrhigaleus, 13
Citrate synthase, 132–33
Cladistic revolution, 8
Cladodus, 47f
Cladoselache, 8, 45, 223
Clasper gland, 449f, 456–57
Claspers, 1, 6–7, 86f, 88, 94t, 105, 276
  structure and functions of, 453–58
Clasper siphon, 105
Clearnose skate. *See Raja eglanteria*
Cleveland Shale, 45
Cloaca, 7, 148–49, 450
Clump-type spermatozeugmata, 463–65
CNP. *See* C-type natriuretic peptide
Coffin rays, 33
Colipase, 157
Collared carpet sharks, 22, 23f
Collecting duct of renal tubule, 392
Commissural vessel, 221
Common mode rejection, 321
Compacta, 249–51
Complex zone, 357, 359f, 385
Compound (multilayered) spermatozeugmata, 463, 464, 465
Compound testes, 445, 446f
Compression phase of respiration, 110
Cones of the eye, 305–6
Constrictor muscles, 98
Conus arteriosus, 252–53
Convection, 189
Copulation, 457–58
Coracoarcualis, 102
Coracobranchial muscles, 102
Coracohyoideus, 102
Coracomandibular muscle, 101
Cornea, 51, 304
Coronary arteries, 204, 255–56
Coronary veins, 256
Corpora lutea, 401, 432, 434
Corpus cavernosum, 185, 186, 187
Corpus cerebelli, 282–84
Corpus luteum, 403, 435
Cow-nose rays, 38f, 40
Cow sharks, 9, 11
Cranial arteries, 221f
Cranial muscles, 107–8, 110
Cranial roof, 71, 76
Cranium, 71–82. *See also* Neurocranium; Splanchnocranium
Crassinarke, 33
Cretaceous era, 48
Cretaceous-Tertiary extinction event, 2
Crocodile sharks, 20f, 21
C17,20-Lyase, 459
C-type natriuretic peptide (CNP), 248, 348
Cucullaris muscle, 100
Cutaneous mechanosensitivity, 316

Cutaneous veins, 215–16
Cyclic adenosine monophosphate (cAMP), 346f, 347, 348
Cyclic guanosine monophosphate (cGMP), 348
Cyclomorial scales, 55
Cyclostomes, 240, 330t
Cystic fibrosis transmembrane conductance regulator (CFTR), 347

Daenia quadrispinosum, 53
Dalatiidae, 13f, 16–17
Dalatiiformes, 9
Dasyatidae, 38f, 39–40
Dasyatis
  circulatory system of, 199, 200, 201f, 202, 204, 208, 211–12
  female reproductive system of, 434f
  muscle fibers of, 134, 139
  nervous system of, 275
  skin of, 50
Dasyatis akajei, 51
Dasyatis americana, 145, 157, 355t, 419–20
Dasyatis guttata, 355t
Dasyatis marinus, 355t
Dasyatis okajei, 355t
Dasyatis pastinacea, 157, 209f, 355t, 357
Dasyatis sabina, 194, 314f
  electrosensory system of, 318f
  female reproductive system of, 399
  kidney studies on, 355t
  male reproductive system of, 460
  rectal gland of, 339
Dasyatis sayi (Dasyatis say), 355t, 399
Dasyatis violacea, 51, 399, 418
Deadwood Formation, 45
Denticles, 3
  dermal. *See* Placoid scales
  stomodaeal, 54–56
Dentine, 54, 55, 57, 58, 60–66
Depressor rostri muscle, 102–3
Dermal denticles. *See* Placoid scales
Dermal fin rays, 56–57
Dermis, 50
Devil rays
  external body form of, 6, 38f, 41
  girdles of, 85
  nervous system of, 291
Devonian era, 2, 8, 48
Diametric testes, 445, 446f
Diastole, 258, 261, 262, 263
Diencephalon, 285, 287–89, 307, 321, 322
Diffusion, 189
Diffusion conductance, 190
Diffusion resistance, 190–92, 193–94
Digestive system, 144–68. *See also* Cloaca; Esophagus; Gallbladder; Gut; Intestine; Liver; Mouth; Pancreas; Pharynx; Rectum; Stomach
3,4-Dihydroxyphenylalanine. *See* Dopa
Diplospondylous caudal vertebrae, 84
Diplospondylous precaudal vertebrae, 84
Distal segment of renal tubule, 370, 390–92
Dogfish sharks, 43. *See also* Squalidae; Squaliformes
  external body form of, 7, 11–17
  heart of, 248
  male reproductive system of, 460
  olfactory system of, 301f
  systematics of, 9
Dopa, 164, 412, 413
Dopachrome, 413
Dopamine, 164, 282
Dorsal aorta, 150, 209–13
Dorsal cutaneous vein, 215–16
Dorsal fins, 1, 7, 88–91

Dorsal funiculi, 276
Dorsal interarcuals, 100
Dorsal mixipodial muscle, 105
Dorsal pterygoideus muscle, 104, 105
Dorsal thalamus, 289
Ductus deferens, 444, 449, 451–52, 465
Duodenum (proximal intestine), 145, 148

Eagle rays, 5, 38f, 40
Ears, 308–11
Echinorhinidae, 12, 13f, 57
Echinorhiniformes, 9
Echinorhinus brucus, 3, 12, 48, 53
Edinger-Westphal nucleus, 277
Efferent branchial arteries, 200–205
Efferent connections
  of cerebellum, 283–84
  of tectum mesencephali, 285–87
Efferent ductules, 444, 450
Egg capsules, 2, 404–15, 436
Egg envelopes, 422, 423f, 425, 428f
Egg jelly. *See* Albumen
Egg predation, 414
Eggs, 399, 403, 415
Egg shells, 409–10
Elasmobranchii, 9, 58
Elasmobranchimorphi, 58
Elasmobranchs. *See also* Rays; Sharks
  blood pressure of, 219t
  checklist of, 471–98
  description of, 2–3
  diversity of, 10
  external body form of, 3–8
  fluid compartments of, 330t
  kidney studies on, 355–57t
  systematics of, 8–10
  teeth of, 58–66
  warm-blooded, 225–29
Electric rays
  external body form of, 33–35
  girdles of, 85
  splanchnocranium of, 82
  systematics of, 9
Electrosensory system, 316–22
Elephant fishes, 1
Ellipsoids, 232
Enameloid, 2, 3, 45, 49, 54, 58, 60–66
Endeavour dogfish. *See Centrophorus scalpratus*
Endings of Poloumordwinoff, 275
Endocardium, 247, 253
Endocrine cells, 154–55
Endorphin, 155
Endoskeleton, 1, 69–91. *See also* Appendicular skeleton; Axial skeleton; Cranium; Vertebral column
Enkephalin, 155
Epaxial lobe, 7, 8
Epaxial muscle, 102, 104
Epibranchial artery, 221f
Epibranchial muscles, 94t, 102–3
Epicardium, 242–43, 247, 253
Epidermal coparticipation hypothesis, 50
Epidermis, 50
Epididymis, 444, 449, 450–51, 461, 465
Epigonal organ, 232, 234, 445
Epinephrine (adrenaline), 163–64, 166, 225, 233
Epiphysis. *See* Pineal
Epithalamus, 289
Epithelium
  of female reproductive system, 419–20, 429–30
  of renal tubules, 380–92
Eptatretus stoutii, 330t
Esophagus, 145

Estradiol, 431–34, 435–36, 459
Estrogen, 404, 432, 460
Estrogen receptors, 460
Ethmoidal region of neurocranium, 71
Ethmopalatine ligament, 106
Etmopteridae, 13f, 14–15
Etmopterus, 14, 50, 118, 120t, 121t, 232
Etmopterus bullisi, 53
Etmopterus spinax, 115, 117, 155, 245, 247, 251
Eumelanins, 51
Euprotomicroides, 16
Euprotomicrus, 50
Euselachii, 2, 9
Evoked potentials, 307, 312, 313, 322
Excretory canal, 340
Exoskeleton, 69, 91
Expansion phase of respiration, 110
External body form
    of elasmobranchs, 3–8
    of rays, 29–41
    of sharks, 10–29
External carotid arteries, 204
External gill filaments, 431
External yolk sac, 410, 411f
Extrabrachials, 82
Extrahilar vessels, 256
Extratubular matrix, 333–35
Extravisceral cartilages (arches), 82
Extrinsic nerves of the gut, 151, 152f
Extrinsic ocular muscles, 94t, 95
Eyeballs, 77–78, 95, 304
Eyelids, 3, 304
Eyes, 3, 51–52, 304–6. *See also specific structures*

FABP. *See* Fatty acid–binding protein
Facial (VII) nerve, 97, 98, 277, 279, 302
False cat sharks, 26f, 27
Fatty acid–binding protein (FABP), 225
Fatty acids, 265–66
Feeding
    electrosensory system and, 319–20
    musculature involved in, 107–10
Female reproductive system, 398–437. *See also* Oviparity; Viviparity
    endocrinology of, 431–37
    gross and microscopic anatomy of, 400–409
    reproductive cycles in, 399–400, 432
Fetoplacental unit, 425
Fick's law of convection, 189
Fick's law of diffusion, 189
Fields induced by moving recipient, 317
Finback cat sharks, 5, 26–27
Fins, 85–91
    anal, 1, 7, 88–91
    caudal, 2, 3, 7–8, 88, 91
    dorsal, 1, 7, 88–91
    paired. *See* Paired fins
    pectoral. *See* Pectoral fins
    pelvic. *See* Pelvic fins
    unpaired. *See* Unpaired fins
Fin spines, 7, 54, 57–58
Flagellar cells, 367f, 380–81, 382
Flagellar ribbons, 367f, 381
Flegestolepis grossi, 46f
Fluid compartments, 329–31
FMRF. *See* Phe-met-arg-phe amide
Follicles, 401–3, 431–34, 435, 436, 445
Frilled sharks
    anatomical features of, 11f
    external body form of, 5, 10
    splanchnocranium of, 81
    systematics of, 9
Furgaleus, 27

GABA. *See* Gamma-aminobutyric acid
Galanin, 158, 159
Galeaspida, 45
Galeocerdo, 28, 47f
Galeocerdo cuvieri, 48, 51, 332
Galeoids, 203
Galeomorphii, 9, 10, 46
Galeomorphs
    circulatory system of, 200–202
    male reproductive system of, 445
    nervous system of, 287, 291
    systematics of, 9
Galeorhinus australis, 260–61f, 262
Galeorhinus galeus, 249
Galeus, 118, 120t, 121t, 132, 255, 401
Galeus canis. See Mustelus canis
Galeus melanostomus, 52, 53f, 115, 117, 245
Gallbladder, 159
Gamma-aminobutyric acid (GABA), 123
Ganglion cells, 245–46
Gape cycle, 108
Gas exchange, 176, 178–82, 186, 189–95
Gastric acid, 148
Gastric artery, 212
Gastric emptying time, 160
Gastric inhibitory peptide (GIP), 155, 159
Gastric secretion, 155–57
Gastric veins, 216
Gastrin, 154t, 155, 156–57, 158, 159, 164
Gastrin-releasing peptide (GRP), 154, 155, 158
Gastroduodenal artery, 212
General cutaneous sense, 307–8
Genital ducts, 444, 449–453
Germ cells, 400, 445, 446–47, 458
Giant megatooth shark, 5
Giant stingarees, 37, 38f
Giant whale shark. *See Rhiniodon typus*
Gill arches (branchial arches), 1, 78, 81–82, 174, 240, 279
Gill filaments, 110, 174, 175f, 176, 431
Gill openings, 1, 3, 5
Gill rakers, 54
Gill rays, 174
Gills, 3, 110, 174–95, 240
    arrangement of, 175f
    circulation in, 182–89
    gas exchange in relation to morphology of, 189–95
    morphology of, 178–82
    osmoregulation and, 332
    ventilation and, 176–78
    ventilation:perfusion ratio regulation in, 220
Gill slits, 110, 174, 176, 178
Ginglymostoma, 225
    epigonal organ of, 232
    external body form of, 24
    nervous system of, 291, 294
    teeth of, 47f
Ginglymostoma cirratum, 157
    circulatory system of, 230
    kidney studies on, 355t
    male reproductive system of, 457
    mechanosensory lateral line of, 313
    teeth of, 48
    visual system of, 306, 307
Ginglymostomatidae, 23f, 24
GIP. *See* Gastric inhibitory peptide
Girdles, 85–88
    pectoral, 3, 72f, 85, 86f
    pelvic, 85, 86f, 87
Glenoid ligament, 106
Glicentin, 155
Glomerular mesangium, 372–73

Glossopharyngeal (IX) nerve, 98, 277, 279, 302
Glucagon, 159, 348
Glucose, 461
Glucuronide, 459
Glycine, 412
Glycogen, 133
Goblin sharks, 19, 20f, 79
Gogolia, 27, 28
Gollum attenuatus, 399
Gonadal cycles, 460–61
Gonadosomatic index (GSI), 460–61
Gonadotropin-releasing hormone (GnRH), 276, 289, 294, 435, 458, 459–60, 461
Granular cells, 371–72
Gravity receptors, 311
Gray reef shark. *See Carcharhinus menisorrah*
Great white shark. *See Carcharodon carcharias*
Ground sharks. *See* Carcharhiniformes
Growth factors, 460
GRP. *See* Gastrin-releasing peptide
GSI. *See* Gonadosomatic index
Guitarfishes
  external body form of, 31–33
  nervous system of, 291
  splanchnocranium of, 82
  systematics of, 9
Gulper sharks, 13f, 14
Gurgesiella, 35
Gustatory system, 300–303
Gut, 146f
  absorption in, 160
  circulation in, 165–67
  endocrine cells and hormones of, 154–55
  innervation of, 151–54
  motility and food transport in, 160–65
  secretion in, 155–59
  vessels of, 150–51
Gut wall, 147f, 149–50
Gymnuridae, 38f, 40

Habenular nuclei, 289
Hair cells, 308, 311, 313
Hammerhead sharks
  electrosensory system of, 318
  external body form of, 26f, 28–29
  nervous system of, 291
  urinary system of, 375f
Head, 3
Headgut, 145
Hearing, 308–13
Heart, 225, 238–68. *See also* Atrium; Conus arteriosus; Sinus venosus; Ventricle
  blood supply of, 253–56
  cardiac cycle and, 258–62
  circulation of, 262
  general design of cardiovascular apparatus, 240–42
  innervation of, 256–58
  location of, 242–44
  metabolism and energetics of, 265–66
  structural design of, 244
  working, 263–65
Helicoprion, 46
Hemal arch pump, 224
Hemigaleidae, 26f, 28, 76
Hemigaleops fosteri, 312
Hemigaleus, 429
Hemipristis, 28
Hemiscylliidae, 23f, 24
Hemitripterus, 265
Hemopoiesis, 229–32, 234
Hepatic artery, 212
Hepatic portal system, 216

Hepatic sinuses, 224, 240
Hepatic vein sphincters, 225, 240
Heptanchus, 174
Heptanchus maculatus, 53, 198, 355t
Heptranchias, 202, 203
Heptranchias perlo, 133
Heterocercal tail, 112–13
Heterodontidae, 18f, 19, 57, 410. *See also* Bullhead sharks; Horn sharks
Heterodontids, 7, 91, 287
Heterodontiformes, 9, 10, 18–19. *See also* Bullhead sharks
Heterodontus
  circulatory system of, 222, 223
  fin spines of, 57
  heart of, 258, 261
  male reproductive system of, 453
  placoid scales of, 52
  thymus of, 231
Heterodontus francisci, 213f, 230, 355t
Heterodontus japonicus, 63, 355t
Heterodontus mexicanus, 405f
Heterodontus portusjacksoni (Port Jackson shark)
  blood pressure of, 219t
  female reproductive system of, 399
  heart of, 242, 260–61f
  male reproductive system of, 450–51, 465
  rectal gland of, 339
  respiratory system of, 176–78, 179f, 195
  teeth of, 46
Heteronarce, 33
Heterostraci, 45
Hexanchidae, 9, 11
Hexanchiformes
  anatomical features of, 11f
  circulatory system of, 199
  digestive system of, 145
  external body form of, 10–11
  systematics of, 9
Hexanchoids
  circulatory system of, 201f, 203
  fins of, 88
  girdles of, 85
  splanchnocranium of, 79, 80, 81, 82
Hexanchus, 47f, 48, 202, 204
Hexanchus griseus, 202f
Hexatrygonidae, 37–39
Hilar vessels, 256
Himantura, 39
Himantura warnak, 125t
Hindgut, 145
Histamine, 155, 156
Histotroph, 431
Holocephalans, 1. *See also* Chimaeroids
Holocephalii, 8, 9, 58
Holorhinus tobijei, 355t
Horizontal skeletogenous septum, 103
Hormones
  of the gut, 154–55
  of the heart (neurohormones), 266–67
  pancreatic, 154
  steroid. *See* Steroid hormones
Horn sharks, 19, 306, 410. *See also* Heterodontidae
Houndsharks, 26f, 27–28
Hybodonts, 9, 46
Hybodus, 45, 47f
Hydrolagus, 277, 279
Hydrolagus collei, 154, 447, 463, 465
17 α-Hydroxylase, 459
Δ5-3β-Hydroxysteroid dehydrogenase, 458, 459
5-Hydroxytryptamine. *See* Serotonin
Hyobranchial region, 3
Hyoid arch, 1, 78, 80

Hyoidean arteries, 204, 205
Hyoid muscles, 98
Hyomandibulo-hyoid ligament, 107
Hyoscine, 164
Hypaxial lobe, 7, 8
Hypaxial musculature, 104
Hypnidae, 33
Hypobranchial arteries, 203–4, 210f
Hypobranchial muscles, 94t, 100–102
Hypobranchial veins, 215
Hypochordal rays, 91
Hypothalamus, 274f, 287–88, 289–91, 435, 459
Hypoxia, 266

IGF-1. *See* Insulin-like growth factor
Ileum, 145
Iliac arteries, 212
Indo-Australian archipelago, 8
Inferior orbital artery, 206
Inferior raphe, 277
Inner ear, 307–8, 309–11
Inner red muscle fibers, 118, 119f, 132
Innervation. *See also* Nervous system
  of the gut, 151–54
  of the heart, 256–58
  of the male reproductive system, 458
  of muscle fibers, 122–23, 138–40
Inner white muscle fibers, 119f, 120, 132, 135
Insulin, 154, 159
Insulin-like growth factor (IGF-1), 155, 158, 159
Integrins, 465
Integumentary system, 43–58. *See also* Scales
Interarcual muscles, 100
Interbasilar muscles, 100
Interbranchial muscle, 100
Interbranchial septum, 186–87, 188
Intercalated disks, 244
Interhyoideus muscle, 97–98
Intermandibularis muscle, 97
Intermedialia, 83
Intermediate fibers, 118, 138f, 139, 140
Intermediate segment of renal tubule, 384–90
Internal carotid arteries, 206, 208–9, 229
Internal yolk sac, 410, 411f
Intestinal secretion, 157
Intestine, 148
Intraintestinal vein, 216
Intrauterine cannibalism, 420–21
Intrinsic nerves of the gut, 151
Ion transport, 345–47
Iris, 277, 304
Isistius, 16, 17, 50
Islet cells, 159
Isthmus, 408
Isurus, 5, 208
Isurus oxyrhynchus
  circulatory system of, 210f, 226, 228f, 229
  heart of, 249, 251
  muscle fibers of, 117f, 132–33

Jagorina, 234
Jaw protrusion, 108–10
Jaws, 1, 2, 79, 80
Jaws (Benchley), 44
JGA. *See* Juxtaglomerular apparatus
Jurassic era, 3, 46, 48
Juxtaglomerular apparatus (JGA), 366–74

Ketone bodies, 132, 133, 265
Kidneys, 331–32, 353–54. *See also* Renal *entries*
  microscopical studies on, 355–57t
  vasculature of, 374–80
Kidney tubules. *See* Renal tubules

Kinetic electric fields, 317
Kitefin sharks, 13f, 16–17
Knifetooth sawfish, 8
Krough's diffusion constant, 190

Labial cartilages, 80
Lactate, 132, 229
Lactate dehydrogenase (LDH), 118, 120
Lagena, 308
Lamiopsis, 28
Lamna cornubica, 228f
Lamna nasus, 226
Lamnidae, 20f, 21–22, 112. *See also* Mackerel sharks
Lamniformes. *See also* Mackerel sharks
  circulatory system of, 204
  external body form of, 19–22
  female reproductive system of, 420
  male reproductive system of, 445
  systematics of, 9, 10
Lamnoids
  circulatory system of, 209
  digestive system of, 162
  fins of, 91
  muscle fibers of, 130–31
  muscular system of, 109
  splanchnocranium of, 79
  vertebral column of, 83
Lanceolate scales, 53
Lantern sharks, 13f, 14–15
Lateral abdominal vein, 223
Lateral artery, 212
Lateral cutaneous veins, 216
Lateral eyes, 51–52
Lateral funiculi, 276
Lateral membrane, 338–39
Lateral reticular zone, 277–78
Lateroventral hypobranchial arteries, 204
LDH. *See* Lactate dehydrogenase
Legged skates, 35–37, 88
Lemon sharks, 303. *See also Hemigaleops fosteri; Negaprion brevirostris*
Lens of the eye, 304
Lens placodes, 50
Leopard shark. *See Triakis semifasciata*
Lepidomorial theory of scales, 55–56
Lepidomorium, 55–56
Leptochariidae, 26f, 27
Leucoraja naevis, 134
Leu-enkephalin, 289
Levator palatoquadrati, 96, 98–100
Levator rostri muscle, 102–3
Leydig cells, 458, 459, 466
Leydig gland, 444, 449–50, 452–53, 465
Leydig gland bodies, 450, 451
Ligaments, 106–7
Lipid metabolism, 133
Little skate. *See Raja erinacea*
Liver, 159, 240, 242
Locomotion, 111–13
Long-snout dogfish, 53
Long-tailed carpet sharks, 23f, 24
Lower jaws, 1, 108
Lower Silurian era, 45
Lungfishes, 198
Lymphatic system, 330

Mackerel sharks. *See also* Lamnidae; Lamniformes
  external body form of, 19–22
  heart of, 240
  locomotion in, 112
  systematics of, 9
Macula densa, 370, 371, 372, 374f, 390
Macula neglecta, 309–11, 312

Magnetic fields, 320–21
Mako sharks, 227
Male reproductive system, 444–66
  copulation and, 457–58
  endocrinology of, 458–60
  gonadal cycles in, 460–61
  gross and microscopic organization of, 444–45
Mandibular arch, 78–79
Mandibulohyoid ligament, 106
Manta, 291
Manta birostris, 51, 319
Maxillary valve, 6
Mechanosensory lateral line, 307–8, 313–16
Mechanosensory systems, 307–8
Meckel's cartilage, 1, 58, 78–79, 108–9
Medial cerebral arteries, 208
Medial reticular zone, 277
Median reticular zone, 277
Medulla oblongata, 274f
Megachasma, 19
Megachasma pelagios, 6, 319
Megachasmidae, 20f, 21
Megamouth sharks, 20f, 21
Melanin, 50–51
Melanocytes, 50
Melanophores, 50–51
Melanophore-stimulating hormone (MSH), 52
Melanosomes, 51
Mesangial cells, 372–73
Mesencepahlic tegmentum, 281
Mesencephalon, 279–80
Mesenteric arteries, 150–51
Mesozoic era, 2, 45, 46
Met-enkephalin, 289
Microvillar cells, 430, 431
Midgut, 145
Midventral hypobranchial artery, 204
Miroscyllium, 14
Mitsukurinidae, 19, 20f
Mixed type hearts, 249, 250t
Mixopterygia. See Claspers
Mobula, 134, 199, 202, 208, 291
Mobula japonica, 275
Mobulidae, 6, 38f, 41
Modern rays, 9
Modern sawfishes. See Pristidae
Modern sharks, 9
Mollisquama, 16
Mononuclear phagocytic system (MPS), 229–32, 233
Monospondylous precaudal vertebrae, 84
Mouth, 3, 145
Mouth floor, 6
MPS. See Mononuclear phagocytic system
MSH. See Melanophore-stimulating hormone
Muscle fibers, 115–42
  of batoids, 115–16, 132–41
  biochemistry of, 132–34
  electrophysiology of, 123–27, 138–39
  mechanical properties of, 127–30, 140–41
  microscopical anatomy of, 116–23
  mitochondrial volume of, 120, 122t
  motor innervation of, 122–23, 138–39
  of rays, 141
  sensory innervation of, 123, 139–40
  of sharks, 116–34
  structure of, 117–22, 136
  ultrastructural features of, 118t, 121t
Muscular system, 93–113. See also Branchiomeric muscles; Epibranchial muscles; Extrinsic ocular muscles; Hypobranchial muscles; Muscle fibers; Trunk muscles
  descriptive anatomy of, 95–107
  functional morphology of, 107–13

Mustelus
  acoustic system of, 310f
  circulatory system of, 198, 222
  female reproductive system of, 401
  heart of, 256, 266
  nervous system of, 277
Mustelus antarcticus, 195, 221f
Mustelus californicus, 306
Mustelus canis (smooth dogfish), 52
  acoustic system of, 312
  blood pressure of, 219t
  circulatory system of, 230
  digestive system of, 155
  electrosensory system of, 319, 320
  female reproductive system of, 399, 408, 415, 416, 422
  gustatory system of, 303
  heart of, 260–61f
  kidney studies on, 355t
  male reproductive system of, 455
  mechanosensory lateral line of, 314f
  urinary system of, 375f, 382f, 384
Mustelus griseus, 460
Mustelus laevis, 301f, 418
Mustelus manazo, 460
Mustelus mustelus, 206f, 355t
Mustelus schmitti, 145, 155
Mustelus vulgaris. See *Mustelus mustelus*
Myliobatidae, 38f, 40
Myliobatiformes, 9, 33, 37–41. See also Stingrays
Myliobatis, 47f, 48, 50, 139, 419
Myliobatis aquila, 355t
Myliobatis californicus, 355t, 399
Myliobatis freminvilli, 418
Myliobatoids
  girdles of, 85
  nervous system of, 291
  neurocranium of, 74
  splanchnocranium of, 80
  vertebral column of, 84
Myocardium, 242, 244, 247, 249, 253
Myomeres, 103–4, 105f, 111–12
Myosin, 447
Myotomal muscle fibers, 116–30, 138
  functional role of different types, 130–32
Myxine, 243

Naked sperm, 463
Narcine brasiliensis, 355t
Narcinidae, 33
Narke, 33
Narke japonica, 248
Narkidae, 33
Narkid electric rays, 74, 84
Nasal capsules, 71, 75
Nasal flaps, 5
Natriuretic peptide (NP), 348
Neck segment of renal tubule, 380–82
Negaprion, 28
Negaprion acutidens, 313
Negaprion brevirostris (lemon shark), 52
  acoustic system of, 311
  ceratotrichia of, 56f, 57
  circulatory system of, 226
  digestive system of, 160
  ligaments of, 106–7
  pit organs of, 314f
  teeth of, 59, 63, 64f, 65f
  visual system of, 306, 307
Negaprion fosteri, 312
Neoraja, 35
Neoselachian revolution, 2

Neoselachians
  endoskeleton of, 70f
  external body form of, 3–8
  fins of, 89f
  neurocranium of, 72f, 73f, 74f
  paired fins and girdles of, 86f
  splanchnocranium of, 80, 81f
  systematics of, 8
Neoselachii, 2, 9, 10, 46–48
Nervous system, 162–64, 273–94. *See also* Brainstem; Cerebellum; Diencephalon; Hypothalamus; Innervation; Spinal cord; Spinal nerves; Tectum mesencephali; Telencephalon
Nervus terminalis, 303
Neural crest cells, 48, 50
Neurocranium, 1, 2, 71–78
Neurohormones, 266–67
Neuropeptides, 151, 153, 154t. *See also specific types*
Neuropeptide Y (NPY), 154, 155, 157–58, 159, 166, 257, 289, 348
Neurotensin, 155, 159
Neurotransmitters, 151–54, 164–65. *See also specific types*
New Zealand legged torpedos, 88
Nictitating membrane, 304
Nidamental gland, 406
Nonflagellar cells, 380, 381
Nonjunctional contacts, 465
Nonrespiratory blood pathway, 183, 186–89
Nonvitellogenic follicles, 435
Norepinephrine (noradrenaline), 166, 225, 233
Nostrils, 2, 3, 5
Notorynchus maculatus, 401
NP. *See* Natriuretic peptide
Nuclei C1, C2, and C3, 281
Nucleus cerebelli, 285
Nucleus dorsalis, 280f, 281
Nucleus funiculi lateralis, 276
Nucleus medialis, 281
Nucleus octavus ascendens, 281
Nucleus octavus descendens, 281
Nucleus ruber, 281
Nucleus ventralis, 281
Numbfishes, 33
Nurse sharks, 23f, 24

Oblique muscles, 95
Occipital region of neurocranium, 71
Occiput, 71, 78
Octavolateralis area, 280–81
Octavolateralis placodes, 50
Oculomotor (III) nerve, 95, 277
Oculomotor nuclei, 283
Odontaspididae, 19, 20f
Odontaspis, 226
Odontoblastic process, 60, 66
Odontoblasts, 54–55, 60, 63, 64
Odontoclasts, 55
Odontoderegulation theory, 56
Olfactory bulb, 291, 302
Olfactory placodes, 50
Olfactory system, 300–303
Oncorhynchus mykiss, 330t
Oocytes, 401
Oophagy, 19, 405, 420–21
Ophiodon elongatus, 219, 220f
Ophthalmica magna, 200f
Ophthalmic vein, 214
Optic arteries, 208
Optic nerve, 274f
Optic nuclei, 283
Optic tectum, 274f, 285–87
Orbital processes, 79

Orbital region, 3, 71, 76–78
Orbitonasal vein, 214
Orbits, 71, 76–78
Ordovician era, 45
Orectolobidae, 22–24
Orectolobids, 74, 91, 108–9
Orectolobiformes, 9, 10, 22–25
Orectolobus japonicus, 51, 356t
Organ of Leydig, 145, 232, 234
Osmoregulation, 331–32
Osteichthyes, 45, 58
Osteostraci, 45
Ostium, 404
Ostracoderms, 45
Otic capsules, 71, 78
Otic region of neurocranium, 71
Otolith organs, 308–9, 311
Outer red muscle fibers, 118, 119f, 131, 132
Outer white muscle fibers, 119f, 120, 135
Ovaries, 400–404
Oviducal arteries, 211
Oviducal gland, 404–8, 412, 413, 415
Ovoviviparity, 2, 401, 403, 404, 406, 407, 408, 416–17, 432
  description of, 409–15
  reproductive cycles in, 399
  steroid hormones and, 432, 433f, 434, 435, 437
Oxynotidae, 13f, 15–16
Oxynotus centrina, 203f
Oxynticopeptic cell, 420

Pacific reef blacktip. *See Carcharhinus melanopterus*
Pain receptors, 275
Paired fins, 1, 6
  girdles and, 85–88
  lateral abdominal vein and, 223
Palate, 6
Palatoquadrates, 1, 58, 79, 80, 108–9
Paleoselachii, 45
Paleozoic era, 8, 45, 48
Pallial tract, 293
Pallium, 291
Palmitoyltransferase, 133
Pancreas, 157–59
Pancreatic hormones, 154
Pancreatic polypeptide, 154, 159
Papillary zone, 408
Paragaleus, 429
Paraplacental uterus, 429
Parascylliidae, 3, 22, 23f
Paraselachians, 9
Paratrygon motoro, 300
Parietal pericardium, 242
Parmaturus xaniurus, 399
Pars lateralis, 283
Pars medialis, 283
Pectoral fins, 1, 6, 85, 86f, 87
  lateral abdominal vein and, 223
  locomotion and, 112–13
  muscle fibers of, 134–36
  musculature of, 104–5
Pectoral girdle (scapulocoracoid), 3, 72f, 85, 86f
Pelvic fins, 1, 6, 85, 86f, 87–88
  lateral abdominal vein and, 223
  muscle fibers of, 134–36
  musculature of, 104–5
Pelvic girdle, 85, 86f, 87
Pelvicobasal muscle, 104
Pentagastrin, 164
Pentanchus, 25
Pepsin, 155
Pepsinogen, 145, 148, 155
Peptide histidine isoleucine (PHI), 155

Peptide YY (PYY), 155
Perfusion conductance, 190
Pericardial cavity, 243–44
Pericardioperitoneal canal, 244
Pericardium, 242–44
Peripheral blood leucocytes, 230
Peripolar cells, 373–74
Peritoneal cavity, 244
Peritubular (peribundular) sheath, 361
Permian-Triassic transition, 45
Peroxidase, 412, 413
Petromyzon, 43
Phaeomelanins, 51
Phagocytosis, 229–32
Pharyngobranchials, 82
Phasic control, 268
Phe-met-arg-phe amide (FMRF), 154t, 155, 159, 289
Phenoloxidase, 412, 413
PHI. *See* Peptide histidine isoleucine
Photophores, 50
Pillar cells, 178–79
Pineal (epiphysis), 51–52, 289
Pineal body, 76
Pirstiurus, 275
Pit organs, 313, 314f
Pituitary-testicular axis, 459
Placenta, 240, 429, 431
Placental viviparity, 421–29, 432
Placodermi, 45
Placoid scales (dermal denticles), 2, 43–44, 91
  diagram of, 55f
  evolution of, 45, 46f
  form and distribution of, 52–54
  histology of, 54–56
Platyrhinidae, 31, 32f, 33
Platyrhinoidis, 282, 285, 291, 313
Platyrhinoidis triseriata, 282, 284, 286f, 290f, 291, 292f
  acoustic system of, 312
  electrosensory system of, 322
  mechanosensory lateral line of, 315, 316
  visual system of, 307
Plesiobatidae, 37, 38f
Plesodic sharks, 87
Pleurocanthodii, 45
Pliotrema, 54
Podocytes, 365f, 366, 368f
Poloumordwinoff, endings of, 275
Polyodon spathula, 232
Porbeagle sharks, 227, 229
Poroderma, 25
Port Jackson shark. *See Heterodontus portusjacksoni*
Posterior cardinal veins, 215
Posterior cerebral arteries, 208
Posterior cerebral vein, 214
Posterior gastropancreaticosplenic artery, 212
Posterior intestinal artery, 212–13
Posterior intestinal veins, 216
Posterior mesenteric artery, 213, 340, 341–43
Posterior oviduct. *See* Uterus
Posterior parencephalon, 288–89
Posterior tubercle, 289
Potamotrygon circularis, 332, 356t
Potamotrygon humerosa, 354, 356t, 370, 376, 380, 384, 390
Potamotrygon hystrix, 330t
Potamotrygonidae, 38f, 39
Potamotrygon magdelanae, 356t
Potamotrygon motoro, 356t
Potassium, 461
Potassium chloride, 461
P-Q interval, 258

Precaudal fins, 1
Precaudal pits, 7
Precaudal tail, 3, 7–8
Predentine, 60–63
Preoptic nucleus, 289
Preoptic region, 287–88, 289–91
Preoral snout, 2
Preorbital muscle, 96, 108, 109
Preovulatory follicles, 435
Primary jaws, 79
Prionace, 28, 47f, 118, 401
Prionace glauca
  circulatory system of, 226
  electrosensory system of, 320
  epigonal organ of, 232
  female reproductive system of, 399, 418–19, 422
  heart of, 249
  kidney studies on, 356t
  male reproductive system of, 460, 465
  muscle fibers of, 117
  skin coloration of, 51
  teeth of, 48
Pristidae, 3, 29, 30f, 54
Pristiformes, 9, 29, 30f. *See also* Sawfishes
Pristiophoridae, 18, 54. *See also* Saw sharks
Pristiophoriformes, 9, 10
Pristiophoroids, 7, 74
Pristiophorus, 54, 401
Pristis, 54
Pristis cuspidatus, 401
Pristis pectinatus, 356t
Pristis perotteti, 356t
Pristiurus melanostomus, 356t
Pristoids, 74
Progesterone, 432, 433f, 434, 435, 436–37, 459
Proscylliidae, 5, 26–27
Proscyllium haberrer, 456f
Protacrodus, 47f
Protoselachians, 2
Protoselachii, 45
Proximal club zone, 408
Proximal intestine. *See* Duodenum
Proximal segment of renal tubule, 382–84, 386f
Psammosteus, 46f
Pseudobranch, 221, 222, 229
Pseudobranchial arteries, 204–5
Pseudobranchial plexus, 229
Pseudocarchariidae, 20f, 21
Pseudoginglymostoma, 24
Pseudotriakidae, 26f, 27
Pseudotriakis, 25
Pteromylaeus bovina, 399
Puboischiadic bar, 85, 86f, 87
Pupil, 304–5
Purkinje cells, 283–84
PYY. *See* Peptide YY

QRS complex, 258, 262
Q-T interval, 261
Quadratohyomandibular ligament, 106
Quadratomandibular ligament, 106

Rachycentron canadum, 376
Radial muscles, 222–23
Radial testes, 445, 446f
Rainbow trout, 190, 192
Raja
  circulatory system of, 199f, 201f, 202, 203, 204, 205, 208, 215
  digestive system of, 149f, 151, 155, 157, 158, 162, 164, 165
  heart of, 244

muscle fibers of, 139
nervous system of, 276, 277, 282, 285, 291, 293
Raja asterias, 356t
Raja batis, 232, 356t
Raja binoculata, 219t, 330t
Raja clavata
   acoustic system of, 309f, 311
   circulatory system of, 208f
   digestive system of, 146f, 147f, 153f, 154t, 155, 156, 157, 160
   electrosensory system of, 319
   female reproductive system of, 408
   heart of, 244, 248
   kidney studies on, 356t
   muscle fibers of, 134, 135–36
   placoid scales of, 53
   respiratory system of, 176, 178f, 186, 187
   skin coloration of, 51
   thymus of, 231
   visual system of, 306
Raja diaphanes, 244, 356t
Raja eglanteria, 315
   female reproductive system of, 399, 408
   locomotion in, 113
   male reproductive system of, 449, 457, 465
   visual system of, 305f
Raja erinacea (little skate)
   circulatory system of, 233
   digestive system of, 154t, 159
   electrosensory system of, 322
   female reproductive system of, 399, 404, 407f, 408, 412, 414, 422, 431, 432, 435, 436
   heart of, 244, 257, 261, 265
   kidney studies on, 356t
   male reproductive system of, 444, 447, 449f, 451, 452, 453, 454f
   mechanosensory lateral line of, 315
   muscle fibers of, 132, 133
   muscular system of, 102f, 103f
   respiratory system of, 187, 195
   urinary system of, 354, 357, 358f, 359f, 362f, 364f, 365f, 366, 367f, 368f, 369f, 376, 380, 381f, 382, 383f, 384, 385, 386f, 388–90, 391f, 392
   visual system of, 307
Raja laevis, 356t
Raja macrorhynchus, 356t
Raja marginata, 356t
Raja microocellata, 134, 154t
Raja miraletus, 356t
Raja montagui, 51, 134, 154t, 179
Raja mosaica, 356t
Raja naevus, 154t, 356t, 399
Raja nasuta, 356t
Raja ocellata, 356t
Raja oxyrhynchus, 356t
Raja productus, 356t
Raja punctata, 356t
Raja radiata
   circulatory system of, 221f, 233
   digestive system of, 154t, 155, 158, 159, 164
   electrosensory system of, 322
   heart of, 257
Raja radula, 356t
Raja rhina, 154t, 257, 330t
Raja stabuliforis, 244, 356t, 357
Raja undulata. See Raja batis
Rajidae, 35, 36f, 412. *See also* Skates
Rajiformes, 9, 29, 35–37. *See also* Skates
Rajomorphii, 9, 10
Ram ventilation, 110, 178
Raphe nuclei, 277

Rays. *See also* Elasmobranchs
   external body form of, 29–41
   female reproductive system of, 399–400
   muscle fibers of, 141
   nervous system of, 277, 287
   systematics of, 9
   teeth of, 48
   vertebral column of, 84
Rectal gland, 148–49, 329, 332–49
   anatomy of, 332–35
   fine structure of, 335–40
   mechanisms of secretion, 343–48
   vasculature of, 340–43
Rectin, 348
Rectum, 7, 148–49
Rectus cervicis muscle, 101–2
Red muscle, 227
Red muscle fibers, 116f, 117–27, 135, 136, 139–40, 229
   biochemistry of, 132, 133
   functional role of in myotome, 130–32
   histogram of, 137f
   inner, 118, 119f, 132
   mechanical properties of, 127–30, 140–41
   membrane constants of, 125t
   outer, 118, 119f, 131, 132
Renal arteries, 212, 374–75
Renal corpuscle, 362–66, 375f
Renal lobules, 354–57
Renal portal veins, 215
Renal portal vessels, 374, 376
Renal tubules, 219, 354
   configuration of, 357–62
   epithelium of, 380–92
   segments and segment subdivisions, 378–79t
Renal zones, 354–57
Renin-angiotensin system, 368, 374, 392
Reproductive cycles, 399–400, 432
Reproductive endocrinology
   of female reproductive system, 431–37
   of male reproductive system, 458–60
Reproductive system. *See* Female reproductive system; Male reproductive system
Requiem sharks, 3, 26f, 28
Respiration, 110–11
Respiratory blood pathway, 183–86
Respiratory system, 174–95. *See also* Gills
   direction of water and blood flow in, 192
   nonrespiratory blood pathway in, 183, 186–89
   respiratory blood pathway in, 183–86
   unequal distribution of water and blood flows in, 194–95
Resting potentials, 138–39
Retial vessels, 227
Retia mirabilia, 226
Reticular formation, 276–79
Retina, 277, 285–87, 291, 305–6
Rhina squatina. See Squatina squatina
Rhincodon, 22, 145
Rhincodontidae, 23f, 25
Rhincodon typus, 6, 404, 417
Rhinidae, 3, 10, 31. *See also* Sharkfin guitarfishes
Rhiniformes, 29–31. *See also* Sharkfin guitarfishes
Rhiniodon typus, 51, 54, 318–19
Rhinobatidae, 31, 32. *See also* Guitarfishes
Rhinobatiformes, 9, 10, 30, 31–33. *See also* Guitarfishes
Rhinobatos, 47f, 48
Rhinobatos annulatus, 225
Rhinobatos productus, 159
Rhinoptera bonasus, 357t, 418, 419
Rhinopteridae, 38f, 40

Rhizoprionodon, 429
Rhizoprionodon actus, 456f
Rhizoprionodon terraenovae (Atlantic sharpnose shark), 247, 253, 390, 393f
  female reproductive system of, 406f, 416, 422, 425, 427f, 429, 430, 431, 434f
  kidney studies on, 357t
  male reproductive system of, 460
Rhyncobatus djiddensis, 401
Ristiophoriformes, 17–18. *See also* Saw sharks
River stingrays, 38f, 39, 359f, 380
Rods of the eye, 305, 306
Rostrum, 3, 71–75
Rough sharks, 13f, 15–16
Rubrospinal tract, 281–82

Sacculus, 308
Saccus glandulosus, 105
Salmo trutta, 231–32
Sandbar shark. *See Carcharinus plumbeus*
Sand tiger sharks, 19, 20f, 457
Sarcosine, 134
Sawback angel shark. *See Squatina aculeata*
Sawfishes
  external body form of, 3, 6, 29
  girdles of, 85
  muscular system of, 103
  neurocranium of, 74, 75
  shape of, 30f
  systematics of, 9
  vertebral column of, 84
Saw sharks. *See also* Pristiophoridae; Ristiophoriformes
  external body form of, 6, 17–18
  neurocranium of, 74, 75
  shape of, 17f
  systematics of, 9, 10
  vertebral column of, 83, 84
Scales, 43–49
  cyclomorial, 55
  evolution of, 44–48
  lanceolate, 53
  placoid. *See* Placoid scales
  synchronomorial, 56
Scapular arteries, 212
Scapulocoracoid, 3, 72f, 85, 86f
Schneiderian folds, 5
Sclera, 78, 304
Scoliodon, 429
Scoliodon laticaudus, 422
Scyliorhinidae, 5, 25–26, 410
Scyliorhinins, 154, 166
Scyliorhinus
  circulatory system of, 215
  digestive system of, 156, 164
  heart of, 245, 247, 255
  muscle fibers of, 117–18, 119f, 120, 121f/t, 122t, 123–30, 131f, 132, 133, 136, 138, 139
  nervous system of, 276, 282, 285, 293
  rectal gland of, 343
  thymus of, 231
Scyliorhinus canicula
  blood pressure of, 219t
  circulatory system of, 219t, 230, 232, 233
  cutaneous mechanosensitivity in, 316
  digestive system of, 154, 160, 161f, 162, 163f, 166
  electrosensory system of, 319, 322
  epiphysis of, 52
  female reproductive system of, 399, 404, 408, 412, 413, 422, 431, 435, 436
  heart of, 245, 246, 247, 248, 251, 252, 255, 257, 258
  kidney studies on, 357t
  male reproductive system of, 444, 446, 447, 458, 459, 460, 466
  muscle fibers of, 115, 116f, 117, 125t
  olfactory system of, 303
  rectal gland of, 332, 348
  respiratory system of, 176, 177f, 178f, 179, 180, 184, 185f, 186, 187, 188f, 192, 195
  teeth of, 59
Scyliorhinus catulus. *See Scyliorhinus canicula*
Scyliorhinus stellaris
  circulatory system of, 219t
  digestive system of, 154, 155
  female reproductive system of, 408
  heart of, 243, 245, 247, 251, 252, 253, 258, 259f, 266
  kidney studies on, 357t
  male reproductive system of, 458, 459, 465, 466
  respiratory system of, 180, 190, 192, 195
Scyllium canicula. *See Scyliorhinus canicula*
Scyllium stellare. *See Scyliorhinus stellaris*
SDH. *See* Succinic dehydrogenase
Secondary blood system, 222–23
Secretin, 157n
Secretory cells, 335–39, 343f
Secretory parenchyma, 333–35, 339, 340, 341, 342f
Selachians
  circulatory system of, 218–19, 221, 227
  olfactory system of, 301f
  placoid scales of, 52
  visual system of, 303–7
Semen, 450, 457, 465
Semicircular canals, 308, 311
Seminal plasma, 461, 462t, 463t
Seminal vesicle, 444, 449, 450, 453, 454f, 461, 465
Serine, 412
Serotonin, 153, 154t, 155, 164, 166, 257, 277, 289, 455–56
Sertoli cell bodies, 450, 451–52
Sertoli cell cytoplasts, 450, 451
Sertoli cells, 445, 446–47, 449, 458, 459, 460, 466
Sevengill sharks, 133, 199
Sharkfin guitarfishes, 3, 29–31, 75
Sharks. *See also* Elasmobranchs
  external body form of, 10–29
  muscle fibers of, 116–34
  siphon sac in, 455–56
  systematics of, 9
Shell gland, 406
Shunts, 192–94
Silurian era, 45
Simple zone, 357, 359f
Single-layered spermatozeugmata, 463, 464, 465
Sinoatrial junction, 246
Sinoatrial valve, 246–47
Sinus venosus, 244–47
Sinus zone, 354, 358, 361f, 376, 380, 382, 385
Siphon sac, 455–56
Sixgill sharks, 81, 199
Sixgill stingrays, 37–39, 81
Skates. *See also* Rajidae; Rajiformes
  external body form of, 35–37
  fins of, 88
  male reproductive system of, 445
  neurocranium of, 74
  systematics of, 9
  teeth of, 48
  visual system of, 305
Skin, 43, 48–52
Skin coloration, 50–52
Sleeper rays, 33
Sleeper sharks, 13f, 15
Smell. *See* Olfactory system
Smooth dogfish. *See Mustelus canis*

Snout, 3, 6
Sodium, 330, 331, 332, 339, 344, 346–47, 416, 422, 461
Sodium chloride, 343, 344, 347, 384, 461
Sodium currents, 126
Softnose skates, 35, 36f
Somatic branches of dorsal aorta, 209–12
Somatic motor nuclei, 277
Somatic sensory nuclei, 279–80
Somatic veins, 213–16
Somatostatin, 154, 157–58, 159, 165, 348
Somniosidae, 13f, 15
Somniosus microcephalus, 148f
Southern stingray. *See Dasyatis americana*
Sperm, 461–65
Spermatids, 445, 447, 458
Spermatoblasts, 445, 466
Spermatocysts, 445–47, 458, 466
Spermatocytes, 458
Spermatogenesis, 445, 447, 448f, 461
Spermatogonia, 458, 466
Spermatophores, 461, 463, 464f, 465
Spermatozeugmata, 453, 461, 463–65
Spermatozoa, 7, 219, 445, 447, 458
Sperm bundle, 447–49
Sperm sac. *See* Alkaline gland
Sphyrna, 209, 401, 429
Sphyrna acanthias, 432
Sphyrna lewini, 52, 313, 357t, 399, 434f, 465
Sphyrna tiburo (bonnethead shark), 275, 357t, 432, 434f
  female reproductive system of, 399, 435
  male reproductive system of, 447, 456f
Sphyrna tudes, 399
Sphyrna zygaena, 357t
Sphyrnidae, 26f, 28–29, 318. *See also* Hammerhead sharks
Spinal artery, 208
Spinal cord, 275–76
Spinal nerves, 275–76
Spinax, 50
Spinax acanthias. *See Squalus acanthias*
Spinax niger, 465
Spiny dogfish. *See Squalus acanthias*
Spiracles, 3–5, 220–22
Spiracular cartilages, 80
Spiracularis muscle, 96
Spiracular organs, 313
Spiral intestine, 145
Splanchnic nerves, 163–64
Splanchnocranium, 72f, 78–82
Spleen, 232
Spongiosa, 249, 250t, 251
Spurdog. *See Squalus acanthias*
Squalene, 159, 242
Squalidae. *See also* Dogfish sharks
  external body form of, 12–14
  fin spines of, 57
Squaliformes. *See also* Dogfish sharks
  external body form of, 11–17
  systematics of, 9
Squaliolus, 11, 91
Squaloids
  circulatory system of, 200, 201f
  external body form of, 7
  fins of, 88, 91
  liver of, 242
  muscular system of, 109–10
Squalomorphii, 9, 46
Squalomorph sharks
  male reproductive system of, 445
  nervous system of, 287, 291
  neurocranium of, 74

splanchnocranium of, 79
systematics of, 9
Squalus, 13
  circulatory system of, 198
  heart of, 244
  male reproductive system of, 450, 456, 460
  muscle fibers of, 118, 120t, 138
  nervous system of, 279, 282, 285, 291, 293
  organ of Leydig in, 232
  rectal gland of, 332, 338, 339, 343, 344t, 347–48
Squalus acanthias (spiny dogfish)
  blood pressure of, 219t
  circulatory system of, 219, 220f, 225
  digestive system of, 150, 151, 153f, 154, 156, 157, 158, 159, 163f, 164, 165, 166
  external morphology of, 44f
  female reproductive system of, 399, 401, 403, 405, 408, 409, 416–17, 434, 435, 436
  fin spines of, 57, 58
  fluid compartments of, 330t
  heart of, 245, 248, 251, 253, 256, 257, 265, 266
  kidney studies on, 355t
  male reproductive system of, 444, 449, 455, 458, 459, 460, 463, 465, 466
  mechanosensory lateral line of, 313, 315
  muscle fibers of, 134
  muscular system of, 93–113
  nervous system of, 274, 276, 277
  olfactory system of, 303
  optic tectum of, 286f
  osmotic activity in fluids of, 331t
  rectal gland of, 332, 333, 334f, 335
  respiratory system of, 175f, 178, 184, 187, 193, 195
  teeth of, 58, 59f, 61f, 62f, 63
  urinary system of, 354, 357, 363f, 364, 370f, 372f, 384, 388, 388f, 390, 393f
Squalus brevirostris, 401
Squalus lebruni, 195
Squalus suckleyi
  circulatory system of, 220
  digestive system of, 157
  heart of, 240, 253, 256, 261
  male reproductive system of, 447
  respiratory system of, 194
Squatina
  circulatory system of, 198, 199, 208
  external body form of, 8
  muscle fibers of, 117
  nervous system of, 274, 282, 285
Squatina aculeata, 53, 154t, 166
Squatina angelus. *See Squatina squatina*
Squatina californica, 399
Squatina guggenheim, 409
Squatina occulata, 409
Squatina squatina, 205f, 232, 357t
Squatina tergocellata, 51
Squatinidae, 18. *See also* Angel sharks
Squatiniformes, 9, 10, 18. *See also* Angel sharks
Squatinoids, 7, 74
Squatinomorphii, 9, 46. *See also* Angel sharks
Stapedial artery, 205–6
Stapedial plexus, 229
Starling's law of the heart, 263, 266, 267, 268
Static lift, 242
Statoacoustic nerve, 281
Stegostoma fasciatum, 7
Stegostomatidae, 23f, 24–25
Stem cells, 445
Steroid hormones
  of female reproductive system, 400, 401, 425, 427, 431–37
  of male reproductive system, 445, 450, 458–60

Stingarees, 38f, 39
Stingrays
    external body form of, 7, 37–41
    female reproductive system of, 418
    muscle fibers of, 115, 133
    scales of, 54
    systematics of, 9
Stomach, 145–48
Stomodaeal denticles, 54–56
Stratum compactum, 50
Stratum vasculare, 50
Stria medullaris, 293–94
Striated segment of renal tubule, 388
Stroke volume, 263, 265, 266, 267
Stroke work, 263–65
Subclavian arteries, 204, 212
Subclavian veins, 215
Subepicardium, 243
Suborbital shelf, 77
Subpallium, 293
Substance P, 154, 155, 164, 166, 245, 289
Substantia nigra, 282
Subterbranchialia, 9
Subterminal notch, 8
Succinic dehydrogenase (SDH), 118, 120, 132
Superficial fibers, 118, 119f, 120, 130, 141
Superficial mixipodial muscle, 105
Superior orbital artery, 205–6
Superior raphe, 277
Suprachiasmatic nucleus, 291
Suprahepatic rete, 226–27, 228f
Suprahepatic veins, 216
Supraorbital crest, 76, 77
Swim bladder, 218, 223, 233–34, 242
Swimming, 303
Swordfish, 229
Symmoriioids, 8
Synchronomorial scales, 56
Synencephalon, 288
Systematics, 8–10
Systole, 258, 261, 263

T_3. See 3,5,3-Triiodothyronine
Tachykinin, 154, 164, 245
Taeniura, 39
Taeniura lymma, 125t, 139
Tail, 3
    heterocercal, 112–13
    precaudal, 3, 7–8
Tanning of egg capsules, 412, 413, 414
Taste, 300–303
Taste buds, 302
Tectum mesencephali, 284–87
Teeth, 2, 5–6, 58–66, 91, 108, 110
    distribution and rate of replacement, 58–60
    embryonic development of, 420
    evolution of, 44–48
    primary function of, 145
Telencephalon, 274f, 275, 290f, 291–94, 307, 321, 322
Teleostomi, 58
Teleosts
    circulatory system of, 198, 218, 222, 223, 229, 232, 233
    female reproductive system of, 427
    fluid compartments of, 330t, 331
    heart of, 242, 248–49, 251, 252, 259, 263, 265
    nervous system of, 275, 279, 287
    scales of, 56
    visual system of, 304
Telok Anson shark, 422
Temera, 33

Temperature
    gastric emptying time and, 160–62
    warm-blooded elasmobranchs and, 225–29
Terminal lobe, 7–8
Terminal zone, 408
Testes, 444, 445, 447–49
Testicular arteries, 211
Testosterone, 431–35, 459
Tetrapods, 198
Thalamic eminence, 289–91
Thebesian, 256
Thelodonti, 45
Thelodus, 46f
Third eye, 51–52
Thornback rays, 32f, 33
Thresher sharks, 7, 20f, 21, 84, 88
Thymulin, 232
Thymus, 231–32
T lymphocytes, 231
TMAO. See Trimethylamine oxide
Tonic control, 268
Torpedinidae, 35
Torpediniformes, 9, 33–35
Torpedo
    circulatory system of, 199, 200, 201f, 202, 203, 205, 208
    female reproductive system of, 408
    heart of, 255
    muscle fibers of, 116, 117, 122, 132, 139
    nervous system of, 277
Torpedo fairchildi, 219, 220f
Torpedo galvani. See *Torpedo marmorata*
Torpedo marmorata, 51
    circulatory system of, 207f
    digestive system of, 154
    female reproductive system of, 400
    heart of, 254f, 266
    kidney studies on, 357t
    male reproductive system of, 446, 460, 465, 466
    respiratory system of, 176
    thymus of, 231
Torpedo nobiliana, 251, 309f
Torpedo ocellata, 322, 357t, 418
Torpedo rays, 3, 35, 75
Torpedo torpedo, 465
Tractus olfactorius, 274f
Transosomes, 403
Trapezius muscle, 100
Triaenodon obesus, 65f, 312
Triakidae, 26f, 27–28
Triakis, 261
Triakis acutipinna, 456f
Triakis scyllia
    heart of, 245–46
    kidney studies on, 357t
    male reproductive system of, 456f, 461, 462t, 463t
    teeth of, 63
Triakis semifasciata, 51
    heart of, 249, 261–62
    kidney studies on, 357t
    locomotion in, 113
    respiratory system of, 178, 182
    skin of, 49f
Trigeminal (V) nerve, 95, 96, 97, 277, 279
Triglycerides, 133
Trigonognathus, 12
3,5,3-Triiodothyronine (T_3), 134
Trimethylamine oxide (TMAO), 134, 330, 331, 332, 353
Trochlear (IV) nerve, 95, 277
Trochlear nuclei, 283
Trophonemata, 417–20

Troponin C, 268
Trout, 262
Trunk, 3, 6
Trunk muscles, 94t, 103–4
Trygon orrhina, 31, 139
Trygon pastinaca, 357t
Trygon violaceus, 357t, 376
Turban organ, 376, 377f
Typhlonarke, 3, 33, 88, 134
Tyrosine, 412, 413

Umbilical cord, 422, 429–31
Unpaired fins, 7
  musculature of, 105–6
  supports and, 88–91
Upper Devonian era, 223
Upper jaws, 1, 2, 108
Urea, 132, 134, 218, 330, 331, 332, 353, 354, 403, 416, 422, 461
Ureter, 453
Urinary space, 366
Urinary system, 353–93. *See also* Kidneys; Renal entries
Urine, 362, 452
Urogenital ducts, 7
Urogenital sinus, 409, 450, 453
Urogymnus, 39
Urolophidae, 37, 38f, 39
Urolophus, 335, 337, 338
Urolophus auranticus, 51
Urolophus halleri, 280f, 320–21, 399, 403, 457, 458
Urolophus jamaicensis, 247, 253
  female reproductive system of, 402–3, 406f, 408
  male reproductive system of, 447, 448f, 450, 456–57
  rectal gland of, 332
Urolophus mucosus, 231f
Uterine fluids, 415–16, 461, 462t, 463t
Uterine milk, 418, 429, 431
Uterolactation, 420
Uterus, 408–9, 414, 416–17
  paraplacental, 429
Utriculus, 308

Vagal branch, 257
Vagus lobe, 279
Vagus (X) nerve, 98, 162–63, 277, 279, 302
Vasculature. *See also* Circulation
  of the gut, 150–51
  of the heart, 253–56
  of the kidneys, 374–80
  of the rectal gland, 340–43
Vasoactive intestinal peptide (VIP), 153f, 154, 155, 158–59, 165, 166–67, 348
Venous capacitance, 223–25
Venous sinuses, 223–25, 240, 241f
Venous system, 213–16
Ventilation, 176–78
Ventilation conductance, 190
Ventilation:perfusion ratio, 220
Ventral aorta, 199–200, 220f
Ventral cutaneous vein, 216
Ventral intestinal artery, 212
Ventral mixipodial muscle, 105
Ventral pterygoideus muscle, 104, 105
Ventral subterminal notch, 7–8

Ventral thalamus, 289
Ventricle, 248–52
Ventrolateral artery, 212
Vertebral column, 1, 83–84
Vertebromuscular branch of dorsal aorta, 211
Vertebrospinal branch of dorsal aorta, 211
Vesicles of Savi, 313
Vestibular system, 308–13
VIP. *See* Vasoactive intestinal peptide
Visceral branches, 211, 212–13, 257
Visceral cardiac branch, 257
Visceral motor nuclei, 277
Visceral pericardium. *See* Epicardium
Visceral sensory nuclei, 279
Visceral veins, 216
Visual evoked potentials, 307
Visual system, 303–7
Vitamin A, 306
Vitellogenic follicles, 435
Vitellogenin, 404, 435
Viviparity, 240, 401, 403, 404, 406, 407, 409, 414, 415–29
  aplacental with oophagy and intrauterine cannibalism, 420–21
  aplacental with uterine villi or trophonemata, 417–20
  aplacental yolk sac (ovoviparity), 403, 416–17, 432
  placental, 421–29, 432
  reproductive cycles in, 399
  steroid hormones and, 433f, 434, 435, 437
von Willebrand's factor, 380
V wave, 258

Warm-blooded elasmobranchs, 225–29
Weasel sharks, 26f, 28
Wedgefishes, 29–31
Whale sharks, 23f, 25
Whiptail stingrays, 38f, 39–40
White muscle fibers, 116–23, 134–35, 136, 138–39, 141
  biochemistry of, 132–33, 134
  functional role of, in myotome, 130–32
  inner, 119f, 120, 132, 135
  mechanical properties of, 127–30, 140
  membrane constants of, 125t
  outer, 119f, 120, 135
  ultrastructure of nerve endings, 124f
Wiggers diagram, 258
Wobbegongs, 22–24
Wunderer's corpuscles, 275

Xenacanthus, 47f
Xenocanths, 9
Xiphius gladius, 229

Yolk, 400–401, 403–4, 418, 420, 425
Yolk platelets, 410, 411f
Yolk sac, 422, 424f, 425, 426f, 429, 445
  external and internal, 410, 411f
  external gill filaments and, 431
Yolk stalk, 410, 416, 418, 423f, 429, 430

Zanobatidae, 33
Zebra sharks, 23f, 24–25
Zygaena, 429
Zygaena malleus, 149f

Library of Congress Cataloging-in-Publication Data

Sharks, skates, and rays: The biology of elasmobranch fishes / edited by William C. Hamlett
    p.   cm.
  Includes bibliographical references (p.) and index.
  ISBN 0-8018-6048-2 (alk. paper)
    1. Chondrichthyes.   I. Hamlett, William C.
QL638.6.B56   1999
597.3—dc21
                                                    98-24605
                                                      CIP